Advances in Sustainable Manufacturing

Günther Seliger • Marwan K. Khraisheh
I.S. Jawahir
Editors

Advances in Sustainable Manufacturing

The International Academy
for Production Engineering

Springer

Editors
Günther Seliger
Technische Universität Berlin
Department of Machine Tools and Factory Management
Pascalstr. 8-9, 10587 Berlin
Germany
seliger@mf.tu-berlin.de

Marwan K. Khraisheh
Masdar Institute of Science and Technology
Masdar City
Abu Dhabi
UAE
mkhraisheh@masdar.ac.ae

I.S. Jawahir
Institute for Sustainable Manufacturing (ISM)
University of Kentucky 414 C
CRMS Building Lexington, KY 40506
USA
jawahir@engr.uky.edu

ISBN 978-3-642-20182-0 e-ISBN 978-3-642-20183-7
DOI 10.1007/978-3-642-20183-7
Springer Heidelberg Dordrecht London New York

Library of Congress Control Number: 2011929783

Cover design: eStudio Calamar S.L.

Printed on acid-free paper

Springer is part of Springer Science+Business Media (www.springer.com)

Preface

The annual Global Conference on Sustainable Manufacturing (GCSM) sponsored by the International Academy for Production Engineering (CIRP) is committed to excellence in the creation of sustainable products and processes. These conserve energy and natural resources, have minimal impact upon the natural environment and society, and adhere to the core principle of considering the needs of the present without compromising the ability of future generations to meet their own needs.

To promote this noble goal, there is a strong need for greater awareness education and training, including dissemination of new knowledge on principles and practices of sustainability applied to manufacturing. The Global Conference on Sustainable Manufacturing grants international colleagues opportunity to build effective relationships, expand knowledge, and improve practice globally.

The conference has previously been held at different counties and locations: at the Indian Institute of Technology Madras, India in December 2009, at the Pusan National University, Korea in October 2008, at the Rochester Institute of Technology, Rochester, USA in September 2007, at the University of Sao Paulo, Brazil in October 2006, at the Jiao Tong University, Shanghai, China in October 2005, at the Technische Universität Berlin, Germany in September 2004, as well as in the form of the Environmentally Benign Manufacturing workshop held in Birmingham, Alabama, USA, in January 2003.

In November 2010, Masdar Institute of Science and Technology and Abu Dhabi University in the United Arab Emirates (UAE) hosted the 8th Global Conference on Sustainable Manufacturing under the patronage of His Excellency Sheikh Nahayan bin Mubarak Al Nahayan, Minister of Higher Education and Scientific Research of the UAE.

The Welcoming Address was given by President of Masdar Institute Prof. Dr. Fred Moavenzadeh, and by the Chancellor of Abu Dhabi University Prof. Dr. Nabil Ibrahim. Participants came from all over the world to share the results of their sustainable engineering research. Fruitful exchanges of ideas ensued on the potential of renewable energy, on value adding by sustainable manufacturing, education for sustainability engineering, green supply chain and transportation, microelectronics and resource efficiency and technology driven startups, and how to cope with the key challenge of achieving greater functionality with fewer resources.

Visits to the Masdar City, Emirates Aluminium (EMAL), Anabeeb, and Emirates Steel Industries presented valuable opportunities to appreciate novel approaches in value creation beyond pure resource extraction.

This book includes revised and extended versions of selected papers presented at the 8th Global Conference on Sustainable Manufacturing. These contributions are well-structured in eight chapters covering topical areas: Engineering Education; Entrepreneurship; Manufacturing Processes; Product Design and Development; Remanufacturing, Reuse and Recycling; Renewables and Resource Utilizations; Sustainability Assessment; and Logistics Management and Green Supply Chain.

The Editors wish to thank all authors for their contributions and the reviewers for their time and effort in realizing this book. We especially thank Ms. Dipl.-Ing. Pinar Bilge and Mr. M.Sc Engg Sadiq Abd-Elall. Both are PhD students at Technische Universität Berlin. Thanks are due also to Ms. Sara El Hage from Masdar Institute of Science and Technology. Their outstanding efforts and tremendous support ensured the success of both the conference itself and the publication of this book.

March 31st 2011

Günther Seliger, Technische Universität Berlin, Berlin, Germany

Marwan Khraisheh, Masdar Institute of Science and Technology, Abu Dhabi, UAE

I.S. Jawahir, University of Kentucky, Lexington, USA

Table of Contents

1 Engineering Education

Higher education institutions are a powerful starting point for the preparation of creative engineers who are able to design and develop sustainable products and processes. In the first section of this chapter Seliger introduces new approaches in engineering education; examples from the international master program Global Production Engineering have been presented. In the second section El Maraghy explores the connection between engineering and other disciplines, to offer a rich mixture that balances technical subject and key socio-technical issues affecting our global village. In the third section Weckenmann and Werner argue how measures of further education have to be chosen in order to allow for an efficient achievement of required additional competences. In the fourth section Herrmann et al. present an educational structure and concept of sustainability in production engineering course considering the environmental, the social and economical impact on the manufacturing systems. In the fifth section Marxer et al. describe an education concept with a blended learning approach to support sustainable education such as lifelong learning, largely independent of the learner's geographical location and the available infrastructure. In the sixth section Fernandes and Rocha explore the role for hands-on programs for the development of sustainable manufacturing mindset in youngsters towards industry and future career development.

Curriculum Design for Sustainable Engineering – Experiences from the International Master Program "Global Production Engineering"

G. Seliger, C. Reise, P. Bilge

Department of Assembly Technology and Factory Management, Technische Universität Berlin, Germany

Abstract

Engineering can be characterized as the exploitation of potentials for useful applications. It is oriented towards solutions within a framework of limited resources. Education is needed to enable engineers to act as "change agents". It is their responsibility to innovate in the development, manufacturing and marketing of products throughout their full life cycle, to fulfill the needs of humankind, protecting public health, welfare, and environment. Productivity of learning and teaching has to be increased by a new curriculum including innovative methods and forms for engineering education. This paper will analyze which competence domains a curriculum towards sustainable development should have and how the education in the international master study program "Global Production Engineering" is accomplished. This best practice from Germany can be a pioneer for further countries in the world that target to construct a knowledge-based economy like the United Arab Emirates. The paper shows possibilities of a master program to motivate students with bachelor degrees and initial work experience to walk paths in their careers combining professional success and global responsibility.

Keywords:

Engineering education, sustainability, curriculum, global production engineering

1 NEED FOR INNOVATION

In the 20th century scientific and technical innovations offered us an enormous increase in prosperity. People from industrialized countries raised their income by a factor of 30, and the productivity in the agriculture and production sector increased by a factor of 35. At the same time the material living standard increased 30 times and coincided with the extension of life expectancy by 37.5 years. The pace and the overcoming of distance in the scope of mobility increased by a factor 100, which enables us today to reach every place on earth within 24 hours [1].

These achievements put a progressive burden on the environment and the biosphere in the context of serious income differences among industrialized and emerging countries. 2.4 billion people do not have clean drinking water, 50% of all employees live under the UN-poverty line of 1.50 € a day. 550 million wage earners live on less than one US-dollar a day. Humankind is responsible for 60 million tons of carbon dioxide being added to the atmosphere, and 55.000 hectares of rain forest are being destroyed every day. Worldwide, the oceans are being robbed of fish by 220,000 tons, the farmland is being decreased by 20,000 hectares, and the fauna and flora loses 100 - 200 species per day [2, 3].

Considering that all areas of life are affected by global change, there is a gap between our understanding of the problems and available global and regional solutions. The actions of governments companies and individuals in general are far from being adjusted to the future challenges and their accomplishments [4].

In the case of the United Arab Emirates (UAE) the government has set the strategic priority in so called "Vision 2021" to create a knowledge-based economy and enhance infrastructure through the implementation of innovative products, processes and solutions in health, education, sustainability, security and lifestyle [5]. These shall significantly contribute to the economic growth of the country of the UAE. The country has been chosen as the location of the International Academy for Production Engineering (CIRP) sponsored 8th Global Conference on Sustainable Manufacturing to present innovations to competent partners from research institutes and industrial partners related to the area of sustainable manufacturing worldwide [6]. Tapping existing global knowledge and adapting international technologies to local requirements and conditions leads to a domestic innovation to produce value added products and processes.

In order to facilitate sustainability all segments of society must be educated to understand economic, ecological and social connections. Technological innovations play an important role in solving this global challenge. A new awareness of the importance of innovations in technology and management is fundamental to a global economic development, the preservation of life, conservation of natural resources and creation of social justice.

Innovation occurs when potentials are developed and problems are solved. The engineer has a special relevance by using new technologies to design innovative products and processes. The market-based economic competition should be a trigger for innovation and for the implementation of sustainable products and processes.

This paper identifies the demands on an engineering education curriculum and describes the implementation at the international Master Study Program "Global Production Engineering".

2 POTENTIAL FOR NEW PROCESSES AND PRODUCTS

Society continuously demands products and processes to satisfy their needs. Limited resources require the increased use of renewable materials and the preservation of nonrenewable materials. Production, usage and recycling of products effect different influences on the economy, ecology and society. That influence is highly complex and requires an analysis and a systematic examination of the environment.

Sustainable products are characterized by providing environmental, social and economic benefits while protecting public health, welfare, environment throughout their full commercial cycle, from the extraction of raw materials via design and manufacturing to usage and recycling [7]. Nonrenewable materials can no longer be disposed once humankind´s ever increasing needs cannot be fulfilled anymore due to limited resources. Reuse and recycling become inevitable requirements for product and process design. Renewable resources must not be consumed in quantities higher than can be regained.

With reference to Maslow's pyramid of the human needs in Figure 1, products can be assessed by their ability to satisfy needs. Products that satisfy physiological needs affect the fundamentals of life: breathing of clean air, food from agricultural production, preservation of bodily functions by medical products and artifacts, which enable surviving in different regions of the world by creating appropriate living environments. Mobility can be an important cross -section function in order to secure needs like food or medical supplies.

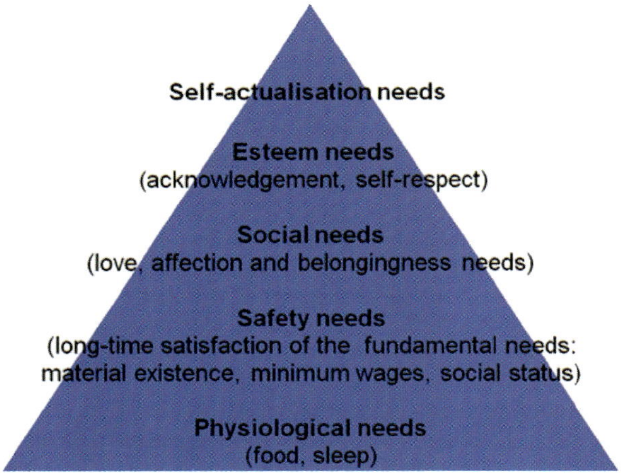

Figure 1: Maslow's hierarchy of needs [8]

Half of the world's population has physiological needs that are not being satisfied. An enormous demand is not settled by organizations. The opposite side builds the oversupply of products in industrialized nations, whose demand is artificially built up and imposes a special obligation to achieve sustainability.

The majority of companies are under constant pressure to optimize costs, time and quality of their products and processes. Abstract goals like sustainability are hard to follow and to integrate into the company goals. Organizations are guided by short-term economic restraints. People in established organizations often do not have innovative mindsets and flexible attitudes, whereas younger people who are not involved in the economic business often do. Organizations, with rigid structures, are slow to recognize new markets and are less inclined to make drastic changes to their processes. Ultimately such organizations face being supplanted by more innovative companies that introduce sustainable products.

The demand for the satisfaction of physiological needs in emerging countries and also the looming demand for sustainable products in industrialized countries creates enormous potentials that are currently not recognized by organizations, and their efforts are not being directed towards the exploitation.

People who are developing solutions for exploiting those potentials strive for a better future. It takes young engineers as entrepreneurs with new business concepts but also in established businesses to exploit these potentials. Education and training should enable humankind to fulfill its primary needs through its own power and to make decisions according to sustainability standards, thereby weakening the inequality of global wealth distribution [9].

It is the educational system's duty to make people aware of sustainability, to help them to conceive the potentials of sustainable products and to enable them to exhaust these potentials through education and training.

3 MARKET MODEL FOR RESEARCH AND EDUCATIONG

Mobility as well as information and communication technology have connected humankind closer than ever before. People from all over the world interact in common markets. This applies also to educational and job markets. Worldwide universities are providers of research and educational services. Students require education for being later able to offer their competence on the job market.

Organizations need resources such as research from universities and a qualified workforce. Figure 2 shows the various participants who meet in the research, educational and job markets. Those markets with its participants build a part of the global society. Cooperation and competition takes place in development and application, purchase and sale of competencies on the societal level [10].

The market for engineering education as well as the job market determines the requirements of the educational curriculum. Participants and their courses of action are described in more detail in the following.

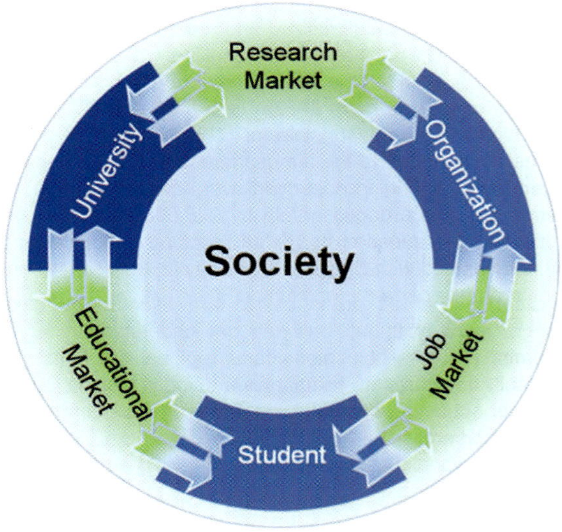

Figure 2: Reference Model for Integrative Competence Management [10]

3.1 Job Market

Organizations are a direct point of demand for qualified workers. Depending on offer and demand, incentives and salaries for engineering-labor are negotiated.

Engineers

Finding technological solutions is a central task for engineers. Therefore people who create and use technology can be considered engineers regardless of their formal job sector. Lifelong learning, market globalization and the changing professional fields determines the engineers' requirements and the composition of their competence portfolio. In response to the increasingly global characteristics of the marketplace, engineers are expected to be more mobile and flexible, to move to different work sites both nationally and abroad, and to change job responsibilities as new needs arise.

Engineers have to act as "change agents" for the important technological, economic and social developments that are challenging our world [11]. The understanding of new socio-technological developments due to global engineering collaboration at the system level of production engineering require broader knowledge of human beings, organizations, and society outside of the scope of traditional engineering education based on natural sciences [11]. The activities have also changed, from developing new technical skills, units, and equipment to advanced problem-solving skills in project planning, implementation, and integration of complex systems of hardware and software. Engineering teams follow the so-called product life cycle: idea generation, product conception, product planning, product development and design, production planning, manufacturing, marketing and distribution, maintenance, repair, and overhaul to recycling. Moreover, engineers must complement their professional and technical expertise with nontechnical competencies such as systematic problem solving, communication, management, and leadership skills [12, 13].

Organizations

Organizations assess the educational portfolios of engineers based on four main competence areas [14]:

- technical and methodological skills,
- personal skills,
- management skills and
- social skills.

Studies about industrial demand for engineers indicate heterogeneity regarding their professional contents, geographic disposition and the degree of demand. Studies suggest that the demand is caused by the economic development, which can be explained with the "Cobweb Model" [15]. Corporations concentrate their search for skilled workers on an operational fulfillment of demands for a current problem.

Regarding their skills companies seek engineers who are technically experienced, culturally aware, and broadly knowledgeable; engineers who exhibit an entrepreneurial spirit and who are innovative and lifelong learners; engineers who understand world markets, who know how to translate technological invention into commercially viable products and services; and engineers who are professionally nimble, flexible, and mobile [12].

Society

The aims of society, which generally have a narrow focus and lack of a global perspective today, must be differentiated from the aims of sustainable development. In general society needs engineers for planning, constructing and operating products and processes, and for increasing or at least holding the prosperity level by satisfying the direct needs of the people.

To meet the aims of sustainable development engineers are also needed to develop products and processes for resource efficiency and effectiveness, supporting qualification and occupational health, being competitive in meeting customer demands. Decision makers and leaders from the innovation side need a norms and values system based on individual responsibility for global concerns. They should be enabled to look at innovations as solutions to the problems of the 21st century, analyze complex problems, choose fields of innovations, in which development should be advanced in order to save the long-term future abilities of society, the economy and the environment, realize innovations through independent action, and successfully market sustainable products on the global market through corporative action.

While defining the curriculum the industrial goals have to be combined with goals for a sustainable development, so that the university educates engineers, who fulfill the company's demands, but at the same time allows them as innovators and entrepreneurs to take new paths without affecting further generations.

3.2 Educational Market

In the educational market people meet their demand by acquiring a competence portfolio that enables them to position themselves in the labor market [10].

Engineers

These days because of increasing demands from markets and innovation, engineers are facing the challenge to continuously and independently plan their competence portfolio. This includes choosing a university, which meets the criteria of expected qualifications. Engineers must also consider the university's reputation, the duration of study and costs. However, a university degree is not the end but a milestone in a life-long learning process. Innovations continuously change markets and the companies acting in them. They have to orient themselves towards new technologies and markets, and adjust their portfolio to it. In organizations with constant pressure to change, life-long employment at the same position will be rare.

Expertise, which ages significantly faster than concept and methodological knowledge, can make an in-demand employee with excellent knowledge today obsolete tomorrow [16]. To keep up -to- date engineers as well as organizations need portfolios for planning competence.

In designing one's competence portfolio a university can merely give advice. It is the responsibility of the student himself or herself to gain competence. This underlies the danger to follow short-term trends proclaimed by a university or industry, instead of gaining longer lasting competencies.

Society

Society exerts influence on the student's choice of university and study program. This influence is related to the reputation of the university or a certain study program. Values and norms of society influence the quantity and quality of people seeking a certain position.

It is an important task of society to sensitize young people early on to the importance of sustainable development. Through such awareness they will be able to assess study programs holistically, and it will be easier for them to identify potentials for their competence portfolio. Graduates of study programs can later become important propagators for creating an awareness of global responsibility in society. Universities are within the society in a unique position to foster understanding, tolerance and informed dialogue regarding globalization.

University

Universities must convince high aptitude students to enroll. Their success on acquiring students is dependent on the ability of the alumni to position them successfully on the job market.

Universities must support young engineers to choose the direction in which they want to innovate. In order to search for technically possible areas, that offer high potentials for innovation, they need a compass for designing their competence portfolio. It should contain values and norms, knowledge about technologies and about ways to use them according to their requirements.

Professional specialization for a certain job profile can be gained at the work place in a temporally restricted adjustment phase, e.g. during a trainee position, an internship or a training on the job. The universities´ assignment is to teach essential competencies that enable a person to innovate and to solve complex problems.

Universities teach traditionally declarative knowledge, and the teaching of implicit knowledge is often taken over by other parts of society and is not being administered as the universities [17]. In order to achieve a universal engineering education, this situation is slowly changing towards a recognition that serves all competence areas. While there is a unilateral lack of education in social, personal and managerial competence, there is no agreement on the elimination of this nonconformance, but rather competing systems of teaching [17].

4 CURRICULUM DESIGN

The design of a curriculum is based on the requirements of the education and the job market. Defining the curriculum involves learning targets and contents as well as learning processes and organization.

Integration of methodological, social and personal competencies, as well as a comprehensive system-view for sustainable questions together with in depth knowledge of technological problems leads to a learning volume which exceeds traditional syllabi and cannot be taught with traditional teaching methods in a normal time frame.

For increasing learning productivity teaching styles will be analyzed, and methods to cope with the needs of increased productivity and necessary technical and methodological contents will be identified. Based on that, key competence domains will be identified and a curriculum for sustainable manufacturing presented.

4.1 Learning and Teaching Styles

"Not the mass of what someone knows or has learned constitutes education, but power and feature, with which someone has acquired education, and how one understands to use the available mass of information with his perception and personnel review. The content does not make the education, but the style" [18].

Felder defines five fields of learning and teaching styles in engineering education [18]. He classifies students according how they receive and process information and classifies instructional methods how they address the proposed learning style components.

The student's learning style can be analyzed by individually answering the following questions: What type of information students preferentially perceive sensory - sights, sounds, physical sensations - or intuitive - possibilities, insights hunches? Can visual – pictures, diagrams, graphs – or auditory input – words, sounds – by the sensory channel be more effectively perceived? Do students find inductive – facts and observations are given, underlying principles are inferred – or deductive – principles are given, consequences and applications are deduced – organization of information most comfortable? Do students prefer to process information actively – through engagement in physical activity or discussion – or reflectively through introspection? How does the student progress towards understanding: sequentially – in continual steps, or globally – in large jumps, holistically?

The teaching styles include: Emphasizes the teacher concrete factual or abstract conceptual information, stresses he visual – pictures, diagrams, films – or verbal – lectures, readings, discussions presentation, and does he organize it inductively – phenomena leading to principles – or deductively – principles leading to phenomena? Is the student participation active, with talking, moving reflecting or passive, where students sit and listen. Is the perspective how information is presented sequential – step- by- step - or global – context and relevance [19]. Figure 3 gives an overview over the possible teaching and learning styles within these dimensions.

Dimensions of Learning and Teaching Styles			
Preferred Learning Style		*Corresponding Teaching Style*	
sensory intuitive	perception	concrete abstract	content
visual auditory	input	visual verbal	presentation
inductive deductive	organization	inductive deductive	organization
active reflective	processing	active passive	student participation
sequential global	understanding	sequential global	perspective

Figure 3: Dimensions of learning and teaching styles [19]

Teachers, who adapt their teaching style at both poles of each of the given dimensions, come close to providing an optimal learning environment for most students in a class. By integrating every teaching style in a training module, it improves the learning outcome for all participants. Nowadays, the usual methods of engineering education address mainly the five categories - intuitive, auditory, deductive, reflective, and sequential - and effective teaching techniques substantially overlap the remaining categories [20].

To reach a balance of teaching styles in individual courses as well as over the whole curriculum, new teaching techniques, which address all learning styles, must be integrated. The teacher should be capable of adopting new teaching styles and adapting them to his individual personality. The main idea is to find a balance between the two teaching dimensions, e.g. provide concrete information and abstract concepts, balance material, which emphasizes practical problem solving with material focusing on a fundamental understanding. Deductive and inductive learners can be addressed by following the scientific method in presenting theoretical material, providing concrete examples, addressing sensing and inductive learning style, developing a theory or formulate a model, for intuitive, inductive and sequential learners, showing how the theory or model can be validated and deducing its consequences, addressing deductive and sequential learners, and presenting applications for sensing, deductive and sequential learners. The combination of using pictures, graphs or sketches during the presentation of verbal material, showing films, providing demonstration and a hands-on approach can support sensory perception, visual input and active progressing [19].

It is a challenge to implement all this in one course, especially inductive organization and opportunities for student activity, while still covering the whole content in the same amount of time. This challenge can be met, if an increased learning productivity can be reached by teachers having new teaching instruments and trying new methods.

4.2 Experience-Based Learning

Currently nontraditional teaching styles such as using inductive organization and active student participation by visual presentation with a global perspective are a minority in classes. It is therefore favorable to promote these styles in a curriculum development process as long as they are not fully integrated into normal teaching practice.

Experience based learning proclaims that only a direct, practical examination of a teaching topic makes effective, inspiring learning [20]. In this model learning presupposes a concrete experience outside of artificial learning environments. Experience-based teaching and learning arrangements are a form of situational learning, whereby the learner stands as a protagonist in the center. Learning always presupposes thereafter an active, reflexive conflict with concrete experiences. Problematic situations, where solving represents a challenge, are the origin of learning processes. Only reflection, the intensive thought about such situations as they arise in everyday's life, leads to instructive experiences and to the expansion of knowledge. Teaching productivity will be increased in formal education situations, if the abstract knowledge, which is obtained there, is attached to concrete individual experiences [21].

This kind of learning basically transforms the learner into the role of the teacher and his learning process is not just to consume input but to create output. The application of scientific principles in learning factories that create useful artifacts has to achieve the special purpose of giving students lasting knowledge with highly practical components.

The educational technical term "Learning by Doing" linked action orientation with experience orientation. "Learning by Doing" is so developed to a project orientation. A subform of experience-based learning is to learn through teaching. The training which is based on these concepts is supported also by the trend from the input to the so called outcome-orientation. The organization of the teachings is not any more at the teaching input, e.g. course contents but rather oriented at their outcome: the learning results.

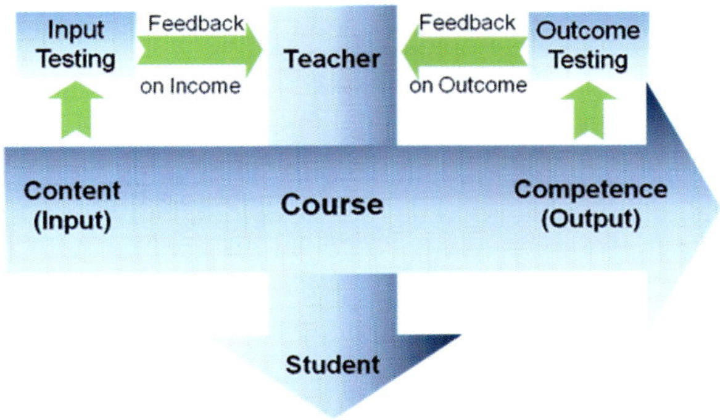

Figure 4: Input- and outcome-orientation of courses [10]

The quality of the graduates is here controlled by product properties. The outcome is oriented towards the ability to create artifacts instead of reproducing knowledge. Active and communicative forms of study and labor will grow in engineering education, and the contents of teaching by presentation of expert knowledge will decline [22].

Experience-based learning, in which the learner becomes the teacher and interaction is in the orientation of learning, can help to increase learning productivity to satisfy the necessities of an increased competency demand in engineering education by entrepreneurial and a higher grade of methodological and social skills.

4.3 Curriculum Design for Sustainable Development

In the process of developing new products and processes students have to recognize and assess the interrelations between economics, technology, environment and society. They face the following questions:

- How can I use technology to create new artifacts with the potential for a sustainable development?
- What kind of technology do I use to create products and processes?
- With what methods and tools do I produce my product?
- How can I motivate others to support technological advance for sustainable development?
- Who will be in action, when and where?

On the basis of these questions a curriculum model was developed in which technical, methodological, social, managerial and personal competences are embedded. Based on the requirements of the markets, four competence domains were identified.

- Technology potentials for sustainability which are showcased as future potentials by applying this technology to products and processes,
- Production for acquiring know-how in development, manufacturing and recycling of products and processes,
- Entrepreneurship, as a knowledge domain for managing technology potentials and organizing the transfer into products and processes.
- In the global responsibility domain contents of the other domains are brought into a global perspective. It promotes awareness of the global challenges in the context of different cultures.

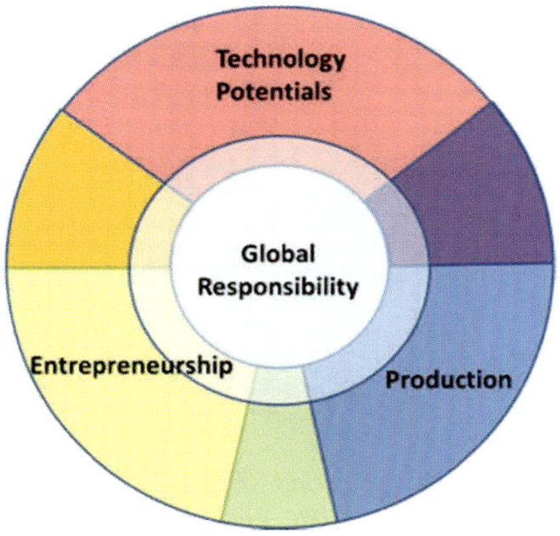

Figure 5: Curriculum for sustainable development

The competence domain, **technology potentials for sustainability** enables the students to deepen their knowledge in technological dimension. Students are introduced to technologies from which they individually must find answers to the question: "How can I use technology to create new artifacts with the potential for a sustainable development?". Students recognize, which technologies are suitable to be transferred into products or processes. Subject areas are technology fields where great potential are expected or have shown partially their potential already, like ecological and social building, energy efficiency of systems and renewable energies, biotechnology and medical technology, water and water-disinfection technology, sustainable mobility and traffic engineering, energy storage techniques or bionics [4].

Design, manufacturing and recycling of product and related processes is the subject of the competence domain production. "With what methods and tools do I produce my product?" is the central question. Students learn with which methods and tools they are able to produce products or create processes sustainably by gaining technical and methodological knowledge. Therefore students learn to view product life cycles holistically. The life cycle steps idea finding, product conception, product planning, product development and design, production planning, manufacturing, marketing and distribution, maintenance, repair, overhaul, and recycling are presented. Students become confident at planning and independently implementing selected processes within the life cycle chain. They learn the relationship between the particular steps in a process and to arrange applications of methods into these steps.

By acting as an entrepreneur with responsibility in intercultural and multidisciplinary groups students become able to independently answer the question: "Who will be in action, when and where?". Decentralized and distributed teams demand from the engineers to perform collaborative actions across various cultural, disciplinary, geographic and temporal boundaries. Through lectures and exercises students learn decision-making, strategic thinking, negotiating skills, and a willingness to take responsibility. Languages promote the motivation to overcome limits and to act globally. Students gain managerial qualifications, assertiveness skills, decision-making skills, the ability to analyze and evaluate, strategic thinking, and negotiating skills.

Handling and developing technology in a responsible manner is at the center of **global responsibility**. "Act accordingly, so that the effects of these actions are reconcilable with the permanency of real human life on earth." So challenges the philosopher Hans Jonas in 1979 in accordance with Kant´s "categorical imperative" from his work "The Responsibility Principle, Trial of Ethic for the Technological Civilization" [23]. Students strengthen values and norms and develop awareness for collaboration on a global scale. System knowledge to recognizing and assessing the interrelations between economics, technology, environment and society needs a global perspective. The shift from egocentricity towards global responsibility, combined with the ability to innovate products today without harming future generations, needs an education process with a multidisciplinary approach which combines humanistic and natural scientific courses in a curriculum.

Within these four competence domains all criteria for comprehensively developing skills are covered. Most courses address all four domains with different importance. Skills are conveyed in different teaching units within one teaching module, such as technical, methodological, social, personal and managerial skills. Productivity of learning and teaching is planned to be increased by innovative methods and forms of engineering education. A variety of teaching dimensions, online learning for methodological skills and web-based learning for specialized knowledge as well as experience centered lectures in realistic environments are instruments to be implemented.

5 GLOBAL PRODUCTION ENGINEERING (GPE)

5.1 Overview and History

The degree program was created in 1998 with the goal to educate international bachelor graduates with first professional experience in the management field and in technology [26].

Engineering graduates learn to work in intercultural and multidisciplinary teams. They are enabled to work as managers on the border between technology and economy and to design and successfully operate sustainable international value creation networks.

The program is arranged into five modules: production, engineering, management, intercultural communication and special profiles. Education focused on the research field in "Recovery of Resources in Product and Material Life Cycles" [24]. With the reorientation of the research at the department to „Sustainable Manufacturing" [25] this reconfiguration is also going to be carried out within the study program.

The guidelines for the degree program are created in the context of the Bologna process to ensure that transparency, comparability and permeability of educational system in Europe are given.

Within the 13 years GPE successfully positioned itself as an international brand in the global competition. Since the beginning 350 engineers graduated successfully. From the students appraised general quality, the close contact to the alumni and university partners worldwide, make a promising broadening of the activities possible. Presently integration of potentials in technology through including solar technologies in the curriculum and the setup of an international alumni network are furthered.

The new challenges are to integrate new teaching methods and interdisciplinary courses without losing technical depth. This will bring competitiveness in the international market for education and reorient education goals towards enabling students to apply technological potentials for future sustainability.

5.2 Internationality

Every intake consists of approx. 30 students from about 15-20 different countries from every continent. The global village truly exists in the classroom. While working together in intercultural groups global thinking, intercultural communication and the recognition of cultural values is integrated into the curriculum. The international orientation of the program seeks to bridge cultural gaps. It brings expanded educational and cultural opportunities to the classrooms, encourages a better understanding of other norms and values, improves communication, and supports the concept of collaboration. The goal is to "meet your global neighbor" and build "cross-cultural bridges". Figure 6 gives an overview of the global distribution of the GPE students from 1998 to 2008.quations should be justified to the left margin and numbered at the right margin. Leave 6 points before and after the equation, as indicated in the Equation style on this template.

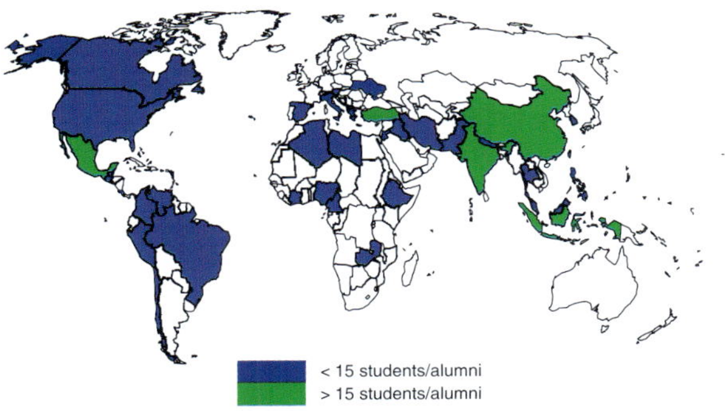

Figure 6: Origin of GPE students

Students learn to develop global products while making adjustments and specifications for regional usage requirements. Development can take place anywhere which is supported by communication technologies and with student specialists from the regions working in international and multidisciplinary teams. The understanding of new socio-technological developments due to global engineering collaboration at the system level of production engineering is part of the education. Knowledge about global interaction between humans, organizations, and society outside the scope of traditional natural sciences based engineering education will be gained automatically. After their study alumni promote and propagate ideas of global responsibility in their home regions and internationally.

5.3 Market-Driven Curriculum

A market oriented curriculum design is derived from the job market and the education market. As first steps towards a market-driven education, an evaluation of the motives of applicants who are interested in the product GPE and of the qualification of alumni was done.

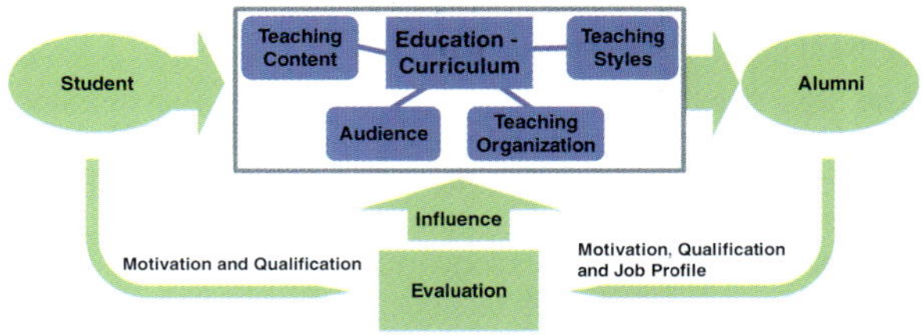

Figure 7: Control-loop for the curriculum design

The study program is in its beginning phase of building up an evaluation directed control loop for the curriculum design. A fine tuning of the measuring instruments within the control loop is planned, which imparts criteria like a holistic outcome oriented competence assessment at the beginning and the end of the study as well as a long-term tracing of the alumni´s career course.

Based on the evaluation, aims and measures for the curriculum design are defined. Motives of students and the alumni´s qualification in the four competence domains were assessed.

Within the curriculum, teaching styles, content and organization have to be adjusted and balanced with offers and demands from the educational and the job market. The curriculum´s outcome will be measured by evaluating students and alumni. Deficiencies will be identified in order to define steps towards a curriculum which gives engineers bright careers while satisfying needs within a sustainable development.

Customer expectations

From the GPE study program students expect a broad variety of education contents, which requires a widespread curriculum. Expectations and the individual qualifications of the applicants in the four knowledge domains vary greatly in part. Creativity, curiosity, international diversity and career orientation are strong general motivations for the study. Within the curriculum these incentives should be encouraged by incorporating current research activities into teaching. An awareness of the potential for technology to assist sustainable development exists and should be encouraged within the curriculum. Through students' curiosity, creative research and global thinking can be utilized to develop technological potentials in entrepreneurial environments. Students become spearheads in development in the society, if this ability is trained purposefully within the competence domains.

To satisfy individual student´s motivation for breadth or depth of knowledge and intercultural and entrepreneurial competence a broad range of educational offers in the competence domains, with the possibility to design the curriculum individually by the student, is needed. In an orientation phase supported by the university students learn to individually select the courses in the curriculum and to act as managers of their own competency portfolio.

Employability

All alumni earned a position in the workplace after their graduation, which suggests the quality in general of the curriculum and the teaching success of the study program. The career of alumni reflects the current range of courses which focus on production technology and the teaching success within the curriculum. Based on the survey, there has not been a high rate of entrepreneurial activity but an assessment of the entrepreneurial qualifications of alumni is not possible at this time without further analysis.

The urge to explore the unknown and use creativity for innovation is supported in the GPE curriculum, and this is reflected in the high rate of alumni working in development and research. This research orientation especially encompasses the fields production technology, automotive, electronics and solar technology. To make technology fields with high potentials within a sustainable development accessible very much depends on the research and teaching orientation of institutes and its partners involved in the education.

Enhancing global responsibility is an important part of the future potentials in an educational oriented curriculum´s development. Even though today´s demand for sustainable orientation is comparatively low in education markets, it is expected to grow within the requirement profile for a study program, due to the fact that visionaries in society propagate the need for change.

6 REFERENCES

[1] Institut für Arbeits- und Berufsforschung: Kurzbericht. 2004, Nr.3.

[2] Guellec, D., Potterie, B.: R&D and Productivity growth; Panel data analysis of 16 OECD countries. OECD Economic Studies, No. 33, 2001/II.

[3] Bundesamt für Umwelt: Nachhaltige Entwicklung in Deutschland. Umwelt Bundesamt, 2002.

[4] Kreibich, R.: Zukunftsfragen und Zukunftswissenschaft. Arbeitsbericht No. 26, Institut für Zukunftsstudien und Technologiebewertung, Berlin, 2006.

[5] UAE Vision 2021 Brochure. http://www.vision2021.ae/downloads/UAE-Vision2021-Brochure-English.pdf, last call 29th Mar. 2011.

[6] 8th Global Conference on Sustainable Manufacturing. www.8th-gcsm-abudhabi.com, last call 29th Mar. 2011.

[7] Sustainable Products Cooperation. www.sustainableproducts.com, last call 29th Mar. 2011.

[8] Kirchler, E.: Arbeits- und Organisationspsychologie. P. 101, Wien, 2008.

[9] Seliger, G., Severengiz, S., Weinert, N.: Sustainable Industrial Value Creation Networks. In: Proceedings of the 15th CIRP International Conference on Life Cycle Engineering, The University of New South Wales, Sydney, Australia, 2008.

[10] Meyer, M. P.: Management von Ingenieurkompetenzen im Spannungsfeld beruflicher Arbeitsteilung. Fraunhofer IRB Verlag, Berlin, 2005.

[11] Lu, S. C.: A scientific foundation of collaborative engineering. In: CIRP Annals - Manufacturing Technology Vol. 56, Issue 2, pp. 605-634, 2007.

[12] Continental AG: In Search of Global Engineering Excellence: Educating the Next Generation of Engineers. Continental AG, Hannover, 2005.

[13] Verband der Elektrotechnik Elektronik Informationstechnik e.V.: VDE-Ingenieurstudie. VDE, Frankfurt, 2005.

[14] Wirtschaftswoche: Ingenieur - der Beruf der Zukunft. No. 48, 2005.

[15] Freeman; R. B.: A Cobweb Model of the Supply and Starting Salary of New Engineers. In: Industrial and Labor Relations Review, ILR Review, ILR School, Cornell University, Vol. 29(2), pp. 236-248, 1976.

[16] Van Weert, T. J.: Education and the Knowledge Society; Information Technology. Kluwer Academic Publishers / Springer Science + Business Media, Inc., New Yourk, 2005.

[17] Coy, W.: Weder vollständig noch widerspruchsfrei. FIfF-Kommunikation, Vol. 4, pp. 45-48, 2001.

[18] Paulsen, F.: Enzyklopädisches Handbuch der Pädagogik. Wien, 1903.

[19] Felder, R. M., Silverman, L. K.: Learning and Teaching Styles in Engineering Education. In: Engr. Education, Vol. 78, No. 7, pp. 674-681, 1988.

[20] Boud, D.; Feletti, G. I.: The Challenge of Problem-Based Learning. Routledge Chapman & Hall, London, 1997.

[21] Dewey, J.: Experience and Education. First published 1938, Touchstone, New York; 1997.

[22] Seliger, G., Meyer, M.: Sustainable Engineering Education. In: G. Seliger, N. Nasr and L. Alting: Proceedings - Global Conference on Sustainable Product Development and Life Cycle Engineering, Uni-edition, Berlin, 2004.

[23] Jonas, H.: Das Prinzip Verantwortung - Versuch einer Ethik für die technologische Zivilisation. Suhrkamp, Frankfurt am Main, 1979.

[24] Seliger, G., Franke, C.: Disassembly Factories for the Recovery of Resources from Product - and Material Cycles. Helsinki, 2002.

[25] Seliger, G.: Global Sustainability – a future Scenario. In: Environment and Science – Concepts and Strategic Goals for the Future, p. 27, 2005.

[26] Global Production Engineering. www.gpe.tu-berlin.de, last call 29th Mar. 2011.

Future Trends in Engineering Education and Research

W.H. El Maraghy
Department of Industrial and Manufacturing Systems Engineering, University of Windsor, Ontario, Canada

Abstract
Engineers play a key role in our societal development, contributing to and enabling initiatives that drive economic progress, enhance social and physical infrastructures, and inspire the changes that improve our quality of life. Simultaneously, industry and manufacturing are facing unprecedented challenges due to globalization and distributed manufacturing. As a result, the business environment of manufacturing enterprises is characterized by continuous change and increasing complexity. The challenges for companies arise not only from the need for flexible technical solutions, but also from managing complex socio-technical systems, and contribute tangibly to the sustainable development of manufacturing and the environment. Researchers and graduates with the ability to understand both complex technological processes and the creative arts and social skills are increasingly sought after in today's industrial and business world in areas of: Manufacturing Management, Health and Service Sectors, Product Engineering and Technical Sales, Transportation and Logistics. Using their strong technical and communication skills, engineering managers oversee a variety of team-based activities. By focusing on the critical role of engineering in solving our most complex global issues, we aspire to make the profession more attractive to both male and female students.

Keywords:
Engineering Education, Grand Challenges for Engineering, Sustainability, Socio-technical Systems

1 INTRODUCTION

The major global challenges we are facing today need to be addressed in the multifaceted context of economy, society, environment and technology. In recent years, the consensus of calling for sustainable development and implementation has emerged. Along with this belief, high added value, knowledge-based, competitive sustainable manufacturing has been widely considered as main enabler.

Engineering in general and engineering design and manufacturing in particular, affects virtually every aspect of our society and engages a substantial set of the population in carrying out engineers' plans and designs. But what is the role of engineering in responding to society's needs as well as in shaping them? This question is being asked with increasing urgency by a society that has benefited from great advances in technology, and at the same time, seen dislocations and experienced fears associated with technology. Often the questions about technology are confused with questions about engineering in the mind of the public despite a growing literature on the relation of technology to the rest of society. The list of impacts and side effects of technology is long and growing and has contributed to society's ambivalence about technology. While it would be wrong to blame the engineer for the apparent lack of interest by large portions of society in understanding the technological process with its constraints and possibilities, engineers can do much to reduce society's ambivalence if they could overcome their own parochialism.

The National Academy of Engineering has announced on 15 February 2008, the "Engineering Grand Challenges". These are: 1) Make Solar Energy Economical; 2) Provide Energy from Fusion; 3) Develop Carbon Sequestration Methods; 4) Manage the Nitrogen Cycle; 5) Provide Access to Clean Water; 6) Engineer Better Medicines; 7) Advance Health Informatics; 8) Secure Cyberspace; 9) Prevent Nuclear Terror; 10) Restore and Improve Urban Infrastructure; 11) Reverse Engineer the Brain; Enhance Virtual Reality; 12) Advance Personalized Learning; 13) Engineer the Tools of Scientific Discovery.

Some of these challenges are imperative for human survival, Some will make us more secure against natural and human threats, and all are intended to improve quality of life. They are of course all complex problems and mostly of a global scale. The other thing that they have in common is they are socio-technical complex systems.

2 ENGINEERING IN SOCIETY

Engineering is an integral part of society. Unfortunately some people, including engineers by training, regard engineering as simply Applied Science. What is needed is an education that emphasizes engineering and society, or better yet, engineering in society or "Engineering Arts", as opposed to the more traditional Engineering Science. In universities, it is something exotic and mysterious that goes on by itself in the Faculty of Engineering. Engineers, in general, have tended to focus on the development of new technologies rather than the social setting - government bureaucracies, school systems, and public service.

As engineering functions inseparably from the society of which it is a part, to operate within that reality, we need to comprehend better than we do what requirements and constraints are put on engineers by the rest of society and what role the engineer realistically can or should play in that society.

2.1 Socio-Technical Systems

The complexity of the interactions between society and engineering is at the root of unrealistic expectations about traditional engineering, as social entities are often inadequately organized to develop and use engineering effectively. It is also at the root of the frustration of engineers unable to bring their capabilities to bear on the solution of social problems or the effective organization of the engineering enterprise.

A more realistic possibility, which engineers should find congenial, is that has been termed the socio-technical system. As Engineers, and particularly Industrial and Manufacturing Systems Engineers, have to deal with systems—technical systems—all the time and are familiar with how they need to be designed, analysed and managed.

In the socio-technical model, the entire society is visualized as a vast integrated system, with the varied social and technical areas of human activity as major interacting subsystems. In this context engineering does appear as one of the subsystems. To analyze the subsystems they must be divided in turn into sub-subsystems and sub-modules and components; these must then be examined individually with an eye towards reassembly of the overall system. Systems engineers are comfortable with such a systems approach, also called System of Systems (SoS).

Engineers, unfortunately, have not had much experience in analyzing even adaptive technical systems; that limited art is only now at the early stages, although significant progress has been made lately within CIRP and other professional organizations by researching Emergent Synthesis, Engineering Design as Collaborative Negotiation [Lu, ElMaraghy, et al. 2007], etc.. .

The image of engineering as an adaptive socio-technical subsystem functioning within the adaptive socio-technical system of society presents an ever greater complex model to implement. It certainly comes closer to reality, however, than the model of engineering and society as distinct and separate entities.

2.2 Why a Socio-Technical System?

It is by now a truism to say that any single technology can be used in multiple, and sometimes unexpected, ways. But we need to add to this observation that, in each different use, the technology is embedded in a complex set of other technologies, physical surroundings, people, procedures, etc. that together make up the socio-technical system. It is only by understanding this system that we can parse out the environmental and societal and ethical issues and impacts. Many of the ethical issues are intimately related to the social and environmental systems. They are socio-technical systems, and the ethical issues associated with them are based in the particular combination of technology and social systems. It is the technology, embedded in the social systems that shapes the ethical issues. The dilemma is to balance society's rights with individual rights and freedoms.

2.3 Trends In The Socio-technical System

Great technologies have over the ages created social revolution. Note how technology made possible the industrial revolution, how the automobile has affected the sprawl of cities and suburbs. In this century the computer-inspired age of information and wireless communication has changed everything all over the globe. The global village seems more a reality than a tired Canadian cliché. We are increasingly dependent on computers as more and more information is coming on-line. When the industrial revolution made the mass production of standard goods possible, it also took away the consumers freedom of choice. Henry Ford, who was striving for utility, simplicity, and low cost, stated: "You can have any colour you want, as long as it is black". Computers in the information age promise to give us back individuality, through flexible, reconfigurable computerized manufacturing that allow large varieties of individualized products.

However, a fundamental residue of the information age is the increase in complexity — complexity of technological systems, of business systems, and of social systems. They seem to demonstrate a form of the second law of thermodynamics. Entropy is always on the rise. We see this particularly in large-scale systems.

Large-scale, interconnected systems include global distributed manufacturing, transportation, the environment and the earth's ecosystem, and the strategic defense and security systems. In his best-selling book Megatrends, John Naisbitt observed that the computer is a tool that manages complexity, and as such, just as highways encourage more cars, the computer invites more complexity into society. The question is whether the ability of computers to manage complexity and information and decision systems can keep up with the continuous increase in complexity. There is considerable hope, because the power of modelling, simulation and availability of supercomputers may be brought to bear on socio-technical problems, giving us new understanding and ability to manage our societal problems.

As the social and business systems have also been adapting to the information age, intellectual property has become an important branch of law, and has contributed its own ambiguities to an increasingly litigious society. The financial system has new problems of stability and control, as exemplified by program trading and the increasing volatility of the market. Moreover, the recent savings and loan crisis has shown the vulnerability of the banking system. The time constraints and turbulence in the economic system have also worked against the development of new science and technology, as business leaders have focused more and more on the short-term profitability rather than the long-term investment required for stable research.

3 SOCIETY AND EDUCATION OF ENGINEERS

To understand how engineering responds to the needs of society, we must examine its social structure and its function. Most people who study engineering in North America have higher physics, biology and mathematics skills and some communication and social ones. This appears to limit their involvement in politics and their success in communicating with the rest of society. Society, in turn, often views the engineer as a narrow, conservative, numbers-driven person, insensitive to subtle societal issues. The systematic study of socio-technical problems is rarely included in the engineering curricula as an important sphere of engineering activity. The curricula usually focus on man-made artifacts to the exclusion, except for specialized cases at the graduate studies level, of biological systems and organisms. This narrow focus has kept engineering away from not only a rich source of inspiration for specific technical feats and lessons offered by systems of great subtlety and complexity, but also a deeper understanding of environmental change.

Most high school students today do not view an engineering education as a path to success and prestige worthy of the sacrifices of a rigorous curriculum. Even bright young engineering students, upon graduation, switch to careers in business management, law, and medicine. On the other hand, engineering continues to be a powerful instrument for social mobility and advancement for immigrants and the poor. But is well recognized by most governments that in order for a country to prosper and compete globally, we need to graduate more scientists and engineering, as they contribute immeasurably to the nations wealth creation.

This situation accentuates the perceived social gap between engineers and other professions in society. In different societies engineering provides most of the same artifacts: shelter, energy and communications, manufacturing, water supply, extraction and use of resources, and disposal of waste. There are societies where engineers carry out broader functions by virtue of the position they hold. In several European and developing countries, they head state organizations and major industry conglomerates, participate in government, and enjoy high social prestige. By contrast, engineers in North America are absent from major positions of societal leadership, and only a handful serve in government, in Congress, in Parliaments, or at the cabinet level.

The profession is, in a sense, handicapped in terms of serving society in a broader way by a pecking order that prizes activities connected with the design of tangible products above the challenges of manufacturing, operations, and maintenance, or public service at large.

3.1 Social Needs And Responsibility

Man-made products, albeit often extensions of our body, have not generally evolved through the gradual process that has shaped man and other biological species. Thus, we constantly face the question of whether the technology we develop enhances the long-range survival of our species. It should be mentioned however, that there is increasing body of research that use biological evolution as a metaphor for developing products and systems, ElMaraghy [2008].

An important determinant of how well engineering satisfies its social purpose is the breadth of engineering. Engineering today continues overwhelmingly to focus on inanimate products or machines, as engineering school curricula worldwide continue to bypass socio-technological. This lopsided orientation grew out of obvious historical origins that have had major consequences for society. The factory environment so single-mindedly rationalized by the engineer F. W. Taylor overlooked the effective integration of the worker - the biological unit - and the machine in the production process. This is the case almost everywhere in the world, with the notable exception of Japan, where a different social ethos has produced a more effective integration with the human, as well as the artificial version, known as intelligent robots.

Another reason for the difficulty engineers encounter in dealing with social issues has to do with the various, and often conflicting, needs of social groups (educational, economic, environmental, health, public service, spiritual, and government) that engineering and technology may be expected to satisfy.

The recurrent conflict between advocates of independent and targeted research is an example and an inevitable result of the tension between short-and long-range needs. If pushed to the extreme, however, such conflicts may cross the boundary between what is socially useful and what is out of control. Most governments that fund research, including in Canada are on the brink with regards to this issue. There must be a balance between short- and long-term needs. They both serve a purpose. You cannot have Strategy without implementation and application. Similarly operating without a long term "discovery" research base can be disastrous in the long run.

The health care system for instance has absorbed an ever-greater portion of our gross national product, regardless of the state of our economic prosperity. At the same time, it has priced itself outside the financial reach of millions of north Americans. Technology has abetted the situation, not only by favouring the higher-cost, high-repair segment of the system, but also by not addressing the structure of the system. Similarly the problem of hunger remains endemic in many parts of the globe despite advances in agricultural technology. Even when production is high, in many countries grain supplies rot for lack of effective storage and distribution systems. It may be argued that engineers need to question their cultural responsibility to society as they contribute to its change. This effort must begin in the universities, in educating future engineers, our researchers, and in professional societies [Duderstadt, 2008], such as CIRP and SME [2008].

In the following paragraphs, Bugliarello [1991] offers five guiding principles, some of which are already deeply embedded in the conscience of engineers.

1) ***Uphold the dignity of man.*** This is a fundamental value of our society that never should be violated by an engineering design. This happens could happen when the design or operation of a technological product fails to recognize the importance of individuality, privacy, diversity, and aesthetics.

2) ***Avoid dangerous or uncontrolled side effects and by-products.*** This demands a rigorous development of a design or a technology considering all the functional requirements and constraints - be they political, economic, popular, or intrinsically technological.

3) ***Make provisions for consequence when technology fails.*** The importance of making provisions for the consequences of failure is self-evident, especially in those systems that are complex, pervasive, and place us at great risk if they fail.

4) ***Avoid buttressing social systems that perform poorly and should be replaced.*** This runs much against the grain of most engineers. Short -run technological fixes can put us at much greater risk in the long term. In the case of energy, for instance, technological or commercial fixes cannot mask the need to rethink globally the impact of consumerism and the interrelationship of energy, environment, and economic development.

5) ***Participate in formulating the "why" of technology.*** At present the engineering profession is poorly equipped to do so both in this country and elsewhere. Few engineers, for instance, have been involved in developing a philosophy of technology. This separation of engineering and philosophy affects our entire society. Engineers, in shaping our future, need to be guided by a clearer sense of the meaning and evolutionary role of technology.

The great social challenges we face require a rethinking of the human-artifact-society interrelationship and the options it offers us to carry out a growing number of social functions using quasi-intelligent products to instruct, manufacture, inspect, control, and so on.

4 ENGINEERING AND RESEARCH INTO THE FUTURE

The current generation of students is much more attuned to global issues and the need for new approaches than their predecessors. Duderstadt [2008] has discussed the future of engineering in detail. By focusing on the critical role of engineering in solving our most complex global issues, we aspire to make the profession more attractive to both male and female students, especially the latter. The new definition of engineering, by the Montreal Engineering Summit [Engineers Canada, 2009] will assist in this regard. The summit defined engineers as: "The enablers of dreams".

Engineers play a key role in our societal development, contributing to and enabling initiatives that drive economic progress, enhance social and physical infrastructures, and inspire the changes that improve our quality of life.

As a profession, we are committed to helping provide the best possible quality of life for all Canadians. "It is our aspiration that engineers will continue to be leaders in the movement towards use of wise, informed and economical, sustainable development. This should begin in our educational institutions and be founded in the basic tenets of the engineering profession and its actions. We also aspire to a future where engineers are prepared to adapt to changes in global forces and trends and to ethically assist the world in creating a balance in standard of living for developing and developed countries alike." [Engineering 20/20, NAE].

The following Montreal declaration [Engineers Canada, 2009] expresses the profession's resolve to help ensure Canada and its citizens thrive and prosper—today and into the future.

1. Deliver Canadian engineering innovation domestically and to the global community
2. Deliver specific engineering capabilities that will be needed in the future to improve health and safety, provide for a cleaner environment, and enable more sustainable development
3. Address areas in which advocacy by the engineering profession can lead to public policy development and directly contribute to Canadians' quality of life
4. Make educational enhancements that will encourage broader participation in the profession by all segments of the population and foster innovation

At a high level, we acknowledge that we must:

- Pursue greater collaboration across disciplines and professions
- Increase engineers' influence in policymaking
- Re-examine our accreditation process
- Transform engineering education and practice
- Encourage the greater participation of underrepresented groups such as Aboriginal Peoples
- Attract and retain women in much greater numbers

It appears that the Faculty of Engineering at McMaster University [2009] has also been paying close attention to the Sustainability and Globalization issues. They recently announced their 2009-2014 Strategic Plan, that includes very similar engineering educational objectives.

"Amongst the most important trends that will help define the future of the Faculty, we identified the following:

- Challenges in developing secure and sustainable forms of resources, including energy and water
- The need to develop more sustainable practices in all branches of engineering
- Increased opportunities for technology to improve human health
- Globalization and its impact on industrial supply chains, education, research and the human condition
- The challenge of demographics that will see an unprecedented wave of retirements in the western industrial world over the next decade."

The Guiding Principles and Core Professional values are also very similar to what was discussed in the proposed University of Windsor "Bachelor of Engineering Arts- Management Engineering" (BEA-ME) articulated as follows: To provide an innovative and stimulating learning environment where students can prepare themselves to excel in life; To achieve the next level in research results and reputation by building on existing and emerging areas of excellence; To build an inclusive community with a shared purpose; To be honest, mutually respectful, fair and inclusive; To foster a collegial, interdisciplinary and innovative work environment; To respect and reflect diversity in our opinions, our recruitment and the community we build; To conduct ourselves according to the highest standards of professionalism, acting ethically and with integrity; and to expect no less of our students; To instill in our students a love of learning ; To inspire our students to see themselves as global engineers; To be stewards of the environment and exercise social responsibility in our research and education.

Goals that were discussed include the following:

1. Develop a vertically and horizontally integrated curriculum following the "Versatilist T Model".
2. Develop clear and achievable learning objectives and student outcomes for each course
3. Offer high level team work and practical experience in all courses.
4. Enhance the multi-disciplinary educational experience of students as well as outreach opportunities
5. Increase the level of student engagement in all engineering programs
6. Increase Faculty involvement formal international exchange (e.g. Stanford University d-Lab in Design)
7. Develop a coordinated approach to Outreach Activities and international networking.
8. Increase our engagement in public policy and debates related to engineering and technology as well as their relation to socio-technical issues.
9. Develop a new undergraduate Minor in Management Engineering with the Faculty of Arts and Social Science.

5 SUMMARY

In the past 25 years, several major trends have emerged that magnify the social impact of engineering and the challenge to engineering to address pivotal social issues.

These trends are too well known and documented to be further underscored here: the sharpening of engineering prowess in the creation of products; the broadening of the social needs that engineering is called to address; geopolitical and economic shifts that are placing new demands on engineering; the coming to the fore of a series of issues of wide societal impact —such as the environment— that stem at least in part from engineering and technology themselves and demand urgent attention.

Engineering has contributed to this situation by its failure to emphasize manufacturing and production in formal engineering education and in the system of professional recognition. That emphasis is being developed, laboriously, only now and engineering has been slow also in responding to the immense challenges of globalization, and of the environment. The greatest challenge that globalization presents engineers and engineering education is how to increase throughout the world the rate of technological, economic, and social progress through the creation of new and more adaptable technologies and better socio-technical integration. Figure 1 illustrates the new paradigm for engineering education and research.

Furthermore, North American engineering has not participated to any major extent in the development of strategies for the reform of the health care and education systems as two key service activities that together absorb well over 15 percent of NA GNP. Engineering also has been absent from the attack on some of the most vexing problems of urban areas. Poverty, drugs, and alienation are all interconnected socio-technological problems of our cities, with their deteriorating infrastructure and the loss of easily accessible jobs in manufacturing.

This creates an added dilemma between societal values and individual values to simultaneously achieve technical excellence, manufacturing competitiveness and quality of life. To address these challenges of the future, the Industrial and Manufacturing Systems Engineering Department (IMSE) at the University of Windsor, Canada is proposing to introduce an innovative, the first in Canada, "Bachelor of Engineering Arts" (BAE) in the field of Management Engineering. Like Industrial Engineering, Management Engineering is concerned with the design, installation, operations and improvement of integrated systems consisting of equipment, materials, information and energy flows, and people. Unlike other engineering disciplines Management Engineering is concerned with the entire system, and especially the role people play in such systems. The educational approach is, therefore, more multi-disciplinary and includes the study of human factors involved in industrial operations, service industries, the health sector, and indeed any business, organization or government. The Structure and themes of the BEA-ME program are illustrated in Figures 1 and 2.

This paper was an opportunity to explore the connection between Engineering and other disciplines, such as the Arts, Sciences, Engineering, Humanities, Social Sciences and Business. The context and objectives from developing a new curriculum and delivery methods for the proposed BEA (Bachelor of Engineering Arts) program have been discussed, with particular emphasize on breadth across engineering subjects and exposure to technology management and communication, as well as the humanities and arts. Our vision is to offer a world-leading integrated, interdisciplinary undergraduate education for students interested in an educational experience that offers a rich mixture that balances technical subject with a deep understanding of the role of an engineer in addressing sustainability and the other grand challenges and key socio-technical issues affecting our global village.

The most obvious applications of the proposed BEA curriculum relates to sustainability and the development of eco-effective designs, human networks, processes and products. The sustainable processes and practices will provide greater applications in the goods and services businesses, as well as in health care, communications technologies, and public service. We aspire to graduate

Leaders who can develop engineering methodologies that can sustain societies anywhere, leading to the concept of a Global Renaissance Engineer.

This is a rare opportunity for educators, researchers and students from everywhere to discuss the merits of a strategic decision to integrate a technical education with the humanities, business, arts, etc.. The timing of this proposal fits well with the University of Windsor construction underway of the new "Centre for Engineering Innovation" (CEI). Such events, activities and projects, as well as the accompanying publicity, will hopefully help in the efforts to increase the breadth and depth of the education of engineers in the future.

6 REFERENCES

Bugliarello, George, 1991, "The Social Function of Engineering," in Engineering as a Social Enterprise (Washington: National Academy Press, 1991) 73- 88.

Duderstadt James J., 2008, Engineering for a Changing World:A Roadmap to the Future of Engineering Practice, Research, and Education. The Millennium Project, The University of Michigan; 131 pages; http://milproj.dc.umich.edu/

ElMaraghy, H., 2008, "Changing and Evolving Products and Systems – Models and Enablers",Changeable and Reconfigurable Manufacturing Systems, Springer-Verlag Publishers, ISBN: 978-1-

Engineers Canada; May 2009a "Leading a Canadian Future: The New Engineer In Society"; A declaration by Canada's engineering profession: The Montreal Declaration Issued at the National Engineering Summit May 21, Montréal, Québec

Lu, S.C-Y.; ElMaraghy, W; Schuh, G; and Wilhelm, R.,2007, "A Scientific Foundation of Collaborative Engineering" (Keynote Paper), CIRP Annals, Vol. 56, No. 2, pp. 605-364, 2007.

McMaster University; 2009 "Engineering a Sustainable Society"; Faculty of Engineering Strategic Plan 2009- 2014; 24 pages.

National Academy of Engineering, 2008, "The Grand Challenges for Engineering", National Academy of Sciences, on behalf of the National Academy of Engineering, 52 pages, http://www.engineeringchallenges.org/cms/challenges.aspx.

SME, 2008 "Manufacturing Education Leadership Conference", Society of Manufacturing Engineers (SME), Robert Morris University, Pittsburgh, PA, 26-27 June 2008.

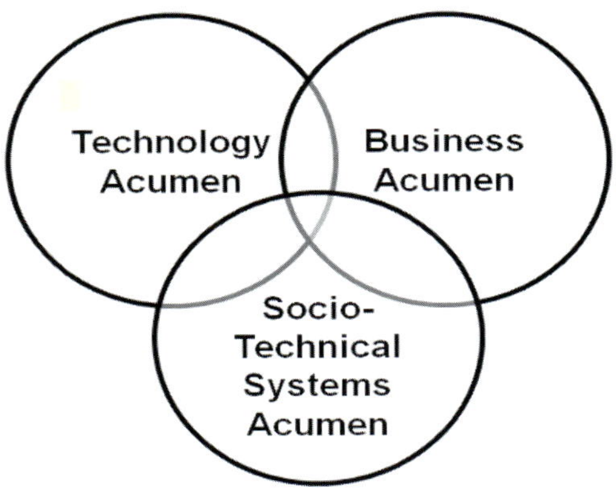

Figure 1 A New Paradigm for Engineering Education and Research

Figure 2 Structure of the BEA-ME Program

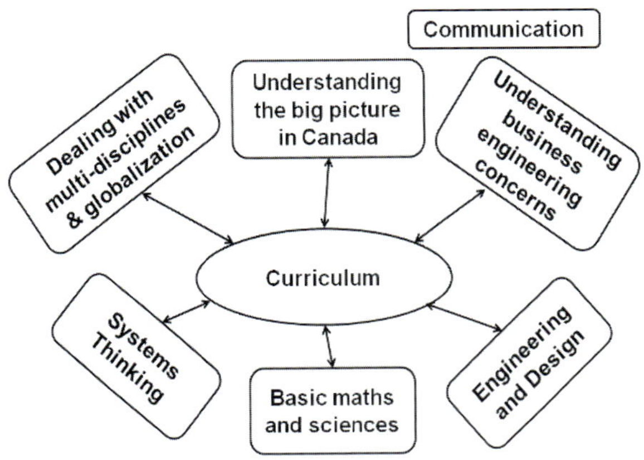

Figure 3 Main Themes for the BEA Curriculum

BIOGRAPHY:

Prof. Waguih ElMaraghy, Ph.D., P.Eng., FCIRP, FASME, FCSME, FEC Professor & Head, Dept. of Industrial and Manufacturing Systems Engineering (IMSE), Director, Intelligent Manufacturing Systems Centre (IMS), Faculty of Engineering, University of Windsor, Windsor, Ontario, Canada

Dr. ElMaraghy research interest are in Engineering Design and Manufacturing, Management of Complexity in Engineering, Industrial and Systems Engineering, Management Engineering, and Global Business Dynamics.

Sustainability in Education

A. Weckenmann, T. Werner
Department of Mechanical Engineering, Friedrich-Alexander-University Erlangen-Nuremberg, Germany

Abstract

Due to the rapid development of methods and technologies in many fields, especially regarding subjects of sustainable engineering, vocational qualification is not finished with basic education. Measures of advanced training have to be chosen adapted to the individual demand, in order to allow for an efficient achievement of required additional competences. But for most interested persons, it is very difficult to find courses suitable for their specific learning aims as well as for their individual state of existing competences. Based on the analysis of possible methods of training, an assistance system is being developed which enables the definition of a training schedule optimized toward the individual demand. For the topical example of manufacturing metrology, a competence net has been defined. Concept and method of the system can easily be adapted to other topics based on the proposed methodology. To prepare the proper application of Life Long Learning, adequate constraints have been established, regarding already basic education.

Keywords:

Advanced vocational training, Assistance system, Competence management, Life Long Learning

1 INTRODUCTION

The aim of vocational education is to enable students to participate successfully in professional life and thus contribute to positive outcome of the employing organisations as well as to sustainable development of society and welfare. The performance of an individual depends on the one hand on his ability to execute his tasks, described by skills and competences, on the other hand on his readiness and willingness. Thus, vocational education has to be oriented towards the establishing of these two conditions with the learners to achieve sustainability regarding their formation.

In a traditional view, it is assumed that employees gather the competences required for the execution of their tasks during education in school and a following specific vocational training for the intended profession in higher education. During vocational life, it is desired that this basic qualification is amended by broadening experiences in the chosen field of work, resulting in a deepened understanding and a honing of necessary skills towards higher efficiency and finally mastery or expertise of the specific area. Yet, nowadays this concept of improvement by growing experiences based upon a once given fundamental qualification is not any more applicable for most areas of work. Rather, especially in areas with strong relation to technology, there is such rapid development of methods and tools that knowledge once gathered during basic vocational training may be obsolete after only a few years. Also, there is a need for all employees to gather additional competences in order to apply newly developed methods correctly in their tasks.

This problem becomes especially evident in techniques and standards related to sustainable engineering or topics of environmental laws and responsibility, as until only few years ago these issues have not been considered on a broad base. Thus, many employees nowadays working in engineering have not been in any contact whatsoever with these issues during their preceding vocational education, but need sufficient competences now for the execution of their tasks in a changed surrounding, considering sustainability with ever increasing importance. But also in a more general view, sustainable engineering requires the application of newly developed methods and technologies in order to achieve an increasingly benign manufacturing, starting with the according development of products [1] [2]. E.g. the reduction of consumption of raw materials by a more efficient manufacturing, or the development and production of a new generation of engines with a more economical use of fuel would have been impossible without a continuous utilization of innovative technologies.

Resulting from this situation, as an indispensable precondition to enable employees to contribute to product development and manufacturing considering the requirements of sustainability, the focus of imparting required vocational competences has to be broadened from a concentration on basic training towards the concept of Life Long Learning and measures of continuous qualification throughout the whole professional life of employees – a change in perspective necessary to strengthen competences gathered during basic and higher vocational education and to enhance the achievement of additionally required competences for sustainable engineering. This demands both adequate offers of further education and the imparting of personal competences required to use these offers.

Also, the sustainability of qualification achieved by educational measures themselves has to be considered, i.e. it is necessary to enhance the long term success of training which will show up in the ability of an employee to transfer the achieved competences to his work and following maintain and enlarge them by continuous learning. If this aspect is neglected, a measure of training will have little effect on the actual competence of the employee. [3]

To support employees and employers in the identification of needs for additional training and the scheduling of an adequate training program assuring a sustainable learning effect, an assistance system is being developed.

2 REQUIREMENTS ON FURTHER EDUCATION

The adequate continuous qualification of all employees has to be considered as a core task in nowadays knowledge based economy, especially in order to qualify employees to contribute to a rapidly developing field such as sustainable engineering. Nevertheless, economical requirements have to be considered as well, regarding costs and required time.

Therefore, each measure of advanced vocational training has to fulfil requirements regarding on the one hand the achievement of certain learning objectives, on the other hand economical aspects as well as operational conditions of industrial working. It is necessary to assure via the chosen method of qualification, that the operator has sufficient knowledge to perform his tasks correctly, and that he is able to apply this knowledge competently on a specific task, considering also the sustainability of the learning effect to assure a well funded base for a further development of competences. Regarding time and cost efficiency, the chosen method has to provide enough flexibility for the learner to enable compatibility with other duties but also adaptation to existing competencies and specific professional experiences or concerns.

In order to enable an efficient and sustainable impartment of competences, the provided measures of qualification have to be of high quality, i.e. they have to fulfil the requirements of customers on the qualification. To express the quality of an educational activity, on the one hand the impact on competences of the participants has to be regarded. They shall on short term be able to apply their newly gathered skills efficiently to their actual tasks, on long term maintain and deepen their knowledge and thus improve the quality of their work. On the other hand, the satisfaction of learners has to be considered which depends not only on the outcome but also on the proceeding of a course regarding contents, implemented methods of teaching and learning and general circumstances. [4]

Yet, the required content and method of any measure of qualification depends strongly on individual constraints. Each learner needs to achieve the competences required for his very specific task, starting from his already existing knowledge. To achieve a high efficiency of qualification, the choice of methods and courses has to be optimally adapted to the individual demands of each concerned employee to close the gaps between desired and available qualification. Here, both the determination of needed competences and of actually available competences as well as the adequate planning of training forms major challenges:

Regarding the *determination of needed competences*, it is often difficult to define the desired state. For this, a good overview of the subject at hand and the intended tasks therein is needed. It can be very difficult for the learner or his superior to identify existing gaps of knowledge and define adequate learning targets – all the more so, if there is only small basic knowledge about the considered area to start from. This situation results in a self-selectivity of employees asking for further training: Only those quite sure of their needs and their standing, being basically well-educated, will ask for additional courses, whereas those with real gaps will rather hide them [5].

Regarding the *estimation of existing knowledge and skills* of an employee, the most common and easy method is to judge based on certificates of previous education. Yet, once gathered knowledge of higher education is lost or obsolete if it has not been used for some time. On the other hand, a person fit for self-driven learning may have achieved a variety of additional competences during professional life which are not certified. Therefore, even if the aspect of insufficient correlation between high marks on a topic and high working performance is neglected, existing competencies of an employee can not at all be judged satisfyingly based on certificates of formal education.

Based on the individual demand, that is derived from the comparison of existing knowledge and required competences, it is necessary to *choose adequate measures of qualification* to fill the identified gap. In order to enable proper fitting, it is possible either to create a personalised training or, to combine ready-made training offers in a modularised way to form a training concept best fitting the actual demand. With individually designed courses, the comparability of gathered competences with common standards and among different employees is reduced considerably and also the arising costs usually are quite high. Yet, regarding the election of standardised training offers to define an individually optimal training concept, it is very difficult for the prospective learner to identify adequate courses and figure out their best combination.

Thus, it is necessary to support the identification of existing need for further education resulting from the gap between current and required competences as well as the according selection of appropriate training offers.

3 METHOD ANALYSIS IN COMPETENCE MANAGEMENT

In order to find suitable possibilities for closing existing gaps in the competence profile of employees and establish adapt measures of advanced qualification, it is necessary to identify the method of competence management most suitable for the given demand. Available methods of competence management have to be classified on the one hand regarding the duration of effect of improved matching of skills and task requirements. On the other hand, the way of implementing the actual measures within operational procedures of working has to be regarded.

Under consideration of these two-dimensions for basic classification of available methods of competence management, the most effective way to set up advanced training in nowadays industrial conditions can be chosen, in order to assure sustainable education of employees for the continuous adaptation to rapidly developing technology-related topics, such as environmental conditions and concepts of sustainable engineering.

3.1 Approaches to competence management

Considering tasks where the required competences exceed the level of skills and knowledge of typical employees charged with the execution, there are two main possibilities to enhance the proper and efficient execution: On short term, the level of demand can be lowered to meet the existing competences, e.g. by partial automation or user guidance; on long term, competences of the employees can be amended by adequate training.

Short-term improvement via assistance facilities

For an immediate improvement of the match between low competences and high requirements, full or part automation of the process to be executed or a guiding through difficult steps can be used, e.g. by so called wizards or other assistance facilities. Thus, the necessary competence level can be significantly lowered. But enabling a performance of complex, knowledge-intensive tasks without qualifying the employee for the underlying procedures and concepts provides only a short-term benefit as it is likely that the employee will not be able to execute the work correctly if even a slight deviation from the usual routine occurs which is not covered by the assistance facility [6]. Even more, there is the risk that an automated process will be applied to a task where it is not appropriate.

Also, a basic understanding of the matter at hand is a necessary requirement to be able to learn self-directed and thus improve performance and transferability of gained skills with increasing experience [7], by critically reviewing one's way to execute tasks and its conjunction to eventually arising problems, by exchange with colleagues or other practitioners or by using other random, unplanned opportunities for informal learning, occurring in every working environment, e.g. maintenance of machines by an external expert or visits to trade fairs.

Still, automation and assistance facilities provide a valuable option to increase efficiency and reduce cost if they are used as an amendment, helping to avoid common mistakes. But to sustainably assure accurate, reliable results and enable a process of continuous self-improvement, it is necessary to provide a solid base for the proper execution of tasks by sufficiently educating the employees [8].

Long-term improvement via qualificational measures

A training adapted to the learning requirements will result in well-funded knowledge as a base for the diligent and well reasoned execution of tasks, thus influencing significantly positive and sustainably the quality of gathered results [9]. But sending an employee to qualification is often time consuming and requires resources. Also, depending on the starting point, it may take some time before the necessary level is reached.

Thus, short term and long term approaches to competence management have to be combined in a suitable way to enable a time and cost efficient, but also sustainable adaptation of employee competences and demands of a given task. Yet, what competences an employee has to gather in addition to existing education, varies strongly with tasks and also with previous education. Thus, the specific demand of a learner will rarely be met by a ready-made training concept as it is adapted to the requirements of an abstract user group described by typical constraints and average abilities. In order to qualify the employees in a way that is both economically sound and sufficient in regard of learning effect, a selection and combination of courses in an individual, demand-oriented training concept is required.

3.2 Categories of qualificational measures

Two main categories of measures to impart the necessary knowledge may be distinguished: Preparative measures of qualification that impart the necessary knowledge in a general way detached from a specific task, and integrative measures of qualification that provide needed information on demand close to its application. [10]

Preparative measures of qualification

By attending adequate courses, as the classical method of preparative qualification, it is possible to achieve comprehensive knowledge regarding both declarative knowledge about fundamentals and procedural knowledge about the handling of machines, devices or software. Preparative measures of qualification exist in a large variety, differing in the used methodology, e.g. face-to-face seminaries, eLearning or combinations, as well as the imparted knowledge.

Yet, to transfer the gathered general information to a special task is often experienced as very difficult and is therefore prone to highly subjective or even erroneous interpretation. This situation can be eased to a certain degree by including into the education problem-based learning with practical tasks. But still the adaptability of knowledge is to be considered a critical aspect. Also, the learner is usually not supported in the maintenance of his competencies after having finished the course. Information often has to be gathered by self conducted inquiries in standards, books or internet. Of course this does not always lead to a satisfactory understanding of complicated subjects. So problems are often caused by misunderstanding or sheer lack of knowledge. Hence, the impact of preparative learning offers on the improvement of labour quality has to be considered as unsatisfying.

Integrative measures of qualification

The most common method of integrative qualification is the traditional on-the-job training by an expert, e.g. a foreman or a senior member of staff. By this method, the learner will have optimal support adapted to his needs and the actual context of work – provided the mentor has enough knowledge and experience regarding both the field of application and didactical aspects. If the expert himself has deficient or obsolescent knowledge, the training cannot lead to the desired qualification. Also, during the period of qualification, the expert will not be available to perform his own tasks.

To ease these problems, holistic information or assistance system may be used instead of experts, such as guidelines or context-sensitive assistance- and information-systems. Here, the operators eventually have to get used to the handling of the system first before using it for learning. For the execution of a task without a profound preknowledge, it would be necessary for the operator to look up a lot of information for each step of the working sequence. This would be very time consuming and also frustrating as accordingly the progress of work will be very slowly.

The knowledge gathered via integrative measures of learning is adapted very well to the specific problem at hand and therefore immediately contributes to a correct performance of the task. Yet, during the phase of basic introduction to the topic, the productivity is decreased considerably due to the efforts necessary to impart the required knowledge beforehand (Fig. 1).

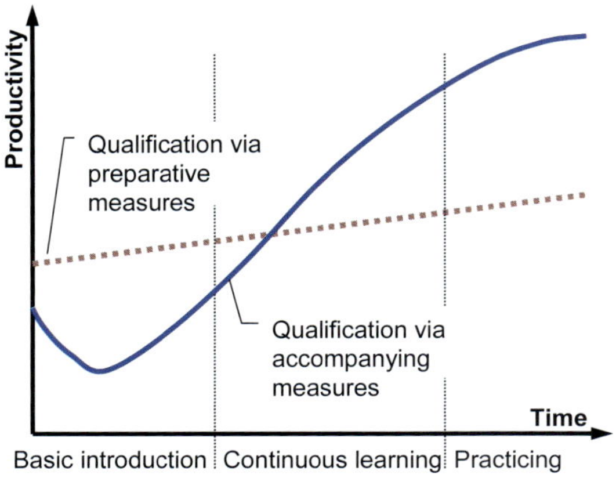

Fig. 1: Progress of productivity with preparative and integrative measures [6]

Thus, pressure of time and cost efficiency in production prohibits a suitable gathering of background information, but enhances a sharp focus on the actually needed information. Therefore it is not recommendable to use integrative measures as a single source of qualification, as this would lead to a merely fragmentary understanding of the area. On long term, this would not enable the operator to solve other tasks on his own.

Taken as a sole source of qualification, neither preparative nor integrative measures provide the required comprehensive, efficient and continuous imparting of the demanded new competences. Therefore, it is necessary to blend both in a holistic concept with a strong view on adaptation to the operator's needs and requirements. In order to augment the respective advantages and extinguish the disadvantages, methods have to be chosen considerately out of the existing offers and combined in a most efficient way.

4 CONCEPT OF AN ASSISTANCE SYSTEM

To support the identification of current needs for further qualification and enable the demand-oriented definition of a schedule with training offers, adapted to the individual requirements of the employee and employer, an assistance tool is being developed which accompanies the necessary steps to define an individualized training concept oriented on the specific demand [11].

4.1 Methodology of training scheduling

Regarding the quality and adequateness of advanced vocational training, there is usually a strict focus on the actual participation in training, by providers of qualification regarding development and quality assurance processes as well as by the customers regarding internal controlling and assessment of training offer. Yet, the sustainable success of a measure of qualification, which finally determines its quality perceived by the participant as customer, does not only depend on the training itself, has to consider also the impact on the actual ability of the employee to apply the newly gathered knowledge in his tasks. Thus, the quality of a training offer has to be assessed also by the change in behaviour achieved by the participant, if sustainability of the learning effect is considered. Only a noticeable effect in performance will enable the desired use of the training, i.e. to qualify the employee for new requirements, e.g. to be able to contribute to benign, environmental-friendly and sustainable product development and manufacturing. To achieve this aim, besides the training itself, also preceding phases of planning and selection of appropriate training offers as well as the subsequent transfer of newly gained knowledge and skills to the performance of actual tasks have to be considered. Regarding advanced vocational qualification, this process has to be envisioned as a staff development cycle rather than as a unique event, considering also the never ending need for continuous improvement. To describe the phases of this development cycle, different models are used [12][13]. Yet, they can be summarised in a seven step model:

1. Analysis of required competences
2. Assessment of available competences
3. Identification of demand for qualification
4. Election of adept training offers
5. Participation in training
6. Utilisation of gained knowledge in professional tasks
7. Assessment of training success

Based on the results of the assessment as well as driven by new requirements, the cycle will start all over again.

In order to develop an assistance tool supporting the planning of qualification with a focus on a sustainable effect of education on increased performance, the required functionalities have to be defined under consideration of the whole cycle of staff development. Thus, based on the seven stepped model, for each phase it has to be analysed, in which way a person trying to find suitable measures of qualification could be supported (Fig. 2). [11][14]

During the first two steps, analyses of competences have to be executed. On the one hand, the target profile of competences has to be identified, focusing on knowledge and skills required to fulfil the given or future tasks correctly and efficiently, but considering also demand of future development. On the other hand, the current competence profile of a specific person intended to perform this task has to be measured. For this, it is necessary to provide a possibility for the comparable description of the intended or available competences, adapted to the considered area of tasks. For both steps, the same scheme for assessment has to be used in order to provide comparability between the two recorded profiles. For the following identification of the actual demand for qualification, targeted and current competence profile have to be compared and occurring deficits have to be highlighted. Depending on the size of the identified gaps, a recommendation has to be provided, if additional, preparative training is necessary or if identified deficits are likely to be easily smoothed out by ways of informal learning or can be more efficiently filled by integrative measures of qualification.

In order to schedule chosen learning offers adequately for the specific situation of the intended learner, besides the content of training specified by the identified deviations of competence profiles also other constraints have to be considered, e.g. requirements regarding disposable time, acceptable costs, formal specifications. Based on this comprehensive analysis of user requirements, possible qualification offers can be identified that fit the identified gaps in competences as well as the learner's constraints. Considering the available time and infrastructure it has to be checked to which extent preparative or integrative methods may be used. Out of the identified possibilities, qualification concepts have to be defined and evaluated according to their fulfilment of the requirements and one or several suitable plans adapted to the individual demands are derived. As a prerequisite for this recommendation, a data base of generally available training offers for the considered area is required, where contents and circumstances are described.

For the actual participation in the qualification program, no support from the tool is required. Yet, afterwards a possibility for assessing the training should be offered. The gathered results can be included in the data base and thus enable a fine-tuning of recommendations. Additionally, the transfer of newly gained competences to the performance of actual tasks has to be supported by suggesting suitable methods to facilitate adaptation of general contents to the specific requirements and enhance activities of knowledge sharing with co-workers. Thus, the effect of training is increased for the actual participant as well as for the whole surrounding.

Finally, after a certain time to assure an unbiased assessment, the now available competences should be measured and compared with the target profile. Here again the same criteria are used as during the first assessment of competences and the definition of target competences. Based on the results, the cycle eventually can be started all over again.

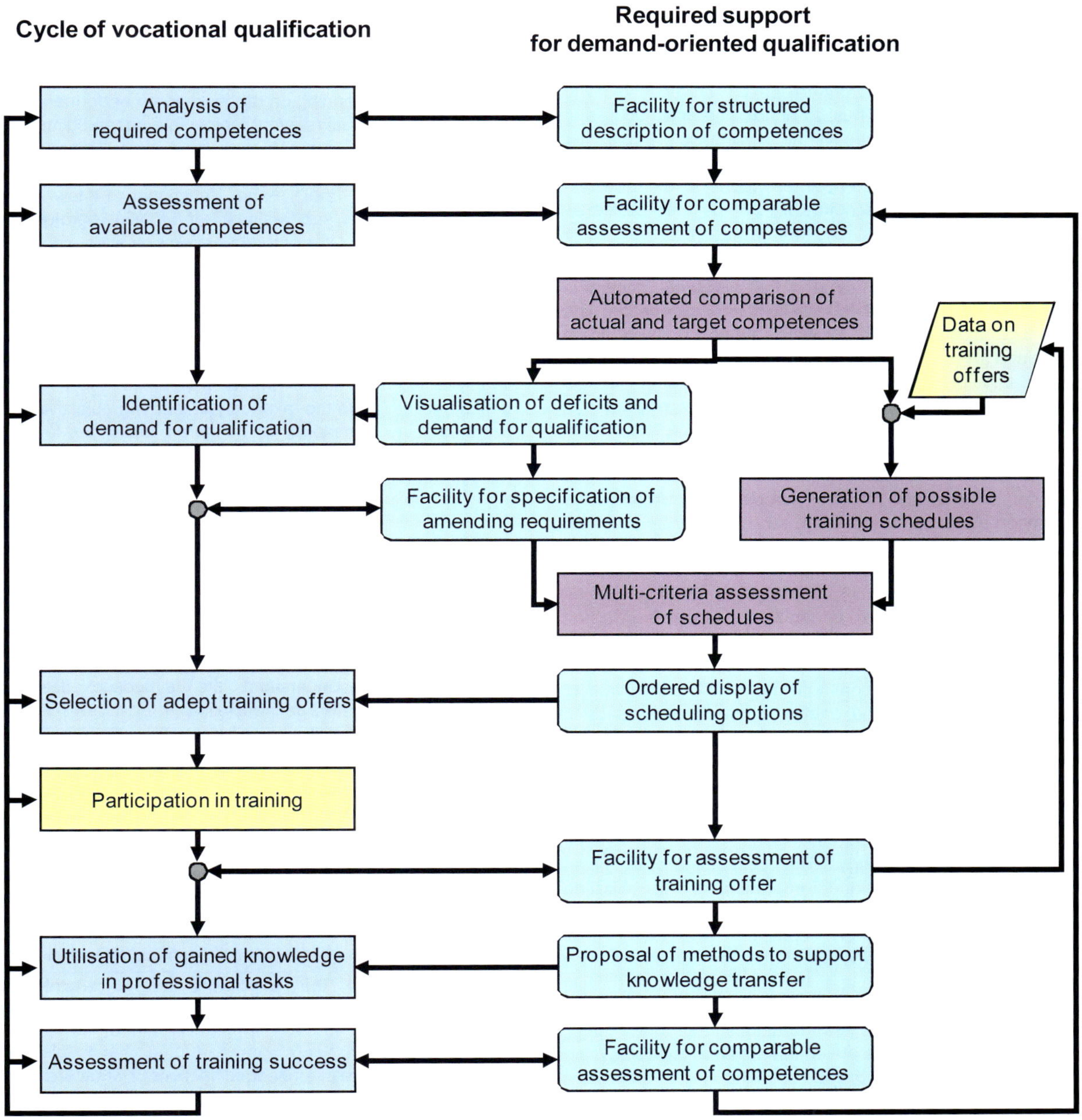

Fig. 2: Functionalities of the assistance system to support the demand-oriented scheduling of continuous training [11]

4.2 Definition of a taxonomical competence net for the Application to Manufacturing Metrology

To actually implement the functionalities of the assistance system, the most crucial step is the definition of a method to express available respectively required competences in the considered subject in a way, that enables a reproducible and comparable description by the user of the system, i.e. by an interested learner or an employer, as well as the processing of the gathered data by the computer.

The required automated generation of training concepts supporting individually adapted learning paths is efficiently enabled by descriptions of training courses focused on the learning outcome [22]. Such a definition of learning objectives, i.e. of intended learning outcomes, is also suitable for the description of the current and desired state of competences of an employee as well as for the comparison of these two profiles in order to identify currently existing needs for improvement. Thus, to enable the description and processing of competences in the assistance system, the concept on a taxonomical competence net was chosen, featuring specified competences as nodes in the network, that can be used as descriptors for competence profile as well as training contents. By the network structure, also interrelations between the single nodes can be represented.

The developed overall structure of the assistance system is completely independent of its specific field of application. The actual topic of qualification, for which support is delivered, is determined by the kind of competences measured for the profiles and the according training offers stored in the database. Considering the area of sustainable engineering, continuous qualification of the employees may be required for many different areas of knowledge, ranging from product development via production technologies to quality control, customer service and product maintenance. Additionally, information about non-engineering subjects may be required, such as laws, environmental policies or society-related topics. Yet, usually it is known in which general area a qualification is required, whereas the identification of proper training facilities within this area forms a major problem. Thus, it is in a first step sufficient to develop competence networks for specific areas of knowledge. For further development, the inclusion of separated networks in a common tool can be considered to increase user comfort and enable a holistic support by a monolithic tool.

As a first example to develop such a taxonomical competence network, the area of manufacturing metrology has been chosen. The demand for a support for the selection of training offers is – among other fields – especially evident in manufacturing metrology. The development of novel measurement devices and improvement of existing techniques is very fast due to the need to provide reliable information on increasingly complex workpieces with increasingly narrow specifications, but at the same time most metrologists do not have a specific vocational training for this field. Resulting from this situation, many activities have been carried out to provide an adaptation of low basic competences of workers in manufacturing metrology on complex measurement tasks, covering both short-term support (e.g. [15],[16]) and long-term qualification (e.g. [17][18], among many others) that are usually designed as preparative measures as well as tools for integrative qualification [19].

In metrology, the operator is held responsible for the delivered measurement result. If an operator has no understanding and control of the performed procedures, the gathered measurement results are not reliable. But due to a scarce basic education, gaps of qualification are hardly noticed. At the same time, superiors often have even less knowledge about this area of work, so that it is virtually impossible for them, to identify existing need of qualification among their employees and choose suitable training. Therefore, manufacturing metrology is considered as a somewhat typical discipline with a high need for adequate support during the scheduling of advanced training and thus is chosen as the first area of application for the assistance system.

A comprehensive competence analysis for a special field of application, such as manufacturing metrology, has to consider in one dimension the various kinds of required competences, depending on the tasks to be performed and in another dimension all relevant topics. In a second dimension, different modes of competences have to be assessed. In context of advanced vocational training, the focus is set on cognitive ("know that") and functional ("know how") competences as subdivisions of topical professional competences [20]. Other modes of competences, social and personal, may be neglected in this specific purpose, as their development – which is necessary for a high overall performance of the employee – cannot be effectively provided via topical training but needs to be included in a more generalised learning culture in the company. Within the cognitive learning aims, various sub-competences can be defined according to the taxonomy of learning objectives [21]. Regarding the complexity of mastered tasks, six levels of learning objectives are distinguished, where each level is relaying on competences of the previous levels (table 1).

Table 1: Exemplary taxonomy of learning objectives

Level of complexity	*Example for the topic "Measurement Uncertainty"*
Knowledge	Know that the GUM provides a guideline for expressing MU
Comprehension	Understand the difference between MU and resolution
Application	Assess the conformity of a measurement result for a given specification under consideration of MU
Analysis	Identify main contributors to MU based on a given budget
Synthesis	Define approaches to reduce an unacceptably high MU
Evaluation	Assess plausibility of a given value of MU for a specified measurement task

GUM: Guide to the Expression of Uncertainty in Measurement, MU: Measurement uncertainty

For performing tasks embedded in overall structures on a mainly operational level, usually competences of application are required. With an increasing degree of freedom and self-responsibility, the required level of competences will rise. To fully master this level, it is necessary that the lower levels are also covered. Otherwise, it will not be possible for the learner to pass on to higher levels of understanding. Therefore, the relevant learning objectives are identified and made transparent in the system together with the learning objectives of levels below. The defined analytical items are also used to describe the learning aims of a specific course in the data base.

This concept also is applied to interdependencies to other topics, e.g. for the performing of measurement tasks in form testing, competences are required in the field of geometrical tolerances to enable the correct interpretation of a given drawing and the specification within. By describing these interrelations between the respective issues and levels of learning objectives, the intended taxonomical competence network is built.

5 PREREQUESITES IN BASIC EDUCATION FOR SUCCESFUL FURTHER QUALIFICATION

Life Long Learning is an essential requirement for the successful participation in vocational life, especially in an area of occupation as fast evolving as sustainable engineering with its demand on the employees to stay up-to-date in the development of their respective area of work. Therefore, the ability and readiness to participate in measures of advanced vocational training of various kinds has to be regarded as a key competence and important factor of employability to start with and of the long-term ability of an employee to contribute to the enhancement of sustainability and benign methods in manufacturing. Most offers of advanced training do not consist of conventional face-to-face training any more, but feature didactical approaches like accompanying or integrated training, that support self-directed learning, as well as new technologies to enable open distance learning.

Yet, compulsory education in universities consists mainly or often exclusively of traditional forms of learning, such as lectures, exercises and limited possibilities of hands-on-training, whereas learning offers featuring new media or approaches are used only for the establishment of additional courses. Thus, the majority of students – aiming at gaining the intended degree with maximum efficiency – leave university with hardly any experiences on learning via new media or self-directed learning. Nevertheless, it is commonly assumed that young engineers can use electronic or web-based offers without any problems for further qualification, due to their experience in the application of computer and internet in general. Also, it is expected that graduates will be able to initiate, organise and control a learning process on their own, as they are responsible for their success during their course of studies as well. But so far, these expectations are still to be proven.

The successful participation in Life Long Learning and thus the sustainment of value for the labour market require ability and motivation to use offered facilities for learning and self-improvement. Virtual courses provide some specific advantages and disadvantages compared to conventional face-to-face teaching to the participant, resulting from the higher degree of freedom. But this choice of possibilities causes also a high self-responsibility for the learning progress and thus demands sufficient organisational and self-motivational skills of the learner. But so far, during university education most students do not experience any courses based on self-controlled learning.

Usually, participants rate the high degree of freedom very positive and thus approve of the provided courses, as observed by the HISBUS study, which was conducted to measure the view of students in German universities on eLearning [22]. But as virtual courses mainly establish an additional offer – amending but not replacing conventional lectures or seminaries – it has to be considered that the positive results are biased by the fact that only those students participate, who are exceptionally motivated and comply with the requirements of a successful participation better than the average student.

This assumption fits well with the observed quite low quote of acceptance and usage of purely virtual lectures, courses or seminaries. The ratio of acceptance describes the percentage of students that use an offer out of those knowing about the offer, and amounts to about 36-48%, depending on the kind of offer and the course of studies. The ratio of participation describes the percentage of students that use an offer out of all possible users, and amounts to about 4-8%. [22]

The data of the HISBUS study suggests strongly that, although it is commonly assumed that graduates are fit to use eLearning based offers of vocational training without problems, most of them do not have any experience with this. By studying the reactions of students obliged to join a virtual course as a compulsory subject, it was gathered, that although all students had been able to participate and rated the course itself quite positively, they experienced severe difficulties to adjust to self-controlled learning [23]. These experiences show that it is necessary to introduce all students to new forms of learning, in order to prepare them for a vocational life requiring Life Long Learning by an increasing number of self-directed, computer-based offers. Therefore it would be useful to emphasize the inclusion of eLearning-based offers in the regular course of study for all engineering students.

To enhance a successful participation in such unfamiliar courses, the students have to be thoroughly prepared and informed. Also, eLearning should not be regarded as a possibility to decrease academic staff, as the opportunity to contact a tutor when required has to be considered as a key component of success for all learning offers.

Conclusively, young engineers have to be educated towards the idea of Life Long Learning, in order to be later on able to continuously maintain their competences and master new technologies, as a fundamental requirement to contribute to sustainable engineering and to an increasingly benign and environmentally sound technical development. It is strongly desirable, that they achieve the required basic competences for this during their education, as currently it can't be assumed that young engineers will be able to participate easily in required measures of advanced training.

6 CONCLUSION AND OUTLOOK

In nowadays knowledge based economy, employees' capability and willingness to perform make up a main influence on success in vocational life. Especially in the area of sustainable engineering, where the application of newly developed technologies and innovative methods in product development and manufacturing is required, the knowledge and skills of an employee have to be adequately maintained and constantly increased, to assure his capability to fulfil the given tasks. For this, an active participation of employees in

Life Long Learning for vocational qualification is necessary. Thus, it can be assured that competences gathered in basic and higher education will be sustainably maintained and amended according to new requirements.

To enhance a proper adjustment to the individual requirements and target competences of the specific learner, a concept for an assistance system has been developed which provides holistic support during the whole staff development cycle. As an exemplary area of application, a taxonomical competence network has been developed for manufacturing metrology as a base for the automated support by the assistance tool. Additionally to these ongoing activities related to employees already in vocational life, it is necessary to prepare students for Life Long Learning during higher education by introducing them to modern learning methods. Thus, competences gathered in higher education later on can be strengthened and amended by adequate further qualification.

By enhancing a process of continuous learning among the employees, the company then can provide a well-funded basis for sustainable development towards the cooperative achievement of high quality products and enable it employees to contribute to a more and more sustainably sound technology development and production, including the whole product life cycle.

7 REFERENCES

[1] Seliger, G., 2009, Sustainable value creation nets – Global Cooperation in Research and Education, In: Shunmugam, M.S., Ramesh Babu, N. (eds.), Proceedings of the 7th Global Conference on Sustainable Manufacturing, , Chennai, IIT Madras, pp. 7-13.

[2] Sakao, T., Lindahl, M., 2009, Introduction to product / service-system design, London, Springer.

[3] Kirkpatrick, D.L., Kirkpatrick, J.D., 2006, Evaluating training programs – the four levels, San Francisco, Berrett-Koehler, 3rd ed.

[4] Weckenmann, A., Werner, T., 2009, Enhancing the quality of advanced vocational qualification by computer-assisted generation of individual training concepts, In: S. Brdarevic (ed.), 6th Research/Expert Conference with international participation Quality 2009 Proceedings, University of Zenica, pp. 685-690.

[5] Gerstenmaier, J., Mandl, H., 2009, Konstruktivistische Ansätze in der Erwachsenenbildung und Weiterbildung, in: Tippelt, R., von Hippel, A. (eds.), Handbuch Erwachsenenbildung/Weiterbildung, Wiesbaden, VS, 169-178.

[6] Severing, E., 1994, Arbeitsplatznahe Weiterbildung – Betriebspädagogische Konzepte und betriebliche Umsetzungsstrategien, Neuwied, Herrmann Luchterhand.

[7] Otto, H.-U., Rauschenbach, T. (eds.), 2004, Die andere Seite der Bildung, Wiesbaden, HWV Fachverlag.

[8] Bullinger, H. J., Wörner, K., Prieto, J., 1998, Wissensmanagement – Modelle und Strategien für die Praxis, In: Bürgel, H.D. (ed.), Wissensmanagement – Schritte zum intelligenten Unternehmen. Berlin: Springer.

[9] forum! Marktforschung GmbH, DGQ (eds.), 2009, Excellence – Erfolgreich in starken Zeiten. ExBa 2009 – Benchmarkingstudie zur Excellence in der deutschen Wirtschaft, Mainz, gzm.

[10] Weckenmann, A., Werner, T., 2007, Reducing Operator Influences on Results in Coordinate Metrology via a Holistic Qualification Concept, Proc. ISMQC, Chennai, India, Nov. 2007, pp. 115-120.

[11] Werner, T., Weckenmann, A., 2010, Computer-assisted generation of individual training concepts for advanced education in manufacturing metrology, Measurement Science and Technology MST 21 (2010) 5, 054018 (6pp).

[12] Weinert, A., Organisationspsychologie, Psychologie Verlagsunion, Weinheim, 1998.

[13] Wood, F.H., McQuarrie, F.O.; Thomson, S.R.: Practitioners and Professors Agree on Effective Staff Development Practices. Educational Leadership, October 1982.

[14] Weckenmann, A., Werner, T., 2010, Holistic qualification in manufacturing metrology by enhancing knowledge exchange among different user groups. Metrology and Measurement Systems M&MS 17 / 1, pp. 17-26.

[15] Germani, M., Mandorli, F., Mengoni, M., Raffaeli, R. 2010, CAD-based environment to bridge the gap between product design and tolerance control, Precision Engineering, vol. 34, 7-15.

[16] Mischo, H.K., Pfeifer, T., Bitte, F., 2000, Model-based optimization of interferometers for testing aspherical surfaces, Proceedings of SPIE, vol. 4101, pp. 497-510.

[17] Samper, D., Santolaria, J., Majarena, A. C., Aguilar, J.J., 2010, Comprehensive simulation software for teaching camera calibration by a constructivist methodology, Measurement, vol. 43, pp. 618-630.

[18] De Sousa, A. R., Gonzáles, R., Conejero, S., 2006, FORMA3D – An educational program for the qualification of technical person involved with coordinate metrology in brazil, Proc. IMEKO, Rio de Janeiro, Brazil, Sep. 2006, p. 10 and CD.

[19] Beetz, S., 2006, Beitrag zur Methode der arbeitsplatz-integrierten Assistenz am Beispiel der Formmesstechnik, Aachen, Shaker-Verlag.

[20] Winterton, J., Delamare – Le Deist, F., Stringfellow, E., 2006, Typology of knowledge, skills and competences - clarification of the concept and prototype, OOPEC, Luxembourg.

[21] Bloom, B., 1971, Taxonomy of Educational Objectives, The Classification of Education Goals, Handbook I: Cognitive Domain, New York, McKay.

[22] Kleimann, B., Weber, S., Willige, J., 2005, E-Learning aus Sicht der Studierenden, HIS GmbH, Hannover.

[23] Weckenmann, A., Werner, T., 2008, Introducing Engineering Students to Open Distance Learning – a case study on the acceptance of a virtual course as compulsory subject, In: Fink, F. (ed.), SEFI 2008 36th Annual Conference, Sense Publishers, Rotterdam, 50 and CD-ROM.

Sustainability in Production Engineering
- Holistic Thinking in Education -

C. Herrmann, G. Bogdanski, M. Winter, T. Heinemann, S. Thiede, A. Zein
Institute of Machine Tools and Production Technology, Technische Universität Braunschweig, Germany

Abstract
Considering the environmental as well as the social impact of manufacturing systems besides economic criteria has become a major topic in research and industrial application in the last years. Production processes demand diverse materials and energy for value creation. Considering the increasing economic and environmental relevance of energy and resource consumption, strategies to improve the efficiency and effectiveness of their usage in manufacturing environments need to be fostered in alignment with the premises of a sustainable development. However, focusing only on research and industrial application is a rather short-term thinking process. While striving towards more sustainability in manufacturing, the integration of this issue into the education of engineering students is a key element on a long-term perspective. Against this background, the paper presents an educational structure and concept of the new TU Braunschweig course *Sustainability in Production Engineering*.
Keywords:
Sustainability; Production engineering; Lecture; Education of university students

1 INTRODUCTION

The manufacturing system is one of the main drivers for the increasing quality of life of the people around the earth but at the same time, is accounted responsible for serious damages and severe impacts to the environment at local and global levels [1]. A sustainable development is targeting these conflicts between the growing economy's goals of bigger throughputs and higher profits and the needs of society and the environment, as noted by the Bruntlandt Commissions in 1987. In 2002, *The World Summit on Sustainable Development* amends in Johannesburg the Agenda 21 with a change of paradigms from defining goals to defining concepts and methods to reach those ambitious goals. Furthermore, the Agenda 21 has identified the need to include educational concepts to consolidate the sustainable development debate. This promotion led to the creation of the *United Nations Decade of Education for a Sustainable Development* (ESD). Essential new skills to ESD are envisioning, critical thinking and reflection, systemic thinking, dialogue and negotiation, collaboration and building of partnerships [2]. Seliger highlights the critical and responsible role of education to foster a sustainable and at the same time competitive market environment [3].Especially engineering students have to enrich their competencies with interdisciplinary views on international und intercultural interdependencies besides their technical knowledge.

The *Technische Universität* (TU) *Braunschweig* is facing these presented requirements by implementing the topics of a sustainable development into the higher education of engineering students on university level.

1.1 Scientific surroundings of the TU Braunschweig

The area around Braunschweig is according to the European Union's statistics the most active research region of Europe [4]. Part of this region are the research institutes of the TU Braunschweig with their academic focus on engineering and natural sciences, closely linked with economics, humanities, education and social sciences. The TU Braunschweig is member of the consortium of the nine leading universities of technology (TU9) in Germany. The *Institute of Machine Tools and Production Technology* is integrated in the department of mechanical engineering and therefore jointly responsible for the educational course system of mechanical engineering as well as industrial engineering students.

1.2 Educational background in the field of ESD

The *Product- and Life-Cycle-Management Research Group* as part of the Institute for Machine Tools and Production Technology has already implemented aspects of ESD in the correspondent course Product and Life Cycle Management (PLM) in German language. PLM is a course concept with a top down approach focusing on the global aspects of a sustainable development and reviewing all aspects of the product's life cycle from cradle to grave [5]. Therefore, the need for an extending lecture on the topic of industrial manufacturing systems arose to complement the present course with its product point of view. The new course would have to be a bottom up approach focusing on the production of goods and services. Having a long lasting experience in the field of engineering education, the Institute of Machine Tools and Production Technology has seen the chance to merge the traditional engineering aspects with the new educational requirements of a sustainable development [6].

Production engineering has a very big lever on the design of industrial systems, and therefore the impact on the environment during the production phase of goods and services. Envisioning the consumed energy and resource flows of present or future production sites the aspect of systemic thinking comes into place as Tilbury and Wortman have noted it to be of high importance [2]. Systemic thinking envisions the production site in a holistic way. Meaning that not only the production machines are regarded when analysing a production site but also the peripheral consumers, like technical air ventilation, tempering devices, compressed air supply and also internal and external effects on the plant's building shell, like machine emissions, heat dissipation and outer climate influences [7]. The Product and Life Cycle Management Research Group was trying to cope with these challenges while setting up the new course concept for *Sustainability in Production Engineering* in English language.

2 SUSTAINABILTY IN PRODUCTION ENGINEERING - EDUCATIONAL OPPORTUNITIES AND CHALLENGES

Engineers are nowadays often in the centre of criticism, that their hitherto work and products were unsustainable and do not fulfil the needs of the current society [8; 9]. To consider this criticism and to turn it into a motivational aspect, the teaching in engineering has to integrate methods and tools for the sustainable development into the curriculum.

An appropriate framework is the *three-pillar-model of sustainability,* formulated at the Copenhagen Summit in 1997, referring to the environmental, social and economic demands. Each dimension has different interests and motivational drivers in the context of a sustainable development. Ashford presents these interests and describes the claims of the three dimensions in respect of a sustainable development [10]. In this context, one important character to handle or solve the challenges is the engineer.

An engineer has usually the standing in society as a "problem solver" as a person "who wants to make things work" [8; 10]. To accomplish this objective an engineer needs clear ideas respectively a precise description of the problem he has to deal with. In the context of teaching sustainable aspects, it is important to identify current unsustainable challenges, presenting methods and tools how to analyse these challenges, deriving possible solutions and making an evaluation of the objective fulfilment.

For teaching contents like sustainability in production engineering, an active encouragement of the students' individual development and excitement is essential. Thus, the lecturer has to motivate them to develop their own thoughts and problem solving strategies. For this reason, the course concept will have to integrate students as early as possible in subjects of current research projects. The educational concept of the single lectures does not only consider the engineering subjects from the point of view of the traditional engineering discipline, but is moreover looking at it from an interdisciplinary perspective. As the tasks, the future graduates will face in their later occupation, will expect them to build up knowledge not restricted to the boundary of the subject's matter. *Sustainability in Production Engineering* is set-up to be integrated by engineering students into their master degree course program, which is oriented to provide the students the necessary apparatus for leading positions and careers in research, development and management.

Gross identifies major obstacles in university education (lack of synergy among other sciences, understanding the concept behind sustainability and lack of analysis of the consequences of unsustainable practices) when trying to integrate the subject of a sustainable development into educational concepts [11]. Filho et al. compare different state of affairs in sustainable development in technology education and indicate qualification goals that are essential to an educational single course concept [12]. The new course Sustainability in Production Engineering is trying to cope with all these challenges. Therefore, it is explained below how the aspect of ESD can be implemented successfully into the universities' curriculums. The integration of ESD into universities' curriculums can be provided on different levels depending on the intensity of the knowledge transfer (see Figure 1).

The range of the intensity of educational training can be distinct on the basis of the used time in the academic teaching. A single lecture in a course is one of the simplest options to introduce innovations or changes of paradigms to a well-known topic. This implementation option does not represent any intensive training because it just allows a shortly overview, but provides the possibility to attract first interest for a new theme. The next possible level is a course that is completely dedicated to the new research topic. This option gives a detailed examination of a new topic and allows the educational training in terms of one specific aspect (economical, social or environmental). A complete course also entails the option for additional practical classes, like case studies, tutorials and laboratories. The most extensive level is a bachelor or master program, which allows the combination of different courses, to cover all aspects of the three-pillar-model. Below, five exemplary higher educational approaches in Germany are illustrated that differ in their level of educational intensity and in the examined aspects of the three dimensions of sustainability.

Single course: "Energy and resource efficient metal processing", Institute for Material Science, Universität Bremen

The Institute for Material Science combines three major scientific disciplines on equal terms: materials science, process engineering, and manufacturing technologies. The course offers a platform for an informed discussion on the environmental dimension of the three-pillar-model. In this context, the course discusses and presents technical solutions like dry machining, coolant management, minimal quantity lubrication, integration of alternative process technologies, integration of the heat treatment into the process chain and the resource-conserving design of process chains. In addition, the technical solutions are evaluated in terms of their impact on the social and economical aspects of the three-pillar-model [13].

Single course: "Sustainability in Production and Logistics", Institute of Automotive and Industrial Production Economics, Professorship for Production and Logistics, Technische Universität Braunschweig

This course focuses on the one hand on production and operations management, in particular contract-related planning, capacity and resource management and production control concepts, on the other hand on logistics and supply chain management, in particular, distribution and market entry strategies and spare parts management. The course examines the economical aspect of the three-pillar-model and illustrates environmental effects of production, environmental legislation, environmental and industrial material flow management, environment controlling, material flow models for the business modelling of industrial material and energy flows and life cycle assessment [14].

Single course „Basic Readings in Environmental Ethics and Philosophy", Institute of Botany and Landscape Ecology, Professorship Environmental Ethics, University of Greifswald.

The Professorship Environmental Ethics focuses on the analytical reconstruction of argument for the environmental ethics with practical intent, in particular dependence arguments, argumentation concept of eudemonia, responsibility to future generations, and a critical appraisal of nonanthropocentric conceptions. In this context, the course examines the social aspect of the three-pillar-model. The intention of the course is a thorough analysis of classic readings in environmental ethics and philosophy. Furthermore, the examination of specific conceptual frames of philosophical text addressing environmental questions for example natural conservation, animal rights, and sustainability issues [15].

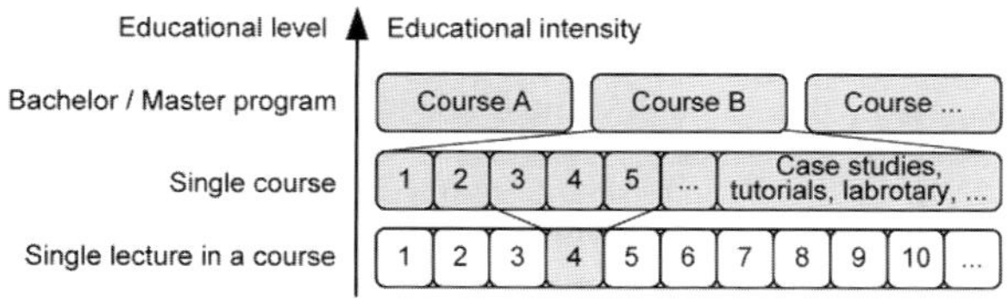

Figure 1: Levels of educational intensity : ESD related issues; white: non-ESD related issues)

Bachelor degree program "Environmental and Resource Management", Department of Environmental Sciences and Process Engineering, Brandenburg University of Technology.

The bachelor degree program "Environmental and Resource Management" has the intention to provide an education in respect to technological, economic and infrastructure determining processes from the perspective of the integrative environmental and resource protection, assessment and monitoring. The students are trained in management of complex economic and technological processes in compliance with the preventive environmental protection, the protection and rational use of natural and economic resources. Moreover, the students should be capable of decision making in the choice of sustainable economic and technological structures of production and material flow management. This intention is served by the availability of multifarious courses in the disciplines of engineering science, science of management, environmental science and social science. This mixture of different courses covers all aspects of the three-pillar-model [16].

Master degree program „Sustainable Development", Faculty of Economics, Universität Leipzig

The master degree program „Sustainable Development" deals with issues of sustainability, environment and development as well as the challenges of transition to a sustainable society. The students consider the possibilities and problems of change from different angles and thus acquire a high and broad expertise in current and future issues of sustainable development. Due to the structure of a master degree program, the students are educated in the disciplines of social science, engineering science, environmental science and science of management [17].

3 EDUCATIONAL CONCEPT

Weizäcker and Schmidt-Bleek discuss that a growth of the worlds gross domestic product can only be achieved by raising the eco-efficiency in order to keep the world's bio-capacity in equilibrium [18; 19]. Radermacher concludes that creating more goods and services from the same amount of natural resources can only result from a gain in engineering progress. Furthermore he notes that a gain in engineering is not a new principle, it has been part of our life ever since. Moreover, it is about the new factor of collective responsibility, implemented by global politics, e.g. through a worldwide limitation of CO_2 emissions [20]. Future engineers are having the great responsibility to act in conformity to the environment and social system while designing new products and services as well as industrial systems to create those products. The new TU Braunschweig course *Sustainability in Production Engineering* comprises an educational approach that has evolved from the problem solving methodologies and tools of two scientific projects in the field of energy and resource efficiency. One of the German Ministry of Education and Research funded research projects (*EnHiPro)* is dealing with the energy and resource efficiency of industrial manufacturing sites, processes and auxiliary materials, spanning various branches of small and medium sized manufacturing companies [21]. The second project (*ProGRess)* is branch specific and focusing on the energy and resource efficiency of aluminium die casting process chains, starting from the alloy creation and transportation, following the supply chain to the casting processes and finally dealing with the mechanical cutting processes for the finished casting products [21]. Both projects have the same methodological approach in common, as depicted in the cyclic problem solving approach in Figure 2.

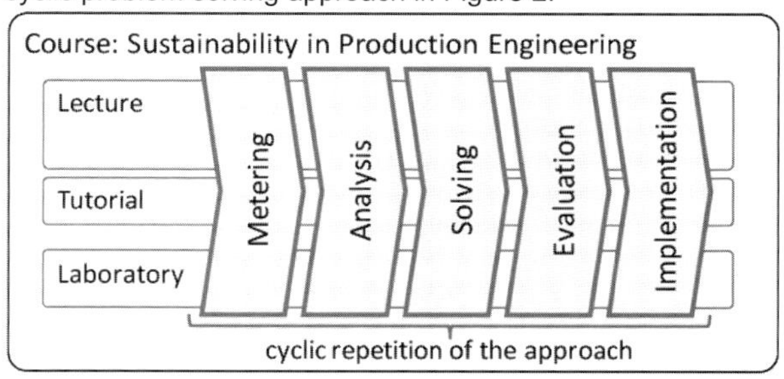

Figure 2: Methodological toolbox inside the course concept

The course concept contains 14 lectures to impart theoretical backgrounds on sustainable development and engineering knowledge. The English course language motivates to work with recent literature in discussions and during self-study times. Accompanying to the lecture a tutorial enriches the didactic concept with hands-on studies and practical knowledge about applying metering strategies on production machines and evaluating dynamic energy and resource consumption behaviour of process chains. The laboratory is an additional course achievement and is an extension to the lecture and the tutorial. It gives a deeper understanding of the technical aspects of data acquisition and data processing in order to enable machine operators to change their operating behaviour towards a more sustainable one.

3.1 Lecture

The lectures of the course extend the traditional lectures of production engineering courses at the TU Braunschweig from a mono-disciplinary-structured course concept to a more inter-disciplinary-structured one that prepares engineering students to live and promote a sustainable development as promoted by the theme of ESD [2].

The lectures promote an understanding how the work of an engineer interacts with society and the environment, locally and globally, in order to identify potential challenges, risks and impacts. The lecturer provides a basic understanding of a sustainable development and a motivation to deal with that topic from an engineer's point of view. Basic theoretical knowledge is imparted starting with general aspects of a sustainable development and soon focuses on the aspects of production engineering and is further on dealing with the mergence and compatibility of both aspects with their often contrary goals [5; 7]. The lecturer prepares the students to approach the sustainability challenges in manufacturing systems with the cyclic problem solving approach of metering, analysing, solving, evaluation and implementation from Figure 2, highlighting the importance for the constant repetition of the approach in order to cope with the changing systemic behaviour and system dynamics.

Metering

The basis of the presented approach is metering. This step describes the allocation of significant data from the physical world. Sufficient metering produces high amounts of time related data, which needs to be analysed in order to derive informational contents from the data. The design of metering strategies is one of the main aspects to aggregate the right amount of data from the right source. An interdisciplinary knowledge of mechanical, electrical and computer system engineering is essential to develop and formulate the right metering strategy.

Analysis

The subsequent step of analysis describes the development of an understanding for the energy and resource consumption behaviour of single machines or systems in interrelation with interlinked factory systems. It is an essential element to understand the systemic consumption behaviour of manufacturing systems in order to resolve deficiencies from a holistic point of view. Depending on the quality of the metered data, the engineer is able to gather a great amount of systemic knowledge.

Solving

In the third step of the approach, existing deficiencies are tackled. The lecturer presents basic measures for problem solving within the strategic and operative layers of production management based on the three strategies of a sustainable development: efficiency, effectiveness, and sufficiency. Solving consists of systemic measures or single process modifications. Both have a distinct impact on the holistic systems, which needs to be assessed within all three dimensional aspects: environmental, economic, and social.

Evaluation and Implementation

In the module of evaluation, the lecturer provides the students with essentials to assess the overall impacts of the addressed changes that have been or are going to be implemented in the existing system. During evaluation phase the students can chose from a toolbox of systemic methods of single key figure calculation, life cycle assessment or life cycle costing and many other well established methods. The last module of implementation indicates the idea of closed loop and systemic thinking. Once a measure is implemented, the cyclic approach is to be repeated.

The lectures deal with the procedural methods of each step of the cyclic approach and finish with a series of presented case studies from the mentioned research projects. The case studies show the practical application of the problem solving approach on a broad field of challenges (e.g. improving the eco-impact of grinding lubricants while conserving the technical properties; improving the energy and resource efficiency of the aluminium die casting process chain while conserving the quality requirements).

Another case study is held by a guest lecturer from our Joint German-Australian Research Group Sustainable Manufacturing and Life Cycle Management (www.sustainable-manufacturing.com) complements the English course concept. Prof. Sami Kara from the University of New South Wales (UNSW) and the students are linked via a video conference classroom. This opportunity allows to integrate practical aspects of global challenges due to global supply chains and regional differences in production environments especially due to other climate conditions, available primary resources and transport distances.

3.2 Lecture accompanied tutorial

The tutorial means to support the understanding of the introduced methodologies and tools and provides the possibility to experience how to approach energy efficiency improvements during hands-on studies in a real manufacturing environment. The students explore the challenges of data acquisition, data processing and data analysis being put into a fictional role-play situation inside a real hardware environment. The given situation puts the students (groups of three or four) into the position of a young engineer in a large manufacturing company whose first job it is to enable the company to win the *green factory award*. The award will be granted to the factory staff, which is able to present the most eco-efficient production line concept. As building blocks for a creative case study, parameter and process alternatives in the given production process chain of an engine starter motor shaft (Figure 3) have to be analyzed and evaluated.

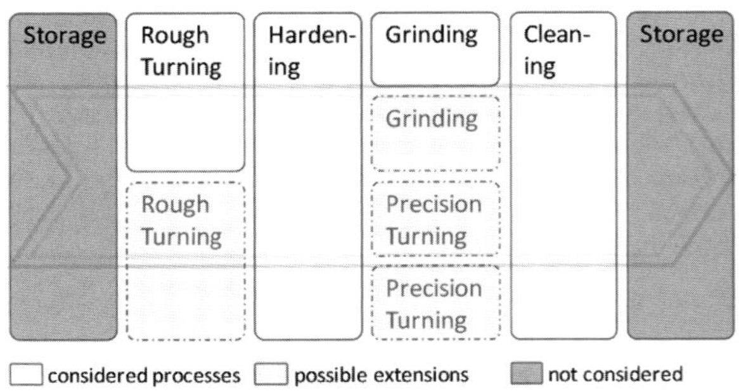

Figure 3: Considered production process chain

Before considering the total process chain, the student groups have to consider the process alternatives of grinding and precision turning. Both machining processes can fulfil the mechanical requirements for the work piece's quality; therefore, the objective is to identify the most energy-efficient process alternative without having to consider the quality requirements. The students' reactions to their self-developed energy metering strategy and the selected approaches to eco-benchmark the alternative processes provide the basis for classroom reflections and discussions.

On the process chain level the student groups focus on the improvement of the existing processes (white boxes in Figure 3) with means of an energy oriented simulation tool for manufacturing systems which was developed at the institute for the last couple of years [7]. The tool leaves the necessary freedom in terms of an alternative design and control of the depicted process chain. The process chain was pre-modelled and made accessible for the students as an intuitive to use Java applet.

At this time, the simulation approach is integrated into the lecture's methodological toolbox to solve complex tasks in virtual models before implementing them into the real world. A simulation tool prerequisites a good metering strategy to parameterize the programmed model. The tool integrates the steps of analysis, solving and evaluation in its simulation sequence. Successful simulation results can be implemented into the physical system.

The given simulation tool has a set of predefined parameters like energy consumption patterns for different operating modes (start-up, idle, machining) and timings (start-up, machining) of selected processes. The simulation moves a predefined amount of semi-finished products from one storage area (on the left) through the parameterized process chain into the finished product storage (on the right) and sums up the embodied energy to derive efficiency key figures as well as the demanded throughput time and energy costs. The students have the freedom to fill in their derived energy consumption parameters (from the benchmarked grinding and precision turning processes). Furthermore, they are free to change parameters like batch sizes for the hardening and cleaning process as well as adding new machines and defining the utilized capacity of the alternative processes in order to manipulate the throughput time, the accumulated energy consumption and energy cost, considering the conflicts in goals between the economic and ecologic dimension.

As a key educational aspect, the students are encouraged to identify shortcomings as well as bottlenecks of the variations in the process chain scenarios considering the resulting output parameters. The emphasis of the tutorial is on both metering and analysis in the real field as well as on evaluation and implementation through virtual simulation.

3.3 Laboratory

In parallel to the lectures and the tutorial, the students have the chance to participate in the complementary laboratory of *Sustainability in Production Engineering*. The laboratory provides the students the opportunity to explore the topic of energy and resource consumption of machine tools and the correlated auxiliary systems depending on the machine operator's behaviour. Following the methodological toolbox from Figure 2, the laboratory is covering the fields of metering, analysis and evaluation, as depicted in the learning path of Figure 4. The machine operator's behaviour is accounted for the field of solving and implementation and is the manipulative target of the students' tasks.

The machine operator's behaviour is significantly affecting the energy and resource consumption of machine tools. Depending on the set-up times, computer numerical control (CNC) coding and usage patterns. The energy and resource demand is highly unpredictable without metering the consumption. Therefore, the student laboratory is equipped with up-to-date production machines and integrated industrial metering equipment. The production machine is a *Studer S120* high precision grinding machine. The grinding process is a very good example for educational aspects, because of its demand for external auxiliary supplies like air purification and coolant filtering that is holding a high potential for disregarded energy and resource consumption [22].

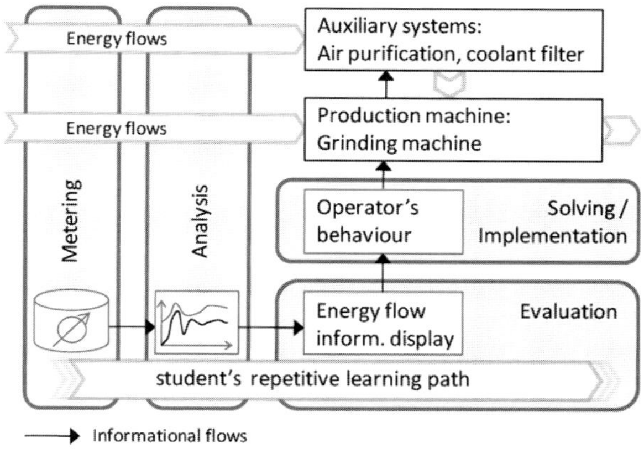

Figure 4: Concept of the complementary laboratory

The grinding machine's operating interface allows the manual and automated control of all auxiliary devices like the air purification system and the coolant filter system. The operator is able to manually switch the devices on if needed and off if not. The auxiliary devices are powered up automatically as soon as the machine's control is started up. Generally, the demand for the auxiliary devices is not given at all times when the machine is powered up, especially during set-up times of the machine. Because the operator's awareness for the wasted energy of auxiliary devices is not given, there will be no change in the operator's behaviour and therefore no rectification of the present deficiencies. Raising the operator's awareness for the presence of energy wasting by means of information and communication technology is the creative key element to the laboratory.

The grinding machine and the auxiliary systems are equipped with energy and resource flow sensors, which are connected to an industrial personal computer (IPC) with a build in software programmable logic controller (PLC). The PLC can be used to acquire the metered sensor data and to analyse the consumption behaviour of each consumer. The students are to program a visualisation of a energy flow information display (EFID) build in the control panel of the grinding machine. The EFID can display any information derived from the PLC.

The students are free to design a creative and meaningful operator's interface providing all necessary information and fields of action to avoid waste of energy and to optimize the resource consumption behaviour of the running machine depending on the change of the production mode of the grinding machine (idle, set-up, machining, etc.). The designed EFID can be tested and evaluated in the field, depending on given usage scenarios.

The laboratory provides the students the necessary knowledge and skills to use information technology as a creative tool for a sustainable development of production processes by means of raising the machine operator's awareness for a more energy sensitive operation of machine tools and their interrelated auxiliary machines. A successful avoidance of unnecessary energy consumption during production processes is the key didactic goal of the student laboratory.

4 STUDENT'S FEEDBACK

In summer term 2010 the course concept was implemented into the curriculum of the TU Braunschweig and approx. 30 students participated in the course. All students had the chance to evaluate the educational quality by means of an anonymous feedback form, which was afterwards charted by the department of mechanical engineering. The course has earned great feedback especially because it was completely held in English language. Furthermore, the students appreciated the lecture accompanied tutorial as an amplifying measure to verify the theoretical background with hands-on practical sessions. The tutorial group students were highly motivated by the competition to win the "green factory award" for their presented improvement scenario, which was secondarily helping to ease up the development of team working and presentational skills. Additionally, the students accredited the course content to be very up-to-date, which was mainly due to lecture accompanying research paper handouts for self-study times.

5 SUMMARY

Sustainability in Production Engineering is part of the newly established ESD contents in the curriculum of the TU Braunschweig and is a single course concept that can be integrated into either bachelor or master programs by engineering students. The course is extending the present course range of the *Product- and Life-Cycle-Management Research Group* from a bottom-up approach in the field of production engineering. The course contains lecture elements that deal with a thorough background of a sustainable development and repetitive elements of the traditional aspects of production engineering. Combining these two elements, the course gives a good understanding of the subsequent challenges and responsibilities. Accompanying practical elements as case studies, tutorials and a laboratory provide the students a better understanding of the theoretical and methodological contents by hands-on studies in the Institute's production environment. The first participating students have shown to develop a critical reflection of the addressed topics in case studies and tutorials. They have become aware of their possible specific contribution to the sustainable development in manufacturing systems based on their different engineering background.

6 REFERENCES

[1] Evans, S., et.al., 2009, Towards a sustainable industrial system, ISBN: 978-1-902546-80-3.

[2] Tilbury, D., Wortman, D., 2004, Engaging people in sustainability, ISBN: 2-8317-0823-0.

[3] Seliger, G., 2004, Global Sustainability, Glob. Conf. on Sustainable Product Development and Life Cycle Engineering, 29-35.

[4] Götzfried, A., 2005, FuE-Aufwendungen und FuE-Personal in den europäischen Regionen, ISSN: 1609-6002.

[5] Herrmann,C., 2010, Ganzheitliches Life Cycle Management, ISBN: 978-3-642-01420-8.

[6] Kastenhofer, K., et al., 2010, The Contribution of University Curricula to Engineering Education for Sustainable Development, GAIA, Vol. 19, No. 1, 44-51.

[7] Herrmann, C.; Thiede, S., 2009, Process chain simulation to foster energy efficiency in manufacturing, CIRP Journal of Manufacturing Science and Technology, ISSN: 1755-5817.

[8] Bras-Klapwijk, R.M., et al., 1999, Training of lecturers to integrate sustainability in engineering curricula, available from: http://www.citg.tudelft.nl/live/binaries /c36dcf0a-335f-4642-81d0-f7fcbe96faf6/doc /luwartikel.pdf [accessed 23 Aug 2010].

[9] Dwyer, J., 2008, Unsustainable measures, the IET-manufacturing, Vol. 86, No. 6, 14-19.

[10] Ashford N.A., 2004, Major Challenges to Engineering Education for Sustainable Development, Int. Conf. on Engineering Education in Sust. Dev., Barcelona.

[11] Gross, D., 2000, Three Questions on Education for Sustainable Development, Human Nature Articles, Vol. 5, No. 2.

[12] Filho, W., et al. 2008, Education for sustainable development: current discourses and practices and their relevance to technology education, Int. Journal of Technology Design Education, DOI 10.1007/s10798-008-9079-z.

[13] Institute for Material Science, Universität Bremen, available from: http://www.iwt-bremen.de [accessed 23 Aug 2010].

[14] Institute of Automotive and Industrial Production Economics, Technische Universität Braunschweig, available from: http://www.tu-braunschweig.de/aip/ [accessed 23 Aug 2010].

[15] Institute of Botany and Landscape Ecology, University of Greifswald, available from: http://www.botanik.uni-greifswald.de/ [accessed 23 Aug 2010].

[16] Brandenburg University of Technology, available from: http://www.tu-cottbus.de/fakultaet4/ [accessed 23 Aug 2010].

[17] Universität Leipzig, available from: http://www.wifa.uni-leipzig.de/ [accessed 23 Aug 2010].

[18] Weizsäcker, et al., 1997, Faktor Vier, ISBN: 3-4267-7286-8.

[19] Schmidt-Bleek, F., 1997, Faktor 10 – Das Maß für ökologisches Wirtschaften, ISBN: 3-4233-0580-0.

[20] Radermacher, F.J., 2003, Die Zukunft der Wirtschaft, Universitätsclub Klagenfurt, Friaul.

[21] Herrmann, C., et al., 2010, Ganzheitliche Ansätze zur Erhöhung der Energie- und Ressourceneffizienz in der Produktion, 10. Karlsruher Arbeitsgespräche Produktionsforschung, ISSN: 1869-9006.

[22] Herrmann, C., et al., 2009, Energy Efficiency of Machine Tools: Extending the Perspective, CIRP Manufacturing Systems, Grenoble.

Sustainable Manufacturing Using a Global Education Concept for Coordinate Metrology with a Blended Learning Approach

M. Marxer[1], C.P. Keferstein[1], A. Weckenmann[2]

[1]Institute for Production Metrology, Materials and Optics, NTB Interstate University of Applied Sciences of Technology Buchs, Switzerland

[2]Chair Quality Management and Manufacturing Metrology, University Erlangen-Nuremberg, Germany

Abstract

This paper describes an education concept with a blended learning approach that takes account of sustainability in terms of environmental, economical and social aspects.

The concept has been developed for coordinate metrology using multi-sensor coordinate measuring devices as an important contribution to zero-fault sustainable manufacturing. In this implementation, combinations of face to face education for theory transfer and practical experience are used as well as online learning using internet possibilities. This approach can be used to support sustainable education such as lifelong learning, largely independent of the learner's geographical location and the available infrastructure.

Keywords

Sustainable Manufacturing; Blended learning; Virtual laboratory; eLearning; Coordinate metrology; Education

1 INTRODUCTION

Nowadays sustainability concepts meet with general acceptance worldwide and are developed and implemented for a wide range of industries, research and development and manufacturing products. In a globalized, rapidly developing world, training and education from primary to tertiary level play an important role especially in respect of social sustainability. In future only those regions and companies that develop modern training concepts with an emphasis on sustainability will continue to be competitive. As regards manufacturing processes and the necessary quality control, a great effort must be made to achieve sustainability: e.g. to optimize material consumption given limited resources.

Generally speaking, sustainability concepts may be described as using a regenerative system in such a way that the essential features of the system are preserved and natural regeneration ensures the continued existence of the system [1].

There are three aspects of sustainability [2], (Fig. 1):

- Environmental sustainability describes the objective of preserving nature and the environment for subsequent generations and includes climate protection. This is relevant for training concepts in that travelling to and from training sites has an effect on the environment by, for example, contributing to pollution.
- Economic sustainability is based on the premise that the economy should function in such a way that it provides a viable basis for employment and prosperity in the long term. As regards training for quality assurance in manufacturing processes, this means, for example, having knowledge resources for lifelong learning in current and future generations that are always available and up-to-date, thus ensuring the earning capacity and prosperity of the individual.
- Social sustainability sees the development of society as a way of ensuring the participation of all members of society [3]. This involves creating a balance between social forces with a view to achieving a liveable society that is sustainable in the long term. As regards training, this means, for example, offering equal opportunities when it comes to accessing learning content irrespective of the geographical location of individual members of society. Manufacturing processes in mass production require workers who have an elementary education and need additional customized training programmes that are independent of specific manufacturers and products.

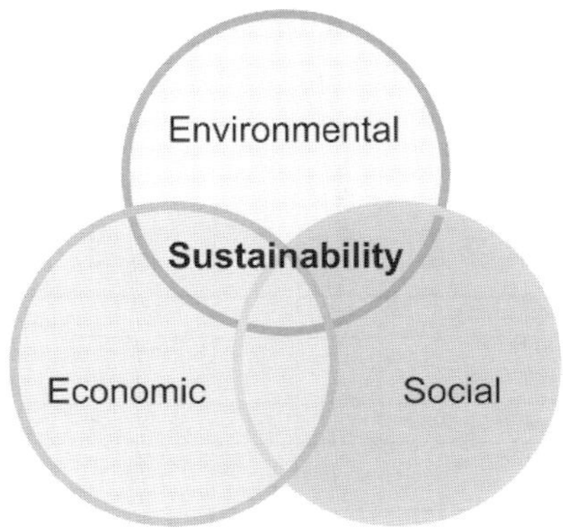

Fig. 1: Sustainability aspects

In this paper, various learning concepts are discussed in terms of their sustainability. One area that is especially affected by globalisation and is consequently especially important in terms of sustainability is production metrology, notably 3D coordinate metrology.

This paper presents a learning concept for coordinate metrology that meets current requirements in terms of sustainability. Individual aspects of sustainability are described in more detail using case studies, and their effectiveness is documented.

2 SUSTAINABILITY OF LEARNING CONCEPTS

2.1 Conventional learning concepts

Nowadays there is a wide variety of different learning concepts ranging from traditional frontal teaching to distance learning. Frontal teaching is one of the most common teaching methods. Curriculum content is taught, exemplified and demonstrated by way of lectures or supervised project work. This teacher-led style of teaching presupposes that course participants are all able to absorb, learn, understand and grasp all learning content at the same time.

In terms of sustainability, this is not the optimal approach. Learners must travel to the place of learning and thus contribute to environmental problems by causing additional emissions. Travel time reduces productivity and economic sustainability. If such a learning approach is taken, only very limited attention can be paid to the needs of individual course participants. This has a negative impact on social sustainability (Tab. 1).

In order to counter the shortcomings of this form of learning, it is preferable to cater for the needs of a heterogeneous group of learners by offering a variety of different learning forms (blended learning) and to work towards learner-centred rather than teacher-centred modes of teaching.

2.2 Blended learning

Blended learning is an alternative integrated learning concept that makes optimal use of an appropriate learning arrangement by using currently available networked learning via the internet together with traditional forms of learning and learning media. Blended learning refers to a form of learning that allows traditional face-to-face teaching to be linked in a didactically appropriate way to modern forms of e-learning.

Tab. 1: Evaluation of features of different forms of learning

Feature	Sustainability			Forms of learning		
	Economic	Environ-mental	Social	Conventional	E-learning	Blended
No time constraints	x		x	-	++	+
No geographical constraints		x	x	-	++	+
Individual learning pace	x		x	--	++	+
Heterogeneous prior knowledge	x		x	-	+	++
Individual. learning behaviour support	x		x	--	--	++
Tutor support	x	x	x	++	--	+
Communicative aspects	x	x	x	+	-	+
Learning success	x		x	+	-	+
Economic aspects	x	x		+	++	+

An appropriate combination of various media and learning methods enhances their advantages and minimizes their disadvantages (Fig. 2). The concept combines the effectiveness and flexibility (economic sustainability) of electronic learning methods with the social aspects of face-to-face communication (social sustainability). Integrating practical learning using a real object plays an important part in the learning process.

3 BLENDED LEARNING FOR COORDINATE METROLOGY IN MANUFACTURING

Sustainability in manufacturing processes includes having a special quality control strategy to avoid producing defective parts. The aim of zero-fault manufacturing is not to waste resources such as materials, time and money. Training in coordinate metrology is one important aspect, because those who use this technology are involved in control strategy and decision processes that have a direct impact on the sustainability of processes and products. The geometric product specification standard (GPS) [4] includes tolerance specifications and measuring system requirements. Measurement technicians and engineers still have a considerable influence on measurement results [5]. In order to take qualified and traceable decisions, it is necessary for them to have basic training that is sustainable and standardized and not dependent on specific measuring equipment. The automotive industry and consumer electronics are important global industries in which coordinate metrology plays an important role. Nowadays these industries have production and assembly facilities located in different countries and cultures and involve a large number of suppliers.

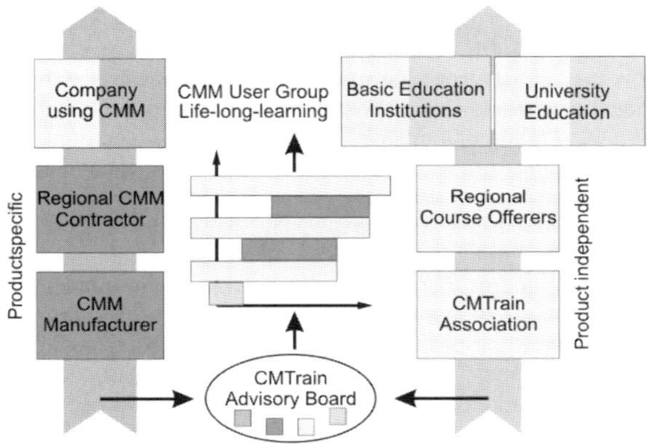

Fig. 2: Sustainable blended learning concept in coordinate metrology involving CMM manufacturers and education establishments

Given this network of different partners, there is a need for standardized quality assurance oriented towards sustainability that goes beyond the boundaries of individual companies and branches of industry; this is essential if international division of labour is to be successful. Only by providing training that is sustainable and not tied to a specific company can demands for high quality and global competition be met. Blended learning particularly lends itself to such training as the needs and learning behaviour of learners in such a network vary greatly. Up until now sustainable training in coordinate metrology has not been offered anywhere in the world; there has been no basic training (apprentice, worker) or advanced training at, for example, universities.

In coordinate metrology it is therefore necessary to include both manufacturers of coordinate measuring machines (CMM) and training institutions that are independent of manufacturers (Fig. 2).

This training concept is based on the very successful training concept of professional apprenticeships used in Germany, Austria and Switzerland, in which training is supported by companies and neutral institutions.

The manufacturer-independent association **C**oordinate **M**etrology **Train**ing (CMTrain) forms the basis of the concept (Fig. 2). The course comprises a training programme at several hierarchical levels (levels 1 ... 3); this includes hands-on training on a real CMM in keeping with the blended learning concept.

Regional competence centres organize individual courses in the respective mother tongue and provide access to modern infrastructure. It is immaterial whether the infrastructure is available on site (practical e-training), at a partner's facilities or at a company / education establishment. This is only possible using the blended learning concept, which increases sustainability in many areas.

After successfully completing their respective course, participants form lifelong learning groups of alumni who remain in touch. All training components (courses and alumni) are overseen by well trained tutors. All activities are coordinated and further developed by CMTrain.

CMM manufacturers are responsible for training on specific machines in the form of training packages (Fig. 2). Manufacturers also have regional competence centres that provide this support for their clients (companies and education establishments) in collaboration with the parent company. They also offer product-specific user groups. These groups may operate on site or via the internet.

3.1 Manufacturer-independent training concept CMTrain

A sustainable learning concept using blended learning components was developed in the course of several projects. Based on the AUKOM project [6], which was a face-to-face concept, EUKOM - the European Training Concept for Coordinate Metrology [7], [8] was developed to include sustainability and new learning methods. In addition, internet-based workshops have been developed. After completion of the EUKOM project, CMTrain was founded (www.cm-train.org) in order to guarantee sustainable use and further development of the project results.

CMTrain training comprises three competency levels (Fig. 3). Level 1 is basic training, level 2 advanced training and level 3 training for specialists. Each training level is followed by further training and hands-on experience using CMM manufacturer specific measuring devices.

Each level of training consists of a combination of several forms of learning (blended learning). Participants work on e-competence and some of the specialist skills in contact hours. This form of learning is also used for supervised intensive hands-on training in small groups (4-5 people per group). Participants use a learning platform to develop their specialist knowledge. At the end of each level of training there is an examination in theory and practice.

In addition to hands-on training in small groups, practical e-training is also offered. For this type of practical training, participants have access to a real CMM via the internet. They have full access to control and programming of the CMM and are in touch with a competent tutor who is on site where the CMM is located. Communication with this tutor also takes place via the internet using information and communication technologies such as video and audio transmission.

This training concept is used for further training in industry and at private and state education establishments up to tertiary level [9]. For use in industrial enterprises, courses are offered as blended learning packages in various languages and include a standardized final examination which is internationally recognized.

For use in universities, individual packages are integrated into conventional lectures and the e-learning component supports self-study. In this way, the concept supports the Bologna Process initiated by the European Union. This process is also aimed at sustainability, standardization and supports student mobility. Up until now, this concept has been used in Bachelor and Master courses in Germany, Switzerland, Italy and Poland.

By the end of 2009, over 400 participants had already completed studies using this training concept [10]. It has already been used very successfully in Europe and South Africa. The training concept is available in English, German, Rumanian and Polish. Versions in French, Spanish, Portuguese and Italian are currently being developed.

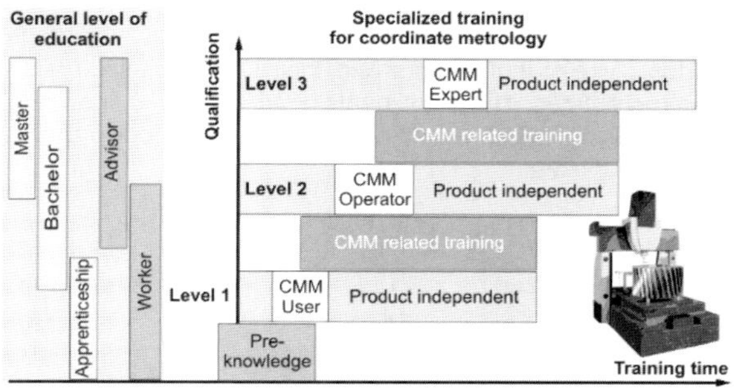

Fig. 3: Sustainable CMTrain training concept in coordinate metrology [11]

The latest technological developments are made available not only to current course participants but also to members of the alumni group (Fig. 2). This takes into account the need for the rapid dissemination of knowledge.

4 SUSTAINABILITY FACTS AND FIGURES IN PRACTICE

4.1 Economic aspects

The new CMTrain blended learning concept has mainly been used in Central Europe up to now. For the purposes of calculation, the following data were taken as typical for this region:

- Average distance between training location for face-to-face activities and place of work: 250 km
- Average workload for CMTrain level 1: 70 hours (Fig. 3)
- Standard rates of pay, travel expenses and course fees

Using the blended learning concept instead of traditional face-to-face teaching means a cost saving of approx. 35%. Both course costs and travel costs are lower (Fig. 4).

If some of the workshops take place as e-learning, the costs of the course increase slightly. However, this increase in costs is offset by lower travel costs. This means a further cost saving potential of approx. 5%. This cost saving potential increases if CMTrain is used in less industrialized countries where distances between place of work and course location are greater.

In addition to these economic advantages, it is possible to greatly enhance economic sustainability using this CMTrain concept.

- With this blended learning concept, participants first undergo comprehensive training amounting to two weeks learning content (CMTrain level 1). Such comprehensive training was not possible in the past for small and medium-sized companies as it meant that employees were not at their place of work.
- While on the course, participants are available on call for urgent measurement jobs at their place of work as they can organize their own learning.
- Well trained tutors and state-of-the art measuring machines are available to participants and guarantee effective training.
- The blended learning concepts were rated by industry, participants and tutors as better than conventional face-to-face training. The criteria taken into consideration were aspects such as learning success and learning efficiency due to much higher motivation.
- CMTrain blended learning also has several advantages for CMM manufacturers and their CMM specific training programmes. The participants share the same knowledge and skills before attending manufacturer specific training and services.

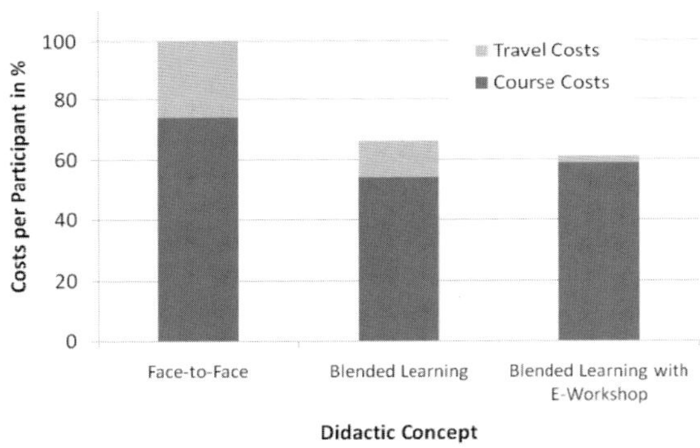

Fig. 4: Comparison of costs between different learning concepts

Furthermore, economic sustainability is documented by measuring results that are based on an equal level of understanding in the global perspective. This means higher quality products, fewer rejects and shorter time to market.

4.2 Environmental sustainability

Using the blended learning concept enables participants to reduce travel. Conservative estimates put the need for training in coordinate metrology at 1000 ... 3000 participants per year.

Assuming that an average distance between course provider and course participant amounts to 250 ... 500 kilometres, this course concept enables total travel to be cut by approx. 1 ... 2 million kilometres per year. The reduced number of overnight stays also constitutes ecological added value.

A further ecological advantage is that paper consumption for course materials is significantly reduced as information is made available electronically. This reduction is estimated at approx. 1...2 tons of paper each year.

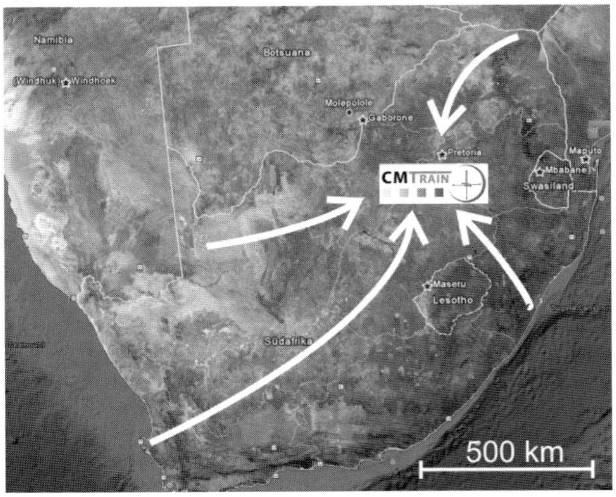

Fig. 5: Travelling distances at the example of a CMTrain course in South Africa

4.3 Social sustainability

The training concept presented significantly increases equality of opportunity when it comes to accessing learning content. Companies and learners in, for example, emerging markets that do not have widespread modern infrastructure can benefit from this training concept.

Initial experience from training courses in South Africa shows the effectiveness of this training concept and the potential for social sustainability. The 15 course participants had an average journey to the course location of approx. 500km. Learning groups that formed during the course of the training sessions continue to bridge the great distances now that the course is over by exchanging ideas in the alumni group via the internet.

A further advantage of this concept is the fact that individual learners determine their speed of learning. This plays a particularly important role in emerging markets, where groups are likely to be very heterogeneous in terms of prior knowledge and speed of learning.

5 FURTHER DEVELOPMENTS

Rapid technological development in coordinate metrology makes it necessary to prepare new subject-specific content didactically and to integrate new learning modules in the learning programme. Current examples are the integration of modules on computer tomography and on articulated measuring arms.

Emphasis is also given to further developing ways of implementing practical training on real CMM via the internet. On the one hand, it is necessary to integrate an increasingly easy-to-use and rapid internet for the parallel transmission of data, images and sound. On the other hand, it is necessary to further develop security and interfaces to companies where CMM are located. In addition, problems with security and interfaces to CMM themselves must be solved. This involves, for example, ensuring that learners cannot damage CMM due to incorrect operation via the internet.

In order to spread CMTrain globally, further competence centres are being established and versions of the learning content are being made available in the respective languages required. The new sustainable learning arrangement presented at the example of coordinate metrology is not only limited for this application. The approach can be of big impact in other fields as well. Using this methodology sustainability advantages in the fields of social, economical and environmental sustainability can be gained.

In a general perspective, the methodology realized can be applied in many other application-fields where the access to special or expensive infrastructure is required for training activities. This is the case for example in sustainable manufacturing technologies. Especially for training on tooling machines, laser shaping or rapid prototyping. It can be of big interest as well in the field of automation where robots and complex handling units are needed for employee training.

In many other fields of applications this approach can support an international cooperation to accelerate sustainable training, reducing travel time for students and tutors causing less pollution and be more efficient at the end.

"In a sustainable society, where people take more responsibility for the consequences of their actions and play a more active role as citizens and workers, they must have access to quality education throughout their lives" [12]

6 ACKNOWLEDGEMENTS

The authors wish to acknowledge the support of the European Union, the Federal Office for Professional Education and Technology OPET and the International University of Lake Constance IBH.

The project results would not have been possible without funding from the European Commission, the Federal Office for Professional Education and Technology OPET and the International University of Lake Constance IBH. This publication reflects only the views of the authors, and the Commission cannot be held responsible for any use which may be made of the information contained therein.

7 REFERENCE

[1] Deutscher Bundestag, 14. Wahlperiode (Hrsg.): Schlussbericht der Enquete-Kommission Globalisierung der Weltwirtschaft – Herausforderungen und Antworten (German Parliament Paper: Final Report of the Commission of Inquiry: Globalization of World Economy – Challenges and Answers); Drucksache 14/9200, 12. Juni 2002.

[2] Wissenschaftliche Dienste des Deutschen Bundestages: Nachhaltigkeit, Der aktuelle Begriff (Scientific Services of the German Parliament: Sustainability Definitions); 06/2004, 6. April 2004.

[3] Weckenmann, A.; Werner, T.: Quality Management for Sustainability in Manufacturing. In: Shunmugam, M. S.; Ramesh Babu, N. (Eds.): Proceedings of the 7th Global Conference on Sustainable Manufacturing (GCSM) "Sustainable Product Development and Life Cycle Engineering" (02.-04.12.2009, Chennai, India), p. 19-25. ISBN 978-81-904235-9-5.

[4] ISO/TR 14638:1995-12: Geometrical product specification (GPS) – Masterplan. Beuth: Berlin: 1995.

[5] Weckenmann, A.; Gawande, B.: Koordinatenmesstechnik – Flexible Strategien für Mass, Form und Lage (Coordinate Measuring Technology – Flexible Strategies for Dimension and Form). 2. Auflage. Carl Hanser Verlag, München Wien, 2007. ISBN 978-3446407398.

[6] Beetz, S.; et al: Ausbildungskonzept Koordinatenmesstechnik – AUKOM. (Education Concept Coordinate Measuring Technology - AUKOM) In: DGQ (Hrsg.): FQS-DGQ-Band: 81-01, 2002 – Project report.

[7] Weckenmann, A.; Beetz, S.: Training in Coordinate Metrology - EUKOM: Made in Europe - Open to the world. IMEKO XVIII World Congress and IV Brazilian Congress of Metrology "Metrology for a Sustainable Development" (17.-22.09.06, Rio de Janeiro, Brazil). In: Book of summaries. IMEKO, 2006, p. 7. - Proceedings on CD-ROM.

[8] Keferstein, C. P.; Engelhardt, W.: Hochschulübergreifende Ausbildung in der Qualitätsprüfung und Fertigungsmesstechnik im Rahmen der Internationalen Bodenseehochschule (IBH); (University-network in Education for Quality Control and Dimensional Measurement within the International University of Lake Constance); Beiträge zum 6. Tag der Lehre. Fachhochschule Ulm 24. November 2005. S. 29-32. Hrsg.: Studienkommission für Hochschuldidaktik an Fachhochschulen in Baden-Württemberg. ISBN 3-00-017324-2.

[9] Keferstein, C.; Marxer, M.; Jakubiec, W.; Weckenmann, A.; Beetz, S.: EUKOM – European Training for Coordinate Metrology. At: 8th ISMQC, October 12th-15th, 2004, Erlangen, Germany.

[10] Keferstein, C.P.: Fertigungsmesstechnik - Praxisorientierte Grundlagen, moderne Messverfahren. (Dimensional Metrology – Practice Oriented Basics and Modern Technologies) 7. Auflage. Teubner Verlag, Berlin, 2011. ISBN: 978-3-8348-0692-5.

[11] Weckenmann, A.; Blunt el.; Beetz, S.: European Training in Coordinate Metrology, 2005. ISSN 1618-9477, ISBN 3-88555-775-6.

[12] http://www.vision21.org.uk/ (August, 30rd, 2010).

Manufacturing Sustainability: Aligning Youth Mindsets

C. Fernandes, L. Rocha
Technological Centre for Metal Working Industry, Porto, Portugal

Abstract
With the present work the authors aim to explore the role of hands-on programs as an important basis for the development of a sustainable manufacturing mindset in youngsters towards industry and future career development.
This project rests on the belief that hands-on programs are important tools for education in consideration of the global citizen and of the sustained society development. The empirical part of this paper will be developed according to the case study approach methodology and grounded on the "Think Industry Project" (TIP), a hands-on program focused on experiencing, and on contradict the myth that manufacturing is not a sustainable economic activity. The authors will analyse and discuss the Technological Centre for the Metal Working Industry' (CATIM) experience as a member of the Association of the Technological Centres of Portugal (RECET) in TIP. CATIM has had this project running since 1995 and it has encompassed more than 60.000 youngsters with ages between 13 and 17 years old. The main objective for TIP is to develop a positive vision of the industry, of employment opportunities and technical careers in the sustainable industrial sector, towards sustainable and active citizenship. In other words to promote "industrial literacy". The authors will centre on one experiencing activity - "This is an idea!" (Tiai) game. Tiai is a pedagogical game designed and implemented in the field with the intention to: (1) foster entrepreneurial spirit focused on innovation and sustainability manufacturing, (2) to show the importance of brands and Industrial Property' (IP) rights and (3) highlighting the opportunities open in engineering careers for future career development.
The importance of hands-on projects (e.g. TIP) and activities (such as Tiai) mixing technology, innovation, sustainability and entrepreneurship, as means for promoting education for sustainable innovation and entrepreneurship.

Keywords:
CIRP International Conference; Industrial Literacy; Manufacturing for Kids; Entrepreneurship; Sustainable Manufacturing;

1 INTRODUCTION

The ambitious objectives established in the Lisbon strategy [1] [2] [3] can only be achieved through sustainable entrepreneurship behaviors aimed at the industry. But, its success will depend upon continuous innovation in products, services and processes sustained in a qualified workforce. Promoting youngsters' global awareness towards lifelong learning, innovation and entrepreneurship are central issues. There are several research groups and associations that focus on these problems (e.g. Manufuture).

As we know, not only in Europe but also worldwide, there is a vocation crisis for the technical careers related to industry. And, in the particular case for the metal working industry, it has been systematically associated with "dirty, hard and difficult" tasks. Having these stereotypes in mind youngsters tent to choose "easier" paths and careers with another public image and recognition. Various studies have issued students' attitudes towards "industry", "science" and "technology". Research shows that youngsters often have stereotypical images and that these affect their attitude toward science, industry and technology [4] [5] [6] [7] [8] [9] [10] [11]. It appears that if a youngster can see himself/herself in a career, then, the likelihood of that person pursuing an educational program to prepare him/her for that career increases [12].

Educational programs (in or out-of-school) in the formal or informal teaching systems should be designed according to the needs shown by the industry and global trends. Only by aligning offer and demand, can countries achieve competitiveness and excellency. It's noticed that global awareness of this problem is a reality nowadays as we can prove in several conferences and seminars on the topic, training programs, papers, literature, programs and projects in educational and industrial fields, among other examples.

The Manufuture agenda focuses on the trends for the new industry and consequently on training and development needs (training and lifelong learning). This agenda [14] focuses on the transformation of industry, research activities and approaches to learning and industry. Manufure addresses the main drivers for change:

1. Competition, especially for emerging economies;
2. Shortening life cycle of enabling technologies;
3. Environmental and sustainable issues;
4. Regulatory climate;
5. Values and public acceptance.

This transformation (personal, societal and global) is seen through five strategic pillars [14] and their associated enabling technologies:

1. New added value products and services;
2. New business models;
3. Advanced industrial engineering;
4. Emerging manufacturing sciences and technologies;
5. Infrastructures and education.

The authors' approach is grounded on the "transformation of research and development" objectives and on pillar five "Infrastructures and education". The strategic goal number three for the Manufuture agenda can be transcript as: "A Europe-wide co-operative Research and Technological Development (RTD) infrastructures with systems and education that favour collaborative research efforts for manufacturing excellence promotes technology transfer and market take-up of R&D results, provides competitive technical assistance, and improves lifelong learning and re-training of a workforce including growing numbers of aging and displaced workers" [14]. It's clear in this definition that lifelong learning and training are central problems for the development of a capable workforce. Nevertheless, research, technological development, innovation and entrepreneurship are central themes for the bases of learning programs.

Starting sensitization by younger people can help globally and individually. Globally by helping with the mentality shift and the promotion of global awareness. Individually as a mean to foster curiosity, entrepreneurship spirit and the adoption of sustainable behaviors.

2 HANDS-ON PROGRAMS

A growing body of literature suggests that variations across countries, in entrepreneurial activity and the spatial structure of economies could potentially be the source of different efficiencies in knowledge spillovers, and ultimately in economic growth [15]. These contexts lead to the need of and desire for designing programs for children and young adults to make the most of time-out-of-school (and in-school) to contribute positively towards a productive and conscientious adulthood. The youth of today will be the working force and the active citizens of tomorrow.

Over the past 20/30 years "scientific literacy" is one of the goals for science education worldwide (in and out of school). The proofs of it are the several educational reports, networks and projects on the subject, e.g. "EU-HOU – Hands-on Universe Europe" [16], "Hands-on science" [17] as European Projects; "Beyond 2000: Science Education for the Future" [18], "Sector Workforce Development Plan for Engineering Manufacture" [19], activities under "Imagineering Foundation" [20] in the United Kingdom; "Cité des metier" [21], "Youngs in Industry" [22], "Bravo Industry" [23] in France; "Science for all American – Project 2061" ([24] [25] and "Benchmarks for Science Literacy" [26] in the United States. Programs like TIP (description and analyses bellow) also intend to promote "industrial literacy" and knowledge of several industrial processes and trends and its links with everyday life contexts.

Through an adequate formal interaction with adults', youngsters (with different ages) may develop important scientific skills [27], attain vocational identity development [28], develop alternative visions of reality and "industrial literacy". According to Johnston and Gray the interaction with adults and other youngsters may [29]:

1. Promote experiences with a specific focus or learning objectives.

2. Function as a role model by observing and expressing ideas themselves.

3. Create an opportunity for asking questions to challenge thinking or develop the experience further.

4. Act as motivation for students to express their ideas and look for other extended, similar or new experiences.

The authors defend that this interaction, specially the one based on TIP activities, also allows youngsters to:

1. Give new meanings and interpretations to reality;

2. Rationalize the impact of their learning in real settings;

3. Promote sustainable consumption and production awareness;

4. Develop new ideas promoting innovation and entrepreneurship;

5. Gain skills concerning divergent thought (e.g. creativity techniques) and viability evaluation (e.g. Tiai game);

6. Promote awareness for technical careers related to industry;

7. Promote awareness of industrial processes and its links to everyday life;

8. Promote and discuss vocational identity development.

Think Industry Project

TIP is a hands-on program designed for the Portuguese reality that encompasses youngsters from 13 to 17 years old. It was based on previously developed and validated models. CATIM has had the project running since 1995 [30] and more than 10.000 youngsters were involved in it just in CATIM.

TIP is a wide project, so as a consequence it encompasses several different stakeholders: general citizen, parents, youngsters, technological centres network, industrial companies, universities, polytechnical schools and research institutes, industry and professional bodies, education and training providers, national and local government, government agencies.

The main objectives for TIP are to develop a positive vision of the industry and of employment opportunities and technical careers in the industrial sector towards a sustainable and active citizenship. In table 1 we present the main problem dimensions, specific targets and general objectives for TIP.

Table 1. Problem dimensions and TIP's specific objectives

Main Problem Dimensions	Specific Targets	Final Goal
Traditional image for the industry	Develop a positive vision of the industry, link the industry to positive values and attractive careers	**Develop a positive vision of the industry and of employment opportunities and technical careers in the industrial sector towards an active citizenship**
Withdrawal among youngsters at a school age and industrial activities and careers	To make youngsters and industry closer (and vice-versa), involve youngsters and industry in mutual approximation processes	
Training choices and market integration heavily influenced by commerce and services	Make youngsters aware of industrial careers in the short term, motivate youngsters to carry on their studies on technological areas	

TIP is also designed to show youngsters new ways of learning, working and living in society e.g. using e-learning and Communities of Practice [31] [32] [33] [34]. There are several activities designed to foster innovative thought development and to promote entrepreneurship and industrial property rights.

3 TIP ACTIVITIES

TIP activities are clustered into three main segments: Lab TI, Distance TI and Tech TI. Each of these segments has its own target audience and specific activities.

The activities developed under the TIP scope, generally tend to promote the understanding of different settings of the industrial value-chain, e.g. "simulation games" that represent management actions and functions, and "technology laboratories" that correspond to the manipulation of equipment related to the industrial activity and the underlying technologies applied in several industrial processes such as robotics, hydraulics, energy consumption, environment, mechanics, milling, lathering, etc..

Lab TI' activities are designed for groups of approximately 15 students from the formal teaching system that come to the technological center every two weeks for a session (global length 32 hours). The activities vary from simulation games (e.g. Tiai) to field trips and to laboratorial activities encompassing several "industrial themes" such as robotics, alternative energies, metrology, and computer numeric control (for all the areas please see table 2).

Table 2. Think Industry activities

Segments	Type	Activities
Lab PI	Simulation games	"This is an Idea"
	Laboratory of Industrial Technologies	Robotics; Alternative energies; Metrology; Computer numeric control; electronics; prototyping; haptical devices; aerodynamics; Mechanics; 3D design software
	Field trips	Industrial companies; Universities; Industrial exhibitions
Distance PI	Online contents	How daily things are manufactured
	Simulation games	"This is an Idea"
	Practical experiments	Assemble a solar energy kit
Tech PI	presentation	Small presentation about the Industrial life cycle
	demo	Short demonstration with some industrial technologies

Distance TI' activities were thought of for all those youngsters that cannot physically go to CATIM (for example due to geographic distance). By using the internet and accessing a Learning Management System (LMS) youngster can learn and explore industrial concepts and do some simulation and experimentation. Distance TI encompasses two presential sessions facilitated by one professional/tutor. The activities vary from simulation games, to on-line contents exploration to practical experiment (e.g. assembling a solar energy kit).

Tech TI' activities are exactly the same as the ones in Lab TI, but it's a more intensive program. There are also groups of 15 students that come to CATIM 3 hours every day for two weeks, normally during the Easter or Summer holidays.

4 TIAI GAME

As an example of the activities developed under the TIP scope, the authors will focus on the activity "This is an idea" a game designed to foster product cycle, innovation and the entrepreneur spirit among youngsters. Tiai is an industrial simulation game to be played by groups. Each team has to come up with an idea for a new product/service. This product/service must have three criteria so that it can go forward:

1. There shouldn't be any product/service alike in the market.
2. It should be feasible and sustainable.
3. And shall have selling potential.

The next step is the constitution of an industrial enterprise. At this point it's necessary to go through every industrial stage, since the birth of the "idea" until the product/service is placed in the market. The objective is to understand the product and/or service industrial cycle and to promote entrepreneurship and innovation among youngsters. The general phases of an industrial cycle are illustrated in figure 1. This is an iterative cycle.

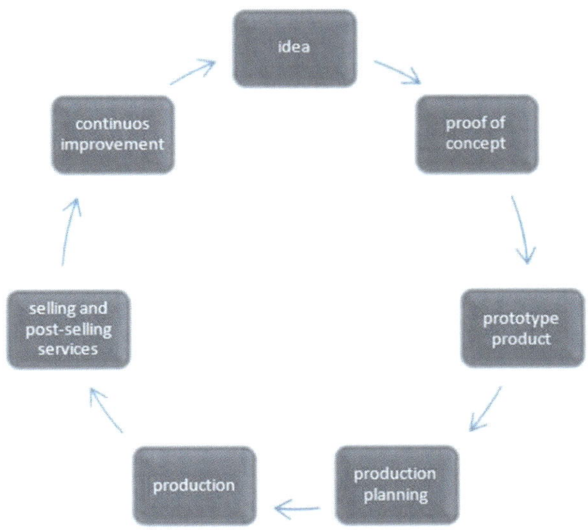

Figure 1. Tiai – Industrial cycle

The game has mainly three phases:

1. Team constitution – the groups should have four to six players between 14 and 15 years hold .

2. Role definition - each player in the team will have a pre-defined role in. Each youngster will have a different role. The process by which the role is chosen is managed by the team, it can be by direct choice, by chance, by appointment. One CEO and one manager/Chief must be appointed by the following areas: Selling and Marketing; Innovation and Development; Finance; Production; Human Resources.

3. Idea operationalization - fill in a form with the main idea assets.

The game begins with a "brainstorming" between all the managers and the CEO, this should take approximately 30 minutes. All the team members give ideas (no matter how silly they might appear) for the development of the new product/service. One idea is chosen based on the commercial potential. The product to be developed and produced by the company is hereby defined. Each "manager" must think about several aspects (see figure 2) and the main conclusions are used to operationalize the idea and to fill in the final form.

The final form contains aspects such as:

- Product/service description (e.g. main characteristics, advantages, disadvantages);
- Selling and marketing (e.g. target public);
- Materials to produce;
- Product/service price;
- Product schema prototype.

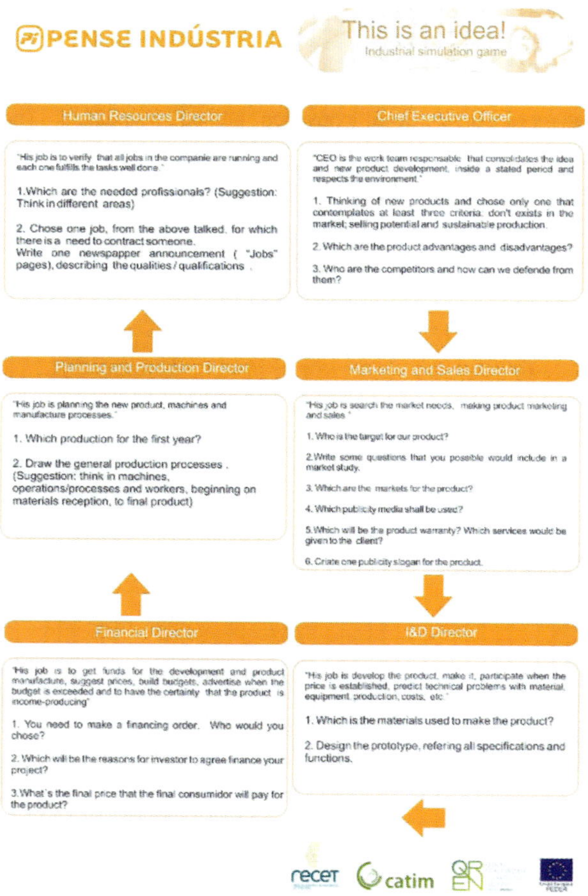

Figure 2. Tiai - Industrial simulation game activities path

For each Technological Centre the three best ideas are chosen according by a one jury. These ideas enter in a National innovation and entrepreneurship contest. In 2008 there were more than 500 ideas from four different Technological Centres at competing.

5 TIAI AN EXAMPLE – "ECOWATERS"

The idea selected by CATIMs' jury for the 2008 year edition of the "Innovation and Entrepreneurship Contest" was named "EcoWaters".

"EcoWaters" was an idea developed by a group of 4 students with ages varying from 14 to 15 years old and in the Portuguese Formal Teaching System. From the first brainstorming session lots of ideas came up, although most of them already existed, e.g. microwave with internet connection, cell phone with solar panel, a pair of tennis shoes that grow with the feet. Then the team started to think of everyday life tasks and arrived at one conclusion: "in society there is a need for products that are both ecological and profitable". This was the pin point when the ideas started to get shape. They figure that when someone wants to take a warm shower, some phases are followed: i. switch on the hot water faucet; ii. Wait for the water to get to the right temperature, and only then they would shower. During in this process good water is wasted, so, having this problem in mind the team decided to solve it. The challenge was pointed out "Why not develop a mechanism that could save water until it reached the right temperature for use? This was the main idea generation process that lead to the birth of "EcoWaters" – a simple idea, with sustainable and environmental concerns, using existing resources, cheap for the final user and that allows real water saving.

The technical system on which "EcoWaters" was based consisted in a temperature sensor and an interconnected valve. The valve and the sensor only allowed the water to flow to the shower when a certain pre-set temperature was reached. The other water was deviated to a parallel system for other domestic uses.

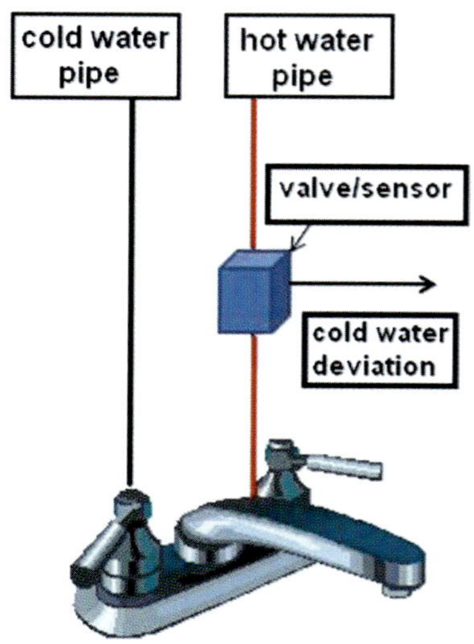

Figure 2. "EcoWaters" idea schema

6 ANALYSES AND DISCUSSION

TIP was not designed to achieve "scientific literacy" but to promote "industrial literacy", although it's based on scientific and engineering principles. By industrial literacy we mean to promote knowledge on several aspects of the industry, industrial product/service cycle, how things work, the application of school contents to industry and industrial processes, IP rights, the perception of industry, etc.. But TIP activities are also designed to attend to specific objectives of developing innovation, creativity and entrepreneur spirit among youngsters. The conjunction of content knowledge, process skills knowledge and sustainable entrepreneurship awareness is added to all of the TIP's activities helping youngsters to give meaning to curricula contents and everyday life activities.

With Tiai game, namely the 2008 winning idea exposed above – "EcoWaters"- the authors want to show that innovation, big breakthroughs, and sustainable entrepreneurship appear anywhere, with people any age! Entrepreneurship spirit can be incited. Intellectual creativity and "different ways" of seeing the same reality are important assets for the development of entrepreneurship capabilities. We can firmly notice that prior to TI' activities youngsters have generic and imprecise images of the industry, careers related to, industrial cycles, etc, but post-TI activities, perceptions were more precise, youngsters showed a bigger awareness of sustainable entrepreneurship issues and made the link between their actions and technology applications, industry and future careers. TIP allows its participants to get in touch with new technologies applied in real settings, to give meaning to several everyday life activities and to curricular contents, along with the rationalization and their application to the industry and industrial settings. TIP also helps youngsters to create mindsets for the industry and technical careers related to engineering and technology and the value of innovation and entrepreneurship. The images that youngsters have of relevant workplaces in the industry, also changed prior and post experiences/field trips, at first they were superficial, unreal and even incorrect. These images might be formed due to several factors e.g. school, TV, movies, news, parents, and teachers. As we have exposed above, when people can see themselves in a career, the likelihood of that person pursuing an educational program to prepare him/her for that career increases. Following that belief, TIP tries to give them new tools so that they can rationalize and make an informed career choice based on correct facts and not on misconceptions. Moreover, these approaches to learning and development processes also provide opportunities for the develop of entrepreneur spirit and social skills, e.g. communicating ideas, taking turns, sharing, helping, working as a group, "selling" ideas to others. This set of skills is transversal to all fields of knowledge and important for an active work life.

7 OVERALL CONCLUSION

Industry, namely manufacturing has huge potential for generating wealth, jobs and better life-quality. The "new" industry must compete by adding value to the processes and products, and not by reducing costs. These assets are made possible by innovation and entrepreneurship. Industry must attract and hold on to capable and qualified people going beyond demographic trends of aging and the unattractiveness of industry as a career. Solutions must encompass a whole lot of areas, such as formal and informal education and life-long learning among many others. The career context crises must be overlapped and youngsters must be motivated to follow technical careers and to be entrepreneurs in a sustainable way.

The authors defend that entrepreneurship can be encouraged and promoted at a young age in a variety of forms e.g. Tiai game. There are contexts that facilitate entrepreneurship, environments that are open to new ideas where there is freedom to research and problematize, sometimes these assets are not present in public schools and it's also important to make youngsters aware of their capabilities and power to innovate and to be entrepreneurs. A society that is open to entrepreneurship and innovation must also deal with failure; good ideas are not always big breakthroughs. On the other hand there are also environments that inhibit entrepreneurship and innovation, e.g. places or schools with excessive regulamentation and control, and not open to new ideas/activities. Youngsters (and generally people all ages) as entrepreneurs need to have a keen eye to understand economic, social, and scientific realities and the capacity to understand evolutionary processes in the future. In other words, they must have the capability to "see the whole setting". Scientific data, common sense and intuition have told us that there is much to be done in the innovation and entrepreneurship fields to achieve economic growth. Government, business people and researchers, are aware and are taking actions. But, programs like TIP and activities like Tiai show that everybody has the power to act. Youngsters have an immense innovative and entrepreneur potential that must be encouraged and fostered so that an entrepreneurial, innovative culture and a future sustainable Industry is possible.

8 ACKNOWLEDGEMENTS

Portuguese and European Funds (Compete/QREN) finance Think Industry Project (project FCOMP-05-0128-FEDER-005248).

This project is promoted by RECET, the Portuguese Technnological Centres Network.

9 REFERENCES

[1] Lisbon Agenda, 2011, Available at http://www.estrategiadelisboa.pt/.

[2] Rodrigues,M. J., 2009, Europe, Globalization and the Lisbon Agenda in collaboration with I. Begg, J. Berghman, R. Boyer, B. Coriat, W. Drechsler, J. Goetschy, B.Å. Lundvall, P.C. Padoan, L. Soete, M. Telò and A. Török, Edward Elgar.

[3] Tausch, A., 2010, Titanic 2010?: The European Union and Its Failed Lisbon Strategy (European Political, Economic and Security Issues Series) Hauppauge, Nova Science Publishers: New York.

[4] Berdslee, D. C., O'Dowd, D. D., 1961, The College-Student Image of the Scientist. Science 133: 997-1001.

[5] Brush, L. R., 1979, Avoidance of Science and Stereotypes of Scientists. Journal of Research in Science Teaching 16(3): 237-241.

[6] Chambers, D. W., 1983, Stereotypic Images of the Scientist: The Draw a Scientist Test. Science Education 67(2): 255-265.

[7] Ferrnandes, C., Rocha, L., 2007, Pedagogy for collaborative learning: Technological and human issues, Proceedings of the IADIS International

Conference on Web Based Communities 2007. IADIS, Salamanca, Spain.

[8] Finson, K. D., Beaver, J. B., Cramond, B., 1995, Development and Field Test of a Checklist for the Draw-a-Scientist Test. School Science and Mathematics 95(4): 195-205.

[9] Flick, L., 1990, Scientist in Resident Program Scientists, School Science and Mathematics 90(3): 204-414.

[10] Mason, C. L.; Kahle, J. B., Gardner, A. L., 1991, Draw a Scientist Test: Future Implications, School Science and Mathematics 91(5): 193-198.

[11] Mead, M., Metraux, R., 1957, Image of the Scientist Among High-School Students. Science 126: 384-390.

[12] Palmer, D. H., 1997, Investigating Students' Private Perceptions of Scientists and their Work. Research in Science & Technological Education 15(2):173-183.

[13] Smith, W., Erb, T., 1986, Effect of Woman Science Career Role Models on Early Adolescents, Journal of Research in Science Teaching 23(8):667-676.

[14] Manufuture, 2006, Strategic Research Agenda: Assuring the Future of Manufacturing in Europe – Manufuture Platform Full Report n.º1/2006, Manufuture, Brussels.

[15] Acs, Z., Varga, A., 2005, Entrepreneurship, agglomeration and technological change, Small Business Economics 24: 323-334.

[16] EU-HOU, 2006, Hands-on Universe Europe. http://www.euhou.net/.

[17] HSCI, 2006, Hands-on Science. http://www.hsci.info/.

[18] Millar, R., Osborne, J. F. (Eds.), 1998, Beyond2000: Science Education for the Future, King's College, London, UK.

[19] Engineering and Marine Training Authority, 2007, Sector Workforce Development Plan for Engineering Manufacture, http://www.mas.dti.gov.uk/content/resources/categories/web-link/LINK_EMTA.html.

[20] Imagineering Foundation, 2010, Imagineerig. http://www.imagineeringweb.co.uk/.

[21] Cité des Metier, 2011, http://www.citedesmetiers.fr/.

[22] Ministere de l'Economie des Finances et de l'Industrie, 2007, Youngs in Industry, http://www.industrie.gouv.fr/index_portail.php.

[23] Union des Industries et Métiers de la Metalurgie, 2007, Bravo Industry, www.metallurgie.org/notremaison/metiers/jeunes.html; www.uimm.fr/fr/index.html.

[24] American Association for the Advancement of Science, 1990, Science for all American: Project 2061. Oxford University Press, New York, USA.

[25] American Association for the Advancement of Science (1993). Benchmarks for Science Literacy. Oxford University Press, New York, USA

[26] National Research Council, 1996, National Science Education Standards, National Academy Press, Washington, USA.

[27] Martins, I., Veiga, L., 2001, Early Science Education: Exploring Familiar Contexts to Improve the Understanding of Some Basic Scientific Concepts, European Early Childhood Education Research Journal 9(2): 69-82.

[28] Taveira, C., Moreno, M.L., 2003, Guidance Theory and Practice: The Status of Career Exploration. British Journal of Guidance and Counseling 31(2): 1897.

[29] Johnston, J., Gray, A., 1999, Enriching Early Scientific Learning, Open University Press, Buckingham.

[30] Rocha, L. (1998). Promover a Inovação e o Empreendorismo Junto dos Jovens Pense Indústria Inovação, In Motor de Innovación Ponencias y Comunicaciones Presentadas en el V Congreso Galego da Calidade, Xunta de Galicia, Santiago de Compostela, Spain.

[31] Fernandes, C., Rocha, L., 2006, The Conceptualization and Analyses of a Value Network: How to Create Value with Inter Organizational Communities of Practice? In José Cordeiro, Vitor Pedrosa, Bruno Encarnação and Joaquim Filipe (Eds), Proceedings on Society, e-Business, e-Government and e-Learning for the Second International Conference on Web Information Systems and Technologies, INSTICC – Institute for Systems and Technologies of Information, Control and Communication, Setúbal, Portugal.

[32] Fernandes, C., Rocha, L., 2006, Value Networks the Source of Collective Community Intelligence: One Case Study, In Piet Kommers, Pedro Isaías and Ambrosio oikoetxea (Eds), Procedings of the International Association for Development of the Information Society – Web Based Communities 2006, IADIS, San Sebastian, Spain.

[33] Rocha, L., Fernandes, C., 2006, eCATIM, In A Implementação do eLearning nas organizações: Práticas de Sucesso [Organizational eLeanring implementation – success practices], PT Inovação, Coimbra, Portugal [Book in Portuguese].

2 Entrepreneurship

Entrepreneurship is a prerequisite for the creation of sustainable products and processes that conserve energy and natural resources; thus adhering to the principle of sustainability which considers the needs of the present without compromising the ability of future generations to meet their own needs. In the first section of this chapter Kohl and Al Hashemi explain the role of the science parks in supporting the knowledge development, as well as intellectual capital management in a rather resources-driven economy like United Arab Emirates. In the second section Abd-Elall et al. present a dynamic model for matching job market qualifications demand and educational market qualifications supply for more productive sustainability engineering education. In the third section Acar and Rother introduces a design thinking approach in engineering education and its adoption in technology-driven startups. In the fourth section Mrozewski et al. explain the technology-oriented university start-ups and a business incubator to enhance the entrepreneurial activities of the students, graduates and researchers. In the fifth section Qureshi and Kratzer propose a framework for marketing capability construct and the various factors influencing the marketing activities in the small technology firm sector.

Science Parks as Main Driver for the Development of National Innovation Systems in Resources-driven Economies! The Importance of Intellectual Capital Management for Sustainable Manufacturing

H. Kohl[1], H. Al Hashemi[2]

[1]Department of Business Excelllence Methods, Fraunhofer IPK, Berlin, Germany

[2]Dubai Institute of Technology, Dubai TechnoPark, UAE

Abstract

A new form of economic development is taking shape worldwide – the so-called “knowledge revolution”. Knowledge is a cornerstone in human development and helps to expand human capabilities. Knowledge has also increasingly become a dynamic factor of sustainable manufacturing and a powerful driver of human capital and productivity. Recent studies about the R&D performance as well as about the R&D capabilities in the Arab region and UAE especially document a serious shortfall in knowledge generation, acquisition, absorption and use as one of the major deficits undermining development. Therefore Science Parks became an important factor of support the knowledge development as well as Intellectual Capital Management in a rather resources-driven economy. This paper will introduce a new approach for Science Park Management by giving an example of UAE.

Keywords:

Intellectual Capital, Knowledge Management, Science Park Management, Sustainable Manufacturing

1 INTRODUCTION

A new form of economic development is taking shape worldwide – the so-called “knowledge revolution”. Knowledge is a cornerstone in human development and helps to expand human capabilities. Knowledge has also increasingly become a dynamic factor of production and a powerful driver of human capital and productivity [3]. The development of the knowledge-based economy puts an increasingly high emphasis on the need to produce, exploit, transfer and apply knowledge [7]. In this context, there seems to be a general agreement about the need to develop and strengthen networking activities between the actors of the so-called triple helix model [15], i.e. the public sector, the business community and the higher education institutions. Intermediaries can facilitate the interaction between these three key groups and liaise with public and private R&D funding and venture capital organizations [18]. Hence, this paper will introduce a new approach of how to manage knowledge intensive organizations and structures like Science Parks by giving an example of the United Arab Emirates.

2 BACKGROUND

Recent studies about the R&D performance as well as about the R&D capabilities in the Arab region and UAE especially documents a serious shortfall in knowledge generation, acquisition, absorption and use as one of the major deficits undermining development [27]. *Therefore Science Parks became an important factor of support the knowledge development as well as Intellectual Capital Management in a rather resources-driven economy, like Dubai and UAE.*

The world devoted app. 1.7 per cent of gross domestic product (GDP) to R&D in recent years. In monetary terms, this translates to app. US$ 830 billion, according to estimates by the UNESCO [26]. These global distributions conceal huge discrepancies. They reflect the enormous divide in terms of development, prosperity, health and participation in the world economy but also in world affairs in general. The question e.g. for resources-driven economies like the GCC-Countries is, what is their role concerning R&D production in the world and what are the challenges e.g. for Dubai and U.A.E. on the way to-wards a value-based knowledge economy? It will take a handful of indicators to answer these questions. Looking at several indicators, there is no doubt that there are wide margins between the scores of regions or countries. Distinctions can be identified between several regions. Only a few countries in the world are producing science and are benefiting from it. To give an overview of the worldwide scientific activities, it is useful to compare the GERD/GDP-ratio for several countries and areas. The shares of North America and Europe in world GERD are on a gently down-wards sloping path. North America was responsible for 38.2 per cent of world GERD in 1997 but 37.0 per cent in recent years. For Europe, the corresponding indicators are 28.8 per cent in 1997 and 27.3 per cent in recent years [26].

The most remarkable trend is to be found in Asia, where GERD has grown from a world share of 27.9 per cent in 1997 to 31.5 per cent in recent years. As for the remaining regions, Latin America and the Caribbean, Oceania and Africa, these each account for just a fraction of the total, at respectively 2.6 per cent (down from 3.1 per cent in 1997), 1.1 per cent (stable) and 0.6 per cent (stable) [26].

Another remarkable fact is all Arab States together were only responsible for 0.2 per cent of world GERD in recent years. This is a first indicator that the Arab States are faced with great challenges to become a value-based knowledge economy.

3 EXAMPLE OF A NATIONAL INNOVATION SYSTEM IN A RESOURCE-DRIVEN ECONOMY

To overcome this shortfall, the Dubai Institute of Technology (DIT) will support Dubai in becoming a leading place in the world to do R&D by setting up a science, engineering and technology base. DIT supports Dubai as well by participating in the current increase in the levels of overseas direct investments for commercial R&D conducted abroad. Such investments are a good indicator of an increased competitiveness of Dubai as a location of choice for the increasing globalization of corporate R&D on the one hand and local R&D on the other hand. International research has consistently demonstrated the positive correlation between R&D investment intensity and company performance measures such as sales growth and share price in the sectors where R&D is important. Countries aiming to become knowledge economies have to embark on generating science and technology advancements and exploiting them. Within this context, in 2008 Dubai Institute of Technology (DIT) was founded as a science park to support Dubai in becoming a leading place in the world for R&D activities. The DIT is a fully owned subsidiary of Economic Zones World (EZW), which belongs to Dubai World, one of the largest Holding Companies worldwide. It symbolizes Dubai's vision to develop a knowledge-based technology-centric sustainable business hub, which will not only support the country's research and development needs in its core sectors but will also support the Emirates' long term economic development and growth [1].

The Mission Statement of DIT is to develop technologies (R&D) focused on a limited number of research areas corresponding to major fields of the progress of knowledge and technology. With "a solving approach" activities answer to market needs and major socio-economic challenges facing the region. Furthermore the delivery of services (R&D) is a cornerstone to maximize the impact of R&D; to pool all the capacities of the regional S&T system, and to integrate science and technology in the economy.

For fulfilling the objectives, DIT sets up an ecosystem which consists of two main substructures, namely the DIT Core and the DIT Strategic Research Areas (Centres of Excellence).

The DIT Core consists of DIT University, DIT R&D-Convention Centre and Science Museum, DIT Services (including Business Development, Policy Advice for R&D, IP Management, Incubator, Investment and Professional Services) as well as the Dubai Institute for Sustainability as DIT's own Research Institute.

The DIT Strategic Research Areas include Health, Energy, Water, Mobility/Logistics, Industrial Engineering and Social Economic Studies. Besides Social Economic Studies, which is part of Dubai Institute for Sustainability, all Strategic Research Areas are established by strategic Joint Ventures and/or Direct Investments with global R&D players respectively R&D intensive industries. To allow the development of a market oriented Innovation System, the overall Ecosystem of DIT is structured as follows:

- The strategic business model of DIT comprises the main objectives, business processes, targeted results as well as the required resources.
- In order to accomplish the strategic objectives representing the complete strategic scope of the DIT Ecosystem three main value adding processes can be distinguished: (1) Research, (2) Technology Development and (3) Service Delivery.

Implementing the business strategy by running the value adding processes successfully leads to the intended business results. The value adding processes themselves are mainly supported by Intellectual Capital Management, namely the human-, structural-, and relational capital management. While human capital management ensures the attraction of best in class knowledge workers through excellent research infrastructure, high quality of life and financial incentives, structural capital management generates sustainable R&D and industry structures through focused funding of R&D projects and incubating entrepreneurs. The attraction of high level R&D organizations and companies to collaborate on strategic research areas is enabled by relational capital management.

4 INTELLECTUAL CAPITAL MANAGEMENT

Following the most frequently used structure to describe intangible assets, the approach for Intellectual Capital Statements divides Intellectual Capital into three dimensions: Human, Structural and Relational capital [2, 5, 12, 13, 16, 23, 24, 25]. Human Capital (HC) includes the staff's competencies, skills, attitudes and the employee's motivation. Human Capital is owned by the employee and can be taken home or onto the next employer. Structural Capital (SC) comprises all structures and processes needed by the employee in order to be productive and innovative. According to a sloppy but useful definition, it "consists of those intangible structures which remain with the organization when the employee leaves". Relational Capital (RC) sums up the organization's relations to customers, suppliers, partners and the public [6, 14].

It is a common ground, that *Intellectual Capital (IC) has become the critical success factor for science parks as well as enterprises operating in a knowledge driven economy.* Especially for science parks it is crucial to utilize and manage their intangible resources efficiently in order to obtain their competitive advantage, since they highly depend on specialized human, structural and relational capital for successful differentiation on the market.

The so-called "new growth theory" is following a similar argumentation by saying that the classic factors of production – labour and capital – do not explain the economic growth and its processes well enough in the globalised world characterized by the importance of knowledge and creativity [11, 22]. Furthermore, the law of diminishing returns of the classic theory is not as useful approach to study growth today as it was before. The new theory emphasizes the process of accumulation of knowledge that results from investment in new technologies and human capital. This, in turn, influences the growth of profits and other benefits. In other words, economic growth is based on the increasing returns associated with knowledge or rather Intellectual Capital. Consequently, growing attention should be paid to all the factors and policies/policy instruments that support and provide incentives for the creation, transfer, introduction and application of knowledge.

4.1 A MANAGEMENT SYSTEM FOR SCIENCE PARK

The fundamental purpose of a management system is to allow for the achievement of the strategic objectives an organization may target through the accomplishment of all related management tasks. This applies irrespective of the scope of management tasks that are addressed by the particular management system in question.

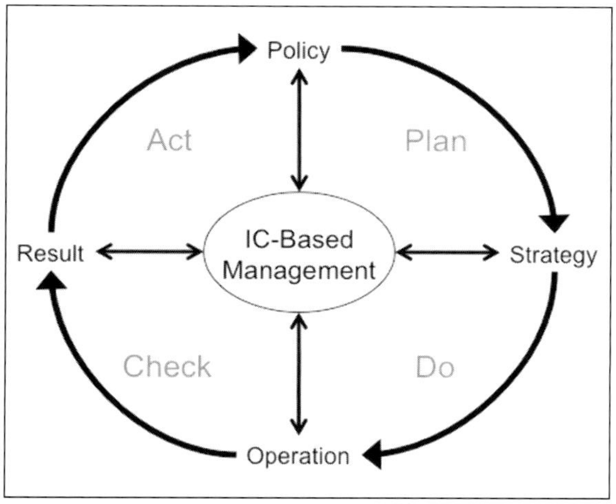

Figure 1: An IC-Based Management System in Terms of the Deming-Cycle [1].

Certainly, the main aim of a Science Park's management is to operate the park economically successful. In order to achieve this overall strategic objective the park management has to accomplish a bunch of specific tasks. Specific management tasks of a Science Park that derive from the Science Park definition are the following [1, 8, 9, 10]:

1. To establish and manage a long-term development plan with dynamic strategic goals including aspects of sustainability.
2. To select or reject which firms enter the park on the basis of the coherence between the firm's business plan and the park's strategic goals.
3. To stimulate and actively manage the knowledge transfer amongst universities, R&D institutions, companies and markets.
4. To stimulate and actively manage the technology transfer amongst universities, R&D institutions, companies and markets.
5. To establish and manage a close relationship with the affiliated science/research institution(s) of higher education.
6. To provide the tenant innovation-led, high-growth, knowledge-based businesses with an environment where they can develop specific and close interactions with a particular centre of knowledge creation and with each other for their mutual benefit / guarantee a stable interface between academia and industry.
7. To provide the tenant ventures and innovation-based companies with business incubation and spin-off support services.
8. To provide the tenant companies and organizations with other value-added services.
9. To market the high valued products and services of the park.
10. To provide the tenant companies and organizations with adequate high quality space and facilities.

The Cabral Dahab Science Park Management Paradigm [8] has been influential in the management of Science Parks around the world and lays down the following conditions for the successful operation of a Science Park [9, 10].

Thus, a Science Park must:

- have access to qualified research and development personnel in the areas of knowledge in which the park has its identity,
- be able to market its high valued products and services,
- have the capability to provide marketing expertise and managerial skills to firms, particularly Small and Medium-sized Enterprises, lacking such a resource,
- be inserted in a society that allows for the protection of product or process secrets, via patents, security or any other means,
- be able to select or reject which firms enter the park. The firm's business plan is expected to be coherent with the Science Park identity,
- have a clear identity, quite often expressed symbolically, as the park's name choice, its logo or the management discourse,
- have a management with established or recognized expertise in financial matters, and which has presented long term economic development plans,
- have the backing of powerful, dynamic and stable economic actors, such as a funding agency, political institution or local university,
- include in its management an active person of vision, with power of decision and with high and visible profile, who is perceived by relevant actors in society as embodying the interface between academia and industry, long-term plans and good management,
- include a prominent percentage of consultancy firms, as well as technical service firms, including laboratories and quality control firms.

A Management System for Science Parks must therefore allow connecting regional/local knowledge generation and knowledge utilization! [1]

5 THE NEXT GENERATION OF STPS MANAGEMENT

Innovation and innovativeness play the key role in national and regional economic growth and national and regional competitiveness. The knowledge-based economy is developing more and more. In this type of economy the basic economic resources are not material, human resources or capital, but knowledge. This economy is global and interlinked. In this process, universities and research and development institutions as resources of knowledge fulfil a basic role. Knowledge creation through research and development (R&D) is the principle initiator in forming this new economy. That is why there is a strong need to combine knowledge theory and business practice, a strong need to strengthen the cooperation between two different environments: academic and business. A Science Park is a widely accepted tool, which is promoting this cooperation.

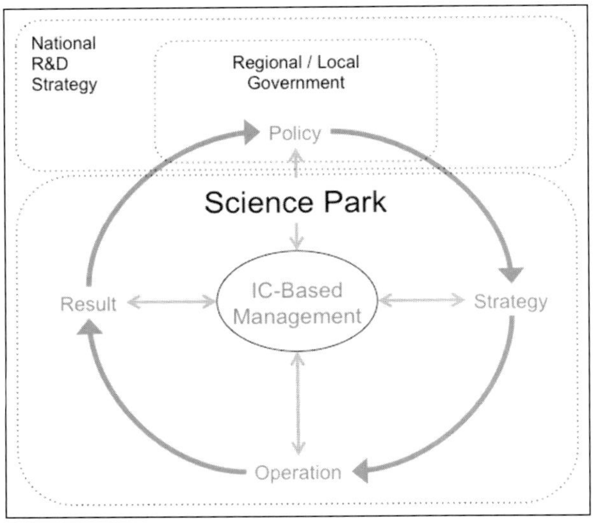

Figure 2: Science Park Management Process – Entities Involved [1, 19]]

Science Parks provide particular conditions for the development of a partnership and collaboration between businesses and public research organizations. While some of these initiatives focus on the formal exchanges of knowledge, and as such have a limited number of target firms, the concept of the Science Park focuses instead on the development of informal links, which are fostered by the physical proximity of producers and users of knowledge. Science Parks do have wider informal contacts that are fostered through the close proximity between new technology-based firms (NTBFs) and universities or other research institutes.

Among the current evolution of Science (and technology) Park generations, a more comprehensive park set-up is being implemented. These "third generation Science Parks" are located within a vibrant urban community and are perceived as the quintessence of science-industry-government relations, increasingly functional and specialized along with their participation in local, regional and even global innovation activities [4]. At the same time, this Science Park generation could become a contradiction in terms, as its management is striving to eradicate the fixed spatial boundaries of the park for it to become truly embedded the urban fabric as a catalyst for innovation. The emerging third generation Science Parks operate interactive models of innovation, embedded in diverse urban environments. In such areas, networks and systems of trust, the development of respective public, private or scientific partners, cultures of interpretation, degrees of public or institutional participation as well as the availability of financial/legal instruments all for an integral part of the innovation environment's global function. Location embeddedness is no longer just a feature, but a key success factor.

However, Science Parks seem to fail to perform as well as might be expected with regard to not only the promotion of Higher Education Institute (HEI) / industry linkages but also the transfer of technology and knowledge from HEIs to Science Park firms.

Science Park key success factors reflected by the literature [8, 9, 10] focus mainly location factors, property management skills and a quality management team. This indicates that although both favourite location factors and firms' integration and interactions are important for the growth of a spontaneous high technology industrial cluster, these have not yet been fully applied in developing Science Parks.

Science Park managers should try to create the highest level of integration and interactions both inside and outside their parks. These include universities, research institutes, and government organizations, but also any sector that may be related with tenant firms in any way. This can lead to easier knowledge transmissions, more innovation, and effective management, which can enhance firms' long-term competitiveness. The success of tenant firms represents the success of a Science Park. The dynamic environment thus formed may attract more actors locating their businesses near the park. In such a way, an industrial cluster may emerge. A Science Park therefore creates the potential development for the region. The best acting Science Parks are in the regions where there are strong network connections and links among the region's knowledge capital, innovation capacity, knowledge economy outputs and knowledge economy outcomes.

Acting as an incubator for innovation, Science Parks are designed to strengthen the link between science and business. The generation and transformation of scientific and technological knowledge within a complex network of different actors is their main activity. Thus, the success of a Science Park highly depends on its ability to utilize and manage its intangible resources efficiently in order to obtain their competitive advantage, since they highly depend on specialized human, structural and relational capital for successful scientific and technological performance [20].

So far, these important aspects have not been addressed by management activities of Science Parks sufficiently. While the number of empirical studies on the performance of Science Parks or the Science Park's tenant businesses is substantial, little advance has been made to systematize this wealth of knowledge and from there generate a theoretical framework which allows for concrete, but flexible, recommendations to managers.

This paper presents a new approach to Science Park management with a special focus on the intangible resources and success factors, the so called Intellectual Capital (IC). It stresses the importance of active IC-based management in complex and knowledge-intensive business environments and presents a process-oriented management system on the basis of human, structural and relation capital management.

Science Parks have to operate as enablers in a three-layer design of strategic stakeholder orientation, with regional/local government and policy being the upper layer, the affiliated HEIs and (public/private) R&D institutes being the middle layer, and the park's tenant companies being the ground layer.

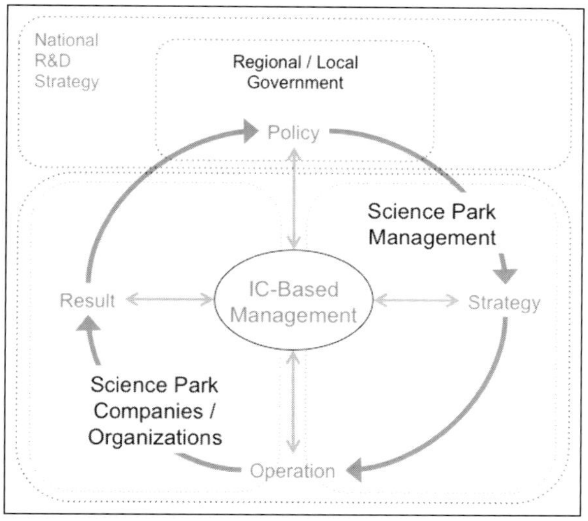

Fig. 3: Science Park Management Process – Actors [1]

A process-oriented and Intellectual Capital-based management system for Science Parks can be seen as critical success factor for a thriving development. Additionally the most important capital of Science Parks is human capital. Currently, there are neither standardized methods nor tools available, which do describe or support the implementation of Intellectual Capital-based management systems in Science Parks. Therefore, for the development of such methods and tools, a standardized process-oriented approach will be developed for helping existing as well as future Science Parks in establishing Intellectual Capital-based management systems on the one hand and enhancing the human capital continuously on the other hand.

Within complex network and knowledge-intensive environments like Science Parks, management has to deal with new and different challenges. Here, Intellectual Capital has comparably strong effects on the value adding processes and therefore is a critical success factor for effective performance and the achievement of strategic objectives.

For the business model of a Science Park nine success factors within three categories of Intellectual Capital (Human, Structural and Relational Capital) can be identified, which play the most decisive role and which have to be managed individually. Therefore, management instruments for Human Capital Management, Relation Capital Management and Structural Capital Management, assessing and managing IC success factors with regard to future value and development potential for a Science Park, have been developed.

- Science Parks are not simply real estate projects, but they are complex organizations. In fact they are organizations with highly differentiated (business) goals/aims --> therefore Science Park management has to become more holistic and should integrate an Intellectual Capital-based management system in order to be able to achieve long term success and sustainability for the Science Park itself as well as for its tenants.

- recent research shows that Science Park management could contribute more efficiently to the value creating process of its tenant firms through fostering active networking

- Science Parks are usually managed in a rather passive way – what is needed, is a more active way of management

- Science Parks have to be managed considering a more holistic management view

Science Parks need a management system that allows them to reach their specific aims concerning their development potential / innovation potential / networking strategies / value added etc.

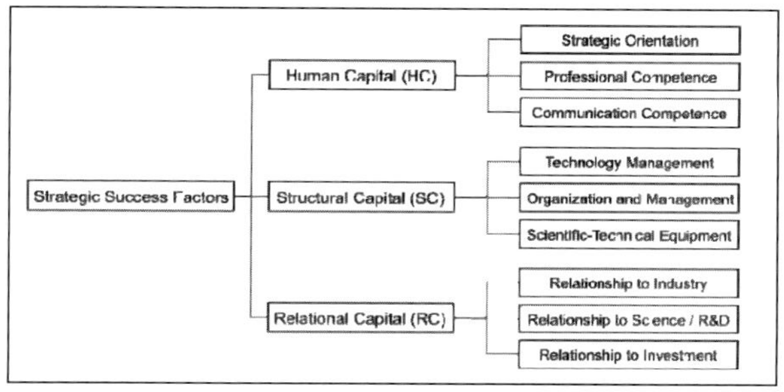

Fig. 4: Strategic Success Factors of IC-Based Science Park Management [1, 17, 21]

6 SUMMARY

Tangible and intangible factors have to be managed effectively in order to support the value adding processes of a Science Park. Due to their complex nature, the management of intangible resources is a great challenge and needs a special focus within a management system designed for a network-like business structure of a Science Park. Especially, the management of IC has to pay attention to the Science Park's business model and strategic objectives. IC management and development has to be implemented on the basis of these general objectives if a positive impact on the value adding processes is to be achieved. Therefore, the IC should be analyzed and managed in regard to its future value for strategic success factors, taking the individual business model and strategy of the Science Park into account.

7 REFERENCES

[1] Al Hashemi, H. (2011) "Development of an Intellectual Capital-Based Management System for Science Parks" TU Berlin.

[2] Alwert, K.; Bornemann, M.; Will, M. (2008) "Wissensbilanz – Made in Germany. Leitfaden 2.0 zur Erstellung einer Wissensbilanz", Guideline Published by the Federal Ministry for Economics and Technology, Berlin.

[3] Alwert, K.; Bornemann, M.; Will, M. (2007) "Does Intellectual Capital Reporting Matter to Financial Analysts?", Paper read at the 8th European Conference on Knowledge Management, Barcelona, Spain.

[4] Annerstedt, J. (2006) "Science Parks and High-Tech Clustering". In: P. Bianchi; S. Labory (eds.), *International Handbook on Industrial Policy*. 279-297.

[5] Austrian Research Centers Seibersdorf (2004) "ARC Wissensbilanz".

[6] Becker, G. S. (1964) "Human Capital"; Chicago, University of Chicago Press.

[7] Botosan, C. A.; Plumlee, M. A. (2000) "Disclosure Level and Expected Cost of Equity Capital: An Examination of Analysts' Rankings of Corporate Disclosure and Alternative Methods of Estimating Expected Cost of Equity Capital".

[8] Cabral, R. (1998) “The Cabral-Dahab Science Park Management Paradigm – An Introduction”. In: International Journal of Technology Management 16, 721-725.

[9] Cabral, R (1998) “Refining the Cabral-Dahab Science Park Management Paradigm”. In: International Journal of Technology Management 16, 813-818.

[10] Cabral, R. / Dahab, S. (1998) “Science Parks in Developing Countries – The Case of BIORIO in Brazil”. In: International Journal of Technology Management 16, 726-739.

[11] Cortright, J. (2001) “New Growth Theory, Technology and Learning: A Practitioners Guide”, Portland, OR: Impresa, Inc.

[12] Danish Ministry of Science, Technology and Innovation (2003) “Intellectual Capital Statements – The New Guideline”.

[13] Del Bello, A. (2003) “Credit Rating and Intangible Assets: Some Observations on Current Practices”.

[14] Edvinsson, L.; Malone, M. (1997) “Intellectual Capital”; New York, Harper Business.

[15] Etzkowitz, H. (2008) “The Triple Helix”. University-Industry-Government Innovation in Action. New York 2008.

[16] InCaS (2009) “Intellectual Capital Statement – Made in Europe. European ICS Guideline”. Ed. by Fraunhofer IPK, Berlin 2009. Available via: www.incas-europe.org.

[17] Kohl, H. (2002) “Performance of Institutes for Applied Research”. In: Mertins, K. (Ed..): Innovation in Indonesia. Assessment of the National Innovation System and Approaches for Improvement. Fraunhofer IRB Verlag, Stuttgart 2002, S. 91-112.

[18] Kohl, H.; Finke, I. (2002) “Design Principles of Institution Building, Reorganization and Support. In: Mertins, K. (Hrsg.): Innovation in Indonesia. Assessment of the National Innovation System and Approaches for Improvement. Fraunhofer IRB Verlag, 146-156.

[19] Kohl, H. (2005) “European benchmarking approaches for SMEs”. Presented at WASME 2005 World SME Convention.

[20] Kohl, H. (2002) “Innovation Management - Bridging the Gap between Academic Research and Industrial Utilization”. German-Serbian Expert Workshop.

[21] Kuhlmann, S.; Holland, D. (1995) “Erfolgsfaktoren der wirtschaftsnahen Forschung”.

[22] Porter, M. E. (1990). The Competitive Advantage of Nations. New York, Free Press

[23] Schultz, T.W. (1961) “Investment in Human Capital”, American Economic Review, No 51.

[24] Steward, T.A. (1997) “Intellectual Capital – the new wealth of organisations”; Doubledays Currency, New York.

[25] Thomas, A. (2006) “A Tale of Two Reports”. In: European Business Forum, Issue 16, 2003/4, pp. 79-81.

[26] UNESCO (2005) “UNESCO Science Report 2005”.

[27] Wagner, C. (2003) “Can the Global Network of Science Contribute to Development?”, presented to the IDRC-UNESCO meeting on Future Directions for National Reviews of Science, Technology and Innovation in Developing Countries, 2003.

A Dynamic Model for Matching Job Market Qualifications Demand and Educational Market Qualifications Supply

S. Abd-Elall, C. Reise, G. Seliger
Department of Machine Tools and Factory Management, Technische Universität Berlin, Germany

Abstract
Matching job market and educational market qualifications will be useful for educational institutions, human resource departments, and students in order to plan the best suitable courses, training programs, and curriculum respectively. In this paper an approach for matching job market qualifications demand and education market qualifications supply will be presented. The data input for this model will be continuously updated and hence provide the users with real-time qualifications demand and supply. Job announcements will be used to represent the job market qualifications demand, and the learning outcomes will be used to represent the education market qualifications supply.

Keywords:
Job market, education market, job announcement, and learning outcomes

1 INTRODUCTION

The global division of the wealth disproportional distributed to the global population. If the lifestyle of the industrialized world and its predominant technologies is to be adopted by the emerging countries, then the resource consumption will exceed every accountable economical, ecological, and social boundary [1]. In order to cope with this challenge, people all over the world must be made aware and enabled to act as agents of change towards sustainable development.

Higher education institutions are a powerful starting point for the preparation of creative engineers who are able to design and develop sustainable products and processes. At the same time engineering students want a quality education with the most relevant skills that will equip them in the best way for a rewarding career of their choice.

Higher education institutions should define a broad sustainable qualification structure for their engineers' occupational careers, considering both the present and future needs of the job market. The success of educational institutions in achieving this role necessitates for them to match their student learning outcomes with the job market qualifications demand, considering the sustainability issues.

2 OVERVIEW

2.1 Job market

Students begin their studies with the hope that a higher education qualification will help them get a job. This hope is mainly influenced by the relationship between their qualifications profile and the employment demand. Engineering field impart certain job-specific skills that are clearly understood in the job market, and hence provide some evidence that these graduates possess 'tools' needed to be productive at work [2].

However, the process of finding a 'suitable' job for some is as a result not easily accomplished. This is partly due to individual circumstances and risk of mismatches between required qualifications and available ones. This risk becomes more serious when the job market demanding certain qualifications that are not available in the workforce market while higher education institutions continue producing workforce with unneeded qualifications. The cost is lost opportunities and recycled qualifications in addition to other negative social disorder [3].

These challenges are increasing as the job market becoming more globally competitive. In order to survive in the current business world, companies are competing for highly qualified manpower. The higher education system on the other hand is traditionally considered conservative and slow in its response to market needs.

In Germany, 15,000 engineering jobs go unfilled every year although 65,000 German engineers are unemployed [4]. A comparative study jointly undertaken by researchers from 13 universities and research institutions in 11 European countries and Japan conclude that, the employment situation of engineering graduates in these countries about four years after graduation, still lags beyond that in the other countries in some respects [5].

2.2 Engineering education

In both Europe and the USA, we are entering an era of intense re-evaluation of engineering education. In the EU, this has, in part, been precipitated by the Bologna Declaration [6], while in the USA it has been encouraged by the transition to the Accreditation Board for Engineering and Technology EC2000 accreditation criteria [7].

As we enter this era of re-evaluation, it is important to understand the goals of engineering education. Ideally, we would agree to a statement of goals that would contain not only a description of the knowledge, skills and attitudes appropriate to university education, but also an indication of the level of proficiency expected of graduating students [8].

The success of educational institutions in supplying the job market by the graduates who possess the required qualifications depends mainly on the level of understanding the job market qualifications need, thus necessitates for them to have a cooperation with the industries supported by a network system which able to collect and evaluate the job market qualifications need in real-time.

3 EDUCATION FOR SUSTAINABILITY

A considerable gap remains between the increasing public awareness about social, ecological and economical challenges and the implementation of sustainability in societal life in different global regions [9]. Often the main concern of the industries is the economical factor, hence shaping the engineering education mainly according to the job market demand is under social and ecological aspects not acceptable. However the sustainability criteria should be considered in the education chain.

A customized framework of sustainability should be developed so that students will have basic concepts about sustainability during their basic studies in the schools. Later in the universities a more advance courses should be integrated in the engineering curriculum, so that the graduates will be able to design and manufacture a more productive products and processes considering the criteria of sustainability.

4 UNIVERSITIES-INDUSTRIES RELATIONSHIP

The relationships between higher education and the world of work have been among the key public debates on higher education over the last four decades, so it does not come as a surprise to note that this theme was among the 12 key themes of the UNESCO World Conference on Higher Education [5, 10].

To understand the relationship between higher education and the world of work, selected models will be discussed. Later an approach for matching engineering graduates' qualifications and employers demand will be presented.

4.1 Supply chain model

In this model the collection of people with different skills and professions available for employment defined as 'workforce market' and the jobs available and open for employment defined as 'job market'. Figure 1, shows the flow of manpower (skilled, semiskilled and unskilled) from schools to higher education institutions to workforce market and ends at the job market. A feedback of manpower from the job market to higher education institutions is shown for continuing education and retraining.

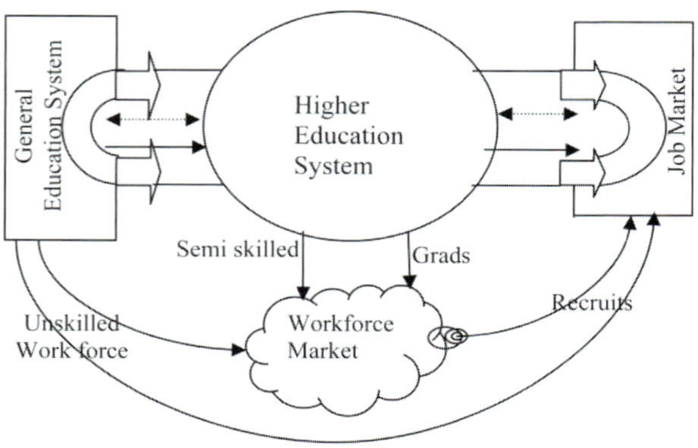

Figure 1: Higher education institutions in the centre of the supply chain [3].

The interaction between educational supply chain components (i.e. educational institutions, workforce market and job market) can be viewed as a chain of suppliers and customers. The raw material in the model is people's minds and the final product is qualified people and hence people are the suppliers and customers and the institutions are service providers. In the supply end, students after finishing high school apply for admission in the universities. Passing through some procedures including entrance exams and interviews, universities admit large number of students and pass them through different ready-made academic programmes and exams for quality assurance.

Graduated students join the workforce market and start searching for jobs. In the demand side companies start searching for the best match of graduates and needed qualifications using CV's and interviews. The flow in the workforce supply chain is analogous to the push production system where higher education institutions 'push' certain qualifications through the chain to load the workforce market with graduates that have various qualifications that might be needed and recruited by employers in the job market [3].

4.2 Reference model

Meyer in his reference model for integrative competence management Figure 2 explains the relationship between universities, industries, and students. The various participants meet in the research "interface with the society", educational, and job markets. The competence portfolio lies in the centre of the reference model with the intention of dynamically specifying and combining competencies.

This subject to the management circle of setting goals, plan, decide, act and control. It is interlinked with the players of the engineering education field, technology based organizations, learning engineers and technology oriented universities, which are connected together within the socio-political framework. The players are connected by the markets they act on.

On the labour market, the competence portfolio is the main criterion to match industrial demand for workforce with the available students. On the educational market, students cover their demand to acquire a competence portfolio that enables them for positioning themselves on the labour market. By applying the mechanisms of free market trade on competence portfolios students become managers of their own competence portfolio.

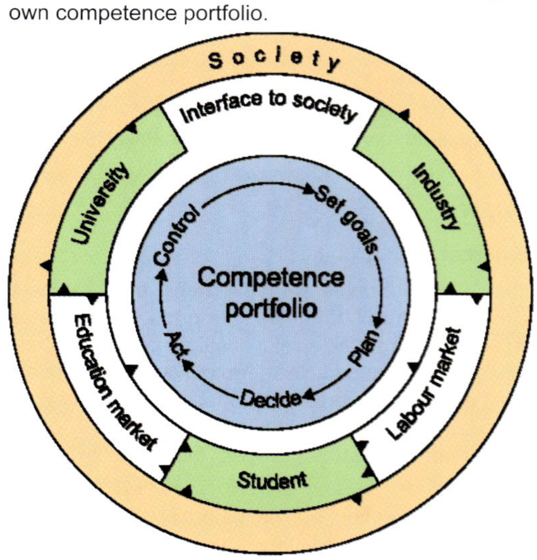

Figure 2: Reference Model for Integrative Competence Management [11].

The markets for engineering education as well as the job market determine the competence requirements to be gained in order to achieve a desired occupation. Competence is the traded value of people's bargaining on the education and labor market. The student, as a main actor in both markets, is largely responsible for his personal learning development. Meyer derives from this model the aim to enable students to create and continuously adapt their individual competence portfolio [11].

4.3 Dynamic model

In this section a dynamic approach for matching engineering graduates' qualification profiles and employment demand will be presnted. The objective of this model implementation is to provide a continously update information from the job market to the students, higher educational institutions and teachers. In this model the information sharing between the industries and educational institution will be thru the use of the job announcements and course outlines contents.

Students will be aware of the job market qualifications demand and hence select the courses and develop themselves with the most relevant qualifications, that will equip them in the best way for a rewarding career of their choice. Higher education institutions will be provided with an efficient communication channels with the employers, hence update the education programs that matches the job market demands. Teachers will be continuously aware of these changes in their area and consider how these can be translated into learning outcomes. These learning outcomes will be offered for the students in the shape of course/courses which they can apply for.

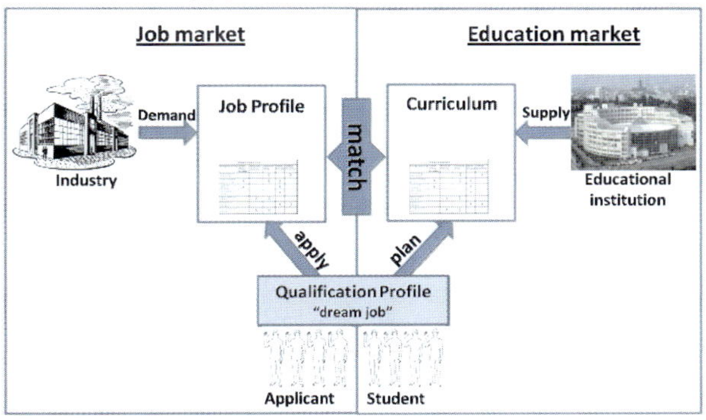

Figure 3: Market perspective on qualifications profile.

Figure 3, composed of job market, and the education market in parallel. Within these markets three major players (student, educational institution and industry). The interaction between these components can be viewed as a chain of suppliers and customers. The qualifications demand of the job market is represented by the job announcements. The qualifications supply from the education market is represented by the curriculum or course outlines offered to the students.

In the job market whenever there is a job vacancy or need for specific skills in any industry, they translate these needs in the form of job announcement "job profile", announce for this job vacancy via the internet, magazine, or any other media to attract the candidates. Candidates read the job announcement and understand the required qualifications, if it is attractive and suit their qualification profiles they apply for the job. Then the industry starts sorting these applications for the best matches with required competences and knowledge as defined earlier in the job profile to select the best candidate.

The job announcements are written carefully to reflect exactly the organization's qualifications demand and to attract as much as possible potential candidates. The large pool of candidate gave the organization a better alternative to select the best available professional for the job vacancy. Job announcements mostly are written in the same format. These formats basically involve the date of the announcement and the deadline for application submission, proposed job title, introduction to the organization, assigned responsibilities, and qualifications needed.

On the other side in the education market; Teachers or instructors whenever they want to offer a course to the students they translate the learning outcomes of this course into course outline "Syllabus", and , give the course objectives, and distributions of the marks for the various course activities. This course outline is considered as a contract between the teacher and the students, i.e. teacher will provide the students the competences and skills stated in the course outline, and students are expected to fulfill the course requirements in order to pass. Students read the course outline, understand it and if it meet their interest they apply for that course.

Teachers who know exactly what the job market qualifications demand and translate these demands into course/s, can easily convince the students to apply for their courses, on the other hand students will be motivated and active in the class as they know the significance of the course. The syllabuses are mostly written in standard format to give a general overview about the expected competence the learner will gain after the completion of the course. Some instructors involve extra information about their courses in the syllabus, but they have to mention the basic information. These basic information are mainly the start and completion date, course description, and course objectives.

In an advance step candidates "graduatess or employees" who want to apply for a specific job "dream job" have to choose the courses and the curriculum which will qualify them for that job. Teachers also have to shape the courses and the curriculum based on the job market qualifications demand, therefore the students are qualified for the stated job. The students have to take responsibility for their competence portfolio by deriving their education demand from the job market. The educational institutions have to detail the competences supply of each course accordingly. If the courses fulfill the demand of the student, they are successful, otherwise the course has to be modified or substituted.

This transparency of the job market and the education market might motivate students to evolve more realistic study strategies. At any time – during the studies, shortly before the end, or even after leaving university – a student can compare in real-time if his education is appropriate for the job market or not, and react by participation in appropriate courses. The demand chain from the dynamic business market to the job market to the education market is closed. Changing demands can be served with little delay. Some students study a subject because of own interest without a perspective on the job market, and are later disappointed as they receive no job [12].

The efficiency of the information sharing between the different components of the system is the major factor for the success of this model. Such cooperation between these stakeholders (student, educational institution and employer) is hard to achieve, especially if we consider the dynamic environment and continuous technological development, but the modern technology like Semantic Web Technology will be used to match the contents of the job announcements and course outlines hence integrating these players together and sharing the information in the real-time.

Semantic web technologies can process distributed data and manage a high degree of complexity. By means of formal ontologies, they can give meaning to the teaching contents, learning processes and performance levels expressed in course descriptions and qualification frameworks on the web. This enables stakeholders in the educational market to analyze the job profile more efficiently, hence matching the demand of the job market with the supply of the educational market.

5 POTENTIALS OF INFORMATION TECHNOLOGY

5.1 Semantic Web Technologies (SWT)

The World Wide Web (WWW) has vast potential to provide information globally. The continuous growth of data published on the WWW creates a challenge to identify desired information. Automated search and data processing solutions try to cope with this challenge. However, unclear semantics, for example the confusion between homonyms and synonyms, often leads to unsatisfactory search results. Software applications need to effectively process data: both the supply of the data and their unambiguous semantics.

Berners-Lee confronted this problem by bringing WWW technologies and methods of artificial intelligence together [13]. Thereby, new web languages, like the Resource Description Framework (RDF) and the Web Ontology Language (OWL), came into existence. RDF information, so-called metadata, can be attached to WWW resources such as documents or pictures. The defined semantic of the documents can, therefore, be processed by machines. With OWL a vocabulary of a knowledge domain can be defined and relationships between entities described [14]. To do this, explicit formal specifications of shared conceptualizations, so-called ontologies, have to be built [15]. Machines, through independent reasoning, can transform existing, but implicit, knowledge into explicit knowledge by using first order logic formalisms based on description logics.

Therefore, the semantic web enables software applications to interpret heterogeneous information. It facilitates precise and unambiguous answers in web queries. The volume and diversity of information can be used more efficiently, and knowledge can be better distributed due to the creation of a vast global knowledge base. New information can be gained through reasoning and used in knowledge intensive applications, such as planning of careers and qualification.

As discussed in section 4.3 the dynamic model mainly consists of two parts. The first part is the job market represented by the job announcements and the second part is the education market represented by the curriculum. In the following case semantic web technology were used to provide high value information about qualifications to all stakeholders in the education market and as a result improve the quality of their decisions in learning, teaching and administrative processes. The approach used so-called Learning Outcomes for the qualification representation and semantic web technologies for the processing of the qualification information. The processes, which are needed to improve the provision of high value information about qualification, embrace the creation, storage and identification of qualification profiles, especially for courses and personal profiles. They also include the analysis of qualification bundles and the synthesis of single qualification statements in order to align them with super ordinate qualification profiles.

5.2 Matching of Qualification Profiles through SWT

Semantic web technology offer promising possibilities for creating a global qualification knowledge base. Figure 4 shows the different layers, real world, intelligent agents, semantic web services and WWW, which together build the architecture to compare qualification profiles. The qualification profiles of job offers, teaching courses and employees build the objects of the real world layer. The lowest layer, the WWW, also exists already. Qualification profiles are already published in websites and online databases. Humans understand the semantics of the published texts, whether they are course descriptions, qualification frameworks or online databases for profiles of employees and job offers. To make the information machine processable a structured knowledge base is needed. It functions together with the WWW resources as “database in the sky” and supplies the information to intelligent agents which provide knowledge for the customer. This knowledge base is built on the basis of semantic web services.

The knowledge base is structured by means of qualification ontology, which is the formalized model of a qualification representation language. The language builds the foundation for the shared concept about qualification. On this basis the communication about qualifications between all agents, human and machines, happens. This language has to be universally valid, generally accepted but still with a high expressiveness. For engineering qualifications, such a qualification representation language was developed at the department for Factory Management and Assembly Technology (MF) [16] and tested in the education market by the International Master Program “Global Production Engineering” at TU Berlin [17]. The language describes qualifications by defining statements of what a learner is expected to know, understand and be able to do at the end of a period of learning, and of how that learning is to be demonstrated. These statements are called Learning Outcomes (LO). To explicitly and formally specify qualifications the concept of learning outcomes can be applied not only to the educational market but also to the job market [1]. The LO language should describe the actual mental and physical knowledge, skills, norms and values achieved. The basics of the LO representation language were developed by Moon, Kennedy und Paquette [18-20]. A verb - the predicate - must be chosen which indicates what the learner - the subject - is expected to do. In addition, LO contain words - the object - that indicate on what and with what the learner is acting, and terms - the attributes - that indicate the nature of the performance. This language and its defined vocabulary was formalized as ontology, named PLOTIN-Ontology, and enriched with a knowledge base [16]. That allows the interoperability between different web sites and software systems. The model enables the storage of instances consisting of very specific LO of single teaching aspects as well as general qualifications of long term learning activities, which is stated in qualification frameworks or job descriptions. The adaptability of ontologies makes the

knowledge base expandable and modifiable, and this, in particular, conforms to the dynamic requirements of the market in a very successful way.
The annotation of qualification statements can be done manually or automatically. For the automated annotation a software application, called module-description-generator, has been developed at MF[21]. For the publication of qualification statements this software application supports the didactically adequate creation of LO. At the same time, while publishing the module or course description on the WWW, metadata about qualification relevant information is annotated. With some extra time investment, a manual annotation that is supported by general annotation tools can also be done by a didactically experienced person after the publication is on the web. Research about software applications for automated annotations of previously published qualification statements has to be performed. The purpose is to automatically identify, analyze and annotate web resources. Such applications can also be based on the PLOTIN-ontology.

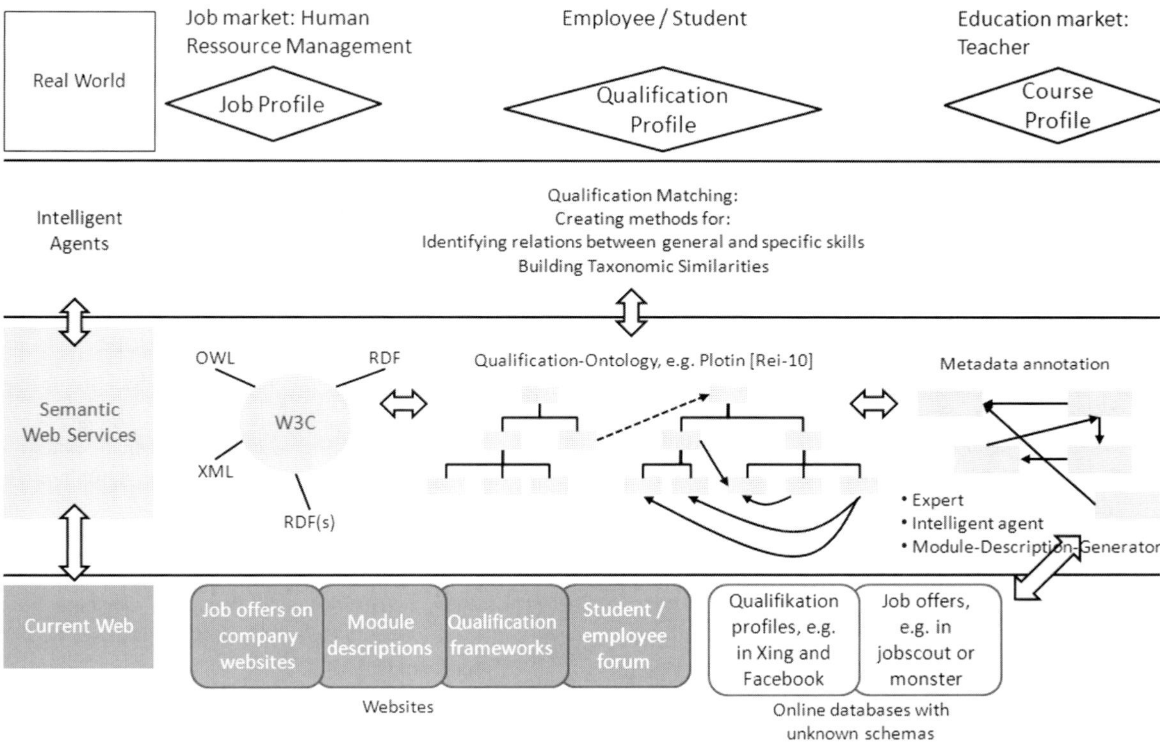

Figure 4: Architecture of Matching Qualification Profiles using SWT.

The matching of profiles between jobs, people and courses is done by knowledge based systems on the intelligent agent layer. They search the knowledge base from different perspectives to identify existing knowledge and to generate new knowledge. The comparison of qualification profiles is based on methods like taxonomic similarity and scaling of qualifications [22, 23]. By scaling qualifications relations between general and specific LO are identified and visualized. This enables agents to support learners in reaching abstract qualification aims or to break qualification profiles down into specific desired LO. With taxonomic similarity subsets of different qualification profiles are compared, and their level of coverage is mapped.

To create efficient knowledge based systems for mapping qualification profiles several questions have to be solved before they will provide ubiquitous services in managing qualification. In the area of metadata annotation there are several qualification models with different expressiveness in usage. Also, the annotation of qualification statements in web documents is not broadly accomplished. Additionally, the creation of relations and algorithms needs further research in order to visualize dependencies and taxonomic similarities of qualification profiles for engineers. Nevertheless the potentials of SWT show the possibilities to increase the management of qualifications tremendously through individual query and provisional knowledge. The idea of knowledge based systems working on a global qualification-database is within reach. First ontologies and its applications in the educational market show promising beginnings.

6 SUMMARY

In this paper, a semantic web framework was proposed, that can be used to improve the effectiveness and efficiency of the information flow between student, educational institution and employer. This enables the students to plan their qualification profile by selecting courses and develop themselves in order to match the job profile they are looking for. On the other hand it helps the teachers to continuously be updated about changes in the job market qualifications demand and hence managing the learning outcome of the courses according to demand. Sustainability issues should be integrated in the engineering education regardless the job market qualifications demand. This demands an open global platform for all stakeholders in the job market to define qualification requirements of future engineers. Semantic Web Technologies have the potential to provide the technical solutions for implementing such a platform.

7 REFERENCES

1. Seliger, G., C. Reise, and R. McFarland, *Outcome-oriented Learning Environment for Sustainable Engineering Education*, in *7th Global Conference on Sustainable Manufacturing: Sustainable Product Development and Life Cycle Engineering*. 2009: Indian Institute of Technology Madras. p. 91-98.
2. Moleke, P., *Finding work: employment experiences of South African graduates* 2005, Cape Town, 8000, South Africa: HSRC Press.
3. Al-Turki, U.M., et al., *Stakeholders integration in higher education: supply chain approach.* European Journal of Engineering Education, 2008. **33**(2): p. 211-219.
4. Continental, *In Search of Global Engineering Excellence Educating the Next Generation of Engineers for the Global Workplace*, A.U. Edited by Tricia Grindel and F.M.G. Layout by Change Communication GmbH, Editors. 2007: Hanover - Germany.
5. Teichler, U., ed. *Careers of University Graduates: Views and Experiences in Comparative Perspectives* 2007.
6. Haug, G. and C. Tauch, *Trends in Learning Structures in Higher Education (II)*, in *Follow-up Report prepared for the Salamanca and Prague Conferences*. 2001, National Board of Education, the Association of European Universities, the European Commission or the European Training Foundation.
7. ABET, *Vision For Change, A Summary Report of the ABET/NSF/Industry Workshops*. 1995.
8. Bankel, J., et al., *The CDIO syllabus: a comparative study of expected student proficiency.* European Journal of Engineering Education, 2003. **28**(3): p. 297-315.
9. Seliger, G., *Sustainability in Manufacturing: Recovery of Resources in Product and Material Cycles*. 2007, Berlin: Springer. 424.
10. Teichler, U., *New Perspectives of the Relationships between Higer Education and Employment.* Tertiary Education and Management, 2000. **6**(2): p. 79-92.
11. Meyer, M., *Management von Ingenieurkompetenzen im Spannungsfeld beruflicher Arbeitsteilung*, in *Department Assembly Technology and Factory Management* 2006, Technische Universität Berlin: Berlin. p. 160.
12. Meyer, M., *A Vision of Sustainable Education with Competence Management*, in *Proceedings of the Global Conference on Sustainable Product Development and Life Cycle Engineering*, G. Seliger, et al., Editors. 2004, uni edition GmbH: Berlin. p. 319-325.
13. Berners-Lee, T., J. Hendler, and O. Lassila, *The Semantic Web.* Scientific American, 2001: p. 29-37.
14. Antoniou, G. and F.v. Harmelen, *A Semantic Web Primer, 2nd Edition (Cooperative Information Systems)*. 2008: The MIT Press. 288.
15. Gruber, T.R., *Toward Principles for the Design of Ontologies Used for Knowledge Sharing.* International Journal of Human Computer Studies, 1993. **43**: p. 907-928.
16. Reise, C. and G. Seliger, *Knowledge Based Systems For Planning Learning Outcomes In Manufacturing Engineering*, in *Diversity unifies - Diversity in Engineering Education*. 2010: Trnava, Slovakia.
17. GPE. *www.gpe.tu-berlin.de download.* 2010 16.08.2010].
18. Moon, J., *Linking Levels, Learning Outcomes and Assessment Criteria.* 2006., University of Bournemouth, UK.
19. Kennedy, D., Á. Hyland, and N. Ryan, *Writing and Using Learning Outcomes: a Practical Guide*. 2007, Quality Promotion Unit, University College Cork.
20. Paquette, G., *An Ontology and a Software Framework for Competency Modeling and Management.* Journal of Educational Technology & Society, 2007. **10**(3): p. 1-21.
21. Reise, C. and G. Seliger, *Analyzing and Planning of Learning Outcomes using Semantic Web Technologies*, in *Interactive Computer Aided Learning Conference*. 2010: Hasselt, Belgium.
22. Falk, T., et al., *Semantic-Web-Technologien in der Arbeitsplatzvermittlung.* Informatik-Spektrum, 2006. **29**(3): p. 201-209.
23. Maillardet, F., *Generic Learning Outcomes for Engineering Programmes*, in *Output Standards Working Group*. 2007, EC^{UK}.

Design Thinking in Engineering Education and its Adoption in Technology-driven Startups

A.E. Açar, D.S. Rother
ingosu – Daniel Rother & Acar GbR, Berlin, Germany

Abstract
As innovation becomes the cornerstone for new problem solving and creation of private and public value, this paper introduces the design thinking approach as a new means of systematic innovation. We introduce the approach with two short examples from programs to graduate students. We illustrate the setup of a design thinking program by using the HPI School of Design Thinking as a case study in order to show how to integrate the approach in engineering education. Finally, we outline how technology-driven startups can benefit by the adoption of the approach and gain a strategic competitive advantage.
Keywords:
Design Thinking; Innovation Process; Innovation Culture; Problem-Based Learning; user-centric innovation; technology-driven startups

1 INTRODUCTION

The general view of design as stated by Evans et al. in 1990 is that "the subject [of design] seems to occupy the top drawer of a Pandora's box of controversial curriculum matters, a box often opened only as accreditation time approaches. Even 'design' faculty—those often segregated from 'analysis' faculty by the courses they teach—have trouble articulating this elusive creature called design" [1].

At the same time, design faculty across the range of educational institutions feel that the leaders of engineering departments and schools do not recognize the intellectual complexities and resources demanded to support good design education [2].

If design is considered to be a distinguishing activity in engineering [3], it can be viewed as part of engineering programs that are able to graduate engineers who can design effective solutions to meet social needs [4].

The goal of this paper is to introduce a recent approach called "design thinking", which reflects a complex process of inquiry and learning that merges engineering with design, by putting a focus on human factors and adopting a systemic view that incorporates economic deliberations [5]. Using case studies, we will depict the results of the design thinking process and present existing institutions educating students in design thinking. Subsequently, we will illustrate the importance of the adoption of this approach by technology-driven start-ups. Finally, we will outline a course of action for cultural and organizational adaptation.

2 THE DESIGN THINKING APPROACH

As the buzz about design thinking is continuing to increase and causing some confusion about the term and its implications, we will present a short overview of its history before illustrating the methodology and the results that can be achieved by its application.

2.1 HISTORY

Design thinking has been introduced as an approach by the California-based design agency IDEO as the result of a quality management initiative to investigate the firm's development process [6]. To develop the design thinking process, IDEO-founder David Kelley participated in a collaboratory setting at the Stanford University with Larry Leifer, head of the Center of Design Research, and Terry Winograd, head of the Human-Computer Interaction Group. With funding by Hasso Plattner, founder of SAP, they founded the Hasso Plattner Institute of Design at Stanford (d.school) [7].

The commercial success of the approach, the high rate of spin-offs from the d.school and the large attention by the business press resulted in the fast spread and adaptation of design thinking by various companies and institutions around the world.

2.2 METHODOLOGY

Design thinking consists of a collection of methods that are common in engineering, ethnologic and anthropologic research, industrial design and business economics. Their use is set by a human-centered innovation process which builds upon further core elements such as the adoption of an integrative mindset, utilization of space, organizational roles and formation of multidisciplinary teams.

THE MINDSET AND CORE ELEMENTS

As Roger Martin states in his various publications, "complex problems that require the integration of various domain knowledge cannot be solved by analytical thinking alone." The goal of analytical thinking is defined as the production of consistent replicable outcomes, a focus on reliability. Analytical solutions can be proven through induction and deduction. He contrasts this to intuitive thinking which he calls "knowing without reasoning". The goal of intuitive thinking is the production of outcomes that meet objectives, a focus on validity. To bridge the fundamental predilection gap between reliability and validity, he proposes an integration of analytical and intuitive thinking and terms this as integrative thinking [8]. The purpose of this mindset is to allow for generative reasoning by integrating exploitation with exploration, analysis and judgment. This new mindset is at the core of design thinking. Bill Moggridge calls people with this mindset "T-shaped" [9].

According to Stanford professor Robert Sutton, design thinking both requires and fosters a "culture of forgiveness, not of permission" [10]. He addresses two core elements of the approach: utilization of space and organizational roles within projects. His research shows that the creation of radical innovation requires a utilization of space and resources that are uncommon in most organizations. Design thinking teams are able to act independently to make changes to their work environment. Dress code, attendance and activities at the office are self-regulated by employees. This freedom to create their own ambiance gives teams the intellectual and emotional safety to pursue the exploration of ideas. Closely related is the assumption of organizational roles. Though design thinking companies do have classical hierarchies, the position within the structure becomes irrelevant within projects as leadership of projects are not determined by seniority [11].

THE PROCESS

The design thinking process is a unified innovation model bringing together managers, engineers and designers. It necessitates the composition of multidisciplinary teams with no organizational hierarchies and no departmentalization within the project [12]. To create a common understanding between the different fields, the process establishes a new vocabulary that characterizes a design thinking "culture". The phrases that have been coined have turned into a lingo that can be understood by design thinking practitioners.

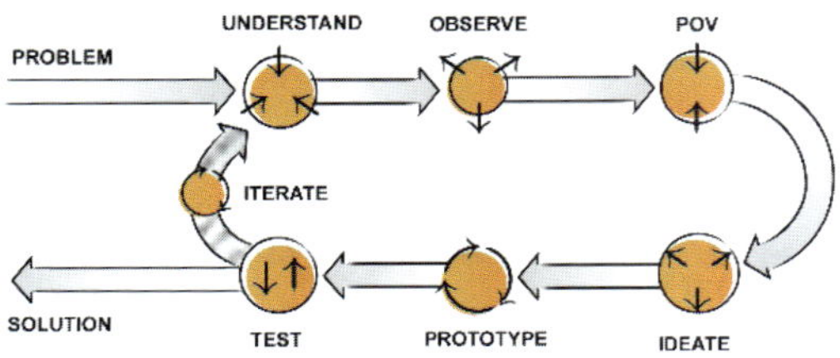

Figure 1: the design thinking process

This nonlinear, iterative process is often equated with the design thinking approach itself, yet it is important to note that it is only one of the core elements [13]. The process begins with a strong focus on the user of a product or a service and places people within a given business process or model into the center of all attempts at solutions.

The **understand** phase of the process constitutes the intensive preoccupation with a problem. Through interviews with experts and extreme users and secondary research, team members are trying to "become experts". This also includes the use of empathy tools [14]. Team members with expertise in the problem field or knowledge of research and interview methods lead this part of the process.

The **observe** phase is used to correlate the findings of the previous step with observations out in the field. The goal is to collect real-life data, using ethnological and sociological methods to uncover hidden needs and unexpressed wants.

After gathering information from a variety of perspectives, the team analyses the collected data and approaches the problem from the **point of view** of the user. The reframing focuses the efforts of problem-solving to finding a solution that is desirable. Problems that seem to be technological or economical in nature now address the human factors.

Based on the new frame, one can **ideate** notions by using a variety of creativity techniques. Though design thinking is mistakenly considered to be a creativity method by some, it only incorporates known techniques such as brainstorming. The important aspect of this step is the focus on quantity, not of quality, of ideas.

A concept that seems to be promising is turned into a **prototype**. Again, the focus here is to create a large number of prototypes using low-cost throwaway prototyping methods. The goal is to build momentum through action and to use the feedback loop for selection.

The real **test** of an idea is in its realization and use. Since concepts that have been visualized in prototype form can be validated by users, the goal here is to get feedback, not to sell ideas. User response is considered to be necessary to **iterate** on ideas, or to trace back to earlier steps, often going back to revise the point of view or even start the process anew by going back to do additional research.

SCIENTIFIC VALIDATION

Currently, the approach has not been validated scientifically but has proven to be successful in economic practice. There is, however, an ongoing collaboratory research program between Stanford University and the Hasso Plattner Institute in Potsdam. The goal of the Design Thinking Research Program [15] is to investigate specific questions such as:

- Why is the design thinking process so successful?
- Which methods or parts of the process contribute most to its success in practice?
- Are design thinking projects more successful than other approaches?
- What are the essential factors for effective application within engineering education?
- How does the approach support interdisciplinary and intercultural collaboration?

2.3 CASE STUDY

To illustrate the different approach to technology-driven business and to demonstrate the wide applicability of design thinking, we have selected the following example:

D.LIGHT

This project began as a Stanford d.school excursion to study the habits and needs of rural villages in Myanmar. Inspired by the candles and kerosene lanterns they saw, the team began their research into the lighting problem.

Figure 2: lighting with kerosene lamp [18]

Research into the problem showed that more than one in five people in the world today use kerosene lanterns as the only option for household lighting. Families spend from 5% to 30% of their monthly income on kerosene oil, a key contributor to indoor air pollution, which claim the lives of 1.5 million people each year, over half of which are under the age of five [16]. As the primary source of greenhouse gas emissions in the developing world, kerosene lamps are responsible for over 100 million tons of CO_2 emissions into the atmosphere [17].

United Nations Development Program studies demonstrate that families with improved lighting have up to a 30% increase in income due to increased productivity at night. The D.light group set out to create LED alternatives and tested their first prototypes, with users reporting an increase in monthly income by as much as 50% from the extended workday.

Further research, however, identified several other non-obvious needs beyond the immediate benefits of the technological solution. Lights needed to be hung easily from the ceiling, from the walls, and still carried comfortably in the hand for outside excursions. Improving on battery size and energy source, LED types for various activities such as children studying or nighttime income generating activities, the team took the context of use into their design considerations.

Figure 3: improved lighting with d.light prototype [18]

By focusing on the reasoning behind the need for lighting, the team was able to come up with a very specific, low-cost solution along with a financial model utilizing microcredits and a distribution model focusing on local production of the product. Today, the project has turned into a start-up [18] that could be seen as technology-driven as they are the leading manufacturers of LED lighting for developing countries. It is worth noticing, however, that the solution was developed out of a human need and incorporates financial and social aspects beyond the technological implementation.

3 DESIGN THINKING EDUCATION

We believe that there are elements in institution teaching design thinking (d-schools) which drive an entrepreneurial spirit. Experience from the d.school shows that alumni of design thinking programs seem to have a disposition towards turning their projects into start-ups, many of which might be traditionally seen as technology-driven.

3.1 DESIGN THINKING PROGRAMS

The common elements at the various programs show a collective understanding of the approach, spreading from the d.school in Stanford.

SPREAD OF EDUCATION

Currently, there are five design thinking programs in institutions around the world:

- Hasso Plattner Institute of Design at Stanford
- Hasso Plattner Institute, School of Design Thinking
- Rotman Desautels Centre For Integrative Thinking
- St. Gallen Design Thinking & Business Innovation
- School of Entrepreneurial Design Thinking, Koblenz

At Stanford, design thinking has now become part of the obligatory program in MBA education. Further programs are being established in Helsinki, Paris and Istanbul.

PHILOSOPHY

The utilization of the human centered process is also discernible in the execution of the programs. “Students first” is a much repeated mantra at the d-schools. In the projects, students also always start off by focusing on users first. Research by observation on the field, to gain insight and inspiration is an extensive part of every design thinking program.

Inviting outsiders to brainstorm or testing solutions with actual users is also an indication for the endeavor to create radical collaboration. Faculty members form teaching teams in order to coach and hold lectures together, paying attention to show the philosophy by their own examples instead of direct explanations.

This also shows a strong bias towards action, which is also recognizable by the effort to be visual. Using graphics and videos, saturating space with sketches and information are all examples of a culture of prototyping. Reflecting upon these actions and creating a mindfulness of the process and interactions themselves is seen as a cornerstone of a good learning experience.

Innovation at the d-schools is not regarded as something that is driven solely by technology or economical models, but rather at an intersection with desirability.

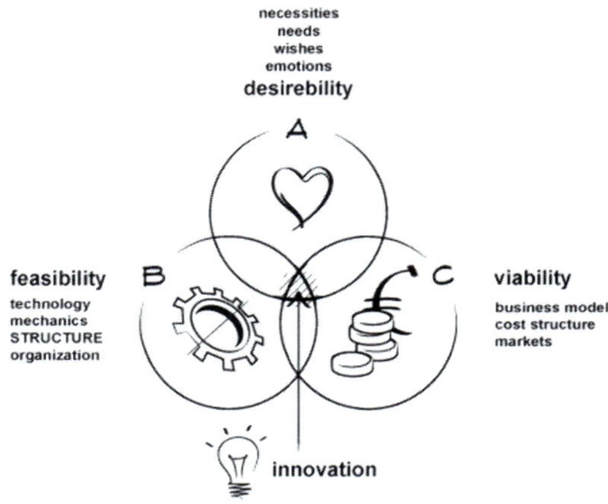

Figure 4: innovation venn diagram

IMPORTANT ELEMENTS

To teach the shift in different modes of thinking, real-life projects are favored over case studies or readings. By expanding on a problem through divergence from the field in question and focusing on the essence through convergence of information, an integrative mindset is being exercised repeatedly [8]. Iteration is a major concept in design thinking education.

As this calls for always shifting group constellations, the process calls for a different perspective on hierarchies within work environments [12]. In combination with the diverse multidisciplinary setup of teams and the ensuing activities and roles [19], there is an emphasis on so-called soft skills. Hence, the communication and information handling within teams is regarded to be a crucial success factor [20], necessitating staff and faculty members trained in psychology with a specialty in group dynamics.

Process requirements, mindset and team interaction can be facilitated by the special design of the space [21]. Existing d-schools pay attention to their interior design, offering a mixture of group and individual workspaces, with spaces for prototyping, socialization and relaxation spread in between.

SYLLABUS

Faculty members teach the process by using projects as teaching frameworks and give lectures on methods from their own specializations. This means that teachers from one class oftentimes attend other classes both in order to deliver content and to gain inspiration for their own courses. The general syllabus at a d-school consists of:

- qualitative research methods taken from anthropology, sociology, journalism and ethnology
- frameworks for point of view generation taken from economics, design and computing sciences
- improvisation and creativity techniques adapted from theatre, generative arts, economics and psychology
- prototyping methods taken from arts and crafts, industrial design and computing sciences

3.2 CASE STUDY: HPI D-SCHOOL

Building on our illustration of the three core elements of design thinking and the results of their application in the previous sections, we will now portray the conveyance of the approach to graduate students with a case study. The following example is an elaboration on the design thinking education as implemented at the HPI School of Design Thinking, based on our own experience as teachers within the program.

HISTORY AND STRUCTURE

The HPI School of Design Thinking (D-School) has been founded in 2007 at the Hasso Plattner Institute (HPI) in Potsdam, Germany. It is an interdisciplinary department within the private institute that is focused on the education of software engineers. The D-School staff consists of an administrative team that is charged with the continuous development of the department and the coordination of teaching activities. Staff members have backgrounds in engineering, design and humanities. The department head is Prof. Uli Weinberg who has a background in film and 3D computer graphics.

The majority of teaching team and students at the D-School are not members of the HPI, being recruited from the universities and colleges in the surrounding states Berlin and Brandenburg instead. Teachers are formally employed as guest lecturers at the HPI and take a leave at their own institution for their part-time commitment. Students are not registered as participants of the program but are required to sign an NDA to participate in commercial projects.

TEACHING CONCEPT

Generally, learners of the design thinking approach are encouraged to act on their own initiative, self-organize their learning experience and evaluate each other in plenary sessions. In contrast to traditional teacher-centered teaching formats [22], which are characterized by passive reception of content and periodic retrieval and testing of understanding, the education at the D-School is activity-oriented. Students are assigned to project teams which work on real-life problems that are derived from observations of both teachers and students, or contributed by companies, NPOs and NGOs. In each case, the goal of the team is to find economically feasible and technically viable solutions, which are based on strong user research to put a focus on desirability.

This practice shows that the D-School has a problem based learning (PBL) approach which is implemented by a teaching concept that is entirely project based. The process steps and methods are introduced by short lectures and are put into practice immediately. By putting theory into application and reflecting on the outcomes, students learn to adapt a “learning by doing” sentiment. To foster the acquisition of an exploratory problem-solving mindset [23], project teams consist of four to six students with different disciplinary backgrounds. By a structured reflection [24] on their experiences and sharing insights with members of other groups, team members are able to learn the different perspectives and approaches of the various disciplines and cultures. At some points within the project, such as the ideation phase, groups are specifically required to exchange members in order to prevent groupthink [25] and to consolidate the goals of PBL as described by [26].

The use of methods as well as the planning and execution of process steps is self-regulated by the students and supported by a teaching team. Every student group is attended by a team of two teachers, consisting of a professor and a research assistant. In a class of forty students, there are twelve teachers from different academic fields, thus extending the idea of interdisciplinarity to the faculty as well. The broad range of disciplines covers engineering, design, economy, natural sciences, humanities and law.

Teachers at the D-School are not of lecturers or project managers but consider themselves to be facilitators and coaches. As such, they are present at a teacher space within the D-School to be approached by students at their own discretion.

CURRICULUM AND SYLLABUS

According to Prof. Uli Weinberg, „there is no curriculum in the conventional sense. That is the strength of the approach in Potsdam: we can pick up on current topics, the latest trends and problems.” [27]

The design thinking education starts with the d-camp, a three-day crash course on the basics of the approach. This period is used to introduce the fundamental concepts with lectures and short exercises, which are also part of an assessment of the student’s experience and personality. Thus, the secondary function of the camp is to select those participants to be accepted as students into the program.

The curriculum is a direct application of the PBL approach, with modules consisting of projects with increasing duration and difficulty. The first modules are short exercises of two to three hours, followed by projects with a length of one day, three days, one week, three weeks, six weeks and a final twelve week project. Students are required to be present at the D-School for two days a week. The total program is spread over two academic terms, consisting of a basic and an advanced track of three months’ time. The advanced track is a project in collaboration with an external commercial partner.

Projects are introduced to the class in a plenary session, with teachers giving presentations with a length of 15, 30 and 60 minutes, depending on the duration of the project. These divisions are also used for lectures on the process and on single methods focusing on user research, creativity and visualization techniques, prototyping, presentation skills, conflict – and project management. Each lecture is followed by a session wherein students are working on their project while focusing on the subject matter of the lecture.

The curriculum also covers optional activities such as short additional workshops on specific topics, led by industry partners and guest speakers [28].

ENVIRONMENT AND CULTURE

The work environment, the interaction between students and faculty and the mode of practice at the D-School are characterized by openness, flexibility and free access to the facilities and informational resources. All these factors form an inspiring and engaging culture that facilitates the formation of the design thinking mindset.

Students are able to create their own workspaces using large mobile whiteboards built from clothes rails common in the fashion industry. These act as notice boards and information spaces, used to gather and structure data that is collected during research and ideation. Mobile bar tables incorporate shelves with writing and crafting materials, as well as digital equipment such as cameras and notebooks. Teams are empowered to design their own area as they see fit; oftentimes including digital projectors, plants, magazines, food and drinks and even office toys into the arrangement. Each workspace is designed according to the needs of the team and the requirements of the project.

Project presentations and lectures are held within a common lecture space with digital projectors and sound equipment. A small library gives access to relevant literature on methodology. A workbench with various power tools and a central material repository are shared for prototyping purposes during projects. Small meeting rooms offer a quiet environment for teleconferences, interviews and meetings with clients and users.

Spread between the team spaces are central areas where groups can interact with each other on the signature red sofas and coffee tables. Gaming consoles, books and magazines are always accessible. Catering with snacks and drinks are available at any time. Students are expected to keep these social areas clean and organized. In the summer, pavilions enable the teams to shift their work sessions out into the surrounding green space. Every person has a keycard to access the building at any time.

At the end of each project, there is an „i like, i wish" session for students and teachers to give feedback about the process and reflect upon the learnings. It is concluded with a party to smoothen conflicts and foster social cohesion.

PROJECT

To illustrate the setup of commercial collaboration, we are presenting an example from the final projects of the academic year 2008/2009.

The project partner is the Future Store Initiative of the Metro Group which is tasked with the development of innovative services. Students were given full access to the processes and data within a supermarket to investigate the question of prevent the waste of fresh produce despite short date of expiry. Project partners offered experts as interview partners and prepared factsheets about internal processes. To gain a deep understanding of the problem source, the team researched central warehouses, logistics centers, interviewed and observed customers and identified these points of interest:

- static reorder of products
- shortness of date of expiry correlates with waste
- insufficient coordination of shelving procedures
- customers pick the longest date of expiry among the same products

Though the project partner saw the problem within its own logistics procedures, the team found that the final insight considering the behavior of customers is a major source of the problem. Based on this insight the students developed a shelving system with a first-in first-out system for fresh produce, preventing customers to select between equal produce with different dates of expiry offering a clear arrangement (Figure 5).

As an additional benefit, the new system released less cold into the immediate surroundings, protecting customers from feeling chilly and saving energy at the same time. A new cartridge system shortened the time necessary for shelving and protected products from being damaged. The team used all these advantages to create a campaign around the new shelving system as well. The combined result incorporated the new shelves, the cartridge packaging, revised procedures for shelving and a campaign to introduce the new solution.

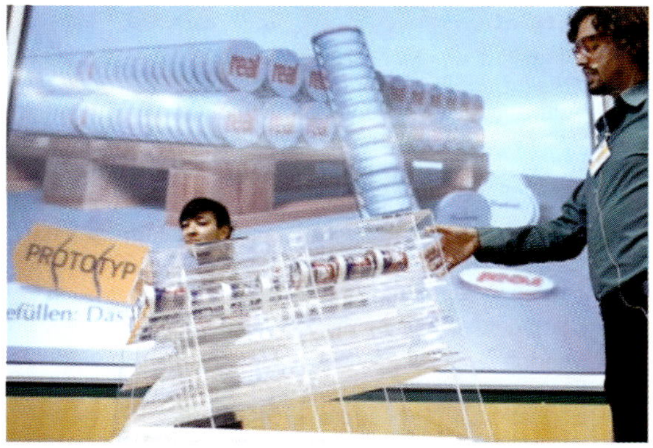

Figure 5: prototype shelving system [29]

4 APPLICATION IN TECHNOLOGY-DRIVEN STARTUPS

The major implication of the design thinking approach for technology-driven start-ups is a shift in self-perception and strategy.

An emphasis on user research brings long-term benefits as products will be desirable and need less marketing efforts. This also limits the danger of the company being enamored with its new technology and limits a natural tendency to overengineer common to technology-driven start-ups [30].

A common complaint in "cool" start-ups is the loss of its culture once the company has matured [31]. An early use of agile development methods creates a culture of prototyping and testing that will prevent the loss of creativity and of an entrepreneurial mode of operation at the later stages of a company [9]. In the same line of thought, a focus on people within the organization as described in 2.2. motivates employees and builds a creative workplace [10].

With institutions increasing efforts to foster entrepreneurial endeavors of their students, design thinking education can deliver a setting that excels in spawning start-ups. Past experience shows that these new companies are built around radical innovations which utilize new technologies or are new combinations of old ones [27]. Thus, the start-ups are not considered to be technology-driven per se, since they are driven by current user needs and future demands.

5 CONCLUSION

Design thinking is a systematic approach to innovation that can produce new products, services and business models. As a best practice, it focuses primarily on the users before generating technological solutions. While its scientific evaluation and validation are still ongoing, the adoption of the approach by high profile institutions around the world confirms its success. Current design thinking programs offer a practical education using PBL as an educational strategy to teach skills essential to start-ups.

We believe that technology-driven start-ups will greatly benefit by the use of design thinking to establish a culture of innovation, as the approach places a strong emphasis on innovation as a result of team effort rather than the achievements of single individuals. Furthermore, offers created by companies using the approach tend to have a high acceptance and desirability among users.

In conclusion, we believe that design thinking education provides a beneficial learning environment to students with entrepreneurial aspiration and delivers a competitive advantage to technology-driven start-ups.

REFERENCES

[1] Evans, D.L., McNeill, B.W., and Beakley, G.C., "Design in Engineering Education: Past Views of Future Directions," Journal of Engineering Education, Vol. 79, No. 4, 1990, pp. 517–522.

[2] Todd, R.H., and Magleby, S.P., "Evaluation and Rewards for Faculty Involved in Engineering Design Education," International Journal of Engineering Education, Vol. 20, No. 3, 2004, pp. 333–340.

[3] Simon, H.A., The Sciences of the Artificial, 3rd ed., Cambridge, Mass.s: MIT Press, 1996.

[4] Sheppard, S.D., "A Description of Engineering: An Essential Backdrop for Interpreting Engineering Education," Proceedings (CD), Mudd Design Workshop IV, Claremont, Cal.: Harvey Mudd College, 2003.

[5] Leifer, L.J. et al., "Engineering Design Thinking, Teaching, and Learning", Journal of Engineering Education, No. 1, 2005, pp. 103 – 120.

[6] Kelley, T. "The Art of Innovation: Lessons in Creativity from IDEO, America's Leading Design Firm", Broadway Business, 2001.

[7] Martin, R.L., "The design of business: why design thinking is the next competitive advantage", Mcgraw-Hill Professional, 2009.

[8] Eris, O., "Manifestation of Divergent-Convergent Thinking in Question Asking and Decision Making Processes of Design Teams: A Performance Dimension," in Human Behavior in Design, Lindemann, U. (editor), p. 142-153, Springer-Verlag, London, 2003.

[9] Brown, T., "Change by Design: How Design Thinking Transforms Organizations and Inspires Innovation", HarperBusiness, 2009.

[10] Sutton, R.I., "Weird Ideas That Work: 11 1/2 Practices for Promoting, Managing, and Sustaining Innovation", Free Press, 2001.

[11] Kelley, T., "The ten Faces of Innovation – Strategies for Heightening Creativity", Profile Books, 2006.

[12] Skogstad, P.L.S. "A Unified Innovation Process Model for Engineering Designers and Managers", Center for Design Research, Stanford University, 2009.

[13] Beckman, S.L., "Reflections on Teaching Design Thinking to MBA Students", University of California, Berkeley, Agent of World Benefit Conference, 2009.

[14] Fulton-Suri, J., "The next 50 years: future challenges and opportunities for empathy in our science". Ergonomics, The Ergonomics Society – the Society Lectures, 1999.

[15] Website of the Design Thinking Research Program http://www.hpi.uni-potsdam.de/forschung/design_thinking/programm.html

[16] United Nations Millenium Development Goals Report 2007, www.un.org/millenniumgoals/pdf/mdg2007.pdf

[17] data gathered by the Lawrence Berkeley National Laboratories, http://www.lbl.gov/

[18] d.light project description at Stanford University http://extreme.stanford.edu/success_stories/dlight_design.html, D.Light http://www.dlightdesign.com

[19] Milne, A., "Analyzing the Activity of Multidisciplinary Design Teams in the Early Stages of Conceptual Design: Method and Measures." Proceedings of Co-Designing 2000 Conference, Coventry, U.K., 2000.

[20] Milne, A., Leifer, L., "Information Handling and Social Interaction of Multi-Disciplinary Design Teams in Conceptual Design: A Classification Scheme Developed from Observed Activity Patterns." Proceedings of the Annual ASME Design Theory & Methodology Conference, Baltimore, USA, 2000.

[21] Tang, J.C., "Toward an understanding of the use of shared workspaces by design teams", Center for Design Research, Stanford University, 1989.

[22] Cuban, L. "How Teachers Taught: Constancy and Change in American Classrooms", 1890-1980. 1st Edition. Research on Teaching Monograph Series.

[23] Hirokawa, R. Y., "Group communication and problem-solving effectiveness II: An exploratory investigation of procedural functions", Western Journal of Communication, Vol 47, Issue 1, 1983, pp. 59 – 74.

[24] Cockayne, W.R., "A Study of the Formation of Innovation Ideas in Informal Networks", Center for Design Research, Stanford University, 2004

[25] Street, M.D., "Groupthink: An Examination of Theoretical Issues, Implications, and Future Research Suggestions", Small Group Research, Florida State University, February 1, 1997, pp. 72-93.

[26] Smith, K.A., Sheppard, S.D., Johnson, D.W., and Johnson, R.T., "Pedagogies of Engagement: Classroom-based Practices," Journal of Engineering Education, Vol. 94, No. 1, 2005, pp. 87–101.

[27] Plattner, H., Meinel, C., Weinberg, U., „design Thinking – Innovation lernen – Ideenwelten öffnen", mi-Wirtschaftsbuch, FinanBuch Verlag GmbH, München, 2009, pp. 89.

[28] Workshop with George Kempel: Innovating for Elementary Education. Documentation http://www.hpi.uni-potsdam.de/hpi/veranstaltungen/d_school_workshop.html?L=1

[29] D-School Jahrbuch 2009, Hasso-Plattner-Institut für Softwaresystemtechnik GmbH, 2009

[30] Wildemann, H., "Kundenorientierte Produktentwicklung in der Automobilindustrie", in: Nachhaltiges Innovations-management, Schwarz, E.J., Gabler, 2004.

[31] Hughes, T.P., „Human-built world: how to think about technology and culture", University of Chicago Press, 2004.

The Impact of Technology-Oriented University Start-Ups on Regional Development: The Case of the Technical University Berlin

M. Mrożewski[1], K. Fajga[2], A. von Matuschka[3]

[1]Chair of Entrepreneurship and Innovation Management, Technische Universität Berlin, Germany

[2]Founder's Service, Technische Universität Berlin, Germany

[3]Centre for Entrepreneurship, Technische Universität Berlin, Germany

Abstract

The emergence of new technology-oriented industries creates opportunities to overcome economic and social problems resulting from the shrinking of traditional heavy industry. For this purpose, a strong presence of research institutions, particularly with technical focus is indispensible because they guarantee technology transfer. Thereby, one of the most relevant channels of technology transfer are entrepreneurship or spin-off activities at research institutions. This article aims at presenting the entrepreneurship support activities of the Technical University Berlin, the largest technical university of the Berlin region. It is furthermore discussing the impact of start-ups deriving from this university on regional development.

Keywords:

academic entrepreneurship, spin-offs, technology transfer

1 INTRODUCTION

The regional economy of Berlin, the capital of Germany, is notably affected by structural transformation, mainly resulting from the shrinking of the traditional heavy industry and the construction sector. Many of these traditional and labor-intensive industries discontinued or relocated their activities to countries with lower production costs. As a consequence Berlin is currently suffering high unemployment at 13.6% which is considerably above the 2010 average rate in Germany of 7.6% (Arbeitsagentur 2010).

At the same time new and more sustainable industries are developing in Berlin, such as renewable energy, biotechnology, medical technology, microsystems technology, information and communication technology as well as traffic and mobility technologies. By 2008 over 145,000 job positions in the Berlin region were created within these future-oriented industry sectors (Technologiestiftung 2008).

These innovation-driven industries are expected to be more sustainable than traditional industry sectors and to bring long-term value in economic, social and environmental terms. They bring about new market opportunities, create jobs for highly qualified employees and are furthermore more eco-friendly than traditional industries as they rely to a lesser extent on natural resources and to a larger extent on human knowledge, innovation and efficient technology. However, for a further development of those industries a strong presence of research institutions, particularly with technical focus is indispensible. In this aspect, Berlin seems to provide favorable conditions as it comes with several universities, universities of applied sciences and research institutions with nearly 140,000 students (German Statistical Office 2010). Furthermore, major research institutions, such as the renowned Fraunhofer Institute for manufacturing lines and construction design have their headquarters in Berlin. Such institutions form the basis of technology development and are sources of technology transfer. Entrepreneurship or spin-off activities are thereby one of the most relevant channels of this process. By founding companies students and the academic staff bring their know-how from science to economy. They furthermore make use of economic opportunities, employ people and thus create economic and social value. It comes as no surprise that university spin-offs and student start-ups have significant implications for local employment creation and revenue generation (Clarysse et al. 2005).

As a consequence, most of the German technical universities are strongly supporting entrepreneurial activities, among them the Technical University Berlin (TU Berlin), the largest technical university in the Berlin region. In 2007 the TU Berlin established the Founder's Service (germ: Gründungsservice) with the objective to enhance entrepreneurial activity and technology transfer and thus to stimulate the regional economy.

This article aims at presenting the Founder's Service as TU Berlin's entrepreneurship support system and discussing the impact of start-ups deriving from TU Berlin on regional development.

The structure of this paper is as follows: Section two provides information on entrepreneurship support at TU Berlin by showing the case of the Founder's Service. In the third section the structure and impact of start-ups deriving from TU Berlin on the regional development are presented. Section four includes the discussion and some conclusions.

2 ENTREPRENEURSHIP SUPPORT AT TU BERLIN

There is a distinct and vivid foundation culture in Berlin with the most company foundations nationwide related to the number of economically existing companies. Thus, in 2009 over 7,500 companies were founded in Berlin which corresponds with 1,310 foundations per 10,000 existing companies (Creditreform 2010). But not only this entrepreneurial friendly environment in Berlin causes that TU Berlin has a large potential to bring forward technology-oriented companies from its own ranks.

TU Berlin is the largest technical university in this region. It comes with 29,000 students, over 2,300 research assistants and over 320 professors, of whom most are stemming from engineering or natural science related areas. It takes an outstanding role in the field of enterprise foundation and technology transfer as it is bringing forward enterprise foundations from its own ranks since its foundation in 1946.

However, it was not until 2007 when the university decided to institutionalize its entrepreneurship support systems by establishing the Founder's Service as a structure within the university. The two main objectives of the Founder's Service are:

- enhancing entrepreneurial thinking of students and researchers and providing them with necessary infrastructure (see section 2.1.)
- providing them with crucial business knowledge and skills (see section 2.2.)

Those goals result from two challenges present at TU Berlin but most probably also at other German technical universities. First, engineering and natural sciences students are less willing to become self-employed than other students which might be explained by the manifold job opportunities for engineers in Germany (Görisch et al. 2002). Second, founders from technical universities possess indeed technological skills but often lack crucial business and management knowledge which is confirmed by a recent internal study of the Founder's Service. Accordingly, only 50% of the polled entrepreneurs related to TU Berlin stated that they had at least good business knowledge at the moment of the company foundation. The others had either no knowledge at all, little or only ordinary knowledge.

In order to address those challenges the Founder's Service developed several services including new sensitization and qualification methods with the goal to establish a successful founder support, especially for technology-oriented founders.

2.1 Instruments for enhancing entrepreneurial thinking at TU Berlin

The Founder's Service is supporting potential founders at every stage of the entrepreneurial process – regardless if there is only an idea or even a fully developed business plan. The services provided by the Founder's Service are based on four main pillars: *orientation*, *consulting, qualification* and *business incubation* for *technology based ventures.*

Orientation - Discover potentials

The first pillar implies the objective to make students, graduates as well as research assistants aware of self-employment as a professional alternative. For this purpose, the Founder's Service provides online tools, e.g. a test in order to reflect entrepreneurial talents. Additionally, potential founders are able to properly evaluate the strengths and weaknesses of being an entrepreneur within a special workshop hosted by a psychologist. There are also presentations of successful start-ups held at TU Berlin showing best practice examples. Every six months an open day offers the opportunity to participate in workshops as well as discussions and to get to know successful entrepreneurs. Finally, students and potential founders are held up-to-date by the quarterly newsletter.

Consulting - Well accompanied

The second pillar includes advice activities aiming at enhancing professional planning and reflecting of the business idea which is essential for the successful presence on the market. Thereby experienced start-up consultants provide any help or advice regarding start-up projects. They offer support and assistance to founding preparations, assistance in the elaboration of a business plan, counselling concerning funding and financing (e.g. federal financial support, such as the EXIST-scholarship for founders or the EXIST- research transfer) as well as contact to business angels and venture capital investors. If a team of potential founders is still missing partners it can make use of the online partnership search engine in order to find them.

Qualification - Well prepared

Another important aspect of the Founder's Service is its focus on skills development. Potential founders are well prepared with a variety of professional and commercial know-how as well as personal and social skills. The Founder's Service offers workshops and seminars to take these exact skills to the next level. Those seminars cover the following topics: first steps into self-employment, conduct of negotiations, corporate design, project management, acquisition or presentation techniques. There is also a lecture called "Entrepreneurship - From idea to market" providing first insight into the establishment process of a start-up. During the annual "Entrepreneurship Academy" the founders have the opportunity to acquire compact, exclusive and practice-oriented start-up knowledge in one week, just before leaving the business incubation centre. The workshop "product-propeller" supports research assistants in generating product ideas and concepts from a market-oriented perspective (see also: section 2.2.).

Gründungswerkstatt - Space to start

The fourth pillar is based on the assumption that the process of generating good ideas is requiring space to develop creatively, professional resources for realization and other start-ups to engage in mutually beneficial exchange. With the Gründungswerkstatt (german for "founding factory") the Founder's Service provides a professional business incubation service including offices with PC, telephone, fax, printer, as well as a consultation room for planning and realizing the business idea. Additionally, the users of the business incubation facilities are eligible for financing their participation at a trade fair to promote the product. The business incubation service is provided for 12 months for free. After the establishment of the business the team is still allowed to use equipment, laboratories and offices at the university within the room-and-device-usage-agreement.

Together with the individual consulting services the training opportunities of the Founder's Service are intensively used by students, graduates as well as research assistants, whereas the number of participants was steadily increasing over the years.

2.2 Instruments for providing crucial business knowledge and skills

A recent internal study of the Founder's Service among academic entrepreneurs of the TU Berlin reveals that 90% based their foundation of enterprises on technical know-how. Management know-how played a subordinated role. In the daily counselling of technology-oriented academic founders it is revealed, that they specifically lack knowledge about market potential, customer needs and financial issues.

To better prepare the potential founders for their future tasks, the Founder's Service established specific instruments for academic entrepreneurs. It developed a special workshop "product-propeller" to qualify the academic staff and is offering special trainings and soft-skills seminars. In 2009 the German ministry of transport, building and urban development (BMVBS) together with the organisation for economic cooperation and development (OECD) awarded these instruments as best practice examples of founder support at universities (BMVBS & OECD 2009).

"Product-propeller" – technology screening

The Founder's Service offers a free technology screening workshop for research assistants, called "product-propeller". The aim of this workshop is the systematic and effective economic realization of research results and technological inventions. The problem is that their potential benefits are not always visible at first glance. That is why this workshop is trying to help identifying and assessing market potential by providing its participants with different perspectives on their research results. Scientific insights are considered from a market-oriented perspective by using creativity techniques. At the same time the workshop serves as a platform for the development of innovative ideas for products and services.

The value adding points of the "product-propeller" workshop are the establishment of realizable ideas, the shift to a market-oriented view, the guidance for further realization as well as the provision of detailed information about potential means of conveyance and financing sources. The basis of the "product-propeller" is an one hour discussion in which the major and common goal of the workshop is elaborated. The concept is then shaped according to the identified goals and contents as well as the potential target groups.

The success of the "product-propeller" workshop results from the assumption that all persons within a team have equal opportunities and that they work and generate ideas together which are then clustered and assessed. The course of action is very structured thanks to a moderator but at the same time it leaves space for innovative ideas about products and services. After narrowing down and evaluating the ideas the whole group decides on the next necessary steps. Finally, the results are documented in a road map.

Soft skills seminars and trainings

The Founder's Service furthermore offers a broad skills training program within the project "Human Venture". This seminar allows persons interested in founding a company to enhance their competences precisely. Besides practice oriented workshops and seminars in soft skills development, there are also regular lectures of TU alumni who successfully founded a company and would like to share their knowledge.

Another training opportunity is the "Entrepreneurship Academy" which is attended by 20 selected founders from the technology-oriented area and is lasting one week. Within this very compact program the participants are trained to act on the market in an entrepreneurial and competent way. They are trained especially in the areas negotiations, pitching, business etiquette, team building and personnel management as well as quality management.

2.3 Participant structure and outcome of service offers

As already mentioned the services of the Founder's Service are not only offered to TU Berlin students but also to TU Berlin graduates, its academic staff as well as to people who already founded a company. Thereby the access to the services of the Founder's Service is also open to persons from other institutions.

All in all, between 2007 and April 2010 1,125 registered participants have joined the service offers of the Founder's Service. During that time, the number of participants triplicated from 160 to 444.

According to an internal study the biggest group of participants were students with 43.2%. 22.7% were graduates and 21.6% were professors or researchers. 12.5% could not be allocated to one of the above categories. In the above time period 15.6% of the participants were coming from other institutions. Table 1 presents the structure.

Table 1: Participant Structure (2007-2010)

Students	43.2 %
Graduates	22.7 %
Academic staff	21.6 %
Others	12.5%

Most participants from TU Berlin came from the faculty of mechanical engineering and transport systems, the faculty of planning, building and environment, the faculty of electrical engineering and computer sciences, as well as the faculty of economics and management. Only a few participants were associated with the faculty of process sciences or the faculty of natural sciences and mathematics. The strong position of technically oriented courses, however, did not correspond with a low share of female participants. Thus, one third of the participants were female which should be considered as a clear success at a technical university.

In general, the most often used service offer was the consulting service followed by the trainings and seminars for soft skills and business know-how. Thereby, those participants who founded an enterprise made use of more services than those who did not decide on founding.

Out of the 1,125 participants, 88 decided to found a company (7.8%). When looking at this specific group it appears that student founders used the open day of the Founder's Service as the first contact opportunity. Graduates, however, decided mostly to make use of the offers related to soft skills and know-how development. Researchers and the academic staff demanded particularly the consulting offer. More than 70% of the founders joined at least two service offers.

3 REGIONAL IMPACT OF TU BERLIN START UPS

The Founder's Service of the TU Berlin is steadily evaluating its impact and the performance of start-ups deriving from TU Berlin. For this purpose a survey of start-up enterprises is carried out on a regular basis (for a detailed discussion of the survey methodology see: Kirchner & von Matuschka 2009). Thereby the questionnaire is not only sent to direct spin-offs from TU Berlin or companies advised by the Founder's Service but to all company foundations which are characterized by a factual, personal and/or timely relation to TU Berlin (in accordance to Lilischkis 2001).

The following analysis is based on the survey conducted by Kirchner & von Matuschka (2009) among 1,682 prospective and existing entrepreneurs. In total 318 persons have responded, whereas the particular sample size is varying dependent on the amount of answers to particular survey questions. Results are presented below.

3.1 Structure of TU Berlin Start-Ups

Location of headquarters

When the location of new start-up enterprises is considered it appears that out of 253 companies who responded to this question (n=253), 74% have chosen the Berlin area as their headquarter. Thereby, the proximity to the university and other research institutions was one of the key factors for choosing this location. 8% established their headquarters in the federal state of Brandenburg which is the region surrounding the city of Berlin, 14% located their headquarters in other German regions and slightly more than 4% were based abroad. This distribution is presented in Table 2.

Table 2: Location of activity of TU Start-Ups

Berlin	74%
Brandenburg	8%
Germany	14%
Abroad	4%

Industry structure and core activities

The industry structure is dominated by the services sector as in 2008 nearly 70% of the respondents (n=200) regarded it as the most important sector for their company. Research & development is second with 14%. Manufacturing is third with 13% and the trade sector accounts for approximately 6%. Table 3 visualizes this breakdown.

Table 3: Start-up structure according to sector

Services	69%
Research & Development	14%
Manufacturing	13%
Trade	6%

When considering the structure according to core business activity (n=187) it appears that consulting services are the most relevant with 22%. 15% of the respondents stated that their activities concern architecture and planning. Further relevant sectors include electrical engineering and telecommunications (10%), software (9%) and internet business models (6%). Table 4 shows this distribution.

Table 4:Start-up structure according to business activity

Consulting	22%
Architecture/Planning	15%
Electr. engineering/Telecommunication	10%
Software	9%
Internet business models	6%
Other industry	6%
Machine building/Automotive	4%
Education/Research	4%
Trade	4%
Environment/Water engineering	3%
Others	17%

The most popular legal form of the start-up enterprises was the form of a German limited liability company (GmbH) with 47%. One third of the respondents were self-employed or freelancers (31%). Approximately 10% of respondents have established a partnership (GbR), about 5% have set up a public limited company (AG), 7% decided on other forms.

3.2 Sustainability and impact of TU Berlin Start Ups

Age of TU Berlin Start-Ups

When it comes to the sustainability of the start-ups it appears that most of the respondents (n=252) were active for quite a long time. Accordingly, 27% of the companies were founded before 1995 and thus active since more than 15 years. 21% were founded in the period of 1995-1999, 29% were founded in the period of 2000-2004, 17% in the period of 2005-2007 and 6% were founded in 2008, the year in which the survey was conducted. This distribution is shown in figure 1:

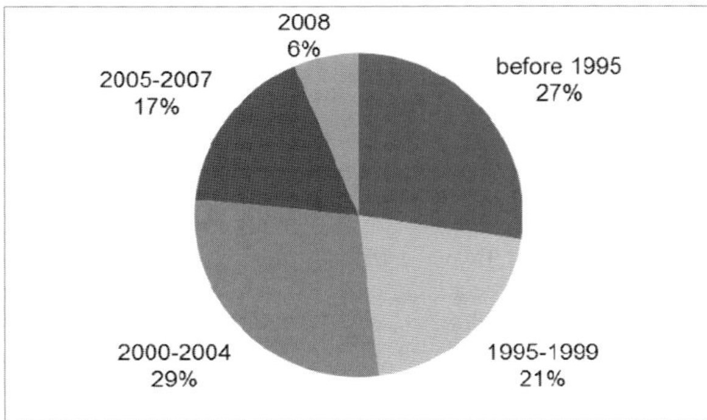

Figure 1: Activity period of TU Berlin Start-Ups.

Economic Impact: Job Market

According to the study of Kirchner & von Matuschka (2009) by 2008 the companies which were founded between 2005 and 2008 and responded to this question (n=151) were employing approximately 5900 people. When adding the still existing enterprises founded in the period 2001-2005 one can act on the assumption that approximately 10,400 job positions were created by TU Berlin start-ups. However, it is necessary to mention that those are only approximated values as they are based on estimates. This is due to the form of the survey which provided respondents with given spans and did not ask for exact figures.

In 2008 nearly 50% of the respondents (n=151) had between 1 and 5 employees. 29% of the polled companies were employing between 6-20 people. 10% had between 21 and 50 job positions and 3% were employing between 51-100 people. 4% of the companies had more than 100 employees and 5% had no employees at all (individual firms). This distribution is shown in figure 2 below.

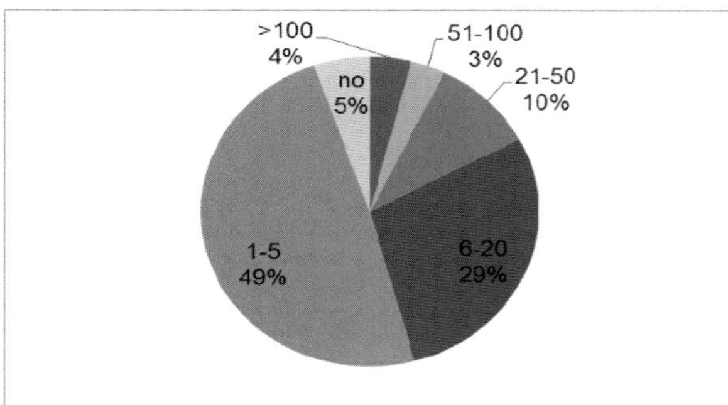

Figure 2: Job positions in TU Berlin companies (2008).

What is moreover important, is that the majority of the job positions were based in Berlin. Thus approximately 5,000 out of 5,900 job positions were created in the Berlin region.

Economic Impact: Generated Turnover

Similar to the amount of created job positions the generated turnover of companies related to TU Berlin can be regarded as an essential indication for their economic impact. However, also in this case no exact figures can be provided due to the same methodological reasons as in the case of job positions.

It is estimated that in 2007 the total turnover of TU Berlin related companies who responded to this question (n=146) was in a range of EUR 713.5m and EUR 1.35bn.

When considering the structure of those companies according to turnover it appears that in 2007 16% of the polled companies (n=146) had a turnover accounting for EUR 50,000 or less. 30% had a turnover in the span of EUR 51,000 and 250,000 and 9% of 251,000 and 500,000. 24% of the start-ups had a turnover between EUR 500,000 and 2 million. Over 21% had a turnover higher than EUR 2 million. This distribution is shown in figure 3 below:

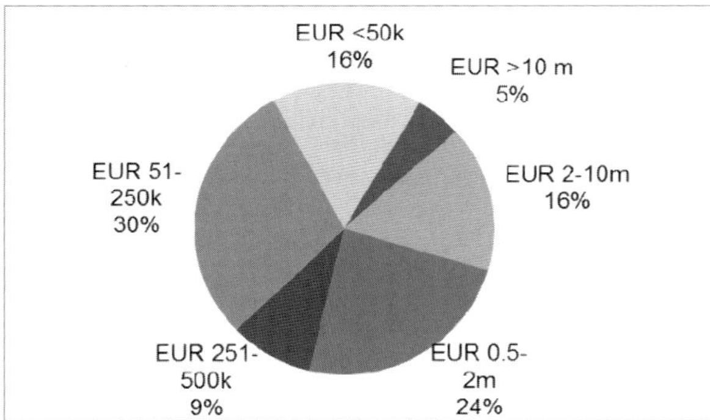

Figure 3: Structure of companies - turnover (2007).

4 DISCUSSION AND CONCLUSION

The regional economy of the city of Berlin is notably affected by structural transformation, mainly resulting from the shrinking of the traditional heavy industry and the construction sector. As a consequence Berlin is currently suffering high unemployment. The emergence of new technology-oriented industries creates opportunities to overcome these problems. For this purpose, a strong presence of research institutions, particularly with technical focus is indispensible because they guarantee technology transfer. Thereby, one of the most relevant channels of technology transfer are entrepreneurship or spin-off activities at research institutions. Science-based start-ups or spin-offs add high value to the city as they create jobs for highly qualified employees, accelerate technology transfer and generate tax income. It is not surprising that TU Berlin is trying to further enhance this development by fostering technology-oriented university entrepreneurship.

The goal of this study was to present the entrepreneurship support activities of the Technical University Berlin and its Founder's Service and to furthermore discuss the impact of start-ups deriving from TU Berlin on regional development.

It was shown, that technical universities in Germany are facing two major challenges in bringing forward technology-oriented start-ups from their own ranks. First, the lack of entrepreneurial spirit at technical universities and second the lack of business and management knowledge of technical university students.

TU Berlin is addressing both challenges through its established Founder's Service which is applying different instruments to cope with them. First, it is providing online orientation tools, workshops hosted by psychologists, consultancy services and seminars on entrepreneurship in order to enhance entrepreneurial thinking at TU Berlin. The second challenge is addressed by two instruments awarded by OECD as best practice, namely the "product propeller" workshop and the soft skills seminar "Human Venture". Both help students from technology-related courses to identify business opportunities and to enhance their soft skills.

After three years of activities it is possible to summarize that the Founder's Service is acting effectively because of three reasons. First, the amount of students and scientists willing to take part in the offer of the Founder's Service is steadily growing which implies a growing entrepreneurial spirit at TU Berlin. Second, most students stem from engineering and technology-related courses which is the natural target group of the Founder's Service' offer. And third, there is already a considerable amount of 88 founders.

It might be expected that these companies will join other companies deriving from TU Berlin which already influence Berlin's economy. As shown in section three, these companies were able to create the considerable amount of approximately 5,000 jobs in recent years in the Berlin area and thus had major impact on economic and social issues of the region. The positive job effect of the companies deriving from TU Berlin becomes particularly evident when it is compared to job creation by spin-offs from other universities. When dividing the total employment of these companies through their total amount it appears that those which derive from TU Berlin on average were employing 14.8 employees. This is a relatively good result, compared to the performance of companies related to other leading universities. Thus, companies related to the University of Twente in the Netherlands were employing on average only 5.6 employees (Van der Sijde et al. 2002), companies related to Swiss ETH Zürich were employing on average 8 employees (Oskarsson & Schläpfer 2008). Even when taking into account that the performance of TU Berlin companies might result from the positive correlation of company age and employment the difference still seems to be relatively high.

Furthermore, the companies deriving from TU Berlin are a big economic power by generating turnover in a range of EUR 713.5m and EUR 1.35bn in 2007. It is expected that this numbers also translate in increasing regional demand and tax income.

What is also important is that there is a high share of companies which were founded before 2004 and are still operating. This is an indication for the sustainability of the companies related to TU Berlin.

However, there is one question regarding the technological-orientation of the founded companies. According to the survey only few were founded in the future-oriented sectors such as renewable energy, biotechnology, medical technology, microsystems technology, information and communication technology as well as traffic and mobility technologies. Most companies were active in the services sector or more specifically in the consulting or IT/Internet segment. It might be assumed that those sectors are less technologically-oriented which would imply that TU Berlin companies do not fully use their technological potential.

Nevertheless, when considering the successful support activities of TU Berlin's Founder Service specifically aligned to the needs of technology-oriented founders, a trend towards founding more technology-oriented companies by TU Berlin students and scientists should be expected.

REFERENCES

Arbeitsagentur, 2010: Arbeitslosenquoten im Jahresdurchschnitt 2010, URL: http://www.pub.arbeitsagentur.de/hst/services/statistik/000000/html/start/karten/aloq_land_jahr.html, accessed on: 10.03.2011.

BMVBS & OECD, 2009: Handbuch: Förderung von Unternehmertum aus Hochschulen: Kriterien und Beispiele guter Praxis, OECD, 2009.

Clarysse, B., Wright, M., Lockett, A., Van de Velde, E., Vohora, A. 2005: Spinning out new ventures: a typology of incubation strategies from European research institutions, Journal of Business Venturing, 20, 183-216.

Creditreform 2010: Insolvenzen, Neugründungen, Löschungen, Jahr 2010, Creditreform Wirtschafts- und Konjunkturforschung, 2010.

German Statistical Office, 2010: Studierende, Insgesamt nach Bundesländern und tiefer gegliederten Angaben, URL:http://www.destatis.de/jetspeed/portal/cms/Sites/destatis/Internet/DE/Content/Statistiken/BildungForschungKultur/Hochschulen/Tabellen/Content50/StudierendeInsgesamtBundeslaender,templateId=renderPrint.psml, accessed on: 10.09.2010.

Görisch, J., 2002: Studierende und Selbstständigkeit. Ergebnisse der EXIST-Studierendenbefragung, Bundesministerium für Bildung und Forschung, 2002.

Kirchner, U., von Matuschka, A., 2009: TU-Gründungsumfrage 2008/2009, TU Berlin, Founder's Service, Berlin, Dezember 2009.

Lilischkis, S., 2001: Förderung von Unternehmensgründungen aus Hochschulen. Eine Fallstudie der University of Washington (Seattle) und der Ruhr-Universität Bochum. Lohmar-Köln.

Oskarsson, I., Schläpfer, A., 2008: The Performance of Spin-Off companies at the Swiss Federal Institute of Technology (ETH) Zürich. Zürich 2008.

TSB Technologiestiftung, 2008: Wie aus Wissen Arbeit wird. Zahlen zur Entwicklung Berliner Kompetenzfelder, TSB Technologiestiftung, Berlin, Oktober 2008.

Van der Sijde, Karnebeek, S., van Benthem, J., 2002: The Impact of a University Spin-Off Programme: The Case of NTBFs Established Through TOP, in: Oakey, R., During, W., Kauser, S. (2002): New Technology-Based Firms in the New Millennium, 2nd edition. Elsevier Science, Oxford.

CONTACT:

Agnes von Matuschka
Head of Center for Entrepreneurship
Phone: +0049 30 314 21456
Technical University Berlin
Hardenbergstr. 38
10623 Berlin
Email: agnes.matuschka@tu-berlin.de
Website: www.gruendung.tu-berlin.de

New Trends for Technology Driven Start-ups: Experiences from Berlin Metropolitan Area: An Empirical Study

S. Qureshi[1], J. Kratzer[2]
[1]Center for Entrepreneurship IBA, Karachi, Pakistan
[2]Chair for Entrepreneurship und Innovations Management, Technische Universität Berlin, Germany

Abstract

The technology driven start up firms operate in a competitive environment. The proper development of skills and capabilities not only enhances the survival rate but also helps these entrepreneurial firms achieve a competitive advantage. The following research proposes an integrative framework built on previous research and tests the various antecedents and outcomes of firm performance capabilities in entrepreneurial firms. Environmental turbulence, entrepreneurial and market orientation of the firm are tested as an antecedent to marketing capabilities of the firm. Firm performance is tested as an outcome of the marketing capabilities construct. A survey was conducted during the period of August 2008 to January 2009 from the CEOs of technology driven start up firms in Germany. A structural equation model using PLS was developed to test the model. Environmental turbulence was found to be having significant impact on the entrepreneurial orientation and market orientation of the firm which in turn were found to be having significant impact on the marketing capabilities. Moreover the marketing capability of the firm was found to be significantly related to firm performance.

Key Words: Marketing Capability, Market Orientation, Entrepreneurial Orientation, Competitive Advantage, Firm Performance

1. INTRODUCTION

The technology driven start up firms operate in an environment consisting of increased risk and uncertainty. This has resulted in a competitive landscape characterised by the forces of change, complexity, chaos and contradiction. The markets are shifting, overlapping and fragmenting and the firms interact as competitors, customers, and collaborators in global knowledge economy. The customers are more knowledgeable and demanding. In this changing context, marketing has emerged to be of great importance for the success of the entrepreneurial ventures. In this changing scenario, entrepreneurial firms use marketing as a path to create competitive advantage, based on differentiating their marketing program by leveraging their superior knowledge of customers, markets and technologies [20].

In an entrepreneurial organization, entrepreneurship and marketing permeates all areas and levels of the organization, with the organization being focused on recognizing and exploiting opportunities. Successful entrepreneurs tend to have a long term orientation to opportunity creation and exploitation that is focused on meeting all the customer's needs by employing creativity and innovation [6]. Morris [37] has defined entrepreneurial marketing as the proactive identification and exploitation of opportunities for acquiring and retaining profitable customers through innovative approaches to risk management, resource leveraging and value creation. This fuses key aspects of marketing thought and practice with those in the entrepreneurship area.

Prior research in the marketing domain has predominantly focused on large, resource abundant corporate organizations and ignored small entrepreneurial organizations [20]. This myopic perspective tends to overlook the more entrepreneurial firms which face resource constraints, capability limits and have different objectives and contexts and the skills and resources employed by entrepreneurs in using marketing as a tool to gain competitive advantage [32].

The research on the interrelation between marketing and entrepreneurship are extremely fragmented so far, and there is no integrated analysis or comprehensive theory yet [29].

This research proposes a framework built on previous research [37] and tests the various antecedents of firm performance in entrepreneurial firms. We examine market and entrepreneurial orientation as key market based assets [36] and firms marketing capabilities as key market relating deployment mechanism. This study investigates the casual linkages of entrepreneurial orientation and market orientation on the marketing capabilities and performance of entrepreneurial firms.

2. THEORETICAL FOUNDATION

There has been considerable interest in the management literature regarding the development of core competencies in order to enhance competitiveness and performance of firms [58],[19]. The resource based view of the firm [58] places a great emphasis on competing on the basis of capabilities, both tangible and intangible [18].

Resource-based theory views heterogeneity among firms in term of resources and assets owned or controlled by the firm that allow its managers to conceive and execute value-creating strategies which are fundamental in explaining firm performance [2]. Theorists have made a number of recent developments, collectively labeled 'dynamic capabilities' theory, addressing limitations in traditional resource-based theory [40], [61]. Dynamic capability theory posits that since marketplaces are dynamic, therefore rather than simple heterogeneity in firms' resource endowments, the capabilities by which firms' resources are acquired and deployed in ways that match the firm's market environment explains inter firm performance variance over time [14], [30], [51]. These capabilities involve complex coordinated patterns of skills and knowledge that, over time, become embedded as organizational routines [17] and are distinguished from other organizational processes by being performed well relative to rivals [3]; [15]. Therefore, a key task for the firm is to identify those capabilities that will provide a sustainable competitive advantage. To be sustainable, these capabilities must be difficult to imitate and should support the organisation's business strategy [1], [12], [11].

The development of marketing capabilities has been identified as one of the important ways firms can achieve a competitive advantage [1], [11], [56]. To develop and sustain a competitive advantage, firms must develop processes that allow them to collect information about market opportunities, develop goods and services to meet the needs of targeted customers in selected markets, price these products according to market information, communicate product advantages to potential customers and distribute products to customers [11].

Marketing capabilities of firms are influenced by both external and internal factors. Environmental turbulence is considered as an external factor in this study. On the other hand entrepreneurial orientation and market orientation are considered as internal factors impacting marketing capabilities. The various resources and capabilities discussed in the study are discussed as follows.

Marketing Capabilities

The extant literature has identified various capabilities which the firm can use to obtain a competitive advantage. [42]; [11]; [1]. Capabilities [17] can be thought of as integrative processes by which knowledge-based resources and tangible resources come together to create valuable outputs. These capabilities come about through integration of knowledge and skills of employees [41], [42]. Marketing capabilities are defined as integrative processes designed to apply the collective knowledge, skills, and resources of the firm to the market related needs of the business, enabling the business to add value to its goods and services and meet competitive demands. Marketing capability [11] is thus developed when the firms marketing employees repeatedly apply their knowledge and skills to fulfil the market related needs of the business. This study investigates the following six marketing areas [54] for evidence of capabilities i.e. marketing research, pricing, product development, channel management, promotion and the marketing management area.

Environmental Turbulence (External Environment)

There are many variables which constitute the external environment i.e. demand and supply heterogeneity, the availability of effective substitutes, competitors, rates of technological change and changing market trends etc. For simplicity such environmental variables are captured by market and technological turbulence.

The external environment that a firm operates in has been shown to impact on many different facets of the organization. Due to the significant influence environment has on the organization in general, it is logical to believe that environmental variation will impact on the various functions of the firm, such as marketing [45] and the strategy of the firm [33].

In this research, more turbulent environments were shown to be related to the development of a strategic orientation that relied on well-developed marketing skills. The development of differentiated products, product innovation and new product development skills enabled firms to outperform other less marketing-oriented firms [33]. When an environment is turbulent, managers need more information to be able to make decisions [10]. An environment is considered turbulent when it produces many rapid changes. The sub dimensions of environmental turbulence include market turbulence – the rate of change in customer composition and their preferences – and the rate of technological change – the degree of technological turbulence [28]. When environments are turbulent, managers have a greater need for market information [31]. In most firms, market intelligence gathering is a key source of the environmental information that manager's need [28], [31]. However, for the information to be useful in the decision-making process it must be disseminated to the right individuals and groups within the organization and these individuals and groups must act on the information [23], [27], [48]. Over time, the organization's employees will routinize these processes by creatively applying their knowledge and skills to the problems and opportunities the environment presents and then these repeated applications of knowledge and skills to the problems and opportunities presented by the environment will evolve into capabilities [16], [17]. Thus, it appears that the development of marketing capabilities will take place to deal with the problems and opportunities created by a turbulent environment.

Higher levels of environmental turbulence directly affects the internal organisation variables i.e. market orientation (MO), entrepreneurial orientation (EO), marketing capability and other organisational factors of the firm. For the sake of simplicity entrepreneurial orientation and market orientation are investigated in this study.

Entrepreneurial Orientation consists of overall level of innovativeness, risk taking and proactiveness within the firm [83], [9], [60]. The entrepreneurial firm is generally distinguished in its ability to innovate, initiate change and rapidly react to change flexibly and adroitly [38]. Based on the firm-behaviour model of entrepreneurship [7], [38], entrepreneurship is conceptualized as a firm behaviour in which the firm displays innovativeness, proactiveness and risk-taking propensity in their strategic decisions.

Innovativeness refers to a corporate environment that promotes and supports novel ideas, experimentation and creative processes that may lead to new products, techniques or technologies. Risk-taking reflects the propensity to devote resources to projects that pose a substantial possibility of failure, along with chances of high returns. Proactiveness implies taking initiative, aggressively pursuing ventures and being at the forefront of efforts to shape the environment in ways that benefit the firm [8]. Entrepreneurship is conceptualized as a continuum using these three attributes that reflects the degree of 'Entrepreneurial Intensity' of the firm.

Market Orientation represents an additional strategic dimension [49] and is a fundamental approach to understanding markets. It represents the implementation of the marketing concept [28], and is a cultural orientation [47] that focuses the firm's efforts on the needs of the market. A market-oriented organisation possesses the ability to generate, disseminate, and respond to information about market forces and market conditions better than less market-oriented rivals [27], [23]. The market-oriented firm thus has a better possibility for building a sustainable competitive advantage. The firm does this by learning what buyers want, building and leveraging the resources and processes necessary to deliver the value they desire [39], [47], and adapting those value-generating processes as market conditions change [48]. Furthermore, the market-oriented organisation looks beyond current customer needs to develop future products that will tap latent needs, thus serving to strengthen the firm's market position over time [49]. To use these processes as the basis for competitive advantage, an organisation needs to develop the capabilities to generate, disseminate, and respond to market intelligence [11] and the processes to act on this information [22].

Firm Performance

Organisational performance is a multidimensional construct, tapping business growth and profitability. Growth reflects increases in sales and is often reflected in market share gains [53]. Growth in sales and market share are important to a business to ensure long-term viability and resource availability [52]. Profitability primarily reflects current performance [53]. Profitability is viewed [22] as the ultimate organisational outcome and is commonly used in strategic management studies.

An Integrative Framework

Figure 1 depicts the integrative model of antecedents of firm performance. Higher levels of environmental turbulence require firms to demonstrate more adaptability and flexibility in approaching competitors and customers, to have high levels of innovation and entrepreneurship as well as demonstrate a high level of strategic orientation. Where firms demonstrate stronger entrepreneurial and market and orientations, they will tend to approach the marketing function differently.

Marketing activities become especially critical under turbulent environmental circumstances. Under normal conditions, firms can concentrate on incremental improvements to their methods of satisfying customer needs. Alternatively, marketers must focus more attention on anticipating and quickly responding to the moves of the competitors. As the environments become fairly turbulent, marketers must take responsibility for introducing greater levels of entrepreneurship into all aspects of the firms marketing efforts. Marketing efforts have to become more customised and unique, with more customer choice in the form of a variety of value packages for different market segments [13]. Finding creative ways to develop customer relationships while discovering new market segments becomes important. In short, firms are incentivized to engage in marketing efforts that are more opportunistic, proactive, risk assumptive, innovative, customer-centric, leveraged, and value creating [37].

The firms approach to marketing is also influenced by internal organisational factors i.e. entrepreneurial orientation, market orientation and other organisational factors.

Entrepreneurial orientation is reported to influence marketing capabilities of innovative firms in a positive way[57]. Bhuian [4] suggests that entrepreneurship provides a filter through which organizations view and direct market intelligence processes. That is, entrepreneurship influences the marketing processes being carried out in the firm. This view is consistent with the dynamic capabilities perspective [51] in which learning, coordination and reconfiguration of, key organizational competencies leads to competitive advantage. Keh [25] reports that entrepreneurial orientation has a direct effect on firm performance.

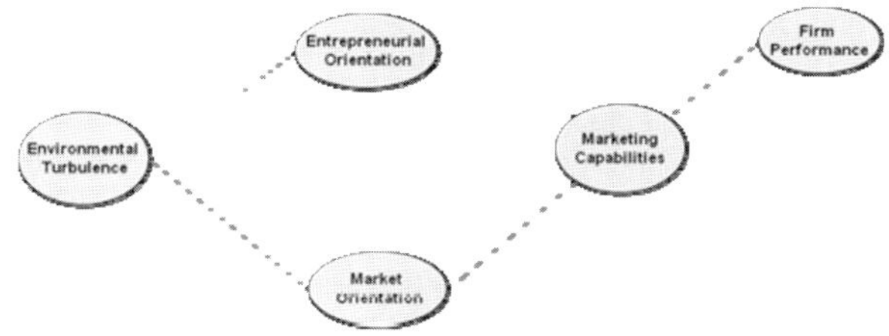

Figure 1: Model of Antecedents of Firm Performance

The market information processing capabilities [31], [26] influence the marketing capabilities of the firm. The traditional literature based on the resource based theory posits that firms with superior firm market orientation achieve superior business performance because they have a greater understanding of customers expressed wants and latent needs, competitors capabilities and strategies, channel requirements and developments, and the broader market environment requirements than their rivals [23].

Due to the important role of marketing capabilities in the selection of product markets and because of their ability to impact on the implementation of market strategies [11,22] the marketing capabilities of the firm have been predicted to positively influence firm performance. Based on the above discussion the following hypotheses are posited.

H1: The higher the level of environmental turbulence, the higher is the entrepreneurial orientation of the firm.
H2: The higher the level of environmental turbulence, the higher is the market orientation of the firm.
H3: The higher the level of entrepreneurial orientation of the firm, the higher is the marketing capabilities of the firm.
H4: The higher the market orientation of the firm, the higher is the marketing capabilities of the firm.
H5: The higher the level of marketing capability, the higher is the firm performance.

3. RESEARCH DESIGN

The research design for this study is a key informant survey designed to collect data from the top marketing decision maker [5, 24]. The top decision maker is selected because he/she would be able to represent accurately his/her organisation's views on the issues covered in this study [24].

The survey was initiated by mailing the questionnaire to the CEOs of (n = 800) small technology firms from the list obtained from Adlerhshof technology park, Berlin and the alumni firms of the Technical University and Humboldt University of Berlin. Only those small firms with at least 2 years of operation and 5 or more employees were included in the survey. The data collection activity was carried out during the period of August 2008 to January 2009 in which follow up telephone calls and email messages served as reminders. The response to the survey was adequate with usable responses received from 143 companies. This yielded an overall response rate of 20 %, which is considered respectable in this type of surveys. It may be noted here that due to the focus on small technology firms, 90% of the respondents were CEOs themselves (who were directly in charge of their company's marketing function), while the remaining were COOs or marketing managers.

A structural equation model using PLS (Partial least squares) was developed to test these relationships.

4. OPERATIONALISATION OF VARIABLES

The measurements of the constructs used in this research are based primarily on previously developed scales. Some amendments were made to the constructs as they were originally designed for large firms. Each of the measures used in the study is discussed briefly.

Environment Turbulence

Two aspects of environmental turbulence were used in this study. Market turbulence refers to the extent to which composition and preferences of the organisations customers change over time [23]. Technological turbulence refers to the extent to which technology in an industry is subject to rapid changes [23]. The respondents rated both of these sub constructs on seven point Likert scale (1 = not at all; 7 = to a great extent).

Entrepreneurial Orientation

The entrepreneurial orientation construct measures the extent to which the firm's leaders are innovative, proactive and risk seeking. High scores on this scale indicate that the firm's key decision makers value innovation and proactiveness and have a high tolerance for risk. The items for the scale are derived from Namen and Slevin [38], which was based on the measure developed by Covin and Slevin [7].

Market Orientation

Market Orientation was measured using the scale developed by [23]. The scale is designed to measure three sub dimensions i.e. generation of market intelligence, dissemination of market intelligence across departments and within the company, and responsiveness to market intelligence. For this research the original 23 item scale was modified to accommodate the small firms. The new scale consisted of 16 items. Seven point Likert scale was used (1 = not at all; 7 = to a great extent).

Marketing Capability

Marketing capability was measured using a scale developed by Vorhies and Harker [54]. Respondents were asked to express their beliefs regarding their business unit's marketing processes in six distinct areas: pricing, promotions, product development, distribution channels, marketing management and planning and marketing research development. Each of these sub constructs were measured with multiple items. To asses the company's marketing capabilities, a seven point Likert scale was used (1 = not at all; 7 = to a great extent).

Firm Performance

In this study, firm success is operationalized by adapting a scale from Spanos and Lioukas [50] and consists of five items (v_113 to v_117) covering different aspects of organizational performance i.e. market-based performance and financially based performance. To operationalize market based performance, two items are used: market share and sales growth. To operationalize financially based performance, three items are used: profitability, return on investment and return on sales. Relative performance on each dimension (item) was measured by asking respondents to assess their firm's performance relative to that of major competitors. Seven point Likert scale was used (1 = not at all; 7 = to a great extent).

5. RESULTS

The respondents were asked to provide their position in the company. This was done to confirm that the individual responding to the survey had the knowledge necessary to answer the questions. In our case more than 90% of the respondents were the CEOs themselves, while the remaining were COOs or marketing managers. Due to the small nature of the companies, the CEO himself was directly in charge of the marketing function. The higher position of the respondent and his direct involvement in the marketing activities indicates that these individuals had the necessary understanding of the issues they were asked to respond to.

All the constructs were tested to confirm the priori loadings of the items on the constructs and to ensure adequate unidimensionality, discriminant validity and convergent validity. Following this analysis, the proposed model was tested using the structural equation modelling software, SmartPLS [43]. The causal model and the empirical results are illustrated in figure 2.

Evaluation of the reflective measurement model

The marketing capability latent variable is being measured by the following indicators i.e. product development, market research, promotions, pricing, channels and market management. The factor-loadings determine the power of the interrelations between marketing capability and its indicators. Assessing the empirical results in fig 2, factor loadings in the reflective measurement model has a very high value of 0,865for the market management indicator followed by market research (factor loading 0,748) and promotions (factor loading 0,745). Besides, the factor loadings of product development (0,638) and channels (0,647) are close to the minimum value demanded in literature. The factor loading for pricing is (0,534) somewhat lower.

Entrepreneurial orientation is measured by the following indicators i.e. innovation, pro activeness and risk taking. Assessing the empirical results, all factor loadings in the reflective measurement model have a value higher than 0,776. Therefore the variable explains the variance of each indicator to a great extent.

Market orientation is measured by the indicators such as intelligence generation, dissemination and response. All the indicators have a loading higher than 0,84. Therefore the variable explains the variance of each indicator to a great extent. Environmental turbulence is measured by the indicators such as market turbulence and technological turbulence. All the indicators have a loading higher than 0,794. Therefore the variable explains the variance of each indicator to a great extent.

Figure 2: The Integrative Model with Path Coefficients

Testing of the overall model

In the SEM (structural equation model) higher environmental turbulence (latent exogenous variable) results in to the highest impact (weight of 0,337) on the latent endogenous variable market orientation, followed by a somewhat lower impact (0,333) on entrepreneurial orientation. In the next stage of the structural equation model, the marketing capability construct is tested. Market orientation results in to the highest impact with a weight of 0,477 followed by entrepreneurial orientation with a weight of 0,283, having a large explanatory share for the latent endogenous variable marketing capabilities with an R^2 of 0,431.

In the final stage of the structural equation model, the impact of marketing capability and entrepreneurial orientation on firm performance is tested. Marketing capability results in to the highest impact with a weight of 0,372, followed by entrepreneurial orientation having a weight of 0,179. Both of these variables have a reasonable exploratory share for the latent endogenous variable firm performance with an R^2 of 0,217. It is worth mentioning that marketing capability has a weight twice of that of entrepreneurial orientation.

These findings thus suggest that the development of marketing capabilities is an important instrument for the new technology based firms to achieve a high level of firm performance. In the long run the higher level of marketing capabilities is determined by the direct and indirect effect of entrepreneurial orientation, market orientation and environmental turbulence.

The total effects (direct and indirect) of various variables on firm performance are mentioned in Table 2. Marketing capabilities have the highest total effect on firm performance (weight of 0,372) followed by entrepreneurial orientation (weight of 0,284) and market orientation (weight of 0,177).

Thus marketing capability has the highest explanatory share for the latent endogenous variable firm performance with an R^2 of 0,213. Therefore a strong focus and development of marketing capabilities and entrepreneurial orientation is a very important instrument for the new technology based firms to achieve high levels of firm performance.

6. DISCUSSION

The hypotheses posited in the framework are statistically significant and thus support the model presented in the paper. Higher levels of environmental turbulence have significant impact on the various capabilities of the firm i.e. entrepreneurial and market orientations. This requires firms to demonstrate more adaptability and flexibility in approaching competitors and customers, to have high levels of innovation and entrepreneurship. Where firms demonstrate stronger entrepreneurial and market orientations, they will tend to approach the marketing function differently. Therefore marketing activities become especially critical under turbulent environmental circumstances.

As the environment becomes fairly turbulent, marketers must take responsibility for introducing greater levels of entrepreneurship into all aspects of the firms marketing efforts. Turbulence evokes fear, uncertainty and doubt among sellers and buyers alike, but also forces firms to make quicker decisions and opens a whole range of new products and market opportunities. Marketing efforts have to become more customised and unique, with more customer choice in the form of a variety of value packages for different market segments [13, 46]. Finding creative ways to develop customer relationships while discovering new market segments becomes important. In short, firms are incentivized to engage in marketing efforts that are more opportunistic, proactive, risk assumptive, innovative, customer-centric, leveraged, and value creating [37].

Further, entrepreneurial and market orientation significantly impact marketing capabilities. Entrepreneurial orientation is reported to influence marketing capabilities of innovative firms in a positive way [57]. The market orientation and market information processing capabilities [31, 26] positively influence the marketing capabilities of the firm. Marketing capability and entrepreneurial orientation in turn have a significant impact on the performance of the firm.

The direct path from market orientations to firm performance was also modelled and tested. However the path strength was very low (weight of 0,027) and it was not significant, thereby confirming the proposed model.

7. MANAGERIAL IMPLICATIONS

This study has important implications for managers and practitioners. It highlights the necessity of firms to develop superior entrepreneurial and market orientation of their members and also to invest on enhancing marketing capabilities as a way of achieving high levels of firm performance. Entrepreneurial orientation based on innovativeness, proactiveness and risk taking has positive impact on other capabilities and the firm performance. Entrepreneurs compete not only to identify promising opportunities, but also for the resources necessary to exploit these opportunities. Entrepreneurs should actively engage in information acquisition as an aid to effective marketing strategy formulation. More importantly, proactive use of such information allows entrepreneurs to predict oncoming trends and enact strategies, supporting the view that the competitive advantage associated with information depends increasingly on whether a firm is able to make the best use of acquired information [35]. Information utilization enables small firms to gain competitive advantage and maintain a stronger position relative to the competition. The information may unveil latent needs, which exist and are unmet but are not apparent to competitors [23]. Being the first to uncover such latent needs provides impetus to develop the marketing capabilities accordingly.

Another implication from the study is that the firms should develop their marketing programs by focusing on developing marketing capabilities. Firms with advanced marketing capabilities should be better able to outperform firm's having lower degree of marketing capability. To enhance marketing capabilities, continued investment in market research, pricing, product development, promotions, channels and market planning and market management capabilities is important. Findings further suggest that market management (ability to segment and target market, to manage the marketing programs, the ability to coordinate various departments and groups to respond to market conditions) , market research and promotion (sales promotions and free samples and trial runs) are the most important marketing capabilities for the small technology firms. Thus the managers are recommended to give priority to these three marketing capabilities at the time of resource allocation.

8. CONCLUSION

This study has provided an insight of the marketing capability construct and the various influencing factors in the context of technology driven start up firms. It is important to understand the limitations inherent in this study. One limiting issue is the use of the key informant approach [5, 24]. Although key informants are frequently used in marketing research, their use presents potential validity problems [41]. Although some researchers advocate multiple informants [21], others have found that CEOs provide data as reliable and valid as multiple informants [59]. One potential problem is that the informant may not be knowledgeable on all of the issues being asked about [48] and this may bias the results.

These finding are significant for the research presented here as the top marketing decision maker for the firm was contacted to fill the questionnaire. As the respondents demographics demonstrated, care in respondent selection yielded responses from knowledgeable top marketing decision makers in most of the cases. However still, due to the importance of this issue caution must be taken in interpreting the study's results. Another limiting issue is the geographical limits of the study. New technology firms in other parts of the country shall also be studied to verify the results in this study.

This study has provided useful insights into the marketing capability construct and the various factors influencing the marketing activities in the small technology firm sector as depicted in the proposed framework. Entrepreneurial orientation and market orientation have a positive impact on the development of marketing capabilities. Furthermore, those firms with highly developed marketing capabilities demonstrated high level of firm performance. As a result, the findings in this research provide important support for many of the recent theories regarding the development of marketing capabilities and the role they play in achieving competitive advantage [12, 55]. This study also adds to the literature on the marketing and entrepreneurship interface and it confirms that higher levels of entrepreneurial and market orientation impacts entrepreneurial marketing capabilities [37].

9. REFERENCES

[1] Barney Jay (1991), Firm Resources and sustained competitive Advantage, Journal of Management, Vol 17, No 1, 99-120.

[2] Barney Jay, Wright Mike, Ketchen, Jr David J. (2001),The resource based view of the firm: Ten years after 1991, 2001 Journal of Management 27: 625-641.

[3] Bingham CB, Eisenhardt KM, Furr NR. 2007, What makes a process a capability? Heuristics, strategy, and effective capture of opportunities. Strategic Entrepreneurship Journal 1(1–2): 27–47.

[4] Bhuian N. Shahid N., Menguc Bulent, Bell Simon J. (2005), Just entrepreneurial enough: the moderating effect of entrepreneurship on the relationship between market orientation and performance. Journal of Business Research 58 (2005) 9-17.

[5] Campbell, D.T. (1955), The informant in quantitative research. American Journal of Sociology 6 (January), 339-42.

[6] Collingson, E., and E. Shaw (2001), Entrepreneurial Marketing: A Historical Perspective on Development and Practice, Management Decision 39(9), 761-766.

[7] Covin, J.G., and Slevin, D.P. (1986), The development and testing of an organization-level entrepreneurship scale. In R., Ronstadt, J.A. Hornaday and K.H. Vesper (eds.), Frontiers of Entrepreneurship Research – 1986, 628-639. Wellesley, MA: Babson College.

[8] Covin, J.G., Slevin, D.P (1989), Strategic management of small firms in hostile and benign environments. Strategic Management Journal 10(1): 75-88.

[9] Covin, J.G., Slevin, D.P (1994), Corporate entrepreneurship in high and low technology industries: A comparison of strategic variables, strategic patterns and performance in global markets. Journal of Euro Marketing, 3(3):99-127.

[10] Daft, R., Lengel, R. and Trevino, L. (1987), Message equivocality, media selection and manager performance: implications for information systems. MIS Quarterly 14 September, 355–66

[11] Day George S. (1994), The Capabilities of Market Driven Organisations, Journal of Marketing Vol. 58 , 37-52.

[12] Day George S. & Robin Wensley (1998), Assessing Advantage: A Framework for Diagnosing Competitive Superiority, Journal of Marketing Vol. 52, 1-20.

[13] Deshpande, R. (1999), Forecasting Marketing, Journal of Marketing, 63(3): 164-167.

[14] Eisenhardt KM, Martin JA (2000), Dynamic capabilities: what are they? Strategic Management Journal, October–November Special Issue 21: 1105–1121.

[15] Ethiraj SK, Kale P, Krishnan MS, Singh JV (2005), Where do capabilities come from and how do they matter? A study in the software services industry. Strategic Management Journal 26(1): 25–45.

[16] Grant, R.M. (1991), The resource-based theory of competitive advantage: Implications for strategy formulation, California Management Review, 33, pp. 114-135.

[17] Grant, R.M. (1996), Toward a knowledge-based theory of the firm, Strategic Management Journal, 17 (Special Issue), pp. 109-122.

[18] Hall, R. (1993), A framework linking intangible resources and capabilities to sustainable competitive advantage, Strategic Management Journal, 14, pp. 607- 618.

[19] Hamel, G. and Prahalad, C.K. (1989) Strategic intent. Harvard Business Review 67, 63–76.

[20] Hills , Hultman , and Morgan P. Miles (2008), The Evolution and Development of Entrepreneurial Marketing. Journal of Small Business Management 2008 46(1), pp. 99-112.

[21] Hogarth and Makridakis, (1981), Management Science, Vol. 27, No. 2, February 1981, pp. 115-138

[22] Hunt, S.D. and Morgan, R.M. (1995), The comparative advantage theory of competition. Journal of Marketing 59, 1–15.

[23] Jaworski Bernard J. & Kohli Ajay K. (1993), Market Orientation: Antecedents and Consequences. Journal of Marketing Vol. 57 , 53-70.

[24] John, G. And Reeve, T. (1982), The reliability and validity of key informant data from dyadic relationships in marketing channels. Journal of Marketing Research 29, 517-24.

[25] Keh, H.T. (2007) et al, The effects of entrepreneurial orientation and marketing information on the performance of SMEs. Journal of Business Venturing 22 (2007) 592-611.

[26] Keller, R.T. (1994), Technology- information processing fit and the performance of R&D project groups: a test of contingency theory. Academy of Management Journal 37, 67–79.

[27] Kohli Ajay K. & Jaworski Bernard J. and Kumar Ajith (1993), MARKOR: A Measure of Market Orientation.. Journal of Marketing Research Vol XXX, 467-77.

[28] Kohli Ajay K. & Jaworski Bernard J. (1990), Market Orientation: The Construct, Research Propositions, and Managerial Implications. Journal of Marketing Vol 54, 1-18.

[29] Kraus et al., (2010), Entrepreneurial Marketing: Moving beyond marketing in new ventures. Int Journal of Entrepreneurship and Innovation Management Vol 11, 19-34.
[30] Makadok R. (2001), Toward a synthesis of the resource based and dynamic-capability views of rent creation. Strategic Management Journal 22(5): 387–401.
[31] Menon, A. and Varadarajan, P.R. (1992), A model of marketing knowledge use within firms. Journal of Marketing 56, 53–71.
[32] Miles, M. P., and J. Darroch (2006). Large Firms, Entrepreneurial Marketing and the Cycle of Competitive Advantage, European Journal of Marketing 40(5/6), 485-501.
[33] Miller, D. (1988), Relating Porter's business strategies to environment and structure: analysis and performance implications. Academy of Management Journal 31(2), 280–308.
[34] Miller, D., Friesen, P.H., (1983), Innovation in Conservative and Entrepreneurial firms: Two models of Strategic Momentum, Strategic Management Journal, 3 (1): 1-25.
[35] Moorman, C., (1995), Organizational market information processes: cultural antecedents and new product outcomes. Journal of Marketing Research 32, 318–335.
[36] Morgan, A. Neil and Vorhies W. Douglas (2009), Market orientation, marketing capabilities, and firm performance. Strategic Management Journal 30: 909-920 (2009).
[37] Morris, Michael, Schindehutte, Minet and LaForge, Raymond (2002), "Entrepreneurial Marketing: A Construct for Integrating Emerging Entrepreneurship and Marketing Perspectives," Journal of Marketing Theory and Practice, 10(4), 1-19.
[38] Naman, J.L and Slevin, D.P (1993), Entrepreneurship and the concept of fit: a model and empirical tests. Strategic Management Journal 14(2), 137–53.
[39] Narver, J.C. and Slater, S.F. (1990), The effect of a market orientation on business profitability, Journal of Marketing, 54, pp. 20-35.
[40] Newbert SL. (2007), Empirical research on the resource based view of the firm: an assessment and suggestions for future research. Strategic Management Journal 28(2): 121–146.
[41] Phillips,L.W. (1981), Assessing measurement error in key informant reports : a methodological note on organisation analysis in marketing. Journal of Marketing Research 28, 395-425.
[42] Prahalad C.K & Hamel Gary (1990), The core competence of the corporation, Harvard Business Review May- June: 3-15.
[43] Ringle, C.M./Wende, S./Will, S.: SmartPLS 2.0 (M3) Beta, Hamburg (2005), http://www.smartpls.de.
[44] Rousseau, D.M. (1985), Issues of level in organisational research: multi level and cross level perspectives. Research in Organisational Behaviour 7, 1-37.
[45] Ruekert, R.W., Walker, O.C. and Roering, K.J. (1985), The organization of marketing activities: a contingency theory of structure and performance. Journal of Marketing 49, 13–25
[46] Sanchez, R (1999), Modular architecture in the Marketing Process, Journal of Marketing, 63(2):92-111.
[47] Slater Stanley F. & Narver John C. (1994), Does Competitive Environment Moderate the Market Orientation-Performance Relationships?, Journal of Marketing, Vol. 58 , 46-55.
[48] Slater, S.F. & J.C Narver (1995), Market Orientation and the Learning organization, Journal of Marketing, 59(3): 63-74.
[49] Slater, S.F., Narver, J.C., (1998), Customer-led and market-oriented: let's not confuse the two. Strategic Management Journal 19, 1001–1006.
[50] Spanos. Y.E. and Lioukas, S. (2001), An examination of the causal logic of rent generation: Contrasting Porter's competitive strategy framework and the resource-based perspective, Strategic Management Journal, 22, pp. 907-934.
[51] Teece DJ, Pisano G, Shuen A. (1997), Dynamic capabilities and strategic management. Strategic Management Journal 18(7): 509–535.
[52] Varadarajan, p and Clark, T (1994), Delineating the Scope of Corporate, Business and Marketing Strategy, Journal of Business Research, 31, pp. 93-105.
[53] Venkatraman, N. (1989), Strategic orientation of business enterprises: The construct, dimensionality and measurement, Management Science, 35, pp. 942-962.
[54] Vorhies Douglas W., Harker Michael (2000), The Capabilities and Performance Advantages of Market-Driven Firms: An Empirical Investigation, Australian Journal of Management Vol 25, No.2..
[55] Vorhies DW, Morgan NA. (2005), Benchmarking marketing capabilities for sustained competitive advantage. Journal of Marketing 69(1): 80–94.
[56] Vorhies DW, Morgan NA. (2005), Benchmarking marketing capabilities for sustained competitive advantage. Journal of Marketing 69(1): 80–94.
[57] Weerawardena, Jay (2003), The role of marketing capability in innovation-based competitive strategy, Journal of Strategic Marketing 11 15–35 (2003).
[58] Wernerfelt, B. (1984), A resource-based view of the firm, Strategic Management Journal, 5, pp. 171-180.
[59] Zahra, S. and Covin, J. (1993), Business strategy, technology policy, and firm performance, Strategic Management Journal, Volume 14 Issue 6, Pages 451 – 47.
[60] Zahra, S. & D. Garvis (2000), International Corporate Entrepreneurship and firm performance: the moderating effect of international environmental hostility, Journal of Business Venturing, 15(5):469-492.
[61] Zott C. (2003), Dynamic capabilities and the emergence of intra industry differential performance: insights from a simulation study. Strategic Management Journal 24(2): 97–125.

3 Manufacturing Processes

Environmental issues and depleting natural resources are among the most important issues in strategic manufacturing decisions. Public awareness and government policies and regulations have forced many industries toward sustainable production that conserve energy and natural resources, and is environmentally friendly and safe for employees, communities, and consumers. In the first section of this chapter Rohrmus et al. present Siemens's concepts to improve products and production at all levels of sustainability, and the energy optimization programs for the manufacturing and supply chain processes. In the second section Schlosser et al. present an approach for the evaluation of energy consumption for manufacturing, the results are discussed and evaluated for drilling operations. In the third section Steingrimsson et al. develop strategies for competition and collaboration which offer a greater economic benefit while promoting sustainability issues. In the fourth section Albakri et al. develop three dimensional models to simulate friction stir processing with in-process cooling from the backing plate using computational fluid dynamics. In the fifth section Krajnik et al. present recent developments in achieving enhanced properties in nanofluids and evaluation of their potential for applications in machining, focusing on their thermal and tribological aspects. In the sixth section Ammouri and Hamade chart quantitative maps of indirect tool-wear of chisel drills undergoing dry machining associated with qualitative descriptions. Based on these tool-wear maps a novel wear criterion is developed. In the seventh section Albakri and Khraisheh develop an optimal variable strain-rate forming path based on a multi-scale stability criterion, aiming at reducing forming time while maintaining the integrity of the formed parts. In the eighth section Mohamed and El Gamal utilize solid waste materials to produce a sustainable sulfur cement product by controlling several parameters for testing and evaluating its impact on the product.

Sustainable Manufacturing at Siemens AG

D. Rohrmus[1], C. Mose[1], J.C. Holst[2], K. Müller[2], M. Schedlbauer[1], U. Raschke[3], N. Weinert
[1]Corporate Technology, Corporate Research and Technologies, Siemens AG, Munich, Germany
[2]Corporate Technology, Corporate Research and Technologies, Siemens AG, Berlin, Germany
[3]Siemens Product Lifecycle Management Software Inc., Ann Arbor, USA

Abstract
Siemens understands sustainability in the holistic context of social responsibility, environment, and economy. In terms of sustainable manufacturing we present concepts that improve products and production at all levels of sustainability. The various company energy efficiency programs focus on key areas of production. It begins with energy aware product development processes supported by the EcoCare Matrix methodology as part of our PLM process. The energy optimization programs for the manufacturing and supply chain processes are implemented by four phases, the Energy Health Check, the analysis, the concept, and the implementation steps. However, sustainable manufacturing is more than just resources flow and consumption optimizations. We discuss an S.M.A.R.T. solution for India exchanging conventional light bulbs by long-life compact fluorescent lamps. Finally, we lead to the road of an integrated consideration of the interactions of resources supply and demand as well as product and production value adding processes to achieve next level sustainability solutions.

Keywords:
Sustainable Manufacturing; Energy Efficiency Program; EcoCare Matrix; S.M.A.R.T. sustainable solutions; Energy measurements related to machine tool states

1 INTRODUCTION

As Siemens 163 year history shows, our understanding of sustainability is closely linked to our company values – responsibility, excellence, and innovation. From the very first beginning, Werner von Siemens insisted on the fulfillment of the company responsibilities for the employees, the society and the nature. Achieving excellence, leading positions in the markets of tomorrow, and developing innovative technologies that help to ensure the future viability of modern civilization is and has always has been our vision and our challenge. Siemens focuses on the mega trends [1]:

- demographic change and healthcare;
- urbanization and sustainable development;
- climate change and energy supply;
- globalization and competiveness.

To challenge these mega trends Siemens has established the company's sustainability principles. They are setup and driven by the company internal sustainability office and controlled by the externally staffed sustainability board. They require a fulfilling of social responsibility and environmental preservation globally, both geared towards long term company economical success, which is a key function of any company [1]. Social responsibility is the Siemens commitment toward the societies in which we are active to help as many people as possible to gain access to essential goods, both today and in the future: food, jobs, security, energy, and affordable healthcare. This is important since companies are mainly successful in societies in which people enjoy a certain level of prosperity. The products and solutions within our environmental program and the greater company portfolio acknowledge the need to preserve the fundamentals of human life in order to remain successful over the long term.

Siemens wants to help maintain a healthy world, one in which natural resources are protected and the needs of future generations are respected by environmental conservation. As one of the important factors for achieving these targets we identify energy consumption and energy efficiency.

The energy consumption of the developed civilizations is one of the main threats for functioning ecosystems globally. Not only green house gas emissions as a consequence of fossil combustion has to be named but also the environmental impairment raised by the extraction of natural resources. Consequently, the excessive use of energy not only threatens the environment but strongly influences the social dimension of the human existence and thus the ability to increase the global living standards. Unfortunately, growing prosperity of societies has always been linked to increased energy consumption. At the same time, the economical challenges due to the increased worldwide demand of the limited available resources become observable.

The industry itself consumes already more than one third of the primary energy worldwide and is at the same time the designer and the manufacturer of the products, which require energy during their use. Thus, improving the energy consumption in manufacturing is as necessary as the development and the production of products using energy efficiently.

For Siemens, as the developer, manufacturer and also user of one-of-a-kind to mass-series products, the starting points are identified throughout the whole product life cycle. In the following sections we present several initiatives and programs related to improving energy efficiency in order to reach our sustainability goals.

2 ENERGY EFFICIENCY PROGRAM

In the Siemens Sectors Industry, Energy, and Healthcare as well as Siemens Corporate Technology energy efficiency programs are established. They are backed by our standardized process definitions of product lifecycle management, supply chain management, and customer relation management:

- The internal Siemens product development processes are carried out including the mandatory target to reduce the energy consumption respectively to optimize the energy output during usage compared to a comparable previous reference product. Examples are the modern energy efficient drive technologies, the latest programmable logic controller developments as well as output optimized energy supply and renewable solutions for wind parks or concentrated solar power plants. We will discuss the underlying process procedures in Section 5 by an illuminants example.
- Production comprises three linked systems that need to be optimized individually as well as in their interdependencies: the manufacturing system itself, the plant, i.e. the building and infrastructure, and the technical building equipment. Additionally, the factory network planning is important for a globally operating company like Siemens. Production site criteria like locally available raw materials and market proximity need to be judged in terms of total lifecycle costs. In Section 3 we present our approach along with an example from a production plant for industry compressors.
- The dynamic measurement of the resources and especially the energy consumptions at all levels of the production is the base for any energy optimization program. The sensor and measurement strategy has to be production adequate in terms of the measurement points, resolution, logging, and the analysis of the acquired data. The strategy is set by the efficiency goals for the specific task. Exemplary, to be able to calculate the required supplies and to balance the energy load of specific manufacturing equipment different kinds of measurements need to be correlated with the process and control states of the machine. In Section 4 we demonstrate consumption comparisons on the machine level and discuss efficiency potentials for existing machine tools.
- The (automated) energy demand response activities as part of our smart grid developments yet focus primarily on the technical building equipment. Smart grid control, which is influenced by the energy suppliers and the consumers, follows the main goal to balance and stabilize the network load. By this exemplarily the need for peak power plants can be lowered and the rising number of renewable energy sources can be better integrated in the existing grids. Several pilot projects in California and South Germany prove currently the high state of the available solutions. In terms of production, the demand response solution is part of a larger picture:
- The integrated consideration of the value stream in the production system and the flow of the required resources is a key for production planning and execution. Exemplarily lean principles, one piece flow, and production leveling may stand in contrast to energy demand reduction and energy leveling of the plant as well as the adjacent regions around the plant. We are developing a methodology that involves an integrated consideration of both aspects similarly as described in [2]. In Section 4 further aspects of energy enhanced value stream mapping are discussed.
- The lifecycle assessment of a product including the production supply chain is extended by the environmental impact dimension (key performance indicators (KPI), like CO_2 equivalent). This includes the current resource investments for the production, the expected life time, and the product phase out as well as the resources requirements at all locations of the supply chain. By this the carbon footprint can be reduced at all involved locations. Our efforts are exemplarily demonstrated in Section 5.

Beside the listed process and technology topics so called S.M.A.R.T. products and solutions (Simple, Maintenance-friendly, Affordable, Reliable, Timely to market, customized to local needs) provide opportunities for emerging markets to reduce their carbon footprint. In Section 6 we present an example how new business models become reality, joining energy efficiency solutions with improved living standards in the emergent world.

3 ENERGY OPTIMIZATION IN PRODUCTION

In this section we demonstrate how our holistic approach to optimize resources consumption in the production is realized in industry. Our standard method is a sequence of four main phases that are usually viewed as quality gates in terms of project management: Energy Health Check, analysis, concept, and implementation phases [7]. The presented results refer to the production of industry compressors, whose production consumption pattern can be viewed as a standard case for semi-automatic production in Europe.

Table 1: Accumulated energy savings in a first analysis phase for a production plant of industry compressors [4].

	Reduction of used		Typical payback periods
	Electricity	Natural gas	
Actions	MWh/a	MWh/a	Years
Reduction of the base load	1.500		4 – 6
Climate control optimization	100		1 – 3
Heat recovery painting		960	
Heat loss over roof-lights		1.820	1 – 3
Heat loss over hall openings		460	3 – 6
Heat recovery compressed air		170	3 – 6
Use of season cool respectively heat	124		1 – 3
Sum	1.724	3.410	

The Energy Health Check targets a first rough estimation of savings potentials [8]. All operational processes relevant to the energy management and the related organizational structures are examined and systematically assessed.

It yields the diagnosis of the plant energy management system (technically and non-technically) and the development of an advisement for improved energy management. The potential savings for the entire plant are derived from a database of more than 2000 analyzed enterprises. For the production of industry compressors in a 100.000m^2 plant in Northern Europe, operating in three-shifts, this high level analysis yield accumulated savings as listed in Table 1. The specified actions are general and point out immediately acquirable potentials. The shown entries are calculated based on data taken from already installed plant-wide supply measurement devices. Figure 1 shows a typical demand run over a week for this type of plant. The 1.500 MWh/a savings potential is derived from the base load run between Monday and Friday.

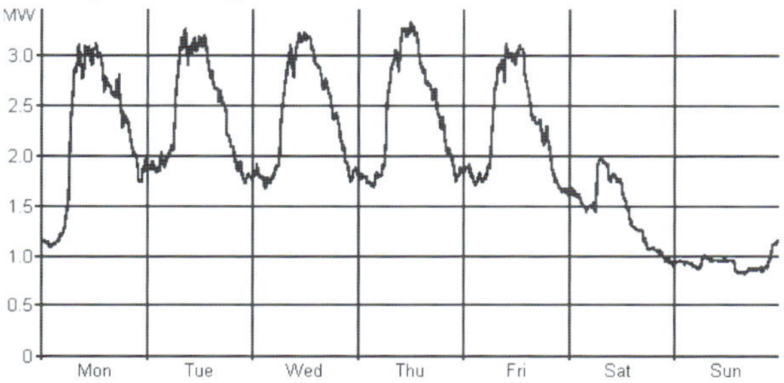

Figure 1 : Demand run for one week for an industry compressor plant.

The technical analysis and review phase actions are derived from classification results, which are based on reference information of an internal data base per industry sector build up from more than 1000 applications. The target is to review the technical and economical feasibility of further measures to achieve permanently improved energy efficiency. Among other goals a complete energy balance is taken into consideration and the required investment is estimated. All forms of energy are considered including electrical, mechanical and thermal energy, and all primary energy sources are involved, such as oil, gas and water. For the mentioned plant one result is a possible base load reduction of 210 MWh/a on weekends and 1.290 MWh/a during the week if a plant-wide load control system is installed. The analysis also showed that certain ventilation and safety systems cannot be shut-down, which explains the difference to the maximal possible 1.500 MWh/a. To be able to judge these actions economically, the investments vs. the energy savings are calculated, which yield a typical four to five years payback period. More typical periods are listed in Table 1.

In addition to the technical analysis of existing installations our layout planning tool IntuPlan® supports the finding of an energy (use and reuse) optimized configuration of the machines and the workshops during the plant planning phase. This helps to understand the relationship between resource efficiency and factory layout (Figure 2). With an intuitive visualization an increased awareness of production flow, machine utilization and energy usage is achieved during the early planning stage. This yields energy improved layouts.

In the concept phase the identified measures are prioritized. The technical efforts are determined in close cooperation with production and plant maintenance and/or the suppliers. For example in about 70% of all machine tool industry applications, the exchange of conventional drives by energy efficient frequency converted drives yields significant energy savings [8]. The actual saving potentials are determined on the basis of measurements and costing studies. An implementation strategy with a detailed technical and economic feasibility study is then developed. In Section 4 we present an example on machine level to demonstrate how detailed the required measurements might be undertaken.

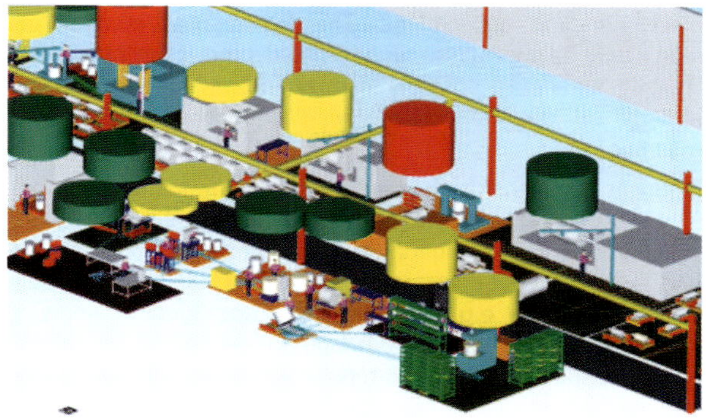

Figure 2 : Energy demand visualization during layout planning with IntuPlan®.

In the final implementation phase the selected measures are realized. Beside the technical and process aspects professional project management is necessary to enable KPI follow up to be able to prove the CO_2 and cost reductions as the base for future optimization steps.

In addition to the four energy optimization phases the factory network planning has a major influence on the global carbon footprint of production and supply chain processes. With our software tool PlantCalc® we plan our global production networks optimized by multiple objectives like local raw materials availability, market proximity, and inter-plant logistic efforts.

4 MEASUREMENTS AND VALUE ADDING

Factory wide the energy efficiency optimization requires the application of quite different measures. Among others these are organizational to level consumption peaks of different production areas or machines, or technically oriented like substituting production processes, equipment or energy carriers. In all cases, a detailed knowledge of the energy consumption behavior of the production system and their components is required. This knowledge has to be acquired by measuring the energy and resource consumptions within the concerned production system.

Any measurement task requires a strategy that depends on the purpose of the measurement. Distinguished can be sequential or simultaneous measurements of single resources up to complex scenarios like a network of machines on a shop floor. Also continuous or temporal measurements of those scenarios have to be considered.

Measuring electrical, mechanical, and thermal energy itself is not enough when the mapping of the energy load on the states of an entire plant or exemplarily the state of a single production machine with respect to the currently produced variant is necessary [9]. These situations appear in reality when energy efficiency targets require two different objectives to be optimized simultaneously: the energy demand balancing of machine networks as well as their value adding functions. On plant-level energy intensive production processes like casting or transport of heavy materials, which might already be energy optimized, can still induce an hazardous peak demand if their processes are not centrally managed in terms of intelligent control, which is able to shift or dim the load within the given production system borders.

Exemplarily we present a simultaneous temporal measurement task for two setups of two axes “xz” pick and place handling applications. This scenario is common in many fully as well as semi automated production machines. By this we detail possible energy improvement potentials already on a small scale, i.e. for functional parts of a machine.

Figure 3 illustrates the results for a fully pneumatically and a fully electrically driven “xz” setup. The different acceleration, velocity, and direction capabilities of the pneumatic pressure and electrical drives become transparent. The mechanically identical setups, e.g. axis bearings and lengths show a significantly lower energy demand for the electrical drives. For comparison reasons the measured pneumatic power was transformed into electrical equivalents based on the electric measurements of a decentralized pneumatic unit.

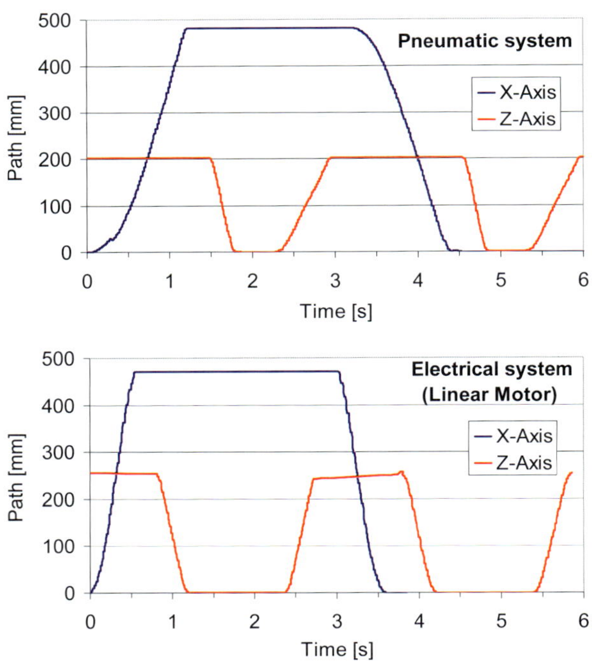

Figure 3 : One cycle of 2-axes pick and place application to compare pneumatic (top) and electric (bottom) setups.

The electrical system can realize a far higher acceleration performance (Figure 3) with a lower energy demand (Figure 4). For deceleration the electrical linear motor offers the potential to re-use shares of the kinetic energy brought into the system during the acceleration process (the same technique that direct motors use for transforming electrical energy into linear motion can be used the other way around when decelerating the masses). The cumulated outcome of our setup (excluding the re-use of kinetic energy) is about 53% energy advantage of the electrical setup compared to the pneumatic one.

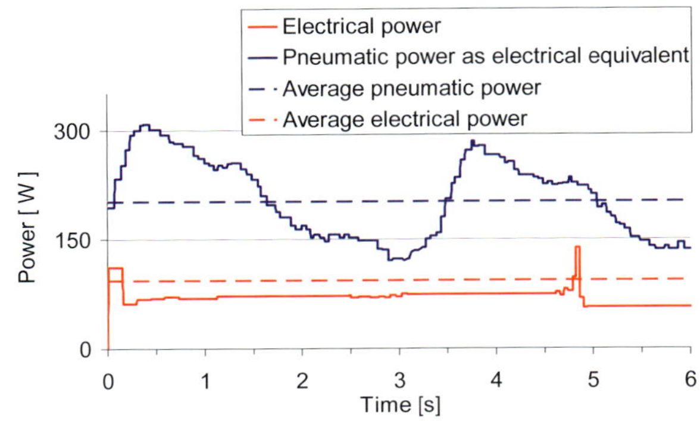

Figure 4 : Power demand for the electrical and pneumatic pick and place system.

Of cause only analyzing the energy demand is not enough to show the whole picture of these very different technologies. We do not discuss the investment aspects in this publication. Following the topic of energy leveling, discussed earlier in this section, another way to reduce the energy consumption is to run the “xz” pick and place system at different speed and acceleration levels controlled by the overall load scheme of the production facility. By this we make use of the advanced controller technology, which is anyway included in an electrical setup.

5 ECOCARE MATRIX

Looking at the existing Product Lifecycle Management (PLM) methodologies used in many companies worldwide, adequate ones to sufficiently support the design of “green” products and solutions are missing. To close this gap in the PLM methodology the Eco Care Matrix (ECM) as shown in Figure 5 has been derived based on a former approach from BASF [3]. The combination of ECM and PLM leads to Green-PLM.

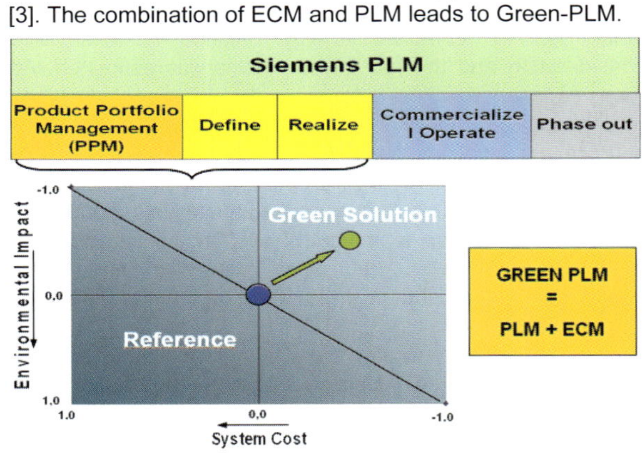

Figure 5 : Green-PLM = PLM + Eco Care Matrix (ECM)

The ECM describes both dimensions of economical performance (horizontal) and environmental impact (vertical). An existing technology, product, or solution is set as a reference in the center of the ECM. The green solutions to be developed should be better in both eco-dimensions, i.e. economical and ecological. To describe the economical dimension it is favorable to use system costs e.g. CAPEX (capital expenditure) and OPEX (operational expenditure). The ecological dimension is described by the Life Cycle Assessment (LCA) methodology [4]. A LCA study aims to provide a holistic overview of the environmental impact of a technological process [5].

Together with OSRAM Opto Semiconductors, a Siemens subsidiary, Siemens Corporate Technology, Berlin conducted such a LCA study [6]. The environmental performance of three different lighting technologies, a conventional light bulb (GLS), a compact fluorescent lamp (CFL), and a light-emitting diode lamp (LED lamp) has been compared. The comparison involved a 40 W GLS, an 8 W CFL Dulux Superstar, and an 8 W Parathom LED lamp.

In order to validate lamps and how they actually deal with energy and resources, more than just the energy consumption during the usage has to be considered. The goal of the LCA was therefore to analyze the environmental impact of a LED lamp over its entire life and to compare it with a CFL and a GLS. The relevant material and energy supplies, including all inputs and outputs, were determined in detail for all LED lamp components and production processes. Beside the detailed analysis of each individual production stage, exemplarily for the LED chips and the lamp housings, these also include all necessary transportation.

In addition to the primary energy and resource consumption, the impact on the environment corresponding to selected environmental categories, i.e. acidification, eutrophication, photochemical ozone depletion, and human toxicity was evaluated. The main findings, additionally proved by three independent experts, are:

- Less than 2% of the total energy demand is needed for the production of the LED lamp: The manufacturing phase is insignificant in comparison to the usage phase for all three different lighting technologies. This study has dismissed any concern that production of LEDs particularly might be very energy-intensive. Merely 0.4 kWh are needed for production of an OSRAM Golden Dragon Plus LED , about 9.9 kWh for the production of the Parathom LED lamp including 6 LEDs.
- LED lamps are competitive to CFL even today: In contrast to the primary energy consumption of incandescent lamps of around 3.305 kWh, CFL and LED lamps use less than 670 kWh. Thus 80% of energy can be saved. The bottom line is that LED lamps are more efficient than conventional incandescent lamps and also ahead in terms of environmental friendliness. Even today, LED lamps show nearly identical impact per hour of use compared to CFL.
- Future improvements of LED lamps will further cut down energy demand: As the efficiency of LEDs continues to increase, LED lamps will be capable of achieving even better LCA results in the future and thus be able to outperform CFL lamps.

This life-cycle assessment proves that LED lamps are among the most environmentally friendly lighting products.

6 ENERGY EFFICIENT HOME LIGHTNING IN INDIA

An example for an S.M.A.R.T. solution by applying an energy efficiently produced lamp in rural areas of India is demonstrated by OSRAM [12]. OSRAM is the first company, which received the approval by the United Nations Framework Convention on Climate Change (UNFCCC) for a new methodology for energy efficient lighting schemes under the Clean Development Mechanism (CDM) of the Kyoto Protocol. The project idea is to replace classic incandescent lamps by energy efficient long-life compact fluorescent lamps (DULUX EL) in defined areas of India (Figure 6). In total 1.7 mio. energy-saving lamps will be distributed door-to-door to households by self-help-groups. Beneficiaries will be the lowest income group of people, which normally cannot afford to purchase these higher priced lamps. Over the entire project duration (Feb. 2009-2019) the energy savings will amount 1.3 mio. MWh. This equals 1 mio. tons of CO_2 emission reduction. To refinance the project OSRAM receives Certified Emission Reductions (CERs), which are continuously sold.

Figure 6 : Currently participating states in India are Andhra Pradesh, Haryana, and Maharastra [12].

7 CONCLUSION

We presented the Siemens understanding of sustainability in a broader manufacturing context. Our energy efficiency programs supported by the Siemens sectors Industry, Energy, and Healthcare as well as Corporate Technology have been introduced. They focus on the key areas of product creation processes. It starts with the energy aware product development process supported by the EcoCare Matrix methodology as part of the PLM process. The energy optimization programs for the manufacturing and supply chain processes consist of four phases: Energy Health Check, analysis, concept, and implementation. Hereby the plant, i.e. the building, the infrastructure, and the technical building equipment are analyzed in terms of their resources flow interactions with the manufacturing processes.

The base for energy optimizations is the measuring of resource flows. The measurements need to be mapped to the plant respectively machine states to be able to derive the actual saving potentials. We point out the important role of the energy suppliers in terms of energy demand response as part of the smart grid developments to level the peak energy load of a plant and its surrounding areas.

All presented concepts are illustrated by results from realized cases: one is taken from the production of industry compressors and one from the evaluation of the environmental impact of conventional light bulbs, fluorescent lamps, and LED lamps over their entire lifecycle.

Sustainable manufacturing is more than just energy optimization. We discuss an S.M.A.R.T. solution in several states of India in which conventional light bulbs are exchanged by long-life compact fluorescent lamps.

The company's holistic understanding of sustainability is the key to improve products and production at all levels. Among new solutions for energy supply and optimized products, integrated considerations of their interactions play an increasing role to achieve next level solutions. Exemplarily production value adding processes in relation to their resource demands are topics that Siemens intensively works on in the future.

8 REFERENCES

[1] **http://www.siemens.com/about/en/index/strategy.htm**

[2] Weinert, N., 2010, Planung energieeffizienter Produktionssysteme, ZWF Zeitschrift für wirtschaftlichen Fabrikbetrieb, 503-507

[3] Saling, P., Kicherer, A., Dittrich-Krämer, B., Wittlinger, R., Zombik, W., Schmidt, I., Schrott, W., Schmidt, S., 2002, Eco-efficiency Analysis by BASF: The Method, Int. Journal LCA (**http://www.basf.com/group/corporate/de/sustainability/eco-efficiency-analysis/publications**).

[4] Wegener, D., Finkbeiner, M., Geiger, D., Olsen, S. I., Walachowicz, F., 2009, Comprehensive Approach to Energy and Environment in the EcoCare Program for Design, Engineering and Operation of Siemens Industry Solutions, RISO International Energy Conference, 14 – 16 September 2009, Copenhagen.

[5] Hauschild, M., 2005, Assessing Environmental Impacts in a Life-Cycle Perspective, Environmental Science & Technology, 81-88.

[6] Osram, 2009, Life Cycle Assessment of Illuminants - A Comparison of Light Bulbs, Compact Fluorescent Lamps and LED Lamps.

[7] **http://www.industry.siemens.de/industrial-services/en/solution_services/simain_bbm_eoa.htm**

[8] **http://www.energy-efficient-production.siemens.com/en/energy-management-process/identify-energy-flows.htm**

[9] **http://www.energy-efficient-production.siemens.com/en/energy-efficient-products.htm**

[10] Raschke, U., Schutte, L., Chaffin, D.B., 2001, Ergonomics in Digital Environments, Handbook of Industrial Engineering, 3rd Edition, G. Salvendy Ed., Wiley.

[11] Raschke, U., Kuhlmann, H., Hollick, M., 2005, On the Design of a Task-Based Human Simulation System, Technical Paper 2005-01-2702, SAE International, Warrendale, PA.

[12] **http://cdm.unfccc.int/Projects/db/tuev-sued1206629154.85/view**

Sustainabilty in Manufacturing – Energy Consumption of Cutting Processes

R. Schlosser, F. Klocke, D. Lung

Laboratory for Machine Tools and Production Engineering, RWTH Aachen University, Germany

Abstract

Rising prices on the international commodity markets as well as the increasing demand on environmental friendly products are generating new needs on product development and process planning. This includes also the development of an energetic product-life-cycle, which does not only comprise the use of a product, but also considers the production and refinement of raw materials at its beginning and possible recycling or end of life scenarios. Therefore within this paper an approach is presented for the evaluation of energy consumption for manufacturing. The results are discussed and evaluated for drilling operations. Final conclusions how cutting processes can be optimised are drawn at the end.

Keywords:

Manufacturing Process Evaluation and Characterisation, Total Process Efficiency, Sustainability

1 INTRODUCTION

Due to the scarcity of resources and the increasing global demand to use these resources for products and power generation the commodity prices on the international energy and commodity market are increasing heavily as shown in Figure 1. Although the prices on the markets crushed as highlighted in the figure during the world economic crisis 2008/2009 they are constantly growing again. By this vast increase in prices on the international energy and commodity market emerges to one of the central problems, which manufacturing industry has to face [1].

Not only in the economic but also in the ecological perspective, resource consumption for products has to be planned more sustainably. Therefore products have to be designed to be resource and energy efficient above the total life cycle. To achieve this aim it is possible to influence the energy- and resource consumption in one or more product life cycle phases. The change of one product life cycle phase might also effect the energy and resource consumption in another phase in a positive or in a negative way. For this reason it is essential to evaluate all changes done in one life cycle phase across the whole life cycle, to guarantee an overall reduction in energy and resource consumption [2].The most important life cycle phase during which product features still can be influenced is the manufacturing phase. [3].

Former studies identified the peripheries of manufacturing processes to be responsible for the majority of the energy and resource consumption and that the process effectiveness is only dependent on the used machine tool and peripheral components. Within this paper it will be also shown, that this assumption is not sufficient. By process parameter variation it is also possible to increase the process efficiency by a huge amount.

Within former projects like PROLIMA it was shown that for an absolute estimation of the energy and resource consumption during the different product life cycle phases of machine tool manufacturing detailed measurements have to be carried out in each of these phases. Additionally these measurements guarantee the identification of optimisation potentials and deduction of optimisation actions for these life cycle phases [4]. Especially the potentials in the production phase are wasted due to missing information about the energy and resource consumption related to the manufactured product.

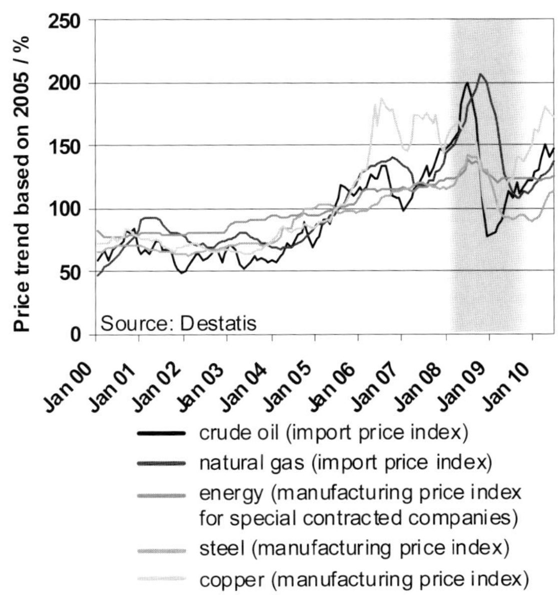

Figure 1: Development of energy and commodity prices

First of all a generally valid proceeding for the evaluation of manufacturing and especially machining operations is proposed. It will be shown, that the total process efficiency might be increased to a large extent by a variation of process parameters.

2 ENERGY RATIOS FOR MANUFACTURING PROCESSES

In this section the energy demand for manufacturing processes is discussed on the example of cutting processes and a proceeding for energy ratios is proposed.

Until now, there is no standardised procedure available, which is capable to evaluate and characterise manufacturing processes or machine tools regarding their efficiency. Only on system and component level efficiency values in form of efficiency categories for electric drives are available. Nevertheless the application of these efficiency categories doesn't provide any opportunity to evaluate the manufacturing process efficiency itself. It rather only enables a qualitative estimation on the energetic performance of the manufacturing system.

The chosen approach for this paper is similar to that of life cycle analysis methods, in which the required amounts of energy and resources are related to a functional unit - in the normal case a product. The life cycle assessment gathers all in- and outgoing energy and resource flows as shown in Figure 2.

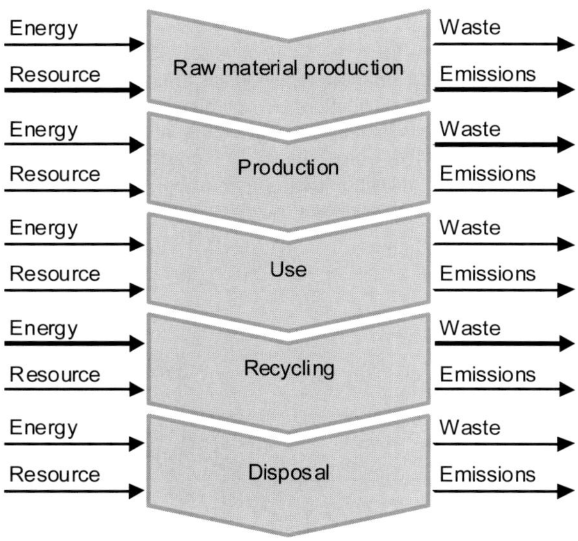

Figure 2: Balance shell for life cycle assessment with in- and outgoing energy and resource flows

In order not to be bound to one distinct product, the following approach will not refer to one single product but the required energy for changing specific product properties will be put into relation to the change itself. This approach also enables to analyse different process and process chain alternatives that are designed to generate the same product. Furthermore, it will be possible to evaluate a specific manufacturing process independent of form or dimension of the workpiece.

Applied to metal cutting processes, the total required amount of energy is put into relation to the amount of removed material. In an analogue manner, the overall power consumption can be put into relation to the material removal rate for constant cutting conditions [5]:

$$E_{spec} = \frac{E}{V_{cut}} = \frac{P}{Q_w} \tag{1}$$

By this ratio it is possible to evaluate the capability of different cutting processes and process variations independent from a distinct product.

2.1 Power demand of machine tools

Figure 3 spreads up the power consumption of a machine tool and its periphery. During one machining cycle several different power demands can be distinguished. On the one hand there is a constant portion containing periphery (Illumination, air conditioning ...) and machine aggregates (basic demand = standby demand). On the other hand the variable portion depends largely on the actual process as well as auxiliary and feed movements [6]. Further on the actual power which is required for the specific cutting operation in the main time is discussed and the results are transferred to idle movements in the non productive time as well.

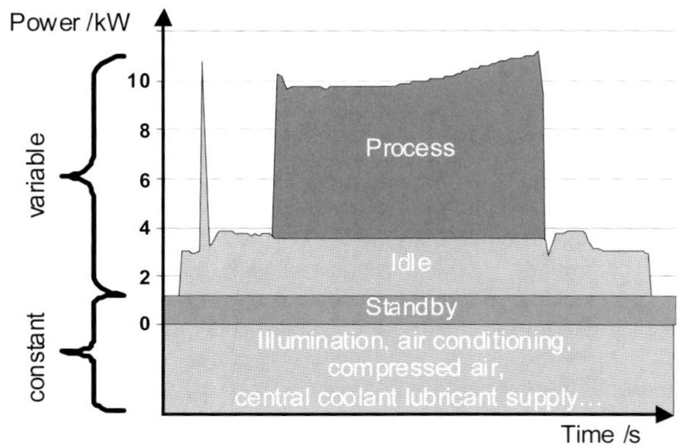

Figure 3: Power consumption in cutting processes

The overall process energy is not only depending on the power level but also on the process times. Therefore different starting points for a reduction of the required energy per volume can be derived from this interrelation. The energy demand cannot be minimised by reducing the specific demands for power and time separately. By a process parameter variation it is possible to influence the power consumption and the time per unit in oppositional directions. With increasing process parameters the process power is increasing as it is derived below. Otherwise with the increase of the process parameters the time per unit is decreased [7].

Due to the fact that these process adaptations are complex, the opposing effects on energy efficiency need a precise investigation on the right parameter selection.

Beside the process parameter optimisation it is possible to decrease the constant basic power demand for a product by extensive additional investments in more efficient machine tools and periphery. As this option has no scientific approach it will not be taken into further consideration.

3 ENERGY CONSUMPTION IN CUTTING PROCESSES

Based on the balancing methodology proposed in section two, a model for power demands and energy demand per volume in cutting operations is established in this section.

3.1 Power demand of cutting processes

Referring to the cutting force model of Kienzle the required cutting force F_c for common cutting processes is calculated using the unit specific cutting force $k_{c1.1}$, the cutting width b, the undeformed chip thickness h, exponent of specific cutting force $1\text{-}m_c$ as well as an equalising correction factor π [8]:

$$F_c = k_{c1.1} \cdot b \cdot h^{1-mc} \cdot \Pi \qquad (2)$$

F_c: cutting force

$k_{c1.1}$: unit specific cutting force

b: width of cut

h: undeformed chip thickness

$1\text{-}m_c$: exponent of specific cutting force

π: equalising correction factor

The unit specific cutting force $k_{c1.1}$ and the exponent of specific cutting force $1\text{-}m_c$ are dependent on the combination of tool and work piece material. The width of cut b and chip thickness h are directly dependent on the process parameters depth of cut a_p and feed per tooth f_z. The equalising correction factor π includes several corrections for process adaption. Commonly it includes corrections for the process, rake angle, friction, cutting velocity and tool wear.

$$h = f_z \cdot \sin \kappa_r \qquad (3)$$

$$b = \frac{a_p}{\sin \kappa_r} \qquad (4)$$

f_z: feed per tooth

κ_r: tool cutting edge angle

a_p: depth of cut

The cutting power P_c is the product of cutting force and cutting velocity [9]:

$$P_c = F_c \cdot v_c \qquad (5)$$

P_c: cutting power

The machining power P_e is the sum of the single power components in cutting, feed and passive direction:

$$P_e = P_c + P_f + P_p \approx P_c \qquad (6)$$

P_e: machining power

P_f: feed power

P_p: passive power

The power level required for feeding is significantly lower than the power demand for cutting, as the feed force and the feed velocity is much lower. As there is no velocity in passive direction in the cutting process, the passive power can be neglected. Therefore the resultant machining power is approximately equal to the cutting power demand.

3.2 Power demand for drilling processes

With respect to the kinematic conditions of the drilling process, the specific cutting energy demand can be calculated as follows:

$$E_{c,spec} = \frac{P_c}{Q_w} = k_{c1.1} \cdot (f_z \cdot \sin \kappa_r)^{-m_c} \cdot \Pi \qquad (7)$$

$E_{c,spec}$: specific cutting energy

Q_w: material removal rate

In this equation the equalising correction factor π comprises, amongst others, several correction factors for the cutting velocity, the applied process and the tool wear.

In addition to the specific cutting energy per volume the constant portion of power demand and portion for auxiliary and feed movements are considered. Due to the fact that in most cases auxiliary and feed movements cannot be influenced directly on the machine tool level, they will simply be assumed as constant. In the following, the balance shell will be drawn around the machine tool for the sake of model simplicity, i.e. the energy consumption of peripheral components will not be taken into further consideration. Yet, in order to evaluate the manufacturing processes correctly, the peripheral power demand has to be recorded and assigned to specific machine tools.

Figure 4 shows the development of the theoretical cutting power for the drilling process dependent on a cutting velocity and a feed per tooth variation. It is indicated that the power demand rises with increasing cutting parameters. Due to the process parameter variation the process power is increased by the factor 13.

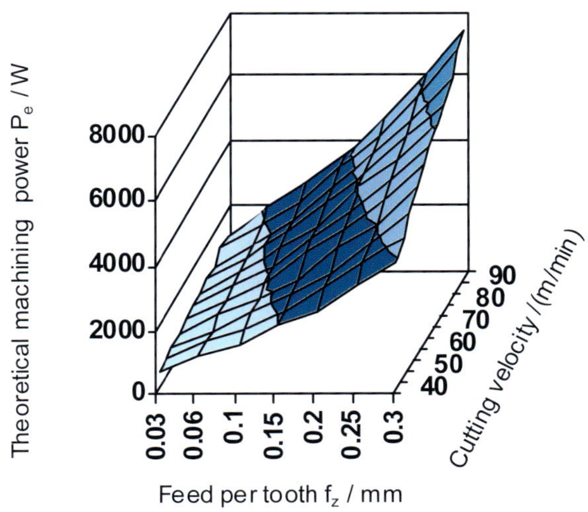

Figure 4: Theoretical cutting power dependent on the process parameters feed and cutting velocity

Figure 5 contrasts this theoretical cutting power with the specific cutting energy for the drilling process. Although the cutting power is increasing the specific cutting energy per volume decreases considerably by a reduction of cycle time.

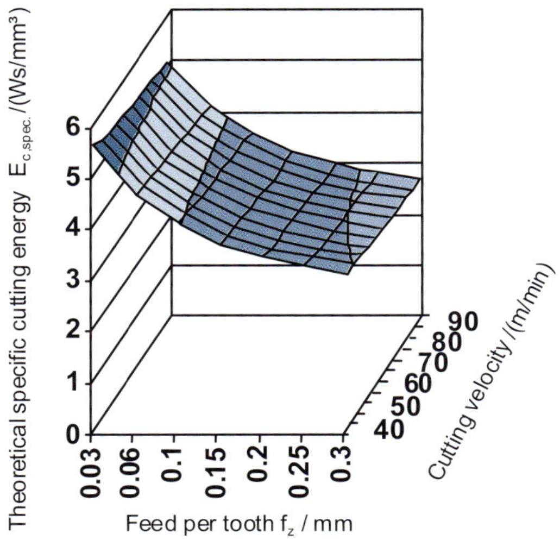

Figure 5: Theoretical specific cutting energy dependent on the process parameters feed and cutting velocity

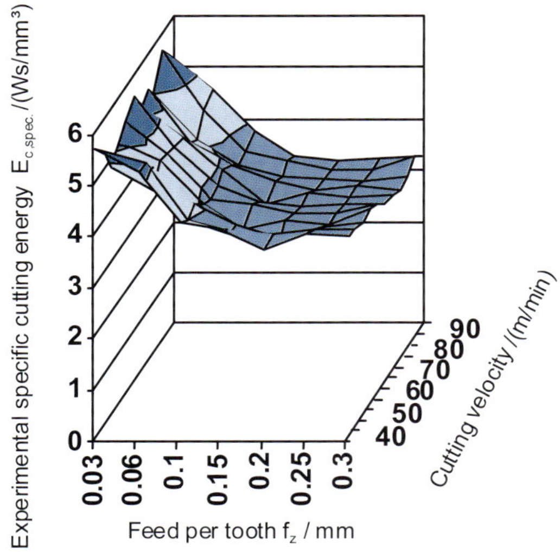

Figure 6: Experimental specific cutting energy per volume dependent on the process parameters

Taking the influence of the constant power consumption in consideration the considerable reduction potential of the directly applied electric energy per volume with higher process parameters is even higher.

4 VERIFICATION OF THE MODEL

For the evaluation of the developed models of section 3 several test runs on drilling with varied process parameters were conducted. Coated carbide drills with the diameter of 10 mm were used for the machining of 16MnCr5. The drilling depth was 50 mm.

The feed per tooth was varied within the range of 0.03 mm up to 0.3 mm while the cutting velocity was varied in the range from 40 to 90 m/min.

As the material removal rate for drilling is calculated by equation (8) it was possible to increase productivity (material removal rate) by a process parameter variation with the factor 22.5.

$$Q_w = A_{Hole} \cdot v_f = \frac{D \cdot z \cdot v_c \cdot f_z}{4} \tag{8}$$

A_{Hole}: cross section of the borehole

v_f: feed velocity

D: diameter of the drill

z: number of cutting edges/flutes

The experimental results showed that as expected the range of constant power consumption is the same for all experimental points. As in the model only the level of variable power demand rises due to the process parameter variation by the factor 13.6 (Figure 4). Still, a close look on the specific cutting energy (Figure 6) clearly reveals the remarkable potential for savings, because processing time and therefore the time of constant power demand can be reduced to approximately 5%.

The results of the model and the experimental results can be compared between Figure 5 and Figure 6, indicating the excellent qualitative accordance between model and experiment.

The detected deviations between model and experiment can be traced back to the measurement principle. The cutting power was measured before the spindle so that the effectiveness of the spindle has to be integrated in the analysis. As no information about the behaviour of the spindle effectiveness over the rotations and torque was existent it was also considered as constant in the equalising factor of the model although normal spindles have a decrease of effectiveness with higher torques. Further investigations will discuss this topic. Giving respect to this failure, the model could be successfully verified. The described problem can simply be solved by an additional calibration and a subsequent adaptation of the correctional factors.

Although this effect was only determined for the primary processing time, it can be transferred to the secondary processing time without any restrictions. Even for faster auxiliary movements that imply a larger power demand, the overall specific energy consumption of the machine tool is lower than for slower motion velocities. However, this is only valid, if due to shorter production times the difference can either be used for shutting down the machine tool earlier or for an increased output of products. In case of an acceleration of the process or the auxiliary movements results in longer standby time respectively, no or only slight savings can be realised, because the continuous standby power demand has to be assigned to the process permanently.

Nevertheless depending on product requirements the process parameters have to be restricted to certain technical boundary conditions, in order to enable the fabrication of a capable product. But products have to be as good as required and not as good as possible. This allows a further optimisation potential for most products.

5 EVALUATION OF ENERGY CONSUMPTION OF CUTTING PROCESSES

The previous sections indicated the positive influence of increased process parameters. However, by changing specific process parameters unwanted secondary effects like increased tool wear or maintenance expenses for the machine tool are triggered. By this the total specific energy demand is also increased. This effect is visualised in Figure 7.

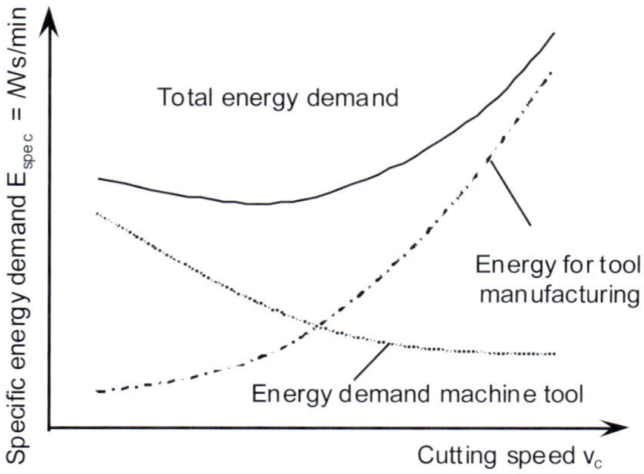

Figure 7: Line-up of the effects on energy per volume

Therefore an energetic trade off can only be reached by changing process parameters thoughtfully and comprehensively. In order to generate energetic improvements, possible positive or even negative synergy effects between distinct processes on the system level - instead of only on the machine tool level - have to be anticipated. Thus, an additional balance of all energetic demands for manufacturing, refinement and maintenance of machine tool and tool has to be set up and taken into consideration.

Until now, this approach is not finally realised within the research project BEAT, which is supported by the German Federal Ministry of Education and Research. However, its final implementation is planned for the future.

The assessment and assignment of maintenance efforts depending on process parameters is currently conducted at the laboratory for machine tools and production engineering within the scope of the BMBF supported project REVISTA.

By combining both research approaches it is going to be possible to evaluate cutting technologies energetically and to define a process-specific minimum of energy demands.

6 SUMMARY

This paper presented and discussed research work within the framework of the BMBF supported project BEAT. It was possible to verify the applicability and suitability of models for the assessment of energy demands for cutting processes. Furthermore, these models are characterised by simple input parameters, which are already available or which are previously defined within the production planning. Therefore the model can be adopted for those areas, where products and processes can still be influenced.

Nevertheless, for a comprehensive balance of manufacturing processes further information are still missing and have to be generated in the future.

7 ACKNOWLEDGMENTS

The authors gratefully acknowledge the financial support of the Federal Ministry of Education and Research (BMBF) and the project supervision of the Project Management Agency Forschungszentrum Karlsruhe, Division Production and Manufacturing Technologies (PTKA-PFT) for the project BEAT - Bewertung der Energieeffizienz alternativer Prozesse und Technologieketten (Assesment of the energy efficiency of alternative processes and process chains).

MANAGED BY

PTKA
Project Management Agency Karlsruhe
within Karlsruhe Institute of Technology

SPONSORED BY THE

Federal Ministry of Education and Research

8 REFERENCES

[1] Klocke, F.; Lung, D.; Schlosser, R.; Nau, B.: Energy- and Resource Effficient Production – a core competence for manufacturers. CIRP sponsored LCE Conference 2009, Cairo, Egypt, 2009

[2] Herrmann, C.; Thiede, S.; Luger, T.; Zein, A.; Stehr, J.; Halubek, P.; Tomey, M.: Automotive Life Cycle Engineering. CIRP sponsored LCE Conference 2009, Cairo, Egypt, 2009

[3] Gerschwiler, K. et al.: Ressourcenschonende Produktion. AWK 2008 – Wettbewerbsfaktor Produktionstechnik, Apprimus Verlag, 2008

[4] PROLIMA BEPM Project Homepage: www.prolima.eu.

[5] Draganescu, F.; Gheorge, M.; Doicin, C. V.: Models of machine tool efficiency and specific consumed energy. Journal of Materials Processing Technology, 141, 2003

[6] Bhattacharya, A.; Das, S.; Majumder, P.; Batish, A.: Estimating the effect of cutting parameters on surface finish and power consumption during high speed machining of AISI 1045 steel using Taguchi design and ANOVA. Production Engineering, Issue 09, 2008

[7] Diaz, N., et al., Strategies for Minimum Energy Operation for Precision Machining, Proceedings of the MTTRF 2009 Annual Meeting, Shanghai, 2009

[8] Kienzle, O.: Die Bestimmung von Kräften und Leistungen an spanenden Werkzeugen und Werkzeugmaschinen. VDI Bd. 94, Nr. 11/12, 1952

[9] Klocke, F.: Fertigungsverfahren 1 – Drehen, Fräsen, Bohren, Springer Berlin, 2008

CONTACT

Dipl.-Ing. Ralf Schlosser

Department of Cutting Technology

Laboratory for Machine Tools and Production Engineering (WZL) at RWTH Aachen University of Technology

Steinbachstr. 19

52074 Aachen, Germany

Tel.: +49 241 80-27466

Fax: +49 241 80-22293

E-Mail: R.Schlosser@wzl.rwth-aachen.de

Business Strategies for Competition and Collaboration for Remanufacturing of Production Equipment

J.G. Steingrímsson, P. Bilge, S. Heyer, G. Seliger
Department for Machine Tools and Factory Management, Technische Universität Berlin, Germany

Abstract
Companies and organizations must cope with the challenge of a demanded sustainability. Collaboration and competition represent a promising approach. The objective of this work is to develop strategies for competition and collaboration which enable increased economical benefit while boosting sustainability issues. The research is focused on value creation strategies, competition, collaboration and their impact as drivers for technological progress. Analysis based on scenarios and forecast reports about value creation in the remanufacturing area of production equipment are carried out. The target is to collect and rank criteria for an evaluation of business strategy regarding sustainability. Existing business strategies for industrial value creation are transferred to the field of remanufacturing analyzed as well as characterized regarding the gathered criteria to support sustainable acting between partners in value creation networks for remanufacturing of production equipment.
Keywords:
Collaboration, competition, strategies, production equipment

1 INTRODUCTION

According to the Organization of United Nations (UNO) the global population increases by approximately 80 million each year. By now around 6.9 billion people are living on the planet and by 2050 this number is expected to be risen to 9.2 billion [1]. As population grows, the demand for consumer goods increases, furthermore, the purchasing power of specific groups within developing countries is growing at high rate. Consequently, global manufacturing increases, which results in a conflict between economy and environment regarding the earth's natural resources. A way out is to introduce periodical life cycles in which resources, materials, components and products are manufactured, used, remanufactured, reused and recycled as seen fitting for each instance. Life cycle management can be seen as a fundamental part of value creation dealing with the issue of an increased need for materials when simultaneously facing dwindling renewable resources.

The challenge for management is to apply chances of economic competitiveness for environmental goals which is linked to the concept of sustainability. Sustainable development is defined by the Brundlandt report of 1987 as development that meets the needs of the present without compromising the ability of future generations to meet their own needs [2]. Sustainability is characterized by the triple-bottom-line of economy, environment and society [3].

Companies cope with the challenge of a demanded sustainability. The question arise which business strategies may promote simultaneously economical, environmental and social benefits in the market. Collaborative competition represents a promising approach. That is why the objective is to address competition and collaboration as drivers for technological progress to achieve profitability and competitive advantages. The paper aims to explore existing business practices for competition and collaboration in relationship with sustainability between actors in the remanufacturing market.

2 BUSINESS STRATEGIES FOR REMANUFACTURING

2.1 Remanufacturing

Remanufacturing focuses on recovering products and parts of products after a usage phase. Unlike repairing products solely, remanufacturing not only extends a product life but also has the goal to bring the product into an “as new” state [4]. Remanufacturing processes include the steps disassembly, cleaning, inspection, reprocessing, storage, testing and reassembly. After conducting remanufacturing, products are reused in another usage phase [5].

In many industries, commercial buyers are more aware about the quality of remanufactured products than buyers of consumer goods. Thus, tire manufacturers offer retreaded tires in their industrial product portfolio but not in their consumer product portfolio. Even though there is little difference in the true quality levels between retreaded and new tires. Bosch Power Tools and Electrolux also follow this strategy, focusing their remanufacturing efforts on the products that are typically purchased by professional contractors [6, 7].

However, remanufacturing has extended to a large number of consumer goods with short life cycles and relatively low values. In widely known remanufacturing cases, the manufacturing is performed by the original producer, integrating new distribution models, e.g. leasing in the case of Xerox with toner cartridges. Other remanufacturing practices are less popular, due to the fact that original equipment manufacturers (OEMs) are rarely involved. Products are not sold through regular retail channels established by OEMs [8]. In general, remanufacturing is accepted as a life cycle tool, but actual application in business of production equipment is still low.

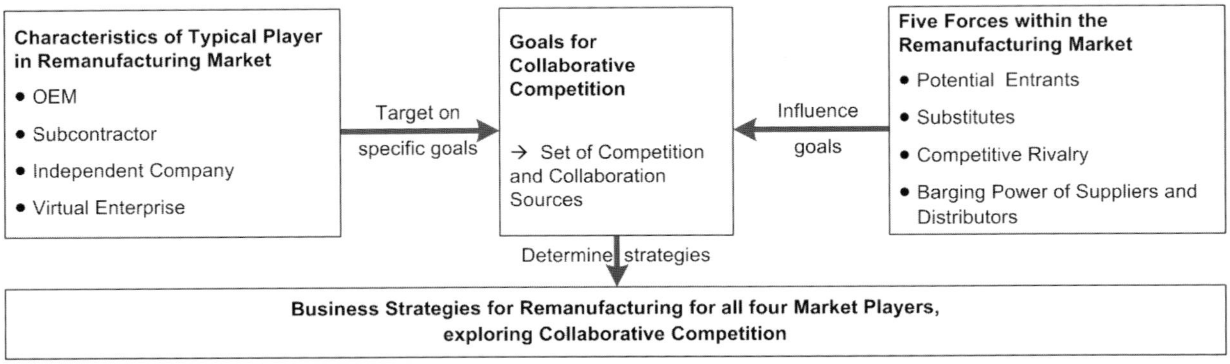

Figure 1: Development of Business Strategies for the Remanufacturing Market

2.2 Business Strategies

A business strategy "[...] is the direction and scope of an organization over the long term, which achieves advantage in a changing environment through its configuration of resources and competences with the aim of fulfilling stakeholder expectations." [10]. A business strategy, applied for a remanufacturing company can be explained in detail regarding the following elements:

- **Direction:** Long term direction of a remanufacturing company indicates strategic aims of the organization in the progress of developing the backwards activities.
- **Markets and scope:** A remanufacturing company of production equipment acts in the area and market of manufacturing and/or remanufacturing. Activities of those companies cover forward and backward activities in core procurement, production, distribution of production equipment.
- **Advantage:** Remanufacturing companies that command a competitive advantage in the market, e.g. access to cores of old machine tools, are likely to perform more successful in their business.
- **Resources:** Resources (skills, assets, finance, relationships, technical competence, and facilities) are required in order to be able to compete. One of the biggest challenges for the remanufacturing industry is finding a way to recycle the production equipment at the end of a usage phase. Depending on the damage and the type of machine, remanufacturing activities are to be adapted to the machine status.
- **Business Environment:** External conditions affect the businesses ability to compete, e.g. regulations and laws have to be taken into account when closing the loop of a life cycle.
- **Stakeholders:** Stakeholders have different values and expectations. They have power in and around the remanufacturing business. Companies like Xerox influence successful growing of the remanufacturing market directly. Others enterprises, which indirectly support the growth of the market, focus on collection of used products.

3 DEVELOPMENT OF BUSINESS STRATEGIES FOR REMANUFACTURING

In the following, the concept for developing business strategies for remanufacturing is presented. In order to come up with proper business strategies, exploiting collaboration to achieve competitive advantages, markets as well as market players have to be known and taken into account. The development process is depicted in Figure 1 and carried out in the following way:

The activities and capabilities of market players are characterized [8]. Porter's five-forces-framework is used to analyze an existing market and possible future markets. Two future scenarios developed in previous work is analyzed in order to identify all sources of competition within the remanufacturing market [11, 12].

Generic goals of collaborative competition can be pursued in order to achieve advantages in competition. The goals are prioritized for each market player in order to come up with individual goal-sets. The goal-sets are linked with the general sources of competition from three scenarios to set basic direction for the business strategies. The three scenarios are the current market scenario and two future scenarios (see Section 6.2 for further details of future scenarios).

4 MARKET PLAYERS IN REMANUFACTURING

Market players primarily vary in their business models. A business model can be defined as architecture for the product, service and information flows laying out the "underlying principles of how a company creates, captures, and delivers value" [13, 14].

The list below depicts the directly and indirectly involved companies in the remanufacturing market, however since the remanufacturing activities, e.g. reuse, repair, remanufacture, refurbish or retrofitting of production equipment, is mainly carried out by few players: OEM, subcontractor, independent company and VE, more detailed description will be provided for them:

- The **customers**: Companies that use remanufactured production equipment. Remanufactured goods can reduce their capital investment expenditures.
- Many **OEMs** in the automotive industry depend heavily on their **subcontractors** in order to decrease production costs, increase production flexibility and ensure that specific expertise are developed [9]. Some OEMs and their subcontractors use also a remanufacturing process as a business strategy to increase profits. The OEMs bare all the production costs and liability for equipment integrity and value, maintaining a continuous chain of custody over a product throughout the entire production process. Typically contracted remanufacturing companies (the subcontractors) have no ownership rights over materials, instead operating within an OEM

facility and utilizing equipment and intellectual property provided by the OEM.

- The **OEM stakeholders** which see greater growth and stability potentials in their investments.
- The **independent companies** are manufacturers of specialized equipment used in the remanufacturing process, such as optical gauges, cleaning and test equipment. These companies are independently operated businesses that purchase machines from former end users or machine brokers, remanufacture them and resell the recovered product afterwards.
- Information technology (IT) based providers such as **virtual enterprises** (VE) create specific IT infrastructures to carry out remanufacturing and distribution activities. Virtual remanufacturing is a new form which is proposed for the adaptation to different markets and environmental aspects.
- The **management consultants** assist new-condition product manufacturers on how to incorporate remanufacturing into their business portfolio.
- The **design engineers** develop and design tools for the remanufacturing processes like disassembly.
- The **investors** provide the capital investment needed to enter the remanufacturing market.
- The **logistics service providers** enable a large increase in reverse logistics activities.

Each class of company has its particular characteristics and goals. The primacy of certain demands, e.g. responsiveness to customer requests and tying existing customers to a product, can be seen as a clear indication of a feature's role in the class of company.

Competitive success can be gained by setting goals for competition based on the organization's business drivers and by understanding the five forces of competition.

5 BUSINESS DRIVERS

Business drivers are internal and external conditions that provoke and shape organizations towards specific forms. Organizations with similar strategic characteristics form strategic groups. Common for these groups are the strategies that they follow and the way that they compete. Comparing one's organization to others can be useful to gain insight in who are currently the biggest rivals and who are most suitable for cooperation. According to Johnson an organization can be classified by the following characteristics [8]:

- The **extent of product and/or service diversity** is a strategy to respond to internal and external changes and opportunities or simply to increase market power.
- The **extent of geographical coverage** is a strategy that represents the geographical widespread of an organization, refers to local, regional, international and global.
- The **number of market segments served** refers to difference in customers. The characteristics for people are: difference in lifestyle, interest or taste, paying capability, age, race, family size, location and life cycle stage. The characteristics for organizations are: difference in industry section, location, size, technology, profitability and management.
- The **distribution channels used** are the paths through which products and services flow from businesses to consumers and to other businesses: trucks, trains, airplanes, freighters, warehouses and retailers.
- The **extent of branding** refers to the promotional activities on specific product and/or services in order to create a uniqueness perception by the consumer with a specific brand.
- The **marketing effort** refers to advertising spread and size of sales forces. It reflects the efforts spent in branding, number of market segments and product and/or service diversity.
- The **extent of vertical integration** refers to the degree to which a firm owns its upstream suppliers and downstream buyers.
- The **product and/or service quality** is the ability to fulfill the customer's needs and expectations.
- The **technological leadership** refers to whether an organization is a technical leader or follower. Influential factors are core competencies of staff, products, processes, programs and knowledge based on cooperation, learning and innovation of the company within the network. This network consists of clients, partners, suppliers, researchers and academia [15].
- The **size of the organization** is measured in the number of employees, turnover, annual balance sheet or in accordance with the micro, small and medium-sized enterprises (SMEs) classification by the European Commission [16].

6 FIVE FORCES IN REMANUFACTURING MARKET

The five forces in markets depicts that rivalry amongst existing firm (the centerpiece) is propelled by buyers and sellers. These have bargaining powers for prices, substitutes and potential entrants that represent replacement threats [10]. The five forces are directly or indirectly affected by external factors such as market drivers.

Market drivers are "[...] factors driving growth, or more specially driving revenues, in an industry" which involve the following [17]:

- **Economic market drivers:** Typically, high costs are a strong driver for every organization. Costs such as for waste disposal can act as a positive incentive for material to be reallocated within the supply chain following the equipment's traditional life cycle.
- **Legal market drivers:** Motivate market allocations to shift toward acting consistent with regulatory requirements. Regulations aimed at controlling the disposal of waste, restriction of the use of certain hazardous substances, and take-back regulation.
- **Social market drivers:** They can both support and undermine the market performance of a product. The public is insufficiently aware of the operative meaning of existing labeling terminology including remanufacturing, refurbishing, recondition, cannibalization, and recycling.

The market drivers are subject to change, such as in taste, in values, in costs for raw materials and how the legal framework is made up. Responsiveness to the changes can be very beneficial to an organization.

6.1 Current Remanufacturing Market

Nowadays many different OEMs, subcontractors, independent and virtual remanufacturers are engaged in the remanufacturing business. Whether a manufacturing company should start to remanufacture or not depends on the structure of the existing as well as forecasted market. It is important to distinguish between independent companies that are only engaged in the remanufacturing business and those that also manufacture original products or components which are OEM and subcontractors. For the production planning and control (PPC) of the second type, it is most of all the high level and variety of uncertainties in return stream that require modifications in the PPC systems [18].

In another case, OEMs sell their production facilities to contracted remanufacturers. The subcontractors achieve high capacity utilization through pooling and supplying many different OEMs. The OEMs focus on innovation in research and development, product design, and marketing. It may result in underinvestment or overinvestment in innovation and capacity, but increases profitability. Further research could examine how this change in industry structure affects investments, and thus profitability [19].

Remanufacturing of production equipment varies in the extent of the remanufacturing activities. Ranging from repairs, where obviously broken parts are replaced to a more extensive remanufacturing and the production equipment is stripped down to the machine bed. Parts and components are then recovered or replaced with the objective to bring the remanufactured production equipment up to a standard that meets or exceeds new equipment [20].

Further impacting factors for the remanufacturing markets are discussed in Section 6.2 (based on prior research), additionally the following two should be considered:

Usage of product-service systems (PSSs) and maintenance, repair and overhaul (MRO) services: In a successful PSS the intangible service augments the tangible product, better fulfilling customer needs. PSS can be seen as means to improve competitiveness and sustainability of a product. MRO services are aimed at the integration of maintenance, repairs and overhaul in order to minimize costs and down time of equipment. MRO services are very useful when applied with use-oriented PSS, since they drive initial costs down for the customers and the lifetime of the equipment is likelier to be prolonged.

Collaborative engineering: An interactive process of working together in order to overcome the complexities in products and organizations. The increased technical complexity set forth by the objective of remanufacturing has to be tackled with effective communication and coordination between stakeholders. Knowledge gained by the remanufacturing activities is handed back into the product design to make the remanufacturing process more efficient [20, 21].

6.2 Possible Futures of the Remanufacturing Market

Two scenarios for possible future markets for production equipment could be identified, using scenario analysis. One consistent future could be that there will be "a paradigm change for sustainable manufacturing" (referred hereinafter as S1), where sustainable development is given a high priority and governmental actions in the form of end-of-life laws in terms of sectoral rules are enforces. High priority is promoted towards the public in order to increase the awareness for the importance of sustainability. As awareness is raised, remanufactured production equipment is more accepted and demand increases. A market with many competing organizations of different shapes and sizes is created. In the other consistent future remanufacturing scenario will "consist of still only a hidden giant" (referred hereinafter as S2), where sustainable development is not promoted and the remanufacturing industry is not stimulated in any way. The remanufacturing processes are handled by few companies or the OEMs themselves [11].

From the scenario analysis the following ten key factors for the possible future market were generated:

1. **End-of-Life laws** enforce a manufacturer's responsibility regarding products and their end-of-life treatment, also referred to as take-back regulation. Take-back regulation encourage remanufacturing since the alternative is to pay for disposal or sell as scrap metal [22]. S1 projects that sectoral rule will be established and manufacturers will be made responsible for the end-of-life treatment. S2 projects that no laws will be established; therefore limited external pressure will exist.
2. **Sustainable development:** Promotion activities of governmental, national or federal bodies include marketing campaigns and incentives that encourage purchasing of remanufactured goods. The aim is thereby to improve societal behavior and attitudes in terms of a targeted search for products and services that exemplify environmentally friendly practices. S1 projects that marketing activities, such as environmental education and green labeling will be undertaken by public bodies. S2 projects that no promotional activities will be undertaken.
3. **Qualification** demonstrates skills and knowledge of employees regarding technical standards and quality criteria of remanufacturing companies. S1 projects that special trainings and informal education will be required and enforced by public and private stakeholders. S2 projects that no focus will be on trainings for remanufacturing processes.
4. **Demand for (re)manufactured machine tools and assembly equipment:** Market volume of remanufactured and reused machines compared to the total remanufacturing market, as well as the shares of most valuable machines. S1 projects that demand increases due to improved quality of remanufactured equipment and components. New standards will be enforced with proper certification to ensure that remanufactured products will be equivalent to new products. S2 projects no change in demand.
5. **Product development** covers the route taken regarding the development of the product and services and includes design decisions that influence product life cycles and functional lifetime. S1 projects that a PSS will be the most significant product development, having a possible eco-efficiency increase by remanufacturing. S2 projects that rapid technological change will be most distinctive for the remanufacturing market.
6. **Reusability** indicates the ability of parts and fittings to be disassembled, inspected, cleaned and repaired for reuse. S1 projects that there will be an increasing reusability of the product itself. Interest of the

manufacturing companies in remanufacturing systems intensifies. Demand increases, diversifies and model change will rise. S2 projects that there will be increasing reusability of components through monitoring. Relevant product and process data will be monitored throughout the product's functional lifetime, by means of prediction systems for maintenance.

7. **Users of (re)manufactured machine tools and assembly equipment:** This factor describes the interests of users in remanufactured tools and equipment. Both S1 and S2 project that customer will primarily lease production equipment, reinforcing PSS-oriented strategies as an important business strategy.
8. **Strategies of manufacturers of machine tools and assembly equipment:** This factor depicts the direction that the major market players follow regarding which entity is responsible for the remanufacturing of the product. S1 projects that the OEMs will handle remanufacturing themselves. S2 projects that the OEMs will accept lower profit margin, manage the remanufacturing of their parts and control the prices of remanufactured parts.
9. **Providers/owners of machine tools and assembly equipment:** This factor describes the interests of providers/owners to offer or request products and/or services such as PSSs. Both S1 and S2 project that the remanufacturing activities will be kept within the owner or supplier companies.
10. **Associations for remanufacturing:** The impact that associations have upon the remanufacturing market. Associations can foster special training programs and fund technical reports, organize public relations efforts for the industry, and participate in lobbying activities in regional and federal levels. S1 projects that associations will have high impact on decision making, causing an increase in demand of production equipment by reduction of technological gap and environmental impacts. S2 projects that associations will have medium impact on decision making, causing an increase in demand of production equipment due to increased quality of products and service.

7 INSTANTIATION IN THE REMANUFACTURING MARKET

As organizations have limitations on their available resources (workers, processes, equipment and raw materials), they cannot serve all customers at once. A good tactic is to do customer market segmentation in order to understand the specific needs of the customers, who share similar purchase criteria. Then the organizations can align their product offerings with their preferred marketing segments, focusing efforts on integrating them better as customers.

As an instantiation, independent company and subcontractor were selected to indicate where the goals for collaborative competition in production systems of mass production and craft production can be located.

Craft production can be seen as fulfilling the requirements of niece markets, where there is a very high degree of personalization. Mass production is seen as the production of standard goods of negligible variety with expected tolerances for defects, for lowest possible cost in large volumes.

The independent companies were selected since they are a common form of entrant threats to OEMs and the subcontractors since they symbolize a dependency role that can be exploited more strategically.

Hormozi identified capabilities for the production systems used in the comparison and quantified them from low to high depending [23]. These values are considered to be the perceived values of the market and compared to values assigned to the independent company and the subcontractor. The factors are following:

Emphasis on elimination of production waste: By focusing on the elimination of waste, the organization produces more value and is therefore more efficient in doing the remanufacturing task. In terms of remanufacturing this would be having a good knowledge of the current status of the part to be remanufactured in order not to have to perform unnecessary or unwanted process steps.

Degree of product leveling shows the extent that the organization takes to ensure that parts, components or machines are remanufactured in order to meet specific customer demand or to utilize the usage of tools and machinery required for the remanufacturing process.

Degree of organizational communication indicates how quick and efficient intra-organization communication is. Low level signals a functional silo setup within the organization, while a high one implies more willingness to aid each other within the organization.

Degree of cooperation between organizations displays to what extent does an organization compare itself and/or share knowledge with other organizations, whether it is competitors, associations or even governmental committees.

Degree of cooperation between organization and customer indicates how quick and efficient cooperation between organization and customer is. High degree shows that the customer is highly involved in the design.

Sensitivity to customer demands: A high sensitivity to customer demands implies that the customer could be on the border of a specific market segment or that the market segment provided with products and services is narrow.

Need for skilled employees: As the remanufacturing activities required differ, the skill sets required differ as well. The higher the levels of customization are the higher levels of skills are required.

Level of automation signifies the level of automation of a process or systems required during remanufacturing activity. Relying on mostly on hand tools represents a low level of automation, while computerized numerical control (CNC) processing would be a higher level of automation.

Degree of flexibility represents the willingness to make changes to product as per customer demand. This includes accommodating changes to the existing parts being remanufactured also to adding of new functionalities.

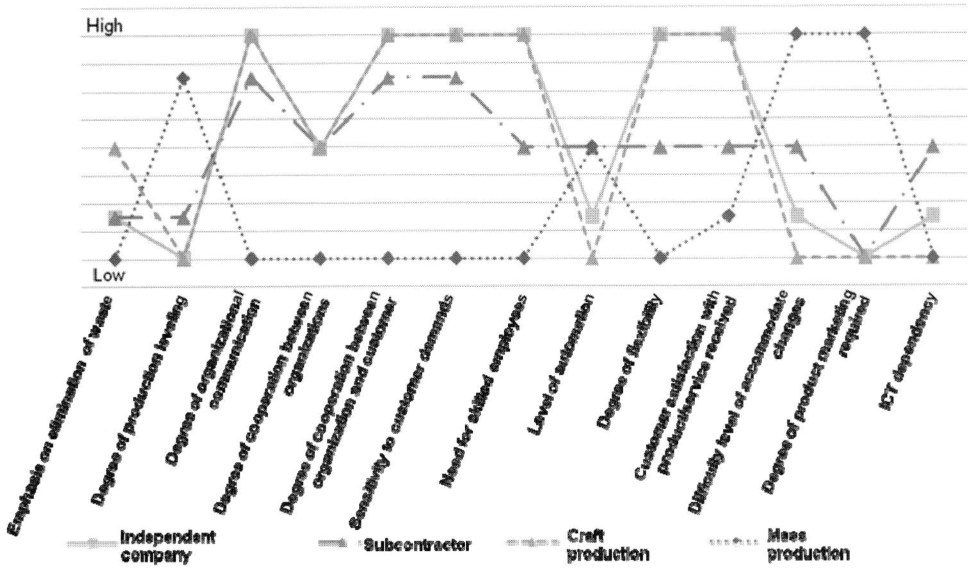

Figure 2: Value Curve for Independent Company and Subcontractor

Satisfaction with product and services received describes how well the organization manages to meet the customer expectation. The customer expectation is driven by the standard practice within a specific market segment regarding technical aspects of products such as defects, lifetime and performance. Aesthetics and ease of use also play an important role in customer expectation.

Difficulty level to accommodate changes (volume, product or process design) indicates how hard it is to implement the changes according to customer demand. As the structure of the organization is more complex the harder it is to accommodate changes, as it is highly affected by the number of different departments and different people involved.

Degree of product marketing required shows to what extent does the organization has to go for marketing their services or products.

Information and communication technology (ICT) dependency presents the dependency towards ICT. Some remanufacturing activities require more intuition than written procedures to perform, due to the nature of part condition at end-of-life and therefore low ICT dependency. On the other hand in order to improve initial design or to be able to handle complex process steps at quick pace the ICT dependency increases.

Figure 2 presents value curve comparison and it shows that independent companies are better suited for craft production than subcontractors. Neither of them is really suited for mass production, but both of them could collaborate in order to be better capable to offering a larger variety of products and services. The assumption is made that the current remanufacturing of production equipment is best described by a craft production system.

As the markets change the production system best suited to serve these markets change as well. The automotive industry is a good example, automotives were first produced with a craft production system, which then changed to a mass production system and is today considered to be best served with lean production systems or mass customization. It is therefore vital for organizations to understand that if the market segments change they might have to change as well.

8 COLLABORATION AND COMPETITION AS DRIVERS FOR TECHNOLOGICAL PROGRESS

Competition can sometimes be seen as traditional approach to achieve competitive advantage. However that advantage can also be achieved by forming inter-organizational alliances. In general, thus alliances between potential competitors, buyers and sellers are likely to be advantageous, e.g. when the combined costs of purchase and buying transactions are lower through collaboration than the cost of operating alone [9, 24]. The intensity of alliances between different companies as well as within one company can vary, e.g. by communication and share of resources. The intensity increases between the layers competition, networking, negotiating, coordination, cooperation and collaboration, where collaboration represents the most intensive form [18].

Forming alliances with other companies or organizations is done to achieve specific goals. The following presents the clustered goals with examples regarding remanufacturing [9]:

- **Increase selling power:** Remanufacturers might build close links with customers to coordinate remanufacturing activities regarding customer demands, reduce stock or generate reputation.
- **Increase buying power:** When offering remanufactured products, cores and knowledge represent the primary resources.
- **Build barriers to entry or avoid substitution:** Faced with threatened entry, organizations in an industry may collaborate to invest in research or marketing. OEMs are able to sell new machine tools while taking back used ones, limiting independent activities.
- **Gain entry:** Organizations seeking to develop beyond their traditional boundaries may need to collaborate with others to gain entry into new arenas. Working with

locals, production equipment can be adapted to specific need of local markets.

- **Share work with customers:** An important trend is a move towards more co-production with customers. Remanufacturing provides the possibility to include used equipment from customers, in order to reduce material costs.

9 SUMMARY

Competition and collaboration are both possibilities to achieve profitability and competitive advantages. For different market players with different characteristics individual business strategies are promising. This paper presented an approach to develop such business strategies for the remanufacturing market. The four generic market players are described as well. The following work will focus on working their goal-sets for collaborative competition which can be achieved by individual business strategies.

10 ACKNOWLEDGEMENT

The results are part of the ongoing research project "Remanufacturing oriented Production Equipment Development". Its goal is to boost remanufacturing of production equipment in an economical as well as technological way. The project is funded by the German Research Foundation DFG. We extend our sincere thanks to Hayden Forrister from the Institute for Sustainable Practice, David Lipscomb University, USA for his valuable contribution.

11 REFERENCES

[1] Deutsche Stiftung Weltbevölkerung, 2010, Weltbevölkerung wächst jährlich um 78 Millionen Menschen, www.weltbevoelkerung.de, last call 30th September 2010.

[2] World Commission on Environment and Development, 1987, Our Common Future, www.un-documents.net/wced-ocf.htm, last call 30th September 2010.

[3] D. Kundisch, P. Herrmann, C. Meier, 2010, Sustainable Process Management - Status Quo and Perspectives, in: W. Dangelmaier, A. Blecken, R. Delius, S. Klöpfer: Advanced manufacturing and sustainable logistics, Springer, 94-106.

[4] R. Steinhilper, 1998, Remanufacturing – The Ultimate Form of Recycling. Fraunhofer IRB Verlag, Stuttgart.

[5] E. Sundin, 2004, Product and Process Design for Successful Remanufacturing, Linköpings Universitet, Linköping.

[6] M. Grismajer, H. -J. Kim, S. Heyer, V. Raichur, G. Seliger, S. J. Skerlos, 2009, Factors Influencing the Success of Product Recovery Business Focusing on Remanufacturing, Proceedings of the 7th Global Conference of Sustainable Manufacturing, Madras.

[7] M.E. Ferguson, L.B. Toktay, 2006, The Effect of Competition on Recovery Strategies, Production and Operations Management, 15/3:351-368.

[8] C. Franke, G. Seliger, E. Yakut, 2004, Mobile Phone Remanufacturing Program Planning in Networks, Proceedings of the Global Conference on Sustainable Product Development and Life Cycle Engineering, Berlin.

[9] R. Kensy, 2001, Keiretsu Economy – New Economy? Japan's Multinational Enterprises from a Postmodern Perspective, Palgrave.

[10] G. Johnson, K. Scholes, R. Whittington, 2004, Exploring Corporate Strategy - Text and Cases, Financial Times, 7th edition, Harlow.

[11] M.E. Porter, 1980, Competitive Strategy: Techniques for Analyzing Industries and Competitors. Free Press.

[12] G. Seliger, S. Severengiz, H. Widera, A.R. Ometto, Barrera, Y. Barrera, T. França, F. Puglieri, F.A. Forcellini, C. Barbosa, A.P. Barquet, 2009, Scenarios for Remanufacturing/Reuse of Machine Tools and Assembly Equipment for Brazil in 2020 – Department of Production Engineering and Deutsche Forschungsgemeinschaft, Report.

[13] P. Timmers, 1998, Business Models for Electronic markets, Electronic Markets - The International Journal of Networked Business, 8/2:3-8.

[14] P. Butala, 2010, Information support of equipment operations – the case of a hydropower plant, 2nd CIRP IPS^2 Conference, Linköping.

[15] H. Willke, 2001, Systemisches Wissensmanagement, 2nd Edition, Lucius & Lucius Verlagsgesellschaft mbH, Stuttgart.

[16] European Commission, 2003, Recommendation 2003/361/EC: SME Definition.

[17] S. Frost, A. Sullivan, 2007, Country Industry Forecast - Social and Infrastructure Analysis for the Brazilian Energy Industry, London.

[18] Jr.V.D.R. Guide, 2000, Production planning and control for remanufacturing: industry practice and research needs, Journal of Operations Management, 18/4:467-483.

[19] E.L. Plambeck, T. A. Taylor, 2005, Sell the plant? The impact of contract manufacturing on innovation, capacity, and profitability, Management Science, 51/1:133-150.

[20] R.B. Aronson, 2003, Remanufactured Machine Tools - Should you fix it or junk it? Manufacturing Engineering, 131/6.

[21] A.K. Kamrani, E. A. Nasr, 2008, Collaborative Engineering - Theory and Practice, Springer, Berlin.

[22] The European Parliament, 2009, The European Council 02.02.2003 – Directive 2002/96/EC of the European.

[23] A.M. Hormozi, 2001, Agile manufacturing: The next local step. Benchmarking: An International Journal, 8/2.

[24] S. Heyer, M. Grismajer, G. Seliger, 2010, Collaboration in Value Creation Networks to improve Material Cycles, 43rd CIRP International Conference on Manufacturing Systems, Vienna.

Modelling of Friction Stir Processing with in Process Cooling Using Computational Fluid Dynamics Analysis

A.N. Albakri[1], S.Z. Aljoaba[2], M.K. Khraisheh[1]
[1]Masdar Institute of Science and Technology, Abu Dhabi, UAE
[2]Department of Mechanical Engineering, University of Kentucky, Lexington, USA

Abstract
Friction Stir processing (FSP) has evolved recently as an energy efficient and green processing technique for enhancing the formability and mechanical properties of some metals. Managing the heat generation during FSP is critical; sufficient heat is needed to soften the material but without melting, which allows for dynamic recrystallization of the grains by the stirring action. Effective cooling was found to improve the resulting microstructure by removing the excess heat that promotes grain growth. In this paper three dimensional models were developed to simulate FSP with in-process cooling from the backing plate using computational fluid dynamics. Various cooling channel geometries, coolants and flow rates are simulated to study their effects on the temperature history, flow stresses, predicted grain size and hardness distributions, where the coolant type was found to have the most significant effect among the different cooling aspects.
Keywords:
Computational Fluid Dynamics, Dynamic Viscosity, Effective Cooling, FSP, Grain Size Distribution, Temperature

1 INTRODUCTION

Friction stir welding (FSW) is a new solid-state joining technique; it was invented in 1991 by The Welding Institute (TWI) to join what has been classified as non-weldable materials as Al alloys. Recently, based on the observed enhancement of the mechanical properties of the friction stir welded zone, which is characterized by equiaxed fine grains and homogenous microstructure, Mishra et al [1] developed a new processing technique which is Friction Stir Processing (FSP).

FSP is a solid state process in which specially designed cylindrical tool made of a pin and a shoulder is plunged into a sheet of metal while rotating, until the shoulder comes in contact with the top surface of the sheet as shown in figure(1). The tool is then traversed along linear path, thus generating heat and causing intense plastic deformation with material flow through the stirring action, which yields dynamically recrystallized fine grains and a defect free homogenous microstructure. Compared to other refining methods like equal channel angular extrusion (ECAE) and thermo-mechanical processing, FSP has the advantages of being green and energy efficient process that uses simple tools and can be easily automated [2].

Heat generation and distribution during the process are critical issues, because they affect the final microstructure and grain size. Different process parameters in terms of tool geometry, rotational and translational speeds affect the thermal history of the sheet. High rotational and low translational speeds can increase the local temperature up to the melting point, or lead to residual thermal cycles and grain growth after FSP which negatively affect the mechanical properties of the processed sheet [3]. For materials of low melting point such as Al and Mg alloys, cooling presents an effective solution to prevent melting during the process and to reduce grain growth after it [2].

In terms of cooling, some research has been carried out to study the effects of FSP or FSW with cooling on the microstructure and mechanical properties. Many cooling techniques were tried; some researchers [4,5] applied cooling by quenching during FSP. While others carried out FSP and FSW with sheet submerged under water [6, 7]. Some researchers concluded that the primary heat loss is made from the bottom of the pin to the backing plate beneath the processed sheet. Chang et al [8] designed a new efficient cooling system through channels made in copper mould backing plate and using liquid Nitrogen as coolant during FSP of AZ31 Mg alloy, they achieved ultra fine grains (100 nm to 300 nm) and significant increase in microhardness from 50 Hv up to 120 Hv. Fratini et al [9] designed a new fixture for FSW process of titanium alloys, they drilled three cooling channels in the backing plate beneath the sheet and used water at room temperature as a coolant in order to maximize the mechanical performance of the weld and keep the integrity of the fixture.

Many experimental works can be found in the literature about effective cooling during FSP and FSW, nevertheless, very few attempted to tackle the problem from a modelling point of view. Although several studies focused on the numerical simulation of FSP and FSW [10-12], limited number of works tried to study the cooling process and connect its various aspects with the resulting thermal histories, mechanical and microstructural characteristics. Fratini et al [13] combined experimental and numerical investigations to study the effect of in process water cooling on the resulting mechanical and metallurgical properties of friction stir welded zone. They simulated the process using finite element code, where the

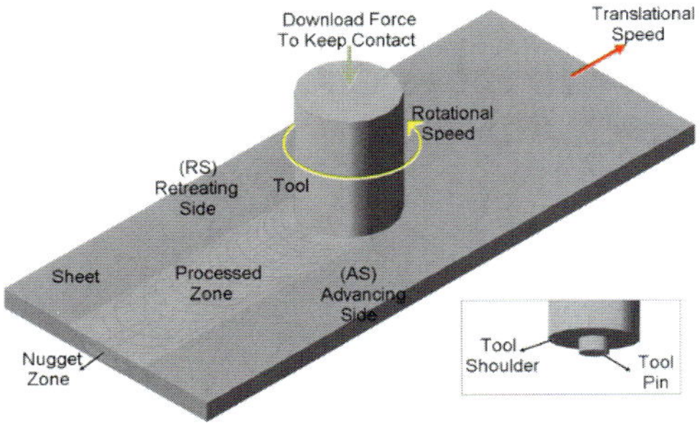

Figure 1: Sketch of Friction Stir Processing

coolant was modeled as water flux behind the tool. The use of water cooling resulted in significant improvement in the mechanical characteristics of the weld by reducing the thermal flux and increasing the micro-hardness. Aljoaba et al [14] used computational fluid dynamics analysis to simulate FSP under different stirring conditions and tool geometries. They created two models, one without cooling and the other with cooling from the backing plate using simple channel with water at room temperature, where cooling proved to enhance the grain size and microhardness of the processed zone.

In this work, FSP is simulated with in process cooling from the backing plate using CFD analysis, where different cooling channel geometries, coolant flow rates and coolant types will be investigated in order to study their effects on the temperature history, flow stresses, predicted grain size and microhardness distributions for the processed sheet.

2 MODEL DESCRIBTION

The CFD commercial code STAR-CCM+ (V.4.06.011) was used in the simulations. STAR-CCM+ is a multi-physics code with Computational Continuum Mechanics (CCM) based algorithms. The simulations were based on solving the continuity, momentum and energy equations applied to finite number of control volumes under steady state conditions. The continuity equation is given by

$$\frac{\partial u_k}{\partial x_k} = 0 \qquad (1)$$

Where u is the velocity of plastic flow that resulted from the sheet material rotation and translation, the steady state, single phase momentum equation is expressed as

$$\rho \frac{u_k \partial u_i}{\partial x_k} = -\frac{\partial P}{\partial x_i} + \frac{\partial}{\partial x_k}\left(\mu \frac{\partial u_i}{\partial x_k} + \mu \frac{\partial u_k}{\partial x_i}\right)$$
(2)

Where ρ is the density, μ is the Non Newtonian viscosity, and P is the pressure. Temperature is calculated from the energy equation as shown below

$$\rho C_p \frac{u_k dT}{dx_k} = \frac{u_k dP}{dx_k} + \nabla^2 kT - \frac{\partial u_k}{\partial x_k} P + \mu\left(\frac{\partial u_i}{\partial x_k}, \frac{\partial u_k}{\partial x_i}\right) \qquad (3)$$

Where C_p is the specific heat, T is the temperature, and k is the thermal conductivity. The convective heat transfer boundary condition is expressed as

$$\nabla T = \frac{h}{k}(T_s - T_e) \qquad (4)$$

Where h is the convective heat transfer coefficient, T_s is the surface temperature, and T_e is the surrounding temperature. The conduction heat transfer q can be modelled as a heat flow through a thermal circuit with a series of resistances as

$$q = \frac{-(\Delta T)}{R}, \text{ where } R = \sum_i^n \frac{\Delta t_i}{k_i A} \qquad (5)$$

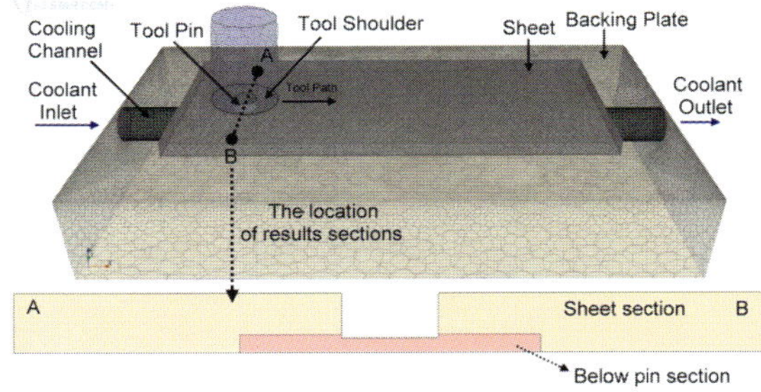

Figure 2: CFD model parts

Where R is the total thermal resistance through the conductive layers, Δt_i is the thickness of the ith layer, k_i is the thermal conductivity of the ith layer, A is the surface area, and n is the number of conducting layers. It will be assumed that contact heat resistance is ignored.

The model parts include the sheet, the rotating tool and the backing plate with cooling channel inside it as shown in Figure (2). The shoulder is assumed to have full sticking conditions with the interface region in the sheet, thus both have the same rotational velocity w (rpm). Three interfaces with perfect contact and negligible heat loss were defined; between the tool and the sheet, the sheet and the backing plate and between the coolant and the channel surface.

2.1 Materials properties

The backing plate and the tool were assumed to be rigid solid bodies; they had the properties of AISI 4142 steel and H13 steel respectively. The sheet was modeled as Non-Newtonian fluid with laminar and incompressible flow with dimensions 120x150x4 mm, while defining the actual sheet material properties which are shown in table (1) and refer for AZ31B Mg alloys. The dynamic viscosity of the sheet is a function of local values of temperature and strain rates as shown in equation (6) [14, 15].

$$\mu = \frac{\sigma_e}{3\dot{\varepsilon}} \qquad (6)$$

Where μ is the dynamic viscosity, σ_e is the flow stress, and $\dot{\varepsilon}$ is the effective strain rate. While STAR CCM+ can find the local strain rate values, a user defined functions were created to find the flow stress, which is a function of Zener Holloman parameter, as shown in equation (7) [14,15].

$$\sigma_e = \frac{1}{\alpha}\sinh^{-1}\left[\left(\frac{Z}{A}\right)^{1/n}\right] \qquad (7)$$

Where α, A and n are material constants shown in table (1), Z is the Zener-Holloman parameter, which is a function of strain rate and temperature as shown in equation (8) [14, 15].

Table 1: Sheet material properties and constants [14]

Property	Value
Density	$1777\ Kg/m^3$
Specific Heat	1000 J/Kg.K
Thermal Conductivity	$96\ W/mK$
Activation Energy Q	130 KJ/mol
Material Property (A)	$2.7 * 10^7\ s^{-1}$

Material Property (n)	1.8
Material Property (α)	0.052 MPa^{-1}

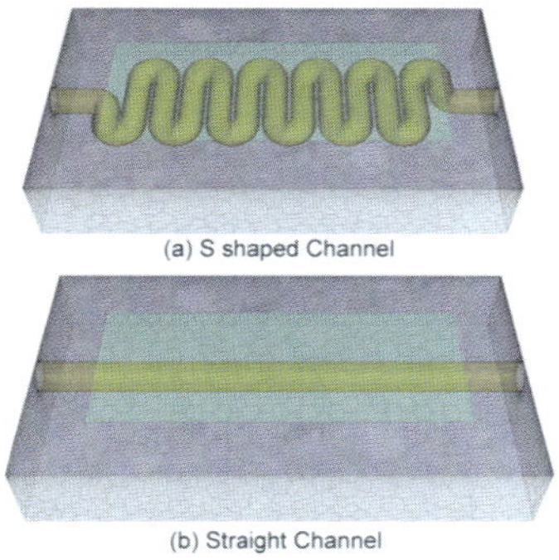

Figure 3: Cooling channel geometries in the backing plate

$$Z = \dot{\varepsilon} \exp\left(\frac{Q}{RT}\right) \tag{8}$$

Where Q is the activation energy of lattice diffusion for the sheet's material with value shown in table (1), R is the universal gas constant, and T is the temperature value.

2.2 Boundary conditions

A full sticking and no slip conditions were assumed for the tool/sheet interface, thus heat was generated from the plastic deformation [11], while frictional heat was ignored. Boundaries of the sheet, backing plate and tool were assumed to lose heat to the ambient (300 K) by convection, while heat was transferred between interfaces though conduction. Convective heat transfer coefficient of 40 W/m^2K was assumed for sheet walls (Mg and air), while a coefficient of 35 W/m^2K was assumed for tool and backing plate walls (steel and air).

3 RESULTS AND DISCUSSION

Three different aspects have been studied and simulated: the effect of cooling channel geometry, the effect of coolant flow rate and the effect of coolant type. Two types of cross sections will be used to show the results as shown in figure (2), the first stands for the whole sheet section, while the second represents the region below the tool pin to show the cooling effects better.

3.1 Effect of cooling channel geometry

Three models have been created, FSP with no cooling, FSP with straight channel cooling and FSP with S shaped channel cooling as shown in figure (3). The same stirring and cooling conditions were assumed in the three cases with rotational velocity of 1000 rpm, translational speed of 12 inch/min, and water as coolant at 10 C° with flow rate of 1 kg/s.

Temperature fields predictions

The calculated temperature contours for the three cases are shown in figures (4) and (5). The maximum temperature values are located at the outer diameter of the shoulder interface with the sheet, as the maximum strain rate is found there, which leads to maximum plastic deformation and heat generation [15]. Both straight and S channels cooling have minor effects on the maximum temperature at the shoulder interface. However, both have significantly reduced temperature below the pin zone as shown in figure (5) with about 30K to 40K. Although S shaped cooling didn't reduce the maximum temperature compared to straight channel one, it has reduced the heat affected zone (HAZ) width, as it provides larger area of heat transfer, where temperature contours are more scattered in the other two cases. Temperature values at the bottom sides of stirring zone are reduced more in case of S shaped cooling (380K) compared to (420K) in straight cooling and (470K) for no cooling.

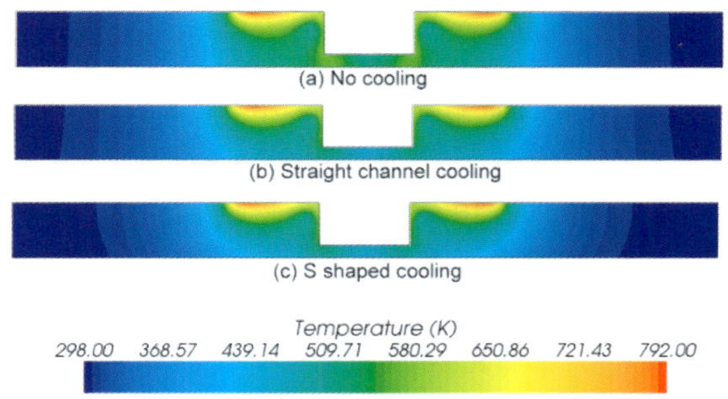

Figure 4: Temperature distribution for different cooling channel geometries

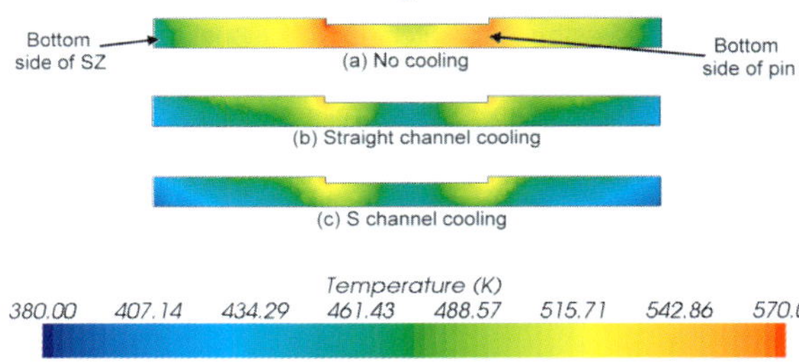

Figure 5: Temperature distribution below pin zone for different cooling channel geometries

Flow stress distributions

Figure (6) shows the flow stress distributions, flow stress is defined as the stress needed to sustain plastic deformation; it separates the elastic and plastic deformation ranges for a material, it is calculated using equation (7). It is noticed that the flow stress values decrease as the temperature values increase; which means that at higher temperatures, lower stresses are needed to achieve plastic deformation. The minimum values of flow stress are located at the shoulder interface; furthermore, its values increase with using cooling options especially below the pin zone. It can be seen that flow stress values at the sides of the sheet are increased in case of S channel compared to straight channel cooling.

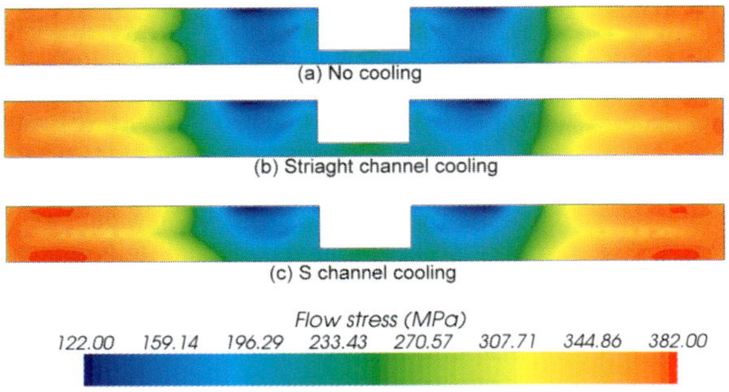

Figure 6: Flow stress distribution for different cooling channel geometries

Grain size and microhardness distributions

Zener-Holloman parameter presents a very useful tool to predict the variations in the grain size and micro-hardness distributions [16]. Darras [17] developed a relation between the Zener-Holloman parameter and the average recrystallized grain size for Mg AZ31B-O alloys for similar stirring conditions as shown in equation (9) where d is diameter in micrometers.

$$\ln(d) = 8.464 - 0.2104\ln(Z) \tag{9}$$

Local evaluation for the predicted microhardness in the stirred zone is possible using Hall-Petch (H_v) relationship [16], which is used to calculate the Vickers scale hardness based on the predicted average grain size d (micrometer) as shown in equation (10)

$$H_v = 40 + 72\,d^{-1/2} \tag{10}$$

Equations (9) and (10) will be applied in the stirring zone only, where temperature is higher than 500 K (temperature of recrystallization of Mg alloys is 523 K [18]) and strain rate higher than 0.5 (1/s). To better show the effects of stirring action on the microstructural evolution, in other zones the average grain size diameter was assumed to be 48 μm, while the microhardness was assumed to be 50 Hv.

Figure (7) shows the predicted distribution for the grain size in the processed zone for different cooling channel geometries, it can be seen that the finest grains are found below the pin zone, where lower heat is generated compared to the shoulder zone. The smallest grains size reached around 1.5 μm compared to 48 μm for the parent material. It is also concluded that as the temperature decreases the grain size decreases, where it has approximately reduced to half with using cooling options. S channel cooling helps in increasing the distribution of the smallest grains, but it didn't produce finer grains compared to the straight channel cooling.

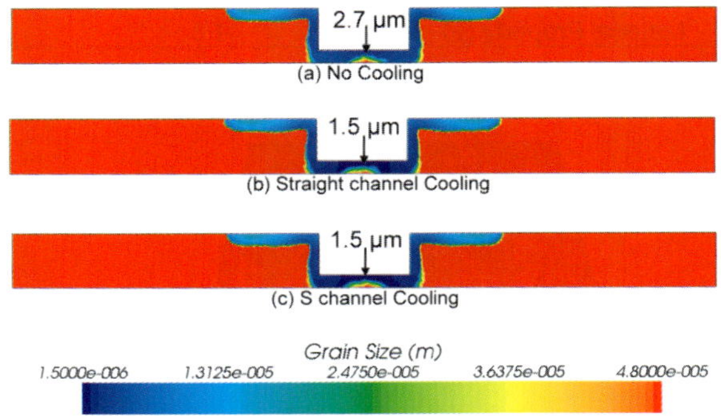

Figure 7: Grain size distribution for different cooling channel geometries

Figure (8) shows the predicted hardness, which has an inverse relationship with grain size. Largest hardness values are located below the pin zone. Cooling has increased the hardness values from about 80Hv in case of no cooling to about 100Hv with using cooling options.

It can be concluded that S shaped geometry didn't have great effect on the grain size and hardness distributions compared to straight channel cooling. However, it has significantly affected the temperature and flow stress values, which can help in reducing the grain growth, residual stresses and thermal cycles in the heat affected zone (HAZ).

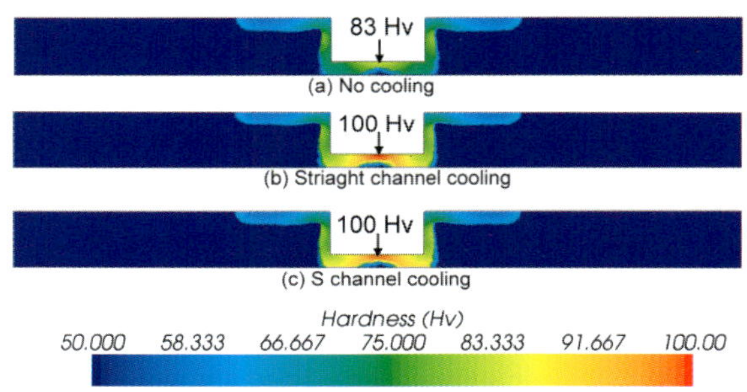

Figure 8: hardness distribution for different cooling channel geometries

3.2 Effect of coolant flow rate

In order to study the effect of coolant flow rate on the predicted temperature fields, grain size and hardness distributions. The straight channel geometry will be used with the same stirring conditions (1000 rpm, 12 inch/min) and the same coolant (water at 10 C°).

Temperature fields predictions

Figure (9) shows the temperature values below the pin zone at different flow rates, as shown the increase in flow rate resulted in lower temperature values, especially in the region close to the bottom of the sheet, where the most significant heat transfer occurs. The temperature has been reduced by about 40K when flow rate is increased from 1 kg/s to 6 kg/s.

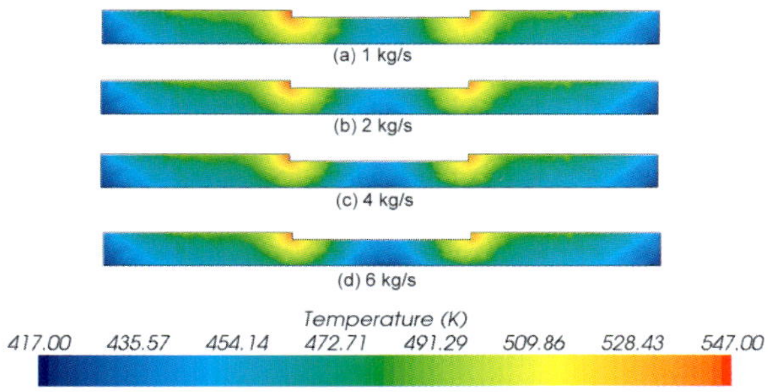

Figure 9: Temperature distribution below the pin zone for different flow rates

Grain size and microhardness distributions

The predicted grain size distribution at different flow rates below the pin region is shown in figure (10). It can be concluded that as the flow rate increases the grain size decreases, until we reach some flow rate beyond which the reduction in grain size becomes insignificant. Increasing the flow rate from 1kg/s to 2kg/s has reduced the grain size more than increasing the rate from 4kg/s to 6kg/s.

The predicted hardness distribution below the pin zone at different flow rates is shown in Figure (11), where increasing the flow rate has improved the hardness distribution especially at lower flow rates.

3.3 Effect of coolant type

Many researchers tried different coolants in their FSP or FSW experiments, such as water [6, 7], mixture of liquid CO2 and

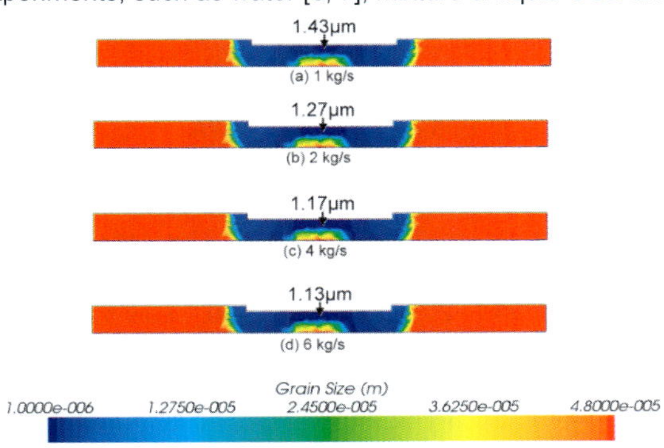

Figure 10: Grain size distribution below the pin zone for different flow rates

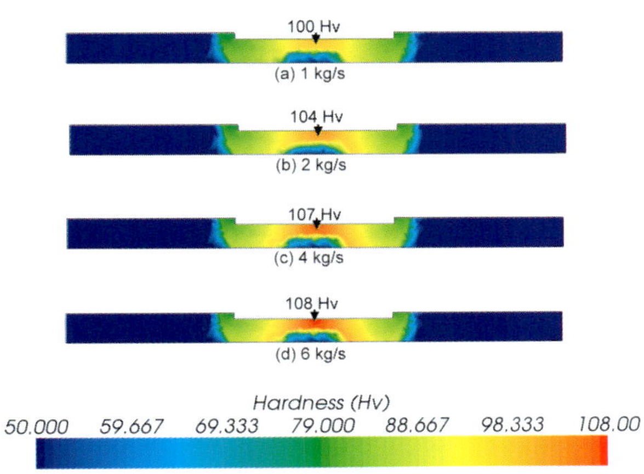

Figure 11: Hardness distribution below the pin zone for different flow rates

methane [5], and liquid Nitrogen [8]. Four types of coolants including water, liquid Propylene (R-1270), liquid Argon (R-740) and liquid Nitrogen (R-728) were tried to study their effects on the results assuming the same stirring conditions (1000 rpm, 12 inch/min, and 6 kg/s) and using the straight channel geometry model. REFPROP, which is an electronic database issued by the National Institute of Standards and Technology (NIST) [19], was used to find the thermodynamic properties of the cooling fluids assuming a working pressure of 1 atm as listed in table (2). The maximum temperature for water was assumed 10 C°, while the boiling temperature was assumed to be the maximum for the other liquids.

Temperature fields predictions

The temperature distributions for different coolants below the pin zone are shown in figure (12). As can be indicated, compared to water cooling, using refrigerants or cryogenic fluids has reduced the temperature significantly, where fluids with lower boiling point have greater effects on the reduced temperature. Using liquid nitrogen instead of water has reduced temperature with around 80 K, while using propylene and liquid argon reduced temperature with around 20 K and 60 K respectively.

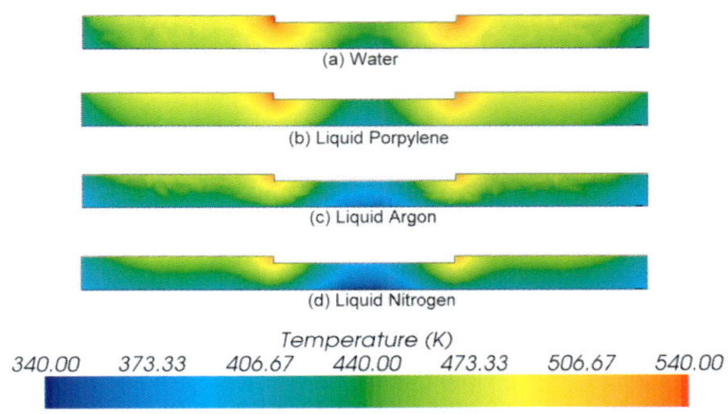

Figure 12: Temperature distribution below the pin zone for different coolants

Grain size and microhardness distributions

The predicted grain size distribution with different coolants is shown in figure (13), as more heat is extracted from the processed sheet, finer grains can be produced. In addition, with cryogenic fluids like liquid nitrogen that has very low boiling point, very fine grains (up to 330 nm) was produced, which is consistent with experimental findings of Chang [8].

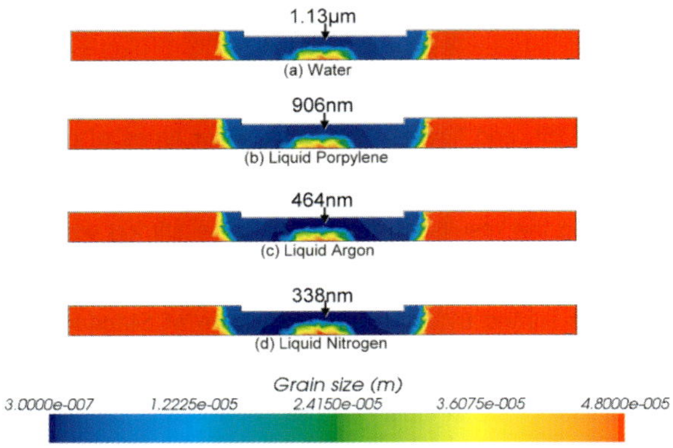

Figure 13: Grain size distribution below the pin zone for different coolants

Figure (14) shows the hardness distribution for different coolants below the pin zone; the highest hardness values were produced by using coolants of very low boiling point. Liquid nitrogen, which has a boiling temperature of about 77.3 K, has produced high hardness values up to 165 Hv.

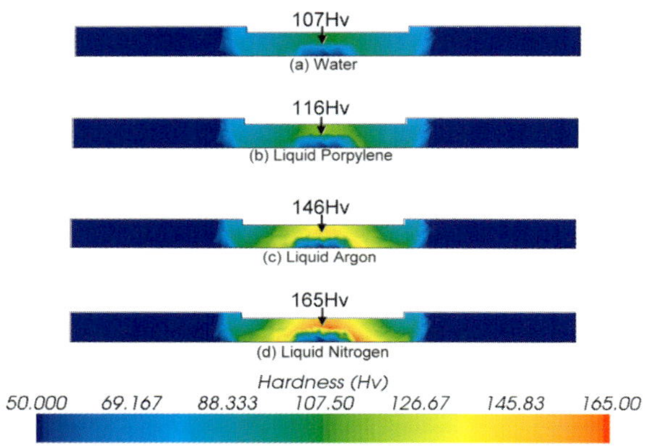

Figure 14: Hardness distribution below the pin zone for different coolants

Table 2: Thermodynamic properties of coolants at 1 Atm [19]

Property	Water	Liquid Propylene	Liquid Argon	Liquid Nitrogen
Density (Kg/m^3)	999.7	610.06	1395.4	806.18
Maximum Temperature (K)	283.15	225.53	87.302	77.335
Specific Heat (J/Kg.K)	4195.2	2192.6	1117.2	2041.4
Thermal Conductivity (W/m.K)	0.5800	0.1493	0.1284	0.1448
Dynamic Viscosity (µPa.s)	1305.9	195.89	260.29	160.76

4 SUMMARY

FSP with in-process cooling through the backing plate has been simulated using CDF analysis, different aspects of the cooling process have been investigated to study their effects on the calculated temperature fields, flow stresses, grain size and hardness distributions. Although using channel geometry of larger area inside the backing plate didn't have significant effects on the resulted grain size and microhardness values, it has reduced the HAZ width by reducing the temperature values around the stirring zone, which could help in reducing grain growth and thermal cycles after the process. In addition, using cryogenic fluids of low boiling point like liquid Nitrogen with high flowing rate has reduced the temperature below the pin zone with about 80 K, and produced very fine grains of about 330 nm and high hardness values up to 165 Hv compared to water cooling. Cooling from backing plate has minor effects on the heat generated at shoulder interface, thus simulating cooling techniques from the top surface of the sheet is needed to try come up with an effective cooling options that cover the whole sheet section.

5 ACKNOWLEDGMENTS

The authors would like to express thanks all who contributed to this paper. This work is financially supported by Masdar Institute, Abu Dhabi, UAE.

6 REFERENCES

[1] Mishra R.S., Mahoney M.W, McFadden S.X., Mara N.A, Mukherjee, 2000, High strain rate superplasticity in a friction stir processed 7075 Al alloy, Scripta Mater,42: 163–68.

[2] Mishra R.S, Ma Z.Y, 2005, Friction Stir Welding and Processing, Mater.Sci.Eng, R50:1-78.

[3] Chen C.M, Kovacevic R, 2003, Finite element modeling of Friction stir welding-thermal and thermomechanical analysis, International Journal of Machine Tools and Manufacture, 40:1319-1326.

[4] Liu F, Ma Z, Chen L, 2990, Low-temperature superplasticity of Al-Mg-Sc alloy produced by friction stir processing, Scripta Materialia,60:968-971.

[5] Su J.Q, Nelson T.W, Sterling C.J, 2003, A new route to bulk nanocrystalline materials, Mater Res Society, 18:1757–1760.

[6] Hofmann D.C, Vecchio K.S, 2007, Thermal history analysis of friction stir processed and submerged friction stir processed aluminium, Materials Science and Engineering, 465:165-175.

[7] Upadhyay P, Reynolds A, 2010, Effects of thermal boundary conditions in friction stir welded AA7050-T7 sheets, Materials Science and Engineering, 527:1537-1543.

[8] Chang C, Du X, Huang J, 2007, Achieving ultrafine grain size in Mg-Al-Zn alloy by friction stir processing, Scripta Materialia, 57: 209-212

[9] Fratini L, Micari F, Buffa G, Ruisi V, 2010, A new fixture for FSW processes of titanium alloys, CIRP Annals Manufacturing Technology, 59: 271-274.

[10] Colegrove P.A, Shercliff H.R, 2005, 3-Dimensional CFD modelling of flow round a threaded friction stir welding tool profile, Journal of Materials Processing Technology, 169:320-327.

[11] Kim D, Badarinarayan H, Kim J.H, Kim C, Okamoto K, Wagoner R, Chung K, 2009, Numerical simulation of friction stir butt welding process for AA5083-H18 sheets, European Journal of Mechanics - A/Solids, 29:204-215.

[12] Aljoaba S, Jawahir I, Dillon O, Ali M, Khraisheh M, 2009, Modeling of friction stir processing using 3d CFD analysis, Material Forming 12th ESAFORM Conference, Netherlands.

[13] Fratini L, Buffa G, Shivpuri R, 2010, Mechanical and metallurgical effects of in process cooling during friction stir welding of AA7075-T6 butt joints, Acta Materialia, 58:2056-2067.

[14] Aljoaba S, 2009, Experimental investigation and modeling of Friction Stir Processing using 3D CFD analysis, Master's Thesis, University of Kentucky, USA.

[15] Ulysse P, 2002, Three-dimensional modeling of the friction stir-welding process, International Journal of Machine Tools and Manufacture, 42:1549-1557.

[16] Chang C, Lee C, Huang J, 2004, Relationship between grain size and Zener-Holloman parameter during friction stir processing in AZ31 Mg alloys, Scripta Materialia, 51:509-514.

[17] Darras B, 2008, Integrated thermo-mechanical investigations of friction stir processing of light weight alloys, PH.D Thesis, University of Kentucky, Lexington, KY, USA.

[18] ZHANG Bing, YUAN Shouqian, WANG Xunhong, 2008, Friction stir welding of AZ31 magnesium alloys processed by equal channel angular pressing, Rare Metals, 27:393-399.

[19] Lemmon, E.W, Huber, M.L, McLinden, M.O, 2007, NIST Standard Reference Database 23: Reference Fluid Thermodynamic and Transport Properties-REFPROP, Version 8.0, National Institute of Standards and Technology, Standard Reference Data Program, Gaithersburg.

Nanofluids: Properties, Applications and Sustainability Aspects in Materials Processing Technologies

P. Krajnik[1,2], F. Pusavec[2], A. Rashid[1]

[1]Department of Production Engineering, KTH Royal Institute of Technology, Stockholm, Sweden

[2]Faculty of Mechanical Engineering, University of Ljubljana, Slovenia

Abstract

Nanofluids could be used to provide cooling and lubrication action and to control thermo-physical and tribo-chemical properties of material processing. It is foreseen that properly designed nanofluids could surpass conventional cutting fluids with respect to thermal conductivity, convective heat transfer coefficient, critical heat flux, viscosity, and wettability. These properties have a promising potential to lead to the development of new coolants and lubricants with applications in a wide variety of materials processing technologies. This paper analyses the developments in research on the properties of nanofluids and evaluates their potential for applications in machining, focusing on their thermal and tribological aspects. The increasing use of nanofluids leads to a need for information on their sustainability in order to recognize and avoid risks. Sustainability is discussed in view of occupational health and safety and toxicity of nanoparticles.

Keywords:

Machining; Cooling; Lubrication; Fluid; Nanotechnology

1 INTRODUCTION

Technological developments in material processing technologies are driven by demands of sustainable manufacturing and higher performances. Changing factors in the manufacturing environment and recent developments in nanotechnology opened a new area for activating technological potential of nanofluids. Tool life, surface integrity, heat evolution (generation and removal), energy consumption etc. are critical concerns in all metal cutting and forming operations. The thermo-physical and tribo-chemical properties of nanofluids can be exploited for effective and efficient control of the above mentioned concerns and improve the sustainability of these operations.

This paper is meant to address some uncertainties in so far the integration of nanofluids into materials processing technologies is concerned and to provide guidelines for future development in this area. More specifically, we are advancing a new idea how to integrate nanofluids into cutting processes that could modify and tailor contact interfaces (i.e. tool-chip and tool workpiece), which can ultimately lead to an enhancement in the: quality of machined surface (e.g. surface integrity, microstructure); reliability of machined components (e.g. durability); tool life; sustainability of machining (e.g. higher fraction of energy spend on useful work, substitution of conventional flood cooling lubrication). We are expecting that these enhancements will be achieved sequentially by:

- Customizing the design of nanofluids to exploit their superior thermo-physical and tribo-chemical properties and scaling-up their fabrication.
- Integration of the nanofluids and machining processes via adapted fluid supply systems.
- Demonstrating the impact of these developments in the machining operations.

In machining processes nearly all the energy spent in the material processing is converted into heat, resulting in high cutting temperatures. Moreover, only 30-50% of the energy is spent for useful work (material removal). The rest of the energy is wasted at the tool-chip and tool-workpiece interfaces [1]. High cutting temperatures, resulting from plastic deformation, chip formation and friction, have detrimental effects on tool wear, surface integrity and machining accuracy [2]. Therefore, the use of cutting fluids is typically required to remove heat from the cutting zone (cooling) and to reduce heat generation by reducing friction (lubrication). Conventional flood cooling lubrication uses large amounts of cutting fluids (usually mineral oil based). These fluids are very often enhanced with anti-wear, anti-corrosion or emulsifying agents [3]. With transitioning to sustainable manufacturing, metalworking industries are systematically looking to reduce costs of procurement, maintaining, and disposing of coolants. These costs, combined with operational health and safety (OHS), and environmental concerns, created a heightened interest in limiting the amount of cooling lubricants applied and in finding alternative solutions (e.g. minimum quantity lubrication).

The conventional way to enhance cooling (heat transfer) is to increase the heat transfer surface area (high volume) and the cutting fluid velocity (high pressure). However, this approach is unsustainable due to high power consumption of the pumps and large volume of coolant required. Considering that the velocity of the cutting fluid affects its cooling ability almost as much as its thermal conductivity [1], it is likely that nanofluids with enhanced heat transfer could meet the cooling challenge at lower cutting fluid velocities. A number of reported experiments show that the dispersal of nanoparticles into a base fluid provides extremely desirable thermal properties, such as higher thermal conductivity and convection heat transfer coefficient. Average heat transfer enhancement for nanofluids is shown to be in the range of 15-40% [4].

Available tribological studies further suggest that the use of nanoparticles in base fluids can enhance properties, such as load-carrying capacity, antiwear and reduced friction between moving surfaces [5]. For example, IrO_2 and ZrO_2 nanoparticles remarkably decrease friction on the surface of 100C6 steel [6]. Taking these findings into consideration, one can expect that the application of nanofluids could result in the reduction of friction related tool wear and reduction of cutting forces (e.g. lower friction at the tool-chip interface).

Cooling and lubrication actions are complex and interrelated to a different extent, depending on the machining operation; therefore it is hard to evaluate their effects separately. In this paper the two mechanisms are not discusses in detail, simply because the quantitative data correlating the properties of nanofluids with machining outputs in different operations are not yet available.

The rapid growth of nanotechnology is exceeding our knowledge of the OHS in terms of risks associated with using nanomaterials. Minimal information is available on exposure routes, potential exposure levels and toxicity. In order to responsibly apply nanofluids to machining operations it is necessary to integrate OHS aspects in the technological development and to take into account risks so that negative impact on sustainability is minimized. It should be noted that there is insufficient data available at present to allow for the identification of any systematic rules that govern the toxicological characteristics of all nanoparticles that have potential applicability in nanofluids. It follows therefore, that risk assessment will be needed on a case by case basis. Engineered nanomaterials comprising both metallic and nonmetallic nanoparticles are quite new and their synthesis and usage occurs mainly on laboratory scale. Consequently, information on risk of exposure to these nanoparticles is limited [7,8]. From the environmental perspective, risk concerns imply the necessity for life cycle evaluation. The discussion on sustainability in this paper does not include the assessment of environmental risk related to disposal of nanofluids.

2 NANOFLUIDS

2.1 Design and synthesis

Nanofluids are defined as suspensions of nanoparticles in base fluids [9]. In the context of this work, however, nanofluids refer to a cooling-lubrication system, regarded as a new class of nano-engineered product, consisting of:

- Nanoparticles: metallic (e.g. Cu, Fe, Au, Ag or Al); nonmetallic (e.g. Al_2O_3, CuO, ZnO, TiO_2, Fe_2O_3, WS_2 or MoS_2); carbon in various forms (e.g. nanotubes, fullerene, graphene).
- Base fluids: water-based (e.g. synthetic cutting fluids); vegetable oils (e.g. coconut, rapeseed or canola); organic liquids (e.g. butanol, ethylene glycol); polymeric solutions.
- Additives: surfactants (e.g. cetrimonium bromide, lecithin); antiwear additives (e.g. phosphorus compounds); corrosion inhibitors (e.g. borate esters, amine carboxylate derivatives); disinfectants; fungicides.
- Scale (volume size-effects).

These four design parameters, shown in Figure 1, are selected by the nanofluids' manufacturers to achieve the desired thermo-physical and tribo-chemical properties.

Several base fluids can be used as dispersing media for the fabrication of nanofluids. Although the nanoparticles with their characteristic Brownian motion tend to be suspended in the solution, the stability of the suspension is enhanced by the addition of suitable surface active agents (surfactants). A variety of surfactants (several types and at a wide range of concentration) can be used to meet the requirements of nanofluid's design objectives. Generally, the properties of nanofluids are determined as a result of the selection of the nanoparticle-surfactant-base fluid system.

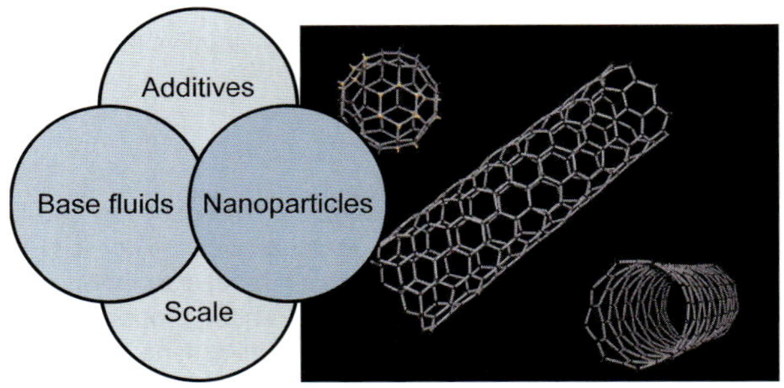

Figure 1: Design of a nanofluid system.

The nanoparticles (1-100 nm in size) can be made of different materials with varying degree of sophistications. Nanoparticles can be of single composition with different morphology (spherical, rod, disks, whiskers, etc). They can also be synthesized as more complex structures with hierarchical architectures such as core-shell structure with single or multi-layers. A typical nanoparticle is composed of the core (ceramic or metallic) and a thin shell, which is often molecular. The core and the shell have various structures and may be composed of more than one entity [10]. Molecular shell has three distinct regions, as schematically shown in Figure 2. Most nanofluids contain less than 1% by volume of nanoparticles.

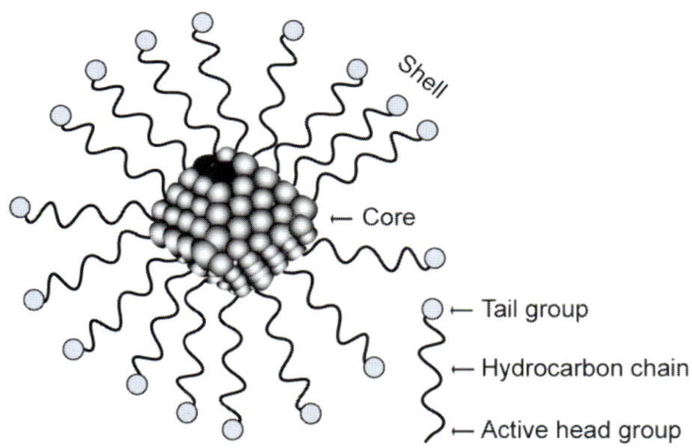

Figure 2: Schematic of a nanoparticle [10].

Nanofluids are fabricated by one- or two-step process. The two-step process first synthesizes nanoparticles and then disperses them into base fluids, using for example ultrasonic devices. This process inherently suffers from particle-agglomeration (sticking together). One way to overcome this is by conditioning nanoparticles through the modification of their surfaces either by electrostatic interaction or the addition of suitable surfactants to create a nanofluid where the particles do not agglomerate. In comparison with the two-step process, the one-step process directly synthesizes and disperses nanoparticles into base fluids, which is cheaper and faster. The one-step process uses several approaches in order to produce uniform particles with a narrow particle size distribution with minimum agglomeration in a base fluid [10].

So far, nanofluids are produced successfully in laboratories only. The challenge is to develop processes for industrial-scale fabrication of the nanofluids. One of the biggest problems refers to the scale of fabrication. As the volume increases, the way the constituents of a nanofluid mix and react changes drastically. This scaling issue is addressed in a design of a nanofluid system.

2.2 Properties of nanofluids

The property that created the most interest in nanofluids is its thermal conductivity; the property that indicates the ability of a nanofluid to conduct heat. The early experimental investigations showed that nanofluids, containing a small amount of oxide nanoparticles, have substantially higher thermal conductivities than the same base fluids without dispersed nanoparticles [11].

For the nanofluids containing the same nanoparticles, the enhanced thermal conductivity ratio is reduced with the increasing thermal conductivity of the base fluid (k_0). On the other hand, the thermal conductivity enhancements of nanofluids using the same base fluid are highly dependent on properties of dispersed nanoparticles; for example, effective conductivity of nanofluids highly depends on the conductivity of the nanoparticles themselves [12]. As shown in Table 1, generally the solids have much higher thermal conductivity than liquids.

Table 1: Thermal conductivity of solids and liquids

Material	Thermal conductivity (W/mK)
Carbon nanotubes	3000
Aluminium oxide	40
Water	0.60
Vegetable oils	0.18

The effect of particle volume concentration on nanofluid thermal conductivity enhancement is straightforward. Thermal conductivity enhancement increases linearly with increased particle volume concentration [13], as shown in Figure 3. Nanofluids containing carbon nanotubes (CNTs) in oil show the largest enhancement in thermal conductivity.

Other distinctive features, such as strong size-dependent [14], temperature-dependent [15], and particle shape-dependent [16] thermal conductivity were discovered during the thermal conductivity measurement of nanofluids. Since direct cooling action of the fluid is due to forced convection, it is essential to outline the importance of increased heat transfer coefficient of nanofluids. The reported experimental convective studies of nanofluids indicate a drastic increase in the convection heat transfer coefficient [11].

Next to thermal conductivity enhancement, any simultaneous rise in viscosity has to be considered. Nanofluid viscosity increases (fluidity is reduced) with nanoparticle volume concentrations, which affects fluid dynamics and causes pressure drop, which is important from the fluid application point of view. In laminar flow of nanofluids the rate of heat removal is a direct function of viscosity [17]. In case oxide nanoparticles are used, volume concentrations are normally below 5% in order to maintain moderate viscosity [10].

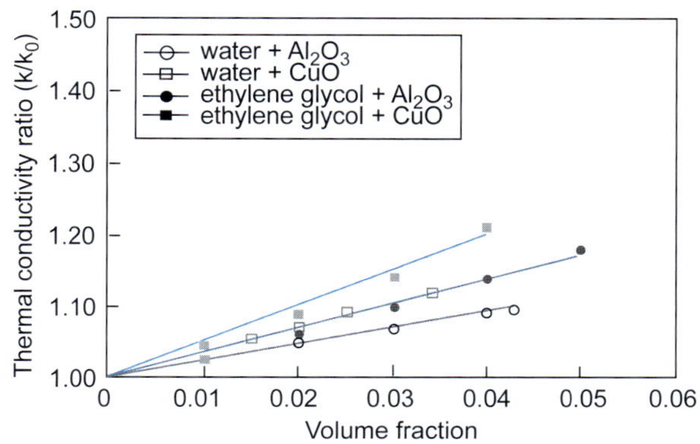

Figure 3: Enhanced thermal conductivity of nanofluids [13].

When nanofluids are used for forced convection, its boiling characteristics must be considered. Nucleate boiling is characterized by a turbulent bubble formation, where high heat transfer rates occur. Nucleate boiling is related to the wetting ability of the fluid. When the wetting is good, the bubbles detach easily. This means that the higher the wetting the higher is the heat transfer [1]. It is believed that nanoparticles increase the wettability of a base fluid [10]. While reported research in nanofluids boiling phenomenon largely differ, the value of the critical heat flux (CHF) is generally enhanced in nanofluids [10].

3 MACHINING

The two most important functions of a cutting fluid are to cool and to lubricate. For cutting fluids, emulsions (solution of oils in water) or straight oils (base oils without active additives) are generally used, depending if a machining operation requires more cooling or more lubrication. Emulsions are characterized by a higher heat transfer characteristics because of their high water content. Straight oils possess a higher degree of lubricity. Transition of the metalworking industry towards sustainable manufacturing is reflected in limiting the use of cutting fluids in machining. This limitation is beneficial for different reasons. First, it reduces OHS risks and environmental impact associated with cutting fluids. Second, it reduces machining costs. It is known that the costs related to the use of cutting fluids range from 7-17% of the total costs of the manufactured workpiece [18]. Most common method of limiting cutting fluid consumption is to use minimum quantity lubrication (MQL) systems.

Cooling or lubrication can be tailored through well designed nanofluids to different extents. In case more cooling (heat removal) is required, nano-coolants can be delivered into the cutting zone through nozzles as flood. When more lubrication (reduction of heat generation) is needed, nano-lubricants can be delivered through MQL system as droplets. In this early development stage, however, the nano-lubricants technology integration based on the principles of MQL seems to have a higher potential of applicability in comparison with the implementation of nano-coolants. The major reasons are high costs of nanofluid fabrication and bigger losses of nanofluids when cooling action is required.

3.1 Tribological aspects

Integration of nano-lubricants in machining operations has a great potential in terms of reduction of heat generation, increased tool life, and reduction of conventional cutting fluid consumption. Turning is the most critical machining operation as far as tribological aspects are concerned. Compared to milling, in turning the large contact stresses are combined with continuous chip that can wipe any lubricant films from the chip-tool contact. Three different types of lubrication circumstances have been observed in continuous chip formation during turning. Cutting fluids can affect friction at low speeds. Thermal softening can ease flow at the chip-tool contact area at high cutting speeds. At intermediate cutting speeds, solid lubricant inclusions can build-up in the chip-tool contact area. At low to intermediate cutting speeds lubricants can penetrate the lower stressed contact regions where the chip leaves the contact [19]. Capillary action, however, is doubted by Astakhov due to high pressure at the tool-chip interface and due to high temperature at the interface causing evaporation of a lubricant [1]. An alternative mechanism of lubrication action has not been suggested though.

Increased wettability (reduction in surface tension of base fluid) is important because nanofluids would be capable of better penetration into the contact interfaces and in the cracks. Friction reducing properties would probably result from flattening of the nanoparticles (non-layered) or an exfoliation of their external layers. In case of layered configuration only a few sheets may be enough to reduce friction. Nevertheless, we are expecting that exfoliated layered nanoparticles such as MoS_2, WS_2 and Graphene dispersed in base fluids would make the most efficient nanolubricants. Each individual layer between the tribo-pairs could enhance the sliding friction resulting in a reduction of cutting forces and improvement in tool life, particularly at low to intermediate cutting speeds.

The potential applications are numerous. Turning of turbine discs, for example, is one of them. Improved lubrication in machining of disc materials (e.g. titanium alloys and nickel-based alloys) is important because these materials are characterized by low thermal conductivity that cause cutting temperature and tool wear to increase even at low cutting speeds and low feed rates [20]. MQL was found to successfully limit friction and reduce tool wear in turning of nickel-based alloys [21]. In this study vegetable oil was used as a lubricant, which is common in MQL applications because of high biodegradability characteristics of vegetable oils. However, it is possible that in some machining cases the boundary film developed with vegetable oils on a tool surface is not strong enough to sustain low friction and to avoid adhesion of work material [22]. In turning with high machining load the friction stress reaches the shear flow stress of the chip material; therefore the existence of boundary film is uncertain. It was shown that the WS_2 nanoparticles appear to form a protective film allowing increased load capacity of the rubbed pairs under severe contact conditions [23]. The roughness of the surface interfaces in machining is typically larger than the size of nanoparticles. Thus we expect that that the added WS_2 (or MoS_2, graphene, IrO_2, ZrO_2) nanoparticles to vegetable oil would yield superior lubricating properties that could remarkably improve the conventional MQL systems and hence tribological properties of machining.

The applications could be extended to milling and drilling, because lubricant can reach the tool face more easily in milling and drilling than in turning. In machining of aluminum and its alloys more effective lubrication in comparison with steels is required, since these materials have highly adhesive characteristics even though they are not so hard [24].

The lubrication function in grinding has a large influence on wheel wear and heat generation. More specifically, lubricants are effective in reducing wheel dulling, in lowering the energy input and in reducing grinding forces. The application of MQL in industrial application of grinding is challenging due to problems related to wheel loading, evacuation of chips and clogging of the machine-tool guideways due to grinding residuals. One possible application could be finish grinding with CBN wheels, characterized by high thermal conductivity of the grits. Nanofluids have been recently used in two reported MQL settings. In the first application, MQL grinding of cast iron employed dispersed MoS_2 nanoparticles in mineral, vegetable, and alcohol based oils that significantly reduced the tangential grinding force and friction between the wear flats and the workpiece, increased G-ratio and improved the overall grinding performance [25]. The results of the second application indicated that graphite nanoplatelets dispersed in isopropyl alcohol significantly reduced the grinding forces as well as specific energy, and improved surface finish during surface grinding of hardened D-2 tool steel [26].

3.2 Thermal aspects

Because the temperature generated in machining with geometrically defined cutting edges (turning, milling, drilling, etc.) affects tool life and limits the productivity, the cooling action is of major interest. Nevertheless, in comparison with these operations, grinding requires much higher energy per unit volume of material removed. Virtually all of this energy is dissipated as heat at the grinding zone, which can cause different types of thermal damage to the workpiece. Thermal damage is one of the main factors which affects workpiece quality and limits the production rates, so it is especially important to understand the role of cooling in grinding.

In order for cooling to lower the grinding zone temperature to any significant degree, it is necessary for heat to be removed from within the grinding zone area. A critical factor in thermal analysis of grinding is the energy partition, which is the fraction of the grinding energy transported as heat to the workpiece. In creep feed grinding (slow workspeeds/large depths of cut) cooling at the grinding zone is effective and is a crucial process requirement. Typically, the energy partition for the creep feed grinding is only about 5% [27]. Such low energy partitions are attributed to cooling by the fluid at the grinding zone. On the other hand, the cooling in most shallow cut grinding operations (fast workspeeds/small depths of cut), by fluids is typically ineffective due to small arc length of the wheel-workpiece contact. For the shallow cut grinding with conventional abrasive wheels, the energy partition is typically 60% to 85% [27]. In this scenario the lubrication effect is more desirable than the cooling and thus creates a more favorable situation for the application of MQL.

The cooling efficiency can be quantified by the convection heat transfer coefficient of the fluid, which is an important variable in determining the energy partition [28]. Here the cooling depends upon whether convective heat transfer occurs by nucleate boiling, in which bubbles nucleated at the heated surface agitate the fluid and promote cooling, or by film boiling, in which a vapor layer that builds up between workpiece and cutting fluid insulates the heated surface and hinders cooling [29]. The transition from nucleate to film boiling is associated with a fluid burnout temperature being reached at a CHF, which is an important thermo-physical property of the fluid. Below the CHF, nucleate boiling prevails and the maximum grinding temperature is low. Above the CHF, film boiling prevails and the temperature rises catastrophically [27].

The application of nano-coolants in grinding is promising, because CHF with nanofluids is much higher than with conventional water-based cutting fluids and straight oils. Up to threefold increase in CHF was reported [30]. Consistently with this study, it was found that a significant enhancement in CHF can be achieved at modest (<0.1% by volume) Al_2O_3, ZrO_2, and SiO_2 nanoparticles concentrations [31]. A porous layer of nanoparticles on the surface interface is likely to buildup during nucleate boiling. This layer can significantly improve the surface wettability and hence cooling effectiveness. The increase in CHF should provide heat dissipation at higher grinding temperatures and thus help to avoid thermal damage where high energy is generated. From the practical point of view, care should be taken while selecting the type of nanofluid, cutting speed, wheel porosity, grain size, and flow rate, so that grinding heat fluxes do not cause film boiling which will leave the workpiece surface temperature much above that expected for the fluid.

4 SUSTAINABILITY

The sustainability concern resides at the centre of nanofluid development and integration efforts. A great deal of critical sustainability issues is focused on process-level improvements. Some of these improvements, including minimizing the use of conventional cutting fluids and reduction of the energy spent in machining do have important environmental ramifications. For the future development, however, the analysis of sustainability of nanofluids has to assess their environmental impact from a system-level perspective, using Life Cycle Assessment. The quantitative comparison between different nanofluids in terms of their potential contribution to e.g. global warming potential, resource depletion, toxicological risk, suspended particulate matter is not yet available.

The integration efforts hence require management of overall risks. In this section, however, only a brief review of potential OHS and toxic effects is given.

4.1 Operational health and safety

Occupational exposure to conventional cutting fluids has a number of well documented adverse health effects including dermatitis, respiratory disorders (aerosols), microbial infections (bacteria, yeast, fungus), and cancer [2]. In machining, aerosols (solid or liquid particles suspended in the air) are typically comprised of metallic dust and cutting fluid mist. Especially, aerosols represent a health concern because they often remain suspended in the working environment for an extended period of time, where they can be inhaled by workers.

Nanoparticles have been identified as one of the main OHS risks and priorities in a review of various national, EU and international regulations and guidelines [32]. The perceived high risk is associated with new industrial applications containing nanoparticles and with the lack of knowledge on toxicity of naoparticles leading to inappropriate protective measures, to poor risk assessment and to unfavorable workplace environment. Health effects of nanoparticles in general have been underestimated so far.

It is difficult to assess the exposure to nanoparticles. No sampler exists to specifically measure the concentration of particles below 100 nm in diameter. The development of monitoring methodologies and equipment that enables measurements for representative exposure to airborne nanoparticles is therefore important. In any case, smaller particles follow airstreams more easily than larger particles, so nanoparticles can be easily evacuated in ventilated enclosures. In addition, they are readily collected by high efficiency particulate air (HEPA) filters. Respirators with HEPA filters should be adequate protection against the inhalation of nanoparticles.

4.2 Toxicity of nanofluids

Much of the toxicology of nanomaterials is uncertain or unknown. Generally, nanosized particles are more toxic than their respective micron sized material due to increased reactivity of their surface area. Therefore, as particles become smaller, their likelihood of causing harm increases. Nanoparticles are considerably smaller than many cells in the body, which means that cells will pick up most of them. There is currently no consensus about the ability of nanoparticles to penetrate through the skin. Currently the most discussed and investigated exposure route is via inhalation. From the MQL application point of view, the lubricant may vaporize completely in machining (due to high temperature) without generating mist. However a small amount of solid nanoparticles may be airborne. Broadly, risks are increased in flood coolant applications, which generate more mist in the form of fine liquid droplets.

As noted previously, micrometer sized metallic dust may be produced by MQL machining. The available studies on toxicity of solid airborne particles point out that exposure to even small concentrations may inflict acute pulmonary effects of the workers. For example, occupational exposure to copper, vanadium, chromium, and zinc has been reported to produce adverse health effects ranging in severity from mild respiratory distress to severe lung dysfunction, and even cancer. Exposure to iron oxide has been found to cause detrimental changes in the human lungs due to deposition of inhaled particles [33]. Knowledge about the toxicity of micrometre particles may be useful in relation to making predictions about the potential toxicity of nanoparticles of the same material. There is an extensive body of evidence, which indicates that for the same material, nanometre particles are more hazardous than micrometer particles in inducing pulmonary toxicity [34]. This has been observed with different types of nanomaterials, including TiO_2, Al_2O_3, and nickel.

Any toxic effects of nanomaterials will be very specific to the type of base material, size and coatings. The risk is that some types of inhaled airborne nanoparticles deposit in the human respiratory tract and may have the ability to translocate and be distributed to other organs, including the central nervous system. Conflicting evidence is present however on the extent of translocation. A significant knowledge gap exists on a complete toxicological profile of metal oxide nanoparticles proposed for future use in machining operations. However, the available studies suggest that ZnO is highly toxic, whereas Al_2O_3 is moderately toxic and Fe_3O_4 exhibits slight toxicity at the high concentration tested. CrO_3 displays no toxicity in relation to biological responses [35]. Nanoscale TiO_2 has shown very different properties from the micron scale material in tests of lung toxicity. Moreover nanoscale TiO_2 is possibly carcinogenic to humans and can travel to the brain by way of olfactory neurons when inhaled [36]. CNTs have been tested in short term animal tests of pulmonary toxicity and the results suggest the potential for lung toxicity, though there are questions about the nature of the toxicity observed and the doses used [37]. Toxicologists point to the need for more realistic exposure methods, namely inhalation, before firm OHS regulations can be put into force. Until toxicologists determine what the safe level is, best laboratory practice would be to prevent all inhalation exposure.

5 SUMMARY

High-performance manufacturing technologies have a strong need for improved cutting fluids that can reduce heat generation or transfer heat more efficiently. In this respect nanofluids have higher potential efficiency than conventional cutting fluids. Although nanofluids offer very promising opportunities, there are still a number of technical issues on the road to their commercialization and cost effective high-volume production. Future research should lead to a development of new cutting fluids for specific applications in machining. Fundamental research is essential because the design of nanofluids can be tailored to optimize their use in each specific application.

In the future, nanofluid properties should be tested under real machining conditions. Combined experimental work and modeling is needed to better understand the underlying physics of heat convection, tribological conditions, and the interactions with workpiece materials. Tribo-chemical aspects of machining have not been addressed in this review. This will be necessary in the future, because nanofluids can drastically modify the machining interfaces, e.g. through the formation of reaction layers. Moreover, models for prediction of cutting temperatures have to consider the effect of nanofluids. This requires modeling not only on the macroscopic, but also the nanoscale level.

There is insufficient knowledge and data concerning nanoparticle detection and measurement to provide for satisfactory OHS. Moreover, there is a lack of validated methods for assessing the effects of nanoparticles on cell/organ toxicity both from the functional and structural view point. This makes it difficult to interpret available experimental results. In this regard it is important for the future development that specific tools addressing these uncertainties as far the inhalation toxicology of nanofluid aerosols is concerned are established and implemented.

6 ACKNOWLEDGMENT

This work was partially supported by the SUSTAIN-MS project funded by the EC within the Seventh Framework Programme (FP7). The authors thank Prof. Mamoun Muhammed from KTH Department of Functional Materials for his consultations on the issues related to design and synthesis of nanofluids.

7 REFERENCES

[1] Astakhov, V. P., 2006, Tribology of Metal Cutting, Tribology and Interface Engineering Series, 52, Elsevier Ltd, Oxford.

[2] Stephenson, D. A., Agapiou, J. S., 2006, Metal Cutting Theory and Practice, CRC Press, Boca Raton.

[3] Merchant, M. E., 1950, Fundamentals of Cutting Fluid Action, Lubrication Engineering, 6:163.

[4] Yu, W., France, D. M., Routbort, J. L., Choi, S. U. S., 2008, Review and Comparison of Nanofluid Thermal Conductivity and Heat Transfer Enhancements, Heat Transfer Engineering, 29/5: 432-460.

[5] Que, Q., Zhang, J., Zhang, Z., 1997, Synthesis, Structure and Lubricating Properties of Dialkyldithiophosphate-modified Mo-S Compound Nanoclusters, Wear, 209:8-12.

[6] Li, J. F., Liao, H., Wang, X. Y., Normand, B., Ji, V., Ding, C. X., Coddet, C., 2004, Improvement in Wear Resistance of Plasma Sprayed Yttria Stabilized Zirconia Coating Using Nanostructured Powder, Tribology International, 37:77-84.

[7] Aitken, R. J., Chaudhry, M. C., Boxall, A. B. A., Hull, M., 2006, Manufacture and Use of Nanomaterials: Current Status in the UK and Global Trends, Occupational Medicine, 56:300-306.

[8] Maynard, A. D., Baron, P. A., Foley, M., Shvedova, A. A., Kisin, E. R., Castranova, V., 2004, Exposure to Carbon Nanotube Material: Aerosol Release During the Handling of Unrefined Singlewalled Carbon Nanotube Material, Journal of Toxicology and Environmental Health, 67:87-107.

[9] Vollath, D., 2008, Nanomaterials: An Introduction to Synthesis, Properties and Applications, Willey-VCH, Weinheim.

[10] Das, S. K., Choi, S. U. S., Yu, W., Pradeep, T., 2008, Nanofluids: Science and Technology, John Willey and Sons, Inc., Hoboken.

[11] Yu, W., France, D. M., Routbort, J. L., Choi, S. U. S., 2008, Review and Comparison of Nanofluid Thermal Conductivity and Heat Transfer Enhancements, Heat Transfer Engineering, 29/5:432-460.

[12] Hamilton, R. L., Crosser, O. K., 1962, Thermal Conductivity of Heterogeneous Two-Component Systems, Industrial and Engineering Chemistry, 1/3:187-191.

[13] Lee, S., Choi, S. U. S., Li, S., Eastman, J. A., 1999, Measuring Thermal Conductivity of Fluids Containing Oxide Nanoparticles, Journal of Heat Transfer, 121: 280-289.

[14] Chon, C. H., Kihm, K. D., Lee, S. P., Choi, S. U. S., 2005, Empirical Correlation Finding the Role of Temperature and Particle Size for Nanofluid (Al_2O_3) Thermal Conductivity Enhancement, Applied Physics Letters, 87:153107.

[15] Das, S. K., Putra, N., Thiesen, P., Roetzel, W., 2003, Temperature Dependence of Thermal Conductivity Enhancement for Nanofluids, Journal of Heat Transfer, 125:567-574.

[16] Xie, H., Wang, J., Xi, T., Liu, Y., 2002, Thermal Conductivity of Suspensions Containing Nanosized SiC Particles, International Journal of Thermophysics, 23:571-580.

[17] Silliman, J. D., 1992, Cutting and Grinding Fluids: Selection and Application, Society of Manufacturing Engineers, Dearborn.

[18] Weinert, K., Inasaki, I., Sutherland, J. W., Wakabayashi, T., 2004, Dry Machining and Minimum Quantity Lubrication, CIRP Annals - Manufacturing Technology, 53/2:511-537.

[19] Childs, T. H. C., 2006, Friction Modelling in Metal Cutting, Wear, 260:310-318.

[20] Kitagawa, T., Kubo, A., Maekawa, K., 1997, Temperature and Wear of Cutting Tools in High-Speed Machining of Inconel 718 and Ti-6A1-6V-2Sn, Wear, 202:142-148.

[21] Settineri, L., Faga, M. G., Lerga, B., 2008, Properties and Performances of Innovative Coated Tools for Turning Inconel, International Journal of Machine Tools and Manufacture, 48:815-823.

[22] Itoigawa, F., Childs, T. H. C., Nakamura, T., Belluco, W., 2006, Effects and Mechanisms in Minimal Quantity Lubrication Machining of an Aluminum Alloy, Wear, 260:339-344.

[23] Rapoport, L., Nepomnyashchy, O., Lapsker, I., Verdyan, A., Moshkovich, A., Feldman, Y., Tenne, R., 2005, Behavior of Fullerene-Like WS_2 Nanoparticles Under Severe Contact Conditions, Wear, 259:703-707.

[24] Min, S., Inasaki, I., Fujimura, S., Wada, T., Suda, S., Wakabayashi, T., A Study on Tribology in Minimal Quantity Lubrication Cutting, 2005, CIRP Annals - Manufacturing Technology, 54/1:105-108.

[25] Shen, B., Minimum Quantity Lubrication Grinding Using Nanofluids (Doctoral dissertation), 2008, University of Michigan.

[26] Alberts, M., Kalaitzidou, K., Melkote, S., 2009, An Investigation of Graphite Nanoplatelets as Lubricant in Grinding, International Journal of Machine Tools and Manufacture, 49:966-970.

[27] Malkin, S., Guo, C., 2007, Thermal Analysis of Grinding, CIRP Annals - Manufacturing Technology, 56/2:760-782.

[28] Jin, T., Stephenson, D. J., 2008, A Study of the Convection Heat Transfer Coefficients of Grinding Fluids, CIRP Annals - Manufacturing Technology, 57/1: 367-370.

[29] Howes, T. D., Neailey, K., Harrison, A. J., 1987, Fluid Film Boiling in Shallow-cut Grinding, CIRP Annals - Manufacturing Technology, 36/1:223-226.

[30] Vassallo, P., Kuman, R., Amico, S. D., 2004, Pool Boiling Heat Transfer Experiments in Silica-Water Nanofluids, International Journal of Heat and Mass Transfer, 47:407-411.

[31] Kim, S. J., Bang, I. C., Buongiorno, J., Hu, L. W., 2007, Surface Wettability Change During Pool Boiling of Nanofluids and its Effect on Critical Heat Flux, International Journal of Heat and Mass Transfer, 50:4105-4116.

[32] Rial-González, E., Copsey, S., Paoli, P., Schneider, E., 2005, Priorities for Occupational Safety and Health Research in the EU-25, European Agency for Safety and Health at Work, Luxembourg.

[33] Sutherland, J. W., Kulur, V. N., King, N. C., von Turkovich, B. F., 2000, An Experimental Investigation of Air Quality in Wet and Dry Turning, CIRP Annals - Manufacturing Technology, 49/1: 61-64.

[34] Health and Safety Executive, Health Effects of Particles Produced for Nanotechnologies, 2004, Retrieved July 1, 2010, from http://www.hse.gov.uk.

[35] Jeng, H. A., Swanson, J., 2006, Toxicity of Metal Oxide Nanoparticles in Mammalian Cells, Journal of Environmental Science and Health, 41/12:2699-2711.

[36] Wang, J., Liu, Y., Jiao, F., Lao, F., Li, W., Gu, Y., Li, Y., Ge, C., Zhou, G., Li, B., Zhao, Y., Chai, Z., Chen, C., 2008, Time-dependent Translocation and Potential Impairment on Central Nervous System by Intranasally Instilled TiO_2 Nanoparticles, Toxicology, 254/1-2:82-90.

[37] Hurt, R. H., Monthioux, M., Kane, A., 2006, Toxicology of Carbon Nanomaterials: Status, Trends, and Perspectives on the Special Issue, Carbon, 44/6: 1028-1033.

Indirect Tool-Wear Maps for Tool Condition Monitoring in Dry Metal Drilling Operations

A.H. Ammouri, R.F. Hamade

Department of Mechanical Engineering, American University of Beirut, Lebanon

Abstract

To avoid damage to work and/or machine, real-time tool condition monitoring is necessary in automatic and sustainable manufacturing operations. In particular, metal machining with NC machine tools can benefit handsomely from the identification of dull tools in real-time so that they can be replaced. This requirement is especially true in dry (sustainable) drilling operations where heat buildup represents a major challenge.

In this work, quantitative maps of indirect tool-wear of chisel drills undergoing dry machining are charted based only on transducers reporting electrical current measurements from machine (spindle and feed drives) motors. Associated with the maps are qualitative descriptions of the various modes of tool-wear afflicting the drill tools. Based on these tool-wear maps, a novel wear criterion is developed that rely on the % increase in motor (spindle and feed drive motors) RMS current values and is dubbed the Current Rise Index (CRI). For verification, this index is found to positively track the corresponding increase in cutting forces. Utilizing this index, an implementation example is presented in this paper by which a real-time tool monitoring of chisel drills is achieved by inputting the CRI along with the cutting parameters to an Artificial Neural Network (ANN) which yielded good tool condition predictions.

Keywords:

Tool condition monitoring, Current sensors, Dry metal drilling, Sustainability, Tool-wear maps, ANN.

1 INTRODUCTION

Wear of cutting tools is an integral part of the machining process since tool condition affects the dimensional accuracy and surface quality of newly machined surfaces. Although most modes of tool wear are progressive in nature, if not replaced, dull tools are likely to fail in a sudden and catastrophic manner causing damage to both work and machine. Therefore, the successful implementation of tool condition monitoring (TCM) becomes critical to the success of fully automated cutting operations. Over the years, many techniques have been used to monitor and detect tool wear in metal cutting in general and, more relevant to this work, in drilling operations. These involve direct as well as indirect measurements of several types of sensors during drilling that correlate to the condition of the tool. While direct measurements rely on visual and computer vision methods to get dimensional values, indirect measurements include a host of techniques including: sonic and ultrasonic vibration/noise detection [1-4], measurements of cutting forces [5-7], and spindle motor / feed drive current measurements [8-11].

For intelligent prediction of tool wear, the artificial neural network, ANN, method is often used. Typically, the network is trained in the first stage where sensor data is fed to the diagnostics software. Once the software has been 'taught' to distinguish a good (sharp) tool from a partially worn one and from a completely dull one, the diagnostic capabilities of the software become automatic resulting in the real-time correct identification of the drilling tool condition by the software. Such an approach has been utilized in [12] where the root mean square (RMS) value of the spindle motor current was used as input to a multilayer neural network.

In this paper, we utilize sensors to instantly measure RMS current of 1) drilling feed drive motor and 2) spindle motor. Further, we establish tracking between these current measurements with corresponding force measurements and, more importantly, with measured tool-wear at various stages of wear. The resulting wear maps are based on a simple measure, dubbed the 'Current Rise Index (CRI)' and which is derived from the increase in RMS current values.

2 EXPERIMENTAL PROCEDURE

In the performed experiments, 2 brands of 10mm HSS classical twist drills were used: TERA (Tera Autotech Corporation, No. 1 Industrial Park 7 Rd., Tachia, Taichung Hsien, Taiwan, R.O.C.) and IRWIN (Irwin Tools, 92 Grant Street, Wilmington, OH, USA). The drilled material was low carbon steel and the 66-holes pattern shown in Figure 1 was used in all drilling trials.

The following cutting conditions were used:

- Drill starts at a height of 3 mm above the workpiece
- Dry drilling depth of 3 mm to ensure full lip engagement
- Feeds and speeds as in Table 1
- Air (not liquid coolant) was used to cool the tool during the cut and during travel

Cutting parameters in the test matrix are chosen so that cutting speeds range from 40 to 55 m/min (1273 to 1750 RPM) and feeds from 0.03 to 0.2 (mm/rev). The resulting test matrix is shown in Table 1 where for example, test cell C2 corresponds to spindle speed of 1750 RPM and feed of 0.06 mm/rev, respectively.

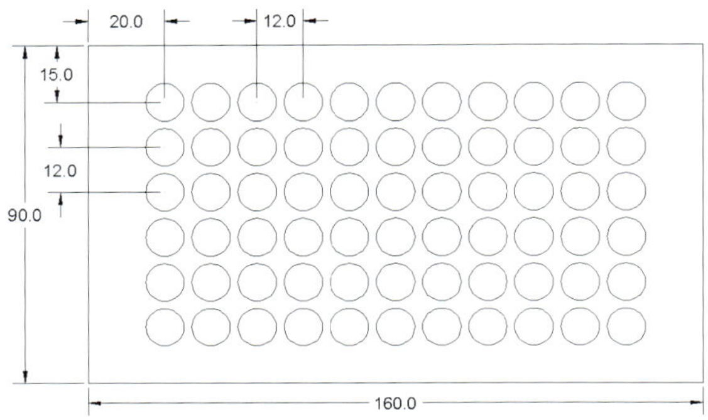

Figure 1: Pattern used in drilling into the low carbon steel

Table 1: Test matrix showing experimental spindle speeds (RPM) and feeds (mm/rev)

		Feed (mm/rev)			
		0.03	0.06	0.1	0.2
Speed (RPM)	1273	A1	A2	A3	A4
	1592	B1	B2	B3	B4
	1750	C1	C2	C3	C4

Current value measurements were carried out by using hall-effect current transducers that were tapped to individual lines of both the spindle motor and the Z-drive motor of a CNC HAAS 5-axis vertical machining center. G-codes were written for each cell in the test matrix. A S22P series current transducer from TAMURA (http://www.tamuracorp.com/clientuploads/pdfs/engineeringdocs/S22PXXXS05.pdf) was used for the spindle motor with a 15A rated current and a LTS6-NP from LEM (http://www.lem.com/hq/en/component/option,com_catalog/task,displaymodel/id,90.37.09.000.0/ was used for the Z-drive motor with a 6A current rating. Both transducers output a 2.5V biased voltage with a 2.5±0.625V full current output voltage.

Signals from the current transducers were acquired by an NI USB-6251 Legacy DAQ device (http://sine.ni.com/nips/cds/view/p/lang/en/nid/209213) into user-friendly LabVIEW software programmed especially for this application. Samples are acquired at a 10K Hz sampling frequency in 2KS batches. The output of the software is raw voltage that represents the cutting cycle current. This raw data is then carefully analyzed to extract the RMS of each cutting trial. Force measurements were collected using Kistler's CompacDyn 3-Component Dynamometer (Type 9254). The Kistler 5070A charge amplifier acquires and amplifies the signal emanating from the dynamometer which is then processed through the Dynamometer's Dynoware data acquisition software (at 1000 Hz).

3 EXPERIMENTAL RESULTS

The results of the experimental study are presented in this Section.

3.1 Flank-wear

As drilling progresses, so do tool wear causing the drill to go from a 'sharp' condition, through a partially worn condition, and ultimately to a dull 'worn out' condition. Catastrophic failures were observed when drilling commenced using dull tools. Wear in drills afflict all regions where metal comes in contact with the drill. However, of the many modes of wear investigated, flank wear, *VB*, is by far the most popular in the metal cutting literature. While many methods exist to quantify flank wear, including taking the average of wear distance at locations along the lip, this work utilizes the maximum *VB*, VB_{max}, at the point where the lip meets the margin and where cutting speeds are largest [13].

Figure 2 shows a TERA twist drill at 3 different stages of progressive wear (a) sharp condition and (b) and (c) at 2 different stages of progressively more worn states of wear with all conditions corresponding to test cell C2. Figure 3 is a typical plot of progressive flank wear of the drill. The three circles on the plot correspond to the 3 stages of wear progression shown in Figure 2.

(a) (b) (c)

Figure 2: TERA twist drill shown at (a) sharp condition (b) and (c) at 2 different stages of progressive wear; test case C2.

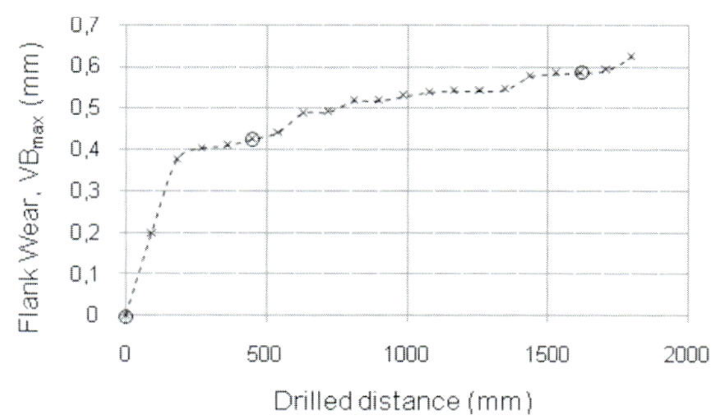

Figure 3: Typical plot of progressive flank-wear of the TERA twist drill; note that circles correspond to images in Figure 2 (test case C2).

3.2 Z-Drive Current

While both torque and thrust force progressively increase as the drill tool undergoes wear [7], thrust force has been found to be more sensitive to drill-wear [14]. To eliminate the usage of the expensive (and machine-tool unnatural) force measurement setup, current measurements of the Z-drive motor line are considered. The LTS6-NP current transducer produces real-time current values with a 400 ns response time. A conditioned calibrated signal of the Z-drive current is shown in Figure 4 superimposed on another signal generated from another drilling test in air (test case B3). The frequency of this particular signal is equal to 3.5 Hz or one full motor cycle = 0.28 s (the value of which depends on the Z-motor driver output for the feed of test case B3 = 0.1 mm/rev). The shown signal corresponds to a tool with drilling history of 7.524 m. Although the two plots are virtually super imposable during the approach stage, significant deviation is recorded during drill engagement. Note that the reason why less current is needed to drive the motor during the drilling stage as compared with during air-drilling is due to the fact that thrust force acts in an opposite direction to tool motion (and gravity). To get a quantitative value of the current, the RMS of the signal is taken into account throughout this paper. In order to remove the bias due to the current consumed from driving the spindle assembly, the RMS of each drilling cycle is subtracted from the current exerted during the air-drill stage for the same drilling conditions.

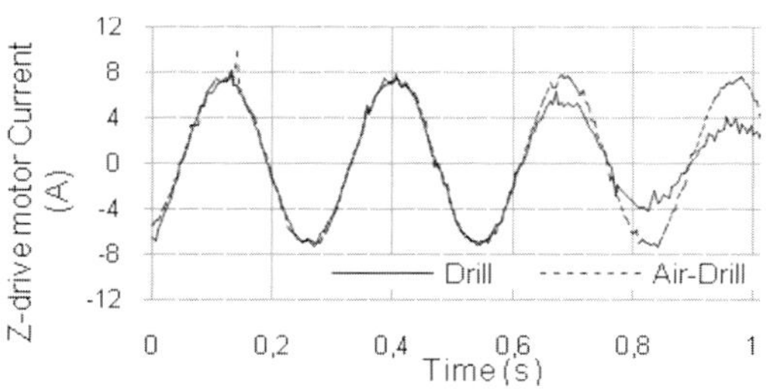

Figure 4: Z-drive motor's true current for a typical signal (one full motor cycle = 0.28 s) generated from an actual drilling test superimposed on another corresponding to an air-drilling test (both for the same parameters: test case B3, drilled distance 7.524 m).

Figure 5 shows the percent increase in measured thrust force values taken over many drilling cycles are found to track fairly accurately that of the percent increase in Z-drive current. Shown in the figure are two superimposed plots of normalized thrust force and Z-drive current over a drilling distance of 1 m (TERA drill; test case C2). Maximum flank wear VB_{max} (mm) is also shown on the secondary axis. The figure indicates that the increase in Z-drive current values is as sensitive as thrust force in tracking flank-wear progression. It can be seen that both sensors (thrust force and current) increase along with progressing tool-wear. The amount of current consumed by the Z-drive motor can be scaled to represent the thrust force.

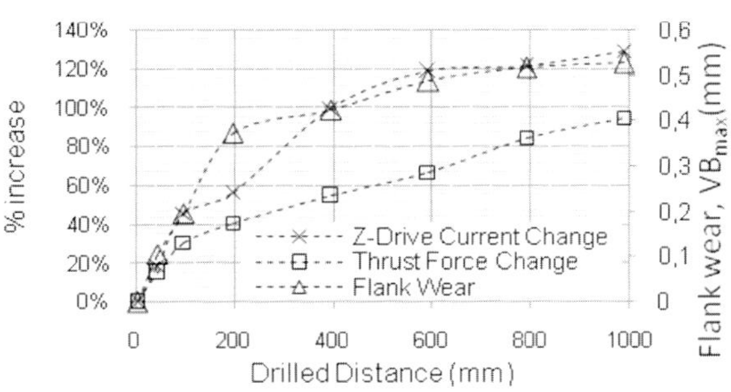

Figure 5: Normalized % change in thrust force and Z-drive current vs. drilled distance for a TERA drill (test case C2).). Progressive flank-wear, VB_{max}, values are also plotted on the secondary axis.

3.3 Spindle-Motor Current

Similar to the Z-drive motor, spindle-motor current values were collected by the S22P transducer at a rate of 10 Ks/s. Figure 6 shows a sample of the collected raw data that corresponds to the same test cell of Figure 4. As wear progresses, the RMS peak value of the spindle-motor current increases given that the motor draws more current to compensate for additional tool-wear.

The effect of drill-wear on the spindle motor current can be observed in Figure 7 which plots the measured progressive VB_{max}, spindle current (RMS), and Z-drive current versus the drilled distance (mm) for a TERA drill (test case C2).

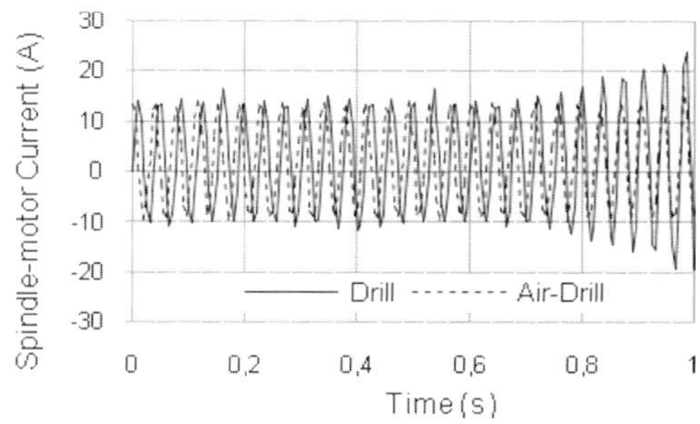

Figure 6: Spindle-motor true current for a typical signal (one full motor cycle = 0.28 s) generated from an actual drilling test superimposed on another corresponding to an air-drilling test (both for the same parameters: test case B3, drilled distance 7.524 m).

3.4 Summary of raw data

For each cell in the test matrix shown in Table 1 a plot that shows the variation of spindle motor current and Z-drive current with the progressive wear is produced. This plot will help in generating the tool-wear maps and the ANN input vector.

Figure 7 is a data summary plot for a TERA drill which shows the increase of the normalized current for Z-drive and spindle motors over drilled distance of 1800 mm (test case C2). Also co-plotted are the maximum flank wear, VB_{max}, values.

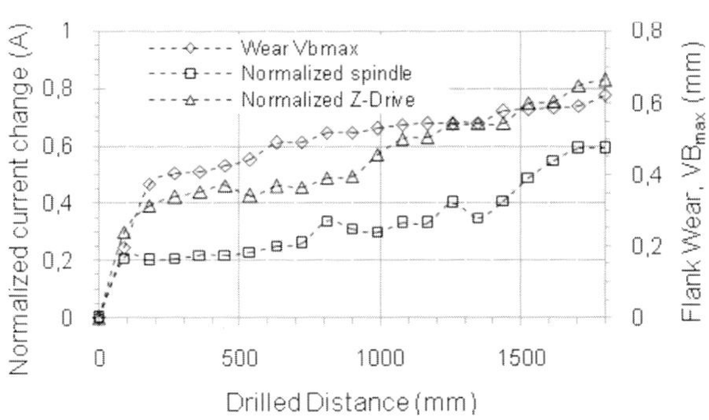

Figure 7: Normalized spindle current (RMS) and Z-drive current vs. drilled distance for a TERA drill (test case C2). Progressive flank-wear, VB_{max}, values are also plotted on the secondary axis.

4 TOOL-WEAR MAPS

A novel representation of progressive tool-wear is presented below based on the acquired data from the current transducer signals. The resulting map is a polar plot based on a measure dubbed here as the Current Rise Index (*CRI*) calculated as the square root of the percent increase in the RMS current values of both transducers. This measure combines both signals in one value that represents the change in current values according to

$$CRI = \sqrt{\left(\Delta I_{Spindle}\right)^2 + \left(\Delta I_{Z-motor}\right)^2} \tag{1}$$

Where:

$\Delta I_{Spindle}$ is the % change in spindle motor current

$\Delta I_{Z-motor}$ is the % change in Z-drive motor current

The percentage change in the current signal is found by subtracting the RMS value of the trial being considered from the air-drill RMS value and dividing the result by the RMS value of the first trial as follows

$$\Delta I_i = \frac{I_i - I_1}{I_1 - I_0} \tag{2}$$

where:

I_i is the RMS value of the test considered

I_0 is the RMS value of the air-drill test

I_1 is the RMS value of the first (wear-free) test

The representation in (2) eliminates the bias of the current signal that comes from the machine dynamics and normalizes the signal such that the first reading is null and the consecutive values are calculated relative to the first trial.

One interesting aspect about *CRI* is revealed by examining (1) which combines 2 complimentary aspects of force generation required for drilling: Z-drive current which is responsible for the feed force action of the drill and spindle current which corresponds to the torque exerted by the spindle. This combination offers wear criteria sensitivity towards both ends of the spectrum where detecting of tool wear affecting either thrust or torque will be possible. It is found in this work that some forms of wear have greater effect on thrust or torque which makes the criteria of practical use in drilling operations irrespective of the specific tool-mode.

Table 2 describes the different ranges of *CRI* and the corresponding tool conditions. For example, a *CRI* value of 1 indicates a 100 percent increase in drilling power-consumption from the starting value. An increase by more than 120 percent means that the tool is dull and should be changed. The values and ranges shown in the table were based on two criteria (1) the drastic increase in power consumption of both motors and (2) the high tool-wear value. Such *CRI* target classifications may equally serve as a measure of wear which lends itself to easy monitoring by human operator, automatic machine tool, or for AI methods such as neural networks as will be shown in Section 5.

Table 2: Wear class criteria, the corresponding CRI values and ANN target value

CRI value			
Min	**Max**	**Tool-Wear Class**	**ANN target**
0	0.29	Sharp – Class 1	1
0.3	0.64	Slightly Worn – Class 2	0.75
0.65	1.19	Workable – Class 3	0.5
1.2	-	Dull – Class 4	0

On a polar wear map, the CRI value and the drilled distance represent the magnitude (ρ) and the angle (θ), respectively, of any point plotted. Figure 8 shows such a tool-wear map corresponding for select cells from the test matrix (Table 1). Both tool brands were tested to show the difference between a low-quality tool and another tool of better quality. A jerk in a wear-map curve indicates a sudden breakage in the tool. The polar wear plot of a desirable tool performance may behave similar to that shown for IRWIN test case B4 where the *CRI* increases smoothly and gradually with low wear rates as the drilling proceeds. Alternately, a bad performance example is the TERA test case B2 where the *CRI* index increased drastically after only a few drills. Although C1 IRWIN (b) and C1 IRWIN (a) are run for the same test case, the jerk seen in he plotted polar curve in C1 IRWIN (a) performance is due to a catastrophic failure in one of the tool flanks.

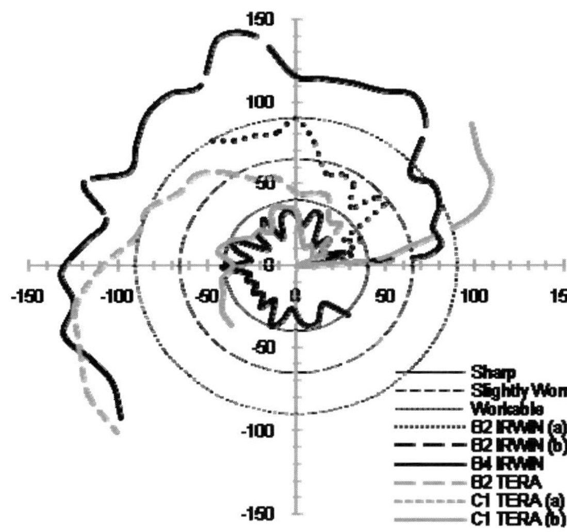

Figure 8: Tool wear maps for select cells of the test matrix performed for drills from both suppliers: TERA and IRWIN.

5 TOOL WEAR PREDICTION WITH ANN

From above, machine tool's motor sensor signals were found to increase as tool wear increased. But the mathematical functional relationship between these signals and tool wear is unclear and is challenging to represent in equation form. In order to combine the signal data from the sensors used and have a functional classification or decision making system, some form of learning or pattern recognition needs to be implemented, in which the machine or program 'learns' how to classify the wear output given the signals inputs. This is done by comparing the values of certain signal inputs to predefined targets. One solution to such a problem is the use of an artificial neural network (ANN) where a neural network emulates a biological neural system in the sense that in a sort of feedback mechanism or learning, optimizes its neural connections or weights to attain a specified target output. In this work, the dual sensor data is integrated into the ANN software module in MATLAB (where a specific type of networks called multiple-layer-perceptron (MLP) network was implemented). Inputs to the network are:

1. Cutting parameters: spindle speed and feed
2. Number of holes (drilled distance)
3. Main spindle motor current (RMS)
4. Z-drive motor current (RMS)

The only desired output is the tool flank wear. The cutting parameters and sensor features represent the input vector according to Equation (3) and the MLP with one hidden layer and an output layer was trained to output the current state or condition of the tool according to Table 2:

$$p = \left(D, N, f, \Delta I_{spindle}, \Delta I_{z-motor}\right) \qquad (3)$$

where

D is the drilled distance

N is the rotational speed

f is the feed rate

This novel wear criterion does not use wear distance as of itself to describe tool condition deterioration but rather uses indirect indicators derived from the machine tool's actual motors current consumption.

MLP network was implemented and tested with 20 neurons in the hidden layer and the 4 'wear class output' criteria. This network used scaled conjugate gradient back-propagation as its learning algorithm. After training this network with data samples corresponding to the experiments in Table 1, the network was given 40 sample inputs and the network outputs were compared to the target classes. The resulting 'confusion (or matching) matrix' which is defined as one that compares the 'wear output class' with 'target class' is shown in Table 3. The results indicate good agreement with the experimental assessment of the state of wear for the cases used to test the network. Out of the 40 test samples, only 3 misclassifications were noticed which corresponds to 92.5% prediction accuracy.

6 SUMMARY

A novel criterion, dubbed the Current Rise Index (CRI), based only on the % increase in the machine tool's spindle and feed motors RMS current values due to tool wear has been presented. Based on this criterion, novel tool wear maps have been introduced which lend themselves to easy decision making by machine operators' visual inspection and monitoring. These maps were demonstrated to be equally useful for automated decision making purposes where artificial neural networks, ANN, were able to make correct determinations about tool (here, drill) condition.

Table 3: Confusion matrix of a test run of 40 samples

Output Class \ Target Class	1	2	3	4	
1	9 22.5%	0 0.0%	0 0.0%	0 0.0%	100% 0.0%
2	1 0.25%	10 25%	2 0.5%	0 0.0%	76.9% 23.1%
3	0 0.0%	0 0.0%	8 20%	0 0.0%	100% 0.0%
4	0 0.0%	0 0.0%	0 0.0%	10 25%	100% 0.0%
	90% 10%	100% 0.0%	80% 20%	100% 0.0%	92.5% 7.5%

7 ACKNOWLEDGEMENTS

The authors wish to acknowledge the financial support of the University Research Board (URB) of the American University of Beirut. The first author also gratefully acknowledges the support of Consolidated Contractors Company through the CCC Doctoral Fellowship in Manufacturing.

8 REFERENCES

[1] M. Mathew, P. Pai, P., and L. Rocha, 2008, An Effective sensor for tool wear monitoring in face milling: acoustic emission. Sadhana, 33/3.

[2] U. Schehl, 1991, Werkzeuguberwachung mit Acoustic-Emission beimDrehen, Fra¨sen und Bohren, Aachen.

[3] S.R. Hayashi, C.E. Thomas, D.G. Wildes, 1998, Tool break detection by monitoring ultrasonic vibrations, Annals of the CIRP 37 (1) 61–64.

[4] K. Kutzner, U. Schehl, 1998, Werkzeuguberwachung von Bohrernkleinen Durchmessers mit Korperschall-sensoren, Industrie Anzeigner 110 (82) 32–33.

[5] K.J. Lee, T.M. Lee, and M.Y. Yang, 2007, Tool wear monitoring system for CNC end milling using a hybrid approach to cutting force regulation, International Journal Advance Manufacturing Technology, 32:8-17.

[6] S. Braun, E. Lenz, C.L. Wu, 1982, Signature analysis applied to drilling, Journal Mechanical Design, Transactions of the ASME, 104: 268–276.

[7] S.C. Lin, C.J. Ting, 1995, Tool wear monitoring in drilling using force signals, Wear 180 (1-2) 53–60.

[8] E. Jantunen, H. Jokinen, 1996, Automated On-line Diagnosis of Cutting Tool Condition (Second version), International Journal of Flexible Automation and Integrated Manufacturing 4 (3-4) 273–287.

[9] Y.-C. Chang, K.-T. Lee, and H.-Y. Chuang, 1995, Cutting force estimation of spindle motor, J. Contr. Syst. Technol., 3/2:145-152.

[10] X. Li, 2005, Development of current sensor for cutting force measurement in turning, IEEE Transactions on Instrumentation and Measurement, 54(1):289-296.

[11] A. Noori-Khajavi, 1992, Frequency and time domain analyses of sensor signals in a drilling process and their correlation with drill wear, PhD Thesis, Oklahoma State University, Stillwater, OK.

[12] K. Patra, S.K. Pal, and K. Bhattacharyya, 2007, Artificial neural network based prediction of drill flank wear from motor current signals, Applied Soft Computing, 7:929-935.

[13] R.F. Hamade, C.Y. Seif, and F. Ismail, 2006, Extracting cutting force coefficients from drilling experiments, International Journal of Machine Tools & Manufacture, 46:387-396

[14] S.E. Oraby and D.R. Hayhurst, 2004, Tool life determination based on the measurement of wear and tool force ratio variation, International Journal of Machine Tools & Manufacture, 44:1261-1269.

Optimization of Superplastic Forming; Effects of Interfacial Friction on Variable Strain Rate Forming Paths

M.I. Albakri, M.K. Khraisheh
Masdar Institute of Science and Technology, Abu Dhabi, UAE

Abstract
Superplastic forming technique provides a unique solution for meeting the growing demand on light-weight materials for transportation applications. However, being a rate controlled process, confined to low strain rates, makes it relatively slow and unfavorable for mass production applications. In this paper, an optimal variable strain rate forming path based on a multi-scale stability criterion is developed, aiming at reducing forming time while maintaining the integrity of the formed parts. This criterion accounts for geometrical instabilities as well as microstructural features. The effects of friction distribution at the die-sheet interface on the deformation stability are also investigated; aiming at identifying the optimal lubrication conditions. These results are demonstrated through finite element simulations of the forming process of bipolar plates flow-field channels used for polymer electrolyte membrane fuel cells.
Keywords
Bipolar Plates; Interfacial Friction; Lightweight Alloys; PEM Fuel Cells; Superplastic Forming; Variable Strain Rate.

1 INTRODUCTION

Environmental issues; global warming and climate change in particular, have been attracting more attention in the last few years. Countries all over the world are requested to cut down their greenhouse gas (GHG) emissions, especially CO_2, in order to meet the international standards. In most of the developed countries, transportation sector has always been a major source of GHG emissions, contributing to about 29% of the total emissions in the European Union and United States [1]. With the growing environmental awareness, these numbers have become a source of a continuously growing pressure on transportation industry, the automotive in particular, to reduce fuel consumption along with the levels of exhaust gas emissions.

To achieve the required emissions reduction, cutting down the mass of the vehicles is one of the most influential and least costly methods. However, customers' increasing demand for safer, more powerful and luxurious vehicles has been adding more weight to the various components of the vehicles, making the realization of lighter cars more difficult and challenging [2]. Thus, the targeted weight reduction will not be feasible on the desired scales without the extensive use of light yet strong-enough materials.

In addition to introducing lighter vehicles, exhaust emissions can be significantly reduced by introducing new, clean and renewable power sources to replace the conventional internal combustion engines. Among the different sources available, fuel cells, proton exchange membrane fuel cells (PEMFC) in particular, are the primary candidates for transportation applications. PEMFCs provide a promising, clean, and efficient energy solution. However, a number of critical issues such as cost, size, weight, and overall efficiency have to be addressed for the widespread use of PEMFCs.

PEMFC consists of three main components; the membrane electrode assembly (MEA), the gas diffusion layer (GDL), and the bipolar plates. Among these different parts, the bipolar plates account for more than 80% of the weight, more than 40% of the cost and occupy a large portion of the volume of the fuel cell [3]. Therefore, introducing new designs and technologies aiming at improving the performance of the bipolar plates and reducing their weight, size and cost are very much needed.

With their low densities, lightweight materials (LWM), such as titanium, aluminium and magnesium alloys are considered as a key solution towards achieving the desired weight reduction. These alloys are generally 43–78% lighter than conventional steels [2]; a fact accentuating the great weight-saving potentials promised by these materials. Unfortunately, due to their limited formability, the use of LWM for transportation applications is pretty much limited. However, several lightweight alloys exhibit extraordinarily enhanced ductility at elevated temperatures; a phenomenon known as superplasticity. This phenomenon has gained a lot of interest over the past few decades, and was put into practice to form several products out of these alloys by means of the Superplastic Forming (SPF) technique [4].

SPF is a near net shape forming process that offers many advantages over conventional forming operations including weight reduction, greater design flexibility, and the ability to form hard-to-form metals into complex shapes. However, low production rate and the non-uniformity of the formed parts are the main obstacles hindering the widespread use of SPF.

To overcome these problems, several studies have been published investigating the possibility of speeding up the forming process by designing variable strain rate forming paths based on different stability criteria [5-7]. Despite of their success, the reduction in forming time achieved with these methods has always been constrained by localized thinning taking place at certain areas of the sheet. Moreover, all of the stability criteria currently available do not account for the effects of frictional forces acting at the die sheet interface. Such forces were found to have crucial effects on the way the metal flows during the forming process [8].

In this paper an optimal variable strain rate forming path based on a multi-scale stability criterion is developed, aiming at reducing forming time while maintaining the integrity of the formed parts. First, a microstructural based constitutive material model that describes the superplastic behavior of the 5083 fine-grained superplastic aluminum alloy is discussed. A multi-scale stability criterion is then developed, and based on it, a variable strain rate forming path is designed. This criterion accounts for geometrical instabilities as well as microstructural features. Moreover, the effects of friction distribution at the die-sheet interface on deformation stability are addressed through a number of FE simulations.

2 MATERIAL CONSTITUTIVE MODEL

A general constitutive model based on the continuum theory of viscoplasticity has been previously applied in modeling different superplastic materials including Al, Mg and Cu alloys [5],[9],[10]. A Similar model is used in this work to describe the superplastic behavior of fine grain AA5083 sheet at 450°C. This model accounts for microstructural features including grain growth and cavitation evolution, as given by the following equation:

$$\dot{\bar{\varepsilon}} = \frac{K}{d^P}\left(\frac{\bar{\sigma}}{1-f_a}\right)^{1/m} \qquad (1)$$

where ($\dot{\bar{\varepsilon}}$) is the effective strain rate, ($\bar{\sigma}$) is the effective flow stress, (m) is the strain rate sensitivity index, (d) is the average grain size, (P) is the grain growth exponent, (f_a) is the area fraction of voids, and (K) is a material parameter defined as a function of the effective strain rate.

To account for the change in microstructure during deformation, evolution equations for grain size (d) and area fraction of voids (f_a) are used. For the former, a simple linear grain growth model similar to that Suggested by Caceres and Wilkinson[11] is used here, as given by Eq.2:

$$d = d_0 + c\bar{\varepsilon} \qquad (2)$$

Where ($\bar{\varepsilon}$) is the effective strain, (d_0) is initial grain size, and (c) is a material constant.

Due to the large deformation associated with superplastic forming, cavitation is primarily controlled by the plastic flow of the surrounding matrix [12]. Such plasticity controlled growth of non-interacting cavities is described by the following equation:

$$f_a = f_{a0}\exp(\psi\bar{\varepsilon}) \qquad (3)$$

Where (f_{a0}) is the initial area fraction of voids, and (ψ) is the void growth parameter. A fitting process is used to obtain the values of the different parameters for AA5083 material model, see table (1). In the resulting single-term phenomenological constitutive model, (P) and (K) are functions of the effective strain rate ($\dot{\bar{\varepsilon}}$), and thus (*m*) is only equal to the strain rate sensitivity at the start of deformation (i.e. $\bar{\varepsilon} = 0$). This model has proved to be capable of predicting the deformation behavior of this superplastic alloy under different loading conditions.

Table 1: Values of the material parameters [9]

Parameter	Value
m	0.5
P	$-3 - 0.43\ln(\dot{\bar{\varepsilon}})$
$\ln(K)$	$-15.251 - 0.202\ln(\dot{\bar{\varepsilon}}) + 0.0346\ln^2(\dot{\bar{\varepsilon}})$
d_0	8.0
c	2.5
f_{a0}	1.25%
ψ	1.5

3 STABILITY ANALYSIS

To design an optimal variable-strain-rate forming path, the effective strain at which the onset of instability takes place needs to be determined as a function of the forming strain rate. To do so, deformation stability of the long cylindrical bar shown in Figure 1 is analyzed. This part contains a local non-uniform region where necking is more likely to occur. The one dimensional modified version of Marciniak and Kuczynski analysis [13], introduced by Hutchinson and Neale [14], is applied here. This stability criterion is coupled with the constitutive model described in the previous section so as to account for the effects of microstructural evolution on the deformation stability. The true stress and strain are defined using the true load bearing area reduced by cavitation. The initial fraction of non-uniformity is defined as:

$$\eta = \frac{A_u^0(1-f_{a,u}^0) - A_n^0(1-f_{a,n}^0)}{A_u^0(1-f_{a,u}^0)} \qquad (4)$$

Where (A^0) and (f_a^0) are the initial cross sectional area and the initial area fraction of voids, respectively. The subscripts (u) and (n) stand for the uniform and non-uniform regions, respectively.

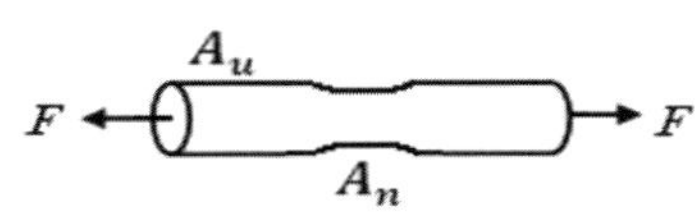

Figure 1: A long cylindrical bar containing a geometric non-uniformity and subjected to a time varying load.

Applying the nonlinear long-wavelength stability analysis, the following equation is obtained:

$$\bar{\sigma}_n e^{\bar{\varepsilon}_n} = \bar{\sigma}_u e^{\bar{\varepsilon}_u}\frac{1}{1-\eta} \qquad (5)$$

Combining the constitutive material model (Equations 1-3) into Eq. 5, a relation between the increments of strain in the local non-uniform and uniform regions is obtained as follows:

$$\left(\frac{\dot{\varepsilon}_n d^P}{k}\right)^m (1-f_{a,n})e^{-\bar{\varepsilon}_n} = \left(\frac{\dot{\varepsilon}_u d^P}{k}\right)^m \frac{(1-f_{a,u})}{1-\eta}e^{-\bar{\varepsilon}_u} \qquad (6)$$

This equation is numerically integrated to calculate the strain ratio between the non-uniform and uniform sections as shown in Figure 2. Instability can be defined as the point at which the slope of the strain ratio curve approaches infinity. At this point, deformation will be localized in the non-uniform region, leading to necking and failure there. A more conservative approach can be adopted to design the variable strain rate forming path by defining a certain value of the strain ratio at which the deformation is considered to be unstable. In this study, a strain ratio of 1.5 is selected to define the critical strain for each forming strain rate shown in Figure 2. These critical-strain, strain-rate pairs form the basis for constructing the optimal variable strain rate paths that are used in the upcoming sections.

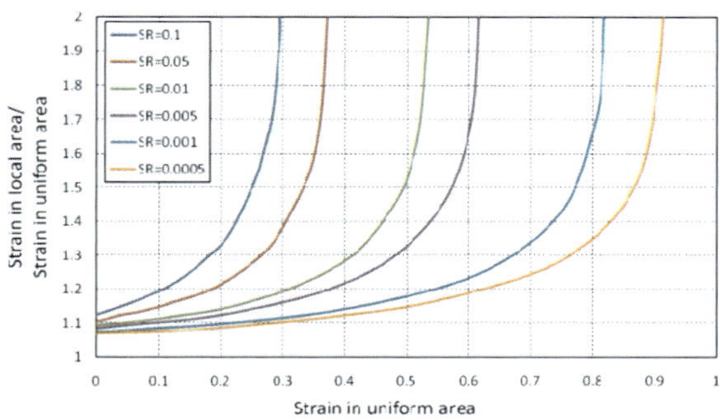

Figure 2: Ratio of strain in local area to the strain in uniform area as a function of strain in uniform area for different strain rates.

4 FINITE ELEMENT MODEL

The forming process of AA5083 sheet into PEMFC bipolar plate is addressed here. This bipolar plate has a serpentine flow field; as shown in Figure 3.a, with the dimensions given in Table 2. Based on a previous study, these dimensions were found to facilitate reactants mass transfer between the bipolar plate's flow field and the adjacent GDL resulting in an enhanced cell performance [15]. For such long channels, a state of plane strain exists in most of the part. This allows simplifying the simulation domain to a two-dimensional geometry by selecting only a section along the minor axis of the flow channel. Due to the shape symmetry only one half of the 2D version of the channel cross section is modeled, as shown in Figure 3.b.

The die is modeled as an analytical rigid surface which allows a better approximation to the physical contact constraints. This is due to the smoother surface description that results from the ability of analytical rigid surfaces to parameterize the surface with curved line segments [16]. On the other hand, the sheet is modeled using four layers of solid plane strain

Table 2: Channel geometry of the bipolar plate flow field

Parameter	Value(mm)
Channel Width	1.5
Channel Height	1
Land Width	1
Entry Radius	0.25
Bottom Radius	0.25

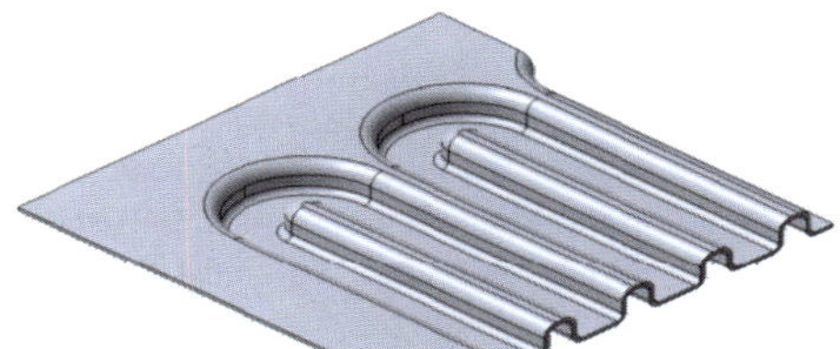

(a)

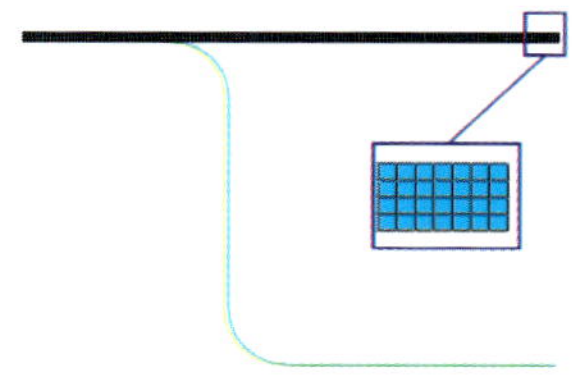

(b)

Figure 3: (a) Section through a serpentine flow field of a PEMFC bipolar plate (b) 2D FE model of the flow channel cross-section showing the layered plane strain elements of the sheet.

quadrilateral elements, with 300 elements in each layer. According to Luckey et al. [17], two-dimensional modeling using layered solid elements is more effective in predicting thinning behavior than three-dimensional modeling with conventional shell and membrane elements.

Frictional forces at the die sheet interface have been defined using the coulomb friction model available in ABAQUS™. With this model, the critical shear stress (τ_{crt}); which is the stress at which sliding between contacting surfaces occurs, is determined by the friction coefficient and the contact pressure, as given by the following equation:

$$\tau_{crt} = \mu . P \tag{7}$$

In blow forming practices, the contact pressure between the die and the sheet surfaces is just the same as the forming pressure applied on the top of the sheet. For all simulations, this forming pressure is controlled so as to maintain a target effective strain rate of 0.01 s^{-1} during the forming process. The averaging algorithm "5-20-75" described by [17],[18] is applied here. With this algorithm, the average of the lower 20% of the top 25% effective stresses found in the sheet is calculated in each step. Using this value, denoted by ($\bar{\sigma}^t_{20\%}$), the forming pressure is adjusted according to the following equation:

$$P_{t+\Delta t} = \left(\frac{\sigma_{tar}}{\bar{\sigma}^t_{20\%}}\right) P_t \tag{8}$$

Where (σ_{tar}) represents the effective stress corresponding to the target strain rate, (P_t) and ($P_{t+\Delta t}$) are the values of the forming pressure at the current and next time increments, respectively.

5 RESULTS AND DISCUSSION

The commercial finite element code, ABAQUS™, is used to carry out the simulations. Three user defined subroutines are developed to define the material viscoplastic behavior and the forming pressure control algorithm. To study the effect of the interfacial friction distribution on the forming process, a homogeneous friction at the die-sheet interface is applied. For the same channel geometry and forming conditions, the coefficient of friction acting at the die-sheet interface is varied from a frictionless condition to 0.5; representing different lubrication conditions. Formed sheet thickness distribution, forming time and forming pressure profile are calculated for each case. This allows the identification of the optimal lubrication conditions for this geometry.

5.1 Frictional Effects on SPF

The effect of the friction coefficient on the formed sheet thickness profile is shown in Figure 4. With excessive lubrication, represented by low friction coefficients, a localized thinning was found to take place at the die entry region. This excessive thinning occurs due to the combined compression and bending acting on this region during the initial stages of the forming process. Furthermore, such localized straining boosts the voids area fraction, given by eq. 3, compromising the load bearing capacity of the die entry region and making it weaker than the rest of the sheet. With small values of interfacial friction coefficients, metal will keep flowing from this weak area, and at later stages of the forming process, this strain peak becomes substantially large and may result in premature failure. On the other hand, with higher friction coefficients, less material is allowed to flow once it gets in contact with the die. This will enhance the deformation stability at the die entry region, and reduce thinning there. However, with such high friction, more thinning is found to take place at the die bottom corner, as it is the last part of the sheet that gets in contact with the die. The thinning taking place at this area doesn't indicate instability, as no strain localization is noticed and the strain is uniformly distributed over the die bottom corner. The friction coefficient at which the location of the minimal thickness moves from the die entry region to the bottom corner represents the optimal case, where the least localized thinning at both regions is obtained. This value will be referred to as the optimal homogeneous friction coefficient, and from Figure 4, it could be noticed that for the channel geometry addressed here, this optimal value is about 0.4.

The shape of the forming pressure profile, shown in Figure 5, is governed by changes in sheet thickness, sheet's radius of curvature, forming strain rate, the die-sheet interaction in addition to the coefficient of friction. During the early stages in the forming process, the pressure required to maintain the target strain rate increases as the rate of change of the sheet's radius of curvature is much higher than the rate of change of its thickness. Thus, higher pressure is required to maintain the target strain rate. However, as the forming process proceeds, the thickness reduction rate increases and the area fraction of voids becomes more significant, causing a slight reduction in the forming pressure. This applies until the

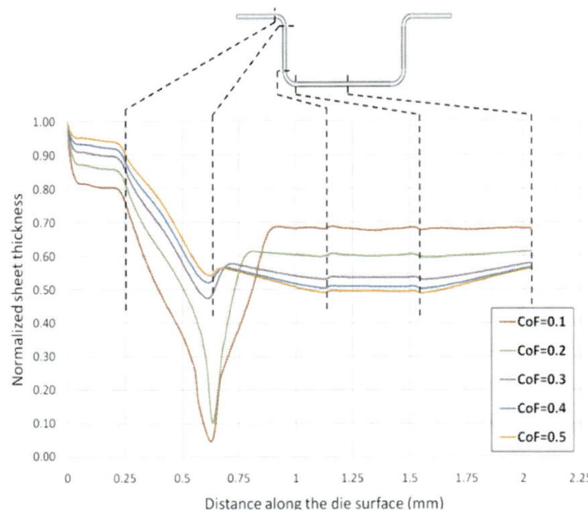

Figure 4: Predicted thickness profile of the formed sheet with different values of friction coefficient

sheet gets in contact with the base of the die. Once that happens, the rate of change of the radius dominates again, resulting in a rapid increase in the forming pressure.

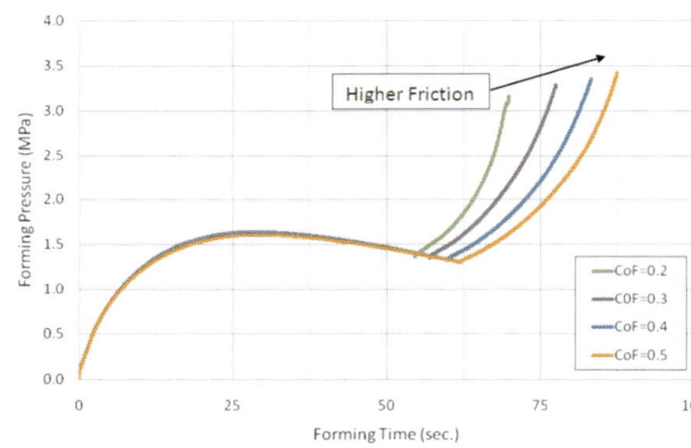

Figure 5: Forming pressure profile with different values of friction coefficient for target strain rate of 0.01 s-1

The previously described trend in the forming pressure profile remains valid regardless the value of the interfacial friction. In the first stage of the forming process, i.e. before the sheets gets in contact with the bottom surface of the die, frictional forces were found to have minimal effects on the forming pressure; as most of the deformation takes place in the free part of the sheet. On the other hand, when the sheet touches the die bottom surface and the sidewalls, surface friction at the die-sheet interface restricts the metal flow, resulting in higher forming pressure and prolonged forming time.

5.2 Variable Strain Rate Forming

Based on the stability analysis discussed in section 3, a variable strain rate forming path, shown in Figure 6, is designed in such a way that the ratio of the strain in the local area to that in the uniform area doesn't exceed 1.5. The smooth pressure control algorithm discussed in section 4 is used here to adjust the forming pressure according to the target strain rate values. The optimal homogenous friction coefficient $(\mu = 0.4)$, found in the previous subsection, is applied at the die sheet interface. The thickness distribution obtained with the variable strain rate path along with that obtained using a constant strain rate of 0.01 s-1 are shown in Figure 7, while Figure 8 shows the forming pressure profiles for the two cases.

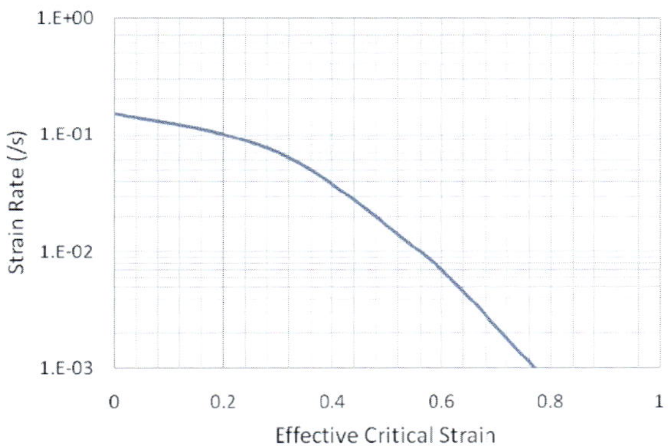

Figure 6: Optimal variable strain rate path.

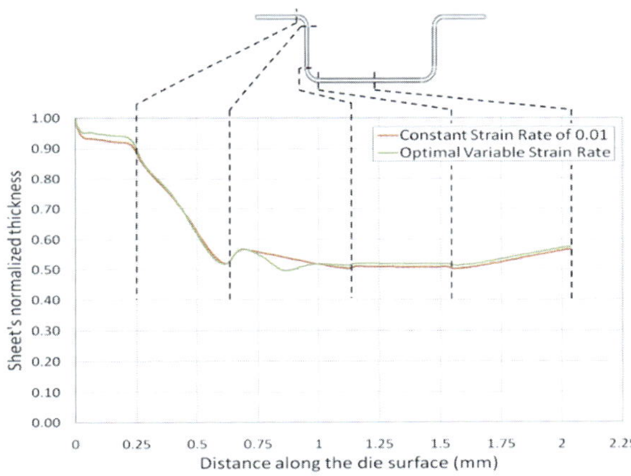

Figure 7: Predicted thickness profile of the formed sheet with constant strain rate of 0.01 s^{-1}, and with the optimal variable strain rate path.

With the variable strain rate path, the forming process is initially carried out at high strain rate values. As the forming process proceeds and more deformation takes place, the edge of instability is approached. Thus, the forming process is slowed down gradually so as to keep the deformation stable, and avoid severe strain localization. With this forming path, it was possible to reduce forming time to 46 seconds, compared to 84 seconds required to form the same part with 0.01 s^{-1} constant strain rate, representing a 45% reduction. Despite of the significant savings in the forming time achieved with the optimal variable strain rate path, the thickness distribution of the formed part is almost comparable with that obtained using the constant strain rate approach.

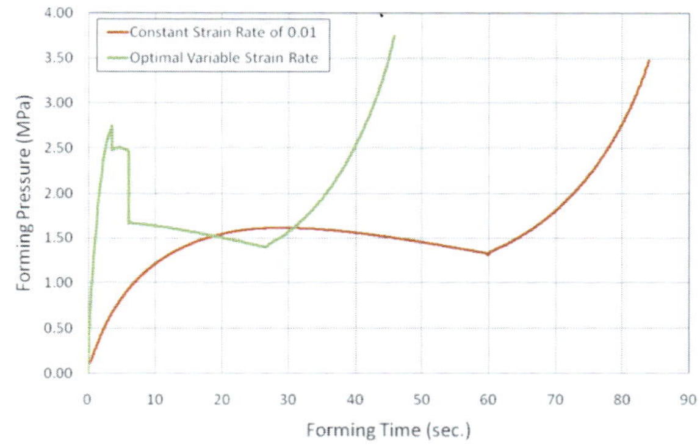

Figure 8: Forming pressure profile for the flow channel formed with constant strain rate of 0.01 s^{-1}, and with the optimal variable strain rate path.

These results show that by carefully designing the forming strain rate path and the friction distribution at the die-sheet interface, significant time savings can be achieved without compromising the integrity of the formed parts.

6 CONCLUSIONS

A nonlinear stability criterion that accounts for geometrical instabilities along with microstructural features has been applied here to design an optimal variable strain rate forming path for a PEMFC bipolar plates' flow channels. The effects of frictional forces acting at the die-sheet interface on the deformation stability were also studied. It was found that larger friction coefficients increase deformation stability, allowing for higher strain rates to be applied.

For the flow channels geometry considered here, a homogeneous friction coefficient of 0.4 was found to be the optimal value. Using this friction coefficient along with the variable strain rate forming path, it was possible to reduce forming time by 45%, without compromising the integrity of the formed part.

References

[1] U.S Department of Transportation, *Transportation Role in Reducing U.S. Greenhouse Gas Emissions*, 2010.

[2] F.K. Abu-Farha and M.K. Khraisheh, "An integrated approach to the Superplastic Forming of lightweight alloys: towards sustainable manufacturing," *International Journal of Sustainable Manufacturing*, vol. 1, 2008, pp. 18–40.

[3] A. Hermann, T. Chaudhuri, and P. Spagnol, "Bipolar plates for PEM fuel cells: A review," *International Journal of Hydrogen Energy*, vol. 30, Sep. 2005, pp. 1297-1302.

[4] D.G. Sanders, "The current state-of-the-art and the future in airframe manufacturing using superplastic forming technologies," *Materials Science Forum*, 2001, pp. 17–22.

[5] M.A. Nazzal, M.K. Khraisheh, and B.M. Darras, "Finite element modeling and optimization of superplastic forming using variable strain rate approach," *Journal of Materials Engineering and Performance*, vol. 13, 2004, pp. 691–699.

[6] M.K. Khraisheh and H.M. Zbib, "Optimum forming

loading paths for Pb-Sn superplastic sheet materials," *Journal of Engineering Materials and Technology*, vol. 121, 1999, p. 341.

[7] L. Carrino, G. Giuliano, and C. Palmieri, "On the optimisation of superplastic forming processes by the finite-element method," *Journal of Materials Processing Technology*, vol. 143, 2003, pp. 373–377.

[8] P.E. Krajewski and J.T. Carter, *Lubrication of magnesium workpieces for hot forming*, Google Patents, 2007.

[9] F.S. Jarrar, F.K. Abu-Farha, L.G. Hector, and M.K. Khraisheh, "Simulation of High-Temperature AA5083 Bulge Forming with a Hardening/Softening Material Model," *Journal of materials engineering and performance*, vol. 18, 2009, pp. 863–870.

[10] M.K. Khraisheh and F.K. Abu-Farha, "Microstructure-Based Modeling of Anisotropic Superplastic Deformation," *TECHNICAL PAPERS-SOCIETY OF MANUFACTURING ENGINEERS-ALL SERIES-*, 2003.

[11] C.H. Caceres and D.S. Wilkinson, "Large strain behaviour of a superplastic copper alloy–I. Deformation," *Acta Metallurgica*, vol. 32, 1984, pp. 415–422.

[12] M.J. Stowell, "Failure of superplastic alloys," *Metal Science*, vol. 17, 1983, pp. 1–11.

[13] Z. Marciniak and K. Kuczynski, "Limit strains in the processes of stretch-forming sheet metal," *International Journal of Mechanical Sciences*, vol. 9, 1967, pp. 609–612.

[14] J.W. Hutchinson, K.W. Neale, and A. Needleman, "Sheet necking—I. Validity of plane stress assumptions of the long-wavelength approximation," *Mechanics of Sheet Metal Forming, Plenum Press, New York*, 1978, pp. 111–126.

[15] Mohammad Albakri and Marwan Khraisheh, "Flow Field Design And Optimization For PEM Fuel Cell Bipolar Plates," 2009.

[16] ABAQUS, *Abaqus Analysis User's Manual*.

[17] S.G. Luckey and P. Friedman, "Aspects of element formulation and strain rate control in the numerical modeling of superplastic forming.," *Advances in Superplasticity and Superplastic Forming as held at the 2004 TMS Annual Meeting*, 2004, pp. 371–380.

[18] F.S. Jarrar, L.G. Hector Jr, M.K. Khraisheh, and A.F. Bower, "New approach to gas pressure profile prediction for high temperature AA5083 sheet forming," *Journal of Materials Processing Technology*, 2010.

A Sustainable Process for the Preparation of Sulfur Cement for use in Public Works

A.M.O. Mohamed, M.M. El Gamal

Department of Civil and Environmental Engineering, UAE University, Al Ain, UAE

Abstract

In this study solid waste materials were utilized to produce a sustainable sulfur cement product. The product employed two complementary stabilizers: (a) bitumen as the chemical stabilizer, i.e. sulfur modifying chemical reagent, and (b) fly ash, as the viscosity increasing surface active stabilizer, acts as thickening agents and generally improve the hardness or strength of the sulfur cement product. Controlling parameters such as the type and percentage of modifiers and the thermal history, including reaction and curing temperatures, curing time, cooling rate, and storage time were tested and evaluated. It is clear that these parameters have a strong impact on product crystallization and grain size development. The amount of the chemical stabilizer required depends upon the end use of the cement and the properties desired. Fly ash have been found to impart an extra measure of durability to the final sulfur cement, independent of its source, and serves the dual function of viscosity increaser and sulfur cement stabilizer.

Keywords:

Bitumen; Fly ash; Modified sulfur; Modified sulfur cement; Microstructures

1 INTRODUCTION

A sustainable approach to integrate waste management would allow the construction industry to use the increasing resources. Sulfur concrete construction materials are used in many specialized applications throughout industry and transportation. They are currently used primarily in areas, such as acidic and saline media, where conventional materials like Portland cement concrete (PCC) failed. Sulfur concrete is a thermoplastic material prepared by hot mixing sulfur cements and mineral aggregates. The molten sulfur solidifies and gains strength rapidly upon cooling, as with other concrete materials.

Construction materials such as sulfur concrete and sulfur asphalt continue to receive more attention since they are environmentally friendly and cost effective [1]. Beginning in the 1970s, successful projects in which sulfur concrete has been used as a construction material have been carried out in different levels. The United States Department of the Interior's Bureau of Mines and The Sulfur Institute (Washington, D.C.) launched a cooperative program in 1971 to investigate and develop new uses of sulfur. At the same time, the Canada Centre for Mineral and Energy Technology (CANMET) and the National Research Council (NRC) of Canada initiated a research program for the development of sulfur concrete [2-4]. A number of investigators and other published number of papers and reports investigated various aspects of sulfur and sulfur concrete [5-15]. All of these activities have led to an increased awareness of the potential use of sulfur as a construction material.

While sulfur concrete materials could be prepared by hot mixing of unmodified sulfur and aggregate, durability of the resulting product was a problem. Unmodified sulfur concretes failed when exposed to repeated cycles of freezing and thawing, humid conditions, or immersion in water. Studies were aimed at establishing reasons for failure of these sulfur concrete materials and determining means of preventing failures. When unmodified sulfur and aggregate are hot mixed, cast and cooled to prepare sulfur concrete products, the sulfur binder, on cooling from the liquid state, first crystallizes as monoclinic sulfur (Sß) at 114°C with a volume decrease of 7%. On further cooling to below 96°C, the Sß starts to transform to orthorhombic sulfur (Sα), which is the stable form of sulfur at ambient temperatures [16] [1]. This transformation is rapid, generally occurring in less than 24 hours. Since Sα form is more dense than Sß, high stress is induced in the material by solid sulfur shrinkage. Thus, the sulfur binder can become highly stressed and can fail prematurely.

It was further necessary to develop an economical means of modifying the sulfur so that the sulfur concrete product would have good durability and reduce the expansion/contraction of sulfur concrete during thermal cycling. Since 1984, the American Concrete Institute (ACI), through its Committee 548, polymers in concrete, and subcommittee 548D, sulfur concrete, has been active in developing guidelines for use of sulfur concrete. There are two main ways for modifying sulfur, both of them trying to control sulfur crystallization chemically or physically. The first one tries to combine chemical substances to sulfur in order to inhibit the transformation to orthorhombic structure; as a result of the chemical reaction with the substance, sulfur remains in monoclinic state after cooling. Several substances have been tried for this methodology; the most common are dicyclopentadiene, or a combination of dicyclopentadiene, cyclopentadiene and dipentene [17, 18]. The second method utilizes a modified sulfur concrete by combining sulfur with olefin hydrocarbon polymers (such as RP220 or RP020 by Exxon Chemical, or Escopol) and a physical stabilizer such as fly ash, or other fine substances [19].

In both ways, the issue is controlling sulfur crystallization, either chemically or physically. Depending on the ultimate use of the produced sulfur concrete, one chooses the method of treatment. Because of the abundance of the physical stabilizers in the United Arab Emirates (UAE), the second treatment method is utilized in this study whereby sulfur, polymer modifier and fly ash were utilized in producing sulfur cement. Each material as well as the mixture was evaluated using thermal analysis techniques, such as thermal gravimetric analysis (TGA) and differential scanning calorimetric (DSC), infrared spectra and scanning electron microscope. The experimental results were evaluated in terms of the potential use of the product for public works and the optimum mixture was quantified.

2 TERMINOLOGY

To prevent confusion, terms found in various publications on the topic are further defined. The ASTM C1159-98 defines sulfur polymer cement as "a polymer product consisting of small amounts of chemical modifying additives dispersed in sulfur." The term "SPC" is generally used as an indicator for sulfur polymer cement. Once SPC is loaded with aggregates or waste products, it becomes sulfur concrete and the term "SC" is used as an indicator. This nomenclature follows the same principle as that for Portland cement. The paste of Portland cement is represented by "PC", while the concrete obtained by the addition of sand and aggregates to PC is designated by "PCC", or Portland cement concrete. Other publications used the terminology "modified sulfur cement" instead of SPC; however, both names refer to the same material.

In the preceding terminologies, the word cement is included without clear indication of the type of additives that will produce this cementious behavior. Generally, chemical additives are added to modify elemental sulfur, nearly all of which fall under the heading of polymeric polysulfide or, alternatively, substances which may react with elemental sulfur to give in situ formation of polymeric polysulfide. Therefore, terms such as "polymerized sulfur" or "modified sulfur" should be used instead.

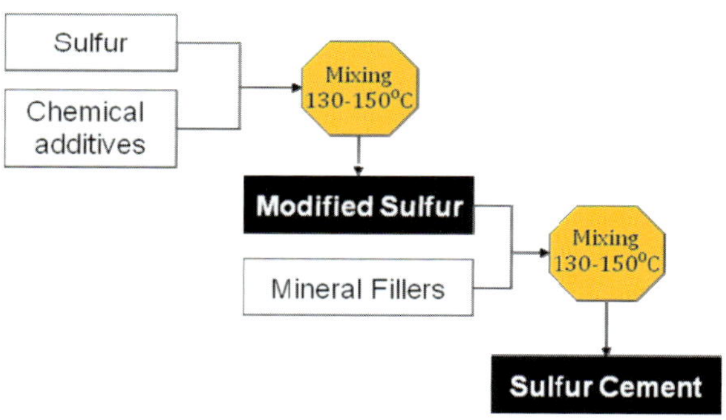

Figure 1: Sulfur terminologies used for sulfur cement production.

For inclusion of the word cement in the definition the mixture should include, in addition to the chemical additives, physical stabilizers or mineral fillers such as fly ash, furnace slag, cement kiln dust, talc, mica, silica, graphite, carbon black, pumice, insoluble salts (e.g., barium carbonate, barium sulfate, calcium carbonate, calcium sulfate, magnesium carbonate, etc.), magnesium oxide, and mixtures. Such fillers typically have a particle size less than 100 mesh (US Standard Testing Sieves) and preferably, less than 200 mesh. Such fillers generally act as thickening agents and generally improve the hardness or strength of the sulfur cement product.

Therefore, throughout this work, the term "modified sulfur" is used when chemical additives are added to the elemental sulfur, and "sulfur cement" is used when mineral fillers are added to the modified sulfur (Figure 1). In addition the term "sulfur concrete" is used when aggregates are added to sulfur cement.

3 MATERIALS

Details of materials used and procedures for preparation of modified sulfur and sulfur cement are described below.

All raw materials used for this study were obtained from UAE. The granular sulfur (99.9% purity) was obtained from Al Ruwais refinery. A notable feature of the sulfur cement of this modification does not require high purity sulfur and can be made with off-grade sulfur containing hydrocarbon impurities, blow dirt, and other "contaminants". Bitumen with a softening point of 48.8°C, specific gravity at 20°C of 1.0289 g/cm3 and kinematics viscosity at 135°C of 431 cSt, with a chemical analysis of C:79, H:10, S:3.3 and N:0.7%, was used. The used physical stabilizer was fly ash, India-97/591. Chemical analyses were performed using ICP; the results are listed in Table 1 while the grain size distribution is shown in Figure 2.

Table 1: Chemical composition of fly ash using ICP analysis

Chemical Composition (Percent by weight)					
Al_2O_3	CaO	Fe_2O_3	K_2O	MgO	SiO_2
32.4	0.46	4.34	0.027	0.66	60.9

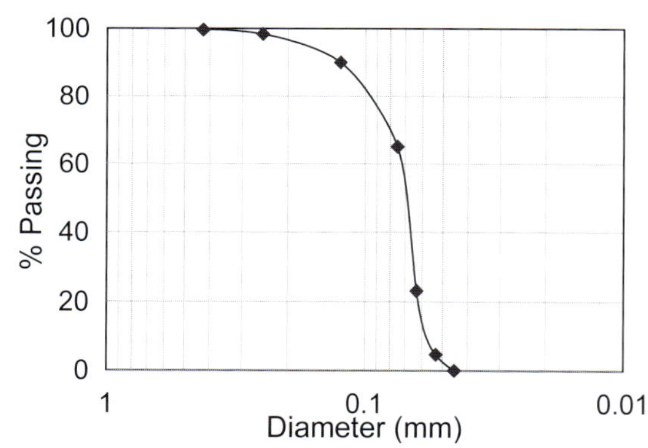

Figure 2: Grain size distributions of Fly ash.

4 MIX DESIGN AND SAMPLE PREPARATION

4.1 Preparation of modified sulfur

In oil bath, 2.5 wt% of hot bitumen, 97.5 wt% molten sulfur and emulsifying agent were mixed and mechanically stirred at 140°C The temperature was maintained at about 135-140°C during the mixing process, which lasted about 45-60 minutes. The progress of the reaction was monitored by the degree of homogeneity of the mixture via careful observation of the temperature and viscosity of the reacting mixture. After mixing, the prepared mixture was allowed to cool at a rate of 8-10°C/min. The final product, which is called sulfur cement, is a sulfur containing polymer which on cooling possessed glass like properties [20-25]. The composition and properties of the MSC were evaluated in terms of chemical analysis, FT-IR spectra, x-ray diffraction, scanning electron microscopy, and thermal processing including TGA and DSC.

4.2 Preparation of sulfur cement

Sulfur cement was prepared through the addition of viscosity increasing surface active stabilizer such as fly ash to the modified sulfur. The amount of chemical stabilizer depends upon the end of use of the cement and the properties desired. It is generally incorporated into the final cement mix by several reaction routes. Preferably, the chemical is pre-reacted at approximately 140°C for about 30 minutes with a smaller proportion of sulfur than is required in the final mix. The resulting concentrate can then be either stored for future use or dissolved in the residual sulfur required for the final mix at the mixing temperature. Controlling parameters such as type and percentage of modifiers, and thermal history including reaction temperature, curing temperature, curing time, cooling rate, and storage time have an impact on product crystallization and grain size development.

4.3 Characterizations of modified sulfur and sulfur cement

The following equipments were used for the characterization;

- Elemental Analyzer Finnigan Flash EA1112 CHN/S, for the determination of C, H, N and S.
- JSM-5600 Joel microscope for microstructure characterization.
- Nicolet FTIR Magno-IR (Model 560) for IR spectrum analysis.
- x-ray powder diffraction Philips PW/1840, with Ni filter, Cu-Kα radiation (λ=1.542 Å) at 40 KV, 30 mA and scanning speed 0.02° / S, for the examination of the interlayer activity in the modified sulfur and sulfur cement.
- JSM-5600 Joel microscope for characterization of the microstructure.
- HPLC Agilent 1100; column PL gel Mixed C, 300*7.5mm*5μm for confirmation of polysulfide molecular weight.
- Thermo Gravimetric Analyzer Perkin Elmer TGA7, for measurements weight changes as a function of temperature.
- Differential scanning calorimeter (Perkin Elmer DSC7), for heat capacity measurements, through phase transitions on heating.

5 RESULTS AND DISCUSSION

5.1 Properties of modified sulfur

Modified sulfur characteristics presented chemical composition and physical properties are listed in Table 2.

Table 2: Chemical composition of the sulfur bitumen mixture

Composition		Properties	
Sulfur	97.4±0.48		
Carbon	1.98±0.08	Sp.grv. g/cm^3	1.89
Hydrogen	0.1±0.067	Viscosity @135ºC	25 cP

The manufacture's certification of sulfur bitumen cement analysis, listed in Table 2, agrees with the requirements for sulfur polymer cement chemical and physical properties [26].

5.2 Chemical reactions

Addition of bitumen in amounts of 2.5% to sulfur initiates chemical reactions whose type depends on the bitumen content, heating temperature and the time of the reaction. Some competing reactions could occur, including those with bitumen incorporation into sulfur molecules or de-hydrogenation with liberation of hydrogen sulfide. As reported in [1], at T < 95°C sulfur exists as a cyclooctasulfane crown with an S-S bond length of 0.206 nm, and S-S-S bond angle of 108 degree; and at T < 119°C sulfur crystallizes. At 119°C (melting point of sulfur), liquid sulfur being thoroughly dispersed in bitumen forming an emulsion (role of emulsifying agent) and cyclooctasulfane turns partly into polymeric zigzag chains (bond length 0.204 nm)

The crystallization features are affected by such factors as chemical reaction of sulfur with bitumen components and its dissolution or dispersion. Sulfur at heating temperature T<140°C, elementary sulfur forms poly-sulfide which initiate formation of a network. Such structures differ considerably in the chemical and thermal stability from unmodified sulfur. At 119 to 159°C, molten sulfur exists essentially as cyclooctasulfane (λ-S). Above 159°C, eight-member ring rapidly break down into bi-radicals. In their turn, bi-radicals recombine to form polymeric chains with the maximal length of up to 10^6 sulfur atoms. However, above 140°C, de-hydrogenation of saturated bitumen components can occur; also linear poly-sulfide can transform into stable cyclic thiophene structures [27]. It is noted that modified sulfur should be produced within its recommended mixing temperature range of 135 to 141°C.

5.3 Composition analysis via FT-IR

The completion of bitumen sulfur reaction was strongly supported by FT-IR spectra of double bonds in bitumen, which is the main modifying bitumen constituent. Figure 3 shows the band at 2975 cm^{-1} which is consistent with CH stretching. Since C-H is associated with C=C double bonds, the disappearance of C-H bonds in the modified sulfur spectra suggested the consumption of aliphatic C=C bonds, which generally leads to the polymerization of sulfur with bitumen. Negative peak in 2400 cm-1 indicates removal of NC=O group indicating the presence of chemical reaction sites for poly-sulfide formation [25].

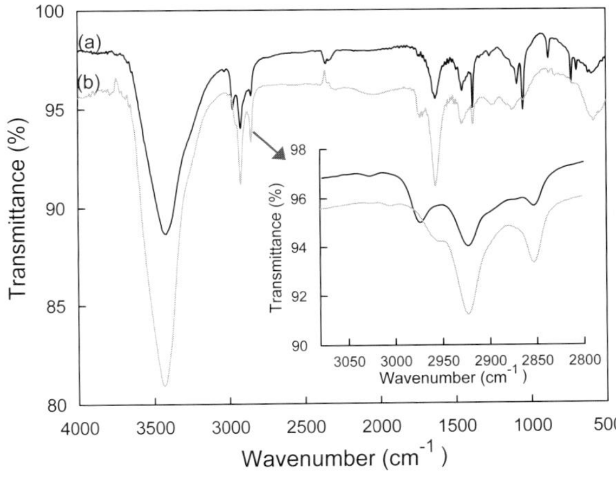

Figure 3: FTIR spectrum of (a) bitumen, (b) modified sulfur.

5.4 Composition analysis via XRD

The XRD diagrams for pure sulfur and sulfur modified with 2.5 wt % bitumen is shown in Figure 4. XRD analysis confirms that the modification of sulfur leads to sulfur crystallizes in a structure different than the unmodified one. Modified sulfur shifted to lower 2-theta, which in turn indicates that the newly formed structure is finer in its microstructure.

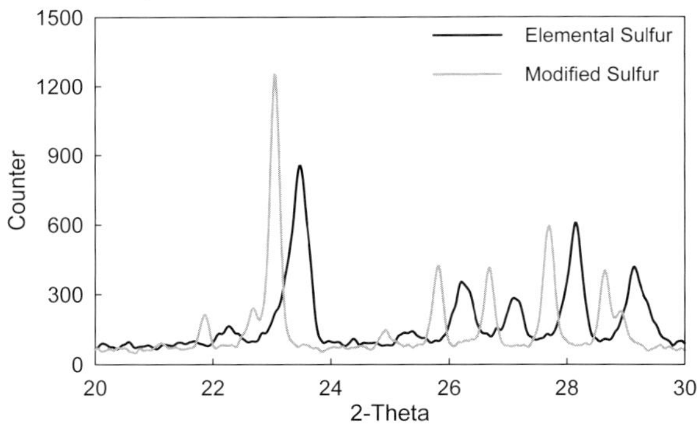

Figure 4: x-ray diffraction of pure sulfur and modified sulfur.

5.5 Microstructure

Distribution of bitumen in sulfur and the free sulfur crystal type (orthorhombic or monoclinic) was evaluated using SEM, which revealed how the bitumen controls the crystallization of sulfur. Pure sulfur crystallizes and forms dense and large alpha sulfur crystals (Sα) with orthorhombic sulfur morphology as shown in Figure 5.

Figure 5: SEM of pure elemental sulfur.

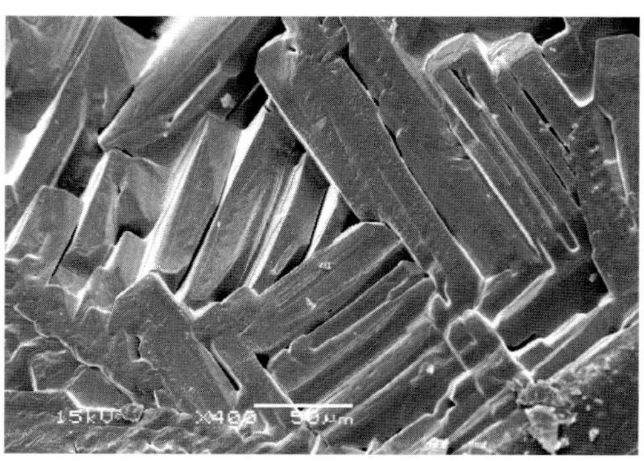

Figure 6: SEM of bitumen modified sulfur.

Orthorhombic sulfur is the stable form of sulfur at room temperature and atmospheric pressure. Orthorhombic sulfur is also called rhombic sulfur. With addition of bitumen, the crystal growth is limited and controlled by the bitumen in such a way that all crystals are plate like of micron dimension as monoclinic sulfur crystal of beta form (Sß) as shown in Figure 6. The formed microstructure confirmed the efficiency of bitumen in inhibiting the formation of orthorhombic sulfur, and the interaction between bitumen and sulfur gave a mixture of monoclinic sulfur and poly-sulfide. The presence of uniformly dispersed polymeric material in sulfur, as coverage of monoclinic sulfur and between plates is evident. The smallest microstructure, plate like, helps to resist cracking and to tolerate any thermal expansion.

5.6 Degree of sulfur polymerization

In the process of producing modified sulfur, bitumen, in combination with a further agent, physically modifies the sulfur by inducing sulfur polymerization. Thus, the resulting modified sulfur cement comprises polymerized sulfur. When polymerized sulfur is present the sulfur phase transformation (β to α) still occurs during cooling, but the polymerized sulfur acts as a compliant layer between the sulfur crystals, and so serves to mitigate the effect of the phase transformation.

Sulfur modified with bitumen analysis test indicate that modified sulfur comprise 55wt % of monoclinic sulfur and 45% wt of polysulfide, based on the total weight of the sulfur component. The structure of polysulfide was confirmed by analyzing the fraction that insoluble in CS2 by column chromatograph HPLC flow rate of 1 ml/min in chloroform, at room temperature 24°C). Analysis data indicates the presence of low and high molecular weight fractions of poly-sulfides with weight average molecular weight of 17417 and average number molecular weight of 344. The polydisperseability index, which is a reflection of the broadness of the product molecular wt., was determined to be 5. As a consequence, sulfur's rheological properties are affected hence, presenting a higher viscosity than unmodified one. This fact has an important effect in the crystallization of sulfur. In a more viscous liquid, and in which the molecules are more polymerized, the growth of the crystals will be more difficult and it will be thus slower.

5.7 Thermal analysis

Figure 7 provides comparison of the TGA curves of pure sulfur and bitumen modified sulfur.

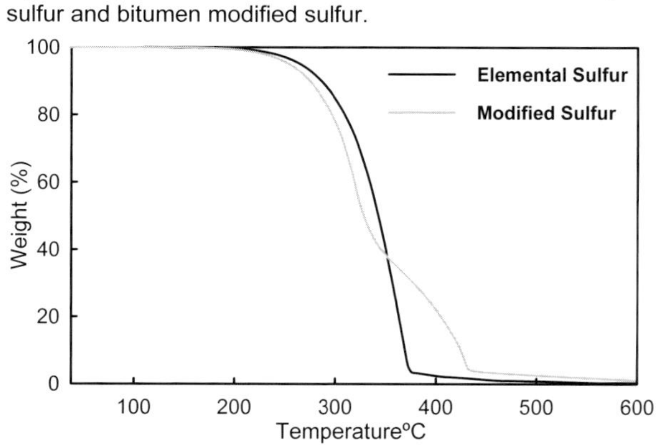

Figure 7: TGA of pure sulfur and sulfur bitumen mixture.

The maximal thermal effect was observed at 180 to 450°C for pure sulfur which was accompanied by active liberation of hydrogen sulfide, while the maximal thermal effect was observed at 200 to 540°C for modified sulfur. This effect reflects the increase of thermal stability of modified sulfur

5.8 Durability of sulfur cement

The molecular weight analysis test of sulfur modified with bitumen indicates that sulfur bitumen cement contained 45% polysulfide and 55% un-reacted sulfur crystallized out as monoclinic. Stability of sulfur bitumen cement was studied by following the thermal stability at different times. The thermal stability of stored modified sulfur cement samples for 12, 24 and 36 months at room temperature was evaluated using the DSC thermo-grams under ambient conditions. The DSC analyses in Figure 8 indicated that sulfur bitumen cement is stable in the monoclinic form (Sβ), and does not transform to orthorhombic sulfur (Sα), after 36 months of storage. The alpha to beta transition is one of the prime sources of failure associated with elemental sulfur cement.

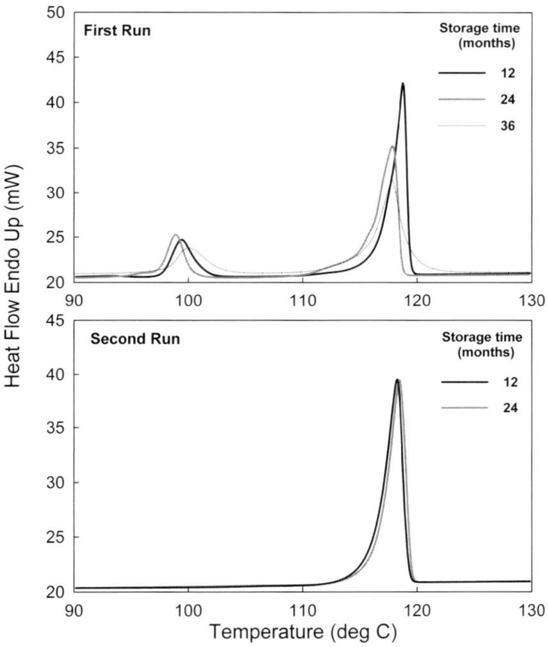

Figure 8: Differential scanning calorimetric results for sulfur bitumen cement, at different storage time.

6 ADVANTAGES AND DISADVANTAGE OF SULFUR CEMENT

6.1 Advantages

1. Sulfur cement has been used as a construction material because of its excellent resistance to acid and salt environments and superior water tightness as compared with Portland cement concrete [28], [29], [25].
2. It has emerged as a possible alternative as a binding and stabilizing agent for the solidification and stabilization hazardous and mixed wastes [16].
3. The thermal expansion coefficient for the modified sulfur is approximately one third that of the unmodified sulfur. Since the modified sulfur does not go through the allotropic transformation upon solidification and has a lower thermal expansion coefficient; it has less shrinkage and hence develops less residual stress upon cooling [16].
4. Sulfur cement operational characteristics have greater flexibility, easier operations compared with Portland cement.

6.2 Disadvantages

1. Sulfur cement material will melt if exposed to temperature above the melting point of 119°C and will lose its integrity [30]
2. The optimum processing temperature range of sulfur cement is between 127and 139°C [30]. Above the temperature range, a sharp rise in viscosity occurs because of additional polymerization within sulfur cement that makes material gummy and un-pourable [31].
3. Sulfur cement is not recommended for use with strong bases and oxidizing agents, aromatic or chlorinated hydrocarbons, or oxygenated solvents because of the risk of chemical corrosion due to H2S formation.

7 SUMMARY AND CONCLUDING REMARKS

In this study, clear distinction was made between sulfur modifiers and sulfur cement. Sulfur modifiers were defined as materials or mixtures of materials which when added to elemental sulfur lower its melting point and increase its crystallization time. Sulfur cement is produced via the addition of viscosity increasing surface active stabilizer such as fly ash, furnace slag; cement kiln dust, talc, mica, silica, graphite, carbon black, pumice, insoluble salts (e.g., barium carbonate, barium sulfate, calcium carbonate, calcium sulfate, magnesium carbonate, etc.), magnesium oxide, and mixtures to modified sulfur. Such fillers typically have a particle size less than 100 mesh and preferably, less than 200 mesh. Such fillers generally act as thickening agents and generally improve the hardness or strength of the sulfur cement product. Studies indicate that, adding the mineral filler after sulfur modification prevents problems of filler dusting balling.

8 ACKNOWLEDGEMENTS

The authors would like to acknowledge the funding provided by the Research Affairs Sector, UAE University, for conducting this study. Also, acknowledgements are in order to the Central laboratories Unit (CLU) and Chemical and Petroleum Engineering Laboratories, UAE University, for the use of their facilities; Al Ain Cement Factory for their help in material characterization; and Al Ain Municipality for providing some of the row materials.

9 REFERENCES

[1] Mohamed, A.M.O., El Gamal, M., 2010, Sulfur Concrete for the Construction Industry: A Sustainable Development Approach, A title in W-F Chen's Civil and Environmental Engineering Series, J. Ross Publishing, Florida, USA.

[2] Malhotra, V.M., 1973, Mechanical Properties and Freeze-thaw Resistance of sulfur concrete. Division Report No. IR 73-18, Energy, Mines and Resources Canada Ottwa, 30.

[3] Malhotra, V.M., 1974, Effect of Specimen Size on compressive strength of sulfur concrete. Division Report No. IR 74-25, Energy, Mines and Resources Canada Ottwa.

[4] Beaudoin, J.J., Sereda, P.J., 1973, The Freeze-Thaw Durability of Sulfur concrete. Building Research Note, Division of Building Research, National Research Council, Ottawa, 53.

[5] Sullivan, T.A., Mc Bee, W.C., Blue, D.D., 1975, Sulfur in Coatings and Structural Materials. Advances in Chemistry Series No. 140, American Chemical Society, Washington, D.C., 55-74.

[6] Vroom, A.H., 1977, Sulfur Cements, Process for Making same and Sulfur Concretes made thereform," U.S. Patent No. 4,058,500.

[7] Vroom, A.H., 1981, Sulfur Cements, Process for Making same and Sulfur Concretes made Thereform. U.S. Patent No. 4,293,463.

[8] Sullivan, T.A., Mc Bee, W.C., 1976, Development and Testing of Superior Sulfur Concretes. BuMines Report No. RI 8160, U.S. Bureau of Mines, Washington, D.C., 30.

[9] Mc Bee, W.C., Sullivan, T.A., 1979, Development of Specialized Sulfur Concretes. Bureau of Mines Report No. RI 8346, U.S. Bureau of Mines, Washington, D.C., 21.

[10] Mc Bee, W.C., Sullivan, T.A., 1982, Assigned to U.S. Department of Commerce, Modified Sulfur Cement. U.S. Patent No. 4, 311, 826.

[11] Funke, R.H.Jr., McBee, W.C., 1982, An industrial Application of Sulfur Concretes. ASC Symposium Series No. 183, American Chemical Society, Washington, D.C., 195-208.

[12] Mc Bee, W.C., Sullivan, T.A., Jong, B.W., 1981, Modified Sulfur Concrete Technology. Proceedings, SULPHUR-81 International Conference on Sulfur, Calgary, 367-388.

[13] Mc Bee, W.C., Sullivan, T.A., Jong, B.W., 1983, Corrosion-Resistant sulfur concretes. Bureau of Mines Report No. 8758, U.S. Bureau of Mines, Washington, D.C., 28.

[14] Mc Bee, W.C., Sullivan, T.A., Fike, H. F., 1986, Corrosion-resistant sulfur concretes. Corrosion and Chemical Resistant Masonry Materials Handbook, Noyes Publications, Park Ridge, 392-417.

[15] Sullivan, T.A., 1986, Corrosion-Resistant sulfur concretes – design mManual. The Sulphur Institute, Wash-ington D.C., 44.

[16] Lin, S.L., Lai, J.S., Chian, E.S.K., 1995, Modification of Sulfur Polymer Cement (SPC) Stabilization and solidification (S/S) process. Waste Management 15 (5/6): 441-447.

[17] Currell, B.R., 1976, The importance of Using Additives in the Development of New Applications for sulfur. Symposium on New Users for Sulfur and Pyrites, 105-103. Madrid.

[18] Bordoloi, B.K., Pierce, E.M., 1978, Plastic Sulfur Stabilization by Copolymerization of Sulfur with Dicyclopentadiene. Advances in Chemistry Series No. 165, American Chemical Society, Washington D.C., 31-53.

[19] Vroom, A.H., 1992, Sulfur Polymer Concrete and its Application. In. proceeding of Seventh International Congress on polymers in concrete: 606-621 Moscow .

[20] Nnabuife, E.C., 1987, Study of Some Variables Affecting the Properties of Sulfur-Reinforced Sugarcane Residue Based Boards. Indian Journal of Technology 25: 363-367.

[21] Mohamed, A.M.O., El Gamal, M., 2009, Hydro-Mechanical Behaviour of a Newly Developed Sulfur Polymer Concrete, Cement and Concrete Composites, Vol. 31, pp. 186-194.

[22] Mohamed, A.M.O., El Gamal, M., 2008, New Use of Surfactant. UK Patent Application No. 0807612.7, filed by J.A.KEMP and Co UK, dated 25 April 2008.

[23] Mohamed, A.M.O., El Gamal, M., 2009, New Use of Surfactant. International Patent Application No. PCT/IB2009/005338, filed by J.A.KEMP and Co UK, dated 21 April 2009.

[24] Mohamed, A.M.O., El Gamal, M., 2009, New Use of Surfactant. GCC Patent Application No. GCC/P/2009/13350, filed by J.A.KEMP and Co., UK, dated 25 April 2009.

[25] Mohamed, A.M.O., El Gamal, M., 2006, Compositional Control on Sulfur Polymer Concrete Production for Public Works, In: Developments in Arid Regions Research, Vol. 3, pp. 27-38.

[26] ASTM C1159-98, 2003, Standard Specification for Sulfur Polymer Cement and Sulfur Modifier for Use in Chemical-Resistant, Rigid Sulfur Concrete.

[27] Syroezhko, A.M., Begak, O.Yu., Fedorov, V.V., Gusarova, E.N., 2003, Modification of Paving Asphalts with Sulfur. Russian Journal of Applied chemistry 76(3): 491-496.

[28] McBee, W.C., Sullivan, T.A., Fike, H.J., 1985, Sulfur Construction Materials, Bulletin 678. U.S. Bureau of Mines.

[29] Mattus, C. H., Mattus, A. J., 1994, Evaluation of Sulfur Polymer Cement as a Waste Form for the Immobilization of low level Radioactive or Mixed Waste, OAK Ridge National Laboratory.

[30] McBee, W.C., Weber, H.H., 1990, Sulfur polymer cement concrete, in proceedings of the 12th Annual Department of Energy Low-Level Waste Management Conference, CONF-9008119, National Low-Level Waste Management Program, Idaho Natl. Engineering Lab., Idaho Falls, Idaho.

[31] Darnell, G.R., Aldrich, W.C., Logan, J.A., 1992, Full-Scale Tests of Sulfur Polymer Cement and Non-Radioactive Waste in Heated and Unheated Prototypical Containers, EGG-W-10109, Idaho Natl. Engineering Lab., Idaho Falls, Idaho.

4 Product design and development

Today's manufactured products have to satisfy a significantly growing requirements providing environmental, societal and economical benefits while protecting public health, welfare, and environment over the entire life, from the extraction of raw materials to final disposal. In the first section of this chapter Musselman and Djurdjanovic design a novel device to consistently excite belt vibrations in the material handling system with greatly reduced variations in belt length and initial condition location. In the second section Hampp and Janajreh outline the development of a drop tube reactor using a simulation-assisted design. In the third section Jamshidi et al. provide a generic algorithm detailing jig installation processes aimed at minimizing the uncertainties associated with the positioning accuracy of the jig components. In the fourth section Alsayyed and Harib present an innovative idea where a FlexDie is designed to manufacture sheet metal parts. The idea is to mimic the representation of graphics and text on a screen using pixels. In the fifth section Bracke and Haller outline the advanced reliability analysis of warranty databases, combine parametric and nonparametric significance tests and therefore facilitate a reliable statistical analysis of small to large sample sizes. In the sixth section Middendorf et al. present the life time monitoring of components. The approach of condition monitoring is then broadened to the so-called innovation cluster "Maintenance, Repair and Overhaul in Energy and Transport". In the seventh section Yang et al. introduce cryogenic processing of biomaterials to modify the surface integrity by severe plastic deformation processes, for significant surface and subsurface property improvement. In the eighth section Kenan et al. show the effect of varying the cross sectional area of the flow channel in the bipolar plates along the length of the fuel cell.

Improvement of Belt Tension Monitoring in a Belt-Drive Automated Material Handling System

M. Musselman, D. Djurdjanovic

Department of Mechanical Engineering, University of Texas, Austin, USA

Abstract

The extension of useful life of manufacturing equipment is one of the keys to sustainable manufacturing. High uptime requirements of the semiconductor industry result in conservative preventive maintenance schemes, leading to replacement of key components before the end of their useful life. This paper presents the results of research toward a more intelligent condition based maintenance scheme for the belt monitoring in a belt driven automated material handling system. An experimental study of belt dynamics showed that transverse belt vibrations were sensitive to changes in belt length, belt tension, belt misalignment, and excitation location. Based on these findings, a novel device was designed to consistently excite belt vibrations in the material handling system with greatly reduced variations in belt length and initial condition location. The standard deviation of tension estimates using the device was dramatically lower than that of a trained technician performing the current standard technique on three different robots.

Keywords:

Automated Material Handling Systems, Belt Tension Monitoring, Design of Experiment, Tension Estimation Variance Reduction

1 INTRODUCTION

Automated Material Handling Systems (AMHS) are crucial to modern manufacturing which requires increased quality and throughput, better inventory control, and a high level of production cell integration. The AMHS benefits of inventory, cycle time, and cost reduction as well as quality and productivity improvement prompt the use of AMHS by the heath care, printing, automotive, wine distribution, food distribution, and apparel industries [1], as well as postal services [2] and semiconductor manufacturing [3]. The top 20 AMHS suppliers' annual revenue grew from $10.7 billion in 1999 [4] to $15.3 billion in 2008 [5], suggesting growing need for automated handling.

The semiconductor industry features the quintessential integration of Automated Storage/Retrieval Systems (AS/RS) with automated material transfer. Wafers are buffered in 'stockers' between sequential processing steps which demands their impeccable reliability because they are a bottleneck for Work in Progress (WIP), leading to immense cost of a potential stocker failure. This cost of downtime has driven extremely conservative preventive maintenance schedules, which are often wasteful because parts are replaced before the end of their useful life. In fact, spending on plant maintenance in the United States rose from $800 billion in 1991 to $1.2 trillion in 2000. Startlingly, it is suggested that as much as one third to one half of this expenditure is wasted due to poor maintenance management [6].

The work presented in this paper focuses on a unitary belt-driven AS/RS used in the semiconductor manufacturing industry. A belt driven central crane robot operates in an aisle to move storage units to and from storage bins. Belt drives are sensitive to tension, suffering from slipping and skipping teeth if tension is too low, or excessive pulley bearing wear and belt fatigue if the tension is too high [7-8]. Currently, belt tension monitoring is performed by a technician manually strumming the belt and recording the ensuing fundamental frequency of belt vibrations. The tension is estimated from the fundamental frequency of transverse belt vibrations based on the one dimensional wave equation and known belt material properties. The process suffers from a high variance in tension estimates and often yields no result as the equipment is unable to isolate the fundamental belt vibration frequency.

Note that the existence of belt drives in the semiconductor industry [9-10], automotive industry [10-14], and robotics in general [15-17] suggest significant implications of the work presented in this paper on fields other than semiconductor manufacturing. The next section briefly reviews relevant literature on this subject.

2 LITERATURE REVIEW

Belt drives play a crucial role for power transmission in many applications. Advantages of belts over gears and chains include weight and noise reduction, improved efficiency, shock absorption, damping characteristics, simplicity, cleanliness, and lack of the need for lubricants [7]. Realization of these advantages require adequate pulley radius, path planning, and correct tensioning [7-8]. Under-tensioning causes slip in flat and V-belt drives and tooth skipping in timing belt drives, which is unacceptable for precision placement operations. Over tensioning causes premature belt failure from abrasion or shearing, as well as overloading of pulley bearings [7-8]. Numerous belt tension monitoring techniques have been developed to assure proper tension.

2.1 Stationary Belt Tension Monitoring

Stationary belt tension monitoring occurs when the belt is not moving and thus requires interruption of the normal machine operation. All stationary methodologies rely on the elasticity of the belt to determine tension either by transverse deflection or by initiating transverse vibration.

The simplest methodology for belt tension estimation assumes linear, elastic material properties. Essentially, the force required for a pre-determined, mid-span belt deflection is measured. Since most belts are thin relative to the span, bending is localized around clamps or pulleys and axial deflection dominates the model. Thus, with knowledge of belt material characteristics, the force required for deflection is separated into tension induced by deflection and the initial tension [18]. This methodology suffers from sensitivity to deflection location, lack of elastic linearity in most composite belts, and neglects the flexural stiffness of belts.

Experimental uniaxial tension distribution in a flat belt measured by optical coating and extensometer was reported in [19]. It was noted that optically measured strain was, on average, 15% higher than the extensometer measured strain. However, the introduction of visual tension measurement cannot be underrated, as it formed the basis for use of the laser Doppler vibrometry in deflection measurements.

Next, the one dimensional wave equation is utilized to estimate belt tension from the fundamental frequency of free vibration [10,12]. The model assumes homogenous, isotropic, linear elastic, undamped material condition with pinned boundary conditions [20]. Clearly, composite belts satisfy none of these conditions since composite construction leads to heterogeneity. Furthermore, parallel tension member design leads to anisotropy. In addition, the use of polymeric matrices in belt construction guarantees non-linear elasticity and the viscoelastic nature of polymers creates frequency dependent damping [21]. Lastly, belt drives suffer from pulley misalignment, among other reasons, for tension distribution across the width of the belt, which was also found to alter the frequency and modes of vibration in [22]. The drawbacks to this methodology are supported by [13], where the damping effect and geometric complexity were noted although no model was formulated. Despite these shortcomings, tension estimation based on fundamental frequency is used almost universally.

2.2 Dynamic Belt Tension Monitoring

This section provides a brief review of advancements made in dynamic belt tension monitoring (i.e. monitoring while the belt is moving). Both longitudinal and transverse vibrations are present in running belt drives. These two phenomena are coupled due to constant belt length. It can be also concluded that transverse vibration frequency is dependent on the belt material, geometry, and tension, while longitudinal vibration is largely material and geometry driven, with much less dependence on tension [23-24]. Friction, whether modeled using a Coulomb friction [25-27] or a tri-linear model [27], is responsible for power transmission in V-belt and flat belt drives [28-29], but is largely replaced by geometric considerations in timing belt drives [30-31]. Next, axial stiffness accounts for the majority of belt modeling accuracy, followed by damping, flexural stiffness, and shear deformation respectively [32-33]. This can be understood by noting that flat and timing belts are thin, making contributions from flexural and shear stiffness negligible. Damping has been modeled linearly, in which case there is coupling between transverse natural frequency and oscillation amplitude [30]. Viscoelastic damping has been modeled as well [34-35], in which case viscoelastic stiffness determines the natural frequency and oscillation amplitude is viscoelastic damping parameter dependent [36]. Parametric excitation of vibration is provided by non-constant pulley velocity [37], pulley eccentricity [30], belt inertia [25-26], and belt-pulley tooth meshing [30,31,35].

2.3 Data-Drive Approach to Belt Tension Monitoring

It is well known that belt relaxation, due to creep and polymeric relaxation [38], and pulley misalignment [39-40] cause temporal and spatial tension distributions in belt drives. However, assumptions used in the current active tension monitoring schemes can be restrictive and overly simplified. This results in false positives, where the belt is calculated to be within tension specification when, in reality, it is either not tensioned properly or is misaligned, as well as false negatives, where the belt is calculated to be out of specification when it is actually within specification. In reality, the power transmission belts exhibit more complex behavior induced by deviation from the assumptions inherent in the string model. These shortcomings, will be tackled by a newly derived data-driven method based on observing and characterizing distributions of features extracted from belt deformations sensed via strain gauge measurement. The most sensitive feature set will be determined by the experiments performed on a specially designed testing device, called the Test Stand in the remainder of this paper. Finally, these experimental results will motivate the design of a novel device for excitation of belt vibrations that lead to more consistent results of the existing belt monitoring procedure.

3 EXPERIMENTS ON TRANSMISSION BELT STAND

3.1 Test Stand

Estimation of belt tension from the fundamental frequency of manually induced transverse vibrations of the belt is the current industrial standard. Fundamental frequency is a function of belt length, tension, and linear mass density. Linear mass density was assumed constant. Furthermore, the position of impulsive excitation for belt vibration was assumed to affect belt vibrations.

To systematically study the effect of these variables on the characteristics of belt vibrations, two sets of experiments were performed on the specially designed test stand on which we could accurately measure and manipulate variables affecting the belt vibrations. The variables and the corresponding mechanisms of control on the test stand, shown in Figure 1, are as follows:

- Belt length, L, is controlled by moving the tensioning clamps.
- Total belt tension, T, is the sum of tensions applied by each of the pistons.
- The ratio of tension on the right side of the belt with respect to the left side of the belt, r, is controlled by the pistons.
- Position of vibration, relative to the belt length, IC_V and width IC_H, is controlled by the mounting platform.

- Excitation of belt vibrations is controlled by the specially designed Initial Velocity Device (IVD). The IVD provides consistent excitation via a solenoid and striking surface.

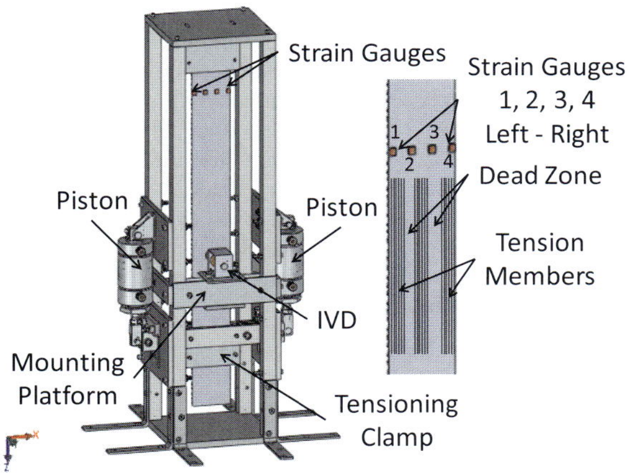

Figure 1: Solid model of test stand

Belt vibrations were measured with 4 strain gauges affixed across the width of the belt. Channels 1 and 4 are placed on stiff tension members, while Channels 2 and 3 were placed in zones with no tension members, as shown in Figure 1.

Lastly, a personal computer based data acquisition system recorded each strain response at 2048 Hz and also controlled the firing time and duration of the IVD strike as illustrated in Figure 2.

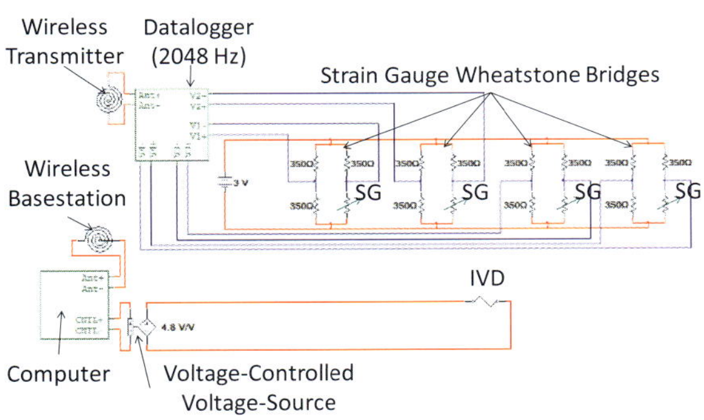

Figure 2: Schematic diagram of data acquisition system

3.2 Experiments

Two three-factor, three-level experiments with a randomized testing order were performed to investigate changes in belt vibration characteristics caused by changes in the independent experimental variables (identified in the previous section).

Specifically, the first experiment, denoted as DOE_1, considered belt length, belt tension, and the ratio of tensions on either side of the belt as the independent variables. The second experiment, denoted as DOE_2, considered belt tension ratio as well as the vertical and horizontal position of the IVD as independent variables.

Dependent variables were drawn from Power Spectral Densities (PSD) evaluated from the Autoregressive (AR) models of the strain gauge readings. An AR model of order 6 was used to study the top three most prevalent modes of the strain gauge measurements. Five features were chosen as a feature vector via Analysis of Variance (ANOVA) based sensitivity analysis [41]. Specifically, linear categorical main-effect models were calculated for each proposed feature, and a one-tailed hypothesis test was run using the following hypotheses

$$H_0: s_{BL}^2 < s_{WL}^2$$
$$H_1: s_{BL}^2 \geq s_{WL}^2 \quad (1)$$

where s_{BL}^2 is the between level sample variance, and s_{WL}^2 is the within level sample variance.

The null hypothesis is rejected if

$$\frac{s_{BL}^2}{s_{WL}^2} > F_{(p,d-1,N-d)} \quad (2)$$

where p is the significance level of the test, d is the number of levels for the factor in question, N is the number of observations, and $F_{(p,d-1,N-d)}$ is the inverse cumulative F-distribution.

Smaller p-values denote higher feature sensitivity to changes in the independent factor. The features with the lowest p-values for both experiments formed the feature vector described below.

It was noted that the use of Channels 1 and 2 retained similar sensitivity as the use of all four channels. The top five features across both Channels 1 and 2 as well as both experiments were retained as a feature vector and are as follows:

- F_1 – Normalized AR(6) PSD peak 1 magnitude
- F_2 – Normalized AR(6) PSD peak 1 frequency variance
- F_3 – Normalized AR(6) PSD peak 1 relative area
- F_4 – Normalized AR(6) PSD peak 2 relative area
- F_5 – Normalized AR(6) PSD peak 1 frequency kurtosis

The feature vector is in good agreement with classical dynamic system descriptors. In dynamic systems theory, each peak in the PSD corresponds do a complex conjugate pair of characteristic roots [42]. Further, each peak is described by its intensity, the corresponding damping ratio, and frequency location [43]. Intensity of the peaks is represented by the relative area features F_3 and F_4, as well as peak 1 magnitude F_1. Damping ratio is indirectly represented by the frequency variance and kurtosis, F_2 and F_5, respectively.

A 3-level, 3-factor experiment is comprised of 9 "classes". Therefore, the degree of dissimilarity between these classes can be seen as a measure of the difference in belt vibration characteristics based on changes in the independent variables. To measure dissimilarity, the overlap [44] of multivariate gaussian distributions characterizing feature vectors within each class was measured by a cosine measure of functional similarity, referred to as the Confidence Value (*CV*) and evaluated as

$$CV(f,z) = \frac{\langle f \cdot z \rangle}{\|f\| \|z\|}, \quad (3)$$

where $\langle f \cdot z \rangle$ is the inner product of two probability density functions f and z, $\|f\|$ is the L_2 norm of a probability density function f, and $\|z\|$ is the L_2 norm of a probability density function z.

In this form, the *CV* value ranges from 0 to 1, where 0 represents no overlap and a value of 1 represents a perfect overlap. Using the CV analysis, it was found that belt vibration characteristics are sensitive to changes in all the independent variables explored.

4 REDUCTION OF TENSION ESTIMATION VARIANCE

4.1 Current, Manual Tension Estimation Technique

The manual technique is performed in the following way. First, the robot is lowered until a belt section of predetermined length is achieved (as measured by a mark on the robot). Next, the technician manually strums each side of the belt, recording the ensuing acoustic emissions with a handheld microphone device. The device returns the fundamental frequency and, with sufficient information about the belt, a tension estimate is derived from the one-dimensional wave equation. This methodology suffers from variance in the belt length, microphone pose relative to the belt, and the location and strength of the initial pluck. These variations are reduced by designing a testing device as described below.

4.2 Device Design

Belt length is controlled through a contact switch on top of the device. Setting belt length is a binary operation where the robot is lowered until the light is illuminated. The IVD is mounted on an X-Y table with a datum arm allowing repeatable placement of the IVD. Two faces of the datum arm are touched to the right side and face of the belt, setting the IVD at half the belt width and ensuring uniform distance from the belt face. The geometry of the device ensures the IVD location is half the length of the belt when contact is registered by the contact sensor. Lastly, the microphone is at a fixed distance from the undeflected belt face and from the site of belt excitation. In this manner, the device effectively reduces variation of all variables considered in this study, with the exception of belt tension and misalignment.

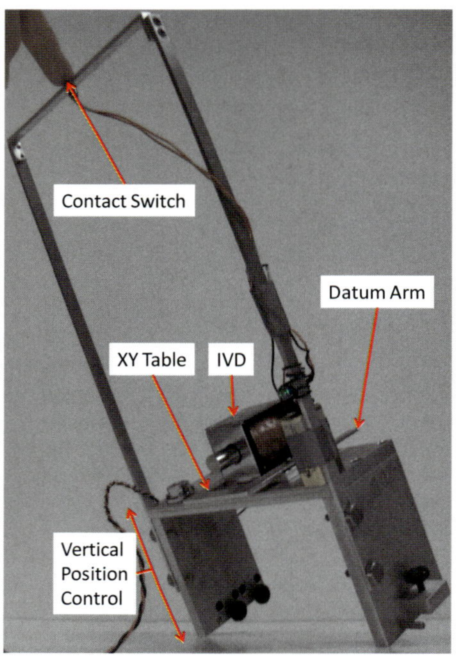

Figure 3: Device to lower tension estimate variance

4.3 Variance Reduction Experiments

Since the manual tension estimation technique relies on the fundamental frequency of belt vibrations to derive the tension estimate, the fundamental frequency will be used as the metric for comparison between the manual technique and the technique based on belt excitations using the newly designed device. In order to validate that the new device lowers the variance of the fundamental frequency, three experiments were run using the fundamental frequency as the dependent variable. In each experiment, the manual tension estimation technique was performed by an experienced technician, and the newly designed device was used to excite and estimate tension from the resulting vibrations. The robots A and B operate in a real production environment while robot operates C in a research laboratory.

The first experiment was run on robot A with technician α performing the manual technique. Belt length for the estimation techniques was reset between each measurement, simulating multiple Preventive Maintenance (PM) sessions. For both the manual technique as well as the one using the new device, 30 tension estimates were gathered.

The second experiment was run on robot B with technician α performing the manual technique. Belt length was held constant for all measurements, simulating multiple estimates within a single PM session. Similarly to the first experiment, 30 tension estimates were gathered for both tension estimation techniques.

The last experiment was run on robot C with technician β performing the manual technique. Belt length followed the pattern of the second experiment. In this experiment, only 20 measurements were gathered.

4.4 Variance Reduction Results

For each of the 3 experiments conducted, statistical hypothesis tests were used to compare the frequency measurement samples. The measurements obtained with excitations from the new device constitute the first sample set and the measurements taken manually constitute the second sample set.

Under the assumption of asymptotic normality of frequency measurements, the one tailed F-test can be used to test the null hypothesis, H_O, that the variance of the sample obtained using the new device, s_1^2, is equal to that obtained using the fully manual technique, s_2^2 (currently used in the industry), against the alternative hypothesis, H_a, that s_1^2 is lower than s_2^2.

$$H_o : s_1^2 = s_2^2 \qquad H_a : s_1^2 < s_2^2 \tag{4}$$

The test statistic is

$$F = \frac{s_1^2}{s_2^2} \tag{5}$$

The null hypothesis is rejected if

$$F < F_{(\alpha, n_1-1, n_2-1)} \tag{6}$$

Table 3 shows the results of the hypothesis tests, at a 95% confidence interval. The variance of the new device is uniformly lower than the manual technique within each robot.

Table 3: The new device significantly reduced the standard deviation of tension estimates in all experiments

		Robot A		Robot B		Robot C	
		New Device	Current Techniq	New Device	Current Techniq	New Device	Current Techniq
Statistics	*s [Hz]*	0.06	0.69	0.14	0.45	0.02	0.03
	n [samples]	30	30	30	30	20	20
F-test		*s1 < s2* (sigificant)		*s1 < s2* (sigificant)		*s1 < s2* (sigificant)	

5 CONCLUSIONS AND FUTURE WORK

In this work, we explored the shortcomings of the standard, fully manual technique for static monitoring of belt tension and misalignment in the belt-driven AMHS systems in modern semiconductor manufacturing systems. The manual technique is theoretically grounded in the one dimensional wave equation and estimation of belt tension from the fundamental frequency of belt vibrations. Belt vibrations are excited by manually plucking the belt and the ensuing sounds emitted by the vibration are recorded. This technique suffers from variance in estimated tension and a method to reduce this variation was sought.

A set of experiments was performed on a specially designed test stand, which measured belt deformations by an array of strain gauges to systematically explore variables influencing the belt vibration behavior. Specifically, the effect of changing the belt length, the belt tension, the belt tension ratio, and the location of excitation of belt vibrations on the belt characteristics were quantified via two 3-factor, 3-level experiments. Features that were descriptive of the belt vibration dynamics were extracted using AR models of belt strain signals and the most descriptive set of features was chosen through an ANOVA based sensitivity analysis.

It was found that the use of two strain gauges placed at the edges of the belt resulted in nearly the same sensitivity to changes in the independent variables as the use of three or four strain gauges placed across the width of the belt. This finding suggests that only two sensors are required for estimation of the belt tension and the difference in the belt tensions on 2 different sides of the belt. Furthermore, this result indicated that the best location for these sensors lies on top of the stiff tension members encapsulated in the polyurethane matrix.

ANOVA sensitivity analysis returned a feature vector that is in good agreement with classical dynamic system descriptors. In addition, the Confidence Value (*CV*) analysis showed that the belt vibration was sensitive to the belt length, the belt tension, the belt tension ratio, and the location of belt vibration excitation. Thus, in order to reduce the variance of belt tension estimates drawn from the current belt monitoring technique, the variance in the belt length and the location of the belt vibration excitation must be reduced.

This finding motivated the design of a novel device to reduce variations in the belt length and the location of belt vibration excitation in a real belt-driven AMHS system widely used in semiconductor manufacturing.

Experiments were run to establish the tension estimation variance when a human completely executed the manual technique (standard approach) and the tension estimation variance when the newly designed device was used for belt excitation and signal collection. The results of these experiments showed a uniform, statistically significant reduction of the standard deviation of tension estimates over all robots tested. This validates the idea that the variance of belt tension estimation can be reduced by better control of the belt length and location of initial excitation of belt vibrations. Lower variation of tension estimates implies that the precision and resolution of tension estimates is higher when the new excitation device is used, as opposed to the results using the standard, fully manual technique.

Despite the improvements in offered by findings of this work, the resulting belt tension monitoring method is still inherently intrusive because the belt-driven AMHS system must be taken out of production for tension estimation. Obviously, estimating the belt tension while the belt-driven AMHS system is in production would represent a large contribution to the CBM in belt drive applications. Belt monitoring using online tension estimation is the subject of future research.

6 WORKS CITED

[1] Daifuku America Corporation, (2008) *Daifuku Material Handling and Beyond.* [Online]. http://www.daifukuamerica.com/case_studies/.

[2] USPS, Material Handling. (2001) *2001 Comprehensive Statement on Postal Operations.* [Online]. http://www.usps.com/history/cs01/c2f-4.htm.

[3] Agrawal, G. and Heragu, S., 2006, A Survey of Automated Material Handling Systems in 300-mm Semiconductor Fabs. IEEE Transactions on Semiconductor Manufacturing, 19/1:112-120.

[4] Feare, T., 2000, The World's Top 20 System Suppliers. Modern Materials Handling. 55/2:68-71.

[5] Rogers, L., 2009, Top 20 Systems Suppliers. Modern Materials Handling. 64/4:23-24.

[6] Ferret, (2004) *Ferret: Australia's Manufacturing and Industrial Directory.* [Online]. http://www.ferret.com.au/n/The-basics-of-predictive-preventive-maintenance-n693449.

[7] Fawcett, J., 1981, Chain and Belt Drives - A Review. Shock and Vibration Digest, 13/5:5-12.

[8] Childs, T., et al., 1998, Automotive Timing Belt Life Laws and a User Design Guide. Proceedings of the Institution of Mechanical Engineers, Part D: Journal of Automobile Engineering, 212/5:409-416.

[9] Tao, W., et al., 2007 Industrial Robot Track Modeling and Vibration Suppression. Industrial Robot: An International Journal, 34/4:317-325.

[10] Tao, W., et al., 2006, Residual Vibration Analysis and Suppression for SCARA Robots in Semiconductor Manufacturing. International Journal of Intelligent Control and Systems, 11/2:97-105.

[11] Fawcett, J., Burdess, J. and Hewit, R., 1989, Belt Natural Frequency as an Indicator of Belt Tension. Proceedings of 1989 International Conference on Power Transmission and Gearing, 1989, 25-29.

[12] Fawcett, J., Burdess, J. and Hewit, R., 1988, Initial Tension Setting for Drive Belts. Automotive Engineering, 13/1:70-71.

[13] Fawcett, J. and Burdess, J., 1980, An Experimental Investigation of the Vibration of Toothed Belts. Proceedings of ASME International Power Transmission and Gearing Conference.

[14] Fujii, A., et al., 2002, Analysis of the Accessory Belt Lateral Vibration in Automotive Engines. JSAE Review, 23:41-47.

[15] Figliolini, G. and Ceccarelli, M., 2002, A Novel Articulated Mechanism Mimicking the Motion of Index Fingers. Robotica, 20:13-22.

[16] O' Halloran, D., Wolf, A. and Choset, H., 2005, Design of a High-Impact Survivable Robot. Mechanism and Machine Theory, 40:1345-1366.

[17] Mitobe, K., et al., 1997, Control of a Biped Walking Robot During the Double Support Phase. Autonomous Robots, 4:287-296.

[18] Beer, F., Johnston, E. and DeWolf, J., *Mechanics of Materials.* 3rd. Boston : McGraw-Hill, Inc, 2002. 0073659355.

[19] Palmer, R. and Jarvis, J., 1977, The Bonding of Birefringment Coatings to Composite Flat Belt Drive Materials. Experimental Mechanics, 17/5:198-200.

[20] Greenberg, M., *Advanced Engineering Mathematics.* 2nd. Upper Saddle River : Prentice-Hall, 1998. pp. 1017-1035.

[21] Capps, R., 1983, Dynamic Young's Moduli of Some Commercially Avalailable Polyurethanes. Journal of Acoustical Society of America, 73/6:2000-2005.

[22] Durvasula, S., 1968, Natural Frequencies and Modes of Skew Membranes. The Journal of the Acoustical Society of America, 44/6:1636-1646.

[23] Beikmann, R., Perkins, N. and Ulsoy, A., 1996, Free Vibration of Serpentine Belt Drve Systems. Journal of Vibration and Acoustics, 118/3:406-413.

[24] Beikmann, R., Perkins, N. and Ulsoy, A., 1996, Nonlinear Coupled Vibration Response of Serpentine Belt Drive Systems. Journal of Vibration and Acoustics, 118/4:567-674.

[25] Alciatore, D. and Traver, A., 1995, Multipulley Belt Drive Mechanics: Creep Theory vs. Shear Theory. Journal of Mechanical Design, 117/4:506-511.

[26] Gerbert, G., 1996, Belt Slip - A Unified Approach. Journal of Mechanical Design, 118:432-438.

[27] Leamy, M. and Wasfy, T., 2002, Analysis of Belt-Drive Mechanics Using a Creep-Rate-Dependent Friction Law. Journal of Applied Mechanics, 69:763-771.

[28] Firbank, T., 1970, Mechanics of the Belt Drive." International Journal of Mechanical Sciences, 12:1053-1063.

[29] Reynolds, O., 1874, Creep Theory of Belt Drive Mechanics. Engineer, 28:396-397.

[30] Callegari, M., Cannella, F. and Ferri, G., 2002, Multi-body Modelling of Timing Belt Dynamics. Proceedings/Institution of Mechanical Engineers, 217/K:63-75.

[31] Leamy, M. and Wasfy, T., 2005, Time-Accurate Finite Element Modelling of the Transient, Steady-State, and Frequency Responses of Serpentine and Timing Belt-Drives. International Journal of Vehicle Design, 39/3:272-297.

[32] Kerkkanen, K., Garcia-Vallejo, D. and Mikkola, A., 2006, Modeling of Belt-Drives Using a Large Deformation Finite Element Formulation. Nonlinear Dynamics, 43:239-256.

[33] Garcia-Vallejo, D., Valverde, J. and Dominguez, J., 2005, An Internal Damping Model for the Absolute Nodal Coordinate Formulation. Nonlinear Dynamics, 42:347-369.

[34] Schiessel, H., et al., 1995, Generalized Viscoelastic Models: Their Fractional Equations with Solutions. Journal of Physics A: Mathematical and Theoretical, 28:6567-6584.

[35] Cepon, G. and Boltezar, M., 2009, Dynamics of a Belt-Drive System Using a Linear Complementary Problem for the Belt-Pulley Contact Description. Journal of Sound and Vibration, 319:1019-1035.

[36] Zhang, L. and Zu, W., 1998, Non-Linear Vibrations of Viscoelastic Belts, Part I: Free Vibration Analysis. Journal of Sound and Vibration, 216/1:75-91.

[37] Karolev, N. and Gold, P., 1995, Load Distribution of Timing Belt Drives Transmitting Variable Torques. Journal of Mechanisms and Machine Theory, 30/4:533-567.

[38] Palmer, R. and Jarvis, J., 1977, Tension Relaxation of Composite Flat Belt Drive Materials Under Simulated Drive Conditions. Strain, 14/1:16-20.

[39] Castellini, P., 2002, Vibration Measurements by Tracking Laser Doppler Vibrometer on Automotive Components. Shock and Vibration, 9:67-89.

[40] Castellini, P. and Mantanini, R., 2002, Automotive Components Vibration Measurements by Tracking Laser Doppler Vibrometery: Advances in Signal Processing. Measurement Science and Technology, 13:1266-1279.

[41] Hinkelmann, K. and Kempthorne, O., *Design and Analysis of Experiments: Volume I Introduction to Experimental Design.* New York : John Wiley & Sons, Inc., 1994. pp. 145-196. Vol. 1. 0471551783.

[42] Pandit, S. and Wu, S., *Time Series and System Analysis With Applications.* Malabar : Krieger Publishing Company, 2001. pp. 11-132. 1575241692.

[43] Franklin, G., Powell, J. and Emami-Naeini, A., *Feedback Control of Dynamic Systems.* 5. Upper Saddle River : Pearson Education, Inc., 2006. pp. 320-325. 0131499300.

[44] Falkowski, B., 1998, On Certain Generalizations of Inner Product Similarity Measures. Journal of the American Society for Information Science, 49/9:854-858.

Development of a Drop Tube Reactor to Test and Assist a Sustainable Manufacturing Process

F. Hampp, I. Janajreh
Department of Mechanical Engineering, Masdar Institute of Science and Technology, Abu Dhabi, UAE

Abstract
This work outlines the development of a Drop Tube Reactor (DTR) following a simulation assisted design. The general purpose of the DTR is to simulate the thermochemical conversion process of solid hydrocarbon feedstock, such as coal, coke, biomass, and industrial waste under controlled reaction conditions. It supports the development of more efficient and flexible conversion devices (i.e. reactors, gasifier, etc.) to accommodate none conventional solid fuel, i.e. biomass and solid waste material. The DTR is extensively used in investigating the thermochemical pathways and in the development of high fidelity reactive flow models. As this device is custom made, the thermochemical loading, i.e. thermal flow/heating and exothermic reactivity requires detailed flow analysis. This work attempts to provide guidelines for the DTR development. It details the functionality of the main device components, investigates the flow conditions and suggests the tube material by considering a variable heating flux and mass flow rate in a conjugate heat flow environment. The results demonstrated how basic analytical calculations, CFD simulation, and conjugate heat analysis influence design decisions. In particular, the effect of the heat flux and mass flow rate and their effect on the flow pattern are investigated. Results have shown that the adjustment of the wall heat flux leads to a more predictable change in temperature whereas the variation in the mass flow rate results in a more predictable change in the velocity profile. The residence time varies linearly with mass flow rate and nearly parabolic with wall heat flux. An increase of the heat flux requires adjustment of the mass flow rate to maintain particle residence time at a constant value.

Keywords:
Gasification, Drop Tube Reactor, Sustainable Product Development, Small Scale Experiments, Simulation Assisted Design

1 INTRODUCTION

Current climate changes demand significant revisions to fossil fuel usage in order to provide a better prospective for energy security. Therefore, one needs to explore potential renewable and sustainable energy resources and increase the efficiency of existing fossil fuel infrastructures. This work supports the improvement of thermochemical conversion devices and the development of more efficient technologies for alternative fuel, i.e. municipal solid and industrial waste, by relying on simulation assisted design.

The thrust of our research at the Waste to Energy Laboratory at the Masdar Institute (MI) is to enhance the gasification efficiency of inhomogeneous waste feedstock. Gasification is a thermochemical conversion process through which hydrocarboneous feedstock is converted into a combustible syngas, essentially a mixture of carbon monoxide (CO) and hydrogen (H_2). Inside a gasifier, the feedstock experiences heating, moisture release, devolatilization, and combustion events. Each event is associated with a different time scale, temperature range and heating rate that steer the chemical kinetics. Information on the limiting temperature and residence time for a given feedstock provides the necessary tuning and design parameters for the conversion device to achieve the maximum efficiency. Comprehensive review on state of the art gasification technologies is given by A.J. Mincher in [1]. Three gasifier technologies are predominant: Entrained Flow Gasifier, Fluidized Bed Gasifier, and Moving Bed Gasifier. For gasification based power generation, the gasifier is, generally, combined with a gas turbine and steam turbine assembly, commonly referred to as Integrated Gasification Combined Cycles (IGCC). An alternative option is a syngas operating fuel cell. Current gasifier technologies however, have several short comings, including: local heating, injector fouling, feedstock inflexibility, slag blockage, scalability etc.

Experimental investigations of gasification are carried out by several [2–6] researchers emphasizing on material characterization i.e. proximate and ultimate analysis as well as calorific values on one hand and investigating the deduction of exothermic and endothermic gasification reactions and their chemical kinetics on another.

The solid feedstock particles experience processes that range from inert heating, moisture release, and swelling to devolatilization of gaseous species and complex surface reactions. Reactions on the particle surface are identified as heterogeneous reactions, whereas reactions of gaseous species, including the oxidizer, volatiles, and gasification products, are identified as homogeneous reactions. The chemical reactions depend on numerous parameters such as feedstock characteristic including elemental fractions, particle size, and porosity, but also on reaction conditions e.g. temperature, heating rate, residence time, and turbulence. These parameters and the onset of the reactions need to be fully investigated in order to design higher efficiency reactors.

Figure 1 illustrates schematically the DTR. Its intention is to simulate the environment of an actual thermochemical conversion device, i.e. gasifier, combustor, furnace, or a general thermochemical reactor. It simplifies the actual gasification process by eliminating two spatial dimensions and applying controlled reaction environment boundary conditions. Controlled conditions enabling the investigation of the conversion influence parameter, i.e. temperature, flow conditions, reaction kinetic, feedstock variations, etc. subjected to the actual reacting environment.

A DRT is an indispensible experimental test facility in gasification and solid combustion research. However, no detailed literature about its development is publically available. S. Kajitani et al. investigated the influence of pressure on the gasification rate of coal within a pressurized DTR [4]. The DTR was developed to gasify coal under the same conditions as a projected 250MW IGCC power plant in Japan and was used to investigate the extent of char reaction while exposed to high pressure, high temperature and different oxidizers. M. Cloke et al. [5] used a DTR to study the gasification behavior of numerous coals, i.e. char burnout, intrinsic reactivity, and char morphology. Shan Ouyang et al. [6] has given a general overview of existing DTR's for different operating conditions. These studies suggest that the DTR can be designed and developed to investigate the different feedstocks and finding their optimal and actual gasification parameters.

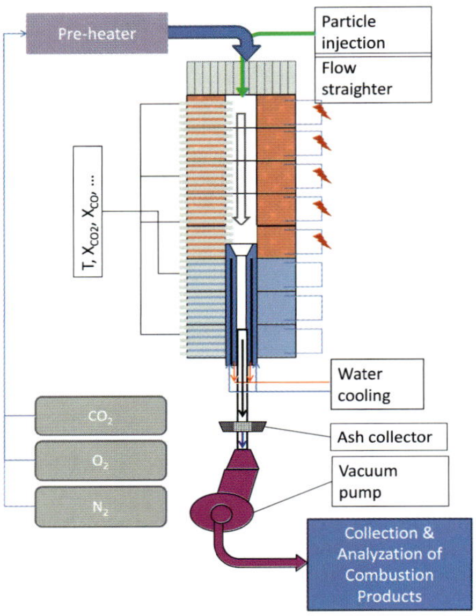

Figure 1: Schematic of a DTR

As the DTR is not readily available on the market, the design and development of the device is undertaken by the Waste to Energy research group at the MI. The intended design operates at atmospheric pressure up to a maximum temperature of 1400K when subjected to 4.6 kW electrical power. The objective of this paper is to detail the development process of this device and to describe the components and operation conditions.

2 DESIGN SPECIFICATIONS

The DTR is designed to accommodate different waste streams i.e. industrial waste, MSW, pet coke, coal, oil shale, and biomass subjected to a wide range of operation conditions as adopted in fluidized bed and entrained flow gasifiers. **Table 1** shows the desired specifications of the reactor furnace.

Table 1: Furnace Specifications

Parameter	Value
Height	1005 mm
Depth	475 mm
Width	390 mm
Heated length	750 mm
Length constant temperature ΔT 10K[1]	250 mm
Power rating	4.6 kW
Weight	105 kg
T_{max}	1100 °C

The DTR is fitted within an outer casing made of stainless steel sheets and is internally insulated with low thermal conductivity vacuum-formed ceramic fiber plates. A hinged tube furnace configuration is used to allow easy exchange of the working tube. Two ceramic semi pipes protect the heating elements from dust, feedstock, and other contaminants. A programmable temperature controller, featuring adjustable heating rates, regulates the temperature of the heating elements and, simultaneously, reactor environment.

For versatility, the DTR is equipped with two tubes; a transparent quartz glass tube, and an Advanced Powder Metallurgy (APM) tube. The quartz glass tube provides a complete optical access and withstands a relatively lower temperature range (up to 1100 °C) than the APM tube. The APM tube is machined with multiple bore holes given probe access to the reactor environment in the mean of thermocouples, velocity meters, pressure, or gas sampling syringes. It can withstand temperatures up to 1250 °C. **Table 2** summarizes the dimensions and specifications of the tubes.

Table 2: Tube Specifications (quartz glass tube – tube A; APM tube – tube B)

Parameter	**Value**
Length tube A	1540 mm
Outer diameter tube A	72 mm
Inner diameter tube A	66 mm
Material tube A	quartz glass
Specialty	transparent
Length tube B	1540 mm
Outer diameter tube B	75 mm
Inner diameter tube B	66 mm
Material tube B	APM
Specialty	measurement access

The carrier gas and oxidizer inlet will be positioned at the top flange, with the gas outlet at the bottom flange. The gas control will operate between 100–1000 l/h given the widest possible range of particle residence time. A simple gas swapping method will be implemented such that just the flow meter requires recalibration with respect to the gas molecular weight. Initially, the flow meter is calibrated for the use of air.
The Lambda Laboratory Instruments powder dosing will be utilized as a particle feeder as it can be fitted to the very top of the tube next the gas supply connection. The particle mass flow rate of this device is calibrated by using Sodium-Chloride (NaCl), which has a density of approximately 2.165 g/cm^3. The feeding rate ranges from 50mg per minute up to 50 g/min for NaCl, and digitally adjustable in 1000 increments. Assuming a coal density of 1.4 g/cm^3, the coal mass flow rate can be adjusted from approximately 33 mg/min up to 33 g/min. The powder dosing device is equipped with a storage capacity as large as 200ml.

Two different gas analysis methods will be used, electrochemical and spectrometry. The EMS five gas analyzer allows online reading of the specie CO, CO_2, HC, NO_x, and O_2, electrochemically, enough to assess the reactions process and main species. The limitation of this

[1] Length of constant temperature ΔT 10K specifies that the temperature varies less than 10 Kelvin within 250 mm. Those 250 mm are located in the middle of the heated length.

device is the relatively large sample size and limited number of detectable species, yet it provides fast and online reading. Thermo Scientific DSQ II GC-MS will be used to analyze collected DTR gas samples. As the gases are inhaled, their reactions are quenched (rapidly cooled), and the sample is stored in a "tedlar bag". This process addresses the reproductively of samples and reduces measurement errors due to ongoing reactions. The samples are transported from the DTR to the GC-MS within the bag, where the trapped gases are then manually injected into the specific GC-MS detection column.

3 ESTIMATION OF DTR CONDITION

Figure 2 depicts the thermochemical conversion of plywood feedstock as subjected to controlled heating environment. The figure shows the evolution of the moisture, volatile, and combustion of char in a series of events. These results, referred to as proximate analysis, are obtained by a thermogravimetric analyzer that also monitors the evaluation of heat release of pyrolysis and char combustion, as depicted in figure 2. Elemental analysis is obtained by the CHNOS thermoscientific Flash analyzer at Masdar Institute. Combined results from proximate and elemental analysis enable to infer the chemical formula of the feedstock to regulate the equivalence ratio as desired. Summary of these tests, applied to plywood and coal as the baseline feedstock, are given in **Table 3**.

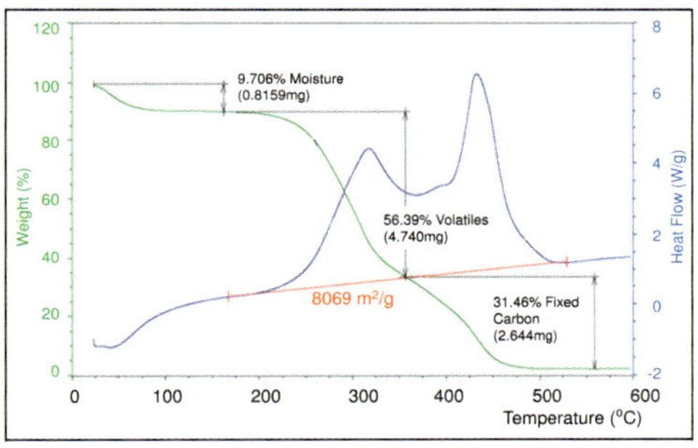

Figure 2: Proximate composition and heat flow for plywood sample subjected to STA

Table 3: Proximate and ultimate analysis for coal and plywood [7]

Proximate analysis	Baseline Coal Mass%	Plywood Mass%
Moisture	2.00	9.71
Volatile	36.66	56.39
Fixed carbon	52.92	31.46
Ash	8.42	2.44
C	79.88	49.59
H	5.80	6.28
O	11.55	43.74
N	1.6	0.39
S	1.10	0.00
Total	100.00	100.00
Heating value [J/kg]	3.30E+07	1.79E+07
Molecular weight	15.04	24.22
Heat of formation [J/kmol]	4.96E+08	4.34E+08
Coal	$C_{1.0000}H_{0.8652}O_{0.1086}N_{0.0179}S_{0.0062}$	
Plywood	$C_{1.0000}H_{1.5087}O_{0.6621}N_{0.0066}S_{0.0000}$	

Boundary Condition for Momentum Conservation:

Estimation of the fluid properties such as density and viscosity is required to evaluate the range of flow conditions. These are estimated under the following assumptions:

- Constant properties in radial and circumferential directions at inlet and outlet
- Boussinesq approximation is applied (density changes only occur due to temperature changes and not due to pressure gradients. This approach is common for low flow velocities and laminar flow regimes)
- Particle are carried by the carrier gas and therefore have the same velocity as the gas
- Outflow temperature is equal to T_{max} of the tube

The lower and upper temperature limits of the furnace are used to calculate the inlet (1) and outlet (2) density of the fluid following the ideal gas law:

$$\rho_{1,2} = \frac{M \cdot p}{R \cdot T_{1,2}} \quad (1)$$

Where R is the gas constant (8.314 [J/K*mol]), M is the molecular weight [kg/kmol], and p is the tube pressure (atmospheric pressure). Values for the viscosity (μ) at T_1 and T_2 can be taken from literature [8]. For a given tube diameter (D) and flow condition Reynolds number (Re) the bulk velocity (u) can be estimated.

$$u_1 = \frac{\mathrm{Re}_1 \cdot \mu_1}{\rho_1 \cdot D} \quad (2)$$

Having a constant cross section area A enables one to estimate the averaged outlet velocity:

$$u_2 = \frac{u_1 \cdot \rho_1}{\rho_2} \quad (3)$$

Boundary Conditions for Energy Conservation:

The given temperature distribution in the DTR is complex due to numerous heat sinks and sources. The supplied electrical power is implemented as a wall heat flux boundary condition and, as the particle combustion event, considered a heat source. The heating of gas and particles, particle moisture release, and devolatilization are considered as heat sinks. **Figure 3** depicts the heat flux that goes in and leaves out of the system.

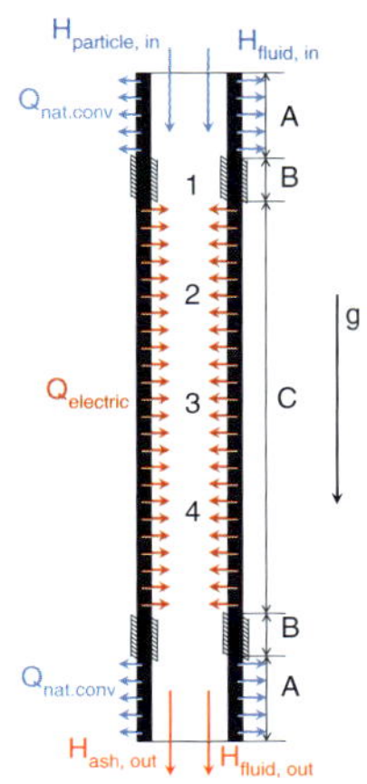

Figure 3: Heat Flux Diagram

Area A is subjected to natural convection $Q_{nat,conv}$ with the surrounding air. The adjacent tube sections (B) are considered to be perfectly insulated/adiabatic. The middle of the tube, section C, is subjected to a positive heat flux due to electric heating. The incoming heat fluxes $H_{fluid,in}$ and $H_{particle,in}$ are due to the enthalpy inflow of gas and particles into the tube and $H_{fluid,out}$ and $H_{ash,out}$ are the corresponding enthalpy flows out of the system, respectively. The two phases, gaseous and discrete particle, experience a change in their enthalpy along the tube, zone 1 through zone 4, as explained below.

- Zone 1: Energy is consumed by the feedstock and fluid through heating. The feedstock is subjected to inert heating, no change in mass or chemical composition of the particles.
- Zone 2: Drying of the feedstock, moisture is released from the particles and added to the gas flow. Energy is consumed due to fluid heating and moisture is released from the particle which is associated with mass change.
- Zone 3: Devolatilization species e.g. CO, CH_4, H_2, and potentially tar in the form of vapor are released from the particle. The particles are reduced into char/solid carbon and the event is associated with endothermic energy and particle mass loss.
- Zone 4: Char gasification and volatile combustion occurs where oxidizer present. The event is associated with exothermic reactions and substantial mass reduction depending on the stoichiometry.

Electric Power Heat Source:

As the maximum electrical power (P_{el}) of the oven is prescribed as 4.6 [kW] a corresponding wall heat flux of approximately 20 [kW/m^2] is applied, given the heater surface area by D 0.1 [m] and L of 0.75 [m]. This is an essential boundary condition for the flow analysis.

Heat Sink for Fluid Flow:

For constant gas properties at the inlet and given inflow velocity from equation (2), the mass flow rate ($\dot{m}_G$) of the gas is calculated as:

$$\dot{m}_G = u_1 \cdot A \cdot \rho_1 \tag{4}$$

Where u_1 is the inlet gas velocity, A is the tube cross section area, and ρ_1 is the inlet gas density. The inflow gas temperature is 300K and the desired outlet gas temperature is 1300K. Energy balance is applied to the tube in a control volume approach under the assumption of zero pressure gradient and constant volume, that is:

$$\dot{Q}_G = \dot{m}_G \cdot c_{p,G} \cdot \Delta T \tag{5}$$

Where $\dot{Q}_G$ is the required heat flow rate, $c_{p,G}$ is the specific heat capacity of the gaseous phase at constant pressure, and ΔT the difference in gas temperature between outlet and inlet. The gas mass flow rate is bounded such that the electrical power multiplied by the furnaces overall heating efficiency η_{eff} has to be higher than the required heat flow rate $\dot{Q}_G$ that is:

$$\dot{Q}_G \le P_{el} \cdot \eta_{eff} \tag{6}$$

Heat Sink for Feedstock:

The evaluation of the feedstock can also be modeled as a simple heat sink/source for design purposes while assuming the following:

- Continuous flow for the particles
- Heating, moisture release, devolatilization, and char gasification correspond to different events
- Time is not the limiting factor
- Homogeneous feedstock composition

As the mass flow rate ($\dot{m}_F$) of the feedstock is user prescribed, the heat flux is estimated for each event as follow:

1. Inert Heating:
This event elapses over the period that raises the particle to the boiling temperature of the moisture content within the solid particle. The required event heat flux ($\dot{Q}_H$) is estimated as:

$$\dot{Q}_H = \dot{m}_F \cdot c_{p,F} \cdot \Delta T \tag{7}$$

The specific heat capacity ($c_{p,F}$) is estimated from the STA data corresponding to the specified constant mass (flat weight evolution) heating event.

2. Latent Heat of Evaporation:
Assuming moisture is released at a constant temperature, the required heat flux ($\dot{Q}_M$) is estimated as:

$$\dot{Q}_M = \dot{m}_F \cdot f_M \cdot h_{evap,M} \tag{8}$$

Where (f_M) is the moisture mass fraction and ($h_{evap,M}$) is the enthalpy of evaporation. As the moisture mass released to the surrounding gas is small compared to the mass of gas, the change in relative humidity is ignored. Moisture content of the feedstock is determined from the STA. For example the moisture mass fraction (f_m) of petroleum coke is smaller than 0.5%, whereas the f_m of plywood is 9.7% (fig. 2) and for general biomass can reach as high as 50%. Francisco V. Tinaut et al. [9] specified the moisture content of their tested biomass feedstock with 10.91%.

3. Devolatilization / Pyrolysis
Feedstock devolatilization is essentially an endothermic process, thereby another heat sink. In principle, it occurs at temperatures above the moisture's boiling temperature. A two step pyrolysis model [10] is used to describe the process such that:

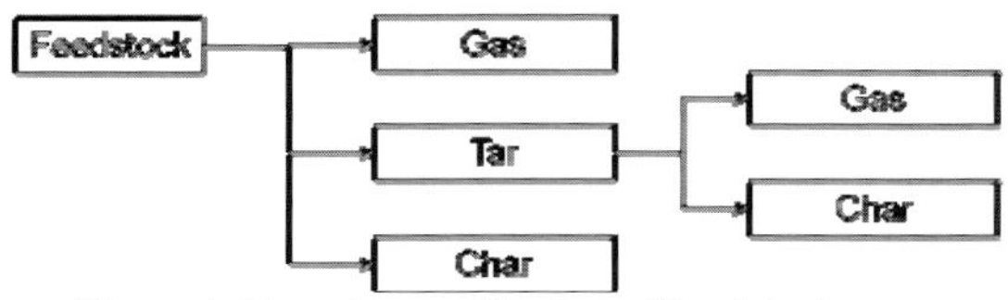

Figure 4: Two step gasification of feedstock

The required heat flux ($\dot{Q}_D$) for the devolatilization process is estimated as:

$$\dot{Q}_D = \dot{m}_F \cdot \left[(1 - f_M) \cdot c_{p,D} + f_M \cdot c_{p,M}\right] \cdot \Delta T + f_V \cdot \dot{m}_F \cdot h_{evap,V} \tag{9}$$

Where $c_{p,M}$, is the specific heat capacity of the water vapor, $c_{p,D}$ is the specific heat capacity of the dry feedstock, f_V is the volatile content fraction, and $h_{evap,V}$ is the enthalpy of evaporation of the mixture of released gases. Using air as carrier gas, the released gases experience an exothermic oxidation process that releases heat to the environment. It is hard to quantify this heat source since it heavily depends on the feedstock. For coal, one can assume that the heat release of the volatile combustion is generally smaller than the heat of char combustion due to their respective fractions.

4. Char Gasification
As the aforementioned processes are primarily endothermic processes, the actual char gasification is an exothermic process. The bomb calorimeter at the MI laboratory can be used to assess the amount of energy released on a dry basis. In general, the energy release due to char combustion is expressed as

$$C_{(s)} + 0.5 \cdot O_2 \rightarrow CO \qquad + 111 \text{ [kJ/kmol]} \tag{10}$$

$$CO + 0.5 \cdot O_2 \rightarrow CO_2 \qquad + 283 \text{ [kJ/kmol]} \tag{11}$$

Thus, the total of heat of combustion (HOC) is +394 [kJ/kmol]. A typical bituminous coal has a fixed carbon fraction near to 53%. Considering a particle feeding rate of 33 [g/min], the coal carbon feeding rate is estimated to be 17.46 [g/min] or 1.5 [mol/min]. The associated heat release from the char combustion is estimated to:

$$\dot{Q}_G = \frac{\dot{m}_{FC}}{\rho_{Carbon}} \cdot HOC \cdot \frac{1}{60} = 9.55W \tag{12}$$

This is a small heating contribution (~2% compared to the applied heating) and even less for other materials such as Plywood, i.e. 5.68W.

The calculations have shown that due to the low mass flow rate of the feedstock in the DTR, one can safely neglect their heat absorbed/released. Given maximum air volume flow rate of 1000 [l/h] and coal mass flow rate of 33 [g/min], the required heat flux is estimated to be approximately 1500 [W] to reach the desired temperature. The 4.6 [kW] furnace capacity appears to be suitable and the excess power accounts for unsteady heating and heat losses.

4 FLOW ANALYSIS

The estimated boundary conditions are used to carry out a conjugated heat transfer analysis. CFD simulation allows better assessment of the flow velocity and temperature field as well as estimating of the heating rate and the conversion residence time. The flow is governed by the mass, momentum, and energy conservation laws that are written as:

$$\frac{\partial \rho}{\partial t} + \vec{\nabla} \cdot (\rho \vec{u}) = 0 \tag{13}$$

$$\frac{\partial (\rho \vec{u})}{\partial t} + \vec{\nabla} \cdot [u \otimes (\rho \vec{u})] + \vec{\nabla} p = 0 \tag{14}$$

$$\frac{\partial E}{\partial t} + \vec{\nabla} \cdot [\vec{u}(E + p)] = 0 \tag{15}$$

Where ρ, t, u, p and E representing density, time, velocity, pressure, and total energy respectively. In reality, the flow is driven by a small pressure gradient in axial direction. This pressure gradient is introduced by the gas supply at the top of the tube. The given geometry lends itself to carry out axisymmetrical analysis. A Dirichlet boundary condition with nil value (no slip) at the wall and a Neumann boundary condition with nil value at the symmetry axis are applied to the momentum equation. An inlet mass flow rate boundary condition is prescribed at the top of the tube and atmospheric pressure is prescribed at the bottom.

Figure 5 shows the thermal resistance circuit for tube furnace components, with the corresponding electrical equivalents. Given the thermal resistance from the outer surface of the tube to the symmetry axis, including intermediate temperatures, T4 – T6, can be evaluated using CFD analysis.

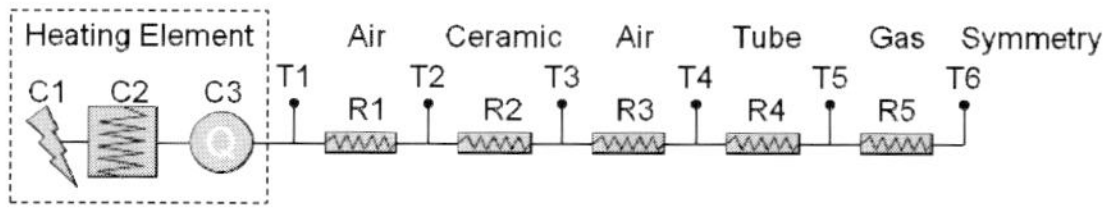

C1: electrical source
C2: heating element
C3: heating element heat source
R1: air resistance between heating elements and ceramic shell – convection, radiation
R2: ceramic shell thermal resistance – conduction
R3: air resistance between ceramic shell and tube wall – convection, radiation
R4: tube wall resistance – conduction
R5: gas resistance – convection (CFD)
T1: Heating element temperature
T2: Ceramic shell outer wall temperature
T3: Ceramic shell inner wall temperature
T4: Outer tube wall temperature
T5: Inner tube wall temperature
T6: Symmetry axis gas temperature

Figure 5: Thermal resistance through the DTR visualized as electric circuit

4.1 Fluid Flow and Conjugated Heat Transfer

The thermal analysis is conducted using the commercial simulation code Fluent 12.1. Due to low Reynolds number, no wall function is implemented and hence the boundary layer is resolved directly with a fine mesh. The flow model is based on a steady state, axisymmetry, viscous, coupled Navier-Stokes and energy equation model. A second order spatial discretization for the convective terms and central for the diffusion term are used achieving momentum and continuity residual as low as 10^{-6}. Gravitational forces are considered and fluid properties are calculated using ideal gas law.

Mesh sensitivity analysis:

To check the grid dependency on the computed results, mesh sensitivity is conducted using three refinement levels as listed in **Table 4**.

Table 4: Mesh Parameter

Parameter	Coarse	Baseline	Fine
Fluid Elements	18750	75000	300000
Axial direction increments	750	1500	3000
Axial cell size [mm]	2.0	1.0	0.5
Radial direction increments	25	50	100
Radial cell size [mm]	1.76	0.88	0.44
Smallest cell in BL [mm]	0.1313	0.0757	0.0387

Resolving the boundary layer demands a high mesh resolution in the near-wall region. The normalized wall distance parameter (y^+) is used to verify the mesh quality within the boundary layer and is defined as:

$$y^+ = \frac{\sqrt{\tau_w / \rho} \cdot y}{\upsilon} \tag{16}$$

Where τ_w is the shear stress at the wall, ρ is the local density, y is the normal distance of the first cell from the wall, and ν is the local kinematic viscosity. Resolving the boundary layer requires $y^+ < 1$ and results of the three mesh levels are verified to satisfy this criterion.

Figure 6 shows a plot of the axial velocity and the temperature at 1/2L downstream for the three refinement levels. The typical laminar velocity profile is not observed due to the buoyancy forces that result from density gradients. These gradients emerge from the wall heat flux and causing a recirculation zone within the tube. The recirculation zone is observed in all meshes and is equally evolved as observed by comparing the vorticity fields. Temperature and axial velocity results show an identical trend for all the mesh levels with an insignificant quantitative difference. No signs of reversed flow at the tube outlet are observed.

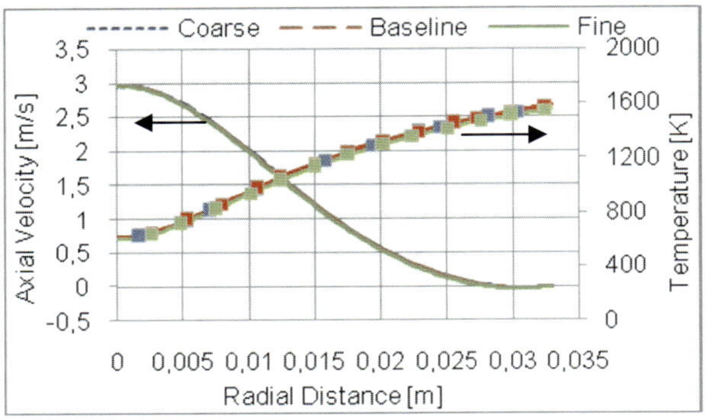

Figure 6: Axial Velocity at 750 mm downstream the Inlet through the domain

The variation in temperature, depicted in **Figure 6**, is in the range of 1 – 2% with the largest deviation near the wall.

Figure 7 shows the velocity and temperature distribution along the symmetry axis. The velocity line plot shows an excellent agreement for all mesh levels along the first 2/3L downstream with marginal discrepancies afterwards caused by the difference in temperature and, therefore fluid properties. The temperature profile shows slight differences, originating from the difference in the resulted wall temperature.

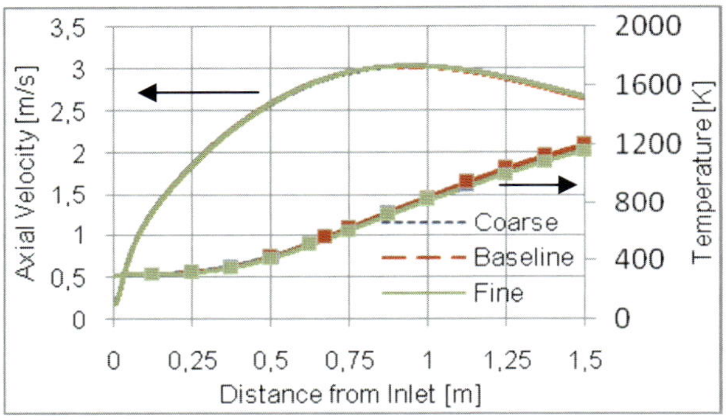

Figure 7: Velocity and temperature profile on the symmetry axis

In general, mesh sensitivity study showed that the baseline mesh is fine enough to carry out a parametric study to the DTR operating conditions. Noting that the high temperature gradients will introduce small, yet acceptable error.

4.2 Parametric Study

A parametric study is conducted to observe the influence of the flow velocity (mass flow rate) and wall heat flux on the steady temperature field and residence time. **Table 5** summarizes the values for the first variation.

Table 5: MFR parameter

	MFR [kg/s]	Volume FR [l/h]	Velocity [m/s]
MFR1	3.4E-05	100	8.1E-03
MFR2	1.0E-04	294	23.9E-03
MFR3	2.0E-04	588	47.7E-03
MFR4	3.4E-04	1000	81.1E-03

The applied wall heat flux is set to 20 [kW/m^2] and results of the axial velocity and temperature along the symmetry axis are depicted in **Figure 8**.

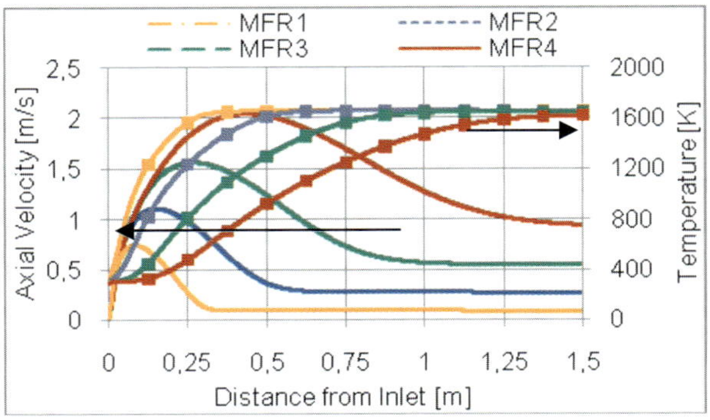

Figure 8: Velocity and temperature profile along symmetry axis

The slopes of the velocity profiles are well correlated. The steep slope of the MFR1 curve which corresponds to low mass flow rate, indicates a smaller recirculation area close to the inlet. This emerges from the heating process which occurs exclusively within the first 1/5L. Downstream from this point, the fluid exhibits constant fluid properties. Considering that the Nusselt number varies less than 15 percent over the whole velocity range, it is assumed that heating of the fluid is preferably a function of temperature difference than of flow velocity. Therefore, as the flow velocity increases, the time for the heating stays constant and the maximum fluid temperature moves downstream as depicted in **Figure 8**. With increasing mass flow rate, the recirculation area stretches due to the longer heating process and continuous change of fluid properties. This is indicated by the smoother and more extended axial velocity profile.

Figure 9 shows the axial velocity and temperature plot at 1/2L downstream along the radial direction. The temperature profile in the cross section is subjected to small variation particularly corresponding to MFR1 and MFR2. The velocity profile, a parabolic function, is typical for laminar flows with constant properties. The MFR3 and MFR4 temperatures continue to change while the velocity can be described by a polynomial function 3rd order.

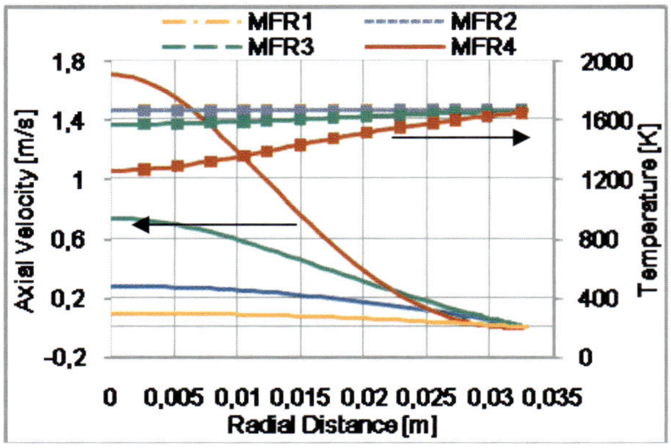

Figure 9: Velocity and Temperature profile 750 mm downstream across domain

The wall heat flux is varied from HF1 5.0, HF2 7.5, HF3 10.0, to HF4 15.0 kW/m^2 to investigate the general temperature profile within the tube. The applied inflow mass flow rate was 2.0E-04 kg/s.

Figure 10 shows the velocity and temperature profile along the symmetry axis for the different wall heat fluxes. A higher wall heat flux results in a higher fluid exit temperature at the cost of lower density. Additionally, constant temperature is evolved further upstream with increasing wall heat flux which emerges from the faster heating of the fluid. Since a high heat flux leads to a fully developed temperature profile at the end of the tube, constant fluid properties can be assumed, and therefore no change in the axial velocity on the symmetry axis is observed downstream 2/3L. Whereas a low heat flux doesn't result in a fully developed temperature profile, and therefore changes in fluid properties still occur. This has a noticeable impact on the velocity profile. A higher wall heat flux, leads to a shorter recirculation area, on the contrary a low heat flux extends the recirculation area.

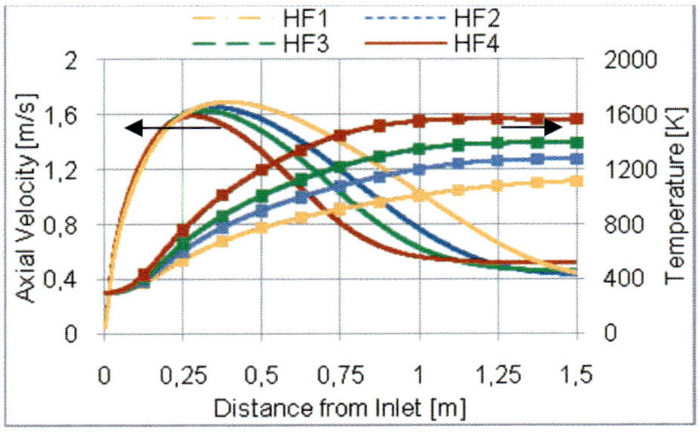

Figure 10: Velocity and temperature profile along the symmetry axis

Figure 11 depicts the velocity and temperature profile in a radial direction at 1/2L distance downstream. The slopes of the temperature profiles are almost equal for all heat fluxes. A slightly steeper slope of the temperature profile with decreasing heat flux is observed that results in a higher temperature difference in radial direction. This is emerged from the fast heating at high heat fluxes and therefore smaller temperature gradients within the domain at given location. The velocity profile shows significant discrepancy for the different wall heat fluxes. This emerges from the different development of the recirculation zone. There is no recirculation at 1/2L downstream distance for high heat fluxes whereas the recirculation is well developed for low heat fluxes. The conservation of mass requires a higher axial velocity at the symmetry axis for a more distinct recirculation zone which leads to a more characteristic velocity profile in radial direction for low heat fluxes.

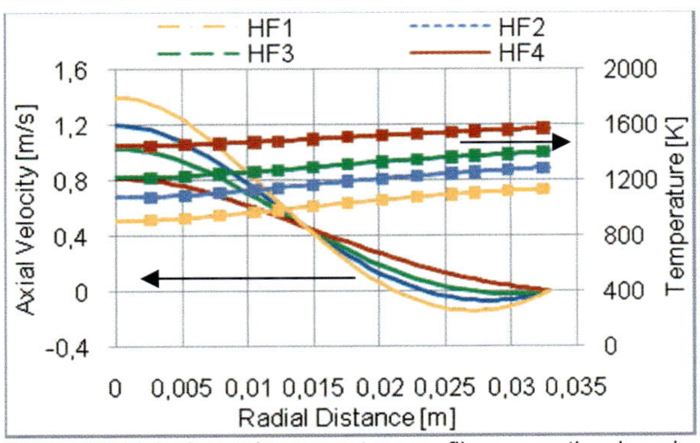

Figure 11: Velocity and temperature profile across the domain

The parameter study shows that the adjustment of the wall heat flux leads to a more predictable change in temperature whereas the variation of the mass flow rate results in a more predictable change in the velocity profile. Both, the change of wall heat flux and mass flow rate, influences the particle residence time in the reactor. **Table 6** shows the influence of the mass flow rate and wall heat flux on this process critical parameter.

Table 6: MFR and HF influence on residence time

MFR1	MFR2	MFR3	MFR4	HF1	HF2	HF3	HF4
8.54	3.33	1.70	1.04	1.30	1.45	1.55	1.66

5 CONCLUSION

To develop a more efficient process, detailed simulations and simplifying experiments are indispensible. It starts with simplification of complex devices and estimation of boundary conditions subjected to complex flow enabling a parametric investigation of an actual gasification process.

In this work simulation assisted design is applied to the DTR. Flow simulation provides information that is difficult to obtain experimentally. The DTR is used to simulate the actual operation of the conversion device and is used during the design and development of these reactors. Investigation of flow pattern, temperature distribution, feedstock consumption, and other quantities have been conducted at different mass flow rate and wall heat flux. The results have shown that the adjustment of the wall heat flux leads to a more predictable change in temperature whereas the variation of the mass flow rate results in a more predictable change in the velocity profile. The residence time varies with mass flow rate and wall heat flux. It is nearly a linear function of the mass flow rate, and a parabolic function of the heat flux. An increase of the heat flux requires adjustment of the mass flow rate to sustain the particle residence time.

Combining computational and experimental investigation, complex processes can be studied and achieved knowledge be used to develop more efficient processes and design a more productive device.

6 ACKNOWLEDGMENTS

The financial support of Masdar Institute for this project is highly acknowledged. The authors wish to thank all members of the waste to energy group (Arnar Valmundsson, Ilham Talab, Syed Shabbar Raza, Rana Qudaih, Mette Thomson) for their support.

7 REFERENCES

[1] Mincher, A. J., 2005, Coal gasification for advanced power generation, *Fuel*, Pages 2222-2235

[2] Ricketts, B., 2002, Technology Status Review of Waste/Biomass Co-Gasification with Coal, *IChemE Fifth European Gasification Conference*

[3] Ciaxia Chen, Masayuki Horio, Toshinori Kojima, 2000, Numerical simulation of entrained flow coal gasifiers, Chemical Engineering Science, Pages 3861-3874 and 3875-3883

[4] S. Kajitani, S. Hara, H. Matsuda, 2002, Gasification rate analysis of coal char with a pressurized drop tube furnace, *Fuel*, 539-546

[5] M. Cloke, E. L., 2002, Combustion characteristics of coals using a drop-tube furnace, *Fuel*, 727-735

[6] Shan Ouyang, H. Y., 1998, A pressurized drop-tube furnace for coal reactivity studies, *Review of scientific instruments*, Volume 69, Number 8

[7] Ilham Talab, Zaki Al-Nahari, Rana Qudaih, and Isam Janajreh, Masdar Institute of Science and Technology, 2011, Solar Assisted Gasification: Implementation and Systematic Analysis, *International Journal of Energy, Environment and Economics*, in Press

[8] Mills A. F., 1999, Heat Transfer, second edition, University of California at Los Angeles, Princeton Hall Inc.

[9] Francisco V. Tinaut, Andres Melgar, Juan F. Perez, Alfonso Horrillo, 2008, Effect of biomass particle size and air superficial velocity on the gasification process in

a downdraft fixed bed gasifier. An experimental and modelling study, *Fuel Processing Technology*, Pages 1076-1089

[10] Colomba Di Blasi, 1996, Heat, Momentum, and Mass Transport Through a Shrinking Biomass Particle Exposed to Thermal Radiation, Chemical Engineering Science, Vol. 51, No. 7, pp. 1121-1132

Methodology for High Accuracy Installation of Sustainable Jigs and Fixtures

J. Jamshidi, P.G. Maropoulos
Department of Mechanical Engineering, University of Bath, UK

Abstract
The ability to accurately measure the components of jigs and fixtures during their installation determines the state of their precision, especially for large size products and applications. This matter is crucial in mass customisation where small batches of products and components with high variety in design are manufactured. Product quality should be in harmony with rapid changeover philosophy as compromising quality for speed is not forgivable for sensitive components and assemblies such as those seen in the aerospace industry. It is necessary for the installation of the jigs and fixtures to be highly accurate in order to minimise the use of tolerance budget due to variations in jigs and fixture positioning. Major overhead costs for jigs and fixtures particularly in the aerospace industry led to the development of the concept of flexible and reconfigurable jigs and fixtures. Reusability of reconfigurable jigs and fixtures makes them attractive for sustainable solutions as their components can be reused for several variant of a product or assembly. The main drawbacks of this type of jigs and fixtures have been their poor accuracy and reliability. In this paper accurate positioning of the key components of sustainable jigs and fixtures is investigated. The factors affecting the performance of the jigs and fixtures are reviewed from the installation stage. The paper introduces a methodology for minimising uncertainties in positioning of the holds and clamps for flexible jigs and fixtures.

Keywords:
Sustainable Jig, Jig installation, Calibration Uncertainty, Jig Monitoring, Metrology, Reusable Jig

1 INTRODUCTION

Factors such as quality and reliability have long converted to implicit characteristics of the new products. Recent market trends have forced manufacturing industries to move towards mass customisation in their products and service range. Increased variation in the design of new products is followed by a second wave of variation with higher amplitude at subassemblies and component level.

State of the art manufacturing systems and technologies have provided more flexibility, enabling designers to think more freely. For instance new large volume measurement systems, developed in the past few years, are capable of measuring several decametre distances. Such technologies facilitate the verification of large size components that used to be manufactured from several assembled components.

The manufacturing of large size products requires specialist jigs and fixtures in order for their components to be held in the desired orientation during build and assembly. This requires major overhead cost that can only be justified by mass production in some cases or otherwise the cost of finished products can be very high. This issue contradicts with the market trends where customers are constantly looking for higher value for their money. In a typical product the variation in the product creates a more sustainable business as it can fulfil the needs of a relatively larger market.

Flexible and reconfigurable jigs and fixture that can be formed in different shapes to support different variation of products is a key solution for the above challenges. The concept of flexible jig existed for several years in the research domain [1]. However, they are not fully utilised to a great extent in real production facilities especially for large size product manufacturers, such as aerospace. This is due to the challenges related to their initial installation, poor calibration, and repeatability that often exceed the tolerance requirement. The manufacturing of these jigs and fixtures from high quality key components as well as their integration with large volume metrology systems can reduce the above limitations.

This paper covers metrology issues related to the installation and calibration of flexible jigs and fixtures as well as their monitoring during service.

2 RELATED WORK

2.1 Manufacturing and assembly of large scale parts

Typically prior to precision manufacturing of mechanical parts it is essential to move the raw material to the machine bench, proceed with rough cutting then fine alignment and clamping. At this stage the part is ready for machining of its high precision key features. However, this is not always possible for large size and/or heavy components. Large scale products refer to those with components that are not economically possible to handle or move around in the factory for fabrication and assembly purposes [2]. The manufacturing and assembly processes of these parts encompass movement of the machines and systems to the desired location and orientation with respect to these parts. Such parts are normally held in their positions using large size jigs and fixtures. If these parts are produced in small batch sizes that is the case for aerospace industries, high overhead cost per product will occur. There have been many attempts to design and manufacture jigs and fixtures so that they can hold a number of variants of components [3, 4]. However, this approach is not feasible for parts with sensitive or key features due to their high accuracy requirements.

Adjustable, reconfigurable jigs and fixtures produce lower repeatability over time compared to fixed ones. Fixed jigs have permanent topology achieved through their permanent joints that are welded or riveted. Mechanical failure of these jigs and fixtures for example due to fatigue and plastic deformation is a main cause of terminating their service and sending them for recycling. With small batch manufacturing requirements it is now common to retire a conforming jig as their service life depends on the life of products. In other words soon after the cease of manufacturing a part's variant, the associated jigs and fixtures become redundant. Even if the jigs are still in working order, they have to be scrapped and sent for recycling. This method brings the burden of high energy consumption for recycling. Even for the fixed jigs and fixtures the drift in the large size parts and jig can affect the accuracy of a large size assembly [5]. Several methods for analysing jig rigidity have been developed [6] to evaluate the impact of vibration on large size jigs. In any case a more sustainable manufacturing can only be achieved by alternative solutions.

Figure 1: Typical components of large scale jig (image courtesy of Electroimpact http://www.electroimpact.com/G150TFIX/gallery.asp)

Extensive lead time to manufacture is another major drawback for fixed jigs and fixtures. These jigs should be ordered well in advance of any manufacturing processes. This can create additional complexity in production planning and product time to market.

Regardless of their type, large scale jigs have a number of common elements including one main frame, one or a number of inner frames, potentially one or a number of moving mechanisms, and smaller components such as clamps, bushings, pickups and adjustable screws (Figure 1).

2.2 Flexible jigs and fixtures

The concept of flexible jigs and fixtures is developed for increased sustainability, rapid changeover as well as low cost. It is now possible to use off the shelf modules and clamps for jigs and fixtures design and assembly. Depending on the requirement only a handful of specialised components for the jigs and fixtures might be needed to be custom designed and manufactured. In this concept the majority of bulk components, joints at the attachments are used in for a specific application. Once the product design variant is fully manufactured it is then possible to disassemble the above components and reassemble them in a new topology to suite the next design variant. This cycle can be repeated over a large number of times resulting in reduced overhead cost for jigs and fixtures. Needless to mention the other factors such as disassembly time, resetting time, operators' time should be considered for evaluating the real cost benefit of using this type of jigs and fixtures. This approach best reduces the cost of jigs and fixtures for the assembly and component ranges that are fairly close in design.

Depending on the level of variations in the components and the type of work required on each a different percentage of the flexible jigs need to be rearranged. This matter is crucial to be considered at design stage in order to increase the benefit of using this type of jigs and fixtures. For example, when possible, the location and 3D positioning of pickups and clamps on different variants of the component or even totally different parts should be in close proximity to increase compatibility and inter-changeability of sub-systems of jigs and fixtures. Having a collection of the key components of the jigs can guarantee the availability of the desired jigs in a short time. In addition to this the storage of the jigs required less space as it is possible to dismantle all the modules that are typically in the form of scaffolding and place them next to each other. Flexible jigs are currently utilised in some of the automotive companies (Figure 2) as their accuracy level is sufficient for this sector. Despite the above benefits there are not many flexible jigs in operation in large size manufacturing facilities such as aerospace factories. Accuracy and uncertainty of positioning pins, repeatability of the clamps and drift of the jig structure are all contributing to the fact that these jigs cannot meet the tolerance requirements of the power generation and aerospace industries. These jigs have high potentials for utilisation in the above industries once their accuracy problems are resolved.

Figure 2: Fixture with reconfigurable components for automotive industry (image courtesy of Witte http://www.horst-witte.de/en/)

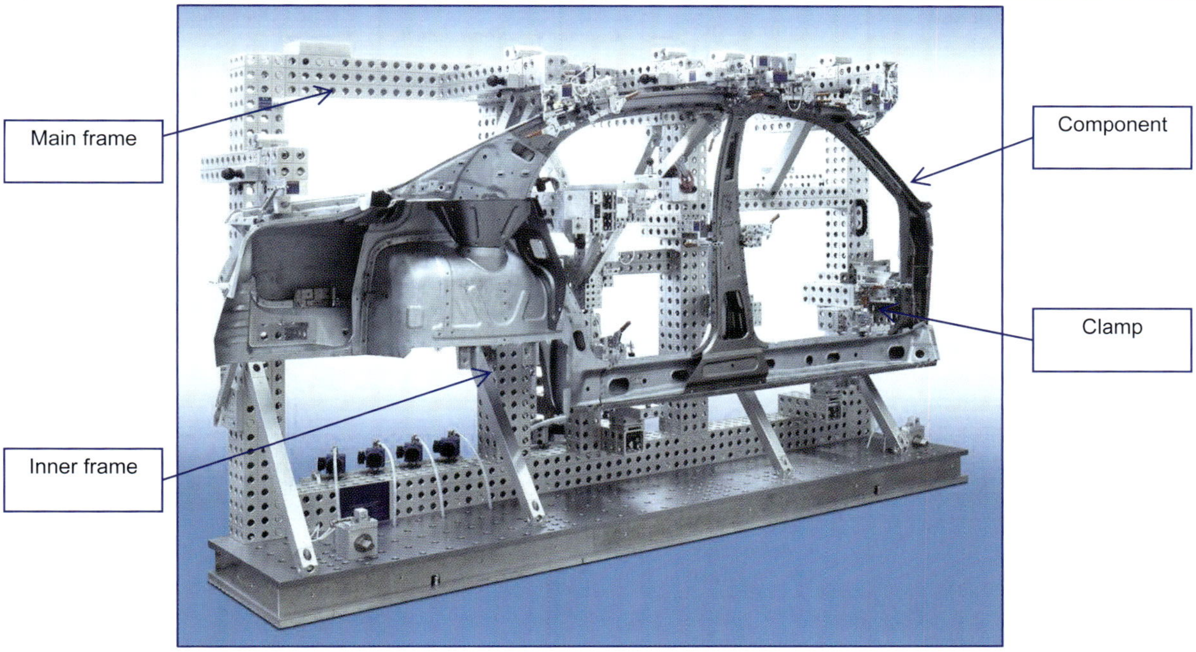

There has been a number of new developments in large volume metrology systems and technologies. Modern laser based metrology systems and technologies are now capable of measuring large size products up to several decametres with acceptable accuracy. These systems can be used to accurately position mountings of the key components of the jig during its installation and initial setup.

The installation of jigs and fixtures typically starts form its base or main frame then large components and gradually to the smaller components such as pickups and clamps. Metrology systems can be used for the installation of flexible jig main frame and its inner frames to guarantee the correct positioning of each and every component. Table 1 shows a few of these large scale measurement systems. For technological review of these systems see [7]. Laser tracker systems capable of measuring reference points, are among the most suitable measurement systems for this purpose. The instrument tracks a Spherically Mounted Retroreflector (SMR) target the position of which can be registered in three dimensional space. SMR can be contacted directly with the target object to provide geometrical positional information or can be used within a mechanically repeatable SMR nest known as Laser tracker target or in short target from this point on. Laser trackers like any other measurement instrument have a level of uncertainty that need to be accounted for during the jig installation process. Also the line of sight issues between the laser tracker and its target point should be considered and if necessary multiple tracker positions should be used. In real measurement activity the result must be accompanied with a statement of uncertainty. Such statement characterises the dispersion of the values that are reasonably attributed to the measurand [8]. This issue is the same for the installation and later verification of any jig or fixture. This knowledge clarifies the jig capability of a given positioning and assembly task. In other word it indicates if a jig can meet the tolerance requirements for its related processes.

2.3 Comparison of jig philosophies

There are a large number of different shape and design jigs and fixtures in different companies for various manufacturing and assembly applications. Some of these jigs and fixtures are readily available in standard forms, while some are designed and manufactured specific to particular parts and tasks. The latter can be very expensive based on the complexity and scale of the products [9].

Regardless of cost and purpose a manufacturing or assembly process can be performed using with no jig, with fixed frame jigs, or with reconfigurable or flexible jigs. Table 2 provides a comparison of these methods with their typical applications.

Fixed frame jigs are typically for heavy duty applications. They are more suitable for applications with a large number of products that can relax the overhead cost of the jig.

There are several advantages in the application of flexible jigs and fixtures for the manufacturing and assembly of large and complex products. In particular for research and development work, as well as for cases where low volume products are manufactured flexible jig and fixture can be very beneficial. In addition to time and money saving benefits the possibility of having a flexible jig gives more freedom to the design, manufacturing and assembly processes due to the low direct and recurrent cost of changing the overall topology of the jig. Reconfigurability and reusability of flexible jig is a main advantage for this type of jig compared to conventional jigs. This is particularly important as it is in line with the industry direction in terms of green manufacturing by recycling components from a used system, reducing project costs with regards to expenses for tooling of associated items.

Table 1: Examples of large volume/portable measurement instruments for jig verification

Instrument	**Auxiliary components**	**Measurement type**		**Image**
		Contact	**Non-contact**	
Laser Tracker	SMR probe	✓		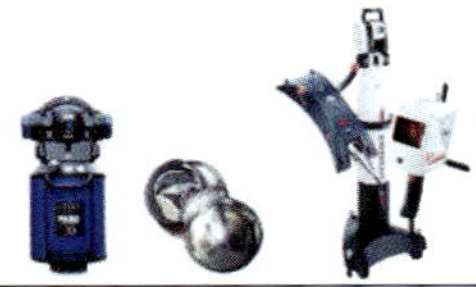
	T-probe	✓		
Laser Radar	Spherical targets		✓	
Photogrammetry	Targets	✓		
	Light projection		✓	
Articulated Arm CMM	Laser based scanning head		✓	
	Contact probe	✓		

3 FLEXIBLE JIG INSTALLATION

The issues and concerns that need to be considered in the jig installation procedure are described in this section. The installation of the jig components in the right position can be a challenging especially when the positioning tolerances are tights. Flexible jigs should also be monitored in order to exploit and compare their rigidity with that of the conventional ones.

Stage by stage measurement instruction for the jig installation based on the results of an initial jig setup in the simulation software environment and practical experiment of a large size jig with dimensions of *5m x 4m x3m* is given in a generic description. This is regardless of whether the jig is in first time installation or it is a change of an existing jig topology into a new shape, for holding a different component.

Depending on the complexity a typical large size jig has between three to five levels of frames. Apart from the base level with normally one main frame, at each level there can be one or several frames. These frames are interrelated with reference to the jig datum in order to facilitate the positioning and functionality required. In an automated, fixed platform, robotic systems carry out several tasks such as part positioning, machining and assembly. The robot working datum therefore is linked with the working frame of the jig. Careful consideration of jig datuming strategy and its subsequent installation can secure achieving the desired tolerance.

3.1 Measurement assisted flexible jig installation

There are several stages for the installation of flexible jigs that can be carried out in first simulation and then real world. The use of simulation exercise can reduce the number of potential errors and rework during this process. The process of measurement assisted installation is similar to tracking objects to position that is common for large size assemblies. In this approach the components of the jig are roughly positioned, within 1mm tolerance from the target position, at first. Then when all of the jig components are attached into their designated positions, within 0.1mm to 0.15mm tolerance, they are tightened using the appropriate torque. The typical stages of metrology assisted flexible jig installation are given below:

1. setting initial reference frames in the factory
2. measurement of initial reference frame
3. installation of base or main frame in its position
4. installation of inner frames offline
5. installation of holding and positioning brackets
6. installation of clamps, bushings and pickups in their rough position on inner frames and main frame
7. installation of inner frame on the base frame
8. fine adjustment and fastening of key locating components
9. verification of reference frames and clamps
10. in service monitoring of key positions on the jig

These stages are related to the complete installation of the jig from the scratch. Needless to mention that in case of slight change in design variation some of the following operations will be omitted.

Table 2: A brief comparison between different jig philosophies

Typical characteristic		**Fixed frame**	**Flexible jig**	**Jig-less**
Application		*Large volume production*	*Low volume production*	*Prototyping*
Pros	Uniqueness	Repeatable	Reconfigurable	Cost effectiveness
	Durability	Very high	High	Low
	Rigidity	Very high	Uncertainty rigidity	Low
Cons	Weight	Heavy	Medium	Low
	Portability	Non-portable	Difficult component positioning in each setup	Difficult to program
	Cost	Very high	Medium	Low
	Manufacturing time	Long	Medium	Short

3.2 Algorithm for flexible jig installation

The installation processes for flexible jigs take the following main stages:

1. the installation of main frame of the jig
2. the assembly of moving units and sub-systems
3. the installation of the jig inner frames on the jig assembly
4. the assembly of pickups and clamps on the jig.

The main frame is the backbone of the jig that is typically fixed for a large number of jig topology and design variations.

Therefore it does not change in shape as regularly as the inner frame or the smaller elements of the jig such as bushings, pickups and clamps. Careful consideration of the manufacturing process can reduce the necessity of rearranging larger elements of the jig components resulting in further time and money saving. Figure 3 in three separate groups of activities shows the processes of flexible jig installation. In this process it is assumed that the standard parts of the jig are selected from the available, off the shelf sections and components. Then in advance of the physical installation a number of tests and trials are carried out to plan the jig installation in such a way that the uncertainty of measurement is reduced. Once the acceptable level of uncertainty is achieved the physical installation can take place.

Figure 3: Measurement assisted installation procedure for flexible jig

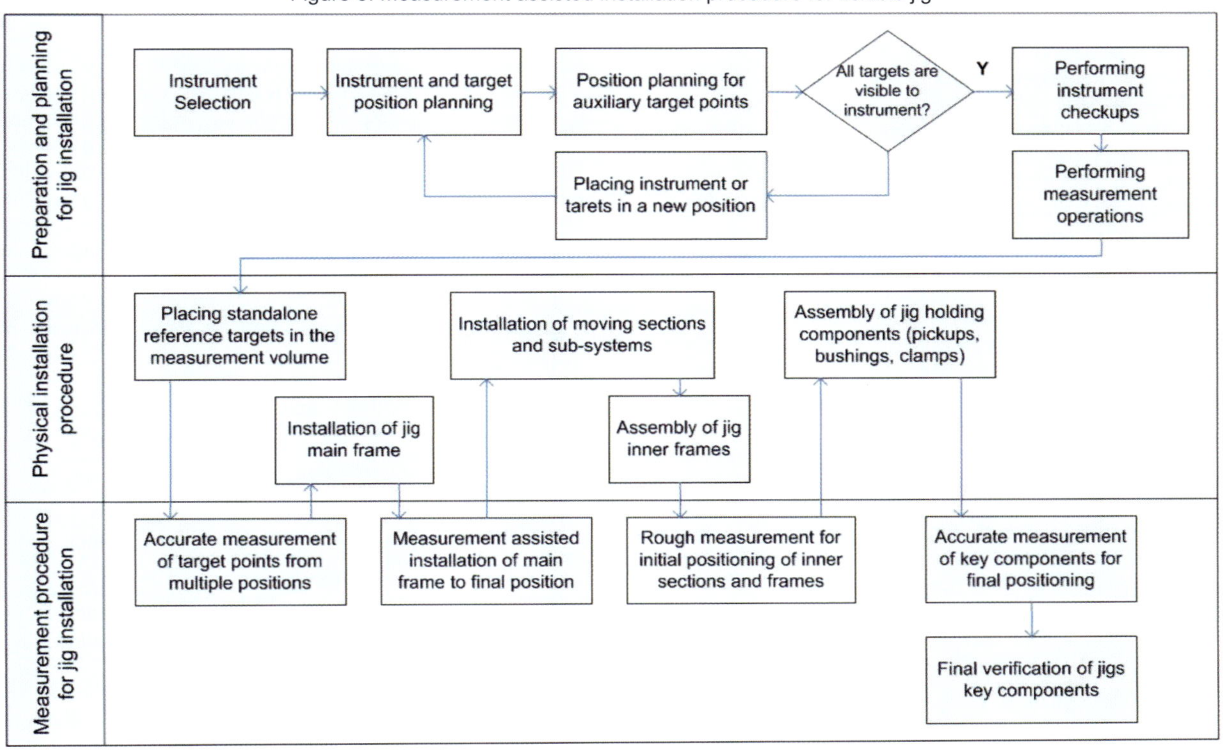

In jig installation process it might be required to use multiple measurement system or a measurement system from several locations to cover the complete set of key points for jig installation. This should include the auxiliary target reference points that are placed on the factory floor and wall for stability and drift check during the jig service.

During the localisation of the key points on the jig the uncertainty of the measurement instrument should be taken into account. As a rule of thumb the accuracy of the jig positioning for each key location should be in an order 10 times better than the required tolerance. In other word if the tolerance on the component is specified as 0.1mm the accuracy of the jig positioning should be at least 0.01mm. The best practice approaches for the flexible jig installation are given below:

1. A planned position for the initial reference points on the factory wall and floor is preferred in order to minimise uncertainty of the measurement.
2. The instrument position should be verified on regular bases using the initial reference points.
3. On each large section of the jig several SMR nests can be attached for better tracking and repeatability.
4. Where the large components of the jig have bend or twist the level measurement should be focussed on the central section of the beams to reduce angular positioning error.
5. The reference points on the inner frames should be selected as distant as possible for creating frame coordinate systems. This can result in smaller uncertainty when in the inner frame coordinate system.

4 UNCERTAINTY

Uncertainty if defined by GUM [8] as the result of the evaluation aimed at characterising the range within which the true value of a measured object is estimated to lie, generally with a given confidence. The Uncertainty here is reviewed from two aspects, the first is related to the measurement process and the second is related to the uncertainty of jig positioning. Measurement uncertainty has a number of contributing factors such as variation in environmental conditions such as dust, gravity temperature, air pressure and humidity, systematic errors within the measurement instrument and its related software, operators' skill, wear of contact probes or SMR. Based on GUM [8] definition a measurement result should be accompanied by its statement of uncertainty. Detailed discussion of these sources of errors is out of the scope of this paper. There have been a number of studies to establish the true uncertainty of some of the measurement instruments [10, 11].

The measurement results of key points on jigs and fixtures should therefore include the related confidence level. In addition to the initial uncertainty rising from the measurement instrument, the jigs and fixtures have other contributing sources of uncertainty. Forces applied to the jig frame due to weight of the product and also by manufacturing processes can potentially create elastic deformation in the jig frame. This matter becomes more complex for jigs that have integrated and automated moving sections.

Careful consideration of the sources of uncertainty at the jig design stage can reduce the risk of overusing the tolerance budget in a given scenario. There is a direct relation between the level of accuracy and cost of jigs and fixtures. However, the overall cost of manufacturing should be considered. The parts that have high number of first pass have longer life. In other words the parts that have quality characteristics closer to their mean values seldom fail during their use. The move towards six sigma [] approach is imperative in the installation and then monitoring of reconfigurable and flexible jigs and fixtures for high precision and sensitive parts.

5 DISCUSSION

Flexible jigs are suitable for assembly and minor machining of a number of parts with different design and geometry. In particular when the design variations between these parts are small this type of jigs proves to be a cost-effective and fast solution. This is because it is possible to adopt the jig to a new product or component variant with minor reconfiguration of the jig. However, when the design of the desired part and assembly is totally different or with different dimensions a complete rearrangement of the jig components may be required. Therefore in designing a new topology for the jig the closeness of the design variant of the parts and sub-assemblies should be considered in order to minimise the time and cost of jig installation and reconfiguration by maximising the reuse of the existing components in the current configuration.

There are a number of ways in which a flexible jig can be assembled simply due to its flexibility. The selection of the right components in the right installation order can maximise time and cost savings. In this selection procedure the potential reuse of the jig for future products should be taken into account. In the case of constant product variation it is important to plan and design jig topology in a collective approach where the mine frame of the jig that is more time consuming and costly can be reused without disassembling for instance. The use of bespoke clamps, fasteners, bushings and other elements of the jig are key issues to the cost reduction.

Extensive use of simulation software and tools can reduce costs associated to rework and waste of jig material. The use of these software help plan for instance the line of sight check between the measurement instrument and its target point. Figure 4 shows the first stage of flexible jig installation in the measurement simulation software. As the jig is being built the benefit of simulation becomes more evident as it can highlight the line of sight issues and uncertainty level of an instrument with respect to the target measurement points.

Figure 4: Simulation for measurement assisted jig installation

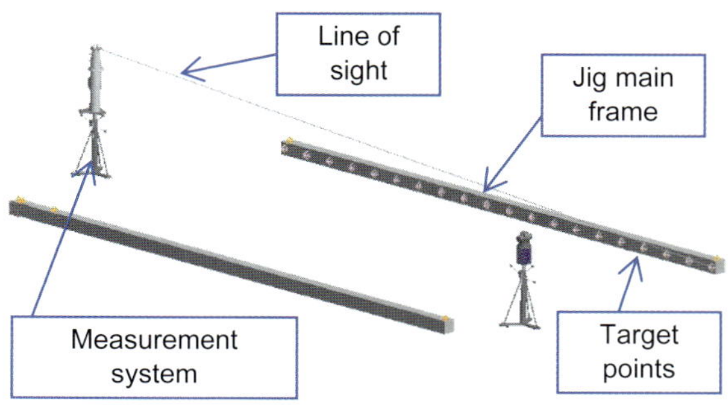

Potential drift and deformation of jig component due to weight and forces related to the jig operation can be analysed in the simulation world well in advance of any financial commitment

with regards to purchasing jig components. Furthermore such tools allow better planning for assembly and manufacturing operations as they can reveal the strengths and weaknesses of the jig prior to its physical setup.

The selection of the right measurement instrument is paramount for successful metrology assisted jig installation. A coherent metrology system with known uncertainty values in its measured results can determine jig accuracy and also expose the capability of the jig and its conformity for the required task within a given tolerance.

6 SUMMARY

Parts and assemblies should be designed in such a way to minimise the cost of jigs and fixtures by accommodating the use of standard components. If thought in advance the speed and cost of changeover of jigs and fixtures for the next variant of products can be increased. The design of jigs and fixtures in their turn can have strong implementation on the total cost, carbon footprint and their sustainability. It is often required to manufacture only a handful of a typical design or a product, subsystem or a component to satisfy variability of products and customer needs. For large scale products where the geometrical dimensions go beyond several meters jigs and fixtures can create major overhead. Furthermore, once the required number of parts is manufactured these jigs and fixtures become redundant. Conventional recycling of redundant jigs and fixtures is not economically viable or green. Therefore the answer should be found in increased flexibility of these jigs and fixtures.

The concept of flexible jigs and fixture has been around for the past two decades. However, their full potentials are not utilised due to uncertainties related to their accuracy and repeatability. High accuracy metrology systems are now available off the shelf that can be used from the initial installation of the jig and throughout its service. Then using the systems it is possible to reconfigure the jig elements accurately to a new topology in order to position components with different geometry, allowing the reusability of the jigs and fixtures for several times.

In this paper a generic algorithm is developed giving stage by stage approach for initial and installation of large scale flexible jigs and fixtures as well as their reconfigurations. The new concept of metrology assisted jig installation proved to be very beneficial for complex settings and installation of the jigs. The conformity of the jig when it is fully assembled can be guaranteed by careful consideration and selection of key points on the jig geometry. This method will be used in the manufacturing and assembly of large size components and products particularly in aerospace and power generation industries.

7 ACKNOWLEDGMENTS

The work has been carried out as part of the LVMA project within IdMRC of the University of Bath under EPSRC grant number EP/E00184X/1 and the Composite Wing Research within the Airbus Composite Structures Development Centre. The authors wish to thank all colleagues, especially Mrs A Miller, Mr A Kayani and Mr Mark Summers, all from Composite Structures Development Centre of Airbus UK, who were involved and supported the work completed within the context of the case study.

8 REFERENCES

[1] Kihlman H, Reconfigurable Tooling for Airframe Assembly",.First International Conference on Agile, Reconfigurable Manufacturing, 21-22 May, 2001, Ann Arbor, MI, USA

[2] Maisano, D.; Jamshidi, J.; Franceschini, F.; Maropoulos, P.; Mastrogiacomo, L.; Mileham, A. and Owen, G. (2008). Indoor GPS: System functionality and initial performance evaluation. International Journal of Manufacturing Research, 3(3):335-349

[3] Jackson M., (2000), An Analysis of Flexible and Reconfigurable Production Systems. An Approach to a Holistic Method for the Development of Flexibility and Reconfigurability, Dissertation

[4] Bi, Z., Lang, S., and Shen, W., (2007), Reconfigurable manufacturing systems: the state of the art, International Journal of Production Research, Vol. 46, No. 4, pp. 967-992

[5] Jamshidi, J., Kayani, A., Iravani, P., Maropoulos, P. G., Summers, M. D., (2009), Manufacturing and assembly automation by integrated metrology systems for aircraft wing fabrication. Proceedings of the Institution of Mechanical Engineers, Part B: Journal of Engineering Manufacture, 224 (1), pp. 25-36.

[6] Nabata, E., and Terasaka, Y., (2006), Jig Rigidity Evaluation Technology by Vibration Analysis, Technical Paper, Kumatsu Technical Report, Vol. 52, No.157, 7 p

[7] Peggs, G. N., Maropoulos, P. G., Hughes, E. B., Forbes, A. B., Robson, S., Ziebart, M., Muralikrishnan, B., (2009), Recent developments in large-scale dimensional metrology. Proceedings of the Institution of Mechanical Engineers, Part B: Journal of Engineering Manufacture, 223 (B6), pp 571-59

[8] GUM, (2003), "General Metrology, Part 3: Guide to the Expression of Uncertainty in Measurement", BSi, British standard, 106 p

[9] Saadat M., Sim R. and Najafi F., (2007), Prediction of geometrical variations in Airbus wingbox assembly, Assembly Automation, 27/4, pp 324-332

[10] Muelaner, J. E., Wang, Z., Martin, O., Jamshidi, J. and Maropoulos, P. G., 2010. Estimation of uncertainty in three dimensional coordinate measurement by comparison with calibrated points. Measurement Science and Technology, 21 (2), 025106

[11] Maisano, D. A., Jamshidi, J., Franceschini, F., Maropoulos, P., Mastrogiacomo, L., Mileham, A. and Owen, G., 2008. Indoor GPS: System functionality and initial performance evaluation. International Journal of Manufacturing Research, 3 (3), pp. 335-349

A New FlexDie Implementation for Sheet Metal Manufacturing

B. Alsayyed, K.H. Harib
Department of Mechanical Engineering, United Arab Emirates University, Al-Ain, UAE

Abstract
As trade competition is getting fiercer, innovation in all areas of product realization is becoming more of a need than a want. In this paper, the authors present an innovative idea where a FlexDie is designed to manufacture sheet metal parts. The idea is to mimic the representation of graphics and text on a screen using pixels. The developed FlexDie in this paper consists of the two halves that are similar in design. It comprises pixel like pins assembled in a two-dimensional array in the XY plane. The pins' heights in the Z-direction are adjustable to yield a desired topology. Each pixel can be altered and locked at a certain Z-coordinate. A prototype is developed to test the concept and produce preliminary results. The developed prototype consists of two-half dies, each having 8x8 square pixels, each of about 10x10 square mm. In the implemented design the Z-coordinates of the pixels are controlled and locked at desired heights via an array of screws. Sample parts consisting of three-dimensional metal sheets were manufactured using the developed FlexDie. Future work to improve the process is also discussed.
Key words:
CAD/CAM integration, FMS, flexible die, group technology, sustainable manufacturing.

1 INTRODUCTION

As the competition is getting sterner in today's market, more pressure is put on all areas of product realization to produce cheaper, better quality in a short time. Manufacturing is not an exception. However, less attention is paid to manufacturing till recent years, where a product lifecycle concept and approach is in practice. Furthermore, sustainable manufacturing is now becoming a need to be able to survive in today's economy that is now more affected by environmentally induced constraints.

New challenges are facing manufacturing as we enter the 21st century. More requirements for variety of configurations of the same product is increasingly pressuring manufacturing to be more innovative in producing different configurations of products in a short time to market and at a low cost. One of the areas where manufacturing was able to fulfill the huge demand of the same product using mass production is slowly phasing out. The trend at the same time is leaning more towards batch production and it is expected that this trend will continue towards lesser quantities of the same product. In this paper, the authors focus on sheet metal manufacturing using a new innovative idea for a configurable and programmable die.Flexible die (FlexDie) is an innovative idea for sheet metal and plastic forming processes that is becoming inline with the current interest in developing sustainable manufacturing processes.

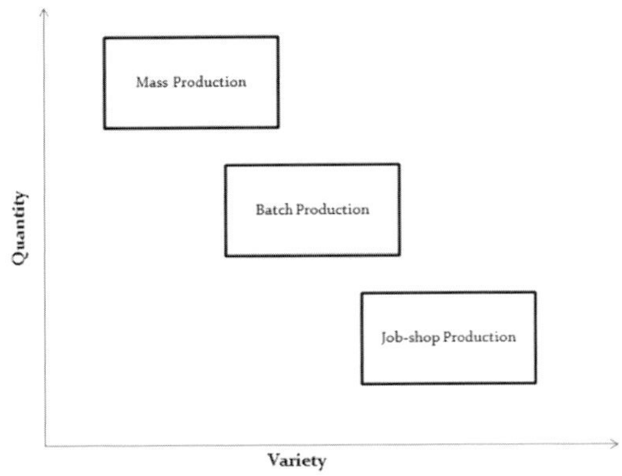

Figure 1, Quantity Vs. Variety

In traditional die manufacturing every part has to have its own die, with very little configuration. Not only it is very costly to produce those dies; it is also more costly to reflect design changes in the die design. It has been one of the evils that industry has to live with. To the contrary of a fixed (traditional) die that is used to produce parts of the same design, with FlexDie various parts with different designs can be produced using the same die. The process can be automated and programmable to form a variety of similar products using the same die. The topology of the die surface could be programmed and constructed via discrete surfaces formed from pixels instead of a fixed topology in traditional sheet metal forming. As such group technology could be used to identify groups of parts that could be manufactured using the same FlexDie.

The benefits of such approach are multi fold. It considerably reduces the cost of the die making as well as all the associated cost of maintaining and storing unused dies. Reducing die development time for newly designed products is also a highly attractive gain of this technology. Currently the general trend in product development is to go from mass production to batch and craft shop quantities with large variety of products [1]. With this trend the die making and maintaining problem is becoming more challenging. Flexible die with adjustable topology is explored by many researchers in recent years. Peng used NURBSS surface extension to design the blend of the surface between the blank holder and the workpiece. Introducing multi point forming [2, 3] is a promising advanced flexible manufacturing approach allowing to handle the variety of products by dividing the surface of the die into array of pins [4]. Sun et al studied the principles of flexible and rigid blank holders.Different types of blank holders resulted in sharply different results [5]. Wrinkles and dimples are the major forming defects in this process [6]. In multi-point forming, adjustable punches are used to form a variable curved surface. This process is taking the place of traditional die forming. There are four types of multi point forming:

1) Relative fixation: where punches are fixed before forming.

2) Passive adjustment: where the opposed punches are driven by the force of the press during the process.

3) Active adjustment: has many types, but main type is where the punches of the upper part or the lower part of the die are formed into the shape before forming and the other part of dies are passively adjusted according to the profile of the first part of the die. This is also called multi-point die forming.

4) Half press forming: the movement of the active punches is controlled according to necessity, at the same time the other half of the die is forced to move by force of the press..

Pins used in previously related work are of circular sections; in this paper we tried square section pins;a method which is meant toutilized the whole effective area of the die. The additional wrinkles resulted from this method yielded unique texture that might be desirable in certain applications. In this paper the behavior of the deformed material is experimentally observed.Further work which is currently underway to refine the design of the pixels and their orientation is also discussed.

2 METHOD DESCRIPTION

In this paper the idea of multi-point forming is introduced using square section pins (pixels). The height of each pixel is controlled independently from the other pixels. A prototype of a FlexDie was manufactured in our lab with an 8x8 array of pixels in both the upper and the lower halves of the die. In this arrangement, the pins, viewed as pixels, are grouped next to each other in the XY-plane. Figure 2 shows the pixels orientation, dimensions and notation.
Each pixel is adjustable in the Z-axis direction independently from the other pixels, and is referred to in terms of its location in the upper die using the notation U(x,y) for the upper half, and L(x,y) for the lower half. An example is depicted in Figure 3 for the upper part. In the implemented design, each pixel represents a 9.8x9.8 mm squared section steel bar with a 40 mm length. Each pixel is supported by an M-8 bolt that controls its Z-elevation.

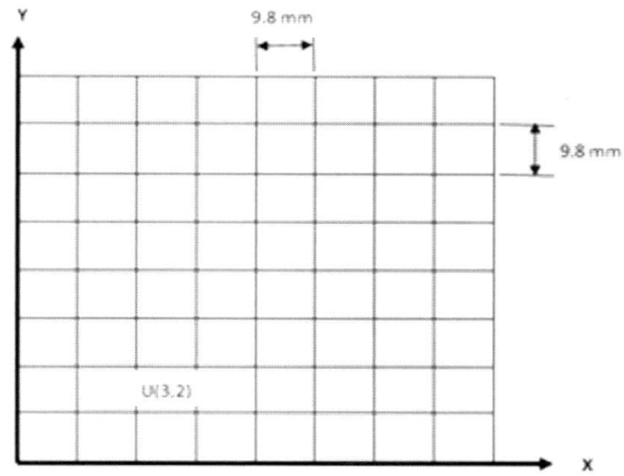

Figure 2, pixels dimensions, orientation and notation

The pixels movement along the Z-axis is coordinated between the lower and upper set of pixels such that each pixel in each set corresponds to another pixel from the other set. The pixels are referred to in terms of their x & y locations. Each pixel in the upper and lower die part has the same magnitude of Z-coordinate but in opposite directions. Hence, the upper and lower parts of the die will take the desired shape of the final part. An example of a concaved forming is used here and is shown in Figure 3.
The die was designed and manufactured in the manufacturing lab at the UAE University. A section drawing of the die is shown in Figure 4. It shows the upper and lower parts of the die as well as the pixels orientation in both sections. The pixels are lined up in a matrix where each pixel is adjusted by an individual bolt. This allows for independent adjustment of the pixels'

heights. This independent adjustment can theoretically produce large number of shapes. The upper and lower pixel's heights will complement each other to the flat level of the pixels.The heights of all pixels in both the upper and lower die parts have to be calculated based on the part to be manufactured profile. The process of calculating the pixels' heights can be automated and integrated with standard Computer Aided Engineering tools.

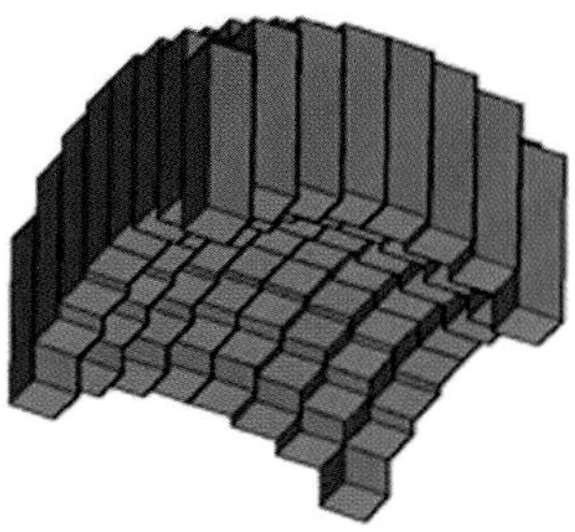

(a)

(b)

Figure 3: Example for pins' arrangements for a concaved part, (a) upper and (b) lower die halves

3 INDUSTRIAL APPLICATIONS

The applications of FlexDie are in all areas of sheet metal forming and plastic and cardboard forming. The value that FlexDie brings to sustainable manufacturing is multifold. One of the most important benefits of FlexDie is the configurability of the die to form different shapes and features. Group technology is a helpful approach to group similar parts into groups that can be formed the same FlexDie. Groups will be based on size, shape topology as well as special features. An example would be in automotive industry, where fenders of all small cars can be formed using one FlexDie. Hoods will all have another die. Another application is for cardboard products, where forming needs less press power; however, variety of shapes could be endless. Recyclable plates are a very good example where a variety of shapes and sizes would be handled by the same FlexDie.

A second major benefit is the cost. While the cost will be reduced tremendously for the fact that less number of dies will be needed for a wide variety of product, the space, storage, and maintenance are of significant cost reduction.

An example of concaved part using a 8x8 FlexDie is employed to demonstrate the idea. The calculated heights using MS Excel are shown in Table 1 below for the upper and lower die halves.

The surface of the part is an upper surface of a sphere with radius *R*, and is defined by the equations below.

$$\begin{aligned} x &= R\cos\theta\cos\phi \\ y &= R\cos\theta\sin\phi \\ z &= R\sin\theta \end{aligned} \tag{1}$$

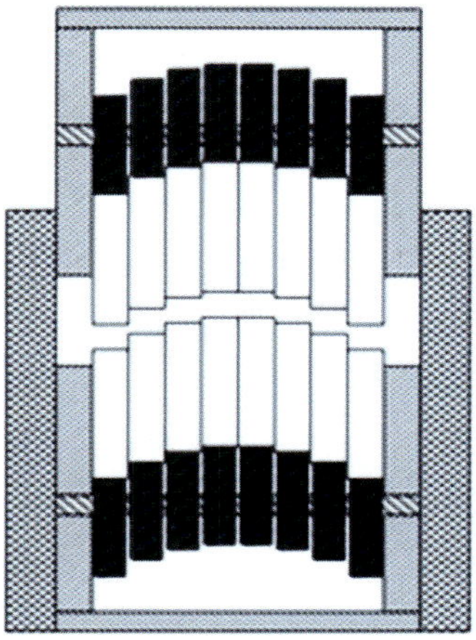

Figure 4: Schematic drawing for a section along the upper and lower die parts

For a given radius R and a set of coordinates (x, y), the first two equations in (1) can be solved for θ, which is then used in the third equation to calculate the pixel height z.

Based on the calculated heights for all pixels, the pins' heights are adjusted individually using the corresponding screws and thus controlling pixels' desired heights. Figure 5 shows the back view of the die.

Figure 5: The screw-side of the die

Setup of the distances is done in reference to the zero level of all the pixels. The upper pixels and lower pixels are adjusted to equal distances in opposite directions. After adjusting the heights of all pixels for the example part, the die upper and lower parts are shown in Figure 6.

Figure 6: lower die for the example part

The pixels are allowed to slide freely as the bolts behind them are adjusted to raise or lower each individual pixel. The setup in Figure 6 was done for a 0.5 mm thick aluminum sheet. The dimensions of the effective area is the summation of the individual pixels in both x and y direction. As the heights are adjusted to different levels, the resulting part will be different.
The number of bolts for this setup is 64 for each part, which make it time consuming to do the setup manual. In the next version of this die, the setup will be automated to reduce the setup time and increase accuracy.

Once the heights of the bolts of both parts of the die are setup and the workpiece is prepared and placed in the die; the die is pressed on the workpiece to form it. A manual press was used to form a 0.5 mm and a 0.8 mm sheet metal parts. Pictures of the resulting parts are shown in Figure 7.
It was noticed, as shown in figure 7, that the

the 0.5 mm part showed some fractures especially near the side pixels,

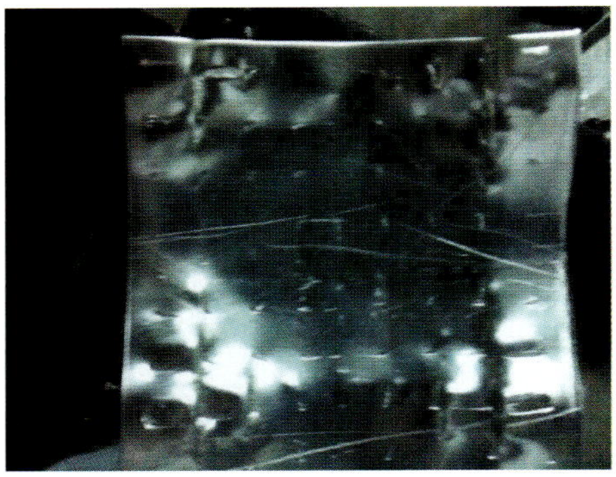

Figure 7: A part manufactured using the FlexDie

Investigation of those fractures revealed that the height of the pixels near the edges needs to be less to reduce the individual pixel drawing distance. The heights were adjusted at one of the edges that showed the max fracture, and the fractures reduced substantially. More investigation is undergoing to optimize on an adjusting factor to the calculated height shown in tables 1. For the 0.8 mm sheet, fractures were much less.

0	7.9	12.4	14.5	14.5	12.4	7.9	0
7.9	14.5	18.5	20.4	20.4	18.5	14.5	7.9
12.4	18.5	22.2	23.9	23.9	22.2	18.5	12.4
14.5	20.4	23.9	25.6	25.6	23.9	20.4	14.5
14.5	20.4	23.9	25.6	25.6	23.9	20.4	14.5
12.4	18.5	22.2	23.9	23.9	22.2	18.5	12.4
7.9	14.5	18.5	20.4	20.4	18.5	14.5	7.9
0	7.9	12.4	14.5	14.5	12.4	7.9	0

Table 1: calculated heights of the pins in upper halve of the die for the considered example

Using a sheet blanket to compensate for the rough pixels' size and sharp edges, showed a reduction in part fracture near the edges.

The blanket used is 0.5 mm thick of textile material. The results show smooth consistent curvature of the part.

4 SECOND GENERATION FLEXDIE

A multi-stage FlexDie is to be used with the automated version of the FlexDie. The die will be divided into course grid pixels shown in red in figure 8 below. Each pixel will have multiple pixels shown in blue, and each level will have more pixels with finer details of the part topology.

Thus, for the first stage 4X4 pixels are active to be adjusted to give the general course topology where they hit the workpiece first. After that, the pixels within each one of the four pixels will be adjusted to give a finer topology.

This new design will help reduce the vertical distance that each individual pixel needs to

Figure 8: orientation of new FlexDie

adjust, thus reduce the possibility of buckling and sticking pixels.

With this design the first trail is done on two stages, however, more stages are considered in order to be able to cover more range of designs and shapes.

5 ARCHETECTURE AND LANSCAPE OF VISIONED SYSTEM

An integrated CAD/CAM system for the automated FlexDie is shown in figure 9 below. It takes a cad model and extracts the features from it. An algorithm will then detects the features and maps them to corresponding manufacturing features. Based on the manufacturing features, the controller of the pixels will set up the pixels heights in relation to each other and with reference to a mapped common reference for the geometry and the FlexDie zero level.

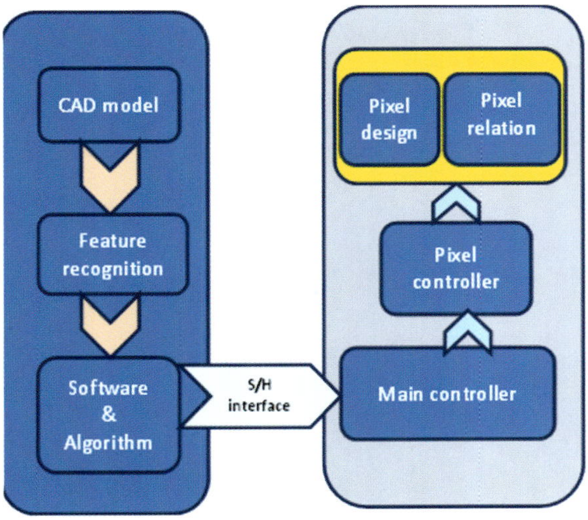

Figure 9: architecture & landscape of vision system

6 DISCUSSION AND FUTURE WORK

In this paper the idea of using a flexible die to manufacture variety of curved metal sheets was examined with squared section pins controlled by a set of screws. Although in the presented design we limited pins shapes to square cross section and flat square face, the used pins could be selected from assortment of square pins with inclined faces with standard angles (e.g. 15°, 30°, 45°, 60°, 75°). This added flexibility will increase the smoothness of the generated curved die topology.

The process is done manually in the presented experiment where the deformation and shape of the workpiece is controlled manually. The preliminary results obtained are promising and work is continuing in our lab to improve the resolution and capability of the process. More applications are understudy where the control of the pixels is done via the CAD system. Work is undergoing for manufacturing different versions of the FlexDie for different applications.

7 AKNOWLEDGEMENT

The authors would like to acknowledge and thank the UAEU research affairs for funding this research as part of the UAEU research affairs activities.

8 REFERENCES

[1] L. Peng, X. Lai, M. Li, Transition surface design for blank holder in muti-point forming, et al. / International Journal of Machine Tools & Manufacture 46 (2006) 1336–1342

[2] M. Li, Y. Liu, et al., Multi-point forming: a flexible manufacturing method for a 3-d surface sheet, J. Mater. Process. Technol. 87 (1999) 277–280.

[3] M.Z. Li, Z.Y. Cai, Z. Sui, Q.G. Yan, Multi-point forming technology for sheet metal parts, J. Mater. Process. Technol. 129 (1–3) (2002) 333–338.

[4] Z.R. Qian et al. / The analyse on the process of multi-point forming for dish head Journal of Materials Processing Technology 187–188 (2007) 471–475

[5] G. Sun et al. / Study of blank-holder technology on multi-point forming of thin sheet metal, Journal of Materials Processing Technology 187–188 (2007) 517–520

[6] M.Z. Li, Z.Y. Cai, Z. Sui, et al., Multi-point forming technology for sheet metal[J], J. Mater. Process. Technol. 129 (2002) 333–338.

Advanced Reliability Analysis of Warranty Databases (RAW) Concept: Contribution to Sustainable Products and Manufacturing

S. Bracke, S. Haller
Institute of Safety Engineering, University of Wuppertal, Germany

Abstract
Sustainable products and manufacturing in the automotive industry become key sales arguments out of the customer focus. Reliable products contribute to sustainability through an increased customer usage phases (lifespan expectancy) and therefore reduced resource usage.
A contribution to achieve reliable products lies in structured field product observation and field failure data analysis. But increasing functionality and complexity of technical products leading to complex damage causes in several ways and therefore aggravate the failure analysis.
Current development trends require early field product observations and therefore lead to small amounts of field data. Industrial parametric statistical reliability methods do not completely fulfil these requirements.
Regarding the outlined requirements the Department of Risk Management and Safety Engineering developed a comprehensive concept for the early, economical and detailed statistical reliability analysis of guaranty and warranty data bases. The advanced reliability analysis of warranty databases (RAW) concept combines industrial established parametrical statistical methods with nonparametric statistical methods in a structured and situational concept. Based on the RAW concept further targeted actions help to increase the product reliability and therefore help to save resources through reliable products.
Keywords:
Reliability, sustainability, field data analysis, prevention damage causes, sustainable design and sustainable products

1 INTRODUCTION

On the one hand sustainable and environmentally friendly products become key sales arguments out of the customer focus and therefore force manufactures to rethink common manufacturing and development concepts. Reliable products – products with a long life cycle – are one way to achieve sustainability and therefore help to save resources in several ways:

- Minimizing replacement parts for maintenance during the product life cycle
- Reducing the number of produced units and therefore using less primary products and materials
- Using reliable products and components in a wide range type series (Company Part – Strategies) and therefore reducing development (e.g. prototype testing) and production (e.g. restructure production) costs

On the other hand growing demands of customers and competitive environments result in increasing components functionality and product complexity and therefore in complex damage symptoms within the customer use phase. In particular in automotive engineering, but also in products of the capital intensive and consumer goods industry, the reasons of complex damage symptoms are often justified in several possible damage causes.

In addition to prototype testing for the preventive detection of existing damage causes during the development, the importance of field product monitoring and field data analysis is steadily increasing. The application of the statistical reliability analysis within the field product monitoring is the evaluation and interpretation of failure data, in detail:

- Comprehensive mapping of the component failure behaviour in terms of damage causes and differences of affected production batches, product optimisations (component changes), climatic influences and regional influences (through customer usage)
- Pre-detection of potential complex damage causes
- Basis of decisions to introduce targeted actions
- Verification of the efficiency of introduced actions for troubleshooting, e.g. in the field or in the current product generation

With the shift from approaches of the reliability analysis during the development to the customer usage phase, a time delay occurs between the failure appearance and failure detection (e.g. construction weak point leads to failure in high customer usage phase [wear out phase of the bathtub model] [1]). This situation requires a change in the statistical field observation.

To reduce the delay, the analysis of field data should be performed the earliest time after market launch, which leads to small available damage cases. In addition, the analysis has to ensure statistical reliable results regarding the component failure behaviour. The industrial method standard of statistical reliability analysis does not completely fulfil these new requirements.

2 GOAL OF THE RESEARCH PROJECT

The industrial method standard of statistical reliability analysis, which is described by Bertsche [1], Ronniger [2] and Linß [3] in detail, is based, in addition to the graphical analysis, on the usage of parametrical statistic (e.g. Weibull distribution, parametrical statistical significance methods).

During the analysis and mapping of simple damage symptoms, the industrial methods ensure reliable and significant analysis results if sufficient large amounts of data are available. Especially Weibull distribution models are intensely used in the automotive industry to describe damage symptoms.

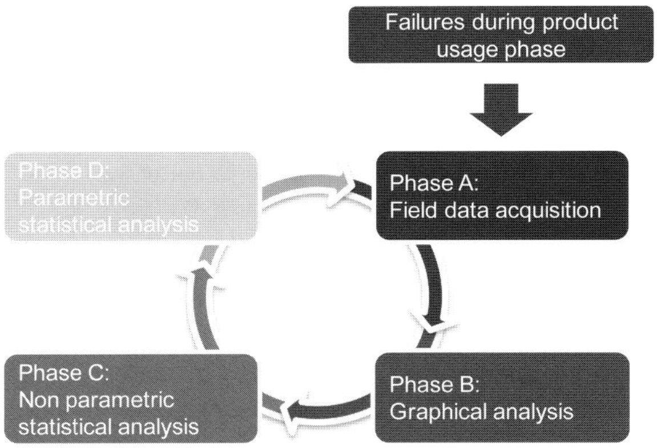

Figure 1: Phases of the RAW concept

Regarding the outlined requirements – early and detailed analysis of failure data – on the statistical reliability analysis the amount of data is reduced significantly and the usage and significance of parametrical statistics is limited.

The goal of the current research project of the Department of Risk Management and Safety Engineering at the University of Wuppertal is the development of a comprehensive concept for the early, economical and detailed statistical reliability analysis of guaranty and warranty databases. The advanced reliability analysis of warranty databases (RAW) concept combines industrial established parametrical statistical methods with nonparametric statistical methods in a structured and situational concept.

Therefore the RAW concept makes a contribution to increase the reliability and sustainability of current and successive product generations. In addition the reflection of the knowledge out of the reliability analysis into the value added network (e.g. Lessons-learned-method) facilitates the long term goal of sustainable product improvement with respect to long lifespan expectancy.

During the product development process, it is preventative to increase the product reliability by using various preventive methods besides the mentioned approach, e.g. FTA, FMEA, Lessons-Learned method, QFD or generally creative techniques (TRIZ, mind-mapping, brainstorming). But these methods are not in the scope of this paper. The mentioned different methods can be combined in different development stages. The purpose of this paper is to contribute to the context of the development of reliable products.

3 ADVANCED RELIABILITY ANALYSIS OF WARRANTY DATABASES (RAW) CONCEPT

The advanced reliability analysis of warranty databases (RAW) concept describes the comprehensive process of the statistical field data analysis from the analysis of small amount of failure data (small sample size) to large amount of failure data (large sample size) (cf. figure 1).

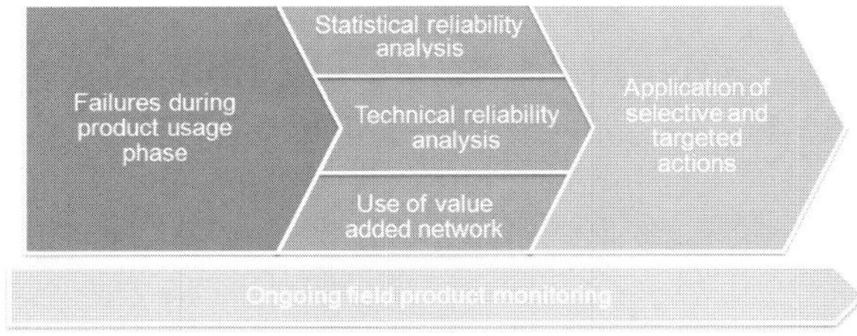

Figure 2: Phases of the FDA concept

The statistical mapping of the product or component failure behaviour or the forecast of further damage cases within the customer usage phase can be the results of the statistical reliability analysis.

Based on the first failure cases of a new damage symptom, a new product generation or a product optimisation (component changes) within the customer usage phase the four phases of the RAW concept can be outlined as follows:

- Phase A: Field data acquisition – Within the warranty and guaranty period the automotive manufacturer (OEM) is provided with sufficient and relevant field failure and vehicle data, gathered by the trade organisation (represented by the dealers)
- Phase B: Graphical analysis – Large amounts of data require a structured graphical analysis to concentrate on data areas – detailed analysed in further phases
- Phase C: Non parametrical statistical analysis – Small amounts of data require methods to ensure reliable analysis results
- Phase D: Parametrical statistical analysis – Growing amounts of data enable the usage of parametrical statistical

Through the continuous data acquisition the four outlined phases are based on each other, complement each other and reiterate with growing amounts of data.

The RAW concept is integrated in the field data analysis (FDA) concept. The FDA concept (cf. figure 2) centres on enhanced methods for a comprehensive reliability analysis of field damage symptoms [4], in detail:

- Organisational aspects (use of the value added network)
- Enhanced statistical reliability analysis methods (e.g. Differentiation of Complex Damage Causes (DCD) algorithm or RAW concept)
- Enhanced technical reliability analysis methods (optimised multi-stage sampling procedures (OMSP) concept)

3.1 RAW Phase A: Field data acquisition

Basis for the statistical mapping of any potential product or component failure cause are two different fundamental sources of data:

- Obtaining damage information during the early product development phase, based on actual prototype tests and validations
- Acquisition of field data (damage data) of previous (current) product generations already used by the customer, based on warranty and guaranty time

databases within the legal warranty and guaranty (for example: Germany 2 years, USA 4 years)

This work only outlines the analysis of field data. But the outlined methods can also be adapted, for example, on the analysis of prototype test data.

Within the warranty database relevant vehicle data of every failure case are collected, e.g. date of production, date of damage, month in service, damage description, kilometrage, motorisation or configuration.

Based on damage cases with the same damage symptom the further phases of the RAW concept are used for a detailed analysis of the damage symptom.

3.2 RAW Phase B – Part I: Graphical analysis – Mapping

Background of the graphical analysis is the visual identification of relevant - specifically to be analysed - data areas and therefore reducing the effort of the statistical and technical analysis.

A well known method of mapping the damage cases is the layer-lines diagram based on specific month in service (MIS) (usual in the automotive industry; one, six, twelve up to 48 months; cf. figure 3). The abscissa shows the individual months of production and the ordinate shows the failure rate, which occurred within a specified customer usage area (MIS). To increase the comparability of individual months, the amount of failure cases are in relation to the amount of produced components.

In addition to the conventional method in the automotive industry using customer usage areas, Bracke and Haller developed a new method using mileage areas to map the failure cases [5]: The layer-lines refer to mileage ranges. This mapping variant is suitable for mileage dependent damage causes, as it avoids inaccuracies between customer usage time (MIS) and mileage.

3.3 RAW Phase B – Part II: Graphical analysis – Selection of relevant data areas

The failure case progress of the realistic case study (electronic instrument panel, in dependence on [4]) is outlined in figure 3, which occurred with the introduction of a new product generation. Based on the shown layer-lines diagram several relevant - specifically to be analysed - data areas can be selected and used to analyse the complex failure symptom with economic efficiency and high detection accuracy of damage causes. Selected areas in detail (cf. figure 3 and 4):

- Area 1 – Abrupt increase of the failure rate: Different damage causes often result in different failure performances and therefore in different failure rates
- Area 2 – Product/component optimisation: Reason for product optimisation is to reduce or prevent damage causes and therefore increase the product reliability and quality
- Area 3 – Fluctuations of failure rates: Based on the layer-lines the fluctuation of failure rates over several production months often results in the dispersion of the failure rate with a constant failure behaviour and therefore with the same damage causes
- Area 4 – Trend of failure rates: Detection of a trend in the failure rates over several production months often results in an increasing amount of damage cases of specific causes and therefore in a constant failure behaviour
- Area 5 – Market or climate specific failure behaviour: Influence of market (customer usage) or climate on failure symptoms (cf. figure 4), e.g. dispersion of the failure behaviour or different damage causes

Depending on the present application cases different statistical significance tests can be used to analyse the drafted areas. Depending on the sampling size, nonparametric (small sampling sizes) and parametric statistical significance tests (large sampling sizes) can be used.

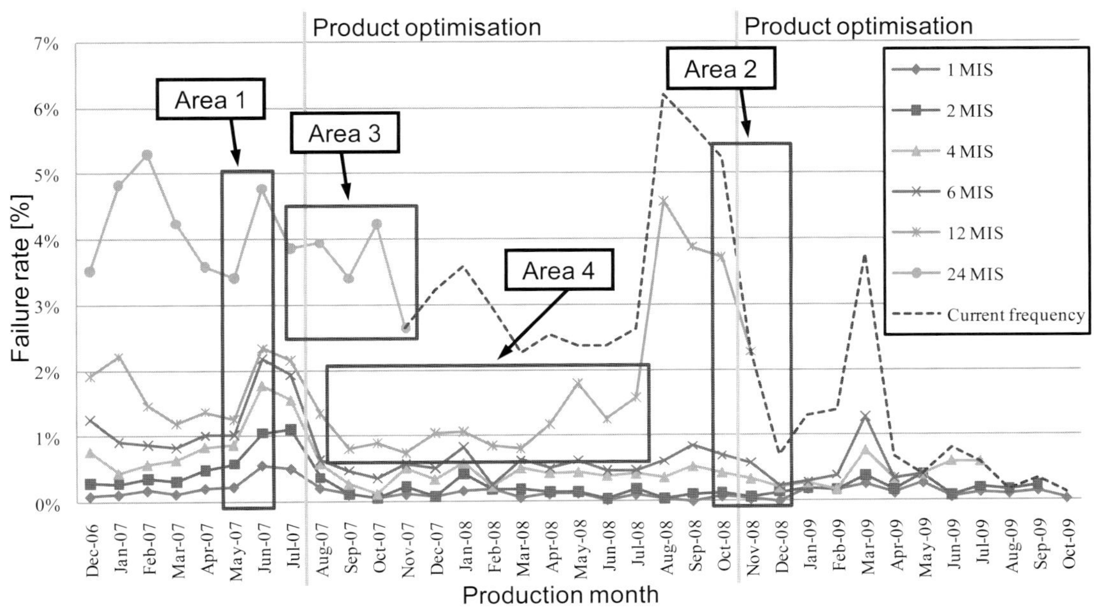

Figure 3: Layer-line diagram of the realistic case study (electronic instrument panel); Area 1 to 4 demonstrate relevant data areas (half synthetic dataset)

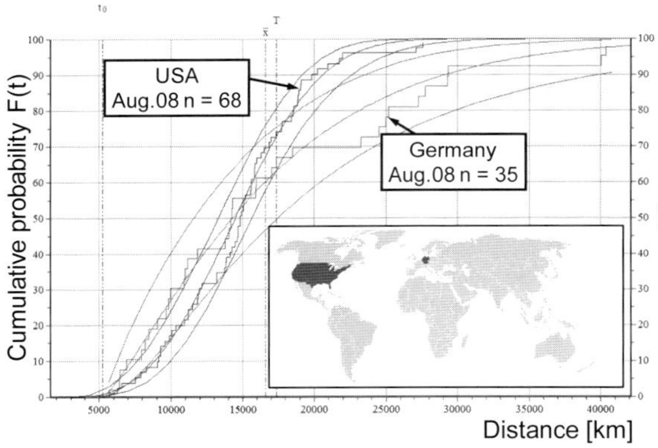

Figure 4: Comparison of the US and German markets (case study electronic instrument panel) using Weibull distribution models and confidence intervals

The following phases C and D outline the application of selected significance tests, to generate additional information for targeted product and process actions.

3.4 RAW Phase C: Non parametrical statistical analysis

As previously described, analysis of field failure data at an early stage is based on small amounts of data (small sampling size). In particular, using layer-lines diagrams result in small amounts of data too, because the occurred faulty products are allocated to the production months, in which they were built.

For small amounts of data (small sampling size) the use of parametric significance tests is not suitable.

In addition, many industrially used parametric tests are generally based on the assumption, that the damage data/population follows the normal distribution. Based on the normally distributed population all required parameter can be determined. But for small sampling sizes, this assumption is only sometimes partially fulfilled.

Because of these limitations of parametric significance tests regarding to an early statistical reliability analysis, other statistical methods have to be used to analyse field data. One approach is in the use of nonparametric significance tests. Nonparametric significance tests do not require dependences to specific distributions, for example they do not need normal distributed data. In comparison to similar application cases of parametric significance tests, they have also further application cases and benefits, in detail:

- Early detection of damage causes with the occurrence of the first field damaged components (small sampling size)
- Application independent of distribution model
- Relative simple calculation
- Early comparison of individual production months/ production batches to each other in relation to
  a. the centre of the failure cases and
  b. the dispersion of the failure cases

  Explanatory note: Both the comprehensive data and specific points in time of the failure occurrence reference (e.g. between six MIS and twelve MIS) can be analysed
- Detection of a trend in the failure rates over several production months/ production batches and within production batches
- Detection of randomness of failure data

A good overview on relevant nonparametric significance tests is described e.g. in Büning [6] and Siegel [7] (cf. figure 5). In the context of the application of nonparametric significance tests for analysing field failure data are the only suitable tests available, which require independent samplings. However, tests, which require dependent samplings, are suitable e.g. for the analysis of in-line and off-line measurements of related components.

This paper focused on selected, representative nonparametric significance tests (two-sided hypotheses), in detail:

- One sampling case:
  a. Sign test: Does the central tendency (median) concur (null hypothesis H_0) or differ (alternative hypothesis H_1) regarding to a hypothetical value:

$$H_0 : \Theta = \Theta_0; \quad H_1 : \Theta \neq \Theta_0$$

- Two samplings case:
  a. Wald-Wolfowitz runs test: Analysis of equality of two distributions F and G (null hypothesis H_0):

$$H_0 : F(z) = G(z); \quad H_1 : F(z) \neq G(z)$$

  b. Wilcoxon rank sum test: Two distributions F and G have the same shape, but differ in a location parameter Θ (alternative hypothesis H_1):

$$H_0 : G(z) = F(z); \quad H_1 : G(z) \neq F(z - \Theta)$$

  c. Sigel-Tukey test: Two distributions F and G differ regarding to variability (alternative hypothesis H_1):

$$H_0 : G(z) = F(z); \quad H_1 : G(z) \neq F(\Theta z)$$

- C samplings case:
  a. Kruskal-Wallis test: C distributions have the same shape, but differ in a location parameter Θ (alternative hypothesis H_1):

$$H_0 : F_1(z) = F_2(z) = \cdots = F_c(z) = F(z)$$
$$H_1 : F_i(z) \neq F(z - \Theta_i)$$

  b. Meyer-Bahlburg test (generalised Sigel-Tukey test): C distributions differ in the variability (alternative hypothesis H_1):

$$H_0 : F_1(z) = F_2(z) = \cdots = F_c(z) = F(z)$$
$$H_1 : F_i(z) \neq F(\Theta_i z)$$

In addition to the outlined two-sided hypotheses – according to the application case – one-sided hypotheses (only if the nonparametric significance test supplies this hypotheses) can be used, e.g. to analyse two samplings if one sampling is based on a basic population with lower central tendency (Boundary conditions for the application of nonparametric statistical significance tests: cf. [6] and [7]).

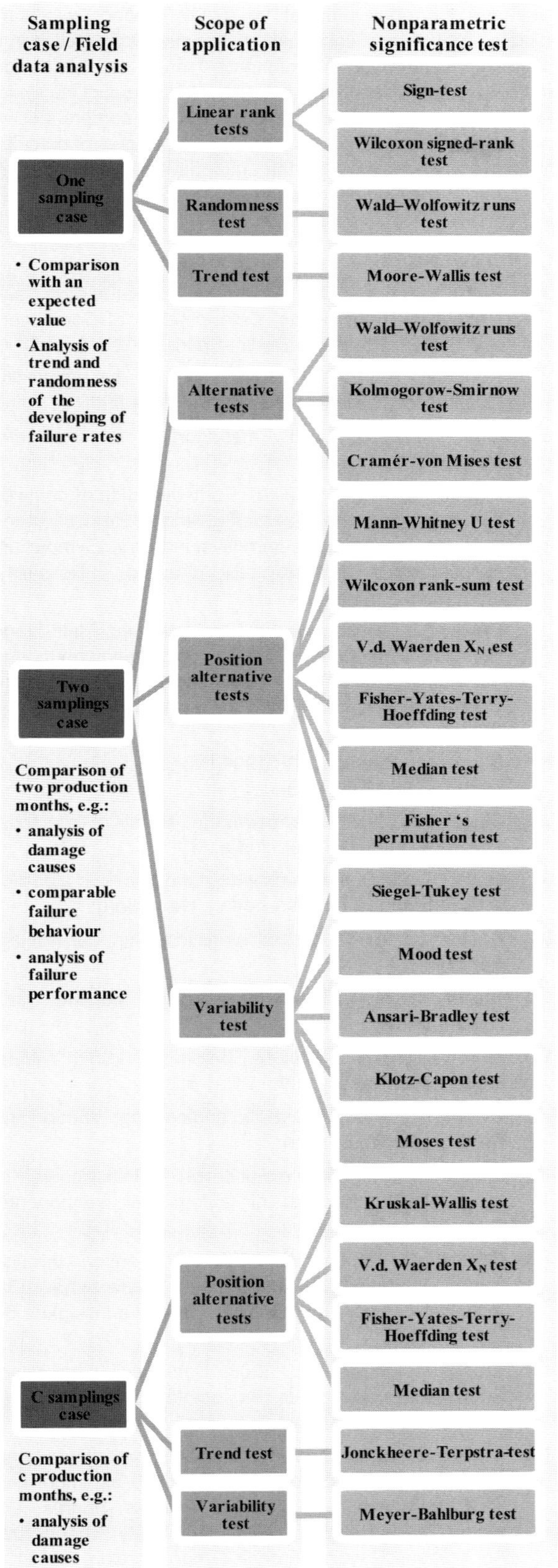

Figure 7: Overview of relevant nonparametric statistical tests (cf. [6] and [7])

With the application of the nonparametric significance tests the outlined five areas of the case study (electronic instrument panel, cf. figure 3, in dependence on [4]) can be analysed. Through the statistical analysis additional information can be generated for targeted product and process actions and therefore help to save resources.

3.5 RAW Phase D – Part I: Parametrical statistical analysis

With the increase of the field failure cases and therefore more data in the warranty database, the application of nonparametric significance tests become limited. Instead, parametric significance tests can be used to analyse and compare individual production months to each other.

Under the assumption of a normal distributed basic population these industrial used parametric significance tests can be performed (testing the assumption of normal distributed basic population: for example Kolmogorov-Smirnov test and Cramér-von Mises test [8]).

Known and industrial used parametric significance tests [9] can be divided in the groups of

- t tests (mean value analysis) and
- F tests (variance analysis), and

require the assumption of a normal distributed population.

Alternative it is feasible to analyse e.g. different damage cases or production batches by comparison of the confidence interval of fitted Weibull distribution models and parameters. Therefore, in addition to statistical significance tests, the graphical mapping of failure cases using the Weibull distribution is a key element in this phase.

3.6 RAW Phase D – Part II: Further statistical analysis

The failure behaviour of products differs; therefore it is often not possible to map damage cases using the normal distribution. For this reason the industrial standard in the automotive industry is the use of flexible Weibull distribution models, because they cover roughly different distribution models (e.g. normal distribution with form parameter b = 3,5; cf. [10], [11] and equation (1)). Therefore the use of the t test and F test is limited. Using the Weibull distribution function $F_{WD}(t)$, cf. equation (1), and the Weibull density function $f_{WD}(t)$, cf. equation (2), it is feasible to form the failure rate function $\lambda(t)$, cf. equation (3). The parameters are t_0 (failure-free time), T (characteristic life time) and b (form parameter). The use of Weibull distribution models allows a description of simple product failures and to indentify different damage phases or behaviours in a product life cycle [12].

$$F_{WD}(t) = 1 - e^{-\left(\frac{t-t_0}{T-t_0}\right)^b} \quad (1)$$

$$f_{WD}(t) = \dot{F}_{WD}(t) \quad (2)$$

$$\lambda(t) = \frac{f_{WD}(t)}{1 - F_{WD}(t)} \quad (3)$$

In principle, it is possible to differentiate the damage behaviour regarding the product life cycle into three superior damage phases: Early failure phase, coincidence failure phase and wear out failure phase. The serial combination of these phases ("bathtub curve") is shown in figure 8, focused on the failure rate λ(t). Subsequently, it is possible to calculate a prognosis regarding the expected damage cases in the field.

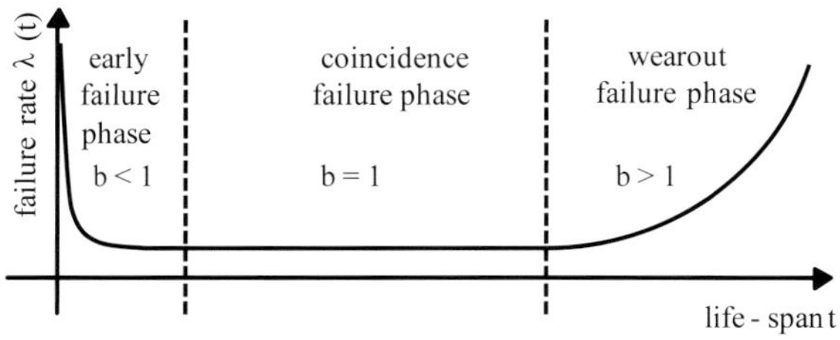

Figure 8: Schematic failure rate λ(t) during the product life cycle including fundamental superior damage phases

Based on the number of produced units it may be determined – e.g. with the sudden death test method or the Eckel candidate method [13] – the time resolved damage probability at each life-span area with respect to the known population in the use phase.

In further steps the performance of a technical damage components analysis can verify the results of the statistical reliability analysis and describe the specific damage causes (not part of this paper).

In addition to the usage of parametric and nonparametric significance tests, for example, correlation methods can be used for further data analysis [12].

4 CONCLUSION AND OUTLOOK

The industrial standard of reliability analysis meets limits especially in the context of early reliability analysis in the custom use phase after product market launch. The outlined RAW concept combines parametric and nonparametric significance tests and therefore facilitates a reliable statistical analysis of small to large sample sizes. In combination with other advanced methods of reliability analysis, such as, for example, the WCF approach, DCD algorithm, and the FDA concept (cf. [14], [15] and [16]), the RAW concept provides a basis for targeted actions and recommendations over the entire product life cycle. The advantages of the statistical reliability analysis using the RAW concept can be outlined as follows:

- Early and focused analysing and evaluating of failure data from small to large sample sizes
- Combining parametric and nonparametric statistical methods in a comprehensive concept
- Using nonparametric significance tests for small sample sizes at an early field observation time, in which parametric procedures are not applicable
- Statistical mapping – using distribution models – of the failure behaviour of field damage cases not needed at an early stage
- Providing the detection of production batches with high loss ratio and therefore supporting the targeted technical damage part analysis
- Supporting the selection of appropriate and targeted actions at an early stage after market launch

Therefore the application of the RAW concept results in reliable products and enhances environmental protection in a long-term goal through environmental-friendly, reliable design and helps to save resources over the whole product life cycle. The long-term development goal is the comprehensive mapping and analysis of complex damage causes of complex components and systems in the automotive engineering at an early date in the field observation after market launch.

5 REFERENCES

[1] Bertsche, B., 2008, Reliability in Automotive and Mechanical Engineering. Determination of Component and System Reliability, Berlin: Springer.

[2] Ronniger, C., 2000, Randomly or systematically, Qualität und Zuverlässigkeit 45(5): 586 – 590.

[3] Linß, G., 2005, Quality management for engineers, München: Carl Hanser.

[4] Bracke, S., Haller, S., 2010, Early detection of complex field damage cases after market launch: Processes and case studies in the automotive industry, Symposium Proceedings International Applied Reliability, Singapore. – in press –

[5] Bracke, S., Haller, S., 2010, 'Optimised Multi-Stage Sampling Procedures (OMSP)' concept, contribution to the technical field failure case analysis, XXII. German-Polishes Seminar, Warsaw University of Technology, Poland. – in press –

[6] Büning, H., Trenkler, G., 1994, Nonparamtric statistical methods, De Gruyter, Berlin.

[7] Siegel, S., 2001, Nonparamtric statistical methods, Klotz, Eschborn bei Frankfurt a. M.

[8] Duller, C. 2008: Introduction to nonparametric statistics with SAS and R, Physica-Verlag, Heidelberg.

[9] Sachs, L., 2002, Applied statistics, Springer Verlag, Berlin.

[10] Pfeifer, T., 2002, Quality Management, München: Carl Hanser Verlag.

[11] Murthy, D. N. P., Xie, M., Jiang. R., 2004, Weibull Models, New Jersay: John Wiley & Sons.

[12] Birolini, A., 2007, Reliability engineering: Theory and Practice, Berlin: Springer.

[13] Eckel, G., 1977, Determination of the initial course of the reliability function of automotive components, Qualität und Zuverlässigkeit 22 (9): 206 – 208.

[14] Bracke, S., Haller, S., 2009, Prevention through precision: Statistics of the complex failure behavior of technical components, Qualität und Zuverlässigkeit 54 (7): 48 – 49.

[15] Bracke, S., Haller, S., 2008, Damaged – but why? Software algorithm analyses complex damage causes, Qualität und Zuverlässigkeit 53 (11): 96 – 97.

[16] Bracke, S., Haller, S., 2010, Analysing multiple damage causes of complex products using DCD algorithm and WCF approach, In: Bris, R., Guedes Soares, C., Martorell, S., Reliability, Risk and Safety: Theory and Applications, Taylor & Francis Group, London.

Electronics Condition Monitoring for Improving Sustainability of Power Electronics

A. Middendorf[1,2], N.F. Nissen[2], S. Guttowski[2], K.D. Lang[1,2]

[1]Research Center for Microperipherics, Technische Universität Berlin, Germany

[2]Fraunhofer IZM, Berlin, Germany

Abstract

Condition monitoring of electronics helps to optimize the use of products, with regards to condition based maintenance, ReUse and refurbishment, and adapting the reliability of a product more precisely to actual use profiles.

One primary area of application is in power electronics, where availability, guaranteed capacity operation and planned maintenance play important roles for the economic and the ecological success of an application. In general, power electronics are crucial for the infrastructure of modern societies, so the improved performance of electronics regarding efficiency and reliability is a building block of emission reductions in energy generation, e-mobility and general transport.

As a specific case, the life time monitoring of IGBT components is presented. The approach of condition monitoring is then broadened to the co-called innovation cluster Maintenance, Repair and Overhaul in Energy and Transport, which has been inaugurated in Berlin last year.

Keywords:

Condition Monitoring, Maintenance, Overhaul, Repair, Remaining lifetime, Life Cycle Unit

1 INTRODUCTION

Maintenance and Repair are significant expense factors for operating a facility, plant or machine. Their efficiency and profitability can be increased by applying intelligent condition monitoring. Expenses caused by unscheduled downtimes, engine breakdowns and consequences thereof are prevented. By applying condition based maintenance, operating time can be increased significantly. However, condition monitoring is nowadays primarily implemented in safety-critical or high value systems or as specific stand-alone solutions, which cannot be adapted one-to-one to other applications.

In order to implement condition monitoring, an extensive analysis of failure causes and fatigue mechanisms for a system is required. These mechanisms depend on various physical quantities that need to be detected and processed. The analysis and the characterization of system behavior during its life time is a fundamental of reliability research. The development in computer science and processing power was the trigger for the finite element analysis that provides the possibility of creating advanced physics of failure models. Based on these models, condition monitoring systems predict the remaining life time or the current state of a device and thus clearing the way for better reliability and sustainability. Therefore, condition monitoring is not only relevant for safety-critical issues, e.g. in aviation or medicine, but implies environmental, economical and technological aspects as well. In the following, two detailed examples of condition monitor systems are given and discussed according to the aspects mentioned above. The first example is a photovoltaic converter provided with a life time prognostic system based on the experienced load cycles. The second introduced application gives an example of an energy self-sufficient radio sensor network that analyzes machine vibrations to draw conclusions of the current condition. Condition monitoring has the potential of increasing the system reliability and functionality in any circumstance, environment and field of application.

2 CONSIDERATIONS FOR DESIGNING CONDITION MONITORING SYSTEMS

There are three different concepts of determining the state of a product during its use phase. One concept is using so called monitor structures. These are designed in the manner that they fail faster than the monitored functional component when exposed to the same external loads. The condition of the monitor structures, working or failed, is periodically evaluated [1]. By combining several monitor structures, as shown in Figure 1, the remaining life time of the functional component can be estimated. The two other approaches are realized in a different way. One method is to measure a set of relevant system parameters of the product, indicating its condition. In electronic systems for example the resistance of a solder joint going to infinity indicates the rupture of that joint.

The other method is to monitor the product's environmental loads and compare these with previously defined physics of failure models. Thus the remaining life time can be predicted [1], [2]. In order to determine the appropriate concept for an existing application the following aspects have to be taken into account for designing condition monitoring systems.

2.1 Technological aspects

The development of micro systems and packaging technologies enables the development and implementation of more complex and reliable sensor devices for condition monitoring. Due to reduced size and improved mechanical robustness the possibilities and the field of application have enlarged. Since condition monitoring is nowadays usually not included in designs of electronic products and systems, the possibility of retrofitting must be given. Therefore, the monitoring system should be small and easy to be integrated into any product. However, for detecting precursor parameters, system parameters or environmental loads for life time estimation an appropriate sensor device and measurement equipment is required. To mount the sensor devices to the monitored system, proper places providing significant data have to be figured out. In order to be applicable to various types of products and thus a broad range of application, monitoring devices must be sufficiently robust to withstand all conceivable environmental impacts.

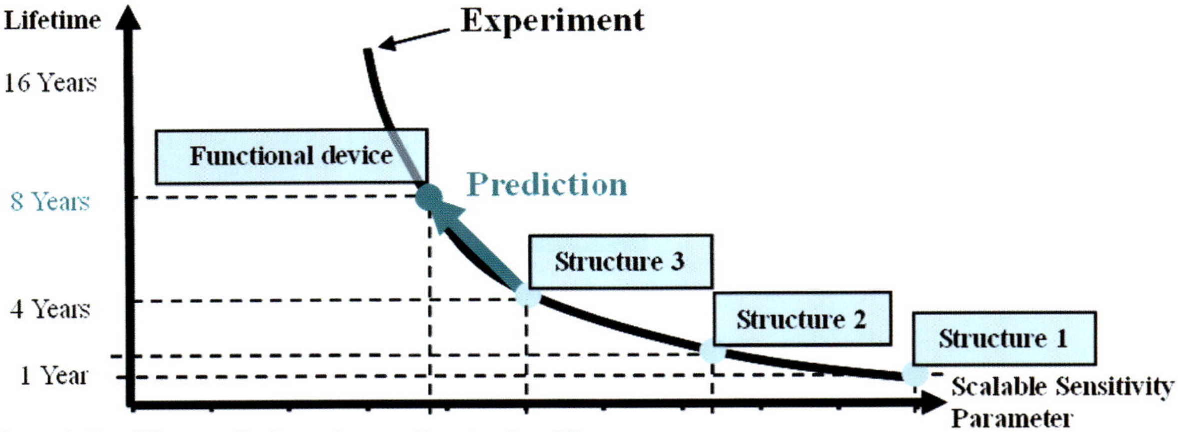

Figure 1: Condition monitoring using monitor structure [4]

By implementing monitor structures no additional sensor devices are required [1]. The energy consumption is marginal and the complexity of generated data lower than for condition monitoring systems that evaluate system parameters and environmental loads. In return, no conclusion can be drawn to the failure cause from monitor structures, in case the monitored structure is very complex.

Figure 2: Device of a silicon based Monitor structure

Fig. 2 shows the second generation of the Monitor structure, which was developed and manufactured at the Fraunhofer Institute for Reliability and Microintegration respectively the Research Centre for Microperipherics at the Technical University Berlin. A silicon wafer is processed with copper lines and lead free solder balls. In a next step the chips are detached. Each chip has a size of 6.4mm x 2.5mm. The devices are a test vehicle to investigate the behaviour of interconnects if various loads are induced. In recent projects we combine thermal with mechanical or thermal with current forces; the singular thermal load was also investigated to generate the basic information for the lifetime modelling. The device in fig. 2 shows the utilization of the physical basics relating to differences in thermal expansion coming from the composition of different materials. There is a strong relation to the distance of the applied forces. Consequence is the earlier fail of the solder ball interconnects in the corners of the device. The not connected solder balls enhance the soldering process because the device is stabilized and allow also the comparison of current and not current loads on one device. According to the particular experiment different printed circuit board designs are available too.

Fig. 3 shows cross-section polishes from the variation of the geometries. The thickness of the solder ball influences the lifetime of the interconnection significantly. With increasing thickness the reliability becomes higher. The process is not used in general because of additional process time and costs; for the research and the understanding of the behaviour of such interconnect this modification is pretty good. We use especially this kind of devices to investigate the electro migration. With this kind of so called Hour glass structures the maximum load is shift into the centre and in the middle of the interconnect.

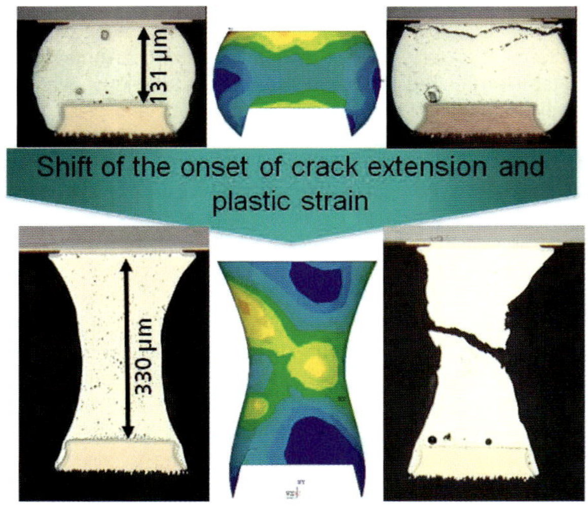

Figure 3: Modification of interconnects for monitor structures

This is also valid for the combination of mechanical and thermal force. Once can see in fig. 3 the shift of the onset of the crack extension from the area close to the termination to the centre and the change of the angle. The simulation is in a very good accordance with the experimental results. The belonging Weibull plot is shown in fig. 4 for the combined force of temperature and vibration. The benefit according to the extension of the life time is moderate. From app. 260 h to 310 h could help to reach the demanded specification for some application, but it is not extraordinary successful. In combination with epoxy underfilling a huge increase of reliability is expected and topic of future research at the institute.

Much more important for the investigation on monitoring structures is the 10% failure probability for the 330 μm structure which is very high compared with the 132 μm one. In other words is the statistic variation very low because the manufacturing processes are very well understand and controlled. This attribute is indispensable for the design and manufacturing of monitor structures for practical applications.

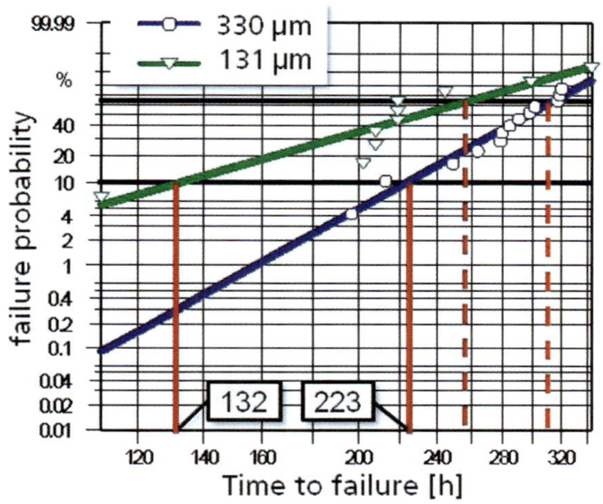

Figure 4: Weibull plot for different indicator structures

Condition monitoring of electronic systems is much more difficult than monitoring most mechanical systems due to more complex structures and nano- or micro-scale effects. A monitoring system should not cause unwanted interference with the monitored system. Especially in electronic systems, the integration of monitoring circuits needs to match various requirements, such as EMC for instance. A further requirement is that the lifespan and the reliability of the monitoring device must necessarily exceed that of the monitored product.

2.2 Economic aspects

Any unexpected failure of systems results in decrease of profits [3]. Especially if complex and expensive manufacturing facilities are stopped or expensive products (e.g. single-units) are faulty, costs can be very high. Loss of comfort, such as cancelled flights and delayed trains can lead to bad reputation, causing falling profits as well. In some cases, e.g. difficult to access machines such as offshore wind turbines, resolving an unexpected failure can take a long time (in that case depending on the weather). During that time no electric energy is produced and thus high financial losses may occur.

In many cases the damage caused by a failure can be minimized if machines are turned off instantly as soon as a problem is detected. For instance, expensive tools, installed in fast running machines, are damaged due to a delayed machine-stop in case of error. Through condition monitoring problems can be detected very fast and appropriate actions can be implemented quickly. Too early replacement of parts can be avoided because the system status is known and the remaining life time can be estimated. Exploiting the entire lifespan of parts increases the operating time and thereby saving money as well. Knowing the time of failure also reduces the amount of spare parts hold in stock because just in time ordering can be done. Another important issue is that the technology has to prove economically feasible. The costs of the monitoring system, including installation and operation, must not be higher than the costs the prevented failure would have caused.

2.3 Environmental aspects

The Re-use of products is a key measure of an overall waste prevention strategy. Used products or their individual components are designated for use in new products. In order to apply appropriate handling at the end-of-(first-)life, precise information on the product's re-use potential is required. By using condition monitoring technologies, the potential for re-use can be determined. By reusing products less material and energy are needed compared to producing new ones [3].

As stated in [3], the quality in reused products is not an issue at all. In fact, studies have shown that second life electronic equipment, if correctly remanufactured, can be more reliable than new [3]. As described, failures in production lines are spotted quickly, therefore none or less faulty products, which in worst case have to be scrapped, are produced. Last but not least, failures in general can be a danger for health, safety, the environment and life. Thus any failure recognized by a monitoring system on time is a great benefit.

3 EXAMPLES FOR IMPLEMENTATIONS OF CONDITION MONITORING SYSTEMS

On the basis of the three concepts of condition monitoring, two examples for the implementation are depicted more in detail. The following applications are examined according to the previously discussed aspects.

3.1 Power Modules in Photovoltaic Systems

The major failure causes of photovoltaic converters are temperature shifts resulting in solder and/or bond-wire fatigues of the IGBT power modules (fig.5) [4].

Figure 5: Wire bond lift-off 1.5 Mio. Thermal Cycle of 60K

Fig. 5 shows the lift off of wire bonds after 1.5 Mio. Cycles with a temperature swing of 60 Kelvin. The load is induced by moderate amperage of 4 Ampere combined with a voltage of 20 Volt. The time for one cycle ranges between 2 and 5 seconds. Therefore one experiment last only weeks or a few month.

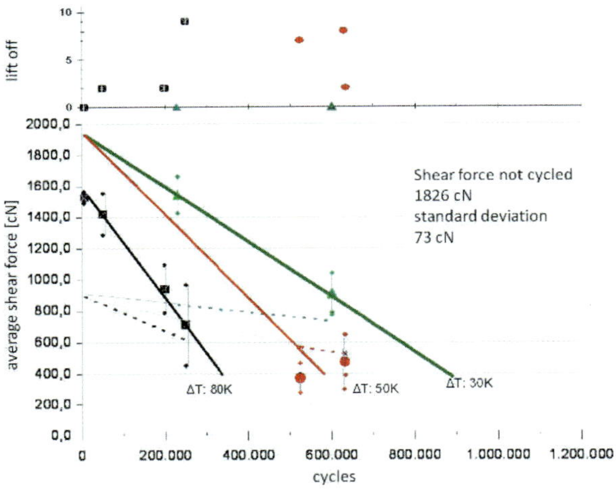

Figure 6: Experimental lifetime data for 400 µm Al-wire-bond

In Fig. 6 different lifetimes dependent on the embossed loads are shown. As higher the temperature swings as shorter the lifetime. The number of wirebond lift off (the total number is 24) and the shear force are the two failure criteria. The different curves are basis for the developed lifetime model which takes also the average temperature into account.

In order to predict the remaining life time, the significant parameters, temperature, electric current and voltage, are measured. The electric current and the voltage, generated by the photovoltaic, are system parameters of the power modules. These parameters are already observed by the system itself and hence no additional sensor devices are needed. An additional temperature sensor detects the temperature of the chip surface, being approximately the interconnection-temperature of the bond wire and the chip [5]. In order to perform a life time prognostic, the measured temperature and temperature shifts are applied to the physics of failure model.

Due to high chip-surface-temperatures up to 70°C, high currents (up to ten or more Amperes) and high voltages (couple of hundred Volts), the robustness of the condition monitoring system is crucial. Furthermore, the environmental temperature, the converter is exposed to, is an important factor. Temperatures can range in some environments, such as deserts, from below zero degrees Celsius in the nighttimes to more than a hundred degrees Celsius during the day. This again shows the requirement of robustness of the monitoring system.

The spare space in converter modules is rare and usually no room has been allocated for monitoring systems in the design-phase. Therefore the later-on installed monitoring system has to be small in size. The lifespan of photovoltaic panels is indicated with around 30 years [6]. This implies that the condition monitoring has to work properly for even more than 30 years. Using condition monitoring systems has two positive economic and environmental consequences, respectively: First of all, the down-time of the system is minimized. In the case of photovoltaic-systems, longer run-time is equal to more electric power produced. The more energy is being inducted into the grid, the higher the return on investment. A grid system based on renewable energy will only be successful if the reliability of the energy production is equal to or higher than conventional systems. To ensure the reliability of smart grids, made up of various (small) power producing systems, the reliability of each single unit has to be ensured. This requirement can be fulfilled by using condition monitoring.

The other advantage is the utilization of a longer lifespan of the alternating-current converters. That means less energy and material are needed for new converters. By using condition monitoring systems, preventive exchange of the converters is not longer needed. They are replaced in case of an upcoming failure only, which reduces the maintenance and acquisition rate, hence the operational costs.

Power electronics are a key technology for regenerative power generation. The example is chosen from the photovoltaic industry, but the results are also transferable to the wind energy as well as for all other electrical engines which are using are electronic frequency converters. The example explained in chapter 3.2 is therefore a holistic approach for the reliability and lifetime consideration by the combination of the mechanical and power electronic system parts.

3.2 Early-warning System for Machine Crash

In order to prevent production facility downtime and machine crash with safety-relevant consequences, a condition monitoring system is applied to the production line of a paper plant. The condition monitoring system comprises numerous energy self-sufficient micro systems and sensor nodes in a wireless sensor network. The sensors record acceleration signals of different machine parts, such as bearings or motor drives (see fig. 7). The basis of machine diagnostic using acoustic signal analysis is state of the art and thousands times implemented. The new approach is the possibility for always online diagnostic because of energy generation from the environment, which is called energy harvesting.

Figure 7: Measurement of Engine Vibrations [7]

The signals are processed, analyzed and compared to vibration patterns. In case of abnormal vibrations the condition monitoring system will trigger a corrective maintenance. Since already slight changes in machine vibration patterns are detectable before wear and tear or cracking becomes a serious problem the maintenance or repair work can be scheduled in advance. An additional feature of this condition monitoring example is the energy self-sufficiency of the sensor components which thereby can be placed in machine parts that are difficult to access without increasing the maintenance effort of the condition monitoring system itself [7]. Due to developments in the area of 'energy harvesting' energy self-sufficient concepts are applicable for condition monitoring and can improve sustainability and reliability [8].

4 SUMMARY

Various reasons for developing and implementing technologies of condition monitoring were depicted. The three different concepts of condition monitoring were briefly introduced and preliminary considerations for designing and implementing these systems were explained more in detail. On the basis of two examples, the feasibility as well as the challenges and boundaries of implementing condition monitoring systems were shown. Besides the demonstrated application areas, many other areas for applying condition monitoring are possible.

Due to the expanding utilization of condition monitoring, large amounts of sensitive data will be generated. The more systems are monitored and data of use is recorded, the easier conclusions to the identity of the consumer and his behaviour can be drawn. An appropriate handling of the gathered information has to be ensured.

It is obvious that condition monitoring has a great potential of increasing reliability, safety and economic profits for a wide range of products and therefore, will play a major role in future.

5 ACKNOWLEDGMENTS

The research topics are sponsored in various projects by the TSB Berlin and EU within EFRE (project CoMoLeFO), the BMBF (project EcoMoS), the Fraunhofer society (projects MRO-Cluster and InCoMo) and the European Commission (ZeroWin). The authors would like to thank for the support.

Contact:

Mail: Andreas.middendorf@izm.fraunhofer.de

Tel.: +49(0)30 46403-135

6 REFERENCES

[1] O. Bochow-Neß, T. Eckert, N. Nissen, J. Jaeschke, A. Middendorf, H. Reichl, "Condition Monitoring of Microsystems Supporting Sustainability", Proc. Electronics Goes Green 2008+, Fraunhofer IRB Verlag, Berlin, Germany, pp. 219-224, 7.-10.09.2008

[2] H. Reichl, A. Middendorf, "Miniaturized Life Cycle Units" in "Günther Seliger: Sustainability in Manufacturing, Springer, 2007

[3] A. Middendorf, W.M. Grimm, H. Griese, O. Ness, and H. Reichl, "Using Life Cycle Information to Improve the Reliability of Electronic Assemblies", Proc. Electronic Circuits World Convention, Cologne, Germany, pp. 61ff, 07.-09.10.2002

[4] M. Errington, and S. Childe, "Is the Reliability of Remanufactured Second Life Electronic Products Better Than That of New?", Proc. Electronic Goes Green 2008+, Fraunhofer IRB, Berlin, Germany, pp. 231-235, 7.-10.09.2008

[5] T. Wernicke, R. Gegusch, S. Dieckerhoff, A. Middendorf, G. Seliger, and H. Reichl, „On-line Condition Monitoring of Power Semiconductors", Proc. Electronic Goes Green 2008+, Fraunhofer IRB Verlag, Berlin, Germany, pp. 225-230, 7.-10.09.2008

[6] A. Middendorf, "Lebensdauerprognostik unter Berücksichtigung realer Belastungen am Beispiel von Bondverbindungen bei thermomechanischen Wechsel-beanspruchungen", Dissertation, Technical University Berlin, urn:nbn:de:kobv:83-opus-26490, 2009

[7] Freistaat Thüringen, Ministerium für Landwirtschaft, Forsten, Umwelt und Naturschutz

[8] M. Niedermayer, J. Hefer, S.Guttowski, „The Dawn of Self-Sufficient Micro Systems", 13th International Forum on Advanced Microsystems for Automotive Applications mst-news 6/08, VDE/VDI/IT, Berlin, p. 21, Dec. 2008

[9] S. Benecke, N. Nissen, H. Reichl, "Environmental Comparison of Energy Scavenging Technologies for Micro System Applications", Proceedings of the 2009 IEEE International Symposium on Sustainable Systems and Technology, Phoenix, Arizona, USA, pp. 1-6, 18.-20.05.2009

Cryogenic Processing of Biomaterials for Improved Surface Integrity and Product Sustainability

S. Yang[1], Z. Pu[1], D.A. Puleo[2], O.W. Dillon, Jr.[1], I.S. Jawahir[1]
[1]Department of Mechanical Engineering, University of Kentucky, Lexington, USA
[2]Center for Biomedical Engineering, University of Kentucky, Lexington, USA

Abstract
Improved functional performance and longer service life of biomedical products offer great sustainability benefits. Surface integrity, which can be modified by severe plastic deformation (SPD) processes, affects the functional performance of materials. Two SPD processes – burnishing and machining – were studied under cryogenic conditions. Cryogenic burnishing of a *Co-Cr-Mo* biomedical alloy using a novel burnishing tool led to significant grain refinement and 80% greater surface microhardness relative to the bulk. Cryogenic burnishing of *AZ31 Mg* alloy led to a more than 2 mm thick SPD surface layer with remarkably refined microstructure (grains <300 nm), where hardness was increased 95%. The SPD layer formed on *AZ31 Mg* alloy after cryogenic machining was about 60% harder than the bulk material. Furthermore, this layer enhanced corrosion resistance during incubation in simulated body fluid. The present results demonstrate significant surface and subsurface property improvement due to cryogenic processes of both alloys, thus providing improved sustainability.
Keywords:
Sustainable manufacturing, Microstructure, Cryogenic burnishing, Cryogenic machining, *Co-Cr-Mo*, *AZ31 Mg*

1 INTRODUCTION

Improved functional performance and longer service life of products, especially biomedical devices, offer great societal and environmental benefits. Surface integrity plays a crucial role in the functional performance of such products.

Severe plastic deformation (SPD) processes have been extensively reported for modifying surface region properties by creating ultrafine or nano-sized grains and grain size gradients in the surface region of many products. These nano-grained materials are due to the high strain/strain rate involved in the processing. Much evidence has shown that SPD induced nano-sized/ultrafine grained materials possess appealing properties compared with their coarse-grained counterparts. Wang et al. [1] fabricated a nanocrystalline surface layer on a low carbon steel plate via SPD process; improvements of wear resistance in parts with the nanocrystalline surface layer were demonstrated. Karimpoor et al. [2] also found that nanocrystalline cobalt exhibits unusual mechanical properties compared to coarse-grained cobalt. Nanocrystalline surface layers generated by SPD process on *Cu*, steels and *Mg* alloy with emphasis on reciprocating sliding wear behavior [3] showed improved wear resistances compared to the surfaces with normal sized grains.

Burnishing and machining are two SPD processes where the strain and strain-rate are usually high [4]. Significant grain refinement has been reported on the processed surface layers. Altenberger et al. [5] created nanocrystalline surface layers on *AISI 304* stainless steel from deep rolling. Their results showed that this nanocrystalline surface region remain stable against cyclic loading, even at high stress amplitudes, and impede dislocation movement and slip band formation. Nalla et al. [6] conducted deep rolling on *Ti-6Al-4V* alloy; nanoscale grains were found on surface layers, which are believed to play a critical role in the enhancement of fatigue life. Nikitin and his co-workers [7] studied deep rolled austenitic stainless steel *AISI 304* and *Ti–6Al–4V* alloy; their results show that deep rolling induced nanocrystalline regions are stable during short time annealing. Isothermal fatigue in the low cycle fatigue regime at high stress amplitudes does not alter the nanocrystalline region up to 600°c. Hot rolled *AZ31 Mg* samples were reported to have a marked reduction in biodegradation rate compared with squeeze cast samples [8]. The reduction was attributed to grain refinement from 450 μm to 20 μm in average diameter. Prevey and co-workers [9-11] showed that low plasticity burnishing can provide a layer of compressive residual stress with sufficient depth to effectively increase the fatigue life of many materials. Iglesias et al. [12] also showed that wear rates of nanostructured copper and titanium specimens created by large strain extrusion machining are significantly lower than that of their coarse-grained counterparts. Nanocrystallized grains of 5-20 nm were reported on the top of the white layer of *AISI 52100* steel after machining [13]. Nanocrystalline layers with grain size about 45 nm were shown to form on the cryogenically machined surface of *AZ31 Mg* alloy [14]; *in vitro* corrosion tests proved that the nano-sized layer improved the corrosion resistance of this alloy.

However, due to the heat generated during SPD processing, refined grains often grow under the high temperatures created. Cryogenic SPD, where liquid nitrogen is applied during processing, has been reported to successfully introduce thicker surface layers consisting of ultrafine/nano-grained structures. By using surface mechanical grinding treatment at cryogenic temperatures, Li et al. [15] synthesized a gradient nano-micro-structure in the surface layer of bulk pure copper. The average grain sizes vary from about 22 nm in the topmost surface to sub-micrometers at 200 µm deep. Ni et al. [16] also reported that by using cooling fluid during machining, grain size in the secondary deformation zone was reduced from 1.2 µm to 360 nm.

In the current study, two SPD processes – burnishing and machining – were studied at cryogenic conditions, where liquid nitrogen was applied so as to reduce the temperature rise created during and following processing. The effects of cooling conditions on microstructure and microhardness changes in two materials are reported herein.

2 EXPERIMENTAL

2.1 Work materials

Co-Cr-Mo alloy

Cobalt-chromium-molybdenum (*Co-Cr-Mo*) alloy has good mechanical properties, wear resistance, and corrosion resistance, which combine to make it biocompatible [17, 18]. It has been extensively used in joint implants such as artificial hips and knees [19, 20]. BioDur Carpenter CCM alloy, which is a high nitrogen, low carbon wrought version of ASTM F75 Cast Alloy with an average initial hardness of 43 HRC is used as the work material in the burnishing experiments reported here. A *Co-Cr-Mo* alloy bar (50.8 mm diameter) is used to prepare disk samples which have a diameter of 50.8 mm and a thickness of 3 mm.

AZ31 Mg alloy

Magnesium alloys are emerging as a new class of biodegradable implant materials for internal bone fixation. They provide good temporary fixation and do not need to be surgically removed after healing occurs, providing relief to the patients and reducing the healthcare costs [14].The particular material studied was the commercial *AZ31 B-O* magnesium alloy. It was obtained in the form of a 3 mm thick sheet. Disc specimens (3 mm in thickness and 130 mm in diameter) were made from the sheet.

2.2 Processing

Burnishing

Burnishing experiments were conducted on a Mazak CNC lathe along with ICEFLY™ cryogenic equipment. Liquid nitrogen as a cryogenic coolant was used. It has the advantages of better surface finish for workpieces, environmentally safer and healthier for the worker. A specially designed and fabricated burnishing tool, with a fixed burnishing roller which was used, is shown in Figure 1(a). The spherical cavity in the tool holder permits the use of rollers with different diameters. Figure 1(b) and (c) show the experiment setup used for cryogenic burnishing. The roller is spring loaded for approximate adjustment of the burnishing force. The forces developed during the processing were measured by a KISTLER 3-Component Tool Dynamometer. A *M2/M7* high-speed tool steel roller with a diameter of 14.3 mm was chosen as the burnishing tool for the current experiments. The hardness and surface roughness of these rollers were measured and found to be 63 HRC and 0.01 µm (R_a), respectively. The roller head is fixed in order to induce enough shear stress and strain to the surface region to cause grain refinement via SPD and possibly dynamic recrystallization (DRX).

The processing conditions used for the burnishing experiments on *Co-Cr-Mo and AZ31 Mg* alloys are shown in Table 1. *Co-Cr-Mo* discs are burnished with and without liquid nitrogen for better study of the effects of cryogenic cooling.

Table 1: Burnishing conditions for *Co-Cr-Mo and AZ31 Mg* alloys

Material	Burnishing time	Burnishing speed	Feed rate	Radial Force
Co-Cr-Mo	20 s	100 m/min	0.05 mm/rev	230 N
AZ31 Mg	60 s	100 m/min	0.05 mm/rev	240 N

The application of liquid nitrogen is expected to effectively suppress grain growth after SPD processing and the occurrence of DRX within the surface region. Due to large strains, high strain-rates and lower temperatures, under the burnishing conditions used, the cryogenic SPD process introduces significant grain refinement to the material surface layer.

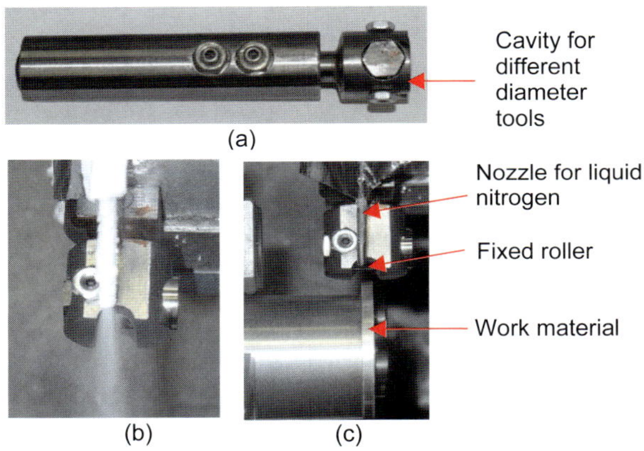

Figure 1: (a) Burnishing tool illustration; (b) application of liquid nitrogen during cryogenic burnishing; (c) experiment setup for cryogenic burnishing

Machining

A Mazak CNC lathe equipped with an Air Products liquid nitrogen delivery system is also used to conduct orthogonal turning of the *AZ31 Mg* discs. As shown in Figure 2, liquid nitrogen was sprayed on the machined surface from the clearance side of the cutting tool for what is being called cryogenic machining. The cutting tools used were Kennametal uncoated carbide *C5/C6* inserts with 68 µm edge radius. *AZ31 Mg* disc was cryogenically machined with 100 m/min cutting speed and 0.01 mm/rev feed rate.

2.3 Material characterization

Co-Cr-Mo alloy

Metallurgical *Co-Cr-Mo* specimens were cut from the processed discs. After hot mounting, grinding, and polishing, the specimens were chemically etched (120 ml 37% hydrochloric acid + 12 g cupric chloride dehydrate, crystalline + 10 ml R.O. Water) [21] for 15 s at 120 °c to reveal their microstructures.

Figure 2: Machining setup with the liquid nitrogen delivery system

AZ31 Mg alloy

Metallurgical *AZ31 Mg* samples were cut from the machined discs. After cold mounting, grinding and polishing, acetic picric solution was used as the etchant to reveal the grain structure as shown below.

Characterization methods

The materials' microhardness and microstructure in the surface region were measured before and after the processing. Microindentation tests were undertaken for *Co-Cr-Mo* specimens by using a Vickers indenter on a CSM Micro-Combi Tester with 300 mN applied load. The hardness of *AZ31 Mg* alloy was measured by a CLARK Digital Micro hardness Tester (CM-700 AT) with 50 mN load and 15 s dwell time. Metallurgical analysis was conducted by using optical and scanning electron microscope (SEM). Chemical compositions were determined from energy dispersive spectroscopy (EDS).

3 RESULTS AND DISCUSSION

3.1 *Co-Cr-Mo* alloy

SEM investigations were conducted to study the burnished workpieces' deformed microstructures. A comparison between SEM photomicrographs of the initial surface and burnished surface using different burnishing conditions is shown in Figure 3. The burnishing direction is parallel to the plane of the figures, and the grains shown are exactly at the edge of the burnished surfaces. Comparisons between the initial microstructure (Figure 3(a)) and the burnished ones (Figure 3(b) and (c)) show that the grains were elongated and more condensed in a thin layer near the surface of the workpiece, grain structures within this layer are not discernable. This layer of indiscernible grain structure was also reported in other materials after burnishing [6] [15], which is defined as the SPD layer. Nano-grains were not observed. When Wu et al. [22] conducted SMAT process on cobalt, a microstructural evolution in the deformed surface layer was observed, which contained recrystallized nano-grains, subgrain subdivisions, elongated subgrains, grains with heavily twins, and equiaxed bulk grains with stacking faults sequentially from the depth of 15 µm down to 180 µm. This is in line with the present investigation. However, recrystallized nano-grains have not been induced by the currently used burnishing conditions.

From Figure 3(b) and (c); it is clearly visible that the depth of the process-influenced layer from the cryogenic burnishing conditions is much larger than that from using dry conditions. This suggests that cryogenic cooling has substantial influence on the surface layer developed during SPD processing. During burnishing, the top surface layer is subjected to the most severe SPD state; the layers beneath the surface are subjected to less severe deformation conditions. For dry burnishing, the effects of plastic deformation on the subsurface layer are compromised by the large amount of heat generated during processing. This layer is hardened by plastic deformation and softened by heat simultaneously; the mechanical and thermal effects oppose each other and finally lead to minor or no influence on microstructure changes. On the other hand, liquid nitrogen application effectively suppresses the heating effect and increases the process influenced depth to a larger extent. As shown in Figure 3(a), a large amount of twinning is present within the initial grain interiors, which may be attributed to pre-existing residual stresses prior to burnishing. This has a substantial influence on the subsequent effects of burnishing.

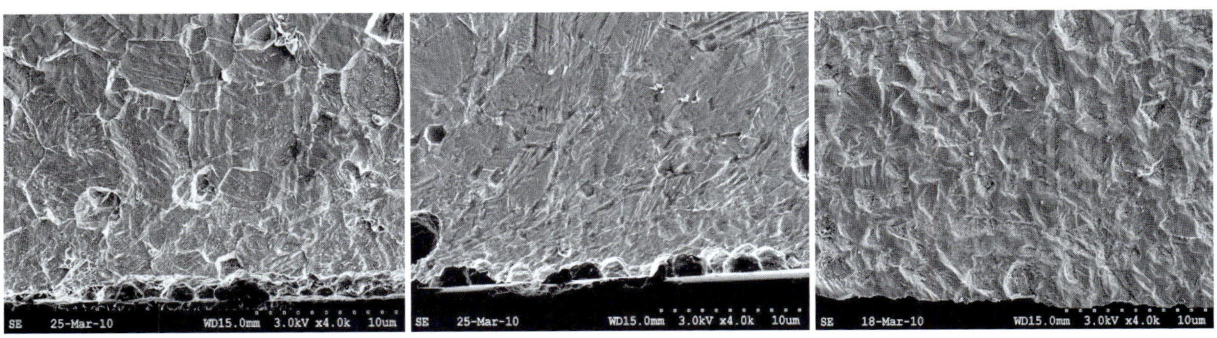

(a) (b) (c)

Figure 3: SEM micrographs of *Co-Cr-Mo* discs before (a) and after burnishing: (b) dry, (c) cryogenic

In order to characterize the hardness variation in the surface region of the processed *Co-Cr-Mo* workpieces, microhardness measurements were made. The minimum measurement depth from the surface is 5 µm in order to avoid the edge softening effect. For each depth, at least three measurements were taken. The microhardness profiles shown in Figure 4 were averaged values from these measurements.

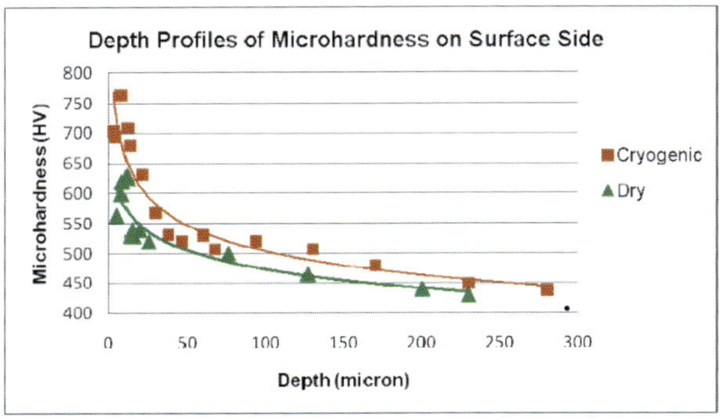

Figure 4: Subsurface microhardness profiles for cryogenic and dry burnished *Co-Cr-Mo* discs

The measured microhardness of the virgin disk is 430 HV on average. Comparing to the initial workpiece hardness, an increase of up to 80% was achieved after cryogenic burnishing. With the same burnishing conditions, the application of liquid nitrogen led to higher hardness value and larger process-influenced depth comparing to dry burnishing. The general trend of the curves shows a gradual decrease in microhardness with distance below the surface until the bulk value is reached.

Based on the well-known Hall-Petch relationship [23]:

$$\sigma_y = \sigma_0 + kD^{-1/2} \tag{1}$$

between yield stress (σ_y) and grain size (*D*) as well as the close interrelations among hardness, yield stress, and residual stresses, high hardness values often indicate fine grain size and large residual stresses. In other word, grain size and residual stresses collectively contribute to the hardness value. In our current work, microstructure changes due to different cooling conditions were only observed within 20 μm depth from the surface. In contrast, microhardness differences were measured to the depth of 250 μm. It is reasonable to state that the variations in microhardness were due to the different residual stresses being generated during processing. The residual stresses of the processed samples will be measured using X-ray diffraction techniques to validate this hypothesis. These results will be published later.

Figure 5 shows the results of the EDS analysis. Measurements were taken on the processed surface and bulk of specimens. The higher oxygen amount on the surface is due the formation of chromium oxide during processing, which is believed to protect the surface from electrochemical degradation and improve corrosion resistance [24].

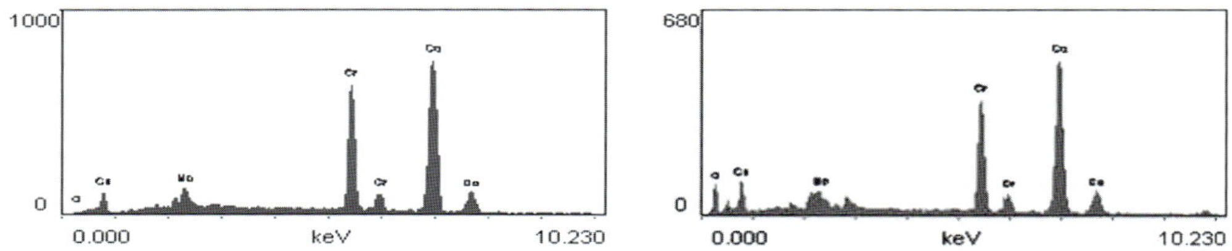

(a) (b)

Figure 5: EDS results of *Co-Cr-Mo* discs from cryogenic burnishing: (a) bulk, (b) surface

3.2 *AZ31 Mg* alloy

The initial microstructure of the *AZ31 Mg* disc is shown in Figure 6. There is no twinning in the bulk material since the as-received material is annealed. However, twinning is visible near the surface of the disc. This is due to the sample preparation in the machine shop where a turning operation is used as the final step in making the disc.

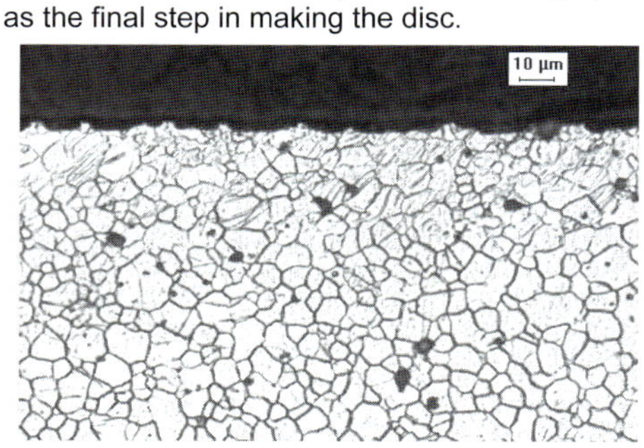

Figure 6: Initial microstructure of *AZ31 Mg* discs before processing

The microstructure obtained from cryogenic machining is shown in Figure 7. Significant changes in microstructure near the processed surface were found; a SPD surface layer of about 8 μm in which grain boundaries were no longer visible (at this magnification) was created by machining.

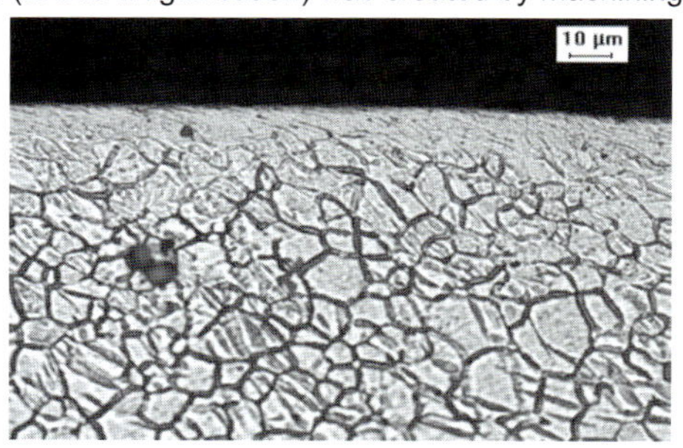

Figure 7: Microstructure of *AZ31 Mg* discs after cryogenic machining

Figure 8(a) shows the microstructure after cryogenic burnishing, in which a 2 mm process-influenced layer was formed, which is similar to the machined surface. The grain structure in the burnished surface region (Figure 8(b)) was no longer discernable compared to the clearly defined microstructure prior to processing (Figure 5). The transition between the initial microstructure and the process-influenced microstructure can clearly be seen in Figure 8(c).

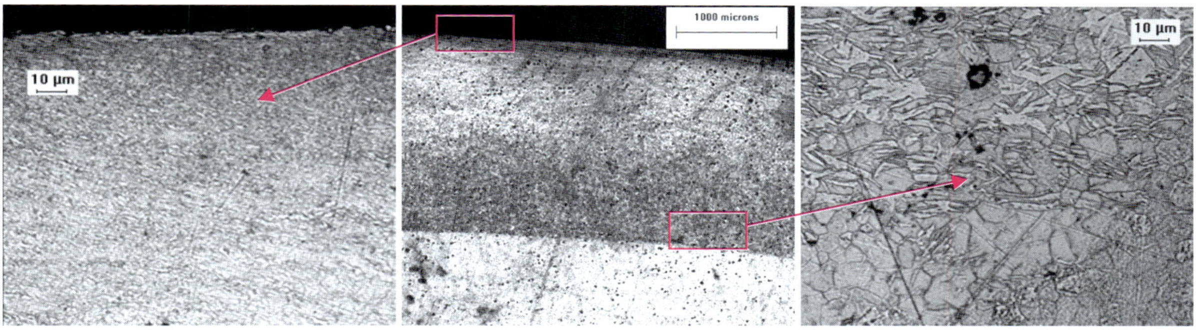

(a) (b) (c)

Figure 8: Microstructure of *AZ31 Mg* discs after cryogenic burnishing: (a) cryogenic burnishing; (b) surface layer in (a); (c) transition layer in (a)

SEM investigations were undertaken for better study of the severe plastic deformed layer from burnishing. As shown in Figure 9, highly uniform grains were present in the burnished surface layers; the grain sizes are generally less than 500 nm, although grains less than 300 nm can be found close to the surface. Comparing to the 10 µm grains before burnishing, the grain size on the burnished surface was more than 20 times smaller.

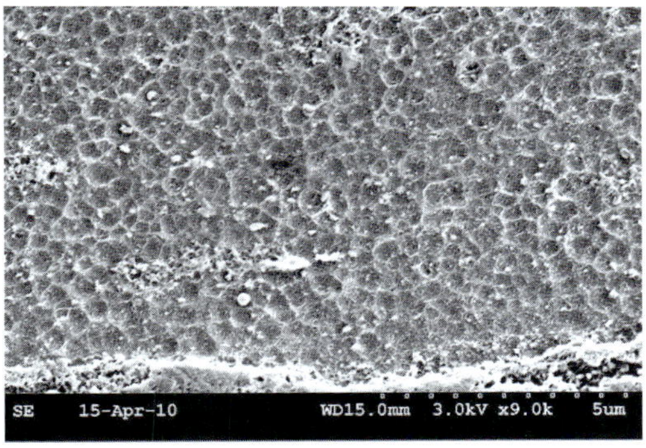

Figure 9: SEM micrograph of *AZ31 Mg* disc surface region from cryogenic burnishing

Figure 10 shows the microhardness profiles of *AZ31 Mg* discs before and after cryogenic processing. The initial hardness of the material was measured to be about 50 HV. The hardness values near the surface on these three profiles (Figure 10) were increased to different extents. The increase of hardness on the edge of the initial disc is due to the disc preparation in the machine shop. Comparing to the bulk material, the measured hardness at about 15 µm from the surfaces was increased about 60% during cryogenic machining and 95% by cryogenic burnishing, which indicate significant grain refinement based on the classical Hall-Petch equation (1).

4 SUMMARY

Two manufacturing processes – machining and burnishing - are shown to be viable SPD routes for introducing ultrafine or nano-sized grains into the surface region of *Co-Cr-Mo* and *AZ31 Mg* alloys. Cryogenic burnishing of *Co-Cr-Mo* alloy experiments resulted in significant grain refinement in the surface region through burnishing-induced SPD. Microhardness in the SPD layer was increased up to 80% relative to the bulk value.

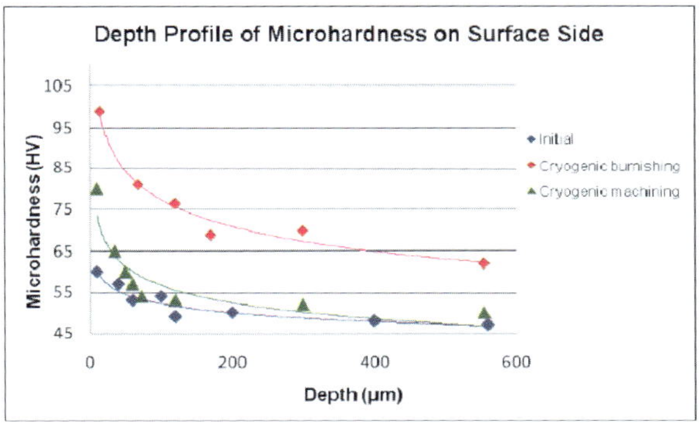

Figure 10: Subsurface microhardness profiles of *AZ31 Mg* discs before and after cryogenic processing

AZ31 Mg alloy was subjected to both cryogenic machining and burnishing processes. The cryogenic burnishing process led to a more than 2 mm thick surface layer with remarkably refined microstructures formed on the burnished surface. A 95% increase in hardness was obtained on the burnished surface, where grains less than 300 nm were observed under scanning electron microscopy. A SPD layer was shown to form on the surface of *AZ31 Mg* disc after cryogenic machining. The hardness of this layer was about 60% larger than the bulk material. It has been reported that the corrosion resistance of the *AZ31 Mg* alloy in simulated body fluid was enhanced due to the formation of this SPD layer [14][25].

The present results demonstrate that both cryogenic processes significantly modify the surface properties of *Co-Cr-Mo* and *AZ31 Mg* alloys and, therefore, may enhance their performances for improved sustainability.

Systematic studies will be done to further investigate the influence of various processing conditions on microstructural changes of *Co-Cr-Mo* and *AZ31 Mg* alloys. Pin-on-disc wear testes will be conducted for studying the relationship between microstructure and wear resistance of *Co-Cr-Mo* alloy.

5 ACKNOWLEDGMENTS

We sincerely thank to Air Products and Chemicals for supplying the equipment for liquid nitrogen application and Dr. Fuqian Yang (Material Science Dept., Univ. of Kentucky) for providing the equipment for microhardness measurements. Additional thanks to our technicians Bill Young and Richard Anderson for their valuable help on conducting the experimental work.

6 REFERENCES

[1] Wang, Z.B. et al., Effect of surface nanocrystallization on friction and wear properties in low carbon steel, 2003, Mater. Sci. Eng., 352:144-149.

[2] Karimpoor, A.A., Erb U., Aust, K.T. and Palumbo, G., High strength nanocrystalline cobalt with high tensile ductility, 2003, Scripta Material, 49: 651–656.

[3] Shi, Y.N. and Han, Z., Tribological behaviors of nanostructured surface layer processed by means of surface mechanical attrition treatment, 2008, Key Eng. Mater., 384: 321-334.

[4] Swaminathan, S., Shankar, M.R., Lee, S., Hwang, J., King, A.H., Kezar, R.F., Rao, B.C., Brown, T.L., Chandrasekar, S., Compton, W.D., and Trumble, K.P., 2005, Large strain deformation and ultra-fine grained materials by machining, Mater. Sci. Eng. A, 410–411: 358–363.

[5] Altenberger I., Scholtes B., Martin U. and Oettel H., Cyclic deformation and near surface microstructures of shot peened or deep rolled austenitic stainless steel AISI 304, 1999, Mater. Sci. Eng. A, 264: 1–16.

[6] Nalla, R.K., Altenberger, I., Noster, U., Liu, G.Y., Scholtes, B., and Ritchie, R.O., On the influence of mechanical surface treatments-deep rolling and laser shock peening on the fatigue behavior of Ti-6Al-4V at ambient and elevated temperatures, 2003, Mater. Sci. Eng. A, 355: 216-230.

[7] Nikitin, I., Altenberger, I., Maier, H.J., Scholtes, B., 2005, Mechanical and thermal stability of mechanically induced near-surface nanostructures, Mater. Sci. Eng. A 403: 318–327

[8] Yang, L. and Zhang, E., 2009, Biocorrosion behavior of magnesium alloy in different simulated fluids for biomedical application, Mater. Sci. Eng. C, 29: 1691–1696.

[9] Prevey, P. and Cammett, J., 2001, Low cost corrosion damage mitigation and improved fatigue performance of low plasticity burnished 7075-T6, J. Mater. Eng. Perform. 10: 548–55.

[10] Prevey, P., Telesman, J., Gabb, T. and Kantzos, P., 2000, FOD resistance and fatigue crack arrest in low plasticity burnished IN718, In: Proceedings of the 5th Nat. Turbine Eng. HCF Conf..

[11] Prevey, P., Hornbach, D., Maston, P., Jacobs, T. and Commett, J., 2001, Low cost surface enhancement method for improved fatigue life of superalloys at engine temperature, NAS3-99116 SBIR, NASA Final Report.

[12] Iglesias, P., Bermudez, M.D., Moscoso, W., Rao, B.C., Shankar, M.R. and Chandrasekar, S., Friction and wear of nanostructured metals created by large strain extrusion machining, 2007, Wear, 263: 636–642.

[13] Ramesh, A., Melkote, S.N., Allard, L.F., Riester, L., Watkins, T.R., 2005, Analysis of white layers formed in hard turning of AISI 52100 steel, Mater. Sci. Eng. A, 390: 88–97.

[14] Pu, Z., Puleo, D.A., Dillon, Jr., O.W. and Jawahir, I.S., controlling the biodegradation rate of magnesium-based implants through surface nanocrystallization induced by cryogenic machining, 2011, TMS, in press.

[15] Li, W.L., Tao, N.R. and Lu, K., Fabrication of a gradient nano-micro-structured surface layer on bulk copper by means of a surface mechanical grinding treatment, 2008, Scripta Material, 59: 546–549.

[16] Ni, H., Elmadagli M. and Alpas, A.T., Mechanical properties and microstructures of 1100 aluminum subjected to dry machining, 2004, Mater. Sci. Eng. A, 385: 267–278.

[17] Disegi, J.A., Kennedy, R.L. and Pilliar, R., Cobalt-based alloys for biomedical applications, 1999, Technology & Engineering, Book Chapter.

[18] Hsu, H.C. and Lian, S.S., Wear properties of Co-Cr-Mo-N plasma-melted surgical implant alloys, 2003, J. Mater. Process. Technol., 138: 231-235.

[19] Nasab, M.B. and Hassan M.R., Metallic biomaterials of knee and hip - A review, 2010, Trends Biomater. Artif. Organs, Vol. 24(1): 69-82.

[20] Yamanaka, K., Mori, M., Kurosu, S, Matsumoto, H., and Chiba, A., Ultrafine grain refinement of biomedical Co-29Cr-6Mo alloy during conventional hot-compression deformation, 2009, Metallurgical and Mater. Trans. A, 40: 1980-1994.

[21] http://www.cartech.com/news.aspx?id=578, Feb. 2008.

[22] Wu, X., Tao, N., Hong, Y., Liu, G., Xu, B. and Lu, J., Strain-induced grain refinement of cobalt during surface mechanical attrition treatment, 2005, J. Acta Mater, 53: 681-691.

[23] Meyers, M. and Chawla, K., Mechanical behavior of materials, 2009, 2nd edition, Cambridge Univ. Press.

[24] Hallab, N.J. and Jacobs, J.J., 2003, Orthopedic Implant Fretting Corrosion, Corrosion Reviews, Vol. 2 J, Nos. 2-3: 184-213.

[25] Pu, Z., Puleo, D.A., Dillon, Jr., O.W. and Jawahir, I.S., 2010, Microstructural changes of AZ31 magnesium alloys induced by cryogenic machining and its influence on corrosion resistance in simulated body fluid for biomedical applications, MSEC2010, in press.

Innovative Bipolar Plates Design for Increasing Fuel Cell Efficiency

N. Kenan, M. Albakri, M. Khraisheh

Department of Mechanical Engineering, Masdar Institute of Science and Technology, Abu Dhabi, UAE

Abstract

Proton Exchange Membrane Fuel Cells (PEMFC's) are a promising technology in the field of clean energy. Bipolar plates are the main components of the PEMFC as they contribute to about 80% of their weight and affect their performance significantly. This study shows the effect of varying the cross sectional area of the flow channel in the bipolar plates along the length of the fuel cell. These ideas were previously oppressed due to manufacturing limitations which will now be overcome with the use of superplastic forming. Results show that for channels of constant height, converging gives the highest current density output while for channels of constant width the diverging channels gives the highest current density ouput at low volumes.

Keywords:

PEMFC, bipolar plates, flow field, superplastic forming, computational fluid dynamics

1 INTRODUCTION

Fuel Cell manufacturers are always investigating ways to improve the efficiency of their products in order to be recognized as a major source of energy for most applications worldwide. The manufacturer's main goal is to come up with an efficient, economical, innovative and handy design for their fuel cells. Fuel Cells are devices used to convert chemical energy directly into electrical energy without eventually being exhausted and are environmentally friendly at the same time. A very promising technology in this field is the Proton Exchange Membrane Fuel Cell (PEMFC) which uses Hydrogen, the most abundant gas in the atmosphere, as its fuel and Oxygen to produce electrical energy and a single side product to be water. The following reactions take place

Reaction at the anode

$$\frac{1}{2}H_2 \rightarrow H^+ + e^-$$

(1)

Reaction at the cathode

$$\frac{1}{2}O_2 + 2H^+ + 2e^- \rightarrow H_2O + Energy$$

(2)

PEMFC's are favored by the automobile industry because they work at relatively low temperatures compared to other fuel cells allowing for faster start-ups and immediate response to changes in power demand [1]. PEMFC's are still not widespread because they are costly and offer some manufacturing limitations when it comes to innovative designs due to the materials they are composed of.

A PEMFC consists of seven layers as follows, two bipolar plates, and two gas diffusion layers (GDL), two catalyst layers, and one membrane as shown in figure 1. The bipolar plates contribute up to 80% of the total weight of the fuel cell and are considered the most important part of the fuel cell because its main functions are distribution of the gas, removal of heat, and conduction of current [3]. The material usually used to manufacture the bipolar plates is graphite which is known to be brittle and difficult to machine and hence will introduce many manufacturing limitations which in turn suppress any innovative ideas regarding the designs of the flow channel and the bipolar plates themselves. The function of the flow channels is to transport gas across the PEMFC. The gas from the flow channel is transported through the GDL to the catalyst layer where the reaction takes place.

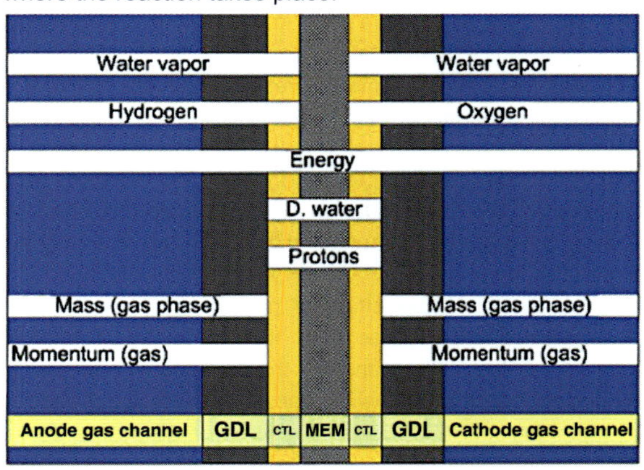

Figure 1 Schematic of a PEMFC indicating regions in which each quantity physically exists [2]

Little attention has been given to flow channel geometry in the PEMFC. The geometry is useful to manipulate the fluid flow properties and accordingly enhance the performance of the PEMFC. Kumar and Reddy [1] tested the effect of varying the flow channel dimensions along with the shape of its cross sectional area on the performance of the fuel cell. Chiang et al. [4] tested the effect of changing the slope of the boundary separating the channel regions and the shoulder region. The channels were varied from converging to straight to diverging. However, two important factors were missed. Firstly, the contact resistance at the interface between the bipolar plates and the GDL was ignored which significantly alters the results of the simulations and secondly, the volumes of the reactants passing through the channels with different configurations are not equal and hence may not provide a consistency in the results. Lee at al. [5] created a CFD model to test the effect of varying different parameters such as channel height ratios between the anode and cathode and the offset ratio between the anode and cathode channels on the current density distribution in the PEMFC. Berning and Djilali [6] developed a three-dimensional, non-isothermal model of a PEMFC. Their model took into consideration the effect of contact resistance at the interface between the bipolar plates and the GDL on the performance of the fuel cell. Li et al. [7] came up with a procedure to design flow channels in PEMFC for proper water management by determining an appropriate pressure drop along the flow channels. Water management is an important issue because it affects the performance of the PEMFC significantly. Al-baghdadi and Al-janabi [8] developed a three dimensional, multiphase, non-isothermal model to optimize the different parameters in the PEMFC. This model allows for the implementation of phase change for water by solving for both gas and liquid in the same domain. Chen et al. [9] developed a two dimensional model for a PEMFC. By solving the resulting ordinary differential equations, the flow channels were designed and the effect of varying different parameters was tested. The parameters tested were inlet velocity, fuel concentration, catalyst activity, channel height, porosity of GDL. Kuo et al. [10] evaluated the convective heat transfer and velocity flow characteristics to enhance the performance of the PEMFC. A series of flow interruptions were introduced in the flow channel in the form of obstacles of different shapes. Wang et al. [11] studied the effect of changing the gas flow channel sizes on the performance of the PEMFC with serpentine flow fields. Liu et al. [12] tested the effect of having a Rib to Channel setup rather than the conventional Rib to Rib setup.

In this work, the effect of having a converging and diverging channel rather than the conventional straight channel on the performance of the fuel cell will be tested. However, this study considers an equal amount of volume to pass through all the channels in comparison in order to have a consistency in the results and also takes into consideration the effect of the contact resistance at the interface between the bipolar plates and the GDL.

2 MODEL DEVELOPMENT

The PEMFC modeled in this work is used to test the effect of having a convergent and/or divergent channel rather than the conventional straight channel on the performance of the fuel cell. For simplicity a single channel model was used. The problem domain consists of the (1) Bipolar Plates; (2) Flow Channels; (3) Gas Diffusion Layers (GDL); (4) Catalyst Layers; and (5) Membrane. The flow channels are the places where the hydrogen and air enter the fuel cell. The GDL's diffuse the gas from the flow channels to the catalyst layers where the chemical reactions take place. The membrane is used to transport the protons produced from the chemical reaction from the anode side to the cathode side. The model was drawn in 3D on the commercial CAD software Gambit® and then imported into the commercial CFD software Fluent® where a special PEMFC add-on module with all equations and subroutines predefined was used.

2.1 Model Assumptions

The assumptions used in this model are the following

1. Steady state conditions
2. Isothermal process (T = 353K)
3. Water only exists in the vapor phase and not in the liquid phase
4. All geometrical properties are isotropic
5. Gases are ideal and the ideal gas equation is used to calculate their density in different areas

2.2 Governing Equations

The equations that govern the fuel cell are the basic transport equations of any fluid flow and the performance equations of the fuel cell itself. The basic transport equations for the fluid flow in the fuel cell are the

1. Mass Conservation Equation (Continuity Equation)
2. Momentum Conservation Equation
3. Energy Conservation Equations

The performance equations are

$$R_{an} = j_{an}^{ref}\left(\frac{[H_2]}{[H_2]_{ref}}\right)^{\gamma_{an}}\left(e^{\frac{\alpha_{an}F\eta_{an}}{RT}} - e^{\frac{-\alpha_{cat}F\eta_{an}}{RT}}\right)$$

(3)

$$R_{cat} = j_{cat}^{ref}\left(\frac{[O_2]}{[O_2]_{ref}}\right)^{\gamma_{cat}}\left(-e^{\frac{\alpha_{an}F\eta_{cat}}{RT}} - e^{\frac{-\alpha_{cat}F\eta_{cat}}{RT}}\right)$$

(4)

Where

j^{ref} = volumetric reference exchange current density (Am^{-3})

$[\ ],[\ ]_{ref}$ = local species concentration, reference value ($kmolm^{-3}$)

γ = concentration dependence (dimensionless)

α = transfer coefficient (dimensionless)

F = Faraday constant (9.65×10^7 C kmol^{-1})

2.3 Working Conditions

The boundary condition imposed on the fuel cell is an ambient pressure of 150 kPa. The mass flow inlets were calculated from the following equation

$$I = \frac{\dot{m} v F}{M}$$

(5)

Where $\dot{m}$ is the consumed mass flow rate, v is the valence of the molecule, F is Faraday's constant M is the molecular weight and I is the current produced by the fuel cell. By entering the desired current output and other parameters, the consumed mass flow rate is calculated. The inlet mass flow rate would be a multiple of the consumed mass flow rate calculated from different trials.

3 RESULTS AND DISCUSSION

The simulations were categorized into three different groups. The first for straight channels in order to have a basis for comparison, the second for varying channel height or width, where one of the dimensions was varied while the other was kept constant, and the third was for varying height and width where both dimensions were varied simultaneously. To ensure that the variation in results depended only on channel geometry, the volumes of all channels were kept constant. The comparisons were made between straight, converging and diverging channels of the same volume.

3.1 Channels of Constant Height

In these simulations the height of the flow channels was kept at a value of 0.5mm. The width of the straight channel was varied from 0.1mm to 0.5mm in increments of 0.1mm. In each case the volume of the straight channel was calculated. The volume of the converging and diverging channels of a certain height was required to be equal to that of the straight channel of the same height. By assuming an inlet width to outlet width ratio of 1:3 for converging and 3:1 for diverging the dimensions were calculated. Figure 2 shows the contours of the current flux density magnitude in the catalyst layer for the cases mentioned.

The contour of the current flux density magnitude for the converging channel shows areas of the highest current density magnitude as compared to the straight and diverging channels. Mainly a higher current flux density magnitude resembles a higher rate of reaction in the corresponding area. Figure 2 also shows the distribution of the oxygen gas along the fuel cell in the catalyst later. The oxygen consumption is highest in the converging channel while it is lowest in the diverging. The higher rate of consumption of oxygen means that the rate of reaction is higher which provides a higher current density in the region.

As the oxygen flows through the converging channel, the cross sectional area of the channel is decreasing. From the continuity equation the velocity of the oxygen gas has to increase with the decrease in area and hence the diffusion of the oxygen through the GDL to the catalyst layer, where the reaction takes place, is enhanced.

3.2 Constant Width

In these simulations the width of the flow channels was kept at a value of 0.5mm. The height of the straight channel was varied from 0.1mm to 0.5mm in increments of 0.1mm.In each case the volume of the straight channel was calculated. The volume of the converging and diverging channels of a certain width was required to be equal to that of the straight channel of the same width. By assuming an inlet height to outlet height ratio of 1:3 for converging and 3:1 for diverging the dimensions were calculated. Figure 3 shows the variation of current density with channel volume for the three cases of straight, converging and diverging channel.

While the straight channel shows that an increase in channel height has no effect on current density the converging and diverging channels exhibit an interesting trend. The current density of the diverging channel decreases significantly with an increase in the channel volume while the current density of the converging channel increases significantly with an increase in the channel volume. A proposed explanation for this trend is the path that the oxygen follows from the flow channel to the GDL. At lower volumes the path the oxygen follows is a short one in the diverging channel, at the entrance of the channel, and hence gets consumed at a higher rate. As the volume increases the length of the path increases and its effect becomes insignificant. While for the converging channel as the volume increases the increase in velocity of the gas along the channel significantly affects its diffusivity through the GDL towards the end of the channel.

3.3 Varying Height and Width

In these simulations both the height and the width of the flow channel were varied simultaneously. The dimensions of the width and height in each case were chosen such that the volume of the channel will be equal to that of the comparable straight channel. Figure 4 shows the variation between current density and channel volume of the converging channel in comparison to a straight channel for different dimensions. The average current density of the varying dimensions

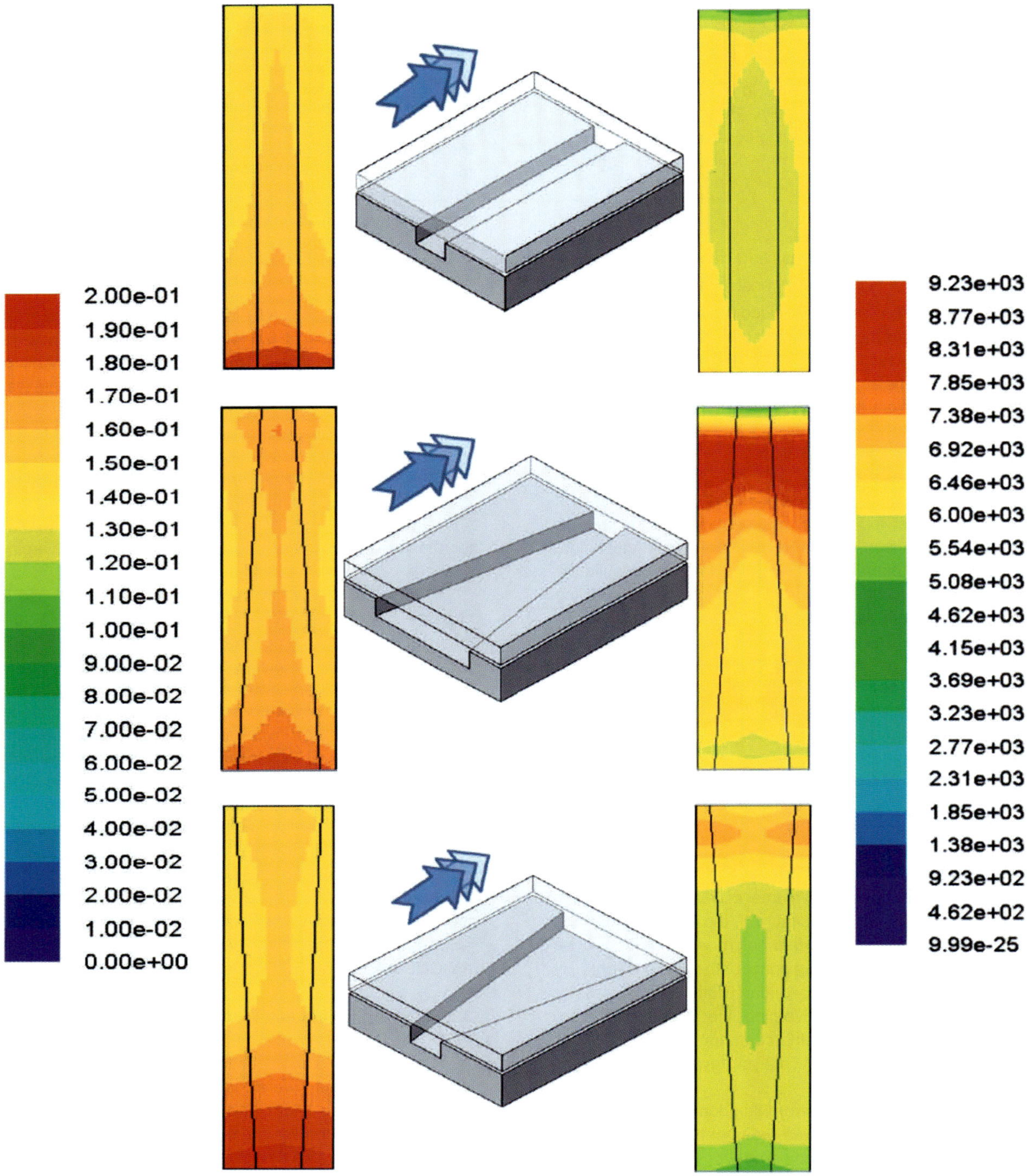

Figure 2 Contours of Current Flux Density Magnitude in A.m-2 (Right) and Oxygen Mass Fraction (Left)

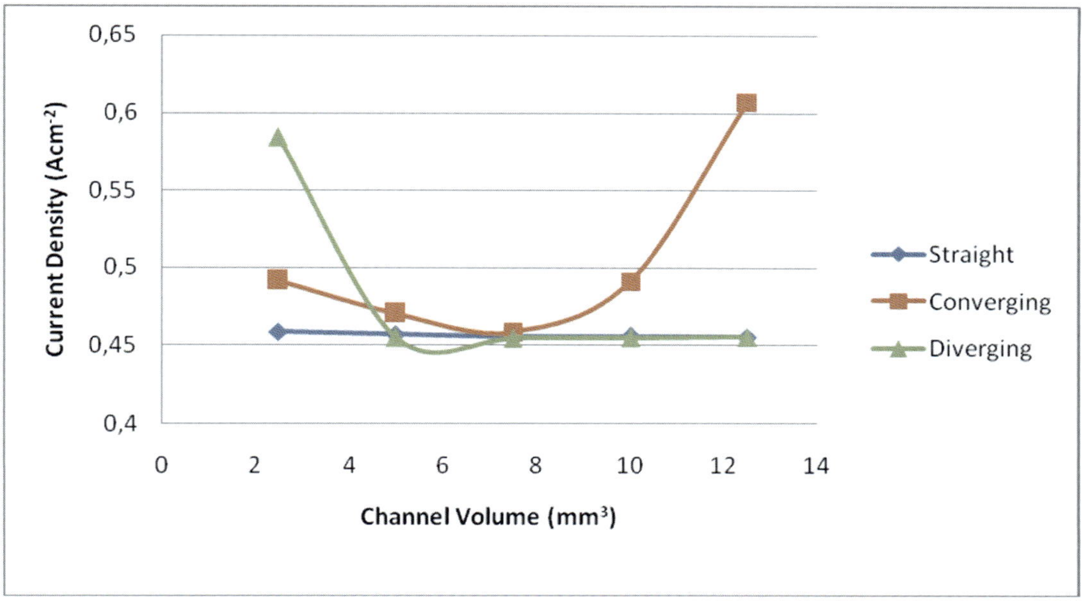

Figure 3 Variation of current density with channel volume at constant width

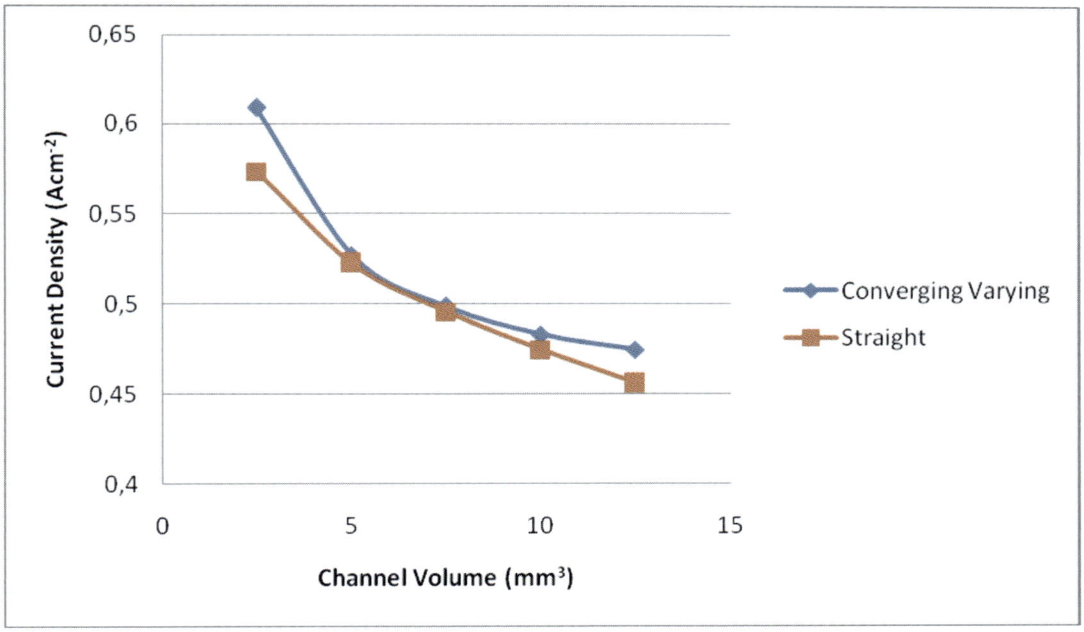

Figure 4 Variation of current density with channel volume for a straight and varying dimensions channel

channel is in all cases at a higher value than that of the straight channel.

The reason in this enhanced performance is the variation of the velocity of the gas in both the x-dimension and the y-dimension simultaneously promotes the diffusivity of the gas through the GDL and into the catalyst layer.

4 CONCLUSION

A three-dimensional, isothermal, one phase, single channel model of the PEMFC was created and a number of simulations were made. It was shown that

1. For a straight channel of constant height a converging flow channel of equivalent volume gave a higher current density output because the diffusion of the gases from the channel to the catalyst layer is enhanced.
2. For a channel of constant width a diverging channel performed better at lower volumes while a converging channel performed better at higher volumes. At lower volumes the diverging channel provides a shorter path for the gases to travel which becomes insignificant at high volumes.
3. Combinations of varying height and width were tested for a converging channel and compared to that of a straight channel. The average current density output for the varying dimensions channel was higher at all volumes.

For the PEMFC market to expand rapidly innovative designs that enhance the fuel cell's performance are continuously needed.

5 REFERENCES

[1] Kumar, A., Reddy, R.G., 2002, Effect of channel dimensions and shape in the flow-field distributor on the performance of polymer electrolyte membrane fuel cells, J. Power Sources, 113, 11–18.

[2] Akbari, M.H., E., Rismanchi, B., 2007, Numerical Investigation of flow field configuration and contact resistance for PEM fuel cell performance, Renewable Energy, 33, 1175–1783.

[3] Middelman, E., Kout, W., Vogelaar, B., Lenssen, J., de Waal, E., 2003, Bipolar plates for PEM fuel cells, J. Power Sources, 118, 44–46.

[4] Chiang, M-S., Chu, H-S., Chen, C-K., Jian, S-R., 2007, Electrochemical reaction and performance of proton exchange membrane fuel cells with a novel cathode flow channel shape, J. Power Sources, 166, 362–375.

[5] Lee, S., Jeong, H., Ahn, B., Lim T., Son, Y., 2008. Parametric study of the channel design at the bipolar plate in PEMFC performances, Int. J. Hydrogen Energy, 33, 5691–5696.

[6] Berning, T., Djilali, N., 2003, Three-dimensional computational analysis of transport phenomena in a PEM fuel cell---a parametric study, J. Power Sources, 124, 440–452.

[7] Li, X., Sabir I., Park, J., 2007, A flow channel design procedure for PEM fuel cells with effective water removal, J. Power Sources, 163, 933–942.

[8] Al-baghdadi, M., Al-janabi, H., 2007, Modeling optimizes PEM fuel cell performance using three-dimensional multi-phase computational fluid dynamics, Energy Conversion and Management, 48, 3102–3119.

[9] Chen, F., Wen, Y-Z., Chu, H-S., Yan, W-M., Soong, C-Y., 2004, Convenient two-dimensional model for design of fuel channels for proton exchange membrane fuel cells, J. Power Sources, 128, 125–134.

[10] Kuo, J-K., Yen T-S., Chen C-K., 2008, Improvement of performance of gas flow channel in PEM fuel cells, Energy Conversion and Management, 49, 2776–2787.

[11] Wang, X-D., Yan, W-M., Duan, Y-Y., Weng, F-B., Jung G-B., Lee C-Y., 2009, Numerical study on channel size effect for proton exchange membrane fuel cell with serpentine flow field, Energy Conversion and Management, 51, 959–968.

[12] Liu, Z., Zhang, H., Wang, C., Mao, Z., 2009, Numerical simulation for rib and channel position effect on PEMFC performances, Int. J. Hydrogen Energy, 35,2802–2806.

5 Remanufacturing, Reuse and Recycling

Remanufacturing, reuse and recycling are important activities involved in recovering residual values in products and materials. Due to the need for resource circulation, these activities with the end-of-life products, have in recent times become significant, and they offer an effective solution for meeting the demands of sustainable value creation. In the first section of this chapter Nasr et al. discuss sustainable production, review the framework and metrics for guiding implementations, and examine the role of remanufacturing and product design to achieve improved sustainable production metrics. In the second section Yamada and Ohta model the three different recovery types: refurbishing, reuse and remanufacturing of the systems by queuing simulation, and compare the three reuse types of the system designs in view of collection rates and economical aspects. In the third section Cunha et al. develop a technology roadmap for remanufacturing oriented production equipment under sustainability criteria. In the fourth section Postawa et al. use automated image based recognition to support faster teaching of humans with different qualification levels by explaining the work task with a universally usable form of description. In the fifth section Fügenschuh et al. apply the scenario technique to examine reuse and its fields of influence for different levels of development, using several reference countries. In the sixth section Jung et al. introduce a diesel engine fuel injector remanufacturing process, analyzes the failure mode, and discusses the performance test methods of the injectors. In the seventh section Qudaih et al. explore the recycling of cross-link polyethylene waste-plastic that generated at a considerable amount from the cable industry. In the sixth section Pigosso et al. develop an ecodesign maturity model to support the application of ecodesign practices in the development of remanufacturable production equipment.

A Framework for Sustainable Production and a Strategic Approach to a Key Enabler: Remanufacturing

N. Nasr, B. Hilton, R. German

Golisano Institute of Sustainability, Rochester Institute of Technology, USA

Abstract

The concern over climate change and related environmental issues has generated a surge of interest and actions towards sustainable development. In response, many manufacturers have begun implementing sustainable production practices. This paper discusses sustainable production, reviews the framework and metrics for guiding implementations, and examines the role of remanufacturing, a key technology within sustainable production. The paper also discusses the important role of product design in implementing remanufacturing to achieve improved sustainable production metrics.

Keywords:

Sustainable Manufacturing; Sustainable Manufacturing Metrics; Remanufacturing; Design for Remanufacturing; End-of-Life (EoL)

1 INTRODUCTION

Studies of climate change in the last decade greatly increased awareness of the myriad complex environmental issues that can negatively affect human progress. Over the past century, industrialization fueled the rapid growth of many countries; today there appears to be a growing consensus for a fundamental re-examination of how industries should develop products for consumers.

It is now widely seen that a major objective for the next two decades will be shifting industrial sectors to sustainable product systems and sustainable production practices. Sustainable production aims to produce better performing products using fewer resources, cause less waste and pollution, and contribute to social progress worldwide. It has been widely regarded as the next industrial revolution and has been touted as inevitable for all production eventually, given the finite resources of the Earth. Sustainable production is a concept and needs a holistic approach. Product developers are moving towards closing the product life cycle to bring about a cradle-to-cradle approach and incorporate different aspects of sustainability at the different stages of a product life cycle. As a result, remanufacturing has become one of the most widely adopted implementation strategies for extending product life cycle, closing the loop on material flows, and reducing total material consumption. This paper will illustrate how remanufacturing related design considerations can help meet sustainable production metrics.

2 A FRAMEWORK AND METRICS FOR SUSTAINABLE PRODUCTION

A key to implementing sustainable production will be application of consistent and comprehensive framework and metrics so that each company can benchmark its process against its competition and the rest of industry as a whole as well as monitoring progress toward more sustainable practices.

The sustainable production framework comprehends the entire life cycle of manufactured products, including consumer use and product end-of-life options, such as disposal, recycling and remanufacturing, as summarized in Appendix 1.

Why are metrics important? There are several compelling reasons. First, industry needs metrics to answer the growing demand by consumers and institutional purchasers for easy-to-use tools to compare and select sustainable products. In addition, industry needs clear, consistent criteria for responding to sustainable requirements not only by consumers, but also to conform to increasingly restrictive international environmental regulations. Properly applied metrics would also provide a market advantage to those companies that can demonstrate that they manufacture sustainable products. Metrics would also promote product design changes that serve to reduce resource depletion, environmental degradation and negative health impacts throughout the product life cycle. Finally, effective metrics will provide a path forward for organizations to advance their systems and measure progress.

There are today a large number of eco-labels, government directives and voluntary metrics focusing on one or multiple aspects of sustainability in products. However, it is often a major challenge for industry to figure out which ones to use. To add to the complexity, many of these metrics can be different around the globe. In response to the clear need for agreement on unified metrics, The Organisation for Economic Co-operation and Development (OECD), which is based in Europe, has made a commitment to research, develop and publish such sustainable production metrics through a dialogue with the OECD countries. These metrics will be based on the following analytical framework:

- *Resources*, including depletion, degradation, efficiency (refer to Appendix 2).
- *Product*, considering design, durability, quality and packaging (refer to Appendix 2).
- *Employment*, relating to health, safety, security, satisfaction and income opportunities of those involved in production.

- *Economic*, including value-added investment in sustainable technology and ethics involved in sustainable production.
- *Society*, including community development and impacts on social balance resulting from implementation of sustainable practices.
- *Environment*, including impact of production on waste, emissions, sound and other components of the environmental footprint (refer to Appendix 2).
- *Supporting infrastructure*, including telecommunications and ease of transport [1].

OECD is currently developing a prototype toolkit of industrial production metrics that comprehend the analytical framework. By design, the toolkit focuses on environmental and financial performance at the facility level, because this level enables a toolkit-user company to review its materials, processes and products in detail to pinpoint which ones can most easily be changed to improve its overall environmental performance. Finalization, release, and distribution of the OECD toolkit should materially advance the adoption of universally applicable sustainability practices that promote business profitability, eco-efficiency, and eco-innovation.

Core indicators included in the OECD prototype toolkit are grouped into the four categories listed below. Indicators that are addressed through remanufacturing (refer to Section 3) are indicated with an * :

1. *Infrastructure indicator*
   - Percent of land occupied that is "natural cover"
2. *Materials indicators*
   - Recycled content of material inputs
   - Renewable proportion of energy consumed*
3. *Process indicators*
   - GHG intensity
   - Energy intensity*
   - Waste intensity (mass balance approach)
   - Waste intensity (total waste approach)
   - Intensity of pollutant releases to air
   - Intensity of pollutant releases to water
   - Water intensity*
   - Non-renewable materials intensity of production
4. *Product indicators*
   - Recycled content of products*
   - Recyclability of products
   - Renewable materials content of products*
   - Restricted substances content of products*
   - Annual energy consumption*
   - Annual GHG production
   - Non-renewable materials intensity over product lifetime (product mix)*

As manufacturing companies utilize the OECD framework to review their production operations, it is important to remember that decisions made during the early phase of the design process can dramatically impact both the sustainable production metrics and the economic viability of the product. For example, Huthwaite [2] revealed that a product design is responsible for only 5% of a product's cost; however, the design is responsible for determining 75% or more of the manufacturing cost. Nevins and Whitney [3] show that 70% of a product's life cycle cost is determined in the design phase. The product development team plays a leading role by impacting how efficiently a design uses material, energy, and controls waste. In 2006, Nasr and Thurston proposed three laws of sustainable product development [4]:

1. Minimize material and energy resources needed to satisfy product function and consumer demand.
2. Maximize usage of expended resources.
3. Minimize or eliminate the adverse impacts of waste and emissions.

This reinforces that the overall viability of achieving improved sustainable production metrics, for a given product or product family, can be greatly influenced by the design criteria for the original product concept. Figure 1 illustrates the concept that decisions made when the least amount of detailed product design knowledge is available often serves to influence the sustainability metrics the most.

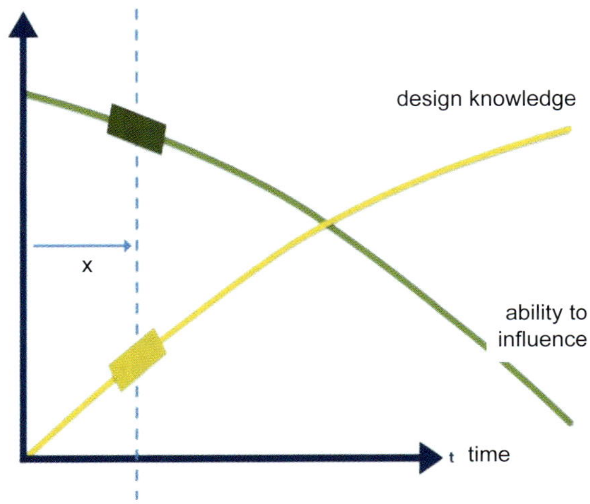

Figure 1: Ability to influence sustainability metrics

3 REMANUFACTURING: A KEY TO ACHIEVING SUSTAINABLE MANUFACTURING

One process that should be considered to minimize the overall life cycle impact is remanufacturing. Remanufacturing (aka 'reman') is a process that restores used durable products to a like-new condition, thereby recovering the value and reducing the environmental impact of the product. Remanufacturing is a production process though which products are systematically disassembled, cleaned, and inspected for wear. Damaged components are replaced, feature upgrades can be incorporated, and the product is reassembled and re-qualified [5]. For this reason, reman differs from other recovery processes in its completeness: a remanufactured machine or component should match the same customer expectations, performance, reliability, and life cycle as new machines.

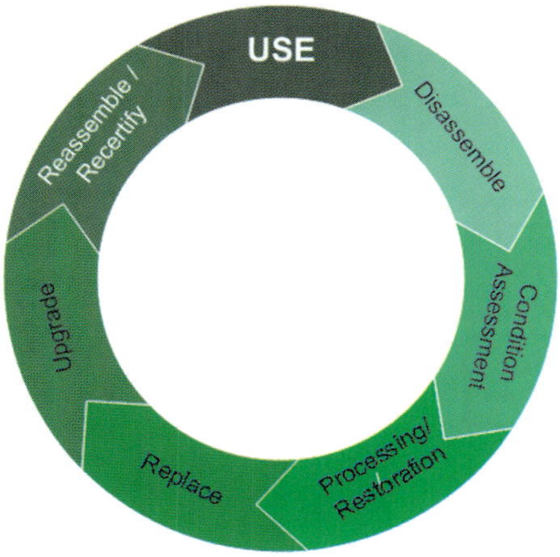

Figure 2: Generic Remanufacturing Process

The discussion around sustainable production practices usually centers on new production that can be maintained with minimal environmental footprint. However, recapturing the value-added component of a product is often both environmentally and economically beneficial since both largely derive from elimination of waste. This reduction of waste contributes directly to sustainable production metrics of a corporation. In fact, remanufacturing is oftentimes referred to as the 'ultimate form of recycling' because it is able to preserve the embodied energy contained in a product [6]. Studies have shown that remanufacturing can recover 85% of the value within the product [7]. In many cases, the ratio of total energy required for original production compared to remanufacture is approximately six to one. This means that for every kilowatt-hour of energy spent in remanufacturing, about six kilowatt-hours are avoided [8].

Remanufacturing can contribute positively towards fulfilling the sustainable production metrics based on the OECD analytical framework in several ways. The direct reuse of components mitigates depletion of nonrenewable resources and reduces environmental impact of material entering the landfill. Corporate remanufacturing programs often contribute to community development and create income opportunities through increased employment. Finally, the economic activity associated with remanufacturing can also contribute to the development of public infrastructure through taxes and community payroll.

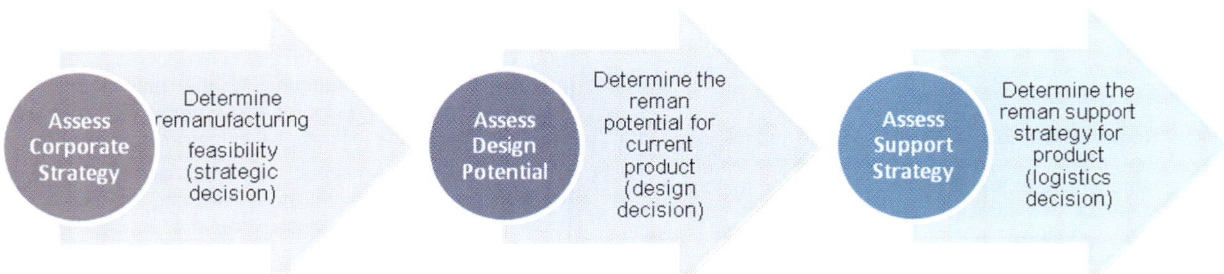

Figure 3: Assessment of Remanufacturing Program Potential

4 REMANUFACTURING PROGRAM STRATEGY

At its heart, remanufacturing is a strategic decision that is not solely based on a single product. A company must decide whether it wants to add remanufacturing into its product support portfolio. It must decide whether it has the knowhow and the infrastructure to support remanufacturing. It must decide whether it can overcome barriers related to policy or market. Once a company can commit to remanufacturing as a global product support strategy, each product it produces can be evaluated to determine the business opportunity.

While remanufacturing can be used to recover the residual value of products at end of life, it must also not conflict with the current business model of the company. In order to better understand the implications of remanufacturing, a company must assess the current corporate strategy in conjunction with market acceptance and customer requirements. In addition, the company must also factor in existing sourcing agreements, institutional knowledge, and future legislation or barriers to the viability of the program.

Next, the company must assess the product portfolio's applicability to using the remanufacturing process. The potential benefit associated with the product must be sufficient to warrant recovery and processing. The product design must be viable into the future long enough to achieve the desired contribution to the business case and the sustainable production metrics.

Finally, the company must determine how to integrate remanufacturing into their global product support strategy. Consideration must be given to the logistical factors associated with the intended remanufacturing deployment model. The remanufacturing program must be structured so that it can best meet the needs of the organization and results in optimized sustainability metrics. The standard framework and metrics from the OECD toolkit could help inform some of this evaluation.

5 REMANUFACTURING DESIGN ENABLERS

The impact that product design can have on sustainable production cannot be understated. As stated earlier, product design only accounts for 5% of the product development cost but can greatly impact the sustainable production metrics. It has been suggested that the decisions made during the product design phase influence more than 80% of the environmental and social impacts of a product [5]. The recovered value and environmental benefits can be increased through the use of Design for Remanufacturing (DfR_{eman}) tools during the development process. Companies that are considering remanufacturing as an implementation mechanism, for improved sustainable production metrics, must establish appropriate design criteria. The following

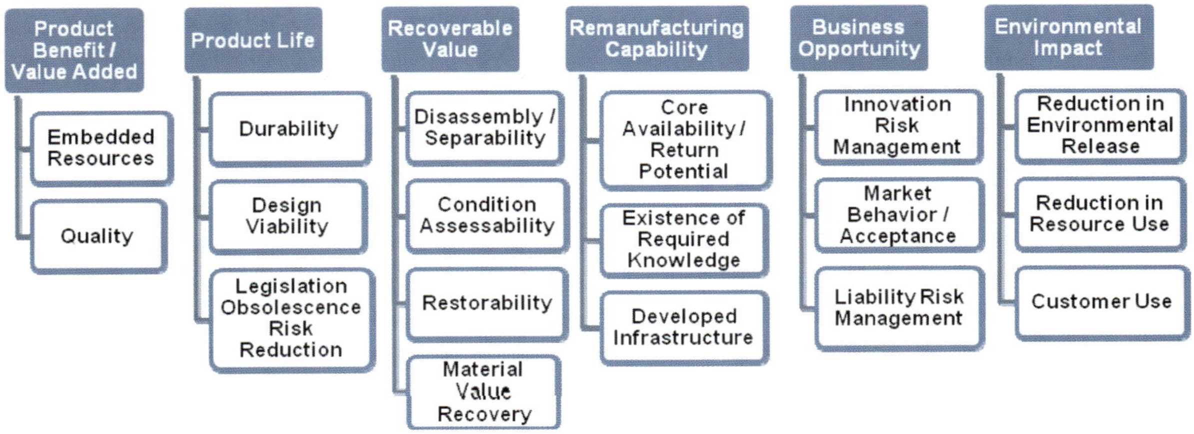

Figure 4: Remanufacturing Design Enablers

criteria are considered the top enablers for a remanufacturable design:

1. **Product Benefit / Value Added** – A remanufacturable design has value worth recovering. Systems and components have investment in tooling, assembly time, and materials; therefore, remanufacturing recovers this embedded value and resources. Additionally, parts are of high enough quality so that part-to-part variation does not become an issue in the remanufacturing process.

2. **Product Life** – A remanufacturable design is useful after one life cycle. Materials are durable to resist degradation, and the design is viable either through timeless design or through upgradeability. The ability to upgrade the product's performance, aesthetics, technology, or software will limit the life cycle risks associated with obsolescence or new regulations.

3. **Recoverable Value** – The value within a remanufacturable design is recoverable through ease of separation, easy assessment of condition, and can be restored to a like-new condition. These factors also help optimize material recovery associated with only the parts or modules that have failed.

Designing a product for remanufacturing must also fit within the product's business case. The following were identified as business enablers.

4. **Remanufacturing Capability** – The manufacturer of the design or contract processor has the capability to remanufacture the product. They have a source of cores or the infrastructure to collect and return the cores, and they have the knowledge and skills required to remanufacture the product.

5. **Business Opportunity** – The manufacturer is able to manage any potential business challenges associated with the opportunity for the remanufactured product. They are able to inform customers on the benefits and performance of the remanufactured product thereby creating market acceptance.

Besides the financial benefit of remanufacturing, there is also an environmental benefit to remanufacturing.

6. **Environmental Impact** – A remanufacturable design recovers more resources than it uses in the process of remanufacturing, and releases less emissions to the environment than the processes used to recover the product. The benefit is calculated on a life-cycle basis.

These six strategic enablers serve to increase the potential for a remanufactured product to positively impact the sustainable production metrics. Figure 4 illustrates the remanufacturing design enabler's relationship to design criteria categories.

6 CONCLUSIONS

This paper discusses the OECD metrics for sustainable production as it relates to product design and manufacturing. These metrics will be supported by a standard toolkit for assessing the production facilities of a corporation. This will serve to improve the environmental profile of an organization and contribute positively to the corresponding sustainable production metrics. Unfortunately the application of the framework, metrics, and toolkit after a product is in production may not achieve the full potential of the possible results.

The paper also presents a strategic approach to developing a remanufacturing strategy. Remanufacturing represents a key strategy to achieving a closed-loop product system and a more sustainable production system.

REFERENCES

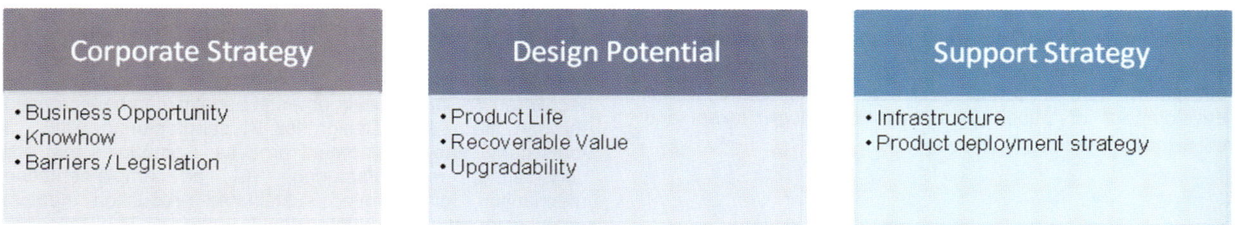

Citations:

[1] Nasr, N. Z., and Bordt, M., 2009, Metrics for Sustainable Manufacturing, 7th Global Conference on Sustainable Manufacturing, Indian Institute of Technology Madras, Chennai, India.

[2] Huthwaite, B., 1988, Designing in Quality. Quality, 27 (11), 34-35.

[3] Nevins, J. L., and Whitney, D.E., 1989, Concurrent Design of Products and Processes (New York: McGraw-Hill), 2-3.

[4] Nasr, N. and Thurston, M., 2006, Remanufacturing: A Key Enabler to Sustainable Product Systems, Proceedings for the 13th CIRP International Conference on Life Cycle Engineering.

[5] Nasr, N., et al., 2002, Closing the Loop: Design Tools for Sustainable Products, EPA Report, Washington, D.C.

[6] Steinhilper R., 1999, Remanufacturing: The Ultimate Form of Recycling, IRB Verlag Stuttgart.

[7] Allen, J., Cradle-to-Cradle: Achieving the Vision Through Remanufacturing, The Organisation for Economic Co-operation and Development, Paris.

[8] Nasr, N., 2004, Remanufacturing from Technology to Applications, 2nd CIRP Global Conference on Sustainable Manufacturing, Technical University, Berlin.

General References:

Nasr, N. and Hilton, B., 2008, Design for Remanufacturing, 15th CIRP International Conference on Life Cycle Engineering, Sydney, Australia.

National Center for Remanufacturing and Resource Recovery 2005, The U.S. Remanufacturing Industry, Terms & Definitions.

Nasr, Berg, Cleveland, Hilton, Roth, Scheldorn and Slocum, 2004, Sustainable Design Tools for Proton Membrane Exchange Fuel Cells, Report to U.S. EPA.

Appendix 2: Metrics for specific environmental measures.

Category	Description	Metric and Unit Measure
Energy use	Any source of useable power—electricity from an outside provider, electric power generated on-site; propane, butane, and natural gas burners; gasoline and kerosene used for transportation or non-transportation.	Kilowatt hours and/or BTUs per unit of production. (Some international protocols, such as WRI, convert kwh and BTUs to megajoules.)
Materials use	Materials used in product and manufacture including: • Ozone depleting substances (e.g., CFC-11) • Packaging materials • Proportion of input materials that were recycled to recovered vs. virgin materials	Percentage input/output materials utilized; percent post-consumer recycled content; tons input materials/year; lbs. or kgs. input materials per unit of production.
Toxic and/or hazardous chemicals used	Use of hazardous or toxic chemicals that are regulated or otherwise of concern (such as listed in RoHS standards and other environmental regulations).	Listing of specific chemicals used; lbs. or kgs. used per year; lbs. or kgs. per unit of production; percentage reduction annually.
Water use	Water from outside sources, such as municipal water supply or wells, for operations and facility use. (Water discharged from operations is recorded in the "water pollution" section below.)	Gallons used annually; percentage recycled water used; lbs. or kgs. of chemicals used for water treatment annually; percentage annual reductions.
Air emissions	Releases of any of the following: Hazardous Air Pollutants (HAPs) described at: http://www.epa.gov/ttn/atw/pollsour.html — Carbon Monoxide, lead, Ozone and its precursors and other Ozone depleting substances, including: Volatile Organic Compounds (VOCs) and Nitrogen Oxides (NOx); Particulate and fine-particulate matter; Sulfur dioxide; Greenhouse gases, including Carbon Dioxide.	Air emissions generated in lbs or kgs. or tons per year; percentage of annual reductions.
Water pollution	Amount of pollutants in wastewater discharged, including: heavy metals (Cu, Pb, hexavalent chromium, cadmium, Zn, Ni, Hg); organic pollutants and pesticides, oil and grease; organic pollutants; suspended solids; pathogens; nitrogen and phosphorous.	Concentration of pollutants discharged in mg/L; lbs. or kgs. or tons per year; percentage of annual reductions.
Solid waste	Wastes other than hazardous wastes.	Gallons or lbs. or kgs. or tons generated or shipped to landfills annually; percentage reductions annually; percentages recycled.

Modeling and Design for Reuse Inverse Manufacturing Systems with Product Recovery Values

T. Yamada, N. Ohta
Department of Environmental and Information Studies, Tokyo City University, Japan

Abstract
Inverse manufacturing systems, consisting of assembly and disassembly, are expected for realizing sustainable manufacturing. Their productivity of these systems is also important for recovering product values and materials. In view of resource circulation, the reuse of EOL (End-of-Life) assembly products is essential in the inverse manufacturing. Inverse manufacturing by reuse is mainly classified into 3 different recovery types: refurbishing, product installation reuse, and remanufacturing. However, inverse manufacturing by reuse has been implemented in Japan in only a few cases. One of the reasons is that the market values for the collected EOL products have been obsolete due to the length of product lifecycle, in addition to the fact that the product recovery lifecycle is different for each product. Therefore, the reuse inverse manufacturing systems should be designed by considering the reuse types, product recovery lifecycle and productivity. This study focuses on the reuse inverse manufacturing systems, models the 3 different recovery types of the systems by queueing simulation, and compares the 3 reuse types of the system designs in view of collection rates and economical aspects.

Keywords:
Environmentally Conscious Manufacturing; Sustainable Manufacturing; Queueing Simulation; Refurbishing; Product installation Reuse; Remanufacturing

1 INTRODUCTION

For sustainable manufacturing [1], inverse manufacturing systems [2] consisting of assembly [3] and disassembly [4] and their productivity are important for recovering product values and materials. Based on [4], there are 2 types of product recovery: reuse and recycling. In view of resource circulation with lower environmental impacts, the reuse of EOL (End-of-Life) assembly products is more important than the recycling. However, in Japan there have been a few cases of reuse inverse manufacturing systems, such as recycling cameras, copiers and computers [5][6]. One of the reasons is that the market values for the collected EOL products have been obsolete due to the long length of product lifecycle, and also the product recovery lifecycle is different for each product [6].

In Japan, many kinds of EOL assembly products are already disassembled in inverse manufacturing and/or disassembly systems in order to recover the parts or materials by reuse and recycling. Based on the Japanese cases [5][6], reuse inverse manufacturing is classified into 3 different types of recovery systems: refurbishing [4], product installation reuse [7] and remanufacturing [4]. To establish the reuse inverse manufacturing economically, the system design should consider the reuse types, product recovery lifecycle and productivity [8].

This study discusses 3 different types of reuse inverse manufacturing systems by queueing simulation in view of productivity and economical aspects. The outline of this study is as follows: Section 2 defines 3 recovery models of the reuse inverse manufacturing systems in view of material flows, and explains their queueing model with reverse blocking.

In Section 3, an example problem for reuse inverse manufacturing is set for evaluating 3 types of systems, and a queueing modeling by a simulator, Plant Simulation, is also explained. Section 4 shows a design example of the 3 different types of reuse inverse manufacturing, and effects of product collection rate, material circulation and economic aspects are also considered. Finally, Section 5 summarizes this study and proposes future studies.

2 MODELING OF REUSE INVERSE MANUFACTURING SYSTEMS

2.1 3 recovery models of reuse inverse manufacturing systems

Reuse inverse manufacturing systems basically consist of assembly and disassembly systems plus consumers (product users) [8]. First, new products are assembled in an assembly system and distributed to the consumers. The consumers buy at sales, use the products and finally finish the usage of the products. Then, they become the EOL products and are collected at a disassembly system for recovery.

The disassembly system mainly has sorting and disassembly processes. By considering the product recovery values such as product/part prices and their lifetime in recovery processes, the collected products arrive at a sorting station and are sorted into each different recovery process such as product refurbishing, disassembly and material recycling. The sorting process in the disassembly system is the first process of the disassembly system, and has reverse blocking [9][10] for decreasing the total productivity of inverse manufacturing. The recovered products/parts in each disassembly process are distributed to consumers again.

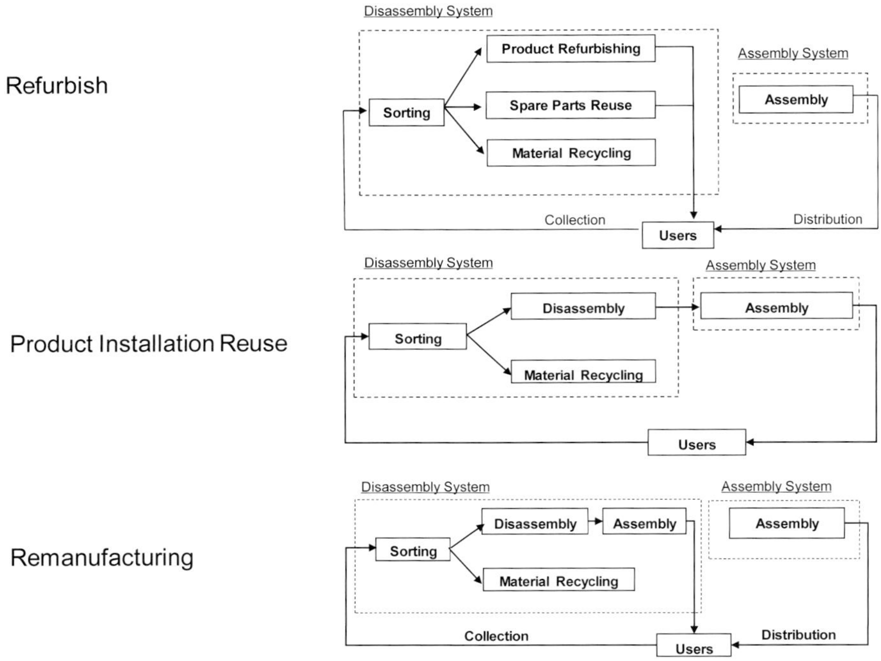

Figure 1: Material flows for 3 types of reuse inverse manufacturing systems

In the disassembly process, reuse of spare parts, reuse product installation reuse or remanufacturing is carried out by the 3 different recovery types of reuse inverse manufacturing. Figure 1 shows the material flows for 3 types of reuse inverse manufacturing systems, and the 3 models of them in this paper are set as follows:

Refurbish

Refurbish is defined to include the processing of an EOL product such that its full functionality is restored [4]. In the disassembly system, the collected products are processed by product refurbishing, spare parts reuse or material recycling. The recovered products/parts with the product refurbishing and spare parts reuse are distributed to the market again. Computers are an example of this reuse type [5][6].

Product Installation Reuse

Reuse is defined as the employment of components and modules obtained from EOL products as spare parts or other items [4]. With reuse, there are 2 types: product installation reuse and spare parts reuse. In product installation reuse [7] by remanufacturing, the collected products are disassembled, and the selected parts are reused for new products and/or recovered products/parts. For the product installation reuse model in this paper, disassembled parts are reused and assembled again into new products at the same assembly systems of the new products. Recycling cameras is a well-known example of this [5][6].

Remanufacturing

Remanufacturing is defined as the composition of reconfigured products with components derived from EOL products [4]. In the models of this paper, product installation reuse is carried out for the reconditioning of (recovered) products by the remanufacturing. The collected products are disassembled, and the disassembled parts are reused and re-assembled for reconditioning products (recovered products), which are not new products, at the different assembly systems of the new products. Some types of copiers are already implemented and recovered by this type of remanufacturing [5][6].

2.2 Assumptions and notation for queueing models for reuse inverse manufacturing systems

This study deals with 3 different recovery types of the reuse inverse manufacturing systems by the queueing simulation modeling. The system consists of assembly and disassembly systems, and users (consumers).

First, product units arrive at the assembly system, where the system configuration is the line type. It is assumed that products arrive at the assembly system according to Poisson arrival with mean arrival rate, λ , and the service times at the assembly and disassembly stations follow the exponential or constant service with mean service rate μ_i $(i=0, 1, 2, \ldots, K)$.

Next, the finished products at the assembly system are distributed to users (consumers). When users finish the usage of the products, they become the EOL products. It is assumed that the EOL products with a collection rate are sent to a disassembly system, and the others are disposed of and removed from the system at this phase.

Finally, the collected products are sent to the disassembly system with the reverse blocking [9][10]. The disassembly system consists of one preceding sorting station and the succeeding disassembly stations for recovery. The collected products from users first arrive at the sorting station. The recovery types and the mixed ratio (routing probability to each succeeding disassembly station) of arrival-collected products are unknown in practice at the disassembly system, because they are used products unlike finished products in assembly systems. In the disassembly system, the mixed ratio of the product *i* (recovery type) is assumed as q_i in modeling this system. After being processed at the sorting station, the recovery types are identified, then they are sorted into each type of product and sent to the appropriate and different succeeding disassembly stations according to the recovery type.

The reverse blocking phenomenon [9][10] occurs in the disassembly system, and is described as follows: It is assumed that the finite buffer capacity at each succeeding station is set as B_i at station i (i =1,...,K). When the buffer capacity at the succeeding station i is full, a new arrival product at the sorting station has just been processed and then identified as a type i recovery. However, the type i recovery product cannot enter the succeeding disassembly station i because the buffer capacity at station i is already full. Therefore, the product stays at the preceding sorting station, and thus it blocks and stops the sorting station until the service at station i is completed and the buffer capacity becomes available. Also, the other succeeding stations starve and stop during the blocking though they are available, and this can decrease the throughput of the sorting process and impact on the total productivity of the reuse inverse manufacturing system.

In the disassembly system, the product refurbish has product refurbishing, spare parts reuse and material recycling. On the other hand, the product installation reuse and the remanufacturing have the disassembly and material recycling. In the product installation reuse, the disassembled products are reused and assembled again at the same assembly system of the new product. However, in the remanufacturing, they are reused and reassembled for the reconditioning products at the assembly line different from that of the new products.

The summary of the notation used in this study is given below:

i	:	station number (i = 0,1,2, ..., K)
K	:	number of succeeding stations
λ	:	mean arrival rate at system
μ_i	:	mean service rate at station i (i = 0,1,2, ..., K)
B_i	:	buffer capacity at station i (i = 1,2, ..., K)
q_i	:	routing probability to station i where $\sum_{i=1}^{K} q_i = 1$ (mixed ratio for product i) (i = 1,2, ..., K)
ρ_i	:	utilization rate at station i (i = 0,1,2, ..., K)
BL_i	:	reverse blocking probability at station i (i = 1,2, ..., K)
BL	:	total reverse blocking probability at system
TH_i	:	throughput at station i (i = 1,2, ..., K)
TH	:	throughput at system
TR	:	total reward during the planning period
TC	:	total cost during the planning period
TP	:	total profit (new reward) during the planning period (= TR - TC)

The assumptions of the models are given below:

1) The system is stationary.
2) The travel time of each product is zero.
3) There is no failure at each station.
4) The dispatching rule at all stations is First Come First Served (FCFS).
5) In case without reverse blocking, the buffer capacity at the sorting station is infinite in the disassembly system.

The objective function of this system is set as a throughput and the total profit. The throughput at station i, TH_i, and at system, TH, are given as equations (1) and (2), respectively.

$$TH_i = \frac{(number\ \ of\ \ products\ \ processed\ \ at\ \ station\ \ i)}{(number\ \ of\ \ input\ \ products\ \ at\ \ station\ \ i)} \quad (1)$$

$$TH = \frac{(number\ \ of\ \ products\ \ processed\ \ at\ \ system)}{(number\ \ of\ \ input\ \ products\ \ at\ \ system)} \quad (2)$$

Also, the reverse blocking probability for a system behavior is observed as busy rates in a graph of the simulator.

The total profit during the planning period TP is the difference between total reward TR and cost TC. The total reward TR and cost TC during the planning period are given by multiplying the number of products/parts processed by coefficients of each price and cost.

3 AN EXAMPLE PROBLEM FOR REUSE INVERSE MANUFACTURING AND QUEUEING SIMULATION

3.1 A design example problem set

Based on actual disassembly/assembly times in an experimental case for a personal computer [11], an example design problem is set for the reuse inverse manufacturing to compare with the 3 models. The mean of total assembly and disassembly times for a product is assumed as 770 [sec] and 472 [sec], respectively. The precedence relations among element assembly and disassembly tasks area also obtained in [11]. The mean of service time at each station in the assembly and disassembly systems is obtained as the sum of the task times of the element task assigned to the station, and has stochastic variations with exponential services in some cases. The reuse inverse manufacturing problem is here set as follows:

Assembly parameter:

Mean of total assembly time 770 [sec]
Exponential service
Each service time 230, 234, 235, 71 [sec]

Disassembly parameter:

Mean of total disassembly time 472 [sec]
Exponential service
Each service time 238, 234 [sec]
Routing probability q_i =0.33 (i=1,2,3), q_i =0.5 (i=1,2)

Number of simulation runs for arrival units 10,000.

3.2 Queueing modeling by plant simulation

In this paper, the 3 types of the reuse inverse manufacturing systems are modeled by using a simulator, Plant Simulation (eM-Plant) provided by Siemens Product Lifecycle Management Software Inc. By using the model with Plant Simulation, the experiments are conducted as shown in Figure 2. Also, the plant simulation has a hierarchical function among queueing simulation models, and the independent models of the asembly and disassembly lines are embedded into one reuse inverse manufacturing model. With the proposed simulation model, each experiment has 10,000 simulation runs for arrival units.

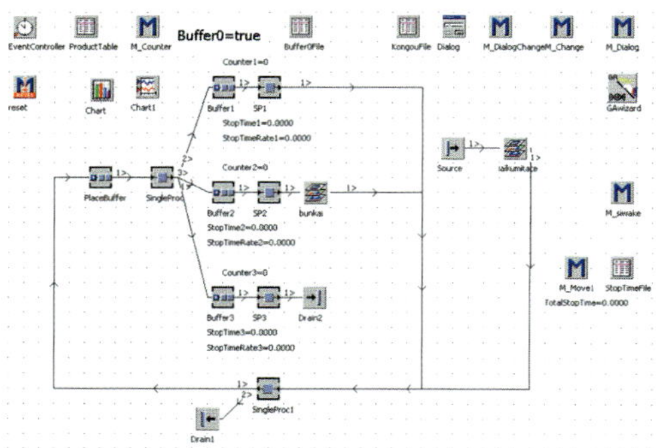

Figure 2 : A queueing simulation model for reuse inverse manufacturing system by Plant Simulation: remaufacturing type

4 DESIGN EXAMPLE OF REUSE INVERSE MANUFACTURING

4.1 Effect of product collection rate

Figure 3 shows the behaviors of blocking probability BL and throughput TH at the system for product collection rates. The product refurbishing has the highest blocking probability BL when the collection rate is 70%. After that, the blocking probability in the respective system decreases as the collection rate decreases. When the collection rate is less than 60%, it is considered that the work stations are almost idle due to no blocking.

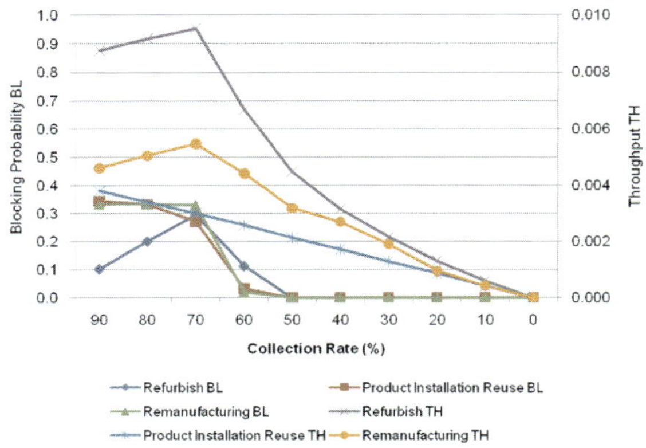

Figure 3: Behaviors of total reverse blocking probability BL and throughput TH at system for product collection rates (B_1=B_2=B_3=0)

The throughput TH at the system in the refurbish is basically higher than the other systems. It is considered that there are 2 types of product refurbish and spare parts reuse for recovery, and the total number of units processed is maintained in the system as time goes by. On the other hand, the product installation reuse and the remanufacturing have the lower throughput TH than one in the product refurbish because the recovered products can only be distributed again to the systems. Therefore, there are a few distributed products in the systems after the second product lifecycle.

4.2 Effect of material circulation and economic aspects for reuse inverse manufacturing

In the experiments, the effect of material circulation and economic aspects are considered. The number of products/parts processed at reuse and recycling implies the effect of material circulation at the reuse inverse manufacturing systems. In addition, the effect of economic aspects on the reuse inverse manufacturing can be obtained by multiplying the number of products/parts processed by coefficients of each cost and price as shown in Table 1.

Table 2 shows the material circulation effect by the number of recovered products/parts at each process in the case without reverse blocking. In the material circulation, the product refurbish has the largest number of the reused products and parts in the 3 systems. However, with the number of the reused products only, the product installation reuse is superior to the other 2 systems. With material recycling, the remanufacturing has the largest number of recycling recovery.

Table 3 shows the economic effect by reward, cost and profit in cases without reverse blocking. The profit in the product installation reuse is the highest in the 3 systems because the total reward in the product installation reuse is also the highest among them.

Table 1: Coefficients of each price and cost for reuse inverse manufacturing in the experiments

Types of Reuse Inverse Manufacturing	Each Cost/ Price	Assembly System	Disassembly System			
		Assembly	Sorting	Product Refurbish (SP1)	Disassembly (SP2)	Material Recycling (SP3)
Refurbish	Price	20	-	2	20	1
	Cost	10	1	1	Spare Parts Reuse 2	1
Product Installation Reuse	Price	20	-	-	20	1
	Cost	10	1	-	Disassembly 2 Re-Assembly 3	1
Remanufacturing	Price	20	-	-	20	1
	Cost	10	1	-	Disassembly 2 Reassembly 3	1

Table 2: Material circulation effect by the number of recovered products/parts: Case without reverse blocking (Collection rate 90%)

Types of Reuse Inverse Manufacturing	Assembly System	Disassembly System			
	Assembly	Sorting	Product Refurbish	Disassembly	Material Recycling
Refurbish	10,000	59,774	18,069	38,580	3,125
Product Installation Reuse	10,000	53,753	-	49,726	4,027
Remanufacturing	10,000	48,677	-	44,086	4,591

Table 3: Economic effect by total reward, cost and profit: Case without reverse blocking (Collection rate 90%)

Types of Reuse Inverse Manufacturing	Total Reward, Cost, Profit	Assembly System	Disassembly System				Total
		Assembly	Sorting	Product Refurbish (SP1)	Disassembly (SP2)	Material Recycling (SP3)	
Refurbish	Reward	200,000	-	36,138	771,600	3,125	1,010,863
	Cost	100,000	59,774	18,069	77,160	3,125	258,128
	Profit	-	-	-	-	-	752,735
Product Installation Reuse	Reward	200,000	-	-	994,520	4,027	1,198,547
	Cost	100,000	53,753	-	Disassembly 99,452 Re-Assembly 149,178	4,027	406,410
	Profit	-	-	-	-	-	792,137
Remanufacturing	Reward	200,000	-	-	881,720	4,591	1,086,311
	Cost	100,000	48,677	-	Disassembly 88,172 Re-Assembly 132,258	4,591	373,698
	Profit	-	-	-	-	-	712,613

The product refurbish is the second highest profit in the 3 systems, while the total reward in the refurbish is lower than one in the remanufacturing. One of the reasons is that the cost in refurbish is minimal among the 3 systems. Therefore, the strategic product recovery values are required to design the reuse inverse manufacturing systems.

5 SUMMARY AND FUTURE STUDIES

This study focused on the reuse inverse manufacturing systems, modeled the 3 different recovery types of the systems by queueing simulation, and compared the 3 reuse types of the system designs in view of collection rates, material circulation and economical aspects.

In view of collection rate, the throughput in the refurbish is basically higher than the other systems because the total number of units processed is maintained by the 2 types of recovery. In view of material circulation, the product refurbish has the largest number of reused products and parts in the 3 systems. However, in view of economical aspects, the profit in the product installation reuse is the highest in the 3 systems because the recovery value (price) of each product is higher than the other recovery types.

Further studies should consider the length of the each product lifetime by users, model a recycle inverse manufacturing system, etc.

6 ACKNOWLEDGMENTS

This research was partially supported by the Ministry of Education, Science, Sports and Culture, a Grant-in-Aid for Young Scientists (B), 20710122 in 2008-2010. Also, the authors would like to thank Mr. Yoshinari Taguchi for the additional experiments.

7 REFERENCES

[1] Seliger, G., 2007, Sustainability in Manufacturing: Recovery of Resources in Product and Material Cycles, Berlin, Heidelberg: Springer-Verlag.

[2] Inverse Manufacturing Forum, 2004, Handbook of inverse manufacturing, Maruzen, Tokyo, Japan (in Japanese)

[3] Nof, S. Y., W. E., Warnecke, H., 1998, Industrial Assembly, Chapman & Hall UK.

[4] Lambert, A. J. D., Gupta, S. M., 2005, Disassembly Modeling for Assembly, Maintenance, Reuse, and Recycling, Florida: CRC Press.

[5] Yamada, T., 2008, Design Issues for Inverse Manufacturing Systems in Japanese Cases, Northeast Decision Sciences Institute, 2008 Annual Meeting, March, Brooklyn, NY, USA, 551-556 on CD-ROM.

[6] Yamada, T., 2008, Inverse Manufacturing Systems with Different Product Recovery Lifecycle: Japanese Cases and A Concept, Proc. of the Global Conference on Sustainable Product Development and Life Cycle Engineering, SUSTAINABILITY and REMANUFACTURING VI, Keynote, Sept., Pusan, Korea, 19-22.

[7] Kobayashi, H., 2003, Product Lifecycle Planning – A Practice for ISO/TR14062 -, Ohmsha, Tokyo, Japan (in Japanese).

[8] Yamada, T., Arakawa, M., 2009, Disassembly System Design Considering Product Recovery Values, 20th International Conference on Production Research (ICPR-20), August, Shanghai, China, 126 (on CD-ROM).

[9] Yamada, T., Matsui, M., 2003, Disassembly Production Systems and Its Design Issues, Reprints of Japan Industrial Management Association, Spring Meeting, May, Chofu, Japan, 144–145 (in Japanese).

[10] Yamada, T., Mizuhara, N., Yamamoto, H., Matsui, M., 2009, A Performance Evaluation of Disassembly Systems with Reverse Blocking, Computers & Industrial Engineering, 56/3, 1113-1125.

[11] Arakawa, M., Yamada, T., 2009, Operation Design for Disassembly Considering a Product Structure, Symposium on Group Technology and Cellular Manufacturing 2009 (GT/CM2009), February, Kitakyushu, Japan, 190-197.

Development of Technology Roadmap for Remanufacturing Oriented Production Equipment

V.P. Cunha[1], I. Balkaya[2], J. Palacios[2], H. Rozenfeld[1], G. Seliger[2]

[1]Department of Production Engineering, University of São Paulo, São Carlos, Brazil

[2]Department for Machine Tools and Factory Management, Technische Universität Berlin, Germany

Abstract

In order to identify sustainable technologies to meet sustainability challenges of 21st century, a technology roadmapping for remanufacturing oriented production equipment is developed. The roadmap was structured under sustainability criteria with consideration for ecological, economic and social aspects, in addition four workshops were conducted with participants from the academic area. In the workshops, timelines were estimated for future technologies, and their likelihoods of development were established. Through the creation of technology paths, possible solutions to achieve sustainable development and potential conflicts for realization remanufacturing oriented production equipment are presented.

Keywords:

Production Equipment; Remanufacturing; Technology Roadmap

1 INTRODUCTION

The concept of remanufacturing is becoming more important for the companies where sustainability and profitability aspects are considered not only in terms of reducing landfills but also reducing energy, raw material and labor costs related to production [1].

Remanufacturing promotes not only the multiple reuse of materials but also enables the improvement in quality by upgrading the functions of products without manufacturing completely new ones and throwing away the used ones. It reduces the production costs via reductions in processing and raw material usage [2].

By remanufacturing the companies can extend the life-cycle of the production equipment, reducing usage of materials and capital investment for the new machines. And considering that the gobal demand for production equipment is expanding and firms require more performance at low cost, there will be an increasing market for remanufactured products with remarkable quality [3].

As remanufacturing of the production equipment gains more importance for the companies, the industry looks for new paths in order to have more knowledge in remanufacturing. Therefore, a technology roadmap is used to support strategic decisions for future investments in the remanufacturing sector.

Technology roadmapping is a needs-driven technology planning process to help identify, select and develop technology alternatives to satisfy a set of product needs. It is an effective technology planning tool to help identify product needs, map them into technology alternatives and develop project plans to ensure that the required technologies will be available when needed. It is an effective tool for providing the linkage between the technology investment decisions and the business requirements [4].

The next session describes the state of the art in remanufacturing as well as a description of the methodology used to develop the technology roadmap. In chapter 3 the results for the development of the technology roadmap oriented for the remanufacturing of production equipment are presented, followed by final considerations, acknowledgements and references sections.

2 STATE-OF-THE-ART

2.1 Remanufacturing

Achieving sustainable production and consumption is only possible with closed 'cyclic' systems in which resources are recovered from the waste stream at the end-of-life of a product. In this sense, remanufacturing is an effective manner to maintain a product in a closed-loop, which also reduces environmental and economical costs of manufacturing and final disposal costs of products [5].

According to Hammond and Bras (1996), "remanufacturing is the practice of disassembling, cleaning, refurbishing, replacing parts (as necessary) and reassembling a product in such a manner that the part is at least as good as, or better than, new" [6].

In remanufacturing, products that are known to be worn, defective, or discarded are brought to a manufacturing environment, where they are disassembled. Then all components are cleaned and checked. Those that can be reused are brought up to specification and those that cannot be used are replaced. When the product is reassembled and tested it is ready for a second life, performing as new with a warranty [7]. Figure 1 describes a product closed-loop considering remanufacturing as end-of-life treatment.

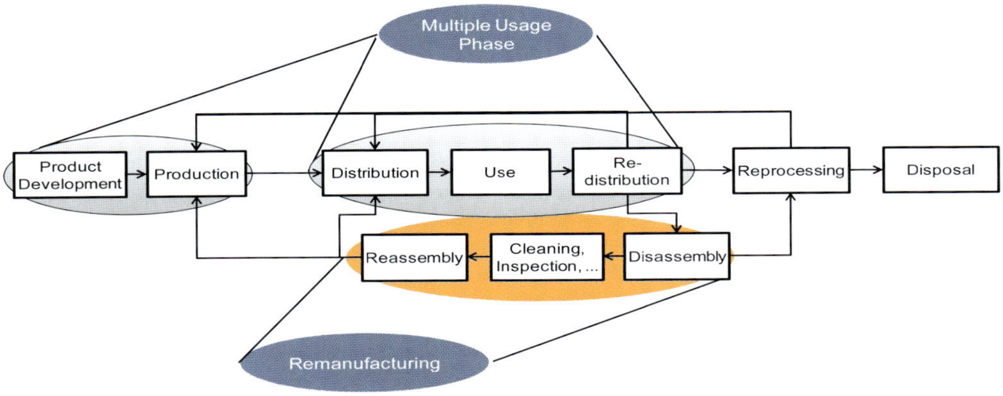

Figure 1: Product closed-loop for remanufacturing.

According to Ijomah et al. [8], up to 85% of the weight of remanufactured products may be obtained from used components with the same quality as new ones and require less energy to produce. One of the most important benefits of remanufacturing is that, it provides 20 % to 80 % cost savings compared to conventional manufacturing via extending the life of used products and avoiding landfilling costs. Thus, companies increasingly require remanufacturing know-how. Remanufacturing integrates the waste back into the production cycle thus, manufacturers can avoid the waste limitation penalties and as a result the companies can maximize their profit [8].

There are some considerations which affect the remanufacturing process and therefore have to be taken into account while designing a product. For example, the products being made up of interchangeable parts which enables ease of the disassembly of the product [1].

In this research, a technology roadmapping technique is used to identify technology solutions that satisfy the needs of remanufactured production equipment [4].

2.2 Technology Roadmapping technique

The technology roadmapping is a flexible technique that describes the market, plans products and processes development, establish technological capacities and analyze resources [9]. The technique shows the interrelations between market, product and technology parameters, and identifies goals related to justify the required efforts [10].

Among the many formats of roadmaps available, the most widely used is proposed by EIRMA (European Industrial Research Management Association) [11], presented in Figure 2.

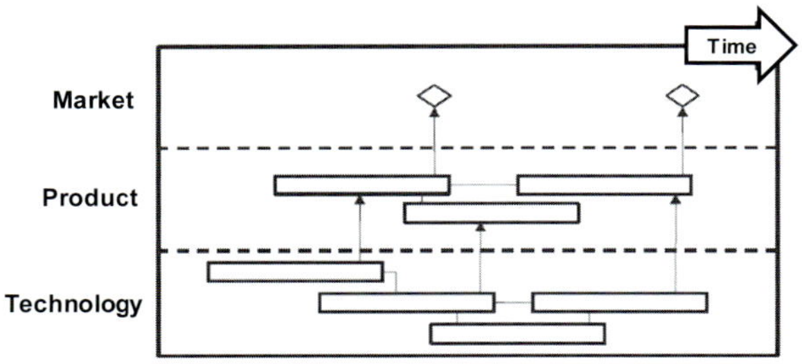

Figure 2: Generic roadmap proposed by EIRMA [11].

The roadmap of EIRMA is formed by three layers on the vertical axis which represent market, product and technology dimensions while the horizontal axis represents the time line.

Best practices regarding the process of implementing the technology roadmapping describe factors that should be considered during the application of the technique. The verification of these factors ensures that the results of the method meet the expectations of the company. Six of these best practices are listed below:

- The success of technology roadmapping depends on support from an influential stakeholder [11, 12, 13];
- The participation of specialists in the process increases the probability of success [12, 13, 14];
- Roadmapping is a multifunctional process; thus, the involvement of participants from different functional areas is essential [12, 15];
- The symbols employed in the roadmap should be known to enable organizational communication [15];
- Customization of roadmapping contributes to the effective application of technology roadmapping, since it can be used in different environments and for distinct purposes [11, 16]; and
- The process is considered responsible for adding value in the technology roadmapping application [9, 11, 16].

The literature describes several processes for the application of technology roadmapping technique. In this work the process for technology planning described in Phaal et al. [17] called the T-Plan was applied.

The T-Plan is a fast start approach for implementing roadmapping, to support products, service and technology planning, as well as general business planning. This process was selected since the activities to develop the roadmap are described in detail. The methodology was developed by the university of Cambridge Centre of Technology Management as part of a three-year research programme sponsored by the UK Engineering and Physical Sciences Research Council.

The process involves a series of four workshops, illustrated on Figure 3, carried out through brainstorming sessions.

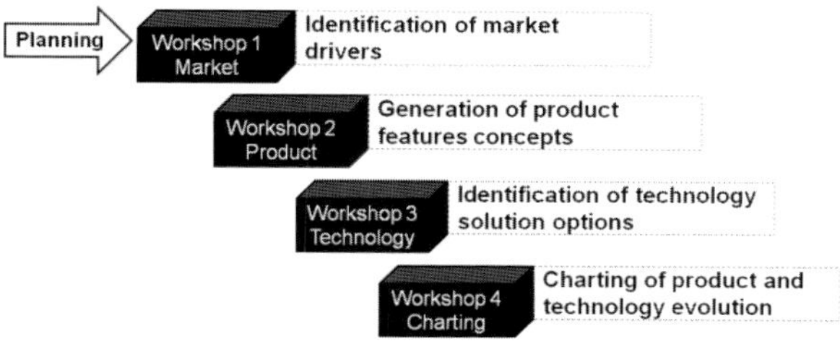

Figure 3: Structure of standard T-Plan process, comprising four workshops [17].

In the first workshop market drivers are identified. The drivers point out the market trends related to the customer, competitors and market segments.

In the product workshop, features related to the market drivers are identified. These features correspond to overall product characteristics, such as design and the environmental aspects.

In the third workshop, technologies solutions capable to achieve the desired product features are defined. Finally, in the charting workshop, the results of the previous workshops are integrated in a multilayer roadmap, where the alignments among the drivers, product features and technology solutions are represented.

3 TECHNOLOGY ROADMAP FOR REMANUFACTURING ORIENTED PRODUCTION EQUIPMENT

3.1 T-Plan workshops overview

Five to ten participants plus two process facilitators responsible for conducting the workshop and ensuring fulfillment of the agenda attended the series of previously described workshops. The participants were all academic specialists familiar with the different areas related to the remanufacturing of production equipment, such as, sustainability, eco-design, production technologies and assembly technology.

During the workshops each participant contributed with his specific knowledge, and when necessary relevant materials were presented as input for the brainstorming sessions.

3.2 Market workshop

The first step of the T-Plan methodology consists of the market workshop, the goals here are introduce the T-Plan process to the participants, brainstorm product performance dimensions related to remanufacturing, as well as identify market drivers and prioritize them.

In this workshop, scenarios for remanufacturing of production equipment for Brazil in 2020, which take into consideration economical, environmental, market-related, political, social, cultural, and technological factors along with particular interests from the different stakeholders were used as basis to identify the market drivers [18]. These scenarios were based on alternative projections of different key factors, which were combined to coherent, consistent, and plausible descriptions of the future.

During this workshop ten market drivers were identified by the participants based on the informations from the scenarios. The drivers are described below:

- Economic viability: Remanufactured products cost less and are at least as productive as new production equipment;
- New market based in lower price equipment: Low cost and environmental benefits of remanufactured products attract customers as the new production equipment prices are getting higher;
- Green labeling: The government is fostering the public discussion on sustainable development. Environmental education through the execution of advertisement in media and television is increasing;
- New demand: The demand for remanufactured machine tools and assembly equipment has increased. This is mainly due to the improved quality of remanufactured equipment and components. Certifications and new standards grant a certain quality level that is equal to that of equivalent new products;
- Product-service system as business strategy: Changing market forces and the recognition that services in combination with products are providing balanced cash flow than products alone motivates companies to invest more and more in PSS. Faced with shrinking markets and increased commoditization of their products, these firms see service provision as a new path towards profits and growth. Due to the reduction of material consumption, PSS are more widely being recognized as an important part of the company`s environmental strategy;
- End-of-life laws: The creation of specific rules leads to an increase in responsibility of the companies for their products end-of-life treatment;
- Increasing reusability of the manufacturing systems: Manufacturing firms have become increasingly interested in manufacturing systems that produce a variety of products. Thus, manufacturing companies have sought methods to rapidly change/deploy/adapt manufacturing systems in economical ways to attend the continuously demand of the customers;
- OEMs remanufacture themselves: Due to expected advantages of remanufacturing, Original Equipment Manufacturers (OEMs) enter this new market through remanufacturing of their own products;
- Remanufacturing inside a user company: New design techniques have increased the interchangeability of components. In addition, the modularity of production equipment makes the remanufacturing operation easier. These operations can therefore be made at the user shop floor. This new practice reduces logistic costs and the time of the overall remanufacturing process, leading to an increased productivity;
- Certifying by associations: Associations support certifications for remanufactured products, which assure the quality of the products.

3.3 Product workshop

Goals of the second workshop are to review information provided by the market workshop, brainstorm product features/concepts and sort them into groups and, finally evaluate the impact of the groups on each driver.

Product and service features or concepts which have the potential of addressing the drivers previously defined were brainstormed during these workshop. The aim here was to identify important features for the remanufacturing of production equipment.

On a second step, the features were sorted into groups according to their attributes and then the overall influence of each group over the different market drivers was ranked through a Pugh diagram. In order to obtain the scores each participant graded each factor with the following scale: (0) No impact, (1) low impact, (2) medium impact, (3) high impact and (4) very high impact. The tool produced a balanced weight that includes the market driver priority provided in the first workshop and the grade for each product feature group. The final score was normalized to 10 to facilitate comparisions of the scores. Table 1 shows the evaluation of the impact of product features groups in the market drivers.

Table 1: Evaluation of the impact of the product feature groups in the drivers.

	Market Drivers											
Weight	5	4	3	4	3	4	3	3	3	2		
Drivers / **Feature groups**	Economic viability	New market	Green labeling	New demand	Product-Service System	End-of-life laws	Increasing reusability	OEMs remanufacturing	Remanufacturing inside user	Certifying by associations	Score	**Final Score**
Design	3	3	2	3	2	1	3	2	3	1	81	**10,0**
Quality and productivity	3	3	1	3	2	1	3	2	3	1	78	**9,6**
Energy efficiency	3	2	3	2	2	1	2	1	1	1	64	**7,9**
Economic efficiency	4	3	1	3	3	1	2	2	2	1	80	**9,9**
Availability of services	2	3	1	3	4	1	2	2	2	1	73	**9,0**
Environmental aspects	2	2	4	2	2	3	2	1	1	2	72	**8,9**

The results shows that the groups Design, Economic efficiency and Quality and productivity had higher influence on the drivers. From this result it is possible to identify that to achieve the tends and motivators for remanufacturing of production equipment special attention should be given to the products features sorted in this groups.

3.4 Technology workshop

The goals defined for this workshop are to review the information provided by the market and product workshops, brainstorm possible technological solutions and sort them into technical areas and, lastly rank the impact of each area on each product feature group.

The definition of technology solution here was kept fairly broad, including component, design, production and information related technology options, together with hard and soft technologies which are capable of satisfying the product features.

The technological solutions were sorted in techical areas and assesed the impact of each area on the product feature groups, for this the same matrix adopted in the product workshop was used, as shown in Table 2. The scores given to the product features groups were used as weight here, indicating the technical areas that make the greatest contribution to the most important groups.

Table 2: Assessing the impact of technical areas on product features.

	Product Features							
Weight	10	9,6	7,9	9,9	9,0	8,9		
Feature groups / **Technical areas**	Design	Quality and productivity	Energy efficiency	Economic efficiency	Availability of services	Environmental aspects	Score	**Final Score**
Eco-efficiency systems	3	3	4	3	1	3	156	**10,0**
Materials	4	3	2	3	1	3	150	**9,6**
Design for disassembly	4	3	1	2	2	3	141	**9,1**
Standardization	4	2	1	3	1	2	124	**7,9**
Modularization	4	2	1	3	2	2	133	**8,5**
Manufacturing control	3	3	2	3	2	1	131	**8,4**
Services	2	3	2	3	4	2	148	**9,5**

From these results it was noticeable that the technical area Eco-efficiency systems represents the greatest impact, followed by the areas Materials, Design for disassembly and Services which are also important. As a final mark, it is possible to conclude that the technological solutions sorted in those areas have higher potential to contribute to the successful performance of the features.

3.5 Charting workshop

The actions intended for this workshop are review the information provided by market, product and technology workshops and develop a first-cut technology roadmap identifying market and strategic milestones, product evolution and technologies responses.

Previously of the charting workshop all the product features and technology solutions were validated through a detailed literature review. In this sense, some of them were changed or adapted and some were removed.

As a first step of the workshop the drivers, as well as the products features and technological solutions were charted on the roadmap within a 10 years timeframe. The next step was to establish links between market, product and technology layers. An important result is the point of view of business objectives and how they can be achieved through identified opportunities for new products and technologies.

In Figure 4 the results of the technology roadmapping is illustrated, it is important to emphasize the presence of three layers and the linkages between them. Due to the complexity to show all the sub-elements and their linkages in a time frame, they were not presented.

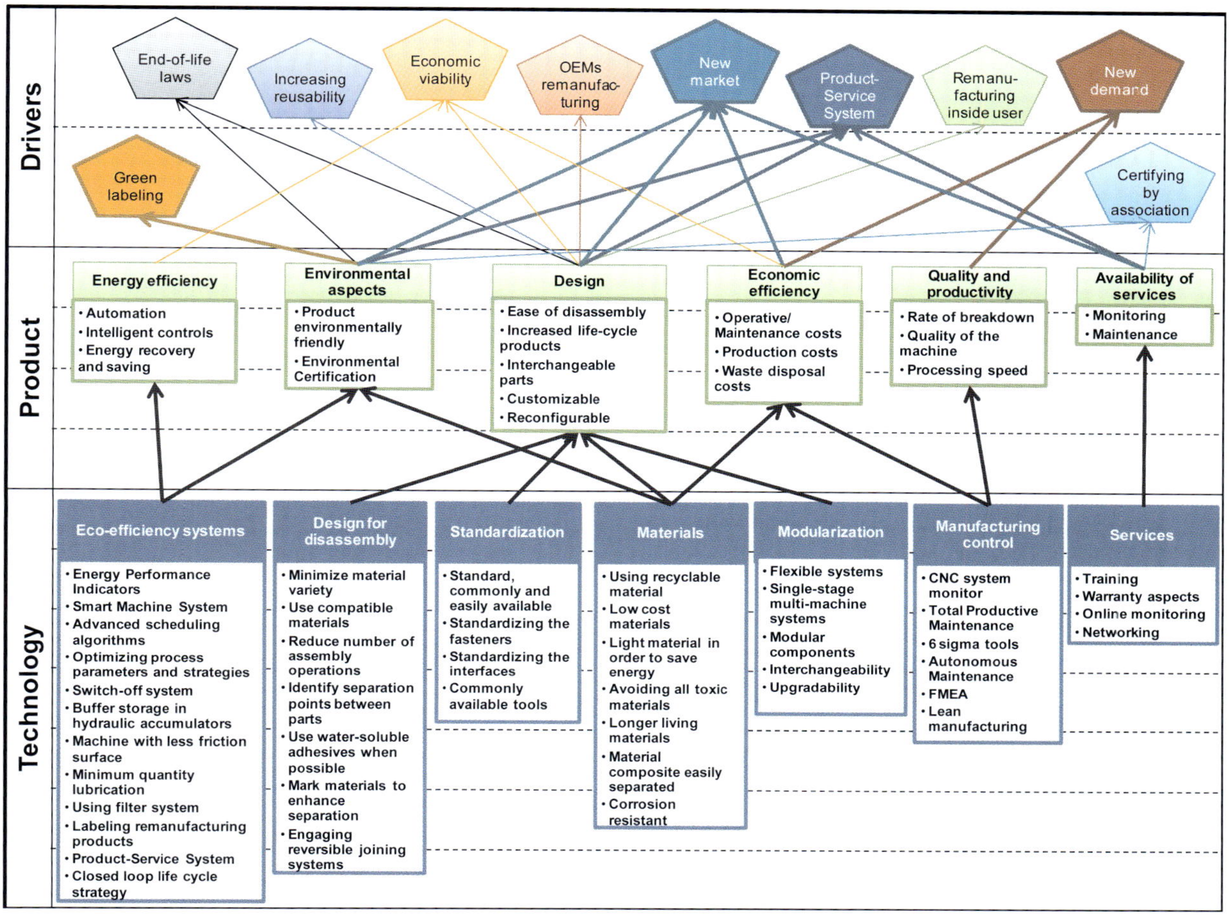

Figure 4: Technology Roadmap for Remanufacturing Oriented Production Equipment.

4 FINAL CONSIDERATION

Remanufacturing of the production equipment is becoming more important for companies in terms of economical and sustainable aspects. This leads to new market opportunities and demands, which requires new paths in order to achieve more knowledge. Therefore, a technology roadmapping is used to support strategic decisions for future investments in the remanufacturing sector. The T-Plan process is used in this case to develop the roadmap, providing the linkage between the technology investment decisions and the business requirements. A series of four facilitated workshops were conducted in order to process the standard T-Plan.

During the first three workshops (market, product and technology) the participants brainstormed key elements for the development of the remanufacturing of production equipment. The defined market drivers have shown that there are several motivators and trends for remanufacturing that go beyond economic benefits and market opportunities. It was also concluded that most of the identified product features and technological solutions have already been or are being developed, but in many cases are not applied oriented for the remanufacturing of production equipment.

Finally, in the last workshop the results of the previous workshops were integrated in the technology roadmap, where the alignments among the market drivers, product features and technology solutions are represented. This information helps identify the paths that should be followed to improve the remanufacturing of production equipment, and plan the development or acquisition of new technologies to satisfy the market drivers.

It is important to consider the limitation of this study, related to the generality of the technology solutions proposed in the roadmap. Thus, seeking continuity of this research, the authors suggest the development of new roadmaps based on specific types of production equipment and machine tools, such as grinding machines.

5 ACKNOWLEDGMENTS

We extend our sincere thanks to CAPES (Coordenação de Aperfeiçoamento de Pessoal de Nível Superior), CNPq (Conselho Nacional de Desenvolvimento Científico e Tecnológico), FINEP (Financiadora de Estudos e Projetos) and DFG (Deutsche Forschungsgemeinschaft) for supporting BRAGECRIM (Brazilian and German Collaborative Research in Sustainable Manufacturing) project intituled "Remanufacturing oriented Production Equipment Development".

6 REFERENCES

[1] Ijomah, W. L., McMahon, C. A., Hammond, G. P., Newman, S. T., 2007, Development of robust design-for-remanufacturing guidelines to further the aims of sustainable development, In:International Journal of Production Research, v. 45, 18-19, p.4513- 4536.

[2] Sundin, E., 2004, Product and process design for successful remanufacturing, Linköping Studies in Science and Technology, Dissetation No. 906, Department of Mechanical Engineering, Linköpings Universitet, SE-581 83 Linköping, Sweden.

[3] Butler, P., 2009, A report on the remanufacture of manual and CNC tooling machinery in the UK, Product Group Report: Machine tools, Centre for Remanufacturing and Reuse, pp. 1-12.

[4] Garcia, M. L., Bray, O.H., 1997, Fundamentals of Technology Roadmapping. Strategic Business Development Department, Sandia National Laboratories.

[5] Kerr, W., Ryan, C., 2001, Eco-efficiency gains from remanufacturing A case study of photocopier remanufacturing at Fuji Xerox Australia. Journal of Cleaner Production, v.9, n.1, p.75-81.

[6] Hammond, R., Bras, B. A., 1996, Design for Remanuacturing Metrics, Proceedings of the 1st International Workshop on Reuse, Eindhoven, The Netherlands, November 11-13, pp. 5-22.

[7] Parkinson, H. J., Thompson G., 2003, Analysis and taxonomy of remanufacturing industry practice. Journal of Process Mechanical Engineering pp 243-256.

[8] Ijomah, W. L., Childe, S., McMahon, C., 2004, Remanufacturing: A Key Strategy for Sustainable Development, Proceedings of the 3rd International Conference on Design & Manufacture for Sustainable Development, UK, 1st-2nd September.

[9] Willyard, C. H., McClees, C. W., 1987, Motorola's Technology Roadmap Process, Research Management, v.30, n.5, p.13-19.

[10] Kappel, T. A., 2001, Perspectives on roadmaps: how organizations talk about the future, Journal of Product Innovation Management, v.18, n.1, p.39-50.

[11] Phaal, R., 2004, Technology roadmapping - a planning framework for evolution and revolution, Technological Forecasting and Social Change, v.71, n.1-2, p.5-26.

[12] Garcia, M. L., Bray, O. H., 1997, Fundamentals of Technology Roadmapping, Sandia National Laboratories, April (report SAND97-0665).

[13] Strauss, J. D., Radnor M., 2004, Roadmapping for dynamic and uncertain environments, Research Technology Management, v.47, n.2, p.51-57.

[14] Albright, R. E., Kappel T. A., 2003, Technology roadmapping: roadmapping the corporation, Research Technology Management, v.46, n.2, p.31-40.

[15] Albright, R. E., Nelson, B., 2004, Product and technology mapping tools for planning and portfolio decision making, in Belliveau, P., A. Griffin and S. Somermeyer, The PDMA Toolbook 2 for new product development, John Wiley & Sons, Chapter 15.

[16] Probert, D., Radnor, M., 2003, Technology roadmapping: frontier experiences from industry-academia consortia, Research Technology Management, v.46, n.2, p. 27-30.

[17] Phaal, R., Farrukh, C., Probert, D., 2001, T-Plan: fast start to technology roadmapping. Cambridge, UK: Cambridge University Press.

[18] Severengiz, S., Widera, H. Seliger, G., Barreira, Y., Ometto, A., Barquet, A. P. B., Forcellini, F. A., 2009, Scenarios for the Remanufacturing of Machine Tools and Assembly Equipment, In:GCSM, 2009, India, http://www.gscm2009.iitm.ac.in/brochure.pdf

Automated Image Based Recognition of Manual Work Steps in the Remanufacturing of Alternators

A.B. Postawa, M. Kleinsorge, J. Krüger, G. Seliger

Department for Machine Tools and Factory Management, Technische Universität Berlin, Germany

Abstract

Automated image based recognition of approved work processes at the manual workplace for remanufacturing of alternators supports faster teaching of humans with different qualification levels by explaining the work task with a universally usable form of description. Automatically classified sequence of work steps, used means of production and work pieces are completed or confirmed by the worker over a user interface. The workers progress during the process may be used for future Man-Machine-Cooperation.

Keywords:

Humans and Technology, Qualification, Image Based Recognition, Remanufacturing

1 INTRODUCTION

Remanufacturing has a wide economical and ecological potential as shown in studies done by Steinhilper [1] and Lund [2]. The remanufacturing of a starter motor for instance requires, compared to production, only ninth of the material and eleventh of energy [1]. In 2009 the automotive supplier Bosch has saved 23000 tons of CO_2 trough the remanufacturing of 2,5 millions of parts instead of their production and is expecting a growth of the remanufacturing market up to 30 millions of products per year till 2015 [3].

Remanufacturing is characterized trough unforeseen products types, conditions and quantity. Due to these variations remanufacturing processes today are mainly performed manually or rather mechanized [4, 5].

If remanufacturing is done within industrial standards economies of scale can be utilized. In the example of Bosch vendor dependent mass remanufacturing can be done, where a large number of similar but own products are refurbished. Vendor independent remanufacturing requires highly flexible tools, production facilities and factory structures. Such systems are been called factories of the next generation. They are adaptable, transformable and intelligent. They are embedded in manufacturing networks to be competitive on turbulent markets [6].

A mini factory for remanufacturing represents such an approach. Through the utilization of potential for value creation in regions with a low infrastructure, mini factories can create local income generation and support regional development. Diverse qualification levels of locally available workers thereby is a challenging aspect to be incorporated in process planning and to be increased through training and qualification measures [7].

2 QUALIFYING FOR REMANUFACTURING

2.1 Information Technology System for Qualification

To enable efficient remanufacturing with small batch sizes, workers with different qualification levels have to be trained on order. An information technology system can support the worker with an automatically generated work description. Interconnected with other workers he has excess to approved work descriptions and is able to compare or exchange them. The worker will be enabled to analyse his working process and qualify himself autonomously.

2.2 Selected work task

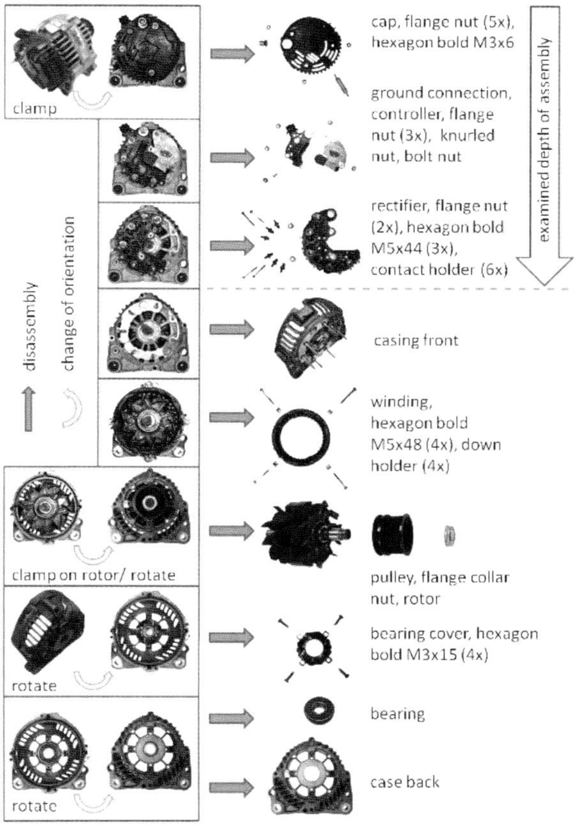

Figure 1: Build up and depth of assembly of the alternator

Remanufacturing includes tasks like disassembling, cleaning, refurbishing, replacing (if necessary), and reassembling parts [8]. For the remanufacturing of an alternator, work tasks disassembly and reassembly have been selected to be recognized by an information technology system.

Exemplarily for the selected tasks, work steps like separation and joining of parts including unsoldering, unscrewing and screwing are been performed in standing position at a manual workplace.

The build up and depth of assembly of the alternator are shown in figure 1.

3 INFORMATION TECHNOLOGY SYSTEM FOR IMAGE BASED RECOGNITION OF MANUAL WORK STEPS

After performance of a sequence of work steps, the positions of the workers hand has been tracked image based and translated into a work description by an information technology system. The automatically generated work description can be used for teaching of workers with different qualification levels.

The information technology system for image based recognition of work steps consisting of a 3D-camera and image processing algorithms for the workers hand tracking as well as a knowledge-based expert system with a user interface for the recognition of work steps (figure 2).

Output of the image based tracking is the case-based knowledge. The expert system screens the trajectory of the workers hand for positions relevant for the work description. The worker completes the knowledge base over a user interface and knowledge acquisition facilities through describing his environment. Expert and worker are teaching the system continuously. Automatically recognised work steps are completed or validated by the worker. A work description is the output of the expert system and comparable with the data base of as well as accessible by other interconnected workers for qualification.

Image based recognition has the advantage to be independent of country languages. Created work descriptions are worldwide applicable.

3.1 Image based hand tracking

Sensors or markers attached on the human body change the touch sensitivity, the freedom of movement and the skin slippage. Marker-less it is very difficult to measure the motion pattern of body parts precisely. Image based motion sequence analysis does not cause these constraints but has the problem of missing data in blind areas. Position of a hidden body part is estimated and underlies spatial uncertainty.

The difficulty level of visual motion analysis increases on interaction of the hand with the environment. The unknown motions of the worker melt together with the tool and work piece during contact of manipulation task. The challenges of the required image processing are the speed, stability and accuracy of the high degree of freedom (DOF) measurement. Self occlusions of the body are not seen from the camera perspective, e.g. some fingers or parts of arm, are estimated and require a very robust modelling of the kinematics. The robust estimation of the full DOF hand pose is still a highly active research field. Inferring the pose and motion of a highly articulated and self-occluding non-rigid 3D objects from images is a mathematical ill-posed problem.

Previous related work of the authors includes implemented 2D hand tracking by skin colour segmentation (figure 3). The skin regions were detected in each image separately. The recognition was done by segment area and position [9]. One extension of that pattern recognition algorithm is done by remembering the hand position to simplify the search to find the hand after short time of disappearing. The proposed approach for analysis of work step sequences achieves more precise hand positions by determining the z-coordinate.

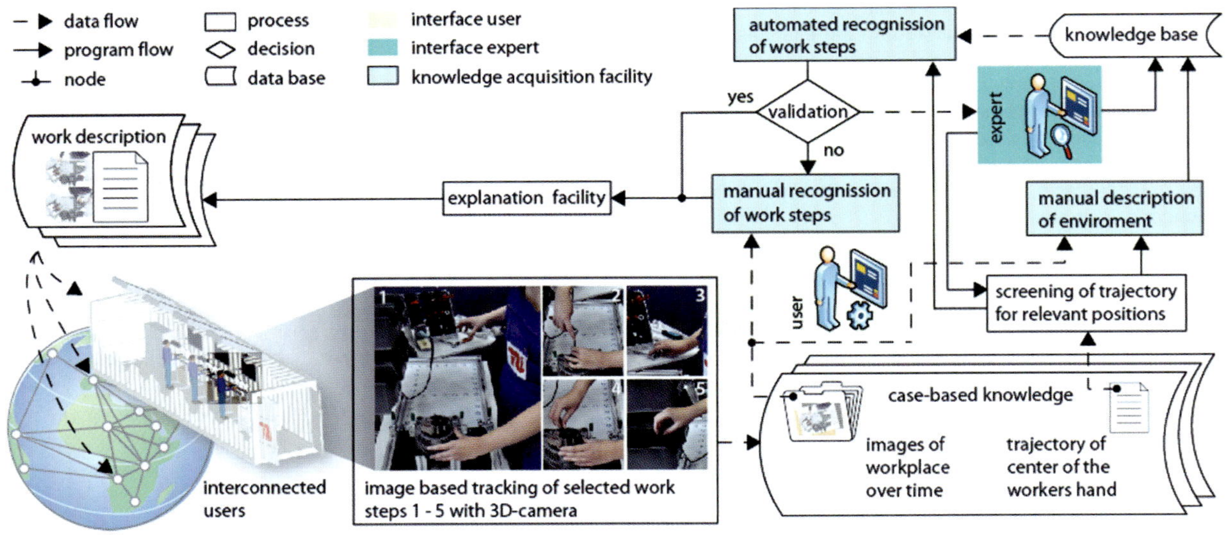

Figure 2: Information technology system for image based recognition of manual work steps – overview

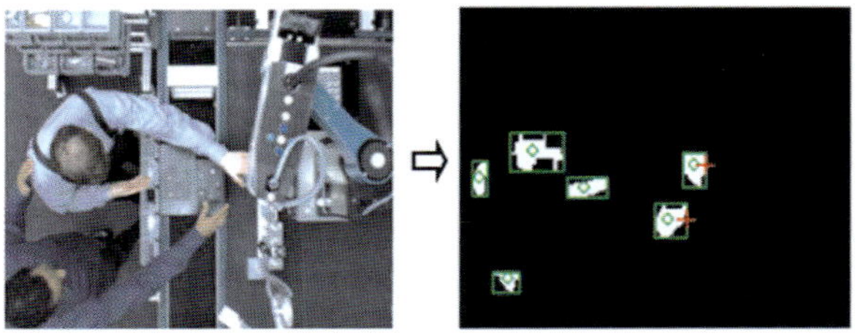

Figure 3: hand detection for 2D collision avoidance

The approach is based on marker-less hand tracking with a 3D Time-of-Flight (TOF) camera which measures the distance by phase shift of the reflected infrared signal. The state of the art electronics are too big and too slow so the emitter and sensor realizes only a resolution of a few centimetres on depth and spatial axes due to time measurement constraints and chip size. The advantage of TOF camera is the distance measurement by hardware which requires no extra computing time. Those Approaches using the computationally intensive stereo reconstruction of two 2D cameras underlie the ambiguous correspondence problem. A Solution for that is structured light but it is too slow. The proposed approach uses one TOF camera.

Each pixel consists of the colour and the distance to the camera. The depth values have an inherent noise and a systematic error which has been reduced by the authors own research [10]. The camera frame rate is 30 Hz. It is fast enough to limit the spatial movement of moving objects between two frames. The image registration is done with the Iterative-Closest-Point (ICP) algorithm that enables the tracking of point sets. ICP minimizes iteratively the sum of squared distances between all points of two sets under transformation of translation and rotation which results in optimal Transformation parameters. The proposed algorithm is as fast as the frame rate on a 2 Ghz PC so real-time hand tracking is achieved.

The implemented image processing program of the proposed approach initializes the hands starting position with a constant value or by the user's mouse pointer. The hand tracking program outputs are the x-, y- and z-coordinates of the centre of the hand. After short time occlusion of the hand its position can be reassigned. It also works for mutual occlusion of both hands. Considering the similarity of both hands this is an outstanding feature (figure 4) [11].

For many work step sequences it is sufficient to evaluate the hand position. The hand pose is not needed for all work steps but later it will be implemented for further work analysis. Figure 5 shows a sequence from the hand tracking and demonstrates the robustness to different hand postures.

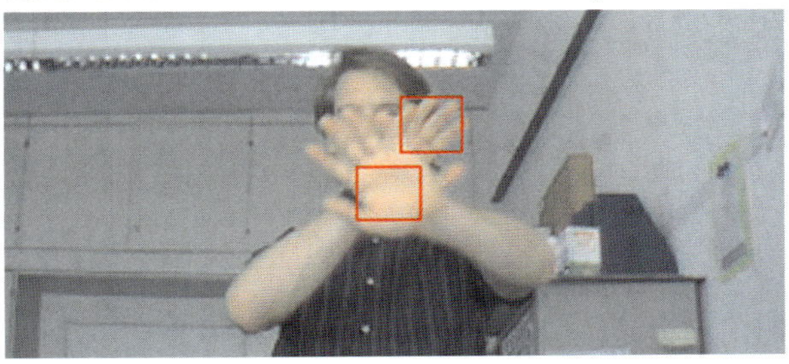

Figure 4: Rectangle marks both 3D tracked hands although mutual occluding

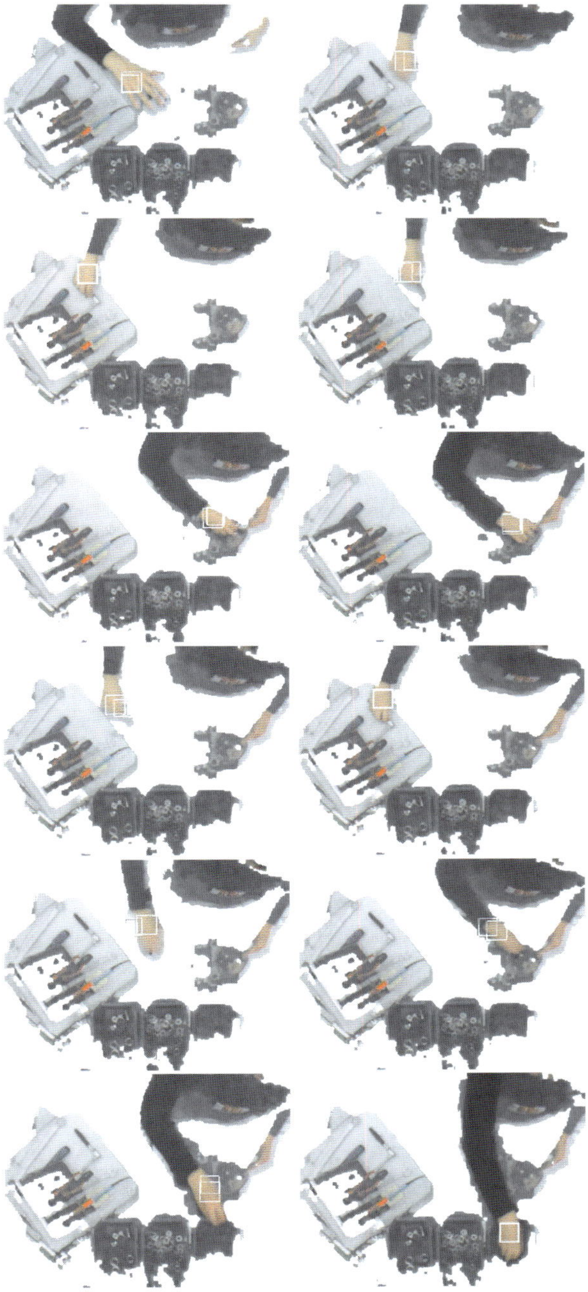

Figure 5: Rectangle marks tracked hand in work sequence

3.2 Data transfer between image based hand tracking and expert system

Initially the expert system is able to recognize sequences performed with only one hand therefore a sequence of work steps within the examined depth of assembly has been performed with the right hand and has been recorded. The coordinates of the recorded hand have been transformed from the camera coordinate system to a centre of reference on the workplace. The x-, y- and z-coordinates of the centre of the hand are stored together with the time of recording in a matrix in a way that each coordinate and the timestamp are assigned to a column.

The matrix contains a time-discrete path of the performing hand. The path underlines a noise of up to three centimetres in every direction and point in time. The hand itself is not punctuated. It is described by a sphere with the radius of the fingers plus the noise. Consequently the trajectory of the hand is represented with a tubular range.

An image of the workspace can be assigned to each timestamp. The matrix saved in a text file and the images are representing the case-based knowledge.

3.3 Expert system for recognition of work steps

The expert system for recognition of work steps consists out of a knowledge acquisition facility, a knowledge base and an interaction component (figure 6).

User and expert are teaching the system continuously. The user is completing or confirming automatically recognised work steps.

After case-based knowledge processing the systems output is a work description that is accessible by as well as comparable with the data base of other interconnected users.

3.3.1 Knowledge base

The Knowledge base represents all information for the inference engine about the environment in particular about various sectors of the manual workplace. In progression with the knowledge acquisition of the expert system, the knowledge base stores all relevant scopes of event E.

Scopes of event are classified with the sections *start*, *workplace*, *bunker* and *tool holder*. Sections are described through a central point and a class related radius. They have class related properties and inference mechanisms. Sectors are stored in tuples within the knowledge base.

Central point of the scope of event classified with *start* is the centre of the hand at t=0. This scope can be used to reset of content data of the tuple *hand*. *Start* is set at the beginning of iteration.

The tuple *hand* describes the condition of the users hand and has its centre point in the moment of investigation. In the scope *start* the hand is expected to be empty and will be reset to its initial condition. Tuple *hand* can insert as well as reinsert means of production and parts like work pieces and standard parts from other scopes of event classified as *workspace*, *bunker* and *tool holder*.

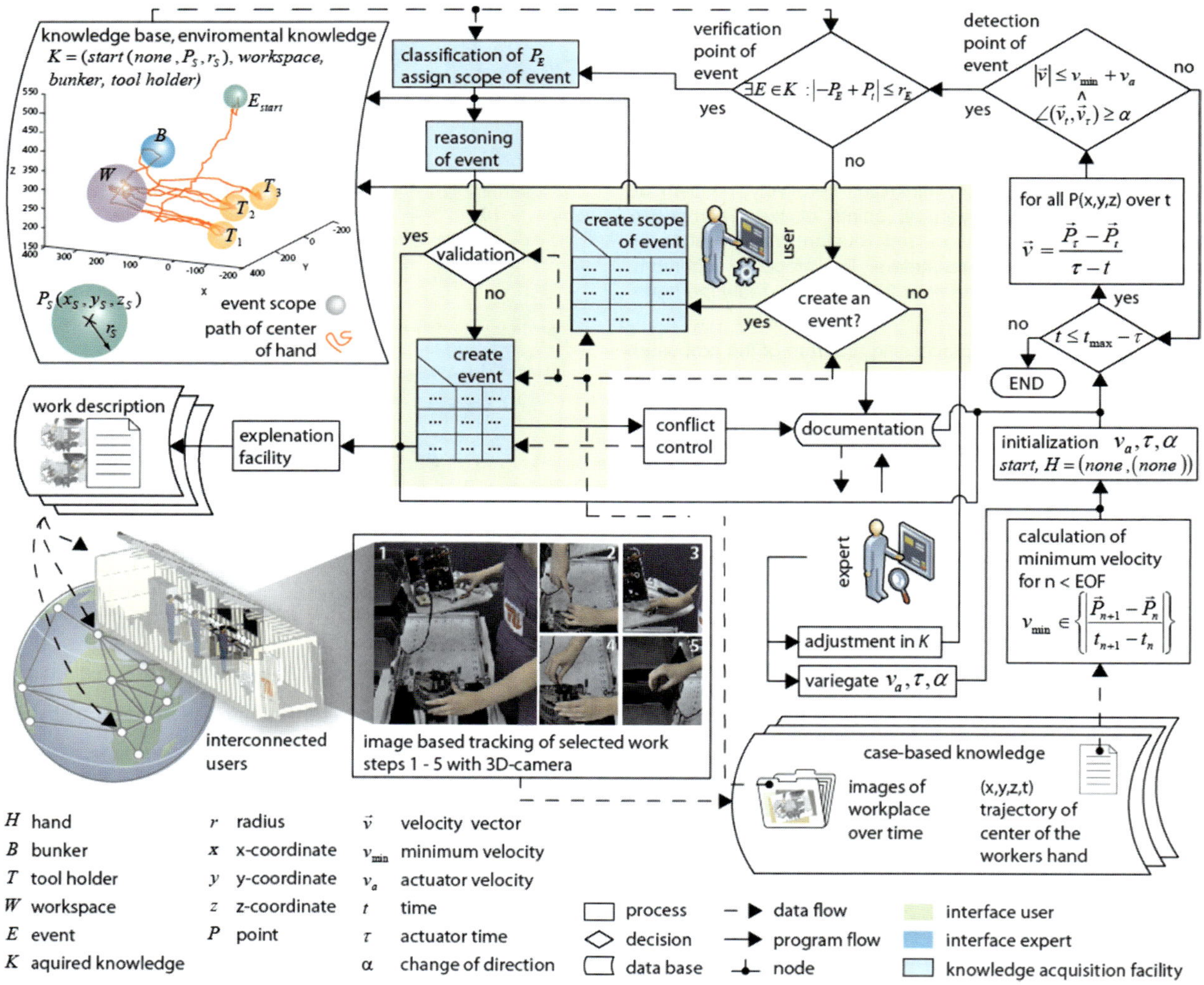

Figure 6: Information technology system for image based recognition of manual work steps – program flow

Work steps of value-creation are performed in scopes of event classified as *workspace*. Here work steps like joining and disjoining of work pieces including desordering, screwing and unscrewing can be performed. All parts as work pieces or standard parts that have been assembled (placed) or disassembled (picked) are stored in tuple *workspace*.

Scopes of event classified as *tool holder* can be assigned with one mean of production. Means of production that have been used here are screwdriver, ratchet and soldering-iron. The scope contains the value true or false depending on the presence of the assigned mean of production. Conflicts created by inserting a tool in the wrong tool holder or reinserting a tool from a tool holder with the value false are stored as information for the expert. Handling of tools (use) is compared to manual operations challenging to detect right because of the higher cover range of tools compared to fingers.

From scopes of event classified as *bunker,* work pieces and standard part can be reinserted (pick) and inserted (place). Parts are listed quantitatively in the tuple *bunker*.

3.3.2 Case-based knowledge processing

After screening the trajectory of the workers hand, relevant points for the work description are detected in which events like *pick*, *place* and *use* could have been performed. A screening algorithm detects positions with an abrupt change in direction and low velocity. The expert system verifies whether the detected point so called point of event P_E is contained in a scope of event E within the knowledge base K.

Is P_E within an E, the point can be classified. Based on classification and contend of *hand* an event can be reasoned through the knowledge acquisition facility *reasoning of event* (figure 7). The expert system requests a validation by the user. The User can validate or complete the reasoned event. He can also decline the validation, skip the detected point or create a new event with the knowledge acquisition facility *create scope of event*.

If no E can be found for a detected P_E, the expert system requests the user to create a scope of event. He can decline the request or access the knowledge acquisition facility *create scope of event*. P_E becomes centre point of the created scope.

Based on the class of scope the expert system reasons an event with the knowledge acquisition facility *reasoning of event*. The user can again validate, complete or create an event.

A control function to detect conflicts within the created work steps reports to the expert. With an increasing knowledge base and decrease of conflicts the validation by the worker can be bypassed.

3.3.3 Interaction component

The interaction component of the expert system is implemented as a switchable interface in MATLAB. The experts interface allows intervention in the knowledge processing of the expert system. It is also possible to adjust the knowledge base from the experts interface. The users interface gives excesses to the knowledge acquisition based return of inferred knowledge as explanation facility, the interface displays also images from the case-based knowledge data base to support the user. So generated work description can be displayed through the interface and are available for other users for qualification facilities (figure 8).

create scope of event

class	*specify*	*Notation*
workspace	"*product*"	*workspace* + ('*product*', P_t, r_W, (*none*), (*none*))
bunker	"*type*"	*bunker* + ('*type*', P_t, r_B, (*none*), (*none*), 0)
tool holder	"*tool*"	*tool holder* + ('*tool*', P_t, r_T, *true*)

explenation facility

object	*work step*	*position*
'*component part*' n	*pick*	P_E, *workspace*[E][0:3]
'*standard part*' n	*place*	P_E, *bunker* [E][0:3]
'*tool*'	*use*	P_E, *tool holder* [E][0:3]

create event

class	*query*	*execute*
workspace	*pick* "*part*" *n* *place* "*part*" *n* *use* '*tool*'	*H* [1] + *n* · '*part*' *workspace* [E][3] + *n* · '*part*' *H* [1] − *n* · '*part*' *workspace* [E][4] + *n* · '*part*'
bunker	*pick* "*part*" *n* *place* "*part*" *n*	*H* [1] + *n* · '*part*' *bunker* [E][5] − *n* *bunker* [E][3] + '*part*' if '*part*' in *bunker* [E][3] = *false* *H* [1] − *n* · '*part*' *bunker* [E][5] + *n* *bunker* [E][4] + '*part*' if '*part*' in *bunker* [E][4] = *false*
tool holder	*pick* *place*	*H* [0] = *tool holder* [E][0] *tool holder* [E][3] = *false* *H* [0] = *none* *tool holder* [E][3] = *true*

reasoning of event

class	*query*		*event*	*execute*
workspace	*none* in *hand* [1]	→ *true*	*pick*	*workspace* [E][3] + '*part*' *hand* [1] + '*part*'
		→ *false*	*place*	*workspace* [E][4] + *hand* [1] *hand* [1] = *none*
	none in *hand* [0]	→ *false*	*use*	
bunker	*none* in *hand* [1]	→ *true*	*pick*	*bunker* [E][5] + 1 *hand* [1] + *bunker* [E][3][0]
		→ *false*	*place*	*bunker* [E][5] − 1 *hand* [1] = *None* *bunker* [E][4] + *hand* [1] if *hand* [1] in *bunker* [E][4] = *false*
tool holder	*none* in *hand* [0]	→ *true*	*pick*	*tool holder* [E][3] = *false* *hand* [0] + *tool holder* [E][0]
		→ *false*	*place*	*tool holder* [E][3] = *true* if *hand* [0] = '*tool*' *hand* [0] = *none*

r radius, *t* time, *P* point, *E* event, *B* bunker, *T* tool holder, *W* workspace, *H* hand

Figure 7: Knowledge acquisition and explanation facility

Classified work steps can be validated or completed with used means of production and work pieces also events can be generated by the user. Besides a text based return of inferred knowledge as explanation facility, the interface displays also images from the case-based knowledge data base to support the user. So generated work description can be displayed through the interface and are available for other users for qualification.

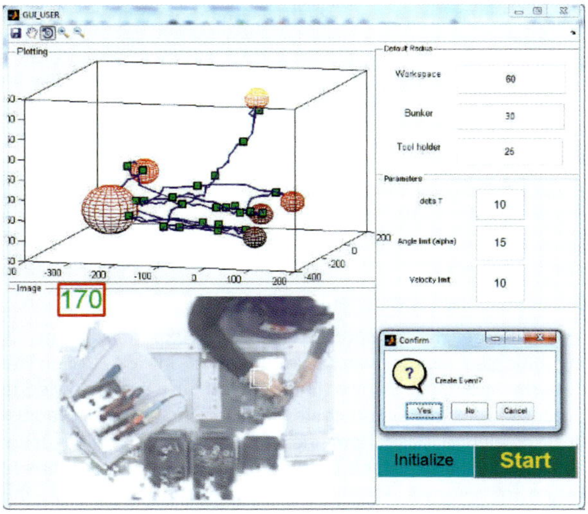

Figure 8: Screenshot of user interface

3.3.3 Output of expert system

After processing of case-based knowledge the output of the expert system is the work description. All events created with the knowledge acquisition facilities are included in the description.

A performance analysis created with basic motions from Methods-Time Measurement (MTM) is a reference for evaluation of the system based work description. Completeness of reasoned motions and the rate of interventions by the user are critical for the quality of the expert system. Continuous improvement process is intended so that the assembly and disassembly performed with both hands can be recognised in higher assembly depth than implemented now. Later on online recognition in a running process shall be implemented.

4 SUMMARY

An Information Technology System for image based recognition of manual work steps consisting of a 3D-camera and image processing algorithms for the workers hand tracking as well as a knowledge-based expert system with a user interface for the recognition of the work steps has been developed.

Through a marker-less hand tracking with a 3D Time-of-Flight (TOF) camera exemplary selected work steps as part of a remanufacturing process of alternators have been recorded. Image registration is done with the Iterative-Closest-Point (ICP) algorithm so that the trajectory of the hand movement could have been stored. Trajectory has been scanned by the expert system for work steps like separation and joining of parts including unsoldering, unscrewing and screwing. Over a user interface the system has been taught continuously, automatically recognised work steps have been validated or completed with used means of production and work pieces by the user. Output is a work description accessible by other interconnected workers for qualification.

Perspectively the developed Information Technology System could be used for intuitive information transmission between worker and machine in man-machine cooperation.

5 REFERENCES

[1] Steinhilper, R., 1998, Remanufacturing- The ultimate form of recycling, Fraunhofer IRB Verlag.

[2] Lund, R., 1996, The Remanufacturing Industry: Hidden Giant. Boston University, Boston, MA.

[3] BOSCH, 11.06.2010 , internet source, http://csr.bosch.com/content/language1/html/6399_DEU_XHTML.aspx.

[4] Kim, H.-J.: "Systemization of a Remanufacturing Process and its general Guidelines with a case study of Automobile Alternators", Journal of the Korean Society of Precision Engineering, 2006, 23/ 7:93-100

[5] National Center for Remanufacturing and Ressource Recovery, 15.11.2008, internet source http://www.reman.rit.edu/.

[6] Wiendahl, H.-P.; ElMaraghy, H.; Nyhuis, P.; Zäh, M. F.; Wiendahl, H.-H.; Duffie, N.; Brieke, M.: Changeable Manufacturing - Classification, Design and Operation. CIRP Annals 56 (2007) no. 2, p. 783-809 .

[7] Postawa, A. ; Brandão, D. ; Rubin, R. S. ; Seliger, G.: Configuration of Value Creation Modules, instantiated in Mini Factories for Disassembly of Printed Circuit Boards, in: Proceedings 7th Global Conference on Sustainable Manufacturing, IIT Madras, Indien, 2.-4. Dez. 2009, 227-233.

[8] Sutherland, J. W.; Adler, D. P.; Haapalaa, K. R.; Kumar, V.: A comparison of manufacturing and remanufacturing energy intensities with application to diesel engine production. In: CIRP Annals - Manufacturing Technology, 2008, 57/1:5-8.

[9] Krüger, J.; Nickolay, B.; Heyer, P. & Seliger, G.: Image based 3D Surveillance for flexible Man-Robot-Cooperation, CIRP Annals - Manufacturing Technology 54 (2005), no. 1, p. 19-22.

[10] Radmer, J.; Fuste, P.M.; Schmidt, H. & Kruger, J. : Incident light related distance error study and calibration of the PMD-range imaging camera, IEEE Computer Society Conference on Computer Vision and Pattern Recognition Workshops, CVPRW '08, Anchorage, 23 - 28 June 2008, p. 1-6.

[11] Radmer, J. & Krüger, J. : Depth data based capture of human movement for biomechanical application in clinical rehabilitation use, Health Informatics and Bioinformatics (HIBIT), 2010 5th International Symposium on Health Informatics and Bioinformatics, Lara Antalya Turkey, 20.-22. April 2010, p. 144-148.

Future Studies for Reuse using Mathematical Optimization of the Scenario Technique

A. Fügenschuh[1], P. Gausemeier[2], R. McFarland[2], G. Seliger[2]

[1]Department of Optimization, Konrad-Zuse-Centre for Information Technology, Berlin, Germany

[2]Department for Machine Tools and Factory Management, Technische Universität Berlin, Germany

Abstract

The reuse of equipment when adapted according to the conditions in the area of application, offers an effective solution for meeting the demands of sustainable value creation. The scenario technique is applied to examine reuse and its fields of influence for different levels of development, using several reference countries. The technique is adapted to observe the economical, ecological and social factors of sustainability. At several stages within the scenario technique we apply linear programming based branch-and-cut methods to solve occurring optimization problems and to prove their global optimality property. This paper includes both a sustainability orientation as well as a methodological development using mathematical optimization applied to sub-problems of the scenario technique.

Keywords:

Scenario technique, mixed-integer linear programs, equipment reuse, sustainability

1 INTRODUCTION

The goal of the developed method is, first of all, to integrate aspects of sustainability in the scenario technique, in order to gain a holistic, long-term understanding of the surrounding environments of different levels of development. This shall facilitate the successful implementation of sustainable technologies. Factors of influence are evaluated in respect to their relevance to sustainability, in order to achieve this goal. Furthermore, an additional development to the method has been undertaken. This development serves to reduce the information loss which occurs when reducing the scope of influence factors to a smaller number of key factors. This is achieved by aggregating the influence factors using mixed-integer linear optimization

1.1 Sustainability

The earth´s resources are limited. With a current world population of about 6.9 billion [1] people – 9.2 [2] billion are prognosed for 2050 – merely one billion people belong currently to the so called industrialized world in Europe, North America, Japan, Australia and a few other islands of wealth. China, for example, with a current population of 1.1 billion people, and other countries as well, are focused on closing the existing gap to the industrialized countries. If the lifestyle of the industrialized world and its predominant technologies, is to be adopted by the emerging countries, then resource consumption will exceed every accountable economical, ecological and social limit [1]. Paradigms of sustainable behavior must be institutionalized and followed, in order to ensure that the needs of future generations can be met. One such paradigm is to no longer use more regenerable resources than can be regenerated and to create cycle economies for non-renewable resources [2]. Resource management is especially a challenge to the producing industries. The reuse of equipment and machine components poses a possible solution approach for this challenge.

Challenges to production are not to be observed in isolation, but rather integrated in a global context. Global challenges are those such as climate change, resource exploitation, unjust distribution of wealth and violent conflicts, to name a few. The specific actions of the globalized population in dealing with these challenges is inadequate to date and demands a holistic understanding of these factors as a part of an interactive system [3].

1.2 Scenario-Technique

The scenario technique provides a useful tool for illustrating the interdependencies between factors and therefore for gaining an understanding of systemic behavior [4]. In order to identify sustainable, strategic courses of action, an understanding of particular environments in their future must be gained. According to Börjesson et al., scenarios can be classified as explorative, prognostic or normative [5]. The analysis of the perspective environment for reuse represents both a prognostic and normative scenario, as possible developments are identified, based on the description of the current situation. The fact that the possibility of several possible developments must be taken into account is ensured in the scenario technique through the principle of multiple futures [6]. The resulting consistent multiple futures allow the provision of strategic courses of action, which meet the demands of a normative scenario.

1.3 Challenges with the traditional Method

The scenario technique and its sub-methods for determining key factors take sustainability into consideration only in so much that influence factors can be chosen, which refer to

[1] Estimation of United Nations.

[2] UNO-forecast.

aspects of sustainability. The selection of key factors is not carried out depending on their relevance to the economical, ecological and social dimension of sustainability. Although a binary evaluation of the relevance to the object of examination takes place, it cannot be determined which factors contribute to the fulfillment of which sustainability dimensions. This information could be used to later derive strategic options in respect to economical, ecological and social sustainability using those factors, which have a great relevance to sustainability.

Information is lost through the reduction of influence factors to key factors. Until now, the reduction of influence factors was the result of the influence analysis and a binary relevance evaluation. Through the aggregation of influence factors to pairs, the information loss itself can be reduced, in that the resulting number of influence factors must only be marginally reduced to the number of key factors.

1.4 Scenarios for the reuse of machine components for different levels of development

Considering the fact that the majority of the world population doesn´t live in the industrialized world, it is not sufficient to examine reuse only on the highest level of technological advance. It is rather of vital importance to adapt technologies for reuse to the level of development in which they are to be utilized, in order to secure this utilization [7]. An understanding of the surrounding environment of the different levels of development must be primarily gained. For this purpose, scenarios for the surrounding environment using three reference countries were developed in the current DFG sponsored project "Szenarien für die Wiederverwendung von Maschinenkomponenten auf unterschiedlichen Entwicklungs-niveaus" (Scenarios for the Reuse of Machine Components on different Levels of Development). Sierra Leone was examined to represent the lowest level of development; it is one of the poorest countries in the world and has been highly shaped by the past civil war. The transition country India was examined for a median level of development and Germany for a high level of development.

2 METHOD FOR DETERMINING THE KEY FACTORS

The developed method for determining key factors can be classified as a sub-method of scenario field analysis according to J. Gausemeier [6]. Figure 1 shows the five step process, in which the sustainability relevance is created and the information loss in reducing the influence factors to key factors is itself reduced through the aggregation of influence factors.

Pair wise comparison of influence factors

In the initial step, influence factors are pair-wise compared and evaluated according to the method of J. Gausemeier so that their systemic characteristics are illustrated with the help of the active and passive sum [6].

Analysis of the relevance to the dimensions of sustainability

Bearing the guiding theme of sustainability in mind, influence factors are not, as in the classic method, weighted against each other using a binary evaluation, but rather the relevance in reference to the dimensions of sustainability is examined. The relevance to economical, ecological, and social sustainability is evaluated for each influence factor. An example for such an evaluation is shown in Section 2.1.

Aggregation of Influence Factors

Influence factors are aggregated in order to minimize the information loss during their reduction to key factors. The aggregation occurs through a pair-wise combination of the factors according to the similarity of their column and row values in the influence matrix. The mathematical procedure as well as the results of the aggregation are described in detail in Section 2.2.

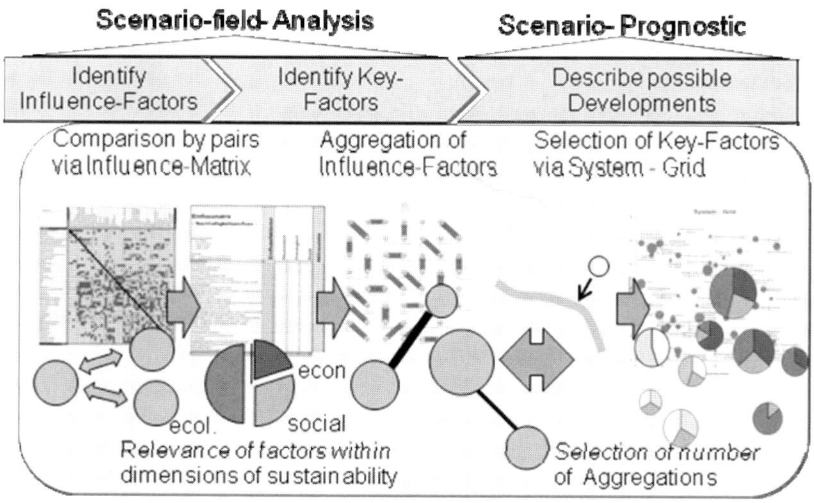

Figure 1: Procedure for determining Key Factors

Selection of the Number of Aggregations

The percentaged similarity of the factor pairs and the number thereof is represented in a diagram. The mean similarity of the factor pairs decreases with the amount of pairs created. It is necessary to determine the point at which the similarity drastically decreases the so-called elbow point. This point determines the number of pairs to be created. Figure 2 illustrates the procedure for determining the number of pairs. If no definite elbow point exists, then the aggregation is stopped at a previously defined minimum similarity (e.g. 80%). The arrival at this threshold then implicitly determines the number of pairs.

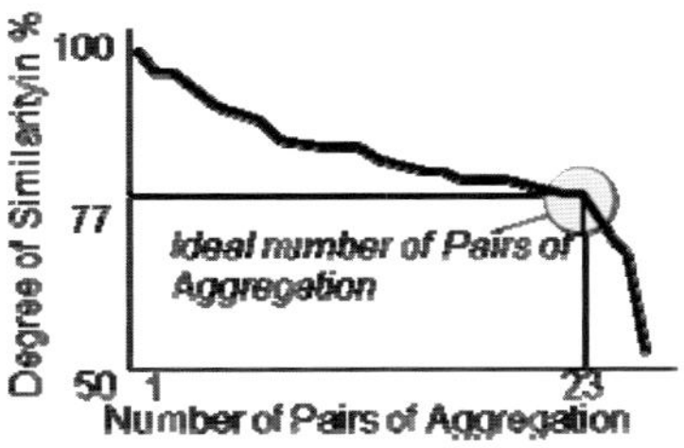

Figure 2: Scree-Diagram

Selection of Key Factors using the System Grid

The information gained in the pair-wise comparison, the relevance analysis and the aggregation are carried over to a system grid in the final stage. This serves as the basis for the selection of key factors in consideration of both the interconnectedness of the factors and their sustainability relevance. The system grid and the results which can be derived from it are described in detail in Section 2.3.

2.1 Relevance of the Factors to the Dimensions of Sustainability

The relevance of singular factors in respect to economical, ecological and social sustainability was evaluated in a workshop. Table 1 shows an excerpt from the results of this evaluation.

Spectrum: 0 no relevance 1 low relevance 2 relevance 3 high relevance		**Dimensions of sustainability**			
		Econ.	**Ecol.**	**Social**	**Sum**
Influence-Factors	**Development GDP**	3	0	2	5
	Influence: Civil War	3	0	3	6
	...				
	Impact of Climate Change	**1**	**3**	**0**	**4**

Table 1: Excerpt from Relevance Analysis

The influence factor *Development of the Gross Domestic Product measured in USD per Capita* has a great economic relevance but a social relevance as well. The social relevance is less than that of the economic. Through economic growth and the associated standard of living, the surrounding environment and the quality of life of the population is influenced.

The influence factor *Effect of Civil War* has both great economic and social relevance. The social structures are significantly shaped by the civil war and its aftermath [UNI10]. The resulting, instable political and economic conditions led to the lack of almost any kind of economy.

The *Transformations through Climate Change* effect mainly the ecological sustainability, but also in part the economical. Global warming, for example, leads to draughts, which effect the agriculture. Agriculture is a main economic source in Sierra Leone, drawback here would have a significant economic impact. The relevance is evaluated as minimal, as these interdependencies are non-direct.

2.2 Aggregation of Influence Factors

In the following the mathematical model, used in the optimization process, is presented followed by a discussion of the obtained results.

Mathematical Model

The central idea of our appraoch is an automatic aggregation of pairs of influence factors based on a similarity measure for the rows and the columns. This measure is defined in such way that it is able to describe the „distance" between the row and column data belonging to two different factors.

Assume that we have a given set of influence factors $F = \{1,2,\ldots,n\}$ and a given influence matrix $A = (a_{i,j})_{1 \le i, j \le n}$. We first define the net similarity $\mathrm{nsim}_{k,l}$ between factor k and factor l as the sum of the column similarity (or active similarity) $\mathrm{asim}_{k,l}$ and the row similarity (or passive similarity) $\mathrm{psim}_{k,l}$:

$$\mathrm{nsim}_{k,l} := \mathrm{csim}_{k,l} + \mathrm{rsim}_{k,l} \tag{1}$$

where the active similarity itself is defined as

$$\mathrm{asim}_{k,l} := \sum_{i \in F} (a_{i,k} - a_{i,l})^2 \qquad (2),$$

and the passive similarity is defined as

$$\mathrm{psim}_{k,l} := \sum_{i \in F} (a_{k,i} - a_{l,i})^2 \qquad (3).$$

By definition the net similarity is a real positive number. The closer to zero its value is, the more similar the two corresponding factors are, that is, the fewer entries in the respective rows and columns are different to each other. Vice versa, if there are a lot of dissimilarities, a high value would be assigned.

We now define a normalization of the net similarity, which will then give the total (relative) similarity. We denote by m, the minimal, and M the maximal net similarity value among all of them. Using the following re-scaling we transfer the net similarity into the total similarity:

$$\mathrm{sim}_{k,l} := 100 \cdot \left(1 - \frac{\mathrm{nsim}_{k,l} - m}{M - m}\right) \quad (4)$$

By definition the similarity $\mathrm{sim}_{k,l}$ is a real number between 0 and 100, where 0 stands for the greatest dissimilarity and 100 corresponds to the greatest similarity, in both cases relative to the entity of all net similarity values.

The so defined similarity measure can easily be extended from pairs of factors to groups of factors having more than two members. In this case it defines a measure for the similarity of all group members among each other. In the case of triples, for instance, we would define it as follows:

$$\mathrm{sim}_{i,j,k} := \tfrac{1}{3}(\mathrm{sim}_{i,j} + \mathrm{sim}_{j,k} + \mathrm{sim}_{i,k}) \quad (5)$$

The final goal of the aggregation is to find pairs of influence factors which have a high similarity. That means, that their influence on other factors is similar, as well as the possibility to be influenced by other factors is also similar. In this case it can be justified to treat the aggregated pair as a new factor in the subsequent analysis of the scenario technique.

In order to find such pairings we make use of methods from the field of discrete, combinatorial optimization, in particular, linear mixed-integer programming. These methods have the advantage that besides a solution they return a rigorous proof of its global optimality. That means, one can be assured that there exists no better pairing other than the one computed by these methods. In order to be able to apply them, we have to first formulate the exact task as an algebraic mathematical model.

In the sequel we restrict ourselves to the case of finding groups of tuples (or pairs) and triples. The presented model can easily be extended to the case of groups with more members (such as quadruples, etc.). But we remark that for such large groups it might be difficult to find a common meaning or interpretation. For this reason we use this model only for tuples.

We introduce binary decision variables $x_i, y_{i,j}, z_{i,j,k} \in \{0,1\}$ for all influence factors $i,j,k \in F$. Here $x_i = 1$ means that factor i is not grouped at all in the solution; $y_{i,j} = 1$ means that the two factors i,j are grouped together; $z_{i,j,k} = 1$ means that the three factors i,j,k are together in one group.

Each factor $i \in F$ must be assigned to exactly one such group, i.e., the factor is either alone, or in a tuple, or in a triple:

$$\forall i \in F: \; x_i + \sum_{j \in F} y_{i,j} + \sum_{j,k \in F} z_{i,j,k} = 1 \quad (6)$$

The number of generated clusters can be controlled by the user. A number N is specified in advance, and the number of clusters has to be equal to that number:

$$\sum_{i \in F} x_i + \sum_{i,j \in F} y_{i,j} + \sum_{i,j,k \in F} z_{i,j,k} = N \quad (7)$$

The ultimate goal is to find pairings that maximize the sum over all individual pairs that are involved in the solution. That means, we want to maximize the following objective function:

$$2 \cdot \sum_{i,j \in F} \mathrm{sim}_{i,j} \cdot y_{i,j} + 3 \cdot \sum_{i,j,k \in F} \mathrm{sim}_{i,j,k} \cdot z_{i,j,k} \to \max \quad (8)$$

Summing it up, we have to deal with the following problem:

$$\begin{array}{ll} \max & (8) \\ \text{s.t.} & (6),(7), \\ & x \in \{0,1\}^F, y \in \{0,1\}^{F \times F}, z \in \{0,1\}^{F \times F \times F}. \end{array} \quad (9)$$

This problem (9) is a mixed-integer linear program. It can be solved efficiently using a linear programming based branch-and-cut-approach. To this end we first relax the integrality condition in (9) and obtain the following linear program:

$$\begin{array}{ll} \max & (8) \\ \text{s.t.} & (6),(7), \\ & x \in [0,1]^F, y \in [0,1]^{F \times F}, z \in [0,1]^{F \times F \times F} \end{array} \quad (10)$$

Such linear program can be solved with Dantzig's simplex algorithm [8]. In general some of the integer variables will assume fractional values in an optimal solution of (10), i.e., they are not 0 or 1, but somewhere in between. Hence we have to re-introduce the integrality condition. On the one hand it is possible to add cutting planes to the relaxation (10). A cutting plane, or cut, is an additional linear inequality which separates the optimal but fractional solution from the set of all feasible integer solutions. On the other hand it is possible to apply branching. Then the problem will be split into two subproblems, where the fractional variable will be round down to 0 in one subproblem, and round up to 1 in the other. Then the fractionality on that particular variable will no longer occur, but at the price of solving now two problems instead of one. Both approaches can be combined into an integral approach known as branch-and-cut. For further details we refer to the literature, for instance [9].

The above outlined branch-and-cut method is implemented in the numerical solver SCIP, which we use to solve problem (9) [10]. For instances with up to 100 influence factors this solver finds optimal solutions within very short CPU time.

Results

From the 53 influence factors, 23 pairs were built in order to reduce the information loss. The degree of similarity used for the aggregation resulted from the evaluations in the influence matrix. The pair-building is not undertaken in steps but globally by solving the optimization problem described in the previous chapter. In this manner, the best possible pairing of the factors can be guaranteed. The degree of similarity lies in this case by 77%. The thickness of the connections, shown in Figure 3, represents the degree of similarity between the evaluations.

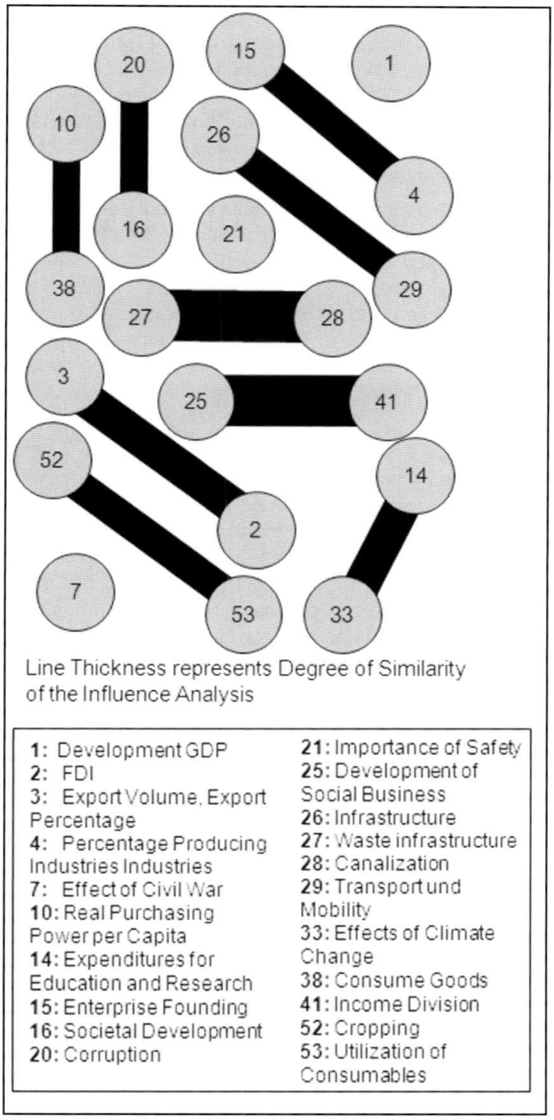

Figure 3: Excerpt from the Aggregation of Influence Factors

Both the factors *Waste Disposal Infrastructure* and *Canalization* have, with 100%, the greatest relative similarity. The similarity results from the strong correlation to further, health related factors. In the mid-field similarity range, with 80% similarity, lies the pairing of *Infrastructure* and *Transport and Mobility*. The similarity results significantly from the effects on the economic factors as well as further, transportation relevant factors such as *Automobile* and *Motorcycle*. Seven factors were not paired, including *Development Gross Domestic Product* and *Effects of Civil War*. Both factors have strong interdependencies with all other factors. This means that they were seldom evaluated in the influence matrix with 0 = “no Influence”. Resulting from this and the large active and passive sums is a strong systemic integration. As the evaluation of these factors greatly deviates from that of the remaining factors, an aggregation of these factors would result in a large information loss and is therefore not reasonable.

In the most cases, factor pairs with a clear correlation could be built through the aggregation. The aggregation therefore presents a meaningful method for illustrating the information included in the factor catalogue and the influence matrix. Some pairs were built with no apparent similarity in terms of the factors themselves, for example *Funding for Research and Education* and *Transformations through Climate Change*. The similarity in this case results from the non-influenceable character of the factors.

2.3 Selection of Key Factors/ Results

The evaluation of influences, the relevance to the dimensions of sustainability and the aggregation of the influence factors are transferred to the system grid, show in Figure 4.The system grid serves as a basis for the selection of key factors in consideration to the dimensions of sustainability.

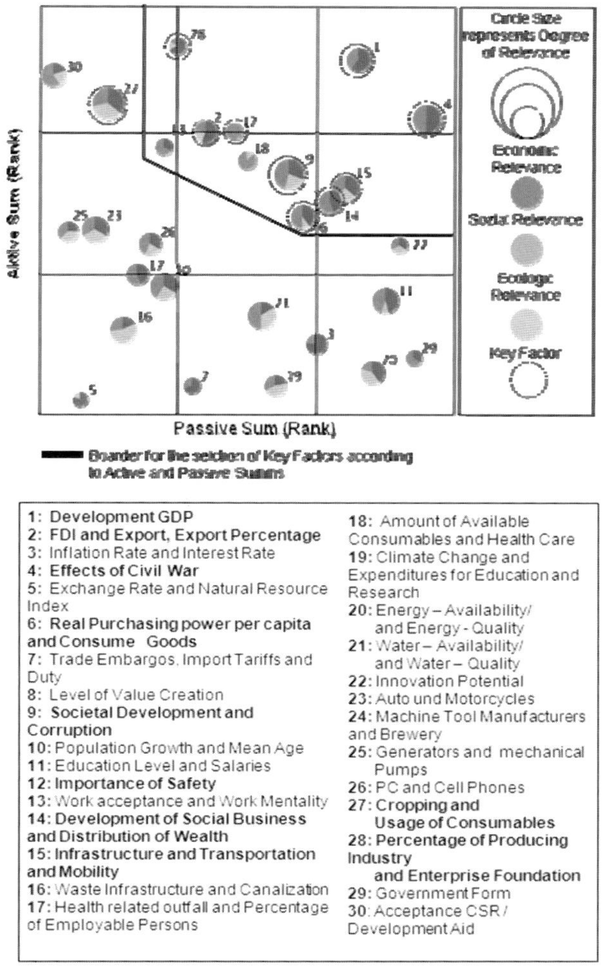

Figure 4: System Grid

The active and passive sums are drawn upon as primary criteria for the selection of key factors. The line in the illustration shows, which of the aggregated influence factors would be key factors, when this criterium alone would be considered. The relevance to the dimensions of sustainability serves as secondary criteria. The great relevance of the factor *Cropping/Consumable Goods Usage* in all three dimensions leads to inclusion in the key factors despite its relative small passive sum. The factors *Work*

Acceptance/Work Mentality was not included as a key factor despite its large active and passive sum, due to its marginal relevance to sustainability.

3 EVALUATION OF RESULTS

In the following chapter, the similarities and differences between key factors resulting from the classical and the proposed method are illustrated.

Three factors, which were not aggregated, were identified as key factors using both methods. These factors are:

- *Development of the Gross Domestic Product*
- *Effect of Civil War*
- *Significance of Safety*

Two key factors resulting from the classical analysis were no longer considered as being key factors though their aggregation and minimal relevance to sustainability. Both of these factors do not show a significant active and passive sum using the classic method, but were combined and chosen as key factors due to their particular relevance to the examined region. An intuitive aggregation took place, which was validated though the mathematical aggregation. These factors are:

- *Energy Supply*
- *Energy Quality*

The new approach allows a larger number of influence factors as key factors through the aggregation. The new key factors are:

- *Real Purchasing Power and Consumable Goods*
- *Cropping/Consumable Goods Usage*

Furthermore, five key factors previously indentified using the classical method were again indentified as key factors and were added to an additional factor through the aggregation:

- *Distribution of Wealth (Development of Social Business* and *Distribution of Wealth)*
- *Transportation and Mobility (Infrastructure* and *Transportation and Mobility)*
- *Societal Development (Societal Development* and *Corruption)*
- *Establishment of Businesses (Proportion of Producing Industry* and *Establishment of Businesses)*
- *Exports, Proportion of Export (FDI* und *Exports, Proportion of Export)*

Noteworthy is that no aggregated key factors are present, which consist of two factors, already identified as key factors using the classical method. No key factors were merged. In the selection process of key factors using the classical method, several influence factors were not chosen, despite their relative large active and passive sums. This is to be interpreted that strong overlapping with existing key factors was intuitively identified. The problem lies in the fact that these factors were not represented using the classical method, resulting in information loss.

4 SUMMARY AND OUTLOOK

The implementation of the newly developed method and the comparison with the results of the existing method, based on the same data basis, has shown that an analysis of the sustainability relevance is beneficial to the selection of key factors. The factor pair *Cropping* and *Consumable Goods Usage* is taken into account due to its large relevance to sustainability. Further factor pairs were omitted, due to a lack in sustainability relevance. The selection of key factors according to their relevance to sustainability was successful.

Furthermore, the information loss could be reduced. Eight key factors from ten, who were identified with the classic method, were also identified using the new approach. The definitive difference lies in that five influence factors, which were previously not taken into account, are represented through the building of factor pairs. These factors were not merely added, but were meaningfully combined with other factors according to their similarity. In addition, two further factor pairs were added.

The analysis of sustainability can be improved based on the result achieved. One possible approach is to perform a separate binary relevance analysis for each sustainability dimension. Three relevance matricies would then be necessary, in which the influence are compared pair-wise factors in respect to their importance to the observed dimension. The central questions for the three matricies would then be: Is the influence factor for the

1) economical,
2) ecological,
3) social sustainability more important than the other?

The method should be applied, in the framework of the existing scenario project for the reuse of machine components, for mean (reference country India) and high level development (reference country Germany) for its further verification.

5 REFERENCES

[1] Seliger, G., 2007, Aufbruch zu nachhaltiger Produktionstechnik, ZWF Zeitschrift für wirtschaftlichen Fabrikbetrieb, Jahrgang: 102. , Nummer: 4/2007.

[2] Seliger, G., 2006, Sustainability in Manufacturing – Recovery of Resources in Product and Material Cycles, Springer, Berlin.

[3] Hermann, C., 2010, Ganzheitliches Life Cycle Management: Nachhaltigkeit und Lebenszyklusorientierung in Unternehmen, Springer, Berlin.

[4] Institut für Zukunftsstudien und Technologiebewertung, 2008, Methoden der Zukunfts- und Szenarioanalyse - Überblick, Bewertung und Auswahlkriterien , WerkstattBericht Nr. 103, (Kosow/ Gaßner), Berlin, September 2008.

[5] Börjeson, L., Dreborg, K., Ekvall, T., Finnveden, G., Höjer, M., 2005, Scenario types and techniques: Towards a user's guide, Futures, 2006.

[6] Gausemeier, J., Plass, C., Wenzelmann, C., 2009, Zukunftsorientierte Unternehmensgestaltung – Strategien, Geschäftsprozesse und IT-Systeme für die Produktion von morgen, Carl Hanser Verlag, München.

[7] Sachs, J. D., 2005, The End of Poverty - Economic Possibilities for Our Time, Penguin Books, New York.

[8] Dantzig, G., Thapa, M., 1997, Linear Programming, Springer Verlag, New York.

[9] Nemhauser, G., Wolsey, L., 1998, Integer and Combinatorial Optimization, J. Wiley & Sons, New York.

[10] Achtenberg, T., 2007, Constraint Integer Programming, PhD Thesis, TU Berlin, 2007.

Remanufacturing Process Issues of Fuel Injectors for Diesel Engines

D.H. Jung, A. Gafurov, Y.K. Seo, C.H. Sung
Korea Automotive Technology Institute, Cheonan, South Korea

Abstract
Korean automobile industry is rapidly growing with annual domestic output already beyond a million vehicles. This fact is closely tied with constant demand on raw materials, energy consumption and, of course, concerns over exhaust emissions to atmosphere. Majority of automobile parts are usually used to be recycled and by this time the list of remanufactured parts are increasing. Remanufacturing is relatively new step in the industry. Several issues still present in remanufacturing of automobile parts such as information of remanufacturing, remanufacturing technology, absence of unified standards. This paper introduces diesel engine fuel injector remanufacturing process. Failure mode of the injectors is analyzed. Performance test methods are discussed by comparing two different OEM remanufacturing specifications. Also, simple process is considered from an independent remanufacturing company.
Keywords:
Remanufacturing, Injector, Standard, Diesel engine.

1 INTRODUCTON

An injector is one of the important parts of an engine. Injectors deliver fuel to cylinders of internal combustion engine. The fuel is sprayed through an injector nozzle, typically at high pressure, to improve the mixing of fuel with air and therefore the combustion efficiency. The most failure prone parts of the injector are nozzle and valves. They may wear off as a result of repetitive contact or damaged by cavitation effect under high pressure [1].

Recent trend of efforts in the manufacturing field are directed to remanufacturing, which is between recycling and reuse processes. Reuse of used parts is the least expensive, time and energy saving method [2]. However, it does not guarantee the reliability of the reused part, since no quality inspection is performed. Recycling is more complex process which involves discarding the used parts and processing them into raw material and then manufacturing a new part from the beginning. Comparing to those two, remanufacturing process is more promising by saving the raw materials, energy and ensuring the reliability of remanufactured parts.

Remanufacturing process includes disassembling, cleaning, testing, reconditioning, and assembling. Nowadays, more than 35 items of the vehicle are being remanufactured in the US, more than 21 items in Japan and more than 25 items in Korea. Remanufacturing procedure of some items have already been certified, while the other procedures are under examination and remanufacturing procedure of some parts like differentials, lower arm, clutches, turbochargers, and injectors are under study.

Market research for the past three years, starting from 2005 to 2007, shows that remanufacturing section also increasing as well as newly manufactured injectors. The global market of new injectors had risen from $39.48 billion in 2005 to $43.412 billion in 2007. Domestic (Korean) market shows increase from $2.16 billion in 2005 to $2.386 billion in 2007. In turn, the domestic market volume of remanufactured injectors had risen from $216 million in 2005 to $238.4 million in 2007.

In this paper two different injectors from BOSCH and DELPHI are considered. In the local market a new Bosch made injector costs $192, whereas remanufactured injector costs $80. Likewise, a new injector made by Delphi costs $200, whereas a remanufactured injector of this brand costs around $105. With expected quality assurance, the price of remanufactured injector can be very competitive. Also, large amount of raw material processing will be avoided.

The main issue with the injector remanufacturing is the absence of certified procedure. Therefore, two different approaches used by local remanufacturing companies are analyzed. Remanufactured injectors are tested for performance using certain procedures. The test procedures are supplied with the test equipment from both Bosh and Delphi. Prior to discussing the remanufacturing process, a failure mode analysis of the injectors will be given in the following chapter.

2 INJECTOR FAILURE MODE ANALYSIS

Nowadays in Korea during replacement of failed part a car repair shop offers compensation for leaving the item while installing a new one. Failed parts are collected by remanufacturing companies for consecutive processing. Currently there is no method for documenting the lifetime of the failed parts. For this reason it appears impossible to predict the usage rate of the part since usage conditions and mileage are unknown. However, manufacturer of new injectors gives three years or 60,000 km of aftermarket warranty, whereas remanufactured injectors are given one year or 20,000 km of aftermarket warranty.

Studied injectors were provided by a remanufacturing company and were randomly chosen. Summary of the Failure Mode and Effect Analysis (FMEA) is given in Table 1. Examination of injectors for the failure mode revealed that mostly damaged parts are solenoid coil, control valve, spring, sealing, needle valve and nozzle. The most often-happening failure is contamination of the nozzle tip. Close examination under Scanning Electron Microscope (Figure 2) shows severely contaminated nozzle tip. This kind of damage leads to spray pattern distortion, which in turn affects fuel-air mixture and uniform spraying of the fuel inside the combustion chamber. Thus, power loss will occur due to inefficient burning process. According to the Hazard rate of FMEA, next frequently happening failures are related with needle valve and control valves. Failure of the solenoid coil also considered being very severe, however this failure is not that frequently happening. So, overall hazard rate is small. Lastly, sealing and spring failures are concluded being least hazardous in this analysis.

Table 1: Summary of FMEA

Item	Function	Potential damage form	Major effects of damage	Severity rate	Main cause of damage/ occurrence mechanism	Occurrence rate	Present process control	Detection rate	Hazard rate
Solenoid coil	A part which moves high pressure valve and control valve.	Short-circuit, fracture.	Poor engine start.	7	Fire phenomenon of solenoid coil. Coil damage related with overvoltage.	3	- Resistance/ insulation test	5	105
Control valve	Flow regulating valve.	Wear, contamination by particles	Engine power deterioration, engine shutdown during acceleration.	8	Appearance of substances in the fuel. Control valve contamination. Poor quality of control valve material. Control valve manufacturing abnormality.	6	-Dimension check, visual inspection; -Dynamic flow test; -Durability test; -Spray opening pressure test.	3	144
Spring	A part which moves the valve.	Deformation.	Engine disharmony, excessive backleak appearance, reduction of dynamic flow rate.	3	Deformation due to spring fatigue. Poor quality of spring material. Poor spring dimensions.	2	-Dimension check, visual inspection; -backleak test; -Spray opening pressure test.	2	12
Sealing	Upkeeps the pressure	Wear, breakage	Poor engine start; Engine output power loss during acceleration.	5	Ingress of contamination to fuel. Appearance of sealing fatigue wear. Sealing dimension defects during manufacturing.	2	-Dimension check, visual inspection; backleak test;	3	30
Needle valve	Preserving and spraying the fuel	Wear, side wear	Poor engine start and irregular condition. Engine power deterioration	8	Contamination development on the needle valve. Side wear appearance on the needle valve.	7	-Dimension check, visual inspection; -Dynamic flow rate test; -Durability test.	4	224
Nozzle	Keeps the spray pattern	Wear, sinking, clogging	Abnormal spray pattern. Engine power deterioration. Faulty engine start. Irregular condition after engine start.	9	Appearance of carbon on the nozzle. Thickening of nozzle tip due to high temperature in the combustion chamber.	9	-Dimension check, visual inspection; -Dynamic flow rate test; -Endurance test; -Spray opening pressure test	5	405

Note: 1) Severity rate is 1~10 / Occurrence rate is 1~10 / Detection rate is 1~5

2) Hazard rate = *Severity rate* × *Occurrence rate* × *Detection rate*

Figure 1: Failure Mode Analysis of the injector.

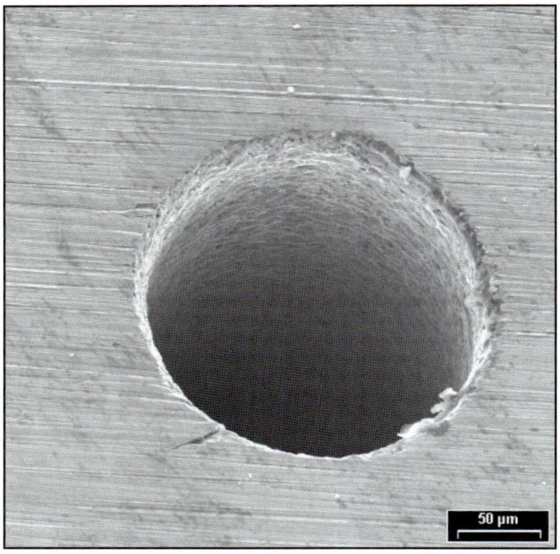

a) New nozzle hole

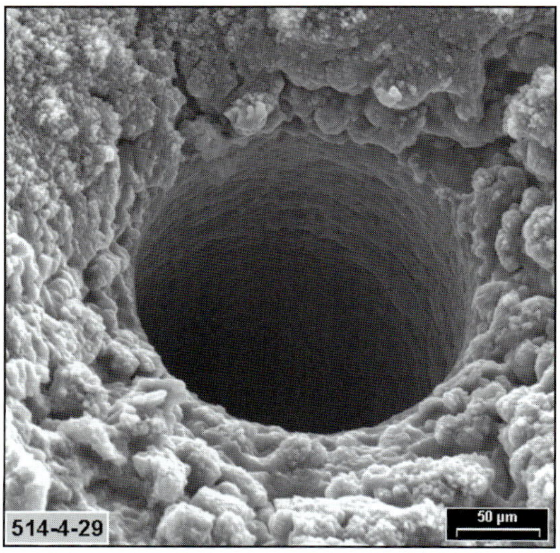

b) Contaminated nozzle hole

Figure 2: SEM analysis of the nozzle tip: *a) new nozzle hole; b) contaminated nozzle hole.*

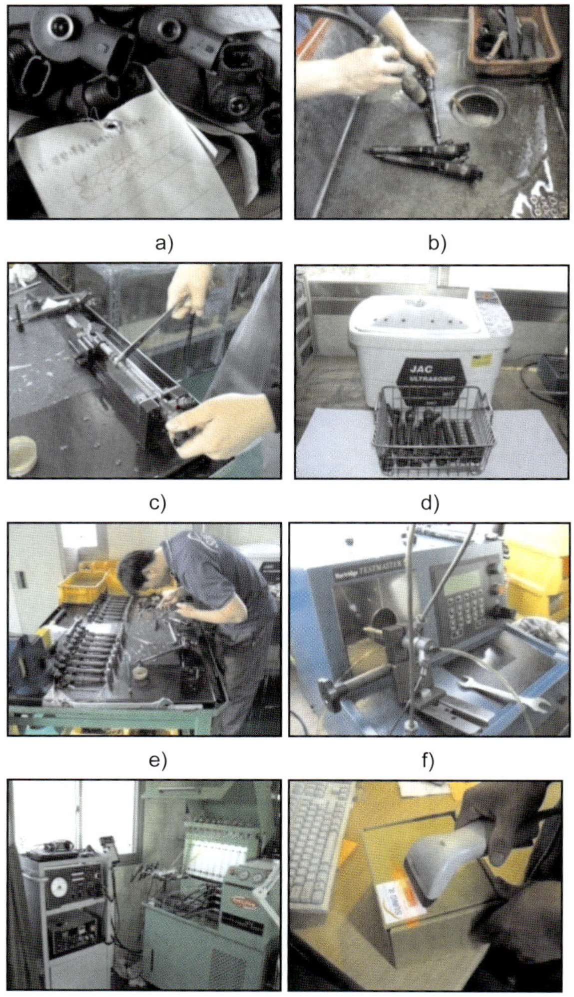

Figure 3: Basic remanufacturing steps of the first company: *a) Collection of injector core; b) washing process; c) disassembling; d) cleansing under high frequency; e) inspection and assembling; f) initial performance test; g) secondary performance test; h) packing and barcode registering.*

3 INJECTOR REMANUFACTURING PROCEDURE

3.1 Approach from Company A

Currently there are two major remanufacturing companies in the local Korean market. They use different approach to remanufacturing and performance testing processes. Approach of the *Company A* starting from injector collection and final packing are shown in Figure 3. Process starts with collection of injector core, washing off the dirt, oil and other contaminants, disassembling, cleansing process under high frequency, inspection of elements for damage, initial and secondary performance tests and finally packing.

Schematic view of the remanufacturing procedure is outlined in Figure 4. According to the procedure, collected injectors are visually checked and sorted out according to their types. Then used injectors are disassembled and observation for failed parts is performed. Failed parts are scrapped or passed for washing step if no visible sign of failure is found. Detailed inspection of parts is performed in the next step and failed components are replaced with either with repaired or brand new component if necessary. Then the components are assembled which is followed by advanced inspection using performance test procedure such as leak test and spray pattern uniformity. If a unit is successfully passed required tests, it is packed by marking with barcode and transferred to storage room for further shipment to end-user customers.

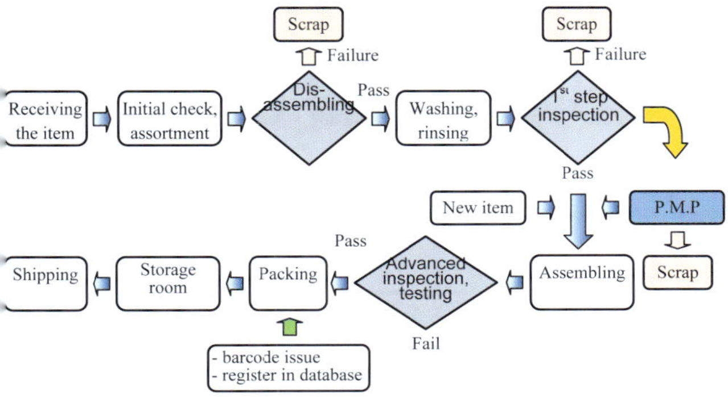

Figure 4: Remanufacturing procedure from *Company A.*

3.2 Approach from Company B

Remanufacturing process from Company B is introduced in Figure 5. Firstly, used injectors are collected and classified depending on types and/or models. Then, exterior washing is performed to get rid of contaminants, oil, dirt, etc. The next step involves visual inspection by measuring resistance of solenoid valve and spray angle. For that purpose, special calibration oil is used under high pressure. If a sample is successfully passes this inspection, it will be handed over to performance test based on test code. In case of abnormality the sample is disassembled and damaged components will be repaired or replaced. After assembling, final performance test is performed based on test code. This code is provided by manufacturer of test equipment. Finally, remanufactured injectors are packed in shipped to customers.

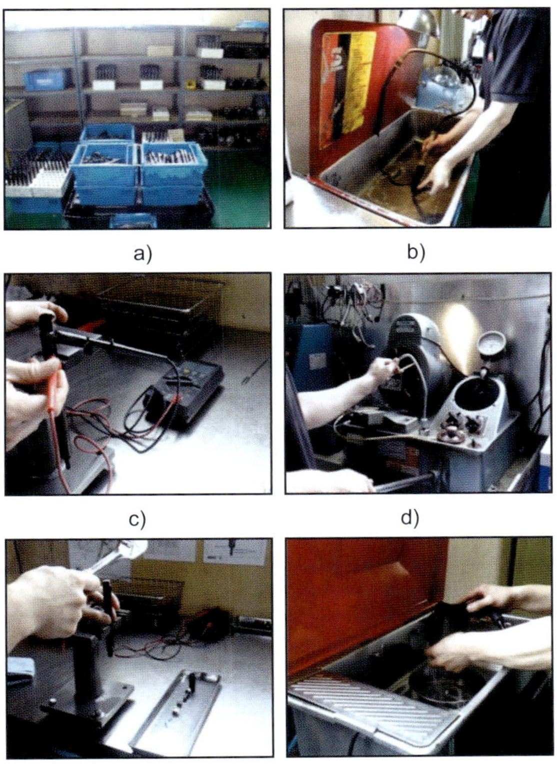

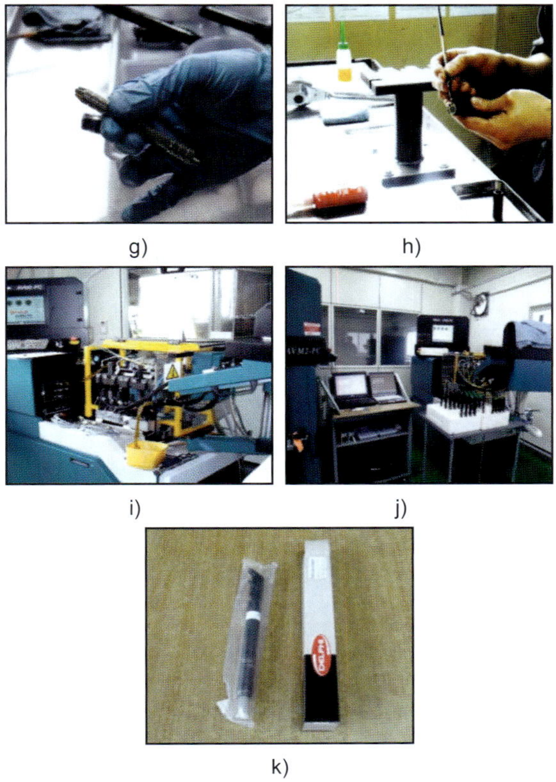

Figure 5: Remanufacturing process of the second company: *a) Collection of injector core; b) washing process; c, d) inspection – solenoid valve resistance measuring, spray pattern observation; e) disassembling; f) washing off the components; g, h) replacement of failed parts and assembling; i, j) performance test; k) packing.*

Remanufacturing procedure is represented in schematic diagram in Figure 6. In contrast to the previously explained procedure from Company A, this procedure has more rigorous visual checking prior to disassembling. As explained, it involves resistance measurement and nozzle hole test for spray uniformity using special calibration oil. Moreover, another distinguishing point is integration of quality control code C21 and C31.

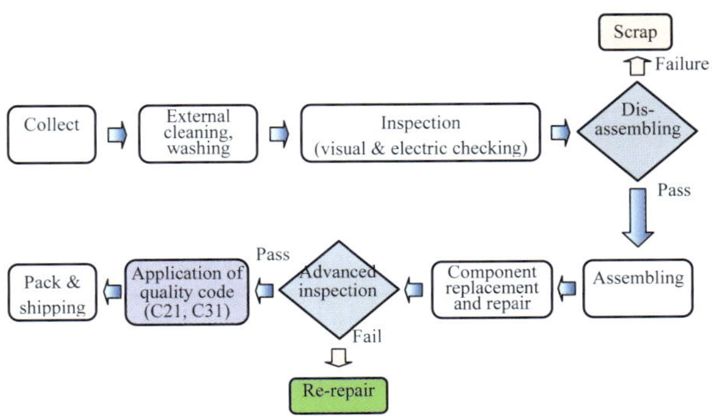

Figure 6: Remanufacturing procedure from *Company B.*

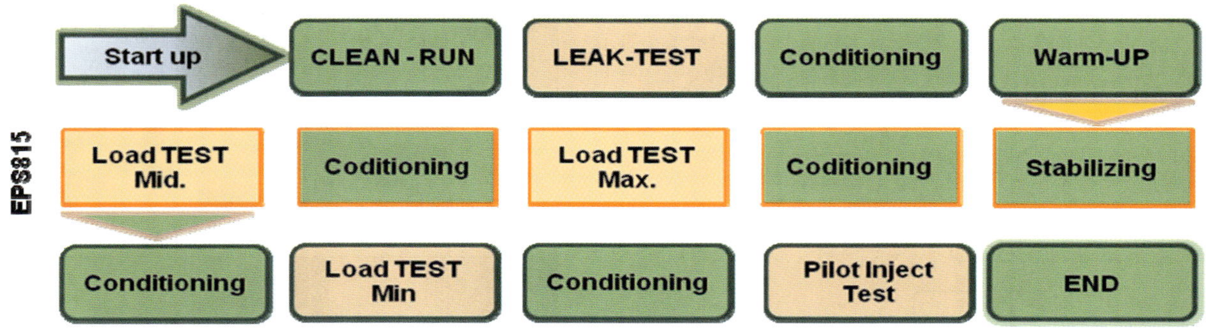

Figure 7: Remanufactured injector test process from Company A

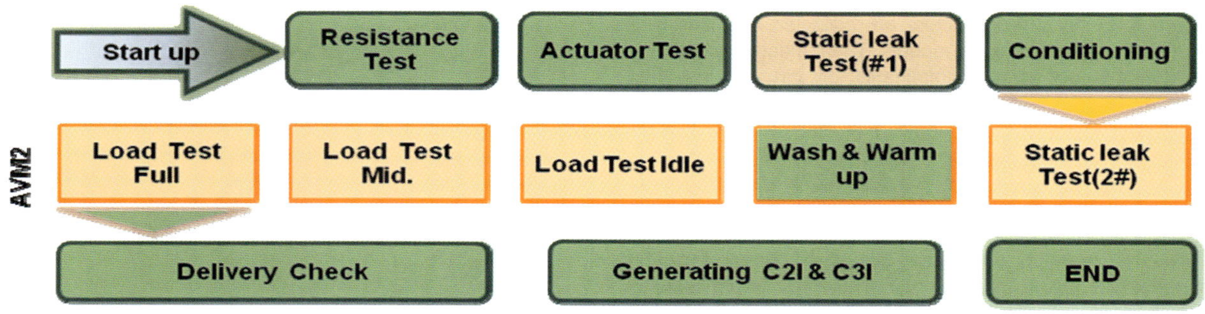

Figure 8: Remanufactured injector test process from Company B.

Table 2: Injector test specification of Company A

1. Leak TEST		2. Load Test Max.		3. Load Test Mid.		4. Load Test Min.		5. Load Test Pilot	
1000rpm/1000bar/ (60s:Wa/300s:Me)		1000rpm/1350bar/850 µs		1000rpm/800bar/600 µs		1000rpm/250bar/600 µs		1000rpm/800bar/160 µs	
Returned Fuel		Injected Fuel		Injected Fuel		Injected Fuel		Injected Fuel	
(㎣/Hr)		(㎣/Hr)		(㎣/Hr)		(㎣/Hr)		(㎣/Hr)	
Setting	Tolerance	Setting	Tolerance	Setting	Tolerance	Setting	Tolerance	Setting	Tolerance

Table 3: Injector test specification of Company B

1. Idle speed						2. Mid Load						3. Full Load						4. Static Backleak			
400rpm/300bar/750 µs						1000pm/1000bar/800 µs						1500pm/1600bar/1000 µs						1000pm/1600bar/0 µs			
Delivery		Backleak		Rest time		Delivery		Backleak		Rest time		Delivery		Backleak		Rest time		Delivery		Backleak	
(㎣/st)		(㎣/st)		(µs)		(㎣/st)		(㎣/st)		(µs)		(㎣/st)		(㎣/st)		(µs)		(㎣/st)		(㎣/sec)	
Min	Max	Min	Max	Min	Max	Min	Max	Min	Max	Min	Max	Min	Max	Min	Max	Min	Max	Min	Max	Min	Max

4 PERFORMANCE TEST

Remanufactured injectors are subjected to performance test following the sequences shown in Figure 7 and Figure 8. Applied test specification by *Company A* is given in Table 2 and specification used by *Company B* is given in Table 3. Introduced test specifications differ in terms of load level sequence, applied pressure, vehicle rpm and measurement method. For example, *Company B* tests the injectors under variable rpm whereas *Company A* does testing under fixed rpm. The test results are given in Table 4 and Table 5.

From the test results of Company A it can seen that the remanufactured injector #1 test on KIA Sorento WGT did not pass the maximum load test. According to the specification it should be 64.7 ㎣/Hr with the tolerance of 4.0 ㎣/Hr. However, the result shows 57.7 ㎣/Hr.

Test results from Company B, given in Table 5, are within acceptable ranges comparing to a new injector test results. Static backleak test is omitted in this report, but it is expected to be well within margins, judging on dynamic backleak data.

Table 4: Injector test results of Company A

Classification		Leak Test (㎣/Hr) Injection vol. (tolerance)	Max. Load Test (㎣/Hr) Injection vol. (tolerance)	Mid Load Test (㎣/Hr) Injection vol. (tolerance)	Min. Load Test (㎣/Hr) Injection vol. (tolerance)	Pilot Load Test (㎣/Hr) Injection vol. (tolerance)
0445110-064 HYUNDAI Santafe WGT	SPEC.	35.0 (35.0)	51.4 (4.0)	24.1 (3.0)	3.5 (1.5)	2.3 (1.3)
	New injector	3.2	52.7	23.4	3.4	2.5
	Reman. injector #1	8.5	51.9	24.9	3.7	3.1
	Reman. injector #2	5.9	52.0	25.5	3.6	3.6
0445110-126 HYUNDAI Santafe VGT	SPEC.	35.0 (35.0)	59.2 (4.0)	22.2 (1.0)	3.3 (2.0)	1.6 (1.3)
	New injector	3.6	59.3	21.5	3.0	1.5
	Reman. injector #1	15.1	59.2	22.9	3.4	2.3
	Reman. injector #2	5.4	60.7	22.7	3.2	1.9
0445110-185 HYUNDAI Porter 2	SPEC.	35.0 (35.0)	55.6 (4.0)	27.5 (3.4)	8.1 (2.3)	1.9 (1.3)
	New injector	2.2	55.1	25.1	7.7	1.5
	Reman. injector #1	5.5	57.7	26.2	7.2	1.8
	Reman. injector #2	8.5	53.8	27.6	7.7	2.1
0445110-092 KIA Sorento WGT	SPEC.	35.0 (35.0)	64.7 (4.0)	15.0 (3.0)	5.0 (2.0)	1.5 (1.2)
	New injector	9.2	61.2	15.9	5.4	1.6
	Reman. injector #1	6.9	**57.7**	15.5	3.2	1.4
	Reman. injector #2	7.9	65.1	19.5	5.1	2.5
0445110-233	SPEC.	35.0 (35.0)	87.6 (3.4)	19.8 (2.8)	4.0 (2.0)	1.3 (1.0)
	New injector	7.6	84.6	21.9	4.0	1.6

Table 5: Injector test results of Company B

Classification		Idle Speed Test			Mid Load Test			Full Load Test		
		400rpm/300bar/750 μs			1000rpm/1000bar/800 μs			1500rpm/1600bar/1000 μs		
		(㎣/st)		(μs)	(㎣/st)		(μs)	(㎣/st)		(μs)
		Delivery	Backleak	Resp.time	Delivery	Backleak	Resp.time	Delivery	Backleak	Resp.time
EJBR02801D KIA Carnival 2 EURO-3	New injector	12.6	8.0	440.0	44.1	12.0	368.0	67.9	24.0	351.0
	Rem. #1	13.1	8.0	435.0	44.5	13.0	359.0	68.9	25.0	356.0
	Rem. #2	12.8	9.0	441.0	43.9	12.0	360.0	69.0	26.0	360.0
EJBR02901D KIA Grand Carnival EURO-3	New injector	13.5	6.0	430.0	46.9	13.0	371.0	70.9	25.0	347.0
	Rem. #1	14.2	10.0	412.0	47.0	15.0	368.0	71.4	26.0	350.0
	Rem. #2	14.1	9.0	415.0	47.9	14.0	359.0	71.9	27.0	344.0
EJBR03001D Bongo 3 EURO-3	New injector	11.9	9.0	452.0	43.0	14.0	366.0	65.9	24.0	360.0
	Rem. #1	12.4	8.0	416.0	43.5	14.0	352.0	66.9	26.0	342.0
	Rem. #2	12.9	7.0	402.0	43.7	13.0	348.0	67.1	25.0	343.0
EJBR03902D KIA Grand Carnival EURO-4	New injector	4.2	4.0	480.0	34.1	8.0	380.0	65.1	18.0	357.0
	Rem. #1	5.2	5.0	472.0	35.2	9.0	374.0	65.4	16.0	347.0
	Rem. #2	5.0	3.0	465.0	35.7	10.0	369.0	66.0	17.0	361.0
EJBR05501D Bongo 3 EURO-4	New injector	4.6	5.0	449.0	31.9	10.0	371.0	50.7	17.0	351.0
	Rem. #1	3.9	3.0	452.0	30.1	9.0	365.0	49.9	18.0	359.0
	Rem. #2	4.0	4.0	449.0	29.8	11.0	377.0	51.0	18.0	365.0
EJBR02601Z SSANGYONG Rexton EURO-3	New injector	10.8	6.0	465.0	36.9	12.0	367.0	55.9	24.0	346.0
	Rem. #1	11.9	7.0	448.0	37.1	14.0	359.0	56.9	26.0	349.0
	Rem. #2	12.0	8.0	451.0	37.9	15.0	351.0	57.1	26.0	342.0
EJBR03401D SSANGYONG Kayron EURO-3	New injector	4.9	7.0	451.0	34.8	12.0	371.0	64.8	22.0	342.0
	Rem. #1	5.2	8.0	442.0	35.9	13.0	364.0	66.5	24.0	342.0
	Rem. #2	4.7	6.0	462.0	35.0	12.0	358.0	65.1	26.0	351.0
EJBR04401D SSANGYONG Action EURO-4	New injector	4.1	4.0	471.0	33.8	10.0	365.0	59.8	19.0	351.0
	Rem. #1	3.9	3.0	480.0	34.0	9.0	359.0	60.8	20.0	348.0
	Rem. #2	3.8	3.0	482.0	33.7	10.0	371.0	61.7	21.0	352.0
EJBR04501D SSANGYONG Action (pickup) EURO-4	New injector	3.9	3.0	480.0	32.9	9.0	375.0	60.2	18.0	344.0
	Rem. #1	4.5	3.0	472.0	33.8	10.0	367.0	62.1	22.0	350.0
	Rem. #2	5.1	4.0	469.0	34.0	10.0	364.0	62.7	23.0	352.0

For in-vehicle exhaust emissions testing, the samples were chosen randomly (Figure 9). Requirements for those vehicles emissions comply with EURO-III and EURO-IV regulations. Prior to test, all the information about injector's OEM number, vehicle model, engine type, engine capacity, number of cylinders, injector reference numbers are stored in database.

Figure 9: Vehicle exhaust emission tests.

5 CONCLUSIONS

Market share of remanufacturing industry is rapidly growing for obvious reasons like increasing restrictions of regulations, reducing raw material processing and more efficient usage of product life-cycle. Benefits of remanufacturing became obvious. However, obstacles of this process still need duly attention, research and development. As reported in the beginning, remanufactured vehicle parts are still not exceeding 30 items particularly in Korea. This leaves great opportunity and substantial reason to concentrate on this issue. Remanufacturing is relatively new to the industry comparing to manufacturing. The manufacturing already passed a long way of development. Almost all requirements are organized according to standards. Currently, remanufacturing process started being organized.

Diesel injector remanufacturing process is preformed based on two different methods. Study involves two local Korean remanufacturing companies. They use their own approach and checking methods. Performance test is carried out on available test bench, which is provided together with testing procedure and specification. Remanufacturing procedure steps are analyzed from both local companies. The quality of the end product and efficiency of remanufacturing approach determines Company B's approach as more suitable.

Performance test specifications also are studied. Based on obtained test data, it is possible to say that there is no conflict in between those specifications even though the mode and variables are noticeable different.

REFERENCES

[1] Osman Asi, 2006, Failure of diesel engine injector nozzle by cavitation damage. Engineering Failure Analysis, 13: 1126-1133.

[2] Giudice, F., La Rosa, G., Risitino, A., 2006, Product Design for the Environment. BocaRaton, CRC Press.

Recycling of Cross-linked Polyethylene Cable Waste via Particulate Infusion

R. Qudaih[1], I. Janajreh[1], S.E. Vukusic[2]

[1]Department of Mechanical Engineering, Masdar Institute of Science and Technology, Abu Dhabi, UAE

[2]Department of Chemical Engineering, The Petroleum Institute, Abu Dhabi, UAE

Abstract

The amount of plastic waste is on the rise each year. It is associated with green house gas emissions and hazardous potential if left untreated at the landfill. Plastic is formed from the polymerization of ethane, essentially fossil fuel based, and with chemical energy content equivalent to diesel fuel, i.e. 43 MJ/kg. Thereby, plastics are excellent candidates/feedstock for energy recovery. This work explores the recycling of cross-linked polyethylene waste-plastic that is generated at a considerable amount from the cable industry. It undertakes thermal analysis of the plastic as the first step using the Differential Scanning Calorimetry (DSC) to observe its thermal conversion and meting temperature. It also undertakes the re-compounding of the waste plastic by infusion the cross-linked polyethylene waste into mini-compounder at different waste-to-virgin polyethylene weight ratio, i. e. 5%, 10%, and 15%. Assessment of the mechanical properties of the compounded product, i.e. recycled mixture, is conducted on standard uniaxial samples using thermo scientific mini extruder, the HAAKE MiniJet II injection molding, and Instron tensile machine. At as high as 15 wt% recyclate to virgin mixture, an increase of 17% in young modulus, 37.2% in yield stress, 22.4% in ultimate stress, and a decrease of 3.3% in the melting point were obtained compared to the virgin/baseline polyethylene. From the resulted stress strain relation Morz bilinear plastic material model constants obtained (Et), that showed 25.6% increase, allowing better investigation of the true stress/strain behavior through FEM numerically simulation.

Keywords:

Cross-linked Polyethylene; compounding waste plastic extrusion; molding; quasai-static tensile, FEM and Morz plastic modeling.

1 INTRODUCTION

The amount of plastic waste that destine to the landfills is increasing each year. Their energy intensity is as high as 43MJ/kg with fossil fuel representing their backbone constituent. Serious recycling effort in plastic will have great impact on reducing carbon emission and slow depletion of fossil resources. In Europe, the amount of plastic accounts for 9.1% of the municipal solid waste MSW [1], and in Abu Dhabi accounts for 19% of the MSW received at Al Dhafra landfill (April 2010) [2]. Polyethylene, polypropylene, polyethylene terephthalate, polystyrene, and PVC accounts for over 95% of the waste plastic of which nearly 69% is low density polyethylene [2]. Polyethylene is essentially created from the polymerization of ethene (CH_2=CH_2); therefore, carbon and hydrogen representing the backbone of its produced waste.

Depending on the branching, crystal structure and the resulting molecular weight, there are several categories of polyethylene, ranging from ultra high molecular weight (UHMWPE) to ultra low molecular weight (ULMWPE) and high (HDPE) to low density (LDPE) as well as cross-linked polyethylene. Cross-linked polyethylene is a thermosetting plastic in which molecules chains are cross-linked by either i) radiation ionization, ii) peroxide or iii) silane compounds that defining the three cross linked polyethylene types. The cross-linking element in the silane cross-linked polyethylene is the siloxane bond (-Si-O-Si-) whereas is the C-C bond in the radiated and peroxide types [3]. Cross-linking of the polymer chain is used to enhance the virgin mechanical and/or thermal properties with the infusion of the additive at small fraction downstream compounder, just prior to the extrusion. Regrettably, most commercial cross-linked plastic are irreversible to the cross-linking due to the thermosetting transform. As the cross-linked PE is subjected to heat, it starts degrading or burning instead of melting. Only when the cross-links bonds are chemically different from those forming the virgin polymer, reversing the cross-links is possible. Therefore, recycling thermosets can be achieved by shredding, grinding, and using it as filler. However, the mixing and compounding need to be optimized depending on the end-user application. In this work, the mixing of recyclate waste is conducted at low weight fraction aiming to achieve an acceptable change in the mechanical and thermal properties, with the overall objective is to attain zero waste production.

Literature Search

Lertrojanachusit et al. [4] studied the effect of active and non-active additives to the recycling of Si-XLPE using single screw extruder at a temperature of 250°C and pressure 16 MPa. They investigated four recycling methods of Si-XLPE and LDPE mixture with: i) Methanol at supercritical conditions, ii) Silica-Alumina catalyst, and iii) Methanol and catalyst, iv) No active additives. Their results suggested that as much as 40% of the virgin LDPE could be saved as they reported that a tolerable 18% increase in the tensile strength and 21% rupture decrease. Toshiharu and Yamazaki [5] investigated the recycling of silane cross-linked polyethylene (Si-XLPE) for potential wire and cable insulation reuse. Their study investigated the continuous process for supercritical alcohol using twin-screw extruder as a compounder to recycle the cross-linked PE at 320 °C and 10 MPa. Their results indicated supercritical alcohol resulted in zero gel fraction and a molecular weight similar to the raw PE. Shigeru et al. [6] attempted to study the recycling of peroxide-XLPE and Si-XLPE industrial waste of Furukawa Electric's via thermal and shearing actions referred as thermoplasticizing technology. The XLPE was fed to the thermoplasticizing device, and the product was converted to pellets of recycled PE. The results showed that cross-linked polyethylene could be thermoplasticized, and was subjected to extrusion and press-molding successfully. It was also found that the chemical composition of the recycled PE was almost the same as that of the virgin LDPE. The above studies have reported successful recycling methodologies while no details given about the material testing and process integration and condition to be implemented in the cable industry. This work provides these details from the testing of the feedstock, preparation, and infusing and mixing, to the final testing and comparing it with the baseline/virgin PE.

Feedstock preparation

The virgin LDPE and silane cross-linked polyethylene Si-XLPE used in this study were obtained from Ducab Inc. Ducab is one of the leading international cable manufacturing companies located in the industrial city of Abu Dhabi. The Si-XLPE is made during the extrusion by infusion of the silane followed by post-extrusion water quenching and curing to be used as cable insulator. The virgin LDPE and Si-XLPE samples for thermal analysis were prepared from a cut out small particles with 1-3 mm size as shown in Figure 1 followed with sieving to 1.25mm. The Si-XLPE-LDPE mixtures were prepared by mixing the virgin LDPE granules with 5%, 10%, and 15% (wt/wt) 1.25 mm sieved granulated waste Si-XLPE as shown in Figure 2. The virgin LDPE granules without cross-linked PE were used as a baseline.

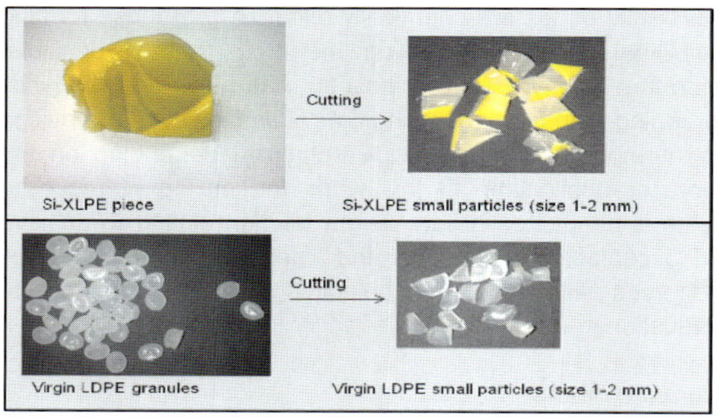

Figure 1: Preparing the PE samples for the STA analysis.

Figure 2: Virgin and the recyclate XLPE.

2 RECYCLING PROCEDURES

2.1 Material Characterization

Thermal analysis, including proximate analysis and melting point, can be performed using Simultaneous Thermal Analyzer. It involves the Thermogravimetric Analysis (TGA), and Differential Scanning Calorimetry (DSC). DSC method involves analyzing the change of the difference in heat flow to the sample and an inert reference against temperature under identical conditions [7]. In this work, the measurement of heat flow of the sample as a function of a controlled temperature profile was determined by using the Simultaneous Thermal Analyzer STA (Q600 SDT TA Instruments) as shown in Figure 3.

Figure 3: Simultaneous Thermal Analyzer STA (Q600 SDT).

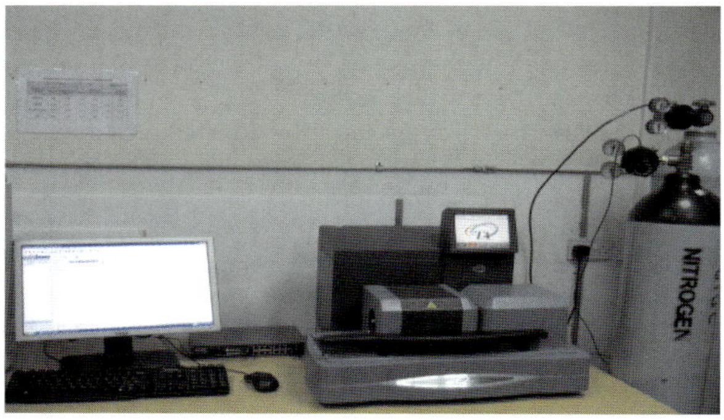

The set up of thermal analysis consists of placing the sample inside a suitable pan in a controlled temperature furnace. Meanwhile, weighting and heat flux sensors measure the desired property change as a function of the applied temperature or rate. Figure 4 shows the STA equipment during the sample loading. There are two separate pans that hold two cups; one cup accommodates the sample and a second accommodates the reference where both are balanced prior adding the sample in the sample cup. A sample of 3 to 12 mg is loaded in the sample cup.

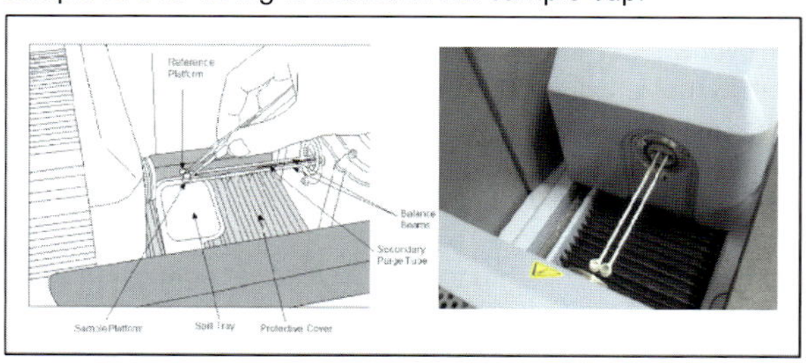

Figure 4: STA sample loading.

Air is used as a cooling gas and nitrogen as a purging gas at a flowrate of 100 ml/min. The samples are subjected to

5°C/min heating rate with sufficient time to reach beyond the sample melting point, i.e. 200 °C.

2.2 Compounding with Conical Twin-Screw Extruder

The Conical Twin-Screw Extruder obtained from Thermo Scientific (HAAKE MiniLabII) as shown in Figure 5 is used to compound small batch of plastic, i. e. 5- 20 g. The system is based on an integrated backflow channel to extrude the filled-in sample in a circulation. The backflow channel is constructed as a rheological slit capillary with two pressure sensors. From the channel and a bypass valve the residence time is defined. The conical twin-screw, shown in Figure 6, is designed to be co-current or counter-rotating screws. Moreover, the compounder can be equipped with an inert gas flush system. At the end of the screw, the bypass valve is opened and the sample is extruded as a strand.

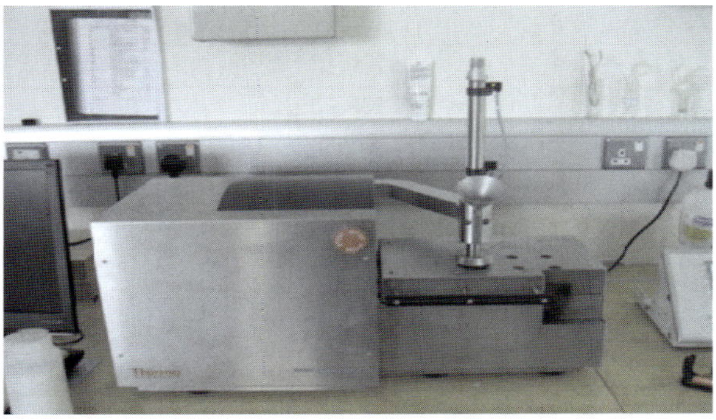

Figure 5: Conical Twin-Screw Extruder obtained (Thermo Scientific HAAKE MiniLabII).

Three different feeding systems are available for the conical twin-screw extruder: a standard pneumatic feeding system, a manual feeder and a force feeder. The feeder is designed for continuous sample feeding (max. pellet size 2 mm). The cooled feeding zone avoids the melting of material in the feeder funnel. Output rates are possible in the range of 2 g of material per minute. The Si-XLPE-LDPE mixture and virgin LDPE are fed manually to the feeder funnel and are passed through the conical twin-screw extruder where they are blended by the co-current rotating screws. The melt blending is carried out at a temperature of 180 °C. The screw torque is set at 550 N cm, while the pressure is set to 210 bars. The circulating time is set to be 10 min, and then the extruded mixtures are pulled and collected as strips using a take-up tool used for molding as shown if Figure 7.

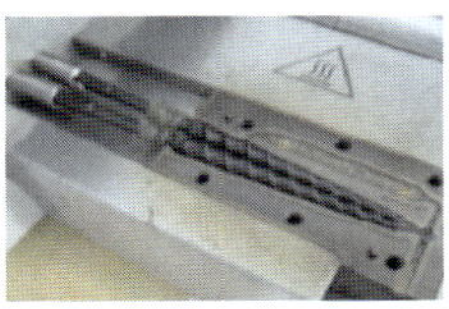

Figure 6: The conical twin-screws.

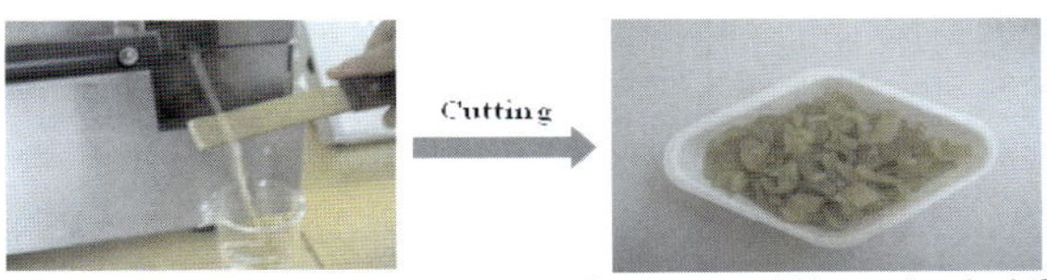

Figure 7: The collected and small pieces of the extruded Si-XLPE-LDPE mixture.

2.3 Injection Molding

The extruded LDPE and Si-XLPE-LDPE compounds are molded into dumbbell test specimens using an injection molding machine (Thermo Scientific HAAKE MiniJet II) as shown in Figure 8. Test specimens can be produced from 5 g powder, pellet or melt material. The main parts of the micro injection molding machine are the retaining pin, cylinder (reservoir), injection nozzle, mould, injection piston, and pneumatic piston. The injection nozzle (needle) is in the corresponding recess in the mould. The injection piston transfers the forces of the pneumatic cylinder to the sample compound inside the injection cylinder.

Figure 8: Injection Molding machine (Thermo Scientific HAAKE MiniJet II type).

The procedure of injection molding caters the subsequent steps: i) Parameter setup for LDPE material molding (the cylinder temperature is set to 200°C and the mold temperature is set to 50°C. The injection pressure is set to 200 bars for 6 sec while the posting pressure is set to 650 bars for 6 sec). ii) Placing the mold into the mold holder inside the machine. The cylinder and its handle bar are hanged out into their stand, and the injection piston is removed as shown in Figure 9. iii) The sample material is filled to the cylinder as shown in Figure 9. The sample is compressed several times during loading in order to avoid the formation of air bubbles in the specimen, and then the injection piston is inserted again into the injection cylinder as shown in Figure 9. iv) Placing the filled cylinder into the retaining pin inside the injection mold machine as shown in Figure 10. v) After placing the cylinder and closing the door, start the set parameters program.

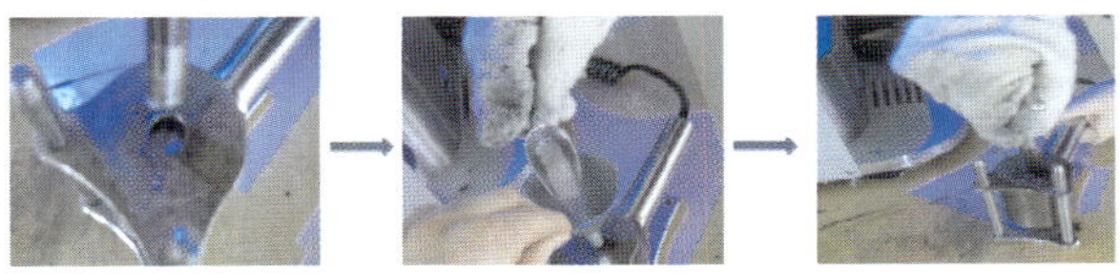

Figure 9: The feeding process into the injection molding cylinder.

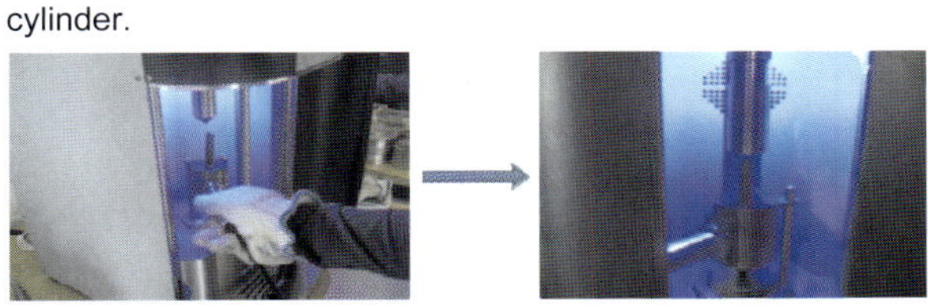

Figure 10: Placing the filled cylinder in the retaining pin inside the injection mold machine.

vi) After the injection ends, the cylinder is removed and the mold is released by the mold remover pliers. The removed mold is then opened with the clamp and the specimen is recovered as shown in Figure 11.

Figure 11: The procedure of removing the specimen from the mold.

2.4 Tensile Test

Plastic mechanical properties such as tensile strength, yield strength, young modulus and maximum elongation (of the virgin and cross-linked mixture) are determined as the sample is subjected to the uniaxial imposed displacement tensile test. The objective of performing the tensile test is to assess the change of mechanical properties of the recycled LDPE. The Instron 5966, shown in Figure 12, is used for performing the tensile test. The specimen is subjected to constant displacement and the resulted force through its loadcell is solicited. The standard molded dumbbell-bars with 25 mm gauge length, 4 mm width, and 2 mm thickness of the virgin LDPE and composites are tested as shown in Figure 13. In this test, the material is subjected to a constant but low displacement rate of 0.05mm/s until it ruptured.

Figure 12: M350-10CT Testometric device.

To obtain the normal stress axial strain plot, the initial sample cross- sectional area and gauge length specimen are recorded prior to performing the test.

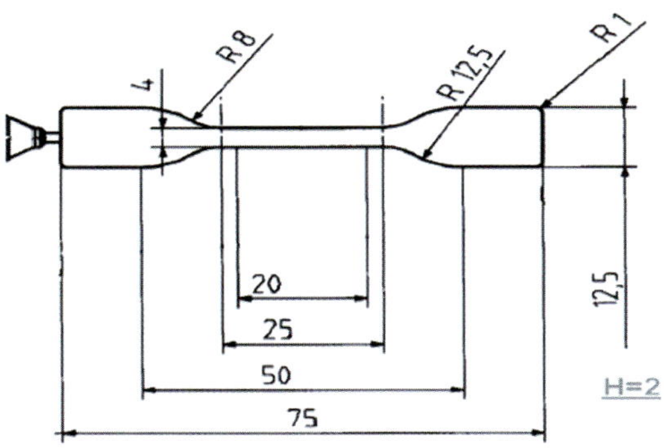

Figure 13: Specimen dimensions for the tensile test.

Following the machine calibration specimen is then placed vertically in the bottom grip and securely tightened to avoid slippage. The test is carried out at room conditions.

3 RESULTS AND DISCUSSION

3.1 DSC results

The results of the DSC test runs are summarized in Table 1 and shown in Figure 14. The values of latent heat of melting are compared to those obtained elsewhere [8].

Table 1: Melting temperature and the latent heat of melting obtained by the DSC

Polyethylene Samples	Melting Temperature (°C)	Δh (J/g)
LDPE (baseline)	109.53	52.87
Si-XLPE	108.17	65.19
Extruded Composites		
15%	105.97	52.10
10%	106.42	53.76
5%	106.90	54.63

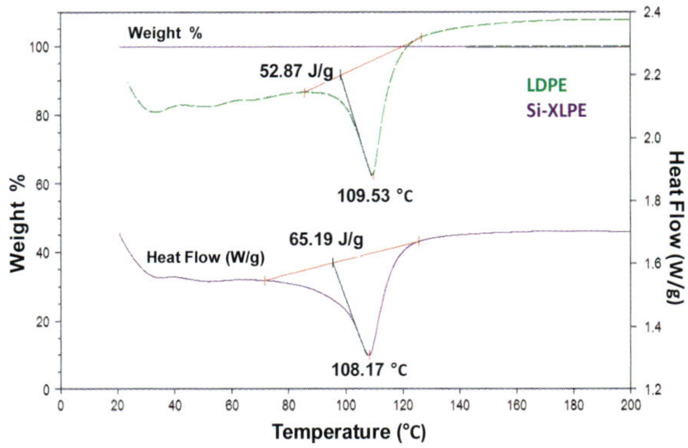

Figure 14: The melting point of the virgin LDPE and Si-XLPE at 5 °C/min.

The DSC results of the virgin LDPE, Si-XLPE and Si-XLPE-LDPE composites show the endothermic behavior of the latent heat of melting. DSC results also show that the melting point of the Si-XLPE mixture is lower than the virgin LDPE melting point. On the other hand, the heat absorbed during Si-XLPE melting is higher than that released during the virgin LDPE melting. It is also observed that as the percentage of the Si-XLPE waste infused in the virgin LDPE increases, the melting point decreases.

3.2 Tensile and material modeling results

Polyethylene, as any plastic, is nonlinear and non-elastic (plastic) material. This elastic-plasticity behavior is modeled following the "flow theory". Von Miseses [9] and Mroz [10] are common formulations used to describe the plastic-bi and –multilinear response. Figure 15 depicts multilinear isotropic hardening and kinematic hardening behaviors.

From the total deformation gradient ${}_0^t x$ at the initial configuration (t=0), one could decompose it into a material rigid body rotation (${}_0^t R$) and a symmetric stretch tensor (${}_0^t U$) written as:

$$ {}_{0}^{t}x = {}_{0}^{t}R \; {}_{0}^{t}U \qquad (1) $$

Where principle diagonalization operation to ${}_{0}^{t}U$ with respect to the global coordinate system can be further decomposed into:

$$ {}_{0}^{t}U = {}_{0}^{t}R_{L} \; {}_{0}^{t}\lambda \; {}_{0}^{t}R_{L}^{T} \qquad (2) $$

Where ${}_{0}^{t}R_{L}$ is the rotation tensor with respect to the fixed global axes and ${}_{0}^{t}\lambda$ is total stretch value for line element of the original configuration in the ${}_{0}^{t}R_{L}$ direction. For inelastic deformation, ${}_{0}^{t}x$ is initially decomposed into elastic and inelastic: ${}_{0}^{t}x = {}_{0}^{t}x^{E} + {}_{0}^{t}x^{INE}$ and each component follows the subsequent decomposition.

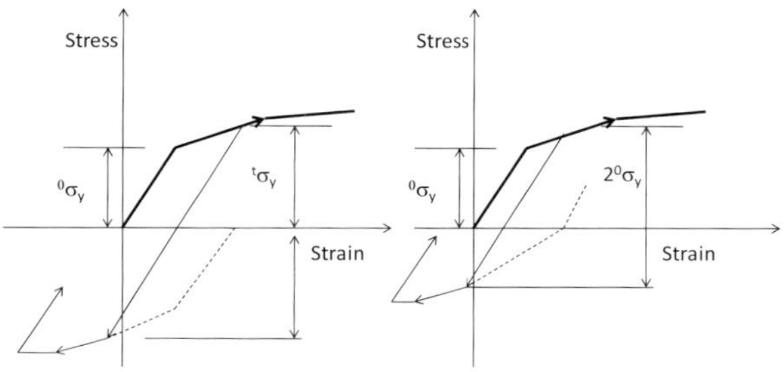

Figure 15: Multilinear isotropic hardening (left) and multilinear kinematic hardening.

In this work Morz bilinear material model is used that based on von Mises yield condition with two associated flow rules one for displacement of the yield surface and another for displacement of the bounding surface. This representation is illustrated in Figure 16. Where plastic deformation starts at ${}^{0}\sigma_{y}$ and continues along line AB reaching the bounding stress σ_{yB} and then continues following the bounding line BC under further stress. As the load is reversed, plastic deformation start at point D following the lines segment DF then bounded with FG line segment under further compressive stress. The yield surface is expressed as:

$$ {}^{t}f_{y} = \tfrac{1}{2}({}^{t}s - {}^{t}\alpha).({}^{t}s - {}^{t}\alpha) - \tfrac{1}{3}{}^{0}\sigma_{y}^{2} = 0 \qquad (3) $$

Where the bounding surface is expressed as:

$$ {}^{t}f_{yB} = \tfrac{1}{2}({}^{t}s_{B} - {}^{t}\beta).({}^{t}s_{B} - {}^{t}\beta) - \tfrac{1}{3}\sigma_{yB}^{2} = 0 \qquad (4) $$

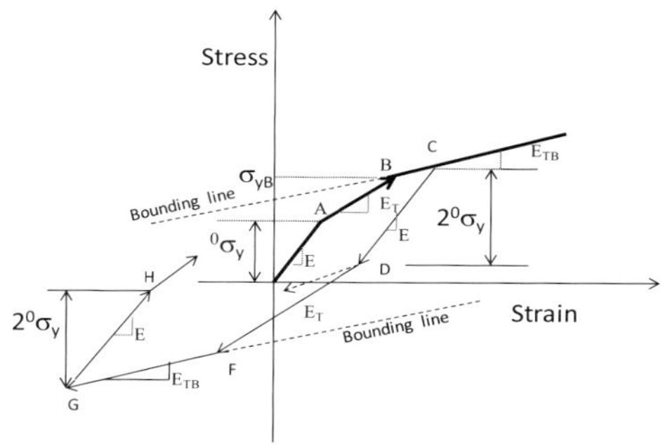

Figure 16: Morz bilinear stress strain model.

The plastic flow rule is written as:

$$ {}^{t}e^{P} = {}^{t}\Lambda({}^{t}s - {}^{t}\alpha) \qquad (5) $$

Where A is the plastic multiplier and the constitutive equation between for the back stress is:

$$ {}^{t}\alpha = {}^{t}C_{p}\sqrt{\tfrac{2}{3}}\frac{d\,{}^{t}e^{P}}{dt}\,{}^{t}m \qquad (6) $$

$$ \text{With } {}^{t}C_{p} = \tfrac{2}{3}\frac{{}^{t}E_{p}}{{}^{t}m.{}^{t}n} \text{ and } {}^{t}E_{p} = \frac{E - {}^{t}E_{T}}{E - E_{T}} \qquad (7) $$

Where the derivative is the rate of effective plastic strain, E elastic modulus and ${}^{t}Ep$ is the plastic modulus and ${}^{t}ET$ is the hardening modulus. m and n are the unit normal. Rupture is modeled when the value of the accumulative plastic strain exceeds a given maximum allowable plastic strain. In numerical FEM program that element can be removed from the model.

The results of the stress-strain measurements are shown in Figure 17 and the inferred values of young modulus, yield stress, ultimate and the tangent modulus (E_t) are summarized in Table 2.The results show that the infusion of XLPE to PE increases the stiffness of the resulted material and decreases its elongation. This proves that the filler material (cross-linked material) is stronger than the virgin PE and the bonding between the filler and the virgins is as strong as or stronger than the virgin bonding.

The goal of material modeling is to simulate the behavior of the infused plastic with minimum experimental solicitations and allow carrying out parametric studies that extremely difficult to obtain experimentally. Contours of stress and strains results of the FEM numerical simulation based on the Bilinear Isotropic Harding Material Model, for the polyethylene and under L/4 displacement are depicted in Figures 18 and 19. Moreover, contours of stress and strains results of the FEM numerical simulation based on the Multilinear Isotropic Harding Material Model, for the polyethylene and under L/4 displacement are depicted in Figures 20 and 21 and summarized in Table 3. Table 4 show the maximum von Mises strain and stress of 15%, 10% and 5% composites. The results based on the bilinear model cannot describe the long range strain behavior as obvious when compared to those obtained experimentally. Multilinear plastic material modeling is more appropriate.

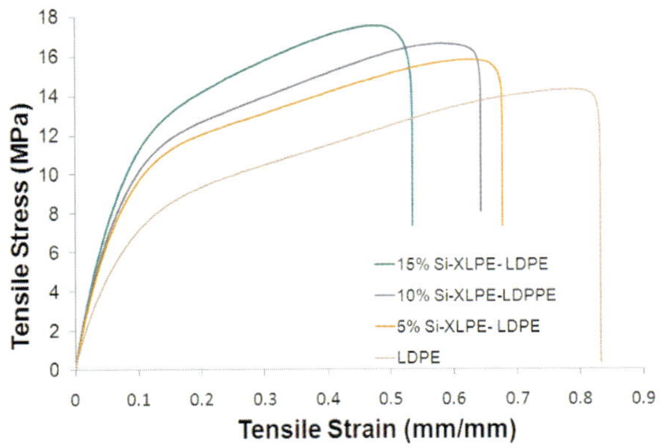

Figure 17: Stress-Strain curve.

Table 2: Summary of the mechanical properties assuming Bilinear Isotropic Harding Material Model

	E (MPa)	ΔE (%)	σ yield (MPa)	Δσ yield (%)	σ ult (MPa)	Δσult (%)	Et (MPa)
LDPE	153.9	-	8.6		14.3		103.5
15%	180	17	11.8	37.2	17.5	22.4	130
10%	171.4	11.4	11	27.9	16.6	16.1	137.5
5%	164	6.6	10.5	22.1	15.7	9.8	143.2

Table 3: Summary of the mechanical properties assuming Multilinear Isotropic Harding Material Model

Young Modulus (MPa)	LDPE	15%	10%	5%
E	153.85	180.00	171.43	164.0
E_t	103.45	130.00	137.50	143.2
E3	73.75	91.43	123.53	92.86
E4	47.44	60.00	100.00	66.67
E5	31.40	28.00	67.50	40.43
E6	16.38	16.36	32.50	16.92
E7	10.67	13.00	20.00	10.00
E8	8.00	4.40	11.61	3.75
E9	4.50	--	4.29	--
E10	0.83	--	--	--

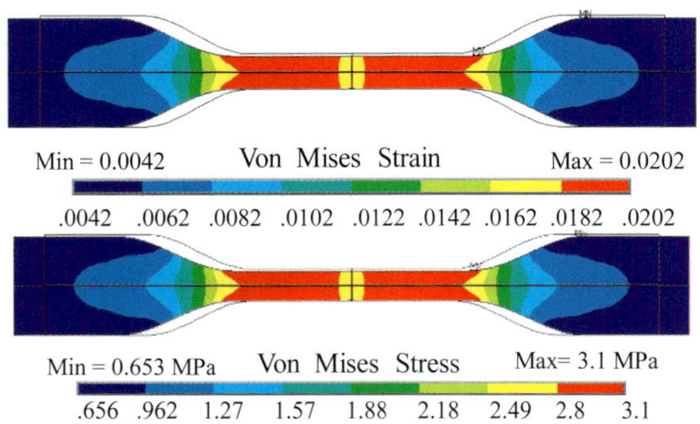

Figure 18: FEM results of von Mises strain and stress using Bilinear Isotropic Harding Material Model for LDPE.

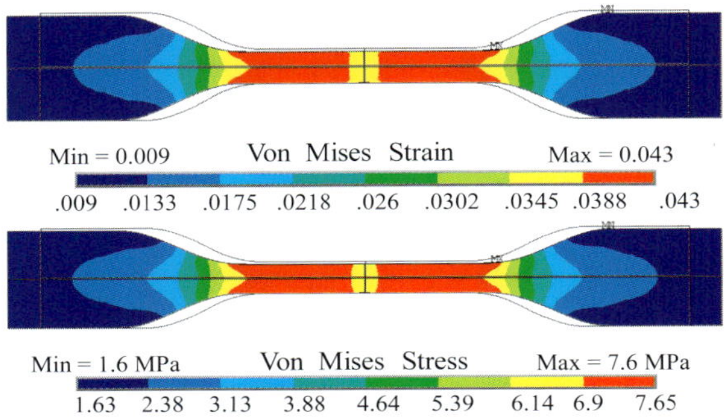

Figure 19: FEM results of von Mises strain and stress using Bilinear Isotropic Harding Material Model for 15% Si-XLPE-LDPE.

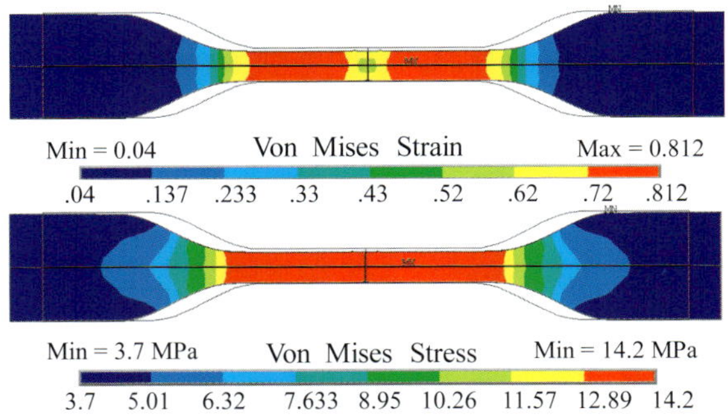

Figure 20: FEM results of von Mises strain and stress using Multilinear Isotropic Harding Material Model for LDPE.

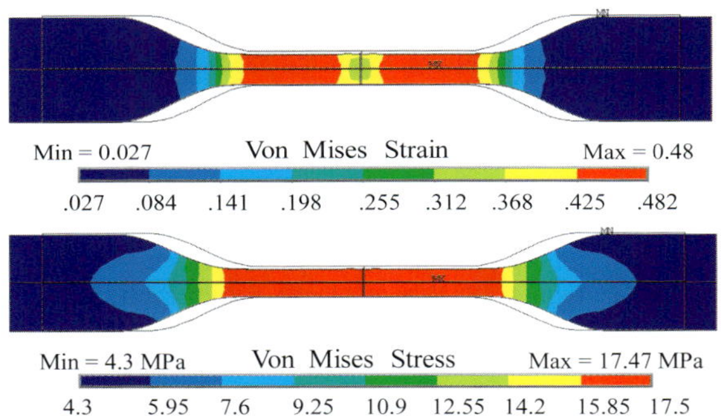

Figure 21: FEM results of von Mises strain and stress using Multilinear Isotropic Harding Material Model for 15% Si-XLPE-LDPE.

Table 4: FEM results of von Mises strain and stress using Isotropic Harding Material Models

	Bilinear Model		Multilinear Model	
FEM results	Max. Strain	Max. Stress (MPa)	Max. Strain	Max. Stress (MPa)
15%	0.043	7.65	0.48	17.5
10%	0.022	3.83	0.59	16.6
5%	0.029	4.77	0.64	15.78

4 CONCLUSION

In this work, the cross-linked polyethylene waste is infused into the virgin LDPE via mini-compounder above the melting point at three waste-to-virgin polyethylene mixing fraction, i. e. 5%, 10%, and 15% weight. Following compounding, they are molded into standard dumbbell-bar and then subjected to tensile test. The waste-to-virgin polyethylene samples are also thermally analyzed by DSC. The DSC results show the heat absorbed during the melting of the Si-XLPE is higher than that released during the melting of the virgin LDPE. It is also observed that as the percentage of the Si-XLPE waste infused in the virgin LDPE increases, the melting point decreases. Finally, at 15% mixing fraction, results have shown an acceptable change in the mechanical properties. i.e. at 15% weight recyclate to virgin mixture an increase of 17% in young modulus, 37.2% in yield stress, 22.4% in ultimate stress, and a decrease of 3.3% in the melting point were obtained compared to the virgin/baseline polyethylene. From the resulted stress strain relation Morz bilinear plastic material model constants obtained (Et), that showed 25.6% increase. Finally, Multilinear Isotropic Harding Material Model shows better representation of the LDPE material model than the bilinear model.

5 ACKNOWLEDGEMENTS

The financial support of Masdar Institute is highly acknowledged. The authors like to extend their gratitude for the help of Salaheddin Abuyahya for the sample preparation and molding.

6 REFERENCES

[1] APME. Good practices guide on waste plastics recycling a guide by and for local and regional authorities. Brussels, Belgium: Association of Plastics Manufacturers in Europe; 2004.
[2] CWM Data Management System, Data set Aug09 to Mar10.
[3] Chong T. Design and process for producing silane-crosslinked polyolefin compound. January 2005.
[4] Lertrojanachusit N, et al. Recycling of Silane-Crosslinked Polyethylene (Si-XLPE) by the Extruder: Effect of Active and Non-active Additives.
[5] Goto T, Yamazaki T. Recycling of Silane Cross-linked Polyethylene for Insulation of Cables using Supercritical Alcohol. Hitachi Cable Review No.23. AUGUST 2004.
[6] Tokuda S. Thermoplasticizing Technology for the Recycling of Cross-linked Polyethylene. Furukawa Review, No. 23. 2003.
[7] Bonilla J, Lobo H. *Handbook of plastic analysis (Plastics Engineering)*. 2003.
[8] Behjat Tajeddin, et al. Thermal Properties of Low Density Polyethylene - Filled Kenaf Cellulose Composites. European Journal of Scientific Research. ISSN 1450-216X Vol.32 No.2 (2009), pp.223-230
[9] K. J. Bathe, Finite Element Procedures, Prentice Hall, Englewood, NJ, 1996
[10] Y. F. Dafalias and E.P. Popov, "Plastic Internal Variables Formulasim of Cyclic Plasticity, "J. Applied. Mechanics, Transaction ASME, Vol.43, pp. 645-651, 1976.

Ecodesign Maturity Model: criteria for methods and tools classification

D.C.A. Pigosso[1], H. Rozenfeld[1], G. Seliger[2]
[1]Department of Production Engineering, University of São Paulo, São Carlos, Brazil
[2]Department for Machine Tools and Factory Management, Technische Universität Berlin, Germany

Abstract
Ecodesign is a strategic design activity established to conceive and develop sustainable solutions in line with the concept of life cycle thinking striving for environmental sustainability. Nowadays, the implementation of ecodesign practices into product-service systems (PSS) development processes, are difficult because the ecodesign practices are not systematized and also because the current maturity level of a company is not considered in the selection of practices to be applied. Depending on the maturity level in ecodesign application, there would exist some ecodesign practices that are more suitable than others. This paper presents the efforts that are being carried out in a cooperative project between Brazil and Germany (BRAGECRIM) to develop an Ecodesign Maturity Model to support the application of ecodesign practices in the development of remanufacturable production equipment. Specifically in this contribution, it will be presented the criteria that were used to classify the ecodesign methods and tools in order to support the selection of the most suitable ones according to the company maturity level and goals. During the systematic literature review, it was identified and classified 105 ecodesign methods and tools according to 13 criteria. The results and the state of the art of the ecodesign methods and tools are presented in this paper.
Keywords:
Classification Criteria; Ecodesign; Methods and Tools; Maturity Model; Product Development; PSS

1 INTRODUCTION AND CONTEXT

In general, production equipment are high value added products which present significant environmental impacts in the end-of-life phases due to the high material intensity. For these products, end-of-life strategies such as remanufacturing, that maintains the value added to the product and reuses product items, is preferable both from an environmental and economical point of view.

In order to successfully implement remanufacturable and more sustainable products, they should have been previously designed for this purpose by the application of ecodesign practices (guidelines, strategies, methods and tools) into the early phases of the development process [1]. Ecodesign is a proactive management approach that directs product development towards environmental impact reductions throughout its lifecycle (from raw material extraction to end-of-life), without compromising other criteria such as performance, functionality, aesthetics, quality and cost [2, 3].

The traditional industrial economy in which value is attributed to material products that are exchanged has shifted towards the new service economy, in which value is more closely related to the performance and real utilization of the products integrated in a system [4]. In this sense, there is a tendency to offer a combination of products and services, called product-service systems (PSS), which are sold in one package to meet the customer's needs. It should be noted, however, that PSS is not intrinsically sustainable [5]. In this sense, it is important to include ecodesign practices into the PSS development in order to minimize its environmental impact, contributing to achieve the PSS sustainability potential [6].

Despite the increasing recognition by companies of the need of applying ecodesign practices into the PSS development process, it is still not clear which strategies, guidelines and methods and tools should be implemented and how to manage the process of integration and continuous improvement. In this sense, it can be argued that the application of ecodesign has not reached companies worldwide over the last decade mainly due to:

(1) There are no systematization of the existing ecodesign practices;

(2) There is an intense development of new ecodesign methods and tools in detriment of the study and improvement of the existing ones [7];

(3) There is a lack of integration between ecodesign and the broad context of the product development process [8][9][10][11];

(4) Ecodesign is poorly integrated into corporate strategy and management [7];

(5) There is not a roadmap for continuous improvement that can be followed by companies in implementing ecodesign;

(6) Companies usually fail in adopting ecodesign because select practices that is not in accordance with their current maturity level.

Consequently there is a need to propose models that help companies to implement ecodesign in their product or PSS development processes in an effective way. Such a model should be based on a systematization of existing ecodesign practices, linking them to the development process and company strategy and showing how to implement them in accordance to the company's maturity level.

In this sense, the main goal of this project is to develop an Ecodesign Maturity Model to guide production equipment developer companies into the effective implementation of ecodesign practices (including remanufacturing) into the product and PSS development process in accordance with their strategic objectives and drivers.

The model is being developed in the context of a cooperative project carried out by Germany and Brazil named "Remanufacturing Oriented Production Equipment Development". This project is one of the initiatives of the BRAGECRIM program (BRAzilian and GErman Collaborative Research in Manufacturing). The Ecodesign Maturity Model, which is being developed in one of the work packages of the project, will be part of a more comprehensive reference model for developing a sustainable remanufacturing oriented product or PSS. It supports the application of ecodesign practices in the development of remanufacturable production equipment and is composed by three main components: Ecodesign Practices (that includes ecodesign operational practices, ecodesign management practices and ecodesign methods and tools), Maturity Levels (it was defined 5 maturity levels, from ignorance on ecodesign to green strategy) and Application Method (for continuous improvement) (Figure 1).

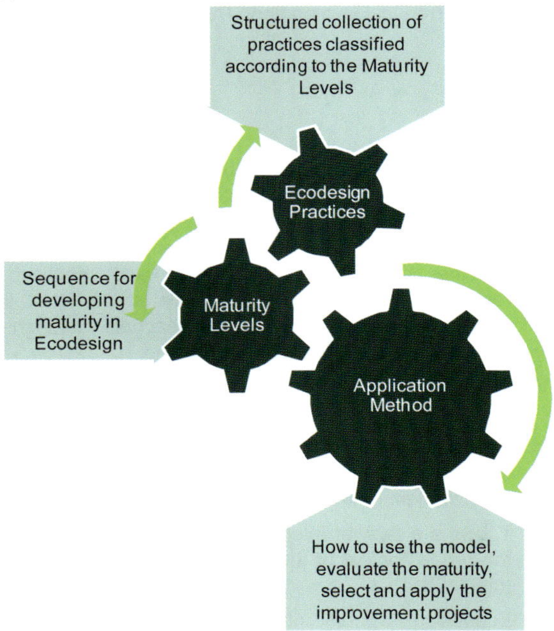

Figure 1: Components of the Ecodesign Maturity Model

In order to achieve the aforementioned goal, the following sub-projects (SPs) are being developed in the Work Package 07: Ecodesign Maturity Model:

SP7.1 Systematization of ecodesign practices on product and PSS development

The ecodesign practices (ecodesign management practices, ecodesign operational practices and ecodesign methods and tools) on product and PSS development that will be addressed in the Ecodesign Maturity Model under development will be obtained and classified by means of a systematic literature review.

SP7.2 Development of the Ecodesign Maturity Model

According to the best practices identified, the maturity levels on ecodesign will be determined. For each maturity level the most appropriated practices will be properly addressed. The correlation and dependencies between practices will also be determined. The model will be assessed by ecodesign experts.

SP7.3 Development of an Application Method

In order to define on which maturity level a company can be classified and which practices should be applied to achieve the next level, it is needed the development of an evaluation diagnoses method. This method will consider also the internal goals and strategies of a company, to filter the best practices on a specific maturity level to be applied by the enterprise.

SP4 Application of the Ecodesign Maturity Model in case studies

Once the initial design of the maturity model is developed, it will be held the application in multiple cases studies to improve and validate the proposal.

The focus of this paper is on the identification and classification of ecodesign methods and tools (one type of ecodesign practice) carried out in SP7.1. The classification into criteria aims to support the selection of the most suitable methods and tools according to the company maturity level and goals. The other ecodesign practices (ecodesign management practices and ecodesign operational practices) identified during the systematic literature review will not be addressed in this paper.

The next session presents the methodology that was adopted to identify and classify the ecodesign methods and tools. Then, it is given the ecodesign methods and tools classification criteria. An analysis of the classification of the identified ecodesign methods and tools is then presented to define the actual state of the art. Finally, the final summary and outlook, acknowledgements and references are presented.

2 METHODOLOGY

The identification and classification of the ecodesign methods and tools in SP7.1 were obtained by means of a systematic literature review.

The systematic review is the way by which the researcher can map the existing and previous developed knowledge and initiatives in a specific research area. Besides the analysis of previous discovery, techniques, ideas and ways to explore topics, the systematic review also allows the evaluation of information relevance to the issue, its synthesis and summarization [12] [13].

The phases of a systematic review correspond to problem formulation (identification of the goal of the review, target and context, beneficed areas and expected results), data collection (identification of the relevant databases, keywords and strings), data evaluation (application of the inclusion/exclusion criteria for the selection of the relevant studies and representation standardization), data analysis and interpretation (synthesis of the studies and definition of the criteria for classification) and presentations and conclusions (registration of the studies and methods and tools, analysis of the classification and determination of the state of the art)[12] [13].

The main goal of the systematic review was the determination of the state of the art in ecodesign methods and tools. The target of the systematic review, according to its context, can be explained by the non-achievement of the ecodesign potential benefits due to the lack of systematization of ecodesign methods and tools. It was obtained 560 studies, including papers, thesis, dissertations, publications, books and books reviews along the systematic review. It was made an initial studies selection, that was then analyzed according to the inclusion/exclusion criteria (studies that presented the development, application or review on ecodesign methods and tools). It was then realized a study review to certify that relevant studies has not been excluded. The valid studies were then analyzed in order to extract the relevant information about the ecodesign methods and tools.

The criteria for classification of ecodesign methods were selected by the understanding of ecodesign methods and tools main functions, characteristics and application possibilities in the development process. After the initial criteria selection, it was done a consult with experts on environmental and product development areas to evaluate the selected criteria. In this occasion, it was suggested new criteria that were added to the method/tool registration form.

The classification of the ecodesign methods and tools using the defined criteria was registered in an online database, which enables the filter according to the desired criteria.

As a result of the systematic literature review, 105 ecodesign methods and tools were identified and classified according to 13 criteria. The results are presented in the following sessions.

3 ECODESIGN METHODS/TOOLS CLASSIFICATION

Despite the existence of a large amount of ecodesign methods and tools in literature, they are still not being used in a systematic way in the product development process. One of the most influencing factors is that companies do not have information to support the selection of the most suitable methods and tools according to their needs. In this sense, the classification proposed in this paper aims to help companies in the selection of the most suitable ecodesign methods and tools according to their needs.

The methods and tools were developed over the last decade by universities and/or organizations, with goals that go from the assessment of environmental impacts of products to the identification of green customer requirements [7][11][14].

The 13 criteria and their alternative values defined in this project to classify the ecodesign methods and tools are presented in following. These criteria represent important aspects that need to be considered by companies for the selection of the ecodesign methods and tools.

Nature of the main goal of the ecodesign method and tool: describe the different types of methods and tools according to their goals

- Prescriptive: methods and tools that present generic guidelines (from a pre-established set of best practices for minimizing the environmental impacts) to increase the environmental performance of products considering environmental impacts that are recurrent for industrial products;
- Comparative: methods and tools that aims to compare the environmental performance of different products, concepts or design alternatives for a given product.
- Analytic: methods and tools that aims to identify improvement potentials in the product performance by means of the assessment of the most relevant environmental aspects. The impact categories can be pre-established according to the method and tool.

Type of the tool used by the ecodesign method and tool: presents the type of tool that is used

- Checklist: tool used to check if a determined parameter related to the product environmental performance was considered during the product development process;
- Guideline: tool that provides general guidelines to be followed during the product development process to improve its environmental performance;
- Matrix: tool that contains a pre-defined scale used to assess the environmental performance of products by correlating two relevant aspects (e.g.: life cycle phase and environmental aspects).
- Software: computational tool used to support the application of an ecodesign method or tool.

Nature of input data: identify the type of input data required by the ecodesign method and tool

- Qualitative: the method/tool requires qualitative data as input;
- Quantitative: the method/tool requires quantitative and numerical data as input;

Nature of output data: identify the type of output data delivered by the ecodesign method and tool

- Qualitative: can support subjective analysis, provide general guidance and do not generate numerical data;
- Quantitative: can support objective analysis, providing quantification and generating numerical data.

Origin knowledge area: describes the research area in which the method/tool were developed

- Ecodesign/environmental management: method and tool which origin is in the ecodesign and/or environmental management field, like methods and tools to assess the environmental impact of products;
- Product development process: method and tool which origin is in the product development field of research;

Lifecycle perspective: presents the phases of product lifecycle considered by the ecodesign method/tool

- Raw material extraction: considers the scarcity of materials, provide guidelines and/or the impact related to the extraction of raw materials;
- Primary Industry: considers the environmental impact and/or provide guidelines related to the raw material processing;
- Manufacturing: considers the environmental impact and/or provide guidelines related to the manufacturing processes;
- Use: considers the environmental impact and/or provide guidelines related to the use phase;

- Reuse: considers the environmental impact and/or provide guidelines related to the reuse of the product, their parts and/or components;
- Recycling: considers the environmental impact and/or provide guidelines related to the recycling of the product, their parts and/or components;
- Remanufacturing: considers the environmental impact and/or provide guidelines related to the remanufacturing of the product, their parts and/or components;
- Treatment and final disposal; considers the environmental impact and/or provide guidelines related to the treatment and final disposal of products;

Environmental Aspect: presents the environmental aspects considered by the ecodesign method/tool

- Energy use;
- Material intensity;
- Chemicals and toxics
- Waste residues;
- Water use;
- Waste water;
- Gas emissions;
- Others:

Current development level: This criterion assess the current development level of the method/tool according to the actual application status

- Theoretical: there are just theoretical academic studies concerning the application of the ecodesign method/tool;
- Experimental: the ecodesign method/tool were already applied in case studies in pilot projects in order to validate them;
- Consolidated: the ecodesign method/tool is already validated and applied regularly in the product development process of companies.

Environment assessment method: this criterion verifies if the ecodesign method/tool presents the existence of an environmental impact assessment method

- Yes;
- No.

Demanded time for use: this criterion assess in a qualitative way the amount of time needed to the application of the method/tool

- Low;
- High.

Application and acquisition cost: this criterion assess in a qualitative way the application and acquisition cost of the method/tool

- Low;
- High.

User specialization level in environmental area: assess the knowledge level on environment required by the designer in the application of the method/tool and interpretation of the results obtained

- Low;
- High.

Information obtained about the method/tool: presents the detail level of the method/tool obtained in the studies

- Superficial: the study presents just general information about the method/tool;
- Succinct: the study presents specify information about the method/tool, but in a succinct manner;

For each maturity level defined in the Ecodesign Maturity Model (it was defined 5 maturity levels in SP7.2 - that is out of the scope of this paper), there is an optimum combination of the proposed criteria to support the selection. In this sense, there are ecodesign methods and tools that are more suitable for application than others according to the current maturity level of the company. For example, the more quantitative methods, as Life Cycle Assessment, are more suitable for relatively mature organizations while more qualitative matrixes as DfE Matrix can successfully be applied by less mature organizations.

4 ECODESIGN METHODS/ TOOLS STATE OF THE ART

The results of the classification of the 105 identified ecodesign methods and tools were analyzed in a way to provide a state of the art view of the current status of ecodesign methods and tools according to the selected criteria. The main conclusions obtained with the classification of the ecodesign methods and tools are:

- The development of ecodesign methods and tools of analytic nature is more expressive in literature than the development of prescriptive and comparative ones. It shows the tendency of using the ecodesign methods and tools to assess the environmental impact of products and PSS;
- There is a tendency in the development of matrixes, tool that contains a pre-defined scale to assess the environmental performance of products by the correlation between two relevant aspects;
- There is a balance among the nature of the input and output data to the application of the ecodesign methods and tools. In this sense, there is equilibrium on the number of methods and tools that uses and delivers qualitative and quantitative data;
- The origin of the ecodesign methods and tools are mainly from the environmental knowledge area. The ecodesign methods and tools which origins from the product development process are predominantly adaptations from the traditional methods and tools of this area (e.g. Qualify Function Deployment (QFD) and Failure Mode and Effect Analysis (FMEA));
- Despite the intense theoretical development of ecodesign methods and tools, their application has not reached companies worldwide yet. The major part of the ecodesign methods and tools are in theoretical or experimental level of application;
- The major part of identified ecodesign methods and tools are presented in a succinct way in the studies. It can be explained by the predominance of papers which presents a limited number of pages in the systematic literature review;
- The most important concerns of ecodesign methods and tools are related to material and energy consumption;

- The majority of the ecodesign methods and tools considers all the phases of the product life cycle (from raw material extraction to end of life);
- 83% of the identified ecodesign methods and tools do not present an environmental impact assessment method;
- There is a need of high level of knowledge on the environmental area to the application of most of the identified ecodesign methods and tools, mainly for the interpretation of the results.

It was also made an analysis concerning the ecodesign methods and tools focused on remanufacturing and end-of-life strategies. It can be addressed that the end-of-life ecodesign methods and tools focused on closing the loop of materials generally include more than one end-of-life strategy. Since product complexity varies substantially, some components, systems or sub-systems are keener to be recycled, reused or remanufactured than others.

5 SUMMARY AND OUTLOOK

This work enabled the mapping of the state of the art and the classification of 105 ecodesign methods and tools. During the systematic literature review on ecodesign, it was also identified 75 management practices for ecodesign (related to the management activities and tasks of product development process that addresses the environmental issues) and more than 450 ecodesign operational practices (related to technical design issues), that was out of the scope of this paper. These practices were interrelated among them and with the ecodesign methods and tools.

The criteria used to classify the ecodesign methods and tools in order to support their selection for application in companies were presented in this paper. It was also performed an analysis of the classification that shows the actual state of the art of the ecodesign methods and tools.

The Ecodesign Maturity Model is inedited and should contribute to the organization of the knowledge in the ecodesign research area and to guide companies in the continuous improvements of their process with the incorporation of the environmental dimension striving for sustainability in their business. It considers the individual needs, drivers, characteristics, maturity levels and strategies of companies to support the decision making process. Further studies will present the other components of the model and the results of application in case studies.

The Ecodesign Maturity Model is part of a more comprehensive reference model for developing a sustainable remanufacturing oriented product or PSS. The goal of the broader project is to open the solution space for a resource-conserving, skills-conductive and competition-compatible development through production equipment for sustainable value creation.

6 ACKNOWLEDGMENTS

We extend our sincere thanks to FAPESP (Fundação de Amparo à Pesquisa do Estado de São Paulo), CAPES (Coordenação de Aperfeiçoamento de Pessoal de Nível Superior) and DFG (Deutsche Forschungsgemeinschaft) for supporting BRAGECRIM (Brazilian and German Collaborative Research in Manufacturing) project intituled "Remanufacturing oriented Production Equipment Development".

7 REFERENCES

[1] Pigosso, D.C.A., Zanetti, E.T., Guelere Filho, A., Ometto, A.R., Rozenfeld, H., 2009, Ecodesign methods focused on remanufacturing. Journal of Cleaner Production, p. 21-31.

[2] Johansson G., 2002, Success factor for integration of Ecodesign in product development: A review of state of the art. Environmental Management and Health: 13(1): 98-107.

[3] Weenrn JC., 1995, Towards sustainable product development. Journal of Cleaner Production: 3(1-2):95-100.

[4] Giarini, O., Stahel, W.R., 1993, The Limits to Certainty: facing risks in the new service economy. Kluwer Academic Publishers.

[5] Tukker, A., Tischner, U., 2006, New Business for Old Europe: Product-Service Development, Competitiveness and Sustainability. Greenleaf Publishing.

[6] Pigosso, D.C.A., Sousa, S.R.; Guelere Filho, A.; Ometto, A. R.; Rozenfeld, H., 2010, Is the Industrial Product-Service System really sustainable?. Industrial Product Service System CIRP Conference.

[7] Baumann H, Boons F, Bragd A., 2002, Mapping the green product development field: engineering, policy and business perspectives. Journal of Cleaner Production:10:409-25.

[8] Bakshi BR, Fiksel J., 2003, The Quest for Sustainability: challenges for process systems engineering. AIChe Journal: 49(6): 1350-8.

[9] Luttropp, C.; Lagerstedt, J., 2006, EcoDesign and The Ten Golden Rules: generic advice for merging environmental aspects into product development. Journal of Cleaner Production, 1-13

[10] Poole, S. et al., 1999, Integrating Environmental Decisions into the Product Development Process: Part 2 The Later Stages. Proceedings First International Symposium On Environmentally Conscious Design and Inverse Manufacturing, p. 334-337.

[11] Guelere Filho, A.; Rozenfeld, H., 2006, Integrating Ecodesign Methods and Tools into a Reference Model for Product Development. IV Global Conference on Sustainable Product Develpment and Life Cycle Engineering. São Carlos: Suprema.

[12] Biolchini, J.; Mian, P. G.; Natali, A. C. C.; Travassos, G. H, 2005, Systematic Review in Software Engineering. Rio de Janeiro.

[13] Brereton, P.; Kitchenham, B. A.; Budgen, D.; Turner, M.; Khalil, M., 2007, Lessons from applying the systematic literature review process within the software engineering domain. In: The Journal of Systems and Software.

[14] BYGGETH, S.; HOCHSCHORNER, E., 2006, Handling trade-offs in Ecodesign tools for sustainable product development and procurement. Journal of Cleaner Production, v. 14, p. 1420-1430.

6 Renewables and Resource Utilizations

In recent years, the demand for achieving better performance and more economical manufacturing process is compounded by the requirement for energy and resource efficient utilization in order to reduce costs while complying with the environmental aspects. In the first section of this chapter Wertheim et al. present examples for the ongoing trend towards more economical process chains in combination with better surface integrity, energy savings and resource efficiency. In the second section Peng et al. show the design of a mechanical grinding pan system to study the grinding mechanism of rubber and provide an effective approach to grind rubber based on the analysis of its mechanical properties. In the third section Herrmann et al. present suitable methods for the evaluation and prediction of energy or resource consumption to integrate energy-related key figures into the decision support systems. In the fourth section Thiede and Herrmann present a simulation approach which is able to realistically consider relevant energy flows in manufacturing systems. In the fifth section Rechenberg et al. introduce a suction pump based on the sap-rising principles of plants by condensing moist air. Distilled water is gained. In the sixth section Lin et al. present the results from an experimental analysis on the sensitivity of CO_2 emissions to renewable energy penetration for regions utilizing power and water cogeneration. In the seventh section Mousa and Diabat present the optimal designs parameters for a hybrid solar-wind power plant.

Energy and Resources Efficiency in the Metal Cutting Industry

R. Neugebauer[1], R. Wertheim[2], C. Harzbecker[1]
[1]Fraunhofer Institute for Machine Tools and Forming Technology, Chemnitz, Germany
[2]Faculty of Mechanical Engineering, Institute for Machine Tools and Forming Technology, Technical University Chemnitz, Germany

Abstract
The demand for a better performance and a more economical manufacturing process is combined today with the requirement for an energy and resource efficient production in order to reduce cost and at the same time to comply with the environmental aspects. Material savings and improvement of component properties in key industries have led to challenges being identified in machining processes.
Process and machine-orientated approaches are emerging from the product manufacturing and utilization phase that illustrate the many and diverse possibilities for overcoming the increasing demand as far as machining operations are concerned. Examples for the ongoing trend towards more economical process chains in combination with better surface integrity, energy saving and resource efficiency is presented with few practical examples from the metal cutting industry.
Keywords:
Cutting; Machine tool; Energy; Surface Integrity

1 INTRODUCTION

The demand for high efficiency in using energy and material resources is influenced by environmental changes and the justified efforts on the part of an increasing proportion of the Earth's population to achieve prosperity. It does force a change of paradigms from "maximum profit generated by minimum funds" to "maximum added value derived from minimum resources" [1]. The long-term trend towards turning away from fossil energy sources represents a fundamental step towards a dramatic reduction in greenhouses gas emissions. The creation of a framework of appropriate political initiatives, e.g. to promote alternative energies and research into electro-mobility, strengthens this process which is supported by relevant international regulations.

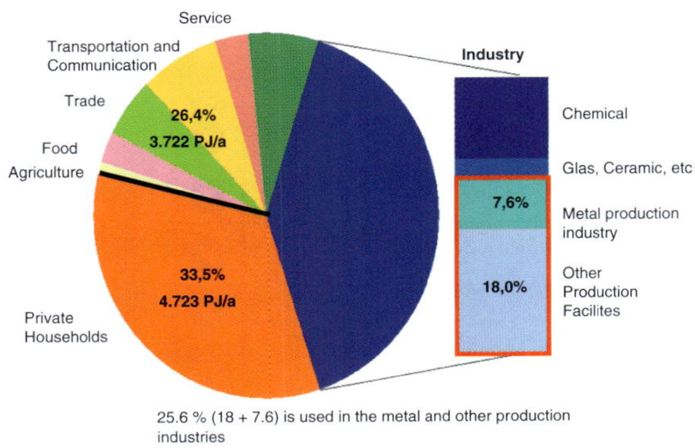

Figure 1: Total Energy Usage in Germany and the Share of the Metal Cutting and other Industries.

What resource efficiency primarily means is sustainable handling of raw materials and energy as well as of time and personnel, not only in the manufacturing phase but also in the product utilisation phase. The investigation results in Figure 1 show that the industry in Germany is responsible for more than 40 % of total energy consumption and the production industries for more than 25 %. The identified potential for energy saving of approx. 30% *[2]* is a convincing proof of this.

The requirement for higher productivity and cost minimization guided and directed the development of the production industries and the metal cutting for many years. Machining conditions and machining performance for every single machining operation as well as for the complete production chain and machine are defined in such a way as to reduce production time in order to ensure higher production rates and a more economical manufacturing process in combination with the demand to minimize production costs.

The required performance and the energy and resources efficiency can be achieved by shortening or optimizing the process chain, by improving and optimizing each process by itself, by the developments and modifications of machine tools and equipment as well as by implementing mechatronic or adaptronic control systems along the complete production chain. The capacity and performance of new machining methods combined with a wider basic knowledge of the cutting process itself is also leading towards more economical production and machining of functionally defined surfaces enabling higher performance and more energy efficient utilisation of the machined parts. Today the more effective R&D process, the elimination or decrease of the phase-in and the prototyping stages and the optimization of the machining processes rely more and more on simulation and modelling methods, which can save time, costly resources, machining energy and long term or expensive investigations.

The challenges that are likely to be encountered in the future as far as efforts to achieve resource efficiency in relation to utilisation and manufacture are concerned, in particular in relation to metal-cutting industry and how these can be influenced is the subject of this paper.

2 DEVELOPMENT TRENDS IN COMPONENT FEATURES AND MATERIAL SPECIFICATIONS

Table 1 shows an overview of significant trends as far as metal-cutting is concerned, for various industries, component features and material specifications.

As far as key industries are concerned, product development activities are conducted on the premise of resource efficiency and characterized by weight reduction. Both in the automotive and the aviation industry, the main topic is how to make use of lightweight construction in order to reduce energy consumption and emissions. The new generations of aircraft demonstrate a clear shift of parts made of aluminium towards composites and titanium *[3]*.

As regards automotive drive systems, the change to alternative forms of operating energy is in sight in the long term; however, in the medium term, the potential of the combustion engine is set to be exploited through improvements of the combustion process, by downsizing in conjunction with supercharging, utilisation of waste heat, reduction in engine friction and the use of alternative fuels. Because of the increased component loads, high-strength lightweight construction materials will be used increasingly. In addition it can be expected that due to the further development and more complex automotive products the number of various components will tend to increase.

In terms of part dimensions and geometric features, the tendency is characterized by more thin-walled features, more complexity, and more requirements as regards to surface integrity e.g. as far as reducing friction is concerned.

As a consequence of an increasing diversity of variants, batch sizes are likely to be smaller. In most of the industries the manufacturing trends depend on the product shape and material while higher complexity, improved quality, smaller batch sizes and more demanding workpiece materials are influencing the developments of metal cutting technologies.

3 CHALLENGES IN THE METAL CUTTING INDUSTRY

Driven by users and manufacturers, increasing challenges will emerge as far as metal-cutting is concerned as a result of changes in the component design and material specifications. Resource efficiency in metal-cutting manufacturing means, in terms of a specific component, primarily a reduction of energy consumption along the complete process chain. In view of the fact that the overall energy consumption of the machine tool and the cooling system exceeds several times the energy used for the metal cutting process *[4]* various essential actions are considered, investigated and implemented:

- Elimination of Process steps; shortening process chain.
- Substitution of process steps.
- Optimisation of process parameters and reducing the primary time (Base time).
- Using hybrid and integrated processes.
- Adaptation of cooling strategies.
- Resource and energy efficiency in product usage.

The developments of machine tools and adaptronic options as well as product utilization are discussed in chapters 6, 7 and 8.

Table 1: Significant Trends in Component Features and Material Specifications

Industry	***Component Fea↘es***	↗	↘	***Material Specifications*** ↗
Type & Products	***Geometry*** ↗ ↗	***Quality*** →	***Batch Size***	***Composition and properties*** ↗ ↑
Automotive and Power-train	Dimensions Complexity		→	High strength and lightweight alloys ↑ Difficult to machine ↓
Aircraft - Airframe	Dimensions NearNetShape preforms	↗ ↗	→ ↘	Composites ↗ Titanium alloys ↗ Aluminium alloys ↗
Aircraft - Engine	Dimensions Complexity	↗		Heat resistance Alloys ↗ Carbon reinforced ↗
Machine tools	Dimensions Complexity	↗	↗ ↘	High strength alloys Specific properties ↗ →
Power and Energy	Dimensions Complexity	↗	↗	Heat resistance alloys ↗ Composites ↗
Toolmaking-Dies & Molds	Dimensions Complexity			High strength alloys Hard material layers ↗
Medical, Micro and IT products	Dimensions Complexity			Titanium, ceramic, High grade steels

3.1 Shortening the Process Chain–Elimination of Steps

Shortening the process chain is an effective way of improving resource efficiency in manufacturing (Fig. 2). Approaches to this are based on reducing of the necessary number of process stages and/or operations by using near net shape forming after primary shaping and subsequent final finish cutting (a), by elimination of heat treatment in combination with finish hard cutting, by using the heat generated during the forming process for integrated hot cutting or by the elimination of separate machining steps replaced by using combined multi-function tools or process substitution, e.g. hard machining with defined cutting edge tool instead of grinding (b).

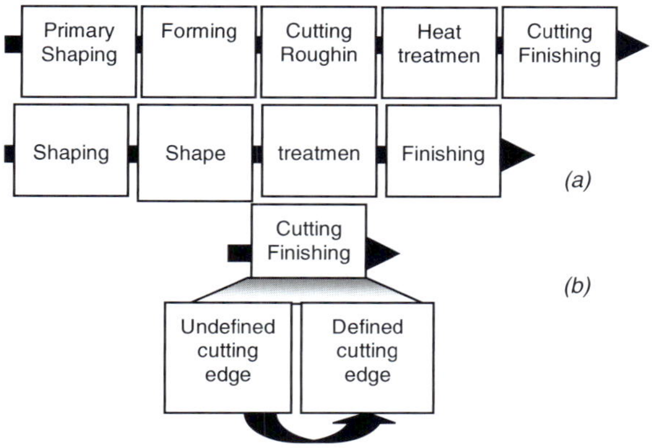

Figure 2: Approaches to shortening process chain by means of a) Process stage elimination, b) Process substitution.

Hot cutting supported by the heat generated during forming is currently at the research stage. In addition to appropriate adaptive process management strategies it requires a highly dynamic change of the cutting position during machining, a problem that can be solved by means of adaptronic systems as well as tool-related secondary-kinematic devices.

One example of shortening the process chain by means of finish hard cutting is a unique turning tool which can produce a perfectly flat finish (Figure 3); The aim of this process is to substitute grinding by hard turning without any characteristic turning marks so that, for example, sealing surfaces on gear shafts can be manufactured faster without being damaged by the helical turning marks typical for hard turning *[5]*. To comply with defined surface characteristics there is for certain applications a specially designed process kinematics linked with a rotation of the unique shaped turning tool around the tool axis providing a mark-free turned surface.

Figure 3: Special shaped turning tool (middle) during machining of a sealing surface [5].

3.2 Substitution of Process Steps

In addition to shortening the process chain, for example by eliminating some machining steps, or by combining several steps together in one continuous step, intensive investigations are directed towards substitution of some expensive operations like EDM or finishing in addition to grinding as shown in Figure 2. The more efficient rough machining or finishing with a defined cutting edge instead of grinding can in many cases provide the same final product quality and performance. A typical example is the production of dies and moulds for various industries shown in Figure 4.

Machining options of hardened steel were limited in the past to long and expensive processes like EDM and Grinding. Requirements set by the complex shapes, deep cavities, larger overcuts, high accuracy and very specific surface integrity properties can be fulfilled today by machining with new designed cutting tools, new cutting tool materials like CBN and PCD and state of the art, machining strategies, machines and adaptronic systems.

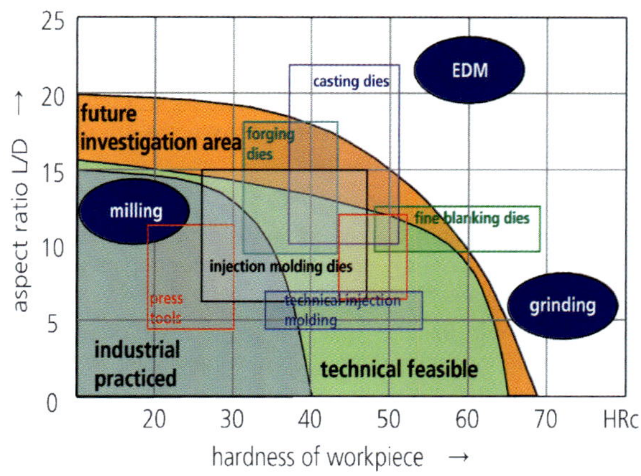

Figure 4: Increase of Resource Efficiency by using new technologies in the Tool and Die Making Industry.

3.3 Optimisation and Reducing the Primary Time

A shorter primary (base) machining time is a fundamental way of improving energy efficiency. On the process side, this generates a need for significantly higher and optimized process parameters like speed, feed and depth of cut which, in connection with the specified material and component design, represents a challenge as far as the performance capability of tool, clamping technology and machine is concerned.

Process-related possibilities for higher machining parameters lie in the use of new tooling systems and materials, hybrid processes, modified machines and specific cooling strategies. Basically it is possible to achieve a reduction in the energy consumption by reducing the required material volume to be removed and increasing the degree of material utilisation e.g. by means of Near Net Shape parts,. The primary time can also be shortened by means of process substitution and/or integration of few process steps together. The increased load on the tools resulting from higher process parameters calls for further developments of cutting materials, coatings, macro- and micro-geometry in order to ensure process reliability and adequate tool life (section 4).

3.4 Using Hybrid and Integrated Processes

Today it is possible to use hybrid processes like drilling or grinding with additional simultaneous ultra-sonic vibration or, by turning and grooving of high temperature alloys to use high pressure flushing which is acting as a "chipbreaker" or "chipformer" in addition to its effects of cooling and reduction of friction for improved cost, resources and energy efficiency.

Investigations are also carried out combining, for example, laser assisted cutting, both in turning and milling applications, thus eliminating more expensive processes like very fine finishing or grinding.

3.5 Adaptation of Cooling Strategies

Machining supported by high pressure flushing is only energy efficient for certain applications and materials. The appropriate use of cooling systems can not only significantly increase productivity but it can also help to decrease energy consumption during cutting operations. This includes dry machining and MQL but, where certain workpiece material compositions are concerned, also cryogenic cooling with liquid nitrogen or dry ice. In connection with the above options, special requirements in regard to the thermal load capability and stability of tools, workpieces and machine tool as well as chip evacuation will in turn need to be met so as to guarantee a reliable process. Examples for cooling strategies are shown in section 5.

3.6 Resource Efficiency in Product Usage

Resource efficiency in the use of machined products means above all reducing energy consumption through lower weights and reducing losses resulting from friction and wear.

Components with thinner walls and/or hollow profiles made from inhomogeneous materials create major challenges for the machining reliability. In addition they require adaptive clamping concepts with improved stability in high temperatures, sharp edge tools with low wear and adaptive machining strategies as well as appropriate machine technology.

The reduction of friction and wear during utilisation calls for development of suitable low-friction and wear-resistant surfaces. One way to achieve this goal is the surface structuring by means of modified metal-cutting or by non-conventional and physical processes.

The use of hard material layers to reduce wear makes extreme demands not only on the cutting tool but also in terms of cooling and hard machining strategy (section 4.4).

Reducing friction also means increasing dimensional accuracy in the operating conditions. This can be achieved by using preventive machining operations to compensate for any deformations that may occur during installation and operation. This solution requires, in addition to appropriate adaptive process management strategies, a highly dynamic in-process change of cutting edge position, a problem that can be solved by means of adaptronic tools as well as an additional, tool-related kinematics.

New demands in relation to machining reliability are also emerging from international regulations for reducing the use of substances that maybe harmful to health, such as lead in workpieces and certain chemicals integrated in the cooling lubricants.

Overall a number of challenges are emerging as regards resource efficiency in machining. Providing solutions is a multi-criteria optimization task which calls for consideration of process and machine tools as a whole, as well as organisational aspects for which simulation represents an appropriate method.

4 INVESTIGATIONS AND DEVELOPMENTS TO IMPROVE ENERGY AND RESOURCE EFFICIENCY

In the following, some research activities and industrial applications to improve energy and resource-efficient machining processes are presented in more detail.

4.1 High Efficiency in the Design Stage Using Modelling and Simulations

Modelling and simulation of the process chain, the machine, the tool, the workpiece and the process itself can save energy, time, material and high investment in real experiments.

During the last few decades more effort has been devoted to the use of analytical models and FEM-finite element methods for the development, application and optimization of cutting tools and the cutting process itself. Designing of cutting tools, insert geometry,chip breakers and coatings has been achieved in the past using a lot of trial and error procedures along with some physical and mechanical basic models. One of the challenges today is to optimize tool design and machining methods even before the tool is actually manufactured. The main objective is to use the finite element models to simulate the cutting process and tool behaviour and the workpiece including stresses, strains, temperatures and all geometrical variables.

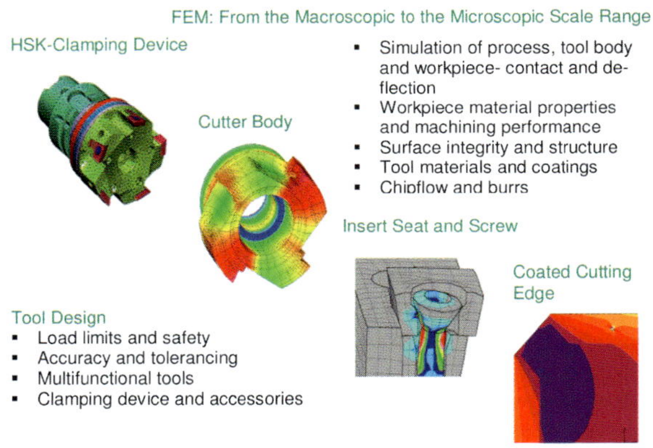

Figure 5: Simulation of clamping device, cutter body insert, micro-geometry and coatings.

Figure 5 is showing the wide range of simulation and optimization options form the clamping device like the HSK, the cutter body, the insert with the clamping screw and effective surfaces up to the coating in the micro scale range.

The various parts are subjected to machining forces and high temperatures in all directions. The cutting forces have a static and dynamic influence on tool behaviour including vibration, frequency, and deflection. The tool body is designed with maximum strength optimal chip gullet, high wear resistance and maximum symmetry to insure balancing, stability and rigidity as well as to minimize deflection. Furthermore, the FEM enables to design the connecting surfaces between the holder and machine, which insure accuracy, stability, and transfer of torque and force to the cutting edge.

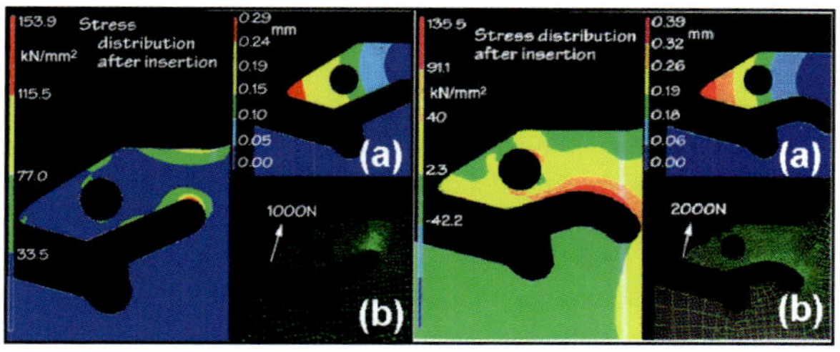

(a) Displacement of the upper jaw after insert insertion.

(b) Clamping forces acting on the upper jaw.

Figure 6: FEM analysis of displacement, stresses and forces of the self-clamping blade with straight and curved slot [6].

Another example, shown in Figure 6, demonstrates the use of the FEM to optimize tool geometry and clamping design of parting and grooving tools for significant improvement of energy and resource efficiency. The challenges of this type of tools are to reduce insert width in order to save workpiece material, to decrease machining time, to reduce forces and to save machine power. Furthermore, the cutting insert should possess optimal frontal and side-clearance surfaces and the tool holder should be able to penetrate to a larger depth with minimum down- and side-deflection. The most significant and effective design is the wedged-shaped or a parallel insert which is elastically clamped between upper and lower prismatic surfaces. The FEM used for simulation and evaluation of the upper jaw deflection upwardly and the actual stresses are shown in Figure 6.

The results are shown for a simple self-clamping device for grooving, using a straight and a curved slot to insure elasticity of the upper jaw and high clamping forces. During insert indexing, a special eccentric key lifts the upper jaw. In the straight slot tool very high stress concentrations are experienced at the rear end which may cause damage even with the 1000 N clamping force and 0,29 mm deflection. During the FEM simulation and the optimization procedure a doubled clamping force of up to 2000 N was assumed for the modified blade with the curved slot. The displacement reached up to 0,39 mm but the stress distribution after insert indexing was much more uniform along the curved slot and the maximum stress was about 16 percent lower than the maximum stress value with the straight slot.

4.2 The Developments of Cutting Tools and Tool Materials

In many industries, especially in the car and other mass production, the development and use of clever, more advanced tool systems, cutting tool materials, insert geometries and combined tools is more and more popular. Cutting tools with non-standard structures, inserts, shapes and material compositions are governing the cutting tool market providing higher performance, more stability, better tool life and easier handling. At the same time the productivity, the economy and the energy efficiency are improved for the design stage of grooving tools as shown for example in section 4.1..

In mass production the use of unique cutting tools combining a large number of single cartridges or inserts can shorten production time and process chains significantly but at the same time they can guarantee high dimensional, shape, and form tolerances. For example drilling, boring, reaming, chamfering and shoulder turning of motor blocks is done with a single very complex tool equipped with many inserts or cartridges doing the complete machining in one single operation. The tool cost is very high but in the case, for example, of an Aluminium motor block the use of many PCD inserts with much higher tool life provide a more effective, lower cost and less energy consumption than using single tools and the conventional machining steps.

The design of the cutting edge, including geometrical features like the rake face shape and rake angles, the clearance face geometry, the corner radius and the edge shape, as well as the various surface integrity features of the tools has a significant influence on the machining stability, cutting forces, power, tool life and energy efficiency.

Using the right parameters can improve not only chip flow and burr formation but can shorten machining time, reduce energy consumption and eliminate the costly finishing steps. For example the wiper type inserts, used in turning (Figure 3), provide excellent surface qualities even when using higher feeds and thereby decreasing significantly the machining time.

A new generation of tangential inserts for milling, drilling and turning applications, which can also be combined with non-continuous, serrated edges, provide better stability, longer tool life as well as improved chip flow and chip control.

Using more positive rake angles does reduce cutting forces and machine power without effecting negatively cutting edge strength due to the improved properties of new cutting tool materials.

One of the most effective and revolutionary developments in metal cutting during the last two decades can be observed in the range of cutting tool materials. It is dictated by the introduction of new workpiece materials and can provide much higher production rates due to an average of 3 to 5 times higher speed and 2 to 3 times higher loads, mainly by using much higher feeds.

With the new materials it is possible to use higher cutting speeds and feeds or to significantly extend tool life. The various coating layers and coating technologies provide a wide range of properties to compete with the demand for higher wear resistance, better chemical stability, higher toughness and edge strengths as well as to lower friction and heat transfer into the tool.

The heat conductivity of various layers differs significantly. In the low temperature range the values for TiN are the lowest ones and for Al_2O_3 ceramic much higher. But while the heat conductivity for most coated layers is increasing with higher cutting speeds or higher cutting temperatures, the heat conductivity for the Al_2O_3 – the ceramic coating, decreases with the increase of temperature. It explains why ceramic layers are considered as a heat barrier avoiding the heat to penetrate into the tool. Due to this unique property the heat flow penetrating into the cutting tool is smaller for Al_2O_3, and higher for the TiCN layers [7].

Therefore the heat transfer with the chips for the same machining conditions is high with ceramic coated inserts, low with TiAlN layers and even smaller with the TiCN coating. It should be mentioned that the largest amount of heat is transferred with the chips reaching between 60 to 70 % of the total heat generated in normal cutting. On the other hand TiC layers are harder; TiN has a lower coefficient of friction while

TiCN and TiAlN or CrAlN has the best average properties in respect of wear, chemical stability and friction required in standard machining operations.

As the result of the various properties mainly multilayer coated inserts are used taking into consideration the advantages of each composition. The TiAlN and TiCN coatings or new modifications like the CrAlN layer are the most applied as the upper coating layers for many machining applications. The coating technology for milling tools and high impact applications is mainly the PVD, due to lower temperatures of the coating process and the compression stresses in the upper layer compared with tensile stresses produced in the upper layers when using CVD coating systems with much higher coating temperatures. An example of a multilayer industrial insert with TiN, Al_2O_3 and TiCN layers is shown in Figure 7.

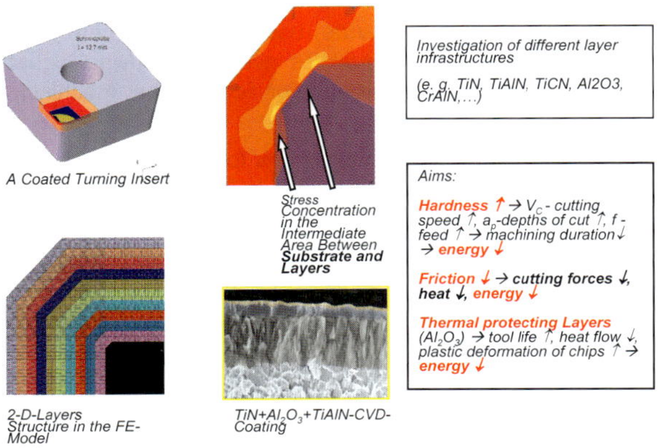

Figure 7: Development directions of coated carbide inserts using FEM.

The use of FEM for the investigation of multilayer coatings is shown in Figure 7. The simulation and analytical approach provides the possibility to evaluate stress concentration in the intermediate area between substrate and the layers, to analyze various phenomena like cracks or delamination, stresses or strain changes. The challenges set for the cutting tool materials are higher hardness and better strength, lower friction, less heat transfer and better tool protection, all to enable higher speeds, larger depth of cuts and feeds and thereby to improve tool life and energy efficiency. In comparison to the coatings the lower edge strengths or lower chipping resistance of solid PCD - Poly Crystalline Diamond, Cermets, solid Ceramics (Al_2O_3 or similar), and of Silicon Nitride (Si_3N_4) does limit the use of inserts and tools made of these materials in high impact operations. The new developments of CBN coatings in comparison with the solid PCBN are very promising but the industrial results are not very successful up to now. Tests with a combination of TiAlN and CBN showed better results and could be a significant contribution in this area. The development of various PCD - Poly Crystalline Diamond, DLC – Diamond Like Carbon and NCD – Nano Crystalline Diamond are offering new options and higher efficiency in many applications today. The thick layer diamond coating was successfully tested in machining new inhomogeneous lightweight materials, (AMC´S), like Aluminium alloys with Si or Al_2O_3 particles in comparison with other diamond coatings described in 4.3 as well as in machining hard carbide layers on forming tools shown in section 4.4.

4.3 Machining of Inhomogeneous Lightweight Materials

The industrial adaption of new workpiece materials means new challenges for production engineering, for process strategy and for the machining parameters. In addition to high strength steels and traditional lightweight construction materials such as titanium, aluminium or magnesium inhomogeneous and lightweight composites are gaining a significant position as construction materials. The inhomogeneous materials are characterized by a relatively soft and lightweight matrix and more solid reinforcing elements.

Composite materials with different properties need to be machined in one process stage under identical conditions, while the material with the higher strength determines the load on the cutting tool. From the performance point of view, the machining operation should generate a defined, uninterrupted surface characteristic on the component without any undesirable removal, or destroying, or changes in the bond between the matrix and the reinforcement material.

For example a material of this type, an aluminium matrix composite (AMC), which is being investigated at the University of Chemnitz is shown in Figure 8. It is possible to identify the structure of the AMC in (a) and a machined surface in (b). The use of CVD-thick diamond coatings with a modified chipformer and sharp cutting edge geometry was the best one to reduce surface damages to avoid the hard particles being pulled out from the matrix and to improve surface roughness [8].

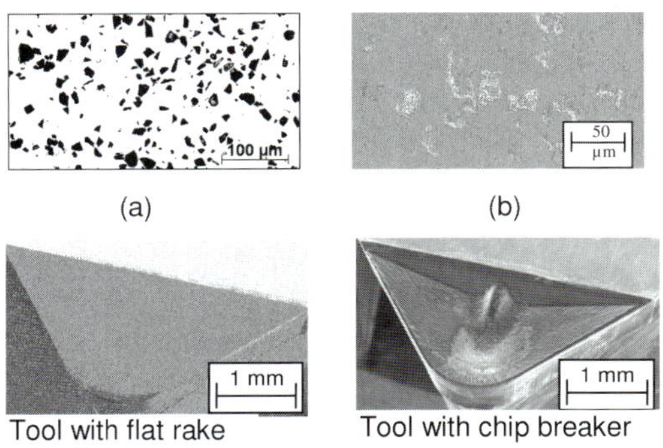

Figure 8: (a) A cross-section of an inhomogeneous AMC (b) Top view on an AMC machined surface *[7]*.

Furthermore it could be found that the homogeneity and damaged free surface can be influenced to a considerable extent by an optimized CVD coated tool geometry, and selection of machining parameters in order to produce an almost damage-free surface without separation of the reinforcement particles on AMC´s.

4.4 Machining of Hard Layers on Steel Components

Coating represents a convenient way of counteracting the higher degree of wear associated as a rule with increasing component loads. Particularly where forming tools or dies and molds and powertrain components are concerned, thick layer coatings in addition to thin layer coatings also have a role to play due to their characteristics such as high wear resistance, heat conductivity, forming properties and defined porosity. The use of hard metal for coating tribological functional areas enables considerably higher operation loads, but also calls for the further development of economic and flexible machining strategies. Up to now machining of hard metals has only been possible using grinding or spark erosion. Both processes are, however, energy, resource and cost intensive and demonstrate environmental, technological and physical disadvantages. Up to this point state-of-the-art technology was such as to permit only the machining of hardened steel, but not the machining of hard metals. The tools and materials used for interrupted cutting of hard metal are exposed to extreme loads and to enormous alternating stresses as regards emerging thermal and dynamic loads. Due to the basically high hardness of hard metals which exceeds the limits for conventional hard cutting, contradictory requirements in terms of compressive strength and toughness are sometimes the result. During the fundamental research at the IWU Fraunhofer, the focus was on the optimization of machining parameters, the efficient cooling strategies and the investigation of super-hard cutting materials such as CBN, PCD and CVD thick layer diamond. The cutting tools were single-edged brazed tips on a ball end mills with a diameter of D=6 mm and the workpiece is a WC+Co layer coated on a steel workpiece. From the test results it could be concluded that it is possible to use interrupted milling of these layers with the PCD tipped tools. The technological boundaries of hard cutting have been pushed outwards to a considerable extent.

Mechanical abrasion on the flank face was observed as the main reason for tool wear. The optimum cutting speed was around 75 m/min. CBN and CVD as tool materials have no advantage due to their tendency to chipping of the cutting edge and the subsequent reduction of tool life (Figure 9).

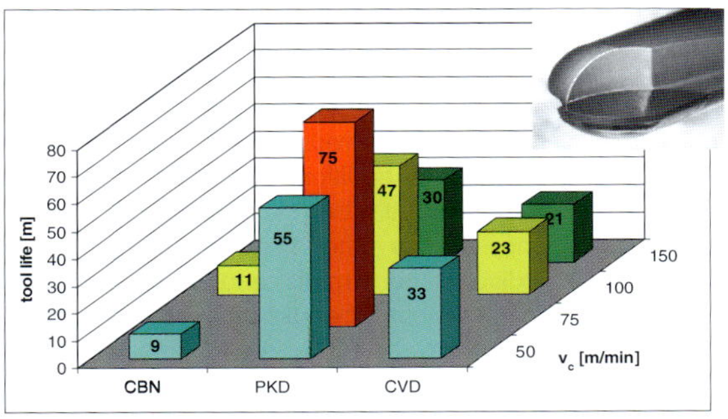

Figure 9: Effect of cutting speed on cutting length (tool life) in machining WC-17Co carbide layers, with fz = 0,05 mm; ap = 0.1 mm; ae = 0.2 mm and various tool materials.

This creates a more economic and more flexible alternative option to conventional grinding and EDM of hard metal. With the implementation of deposition welding of metal matrix composites, an improvement of the bonding strength between substrate and layer can be achieved. However the machinability of material with approx. 50 % matrix material is still very complicated. The use of these particle-reinforced materials represents a further technological challenge and calls for further extensive machining investigations.

5 COOLING AND FLUSHING STRATEGIES

5.1 Alternative Cooling Concepts

Alongside the general trend towards dry cutting or the MQL (Minimum Quantity Lubrication), alternative cooling lubricant concepts, such as cryogenic cooling with dry ice, liquid nitrogen or the use of high pressure cooling can definitely contribute towards resource-saving and economic manufacturing. These types of cooling are characterized by the fact that their range of application is material-specific. Particularly effective is the thermal effect of CO_2 and N_2 on materials that become brittle as the temperature decreases, such as e.g. 16MnCr5.

Standard cooling lubricants, used at high pressure of >100bar, are especially suitable for machining titanium and high temperature alloys. In general, the benefit lies in the improvement of process reliability and process effectiveness. A significant increase in the metal removal rate by a factor of 2 to 3 can be made possible by improving chip breaking, reducing cutting forces and improving tool wear. As a result, the machining time will decrease, which means a more economical manufacturing process despite increased expenditure on cooling.

5.2 Dry and MQL in Grinding

The final-finish cutting of precision components by grinding is frequently necessary due to the requirements in respect of workpiece accuracy as well as the surface integrity and surface roughness. At the same time, the specific grinding energy required for material removal is, at 100...300 Ws/mm3, considerably higher than, for example, for turning (1...20 Ws/mm3) and thus amounts to a multiple of the specific melting energy of steel. During grinding almost 90% of the applied machining energy is converted through friction into heat. In order to guarantee the required shape, form and dimensional tolerances, it is essential to use cooling lubricants. For this reason, as regards to the introduction of dry machining and MQL, the grinding process, by contrast to turning and milling, has not been assigned priority due to the much smaller prospects of success *[9]*. The peripheral equipment of grinding machines requires a considerable amount of energy for the supply, filtering and cooling of the cutting fluids. In practice, the electrical energy used for the peripheral equipment is considerably higher than for the actual grinding process. In view of the rising proportion of energy costs as a part of the overall production costs, the achievement of significant reductions in the grinding energy as a result of substituting the conventional flood cooling lubrication system by using minimum quantity lubrication have become an important aim in current investigations. The energy saving potential in this case can reach more than 80%. In a conventional grinding operation with a cooling lubricant a major proportion of the thermal energy produced is carried away from the working area by the cooling lubricant.

The basic precondition for dry machining with minimum quantity lubrication (MQL) is a clear reduction of the machining energy and thereby frictional heat. A further primary objective is to increase significantly the thermal energy discharged via the chips. Using appropriate process parameters, short chips with a large cross-section can be generated that are capable of absorbing a high degree of thermal energy. This is a basic requirement aimed at avoiding thermal damage to the workpiece.

Possible solutions are the use of grinding wheels with structured surfaces (Figure 10). Here conventional produced grinding wheels with larger pore volumes (Winterthur), wheels with laser drilled microstructures (Tesch) or even tools with several millimetres large slots (Lapport) could be of advantage. Significant advantages include reduced contact surface between workpiece and tool, improved chip evacuation from the contact zone as well as optimized flushing conditions directing the coolant into the contact zone when using minimum quantity lubrication.

In the current investigation it was possible to identify clear benefits of the micro-structured grinding wheels, even for grinding high surface quality.

Since an increase in heat transfer into the workpiece is, generally speaking, unavoidable when using dry machining or minimum quantity lubrication, it is necessary to identify further methods for compensation to ensure that the required form and position tolerances will be achieved.

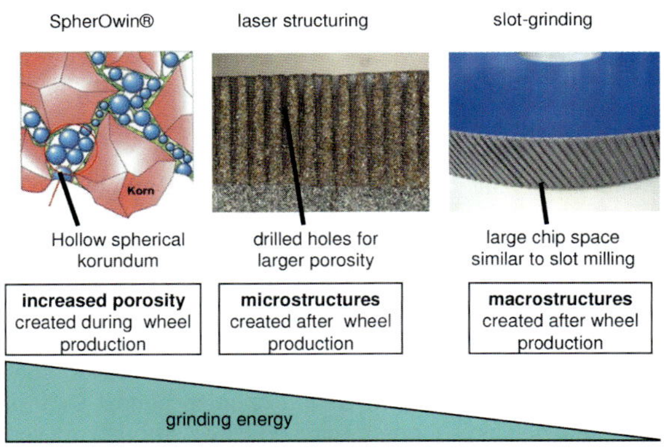

Figure 10: Structured wheels for reduction of grinding energy.

6 DEVELOPMENTS IN MACHINE TECHNOLOGY

Fulfilment of the requirements derived from process and material trends in the various industries (Table 1) will bring not only the cutting tools and the process but also the machine tool and equipment closer to their limit values. The machine must fulfil the contradictory opposed requirements of high rigidity, good damping and increased dynamics. To correspond with smaller batch sizes, flexibility is required, not only in terms of produced quantities but also in terms of higher process flexibility. In order to improve energy and resource efficiency and to diminish the gap between current limits of the machine tools to the desired and possible new limits, some basic developments of the machine tool are required. It is needed to enhance flexibility, productivity, performance, multi-function options and system reliability. Following these directions some requirements can be set towards the machine tool development including implementation of reconfiguration, modular structure, mobility, self-optimization and very high dynamics of the complete machine tool.

In this paper the aspect of machine concepts featuring a high level of dynamics is examined as an example for the above-mentioned challenges. The machine dynamics is always viewed in the sense of acceleration capability. A high level of acceleration capability includes *[10]:*

Rapid traverse: The rapid traverse speed can actually and effectively be utilized for shorter positioning times and faster tool changes.

Machining: A high acceleration capability will permit smaller contour deviations in rapid changes of cutting direction, such as e.g. in corners. The quality, process reliability and the tool life is enhanced, because machining can take place with constant process parameters and hardly any reduction in the reference feed value is necessary.

Various measures, such as e.g. the use of linear drive technology, have led to a clear improvement of machine tool dynamics over the past few years. However, in particular where large machines for example for dies and molds are concerned, these solutions are in principle reaching their limits due to the considerable weights to be moved and/or the construction space required.

From a mechanical engineering point of view, the design of larger machines is linked with the contradiction that, whereas on the one hand long travel paths are necessary, on the other hand there is the need to strive for high machine dynamics and stiffness.

For physical reasons the above specified criteria contradict one another and from mechanical studies it emerges that large machines need to have smaller natural frequencies than small machines i.e. they are dynamically speaking less convenient.

In order to overcome this contradiction two important fields of action may be contemplated. The first one is the use of lightweight construction with fibre composite elements for the mobile machine components. Another basic solution lies in integrating drives and kinematics with distinctly different characteristics into the machine tool, in conjunction with function sharing adapted to the machining task. For example, drives featuring long travel paths and poor dynamics may be combined with small drives and higher dynamic that operate in the same direction. This gives a hierarchical distribution of the overall dynamics over a number of function levels and, linked with this, resulting in the formation of redundant structures, in addition to an improvement of natural frequencies.

To introduce redundancy in the machine structure two basic principles can be used. Motion redundancy is presented in serial structures generating additional degrees of freedom giving the possibility of optimizing kinematic parameters. Actuation redundancy on the other hand is used in closed chains causing additional constraints and enables internal preloading. These effects can be used to optimize force-related parameters in terms of stiffness and singularity control.

The analogies of serial and parallel kinematics suggest the following breakdown of functions using motion redundancy [11]: serial structure for increased travel paths and speeds, parallel structure for high acceleration and jerk. A number of basic studies in this field have led to the concept of “scissor” kinematics, which links the broad operating range of serial kinematics with the high level dynamics of the parallel kinematic machines - PKM (Figure 11).

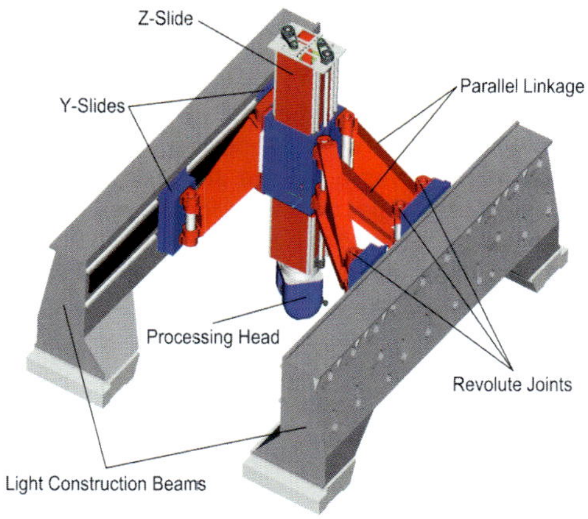

Figure 11: CAD Illustration of the 2 DOF “scissor” PKM with the four y-slides for improved dynamics

For the evaluation of the machining results that can be achieved with the developed machine, specific workpiece geometry was used and the results compared with a conventional machining centre. The machining time of the test-piece with the new machine was approx. 18 % shorter than with the conventional one. The time saving resulted primarily from the reduction in cutting time and illustrates the potential inherent in the new machine concept. In addition to the higher productivity the machining quality was also improved, something which could be demonstrated by the significantly higher contour accuracy compared to the profile achieved with the conventional machine.

7 ADAPTRONIC SYSTEMS FOR HIGHER PROCESS EFFICIENCY AND IMPROVED PRODUCT UTILIZATION

Mechatronic and adaptronic systems can provide significant improvements in resource and energy efficiency. It can optimize each process by itself, shorten the machining time, save expensive finishing operations and provide better component quality. Better quality or targeted shape modifications can save significant energy during utilisation of the component and extend product life. In order to demonstrate the various advantages a recent development from the automotive industry will be presented. Where combustion engines are concerned, deviations in cylinder form have a decisive influence on the tribological system of piston, piston ring and cylinder liner. It was found that 70 – 80% of the unwanted oil consumption originates from the piston system.

The friction proportion of the piston system may amount to approx. half of the overall engine friction. Thus it is particularly important to reduce friction in this area. From an environmental and efficiency viewpoint an important development objective lies in minimizing cylindricity deviations of the liner bores in the crankshaft housing when the combustion engine is in operating condition. Ultimately a reduction in fuel and oil consumption is to be anticipated, assuming the same engine output as well as reduction of CO_2 emission to the environment. The deviations from the ideal cylinder form arise as a result of static distortion e.g. as a result of cylinderhead mounting, because of quasi-static and dynamic distortion as a consequence of temperature differences in operation, variable thermal expansion, gas pressure and forces transmitted from the piston on to the cylinder face.

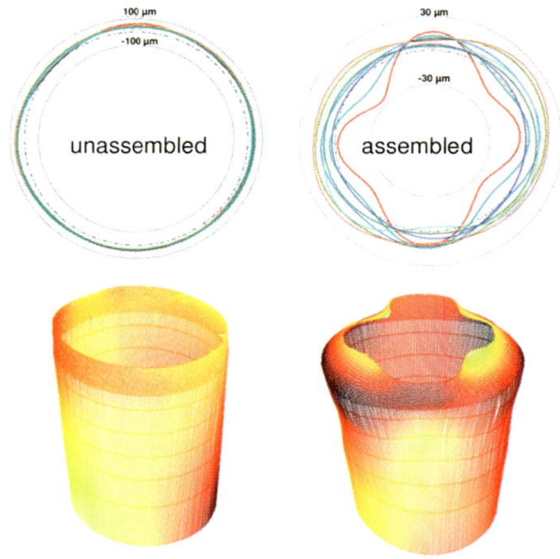

Figure 12: Assembly-related form deviations of the cylinder liner (IWU Chemnitz).

In Figure 12 a maximum cylindricity deviation from the ideal cylindrical of around 50 µm can be observed. Here the largest deviation occurs in the upper half of the liner close to the cylinder head.

The clover leaf-shaped fourth order distortion of the cylinder surface with a superimposed oval distortion in the upper section of the liner is clearly visible and occurs as a result of the introduction of force via the four cylinder head screws.

At present the static distortion resulting from the cylinder head mounting can be compensated for by the complex and expensive use of dummy heads in the final honing process. Quasi-static and dynamic cylinder distortions cannot be entirely prevented or compensated for at the present time, either by means of design or by adopting a production-orientated approach.

The challenge as far as production techniques are concerned lies in developing effective finishing processes that will make flexible and productive out-of-round machining possible.

The necessary highly dynamic relative movement between cutting tool and workpiece can be achieved here by applying a range of process principles.

Where processes with a defined cutting edge tool are concerned, various approaches from tool-based solutions right through to highly integrated mechatronic solutions can be identified (Figure 13).

For a particular problem, the initial decision is whether a single-axis movement or a multi-axis super-imposed system movement should be used. With a single-axis movement, a tool-integrated adaptronic device plays a major role *[12]*.

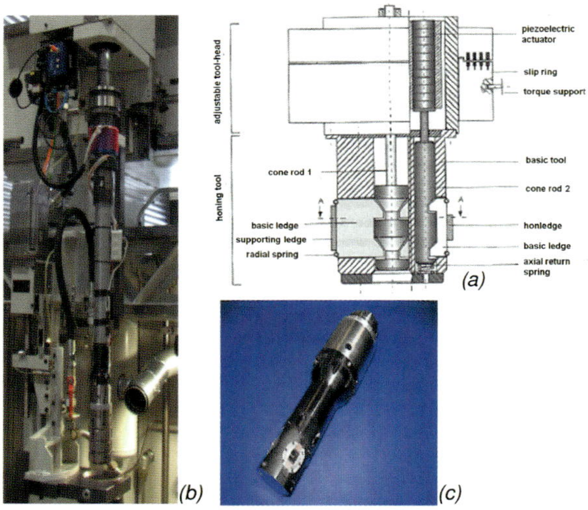

(a) Honing tool. (Gehring), (b) Test bench form honing Fh- IWU, (c) Intelligent drilling device Fh- IPT

Figure 13: Adaptronic tools for highly dynamic out of round-machining.

For multi-axis movements, solutions based on spindles with magnetic bearings and sub-kinematic devices integrated in the machine are used. This leads – similar to large machines – to a redundancy. Nevertheless, the frequencies required here are considerably higher and the space for movement distinctly smaller (< 100μm).

One example of using a sub-kinematic device is the hexapod with piezo-actuators developed at IWU (Figure 14).

The device is designed as an interface between motor spindle and machine structure and allows for free movement in all directions and, thereby for high technological flexibility.

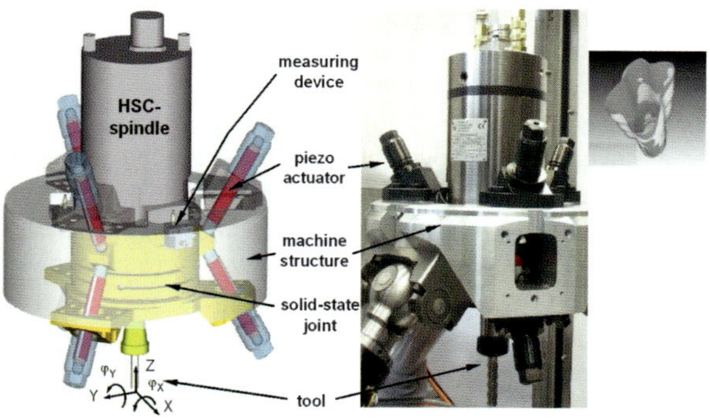

Figure 14: The Principle of a Hexapod Sub-kinematics System for Highly Dynamic Spindle and the achieved Profiles.

Figure 14 also shows the actual honed contour by comparison to the reference geometry. The machining task is the generation of the 4th order form arrangement with a radial value of 20 μm for a liner diameter of 81 mm. The measured difference between the actual and the reference values is in the range of +2 to -3 μm. Transfer of the results into an economic and process-reliable production process is the aim of future activities.

8 MICRO STRUCTURING OF SURFACES FOR REDUCTION OF FRICTION AND WEAR DURING PRODUCT UTILISATION

A further aim of fine machining is to improve the friction and wear characteristics of surfaces moving components during utilization. This involves a defined structuring of the surface with suitable technologies as shown for example in Figure 15. There are various possibilities as regards to surface structuring, including targeted arrangement of the microstructures which can influence friction and wear [13]

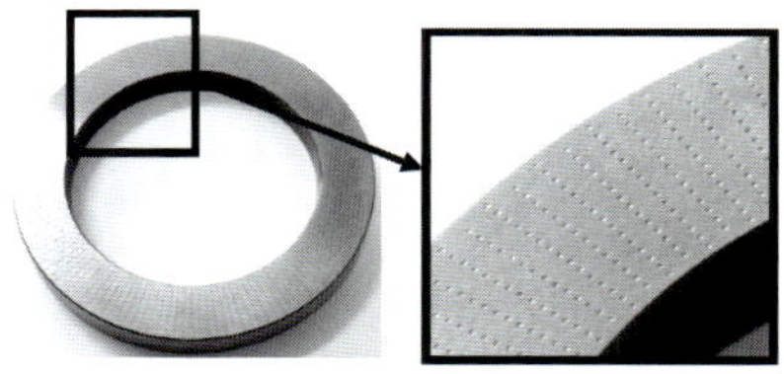

which can influence friction and wear [13]

Figure 15: Targeted structuring of rings using Jet-ECM for minimizes friction.

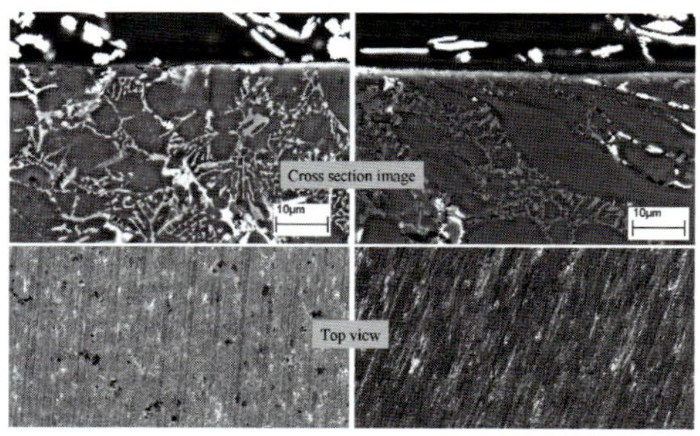

(a) (b)

Figure 16: (a) Machined surface without any modification (left); (b) Surface with nano-structured boundary layer following fine machining;(right) [13].

A further approach aimed towards reducing friction and wear is the specific configuration of the surface, as a function of the machining parameters and the edge geometry of the cutting tool.

In the Project "TriboMan– Production-integrated Reduction of Friction and Wear in Combustion Engines", a modified cutter geometry is used to investigate the surface of a piston bearing of an engine, machined with varying process parameters. By adjusting stresses in the surface layer and thus influencing the boundary layers, it is possible to achieve a "preliminary training" of the bearing surfaces so that, in subsequent operation, friction and wear can be minimized (Figure 16).

In addition to minimizing friction and wear micro structuring through a modified cutting process is also an option for data storage and for the reduction of fluidic losses [14].

9 SUMMARY

Limitations of natural resources and increasing awareness of climate changes are forcing a change of paradigms to the sustainable use of raw materials and energy. At the same time the new challenges need to give consideration to both product usage and the manufacturing phase. Product design, which has a significant effect on resource efficiency, is therefore opting for a reduction in consumption based on lightweight construction and a reduction in friction. Therefore the production and application of thin walls, hollow, geometrically complex components with surface integrity higher in demand and structure using high strength, materials inhomogeneous and, as a result, difficult to machine is continuously growing with a tendency towards smaller batch sizes.

With the transition to shortening of the process chain, an excellent opportunity for improving resource efficiency in manufacturing is emerging, based on elimination or combination of process steps or implementing near net shape preforms as well as using final finishing cutting options.

Metal-cutting itself is likely to maintain its importance in the medium term and will need to find answers for a growing number of requirements. One important route lies in reducing to the primary or basic time used for part machining while assuring process reliability and quality. For this a coordinated increase in the performance capabilities of process, tool and machine will be required.

Considerable potential also lies in the effective use of cooling lubrication system. On the one hand this means wherever possible to use dry cutting or minimum quantity lubrication (MQL) whilst, on the other hand it means a material-specific use of high pressure cooling or the use of alternative cooling media such as liquid nitrogen and dry ice.

The targeted creation of functional surfaces by means of structuring, the preventive deformation compensation through metal-cutting and the machining of hard layers are expected to gain importance in the future.

The implementation of enhanced-performance and/or enhanced-function processes calls for developments in relation to tools (cutting materials, coatings, geometry, actuators in the tool for cutting edge adjustment) as well as for highly dynamic machine technology that can be adapted to the various components that feature good damping and high stiffness. They all have to operate in a resource-efficient manner.

Basically, resource and energy-efficient metal-cutting machining represents a multi-criteria optimization task, the solution to which calls for the optimization of the overall system of process, tool, workpiece, fixture and machine.

10 REFERENCES

[1] Neugebauer, R; Sterzing, A, Koriath, H-J.: Vision einer enrgieautarken Fabrik-Beitrag des Spitzentechnologieclusters eniPROD. In: Neugebauer,R., Energieeffiziente Produkt- und Prozessinnovationen in der Produktionstechnik, Verlag wissenschaftliche Scripten, 2010.

[2] Neugebauer, R.: Ressourceneffizienz in der Produktion –JETZT! Abschlußpräsentation der BmBF-Studie EFFPRO-Untersuchung zur Energieeffizienz in der Produktion, Fraunhofer-Gesellschaft, Karlsruhe 2008.

[3] Danielczik, M.: Prozesssichere Lösungen für das Bohren von homogenen und heterogenen CFK-Stacks. In: Neue Fertigungstechnologien in der Luft- und Raumfahrt, Hannover 24./25.11.2009.

[4] Ablele, E. ; Kuhrke, B.; Rothenbücher, S.: Entwicklungstrends zur Erhöhung und Bewertung der Energieeffizienz spanender Werkzeugmaschinen.In: Neugebauer,R., Energieeffiziente Produkt- und Prozessinnovationen in der Produktionstechnik, Verlag wissenschaftliche Scripten, 2010.

[5] N. Kummer; B. Vogt; Drallfreies Drehen ersetzt aufwändige Schleifprozesse, Industrie Diamanten Rundschau 38 Nr. IV, 2004.

[6] Wertheim, R.& Satran, A. (2000). Vyuziti metody FEM pri konstrukci nastroju, [In Polish], Technicky, Tydenik, 25/2000, Rocnik XLVIII.

[7] Wertheim, R., Ber, A., Porat, R. & Sivan, R. . "Characterization of CVD Coated Carbide Layers and their Thermal Properties". Annals of the CIRP, Vol. 31/1,(1982) 7-11.

[8] Bode, H.-O.: Einfluss einer energieeffizienten Produktion auf Planungs- und Produktprämissen am Beispiel der Motorenfertigung, XII. Internationales Produktionstechnisches Kolloquium PTK 2007, 11./12.10.2007, Berlin.

[9] Klocke, F.; Beck: Technologie des Hochleistungsschleifens mit reduziertem Kühlschmierstoffvolumenstrom, Tagungsband Schleiftechnisches Kolloquium 2000, S.25-39.

[10] Kroll, L.; Blau, P.; Wabner, M.; Frieß, U.; Eulitz, J.; Klärner, M.: Leichtbaukomponenten für energieeffiziente Werkzeugmaschinen. In: Neugebauer,R., Energieeffiziente Produkt- und Prozessinnovationen in der Produktionstechnik, Verlag wissenschaftliche Scripten, 2010.

[11] Neugebauer, R.; Drossel, W. G.; Ihlenfeldt, S.; Harzbecker, Ch.: Design method for machine tools with bionic inspired kinematics. Annals of the CIRP, Vol. 58/1, (2009) 371 - 374 gilt für Fig.7, 8, 9.

[12] Wiens, A.; Dr. Lahres, M.; Dr. Hoffmeister, H.-W.; Flores, G.; Fertigungstechnischer Ansatz zur Kompensation von Zylinderverzügen mittels Formhonen; 5. VDI-Fachtagung Zylinderlaufbahn, Kolben, Pleuel; 07./08.06. 2010 Baden-Baden.

[13] Schubert, A.; Neugebauer, R.; Avila, D.; Sylla, D.: Fertigungs- und Oberflächentechnologien für anforderungsgerechte, tribologisch effiziente Oberflächen von Umformwerkzeugen und Powertrain-Bauteilen, In: Neugebauer,R., Energieeffiziente Produkt- und Prozessinnovationen in der Produktionstechnik, Verlag wissenschaftliche Scripten, 2010.

[14] Denkena, B., Kästner, J., Wang, B.; A, Advanced microstructures and its production through cutting and grinding, CIRP Annals- Manufacturing Technology 59, 2010.

Study on Mechanical Grinding Characteristic and Mechanism of Renewable Rubber

M. Peng, S. Jianbo, X. Dong, D. Guanghong

Department of Precision Instruments and Mechanology, Tsinghua University, Beijing, P.R. China

Abstract

Rubber is a kind of very important strategic natural resource. In China, the consumption growth rate of rubber in 2009 was 6.9% and the total consumption has reached to 5.88 million tons. China has become the largest rubber consumer in the last 8 consecutive years. However, the discarded rubber products also increased greatly and have reached around 7 million tons in 2009. Since main rubber consumption depends on imports, the regeneration of rubber products becomes a critical strategy to reduce rubber supply gap. In order to regenerate used rubber products, such as tires, the grinding process is the essential mechanical process to decompose of rubber and make the rubber particles suitable for regenerating. Based the analysis of the mechanical properties of rubber, a mechanical grinding pan system is designed to study the grinding mechanism of rubber. The rubber could be grinded around 0.2mm and some even reached to Nano-scale. The grinding process is optimized by studying of the rubber particle concentration, morphology features, TGA and mechanical properties of regenerated products. The study provides an effective approach to grind rubber and could save natural rubber resources as well as reducing the potential threats to the environment.

Keywords:

Renewable Rubber; Mechanical Grinding; Regenerating

1 INTRODUCTION

According to the forecast of International Rubber Study Group (IRSG), the global rubber consumption in 2010, including natural rubber and synthetic rubber, will reach to 23.7 million tons. Though the uncertainty of global economy has brought negative effect to the demand for rubber consumption, the stimulation for car consumption in China provides strong support for the steady global growth of rubber demand. China has become the largest rubber consumer in the last 8 consecutive years. For various rubber products, the production of tyre consumes about 60% rubber. With the rapid increase of car consumption, the current annual yield of tyres has reached to 0.4 billion, which shares about the 1/4 output of the global market.

However, the annual amount of waste tyres is near 0.2 billion, and the weight of these tyres has exceed 3 million tons. Besides waste tyres, the discarded rubber products also increased greatly and have reached around 7 million tons in 2009. The amount of waste rubber products accounts for 1% of the total industrial solid waste. Such black pollution has caused severe attention of the world. Rubber products are elastomers which are very hard to degrade. For many countries, such as USA, Canada and China, large amount of stacked waste tyres have caused great threats to the environment.

Since main rubber consumption in China depends on imports, the regeneration of rubber products becomes a critical strategy to reduce rubber supply gap. Take waste tyres for example, the composition and potential usage of waste tyres are show in table 1.

Table 1: Composition and usage of waste tyres [1]

Composition	Weight Ratio	Usage
Synthetic Fibre (nylon)	22~24%	Plastic products
Steel Wire	16~18%	Spring steel
Rubber mixture	58~60%	Reclaimed rubber, rubber powder

Table 1 shows that waste tyres are potential misplaced high-value resource. Almost all parts of the tyre could be recycled and reused. Materials reuse and chemical treatment are the two kind of typical disposing method for waste tyres.

Table 2: Disposing of waste tyres

Composition	Usage
Materials Reuse	1. Tire Retreading; 2. Direct reuse as infrastructure (Dikes, traffic shield, etc.); 3. Grinding (Rubber powder, reclaimed rubber);
Chemical Treatment	1. Pyrolysis (fuel gas/oil, oligomer, carbon black); 2. Combustion (Fuel for cement plant, etc.)

Considering the less negative environmental effect, this paper studied rubber material reuse which concentrates on the mechanical regenerating process.

2 MECHANICAIL TREATMENT PROCESS

2.1 Overview

The aim of mechanical treatment process is to grind waste tyres into rubber powder, which could increase the surface area of the rubber particles and facilitate its vulcanization mixing with matrix rubber to make more stable products [2-5]. According to the disposing temperature, the treatment process could be divided into the following two typical types:

1. Room temperature grinding: The rubber powder acquired at room temperature has irregular shape, larger surface roughness and area [6-7]. The size of most rubber particles ranges from 0.075mm to 0.5mm.
2. Cryogenic grinding: The waste tyres or rubber products are frozen below glass transition temperature with liquid nitrogen. The shape of the rubber powder is spherical particles with smooth surface after grinding.

The efficiency of cryogenic grinding is higher than room temperature grinding, but the cost of the former one is lower and the rubber particles with rough surface are easier to mix with matrix rubber. This paper concentrates on the room temperature grinding.

2.2 Grinding process design

The whole grinding processes are divided into three stages: rough grinding, intermediate grinding and fine grinding. The flow chart of grinding process is shown in Fig.1.

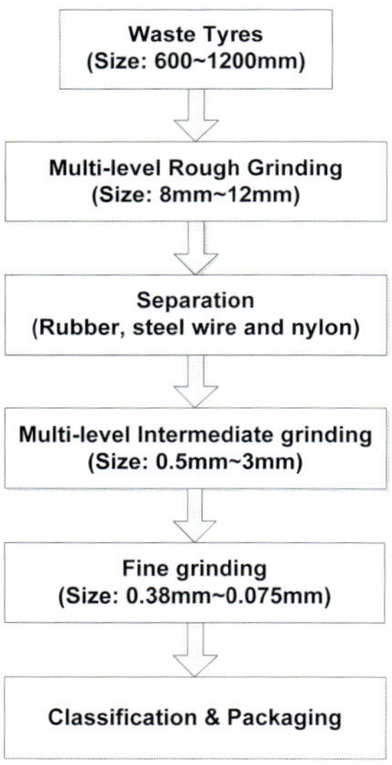

Figure 1: Flow chart of grinding process.

In order to avoid severe temperature rise and prevent rubber particles from getting burn during the grinding process, an effective cooling system is necessary. Usually, water cycle system and air vortex are the main cooling medium. Moreover, the active agent is added during the grinding process to increase the surfactant.

3 EXPERIMENT DESIGN

As for the whole rubber grinding process, fine grinding is the most challenging technology, which decides the efficiency and quality of rubber powder. The experiment design concentrates on the optimization of fine grinding to improve the processing parameters. The key parameters of fine grinding include velocity, temperature and gap between two abrasive discs. The structure of abrasive discs is shown in Fig.2.

Figure 2: Structure of abrasive discs.

The abrasive discs are welded with high speed steel blades around the disc. The gap between the two discs could be adjusted to control the grinding process. The grinding efficiency, operating temperture and granularity of rubber powder are directly decide by the disc gap. During the grinding process one abrasive disc is mounted to the grinding machine without rotating, and the other disc rotates at very high velocity, about 1440rev/min. At the same time, the coarse rubber partivles are sent between the two discs through the center hole of the fixed disc.

3.1 Material and formula

The waste rubbers used for the grinding experiment are cut from the tread rubber to keep the materials more uniform. Other supplementary materials are SBR1502, zinc oxide, stearic acid, sulphur, accelerating agent, carbon black and toluene. The characteristics of rubber powder could be identified by mixing and vulcanization. The composition is shown in table 3.

Table 3: Formula of mixing and vulcanization

Material	Dosage (Unit)
SBR 1502	100
Rubber particles	100
Sulphur	2.4
Stearic acid	3
Zinc oxide	6
Carbon black	55
Accelerating agent	2

3.2 Process conditions of sample

The rubber powder is blended to make test sample blocks according to the formula listed in table 3. The mixing roll is open mill and the velocity ratio of the rollers is 1.2. The mixing time is 10 minutes. The mixing rubber, about 35g, is used to make test block after saving for 24 hours. The curing conditions are limited to 150 degree centigrade, 10MPa and 8 minutes.

3.3 Testing and Characterization

The main experiments are concentrated on the following 4 aspects:

1. Feature of rubber particles: SEM is used to observe and compare the shape and appearance of the rubber particles. In order to increase the conductivity, the surface of rubber particles should be sputtered with gold.
2. Mechanical properties: The tensile strength, tensile modulus and elongation of vulcanized rubber blocks are tested by universal testing machine. The original length of rubber blocks are 25mm and the drawing speed is 500mm/min.
3. Crosslinking density: Taking toluene as the solvent, the crosslinking density is tested by swelling method and calculated by Flory-Rehner. Since tread rubber contained carbon black, the calculated crosslinking density should be improved by formula Kraus[8].
4. Thermo-gravimetric analysis (TGA): TGA is used to compare the thermal stability.

4 RESULTS AND DISCUSSION

4.1 Morphology feature of rubber powder

The size of rubber particles before fine grinding ranges from 0.5mm~3mm. The surface topography of the particles is observed by SEM. Fig.3 shows the surface character of rubber particles before fine grinding.

(a) SEM of rubber particles before fine grinding

(b) SEM of rubber particles after fine grinding

Figure 3: SEM of rubber particles before/after fine grinding.

Fig.3a and 3b show that the size of rubber particle reduced from 1~3mm to around 0.1mm. More importantly, the surface of rubber particle after grinding becomes coarser and has larger specific area, which makes them easier to mix with matrix rubber.

In order to increase the surface activity of rubber powder, the fine grinding process is divided into three stages:

1. Pre-grinding: After multi-level intermediate grinding, the coarse rubber particles are sent to disk mill for grinding without adding any additives. The final size of output rubber powder concentrates around 0.1mm.
2. Surface activation. After pre-grinding, RX-80, a special synthetic activator, is used as the additive. The main components of RX-80 are xylene, formaldehyde and rosin. RX-80 is added to the rubber powder and mixed for about 30 minutes under 125 degrees centigrade. The basic chemical constitution of RX-80 is shown in Fig.4.

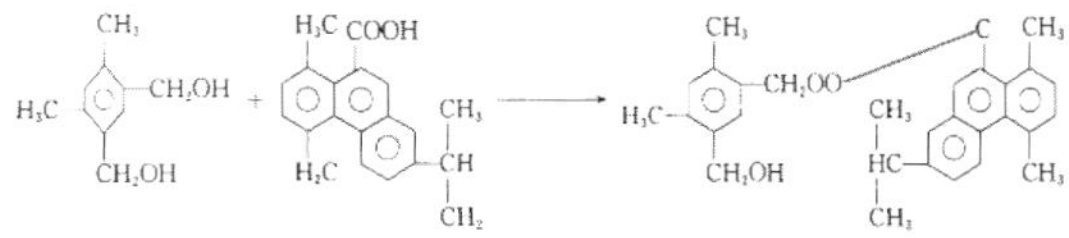

Figure 4: Chemical constitution of RX-80.

3. Re-grinding. After surface activation, the activated rubber powder is sent back to disk mill for second time grinding. The surface activation performance is improved by re-grinding.

The comparison of the above three kinds of rubber powder are shown in Fig.5.

(a) Rubber particles of fine pre-grinding

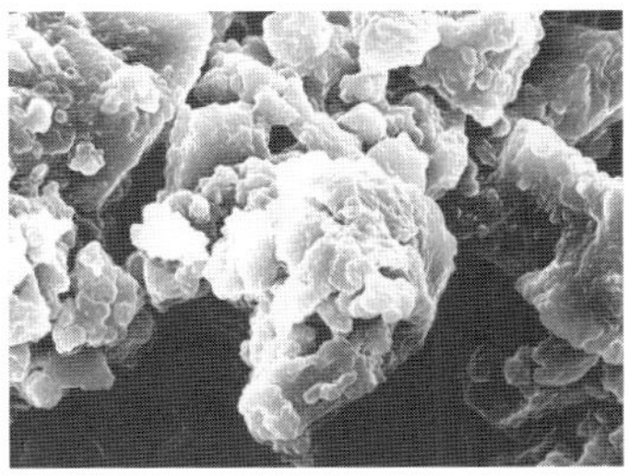

(b) Rubber particles after surface activation

(c) Rubber particles after re-grinding

Figure 5: SEM of rubber particles after fine grinding (The figures in a/b/c share the same SEM scale).

Based on the SEM observation, the actual testing also show that the rubber powder is getting softer after surface activation and re-grinding. The size distribution and specific surface area of re-grinded rubber powder increased greatly.

During the re-grinding process, RX-80 could be completely mixed with rubber powder and the grinding force could cause mechanochemical reaction to improve the surface performance and so as to improve the binding ability with matrix rubber.

4.2 Mechanical properties

The mechanical properties are the most important index to evaluate the quality or performance of rubber powder. The different rubber powder, before and after fine grinding, are mixed with styrene-butadiene rubber (SBR) and then vulcanized to make standard test blocks. The results are shown in table 4.

Table 4: Strength comparison before and after grinding

Test blocks	Fracture elongation (%)	Tensile Strength (MPa)	Hardness
Before grinding	302	18.3	72
After grinding	447.6	17	67

Table 4 indicates that the rubber powder after fine grinding has larger fracture elongation, which has been improved at 48.2%. The experiment shows that the size of rubber particle has great impact to the fracture elongation. Though the tensile strength and hardness of test blocks made by the rubber powder after grinding are lower than the coarse one, the reuslts are acceptable.

Taking velocity and disc gap as the main influencing factors, the tensile strength and elongation at break are compared to optimize the grinding process. The rubber powders under different grinding condition are mixed with SBR for comparison. The mechanical properties at different velocity and disc gap are shown in table 5.

Table 5: Strength comparison at different velocity and disc gap

Velocity (rev/min)	Disc gap (mm)	Tensile strength (MPa)	Elongation (%)
1200	0.57	16.9	427.2
1440	0.57	17	447.6
1540	0.57	17	460.6
1200	0.51	13.6	431
1440	0.51	14.6	453.2
1540	0.51	16.8	471.2
1440	0.61	12.4	367.8

Table 5 shows that the tensile strength and elongation is increasing with the velocity. The velocity plays positive effect to the regeneration of rubber powder. The effect of disc gap is more complicated than velocity. The data show that disc gap has significant impact to the tensile strength and elongation, but the impact is nonlinear and optimal gap should be find out to optimizing the grinding process.

The mechanical property of rubber powder got from surface activation and re-grinding are different and it provides the chance to improve the fine grinding process. The strength of the above two kinds of rubber powder are list in table 6.

Table 6: Comparison of tensile strength at different temperature and additive

Temper ature	Additive	Tensile strength (MPa)	
		Surface activation	Re-grinding
115	5	16.1	20
125	5	17	21
135	5	17.1	17.9
125	10	16.4	18.8
135	10	14.9	17.7
125	15	15.3	18.8
135	15	15	17.6

Table 6 shows that the tensile strength of re-grinding is larger than surface activation, but the temperature and additive did not impact the strength too much. Compared with the testing sample without re-grinding, the tensile strength has been improved app. 23% after re-grinding.

4.3 Crosslinking density

Crosslinking density is the mole number of sulfur cross-linked bond contained in one cubic rubber, and it is an important structural parameter, which indicates the degree of crosslinking of rubber. The comparison of crosslinking density of rubber powder, before and after fine grinding, is listed in table 7.

Table 7: Crosslinking density

Rubber powder	Crosslinking density (10^{-5}/cm^3)
Before grinding	17.204
After grinding	6.111

The table 7 shows that the crosslinking density of rubber powder decreased for about 65% after grinding, which indicates that the cross-linked bond have been cut off greatly. The reduction can verify that the structure of rubber is destroyed to some extent during the grinding process, which is called as mechanochemical effect. Combined with FTIR spectrum and TGA curves, it can be speculated that the mainly destroyed bonds are S-S bonds and/or S-C bonds rather than the C-C bonds which constitute of the backbones of rubber.

4.4 TGA

In order to compare the thermal stability of rubber powder before and after grinding, the TGA analysis is conducted by using TGA-Q600. Thermal stability of different rubber particles is shown in Fig.6. The data of TGA before and after grinding are marked as TGA (A) and TGA (B).

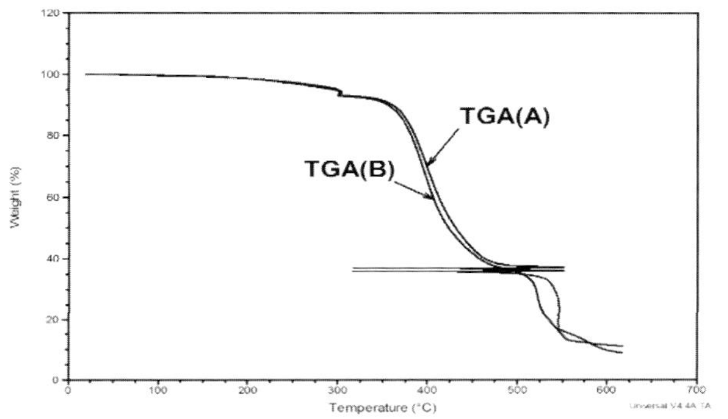

Figure 6: TGA of rubber powder.

The results show that the changing trend of different rubber powder is identical, but the rate of change is different. The reaction of grinded rubber powder is more active, which indicates its lower thermal stability after cutting some cross-link bonds.

5 SUMMARY

Based on the analysis of rubber powder morphology features, mechanical properties, crosslinking-density and TGA, this paper studies the grinding mechanism of waste rubber. The study provides an effective approach to control and optimize the grinding process of waste tyres, which save natural rubber resources as well as reducing the potential threats to the environment.

6 ACKNOWLEDGMENTS

The project is supported by National Natural Science Foundation of China (Grant number: 50875142).

7 REFERENCES

[1] Dong Chengchun, 2008, Recycle of Waste Tyres, Chemical Industry Press, Beijing.

[2] Adhikari, B., De, D.; Maiti, S., 2000, Prog Polym Sci, 25:909.

[3] Klingensmith, W., Baranwal, K., 1998, Rubber World, 41:218.

[4] Stevenson, K., Stallwood, B., Hart, A., 2008, Bioremediation Journal, 1.

[5] Bandyopadhyay, S., Agrawal, S. L., Ameta, R.; Dasgupta, S., Mukhopadhyay, R.; Deuri, A. S., Ameta, S. C.; Ameta, R., 2008, Progress in Rubber, Plastics and Recycling Technology, 24:73.

[6] Samet, H., 1990, Applications of Spatial Data Structure, Addison-Wesley, Reading, MA.

[7] Bilgili, E., Arastoopour, H., Bernstein, B., 2001, Powder Technol, 115-277.

[8] E.Bilgili, H.A.B.B., 2001, Pulverization of rubber granulates using the solid state shear extrusion process Part II: Powder characterization copy, Powder Technology, 115:277-289.

A Holistic Framework for Increasing Energy and Resource Efficiency in Manufacturing

C. Herrmann, S. Thiede, T. Heinemann

Institute of Machine Tools and Production Technology, Technische Universität Braunschweig, Germany

Abstract

The derivation of promising measures to improve energy and resource efficiency demands an integrated, system-oriented perspective in order to avoid conflicts of goals and problem shifting. This involves an extended understanding of production processes, process chains and factories, including all relevant input- and output flows and their dynamic consumption respectively emission patterns. Against this background the topic energy and resource efficiency calls for different fields of action which are presented within this paper: based on different measurement strategies, the deep understanding of single processes and interdependencies within process chains as well as suitable methods for the evaluation and prediction of energy or resource consumption patterns (e.g. through energy oriented simulation), it is finally a major goal to integrate energy-related key figures into conventional industrial decision support systems within production management.

Keywords:

Energy efficiency, energy oriented simulation, resource efficiency, sustainability in manufacturing

1 INTRODUCTION

Production can be defined as combination of production factors labour, material and technical equipment for the purpose of value creation in form of products [1]. Therefore a significant amount of resources and energy is necessarily involved whereas a certain part of the input is used for creating value, another part is wasted in terms of losses, heat, and emissions. For instance, producing companies in Germany consume 47% of electricity and just with that responsible for 18% of the national CO_2-emissions (plus about 20% by direct emissions, e.g. by combustion) [2].

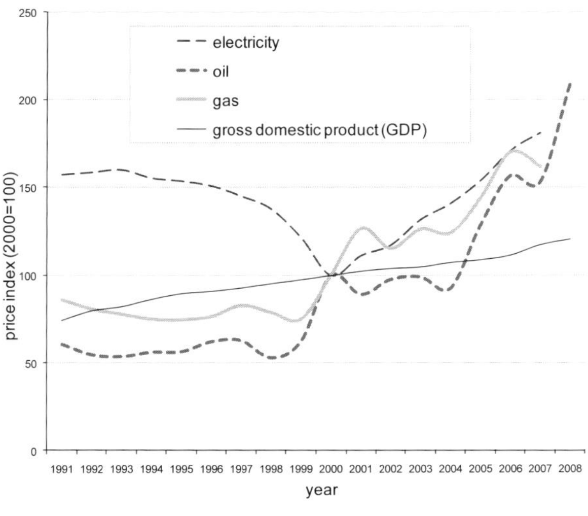

Figure 1: Development of energy prices in Germany (compared to GDP) [2].

For various reasons the consideration of energy and therewith resource consumption is more and more focused by manufacturing companies nowadays [e.g. possible future shortage of strategically relevant natural resources, environmental regulations and regulative incentives (e.g. introduction of CO_2 certificates), rising public awareness/global warming discussion, availability of new and more eco-efficient technologies (e.g. efficient electric drives, biomaterials) and (as a result) necessary investment for environmentally sound technologies]. Whereas all these issues naturally involve an economic dimension as well, a major driver are certainly rising energy costs. Figure 1 shows the progression of energy prices in Germany over the last years. Despite the meanwhile relaxed price level due to the global macroeconomic situation, having in mind global developments (e.g. population and economy growth, resource depletion) it can be assumed that there will be further increases in future.

2 HOLISTIC PROCESS AND SYSTEM COMPREHENSION

While considering energy and resource efficiency is an important issue, a realistic and goal-oriented identification of improvement potentials and actual measures demand a comprehensive, system-oriented view. Therefore, based on the current state of research as well as industry conditions the following preconditions can be defined [3]:

- **Extended process comprehension:** In order to avoid focusing on minor relevant issues (while neglecting major challenges) and local optimization with problem shifting, explicitly all relevant input- and output flows of production processes must be considered. This includes all energetic (e.g. compressed air, electrical power, waste heat) and material (e.g. auxiliary materials as cooling lubricants) flows, which lead either directly or indirectly to additional energy and/or resource consumption. Against this background Figure 2 shows an extended process model which integrates previously rather separately considered ecological and economic process perspectives [4][5].

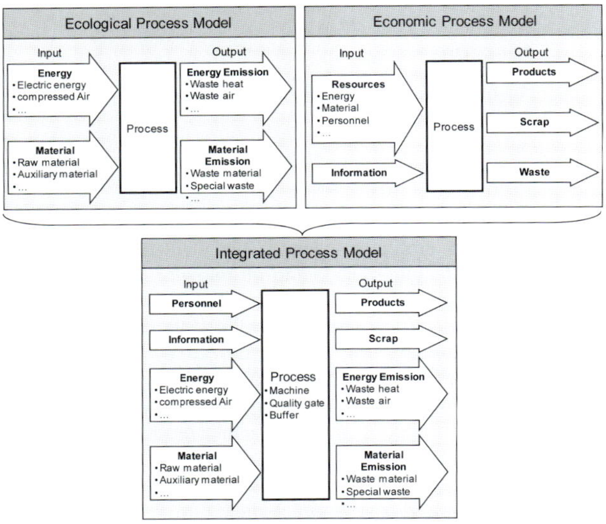

Figure 2: Integrated Process Model [3] [5].

- **Holistic system definition of the factory:** manifold interdependencies between the constituting elements of a factory also demand for an enhanced comprehension of the factory system as a whole. This system can basically be divided into the three sub-systems namely production (machines and employees, coordinated by production planning and –control), technical building services (TBS) and building shell which result in a complex control system with dynamic interdependencies between different internal and external influencing variables [4]. As shown in Figure 3, one major task of technical building services is to ensure the needed production conditions in terms of temperature, moisture and purity through cooling / heating and conditioning of the air. The essential influencing variables are the local climate at the production site (seasonal influences) and the exhaust air and waste heat that is primarily emitted by production machines but also by other production factors like transportation equipment or even personnel. Besides that, production machines need energy (mostly electricity) and also diverse different media like compressed air, steam or cooling water to fulfill their designated processes. Technical building services are also responsible for the supply with these essential media whereas this involves their generation and (mostly) circuitry as well as the required conditioning (e.g. temperatures, pressures, purity). Altogether an evaluation of a factory's energy consumption must consider all non-regenerative energy flows that are externally supplied (e.g. electrical power, oil, gas) for running technical equipment for both production and TBS [4].

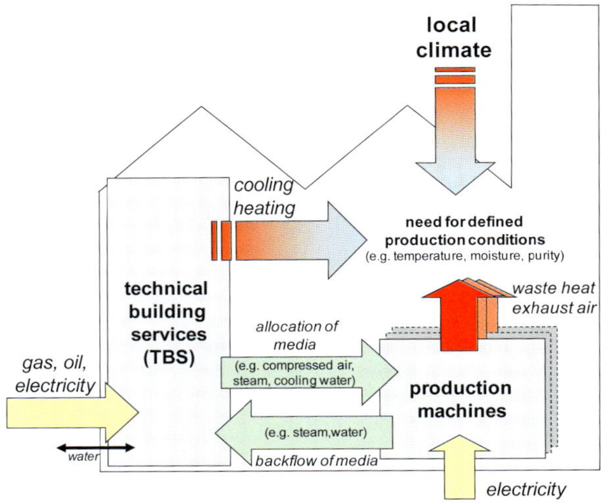

Figure 3: Holistic definition of factory [4].

- **Dynamics of consumption-/emission behavior and reciprocal effects:** All relevant input- and output flows are typically not static values but highly dynamic depending on the operating conditions of the processes and the machines. These profiles add up to cumulative load profiles on the factory level. In the end these dynamic cumulative load profiles (e.g. process heat demand, compressed air demand, heat flow into the factory building, electrical power demand) are decisive for design and control of the technical equipment (e.g. dimensioning of compressed air system) as well as for billing (e.g. energy supplier) [4] [6] [7] [8].
- **Thinking in process chains:** final products are usually not the result of a single production processes, but are rather manufactured in several steps on different production lines in the sense of production process chains. Against the background of energy and resource efficiency, the process chain has to be regarded and evaluated as a whole, as it may involve further potentials (e.g. combination of processes). Moreover, problem shifting might occur while improving measures in one process can possibly lead to worse performance of others.
- **Life-cycle-oriented perspective**: Analogous to the thinking in process chains, all life phases of products (this includes also all the technical equipment within the factory itself) have to be considered when it comes to deriving measures concerning the energy and resource efficiency. Thus, the decisive factor for increasing the energy efficiency of a machine tool, for instance, is less the improvement of single parameters of a specific process than rather the development of the machine itself. Moreover, the choice of specific processes (e.g. joining techniques) has direct effects on the use- and disposal phase which could lead to increased efforts in those phases.
- **Consideration of all sustainability dimensions and integrated evaluation:** In order to deduce advantageous solutions, several relevant target dimensions have to be considered simultaneously. Besides an ecological evaluation (with a correct balance of the different input- and output parameters, e.g. environmental effects of electricity- and gas

consumption), this includes a realistic economic (on the basis of a suitable cost model which integrates real contract conditions) and technical evaluation (e.g. effects on product quality). Possible conflicts of goals must be disclosed and decision support to their solution must be offered.

Against the background of the explanations above, Figure 4 shows a phase based procedure that incorporates all fields of actions to systematically improve energy and resource efficiency in manufacturing. It consists of five different phases whereas one can differentiate between the machine- or process perspective and the view on process chains or factory system. As described and depicted above both are directly connected and need to be considered simultaneously [3].

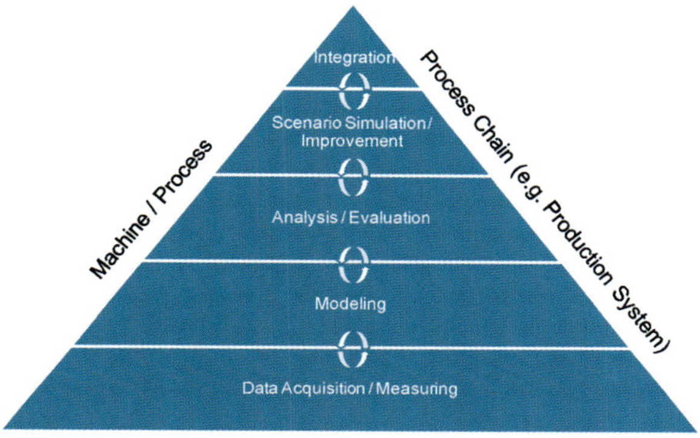

Figure 4: Fields of action for energy and resource efficiency in production [3].

Data collection/Measurement - The basis for the elaboration of improvement measures is a severe data collection respectively measurements that can be done singular or continuously depending on the individual purpose and technical device. The data collection can be done time- or condition-dependent and needs to capture all relevant input and output flows with a sufficient resolution. Based on more aggregate data a prioritization needs to be done before to ensure that the actually relevant measuring points are focussed.

Modeling - Afterwards the collected data gets combined with information about the interdependencies of the focused processes and their individual operation modes and working conditions in order to build up and describe a model of the focused production system that serves as a basis for a thorough understanding of the system that needs to be improved. This understanding of the interdependencies of the single system elements and the combination with metered or calculated data enables the development of individual solutions for improving the energy and resource efficiency in the production system.

Analysis/Evaluation - Important for the long term success of the implementation of such measures is a prior analysis of the actual energetic hotspots in the production system that can be based on the described model. Also it is important to evaluate proposed measures for improving the efficiency of this system before they get implemented. Evaluation criteria may base on economic, ecological, technical or also social objectives whereas conflicts of goals might occur [6].

Simulation of scenarios/Derivation of improvement measures – Interactively connected with the evaluation for the actual assessment of possible actions or strategies the usage of an energy oriented process chain simulation (that also considers material flow oriented key figures) is proposed [6]. While considering dynamic consumption behaviour and interdependencies of all technical equipment involved (production processes/machines as well as technical building services) the simulation approach allows to assess the effects of measures on the process/machine level for the whole production system on a realistic base. Furthermore measures on the process chain/factory level can be derived and assessed which may focus on PPC (production planning and control, e.g. production program, lot sizes) or dimensioning/control of connected technical building services (e.g. compressed air system) [6] [9].

Integration - Finally the knowledge that is generated by the prior data collection and measurement as well as through the simulation of different improvement scenarios needs to be integrated into operational databases and planning

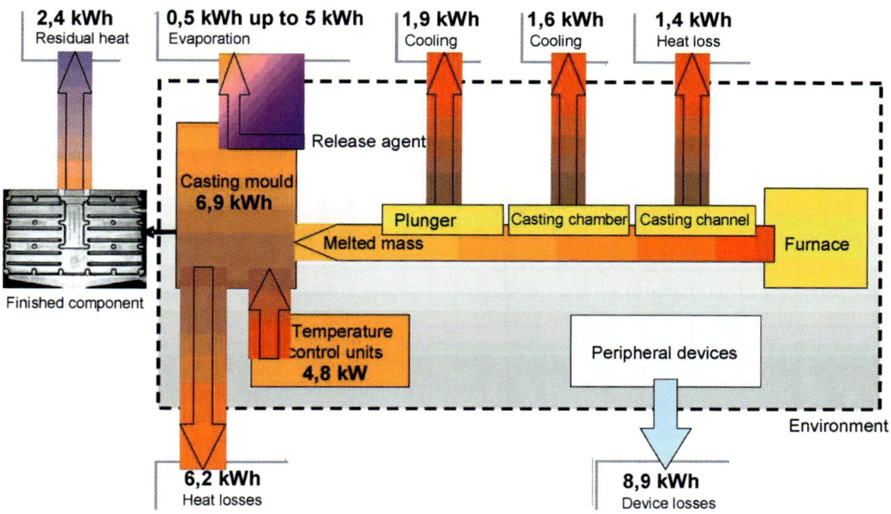

Figure 5: Energetic flows in the die casting process of aluminium [10]

procedures in order to enable a continuous and real time improvement process at the production site (e.g. integration in ERP processes).

In the following, one example of a process chain in which this procedure can be applied and selected steps of the proposed procedure are presented more in detail based on examples.

3 EVALUATION OF ENERGY- AND RESOURCE EFFICIENT PROCESS CHAINS USING THE EXAMPLE OF ALUMINIUM DIE CASTING

Due to its high industry-sector-specific energy consumptions, the competitiveness of the German aluminium die casting industry is closely linked to the current high energy prices. The necessary energy is largely needed for the heating and melting of the aluminium, for the die casting machine with its peripheral devices and for the die temperature control. The heat balance of the entire process is characterised by high losses of energy, which are depicted in Figure 5. The energetic losses were determined in the scope of a project for the energy balance of the die casting process funded by the Deutsche Bundesstiftung Umwelt (DBU) [10]. It is obvious that a large part of the energy input is lost during the process in the shape of heat or by cooling processes. On the whole, the high heat losses in the die casting process are regarded as extremely unsatisfying in terms of environment protection.

Besides the consideration of the efficient use of energy and auxiliary materials in the production of aluminium castings, also the efficient supply and use of the aluminium itself is of great importance. According to the range of parts, amount of the used alloys and available infrastructure, the supply of the melted mass from ingots/recycled material and the reprocessing is carried out in the casting house or the liquid alloy is directly delivered in thermal containers. This involves diverse material cycles, in which material of different quality is melted and reprocessed over and over again. However, it is exactly the poor exploitation of the raw material aluminium that is problematic in the process. Spills and sprue (which can add up to 50% of the cast form) as well as finished components, which do not fulfill the required quality, or parts from the start-up process are re-melted to ingots as cycle material and have to run through the whole energy-intensive process again. According to production parameters this applies to 30 – 70% of the originally used material. Additonally, there are 2 – 5% material losses, where the material cannot be re-processed again (e.g. due to heavy contamination) and is therefore lost [11]. Whereas from the companies' point of view those losses are mainly severe due to the material costs, it is the extremely energy-intensive and environmentally harmful recovery of aluminium that is critical from the environmental perspective (waste that pollutes the environment, big areas necessary for the exploitation of bauxite, electrolysis for recovery of aluminium). Thus, from a global perspective, an improvement of material efficiency and therefore a reduction of the aluminium consumption leads „indirectly" to a significant reduction of the energy demand.

The project ProGRess (design of resource-efficient process chains using the example of aluminium die casting, http://www.progress-aluminium.de) funded by the Federal Ministry of Education and Research (BMBF) pursuits a holistic approach which is in accordance with the enhanced process and system understanding (see above). Besides an analysis of material- and energy consumptions and losses of single processes, ProGRess aims at an evaluation and design of the entire aluminium die casting process chain in regard to its energy and material input. Therefore, it is necessary to regard the energy consumption and material flows from the level of single processes over single production systems up to cross-company supply chains along the process chain aluminium die casting.

Based on company key data, characteristic consumption values of the corporate user are generated first of all and used as efficiency measures for the individual production system. In that way, first interrelations between production volume and production systems in relation to the resource consumption and energy input per finished part can be identified. For the intra-corporate process chain from the melting of the material over keeping the melted mass at temperature and dosing at the die casting installation and the die casting process itself up to the mechanical post-processing, extremely differing consumption values as a result of different grades of automation, degrees of utilization and batch sizes can be documented in regard to the used energy sources. Figure 6 shows the proportion of the energy sources gas and electricity of the energy demand per kilogramme of finished cast parts over the intra-corporate process chain comparing two selected casting houses. Company 1 represents a company with highly standardised and automated processes with a high level of utilization. Company 2, on the other hand, represents a company that is characterized by a high diversity of variants and relatively small batch sizes as well as by a high amount of manual work. Different concepts for the supply and dosing of the melted mass due to the demands of the different production programmes lead to a varying percental proportion of the different energy sources.

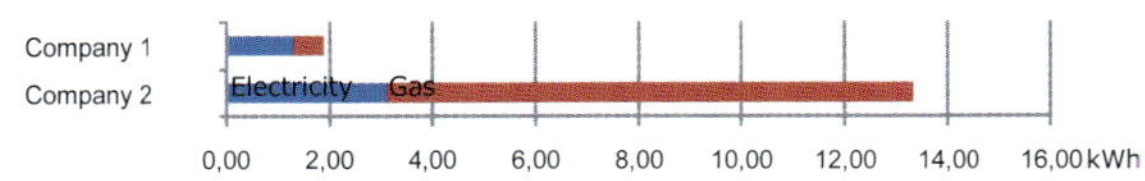

Figure 6: Energy consumption per kg of cast parts in the intra-corporate process chain.

In order to increase the utilization efficiency of the single energy sources in this production system, it is especially necessary to analyse the single system components. In die casting foundries die casting cells are a meaningful sub-system. They unite the die casting machine itself, the dosing furnace, temperature control units as well as diverse further peripheral devices. If one focuses on the consumption of electric energy in this subsystem, different load profiles of the single consumers can be observed. They add up to a total load profile, which in turn reflects the contribution of the die casting cell to the power requirement of the entire production system. As can be seen in Figure 7 characteristic load profiles (regarding standby-consumption, load peaks, load-dependent levels, etc.) can be documented for selected users. In the scope of the project ProGRess those load profiles can also be observed in other production environments of the aluminium die casting sector and which partially only differentiate by a scaling factor. They are useful for the planning of a new die casting line or can be considered in the scope of a dynamic simulation of aluminium die casting. In that way, the behaviour of the process chain can be pre-determined already before new production systems are started or when process parameters are changed and can be designed efficiently in regard to energy consumption. Moreover, the recorded load profiles can also be used for increasing the efficiency in already existing die casting cells in order to identify technological as well as organisational potentials and to derive suitable measures. By doing so, factors for increasing the energy and resource efficiency can be identified on a single process level as well as company-wide in order to reduce emissions from climate gases permanently and at the same time secure the competitiveness of the German aluminium die casting sector.

4 ENERGY ORIENTED SIMULATION APPROACH

As already mentioned the consideration of consumption- and emission dynamics is of high importance for a realistic, energy and resource efficient design of systems with several machines. In that way, non-value adding losses need to be identified and measures to stop them can be deduced. Furthermore, load peaks are important from an economic (e.g. load oriented price components in energy supply contracts) as well as from a technical perspective (e.g. design of the electrical network and technical equipment). As statical approaches are not sufficient for this purpose, a suitable simulation approach for the energy oriented analysis and evaluation of process chains or factory systems was developed [6] [12]. With that approach production systems with all their equipment and the relevant energetic and material flows can be modelled. Further on, it considers reciprocal effects with the technical equipment, as for example the necessary energy for the supply of compressed air. Figure 8 exemplarily shows the result of a case study: On the basis of a modelled production system, scenarios (in this case variations in the range of production planning and control) were simulated and an integrated evaluation considering ecological (power consumption), economic (calculation of electricity costs on the basis of a real contract model) and technical aspects (production time) was carried out. The diagrams show the resulting load profile of different scenarios. It becomes obvious that the energetic behaviour of the production system can be significantly influenced by production planning and -control. Furthermore, target conflicts between the different evaluation dimensions emerge. As depicted, an integrated evaluation and the deduction of promising measures from a comprehensive point of view now becomes possible (in the given example, scenario 3 is to be preferred).

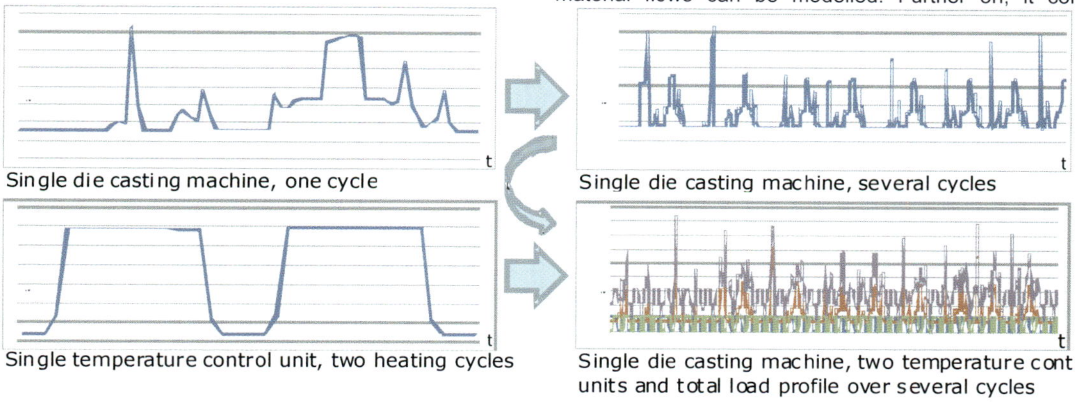

Figure 7: Overlap of load profiles in die casting cells.

Scenario	Description	Power consumption [kWh]	Electricity costs [€] (per month)	Production time [min]
S1	Standard (both lines 50%, syncronous)	94,60	2.747,20	130,92
S2	P2/P5 blocked against each other	105,14	2.744,08	139,00
S3	Start postponed at line 2	100,20	2.370,75	145,50
S4	Prod. line 1: 80%, line 2: 20% + S3	108,40	2.082,64	189,23
S5	Prod. line 1: 100%, line 2 off	95,09	1.476,34	247,55
S6	Prod. line 1: 80%, line 2: 20%	108,40	2.387,57	189,27
S1*	S1, but night-time prodution (base rate)	94,60	2.547,28	130,92

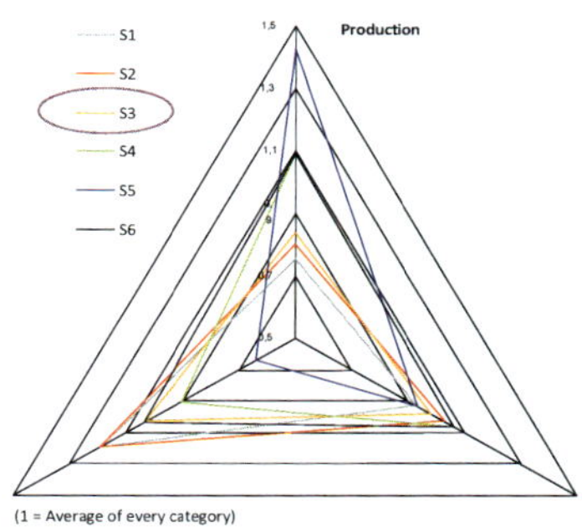

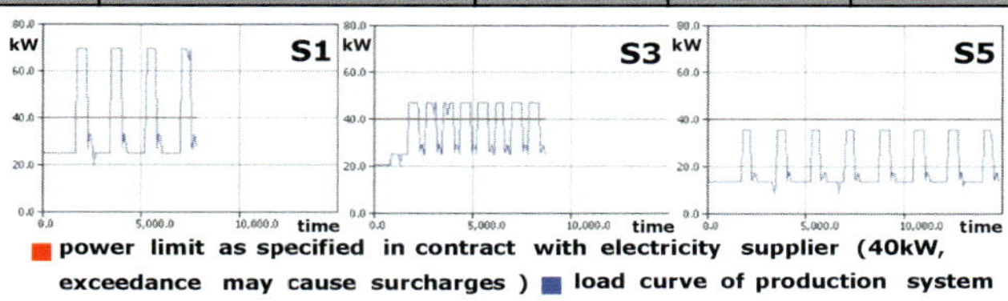

Figure 8: Exemplary results of an energy-oriented process chain simulation and integrated evaluation [12].

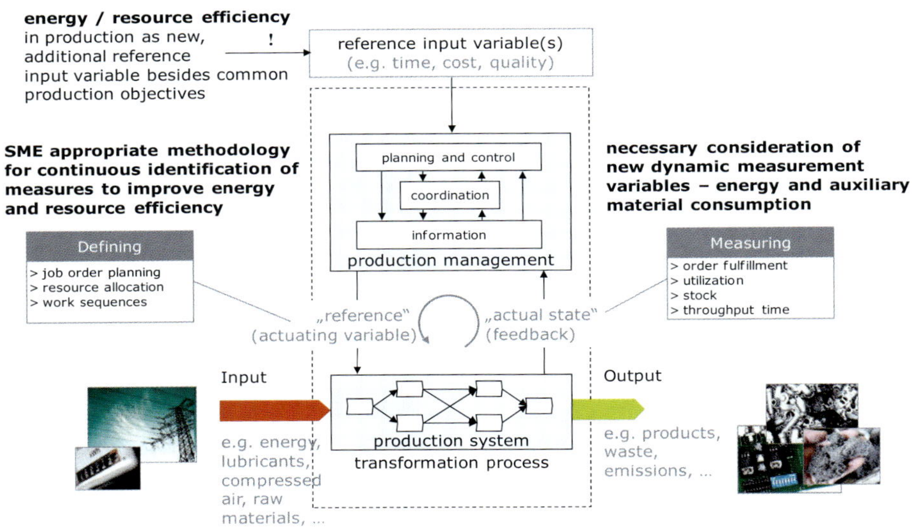

Figure 9: Integration of energy- and resource consumptions into the operational life cycle of production

5 INTEGRATION IN PRODUCTION MANAGEMENT

In research and industrial practice a variety of measures for increasing energy and resource efficieny are generally known by now [13]. However, especially SMEs lack decision support regarding the precise applicability and effectivity of such approaches in specific cases. The project EnHiPro (Improving energy and auxiliary material efficiency in production, http://www.enhipro.de), funded by the Federal Ministry of Education and Research (BMBF) and managed by the project management agency Karlsruhe (PTKA), enables producing SMEs continuously to derive and evaluate such organizational and technical measures for increasing the efficiency. Besides electrical power, gas or oil consumption also other forms of energy and auxiliary materials are explicitly considered, as for instance compressed air, process heat or cooling lubricants. As depicted in Figure 9, the aim is the necessary integration of energy and auxiliary material consumption into the operations/production management. By means of a suitable combination of measurement technology and IT-applications for metering and data processing, actual consumption values are being gathered and transferred into production management applications (e.g. enterprise resource planning system, ERP). This data is the basis for predicting and visualising effects and potentials of improvement measures. Altogether with implementing such kind of a system an important contribution to closing a gap between knowledge and actual is made.

6 CONCLUSION

The paper stresses the necessity for a holistic view on production processes and factories as promising base for deriving measures to improve energy and resource efficiency. Specifically the conscious consideration of process chains as a whole with all relevant input and output flows as well as their independencies and time-dependent behavior is an important element. Against this background fields of action are described. Selected examples show the potentials of involved methods and tools towards more energy and resource efficient production processes and factories. However, in a long term perspective the integration of these aspects into actual planning processes of production companies is a key element to enable continuous application and improvement.

7 REFERENCES

[1] Gutenberg, E., Einführung in die Betriebswirtschaftslehre. Band 1: die Produktion, Springer, Berlin/Heidelberg, 1983.

[2] German Federal Ministry of Economics and Technology, 2009, Energy Statistics.

[3] Herrmann, C., Thiede, S., Heinemann, T., Ganzheitliche Ansätze zur Erhöhung der Energie- und Ressourceneffizienz in der Produktion, 10. Karlsruher Arbeitsgespräche Produktionsforschung, Karlsruhe, Germany, 2010.

[4] Hesselbach, J., Herrmann, C., Detzer, R., Martin, L., Thiede, S., Lüdemann, B., 2008, Energy Efficiency through optimized coordination of production and technical building services; Proceeding of the 15th Conference on Life Cycle Engineering, Sydney.

[5] Schultz, A., 2002, Methode zur integrierten ökologischen und ökonomischen Bewertung von Produktionsprozessen und -technologien, Dissertation, Magdeburg.

[6] Herrmann, C.; Thiede, S., 2009, Process chain simulation to foster energy efficiency in manufacturing, In: CIRP Journal of Manufacturing Science and Technology, Elsevier, ISSN 1755-5817

[7] Gutowski, T., Dahmus, J., Thiriez, A., 2006, Electrical Energy Requirements for Manufacturing Processes, in:

Proceedings of 13th CIRP International Conference on Life Cycle Engineering, (Leuven, Belgium, May 31–June 2, 2006).

[8] Devoldere, T., Dewulf, W., Deprez, W.,Willems, B., Duflou, J.R., 2007, Improvement Potential for Energy Consumption in Discrete Part Production Machines, in: Proceedings of the 14th CIRP Conference on Life Cycle Engineering, (Tokyo, Japan).

[9] Heilala, J., Vatanen, S., Montonen, J., Tonteri, H., Johansson, B., Stahre, J., Lind, S., 2008, Simulation-Based Sustainable Manufacturing System Design, in Mason SJ, Hill RR, Mo¨nch L, Rose O, Jefferson T, Fowler JW, (Eds.) in: Proceedings of the 2008 Winter Simulation Conference.

[10] Optimierung der Energiebilanz beim Aluminium-Druckgießprozess, Final report, DBU, Institut für Füge- und Schweißtechnik, TU Braunschweig

[11] Institute of joining and welding technique, TU Braunschweig. www.ifs.ing.tu-bs.de, 2010.

[12] Herrmann, C.; Thiede, S., 2008, Increasing Energy Efficiency in Manufacturing Companies through Process Chain Simulation, In: Sustainability and Remanufacturing VI, Pusan, South Korea, 2008, pp. 52-57.

[13] Seefeldt, F.; Wünsch, M., 2007, Potenziale für Energieeinsparung und Energieeffizienz im Lichte aktueller Preisentwicklungen, Prognos AG.

Energy Flow Simulation for Manufacturing Systems

S.Thiede, C. Herrmann
Institute of Machine Tools and Production Technology, Technische Universität Braunschweig, Germany

Abstract
Producing companies externally demand at least primary energy sources like gas or oil and typically electricity for running their production. Additionally, transformed from these sources, internally energy carriers like compressed air or process heat (e.g. steam) are of major importance. Thereby the actual cumulative demand of all types of energy for the whole manufacturing system is naturally not static but rather highly dynamic depending on the actual state and interactions of production machines as well as technical building services. Against this background this paper presents a simulation approach which is able to realistically consider relevant energy flows in manufacturing systems. This enables an integrated evaluation of typical production target criteria (e.g. throughput times, utilization rates) and energy driven variables based on appropriate models for realistic cost and environmental impact evaluation.
Keywords:
energy efficiency, sustainable manufacturing, production management

1 INTRODUCTION

Per popular definition, "energy is the capacity to do work" (e.g. [1]) so it is necessary to execute any kind of designated tasks. Fundamental physics distinguish between only two basic types of energy: potential (stored) and kinetic (working) energy. However when detailing further on with mechanical, thermal, chemical, electric, electromagnetic and nuclear energy more forms can be differentiated (e.g. [2] [3]). Conversion between different energy forms is basically possible. Referring to the two basic laws of thermodynamics within a system the sum of energy stays constant but every conversion is connected with losses because not the whole amount of energy ends in the designated form. As an example the conversion of electrical energy into mechanical energy through an electric drive also generates heat (losses).

Table 1 shows the necessary forms of energy for industrial (manufacturing) purposes in the case of Germany. According to that, (space and process) heat and mechanical energy are mainly needed [4] which are getting converted from energy sources like electricity (electrical energy), gas, oil or coal (chemical energy). As the study also underlines the actual composition of energy form and sources significantly differs between different branches. Whereas coal is mainly used by metal founding, cement or chemical industry (almost 90% of coal is used from these branches), oil and especially electricity and gas are far more common through all industries. In machinery and automotive industry for example, electricity counts up for over 50% of total energy consumption [4]. Altogether the most typical energy conversions are from gas to process heat (e.g. generation of steam) and from electricity to mechanical energy (e.g. electric drives or generation of compressed air).

On a national scale, industry is a major consumer of energy – e.g. German industry is responsible for 47% of the national electricity and 36% of the national gas consumption [5]. Considering energy consumption has a very strong relevance from both economic as well as environmental perspective.

The supply with energy is directly connected with environmental impacts like green house gas (GHG) emissions with certain contribution to global warming or depletion of diverse non-renewable resources (e.g. oil, gas, coal). As a result based on currently known securely mineable deposits and demand the statically estimated supply range is approx. 40 (oil) respectively 60 (gas) years [5]. Having in mind global warming, only through its electricity demand industry is responsible for approx. 18% of CO_2 emissions (plus approx. 20% through direct industrial emissions) in Germany [5].

	space heat	process heat	mechanical energy	lighting	total
total	**345.6**	**1589.3**	**522.3**	**72.1**	**2529.3**
electricity	21.8	234.8	490.4	72.1	**819.1**
gas	179.6	792.1	2.0	0.0	**973.7**
oil	97.7	129.6	3.9	0.0	**231.2**
coal	10.3	397.5	0.0	0.0	**407.7**
district heat	27.8	27.9	0.0	0.0	**55.7**
renewable	8.4	7.4	0.0	0.0	**15.7**
fuel	0.0	0.0	26.1	0.0	**26.1**

Table 1: Energy consumption for German producing industry with respect to energy forms and sources (based on data from 2002, in Petajoule) [4].

Thereby the pure energetic view as shown in Table 1 is certainly just one perspective; striving towards sustainability in manufacturing demands a more detailed analysis of connected economic as well as environmental impacts (here depicted with related CO_2 emissions). Therefore (based on the data from Table 1) Figure 1 shows the estimated energy costs and CO_2 emissions for the German producing industry for the main energy sources. The calculation bases on the average prices for 2000/2008 and the emitted CO_2 for generating electricity (energy source mix for Germany) or directly burning oil, gas or coal. The calculations underline the major importance of considering electricity in comparison to primary energy sources (due to upstream chain).

Besides, the calculation also stresses the very strong economic relevance of industrial energy consumption. Energy prices for electricity, gas and oil are steadily increasing for a couple of years [5] – as shown in Figure 1 from the year 2000 to 2008 energy costs for producing companies more than doubled.

	energy consumption [in PJ]	energy costs (2000) [in €]	energy costs (2008) [in €]	related CO2 emissions [in t]
electricity	819,1	10.012.650.793 €	20.073.221.336 €	130.933.135
gas	973,7	4.577.253.331 €	9.094.440.438 €	38.745.623
oil	231,2	10.558.553 €	22.046.590 €	10.395.556
coal	407,8	586.200.977 €	1.566.545.164 €	37.185.949
total	**2431,8**	**15.186.663.655 €**	**30.756.253.528 €**	**217.260.264**

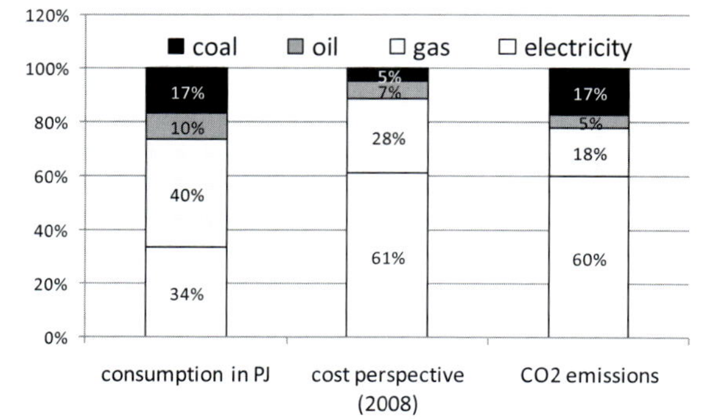

Figure 1: Estimation of costs and CO_2 emission related to energy consumption of German producing companies.

Against the background of these urging environmental as well as economic challenges, the paper presents an approach to foster energy efficiency in manufacturing companies. Hereby energy efficiency is the ratio of the production output (e.g. in terms of quantities with defined quality) to the total energy input (e.g. electricity, gas, oil) for the operation of the whole factory system. Manifold possible fields of action with certain potential to improve efficiency are known nowadays (e.g. Figure 2).

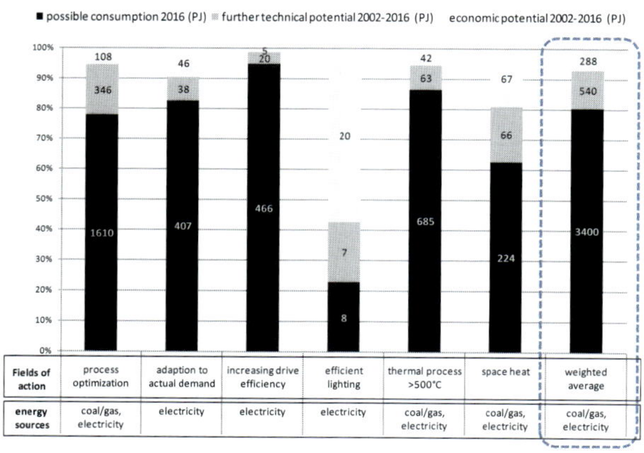

Figure 2: Potentials to improve energy efficiency in manufacturing companies (selection, based on [4]).

Considering these opportunities and the relevance of different energy input flows it becomes clear that a holistic perspective on all energy sources and forms is necessary to identify and tab the most worthwhile potentials for the individual company case. However still energy is just one of several input factors for a production process and minimizing energy consumption is just one of the target objectives of a company (besides e.g. material and personnel costs, production time, quality). While different measures may also cause conflicts of goals an appropriate approach shall be able to consider these different perspectives.

2 THEORETICAL BACKGROUND

System definition

The consideration of all relevant energy flows necessarily requires a holistic factory definition with three partial systems: the production system itself (with machines and controlled through production management), the technical building services (TBS) and the building shell (Figure 3) [6]. These partial systems interact as complex control system with dynamic interdependencies between different internal and external influencing variables. Production machines which execute or support the actual value creating processes directly need energy (typically electricity) to fulfil their designated processes. However they also need diverse other energy forms/media like compressed air, steam or cooling water which are provided by technical building services. Another task of technical building services is to ensure the needed production conditions in terms of temperature, moisture and purity through cooling/heating and conditioning of the air. The essential influencing variables are the local climate at the production site (e.g. seasonal influences) and also the exhaust air and waste heat that is primarily emitted by production machines or personnel. Altogether, besides direct energy consumption through production equipment, TBS need further energy to fulfil their tasks and enable factory operation. Referring to a study of the European Union this consumption counts up for a major part of industry energy consumption [7], additionally high potentials for energy related improvement in that field were identified [4].

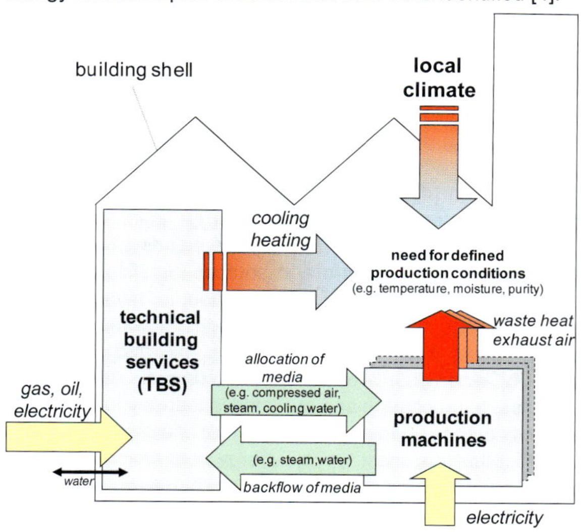

Figure 3: Holistic system definition on energy consumption of manufacturing systems [6].

A typical example for internal energy flows involving TBS is the generation of compressed air. Because of its advantages compressed air is broadly used in manufacturing companies for different purposes. It is basically a conversion of electrical energy to mechanical energy. Altogether about 10% of total industrial electricity consumption is caused by generation of compressed air (which means 80 TWh or 55 million tons CO_2) [8]. As one big disadvantage compressed air usage is often connected with very high system losses [9]. Studies show that typically not even 10% of inserted energy ends up as actual usable mechanical energy at the end use device. As a result compressed air is actually one of the most expensive forms of energy in industry [10] [11]. Studies reveal that saving opportunities are not used yet; saving potentials are estimated with 5-50% (average approx. 33%) for the next 15 years [8].

Defining fields of action

In order to increase the energy and resource efficiency of production processes and process chains, a systematic approach against the background of the aforementioned holistic system understanding is necessary. It is not sufficient to focus on only single selected system elements of the production. To avoid focusing on minor relevant aspects and local optimisation as well as problem shifting the whole production system including all relevant input and output flows needs to be taken into consideration. Against this background, Figure 4 shows fields of action in context of energy and resource efficiency in production [12]. One can differentiate between the machine or process perspective and the view on process chains or production/factory system, yet both are directly connected as depicted in the figure. Based on either singular or permanent data collection, the understanding of interrelations through modelling as well as on suitable methods for the evaluation and prediction of operating behaviour through simulation, it is the ultimate objective to integrate energy and resource consumption as a further dimension into operational decisions/production management (besides conventional objectives like for instance utilization, cycle times, quality rates), e.g. within ERP-systems.

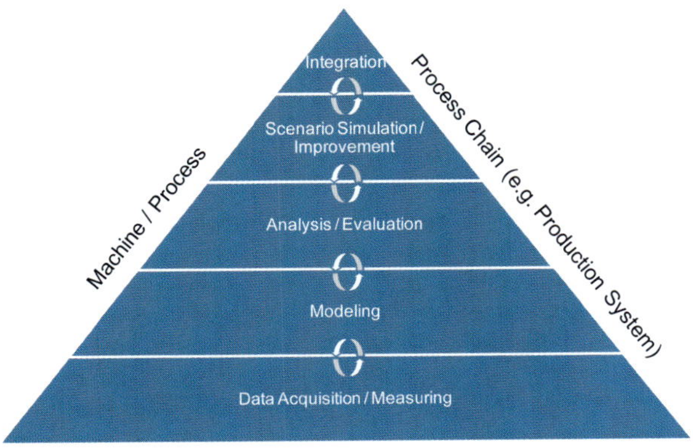

Figure 4: Fields of action for energy and resource efficiency in production [12].

3 SOLUTION APPROACH

Requirements

The previous sections pointed out the necessity of a holistic view on the whole factory with all energy flows in order to be able to tab many energy efficiency potentials. The power consumption (as well as media, materials, emissions) from single machines/processes is naturally not static but directly depends on the actual operating state. Of course it is crucial to understand this behavior and minimize not value creating (energy) consumption (e.g. idle losses or non-process related consumption) [13] [14]. However, depending on production system structure and management single machines consumption profiles lead to cumulative load profiles for the whole factory. Therefore this is the relevant layer for 1) dimensioning and control of TBS (e.g. compressed air) and 2) energy cost calculation (including e.g. peak surcharges) [15]. The other way around resulting states on a system layer may have effect on single production machines again (e.g. functional failures because necessary compressed air flow cannot be provided, overcharge of electrical system) Thus, considering single machines/processes is certainly a necessary but not sufficient prerequisite. A promising solution approach must be able to cope with the dynamic interdependencies of all relevant partial systems of a factory and enable an integrated evaluation (technical, economic, ecological) of energy efficiency measures on this factory layer [6].

Simulation Concept

Against the background of the holistic system definition, the defined fields of action and identified requirements a simulation approach is proposed. Referring to previous publications (e.g. [15] [16]), the architecture (input, logic, user and evaluation layer) of the simulation shall not be explained in detail here. Basically, it is a modular, flexible approach, which allows a realistic representation of the production system with all the interdependencies and dynamics of involved technical equipment. As an extension of popular material flow simulators all energy related input and output flows are explicitly considered. For the case of electricity, Figure 5 shows a simplified illustration of dynamic consumption data handling of production machines within the simulation. Machine behaviour can be depicted with state charts – each operating state has a definable duration (e.g. based on certain time or trigger events) and is connected with a certain consumption of a resource (described as value or equation, e.g. depending on process parameters). Thus, with this technique the dynamic consumption behaviour of e.g. all forms of energy, any (auxiliary) materials or even emissions can be modelled.

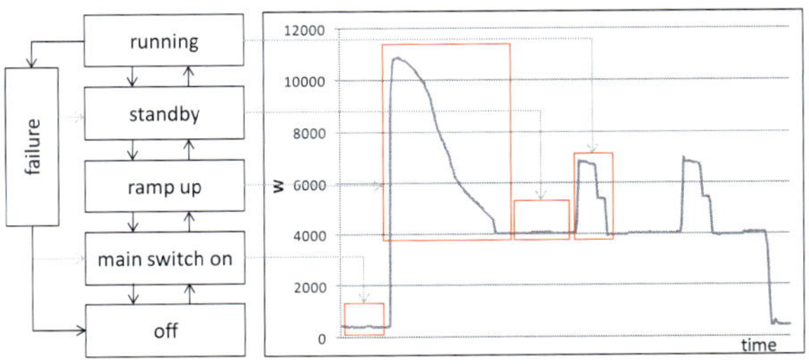

Figure 5: Modelling energy consumption of machines based on state charts (simplified example of power consumption in Watt over time) [17].

As described before single consumption profiles lead to cumulative load curves which are is being simulated based on production structure (e.g. material flow, cycle times) and management (e.g. production program). TBS-related energy demand of the actual production equipment (e.g. compressed air) serves as input for appropriate partial TBS-models (e.g. for generation of compressed air). Herewith additional energy consumption (e.g. electricity needed to generate compressed air) of TBS is calculated – together with certain direct energy consumption of production equipment this leads to the total energy demand of the production site (Figure 6). Additionally TBS models simulate the possible supply with energy or media – interacting with the production system, a lack of e.g. compressed air (air pressure to low) leads to failures of production machines.

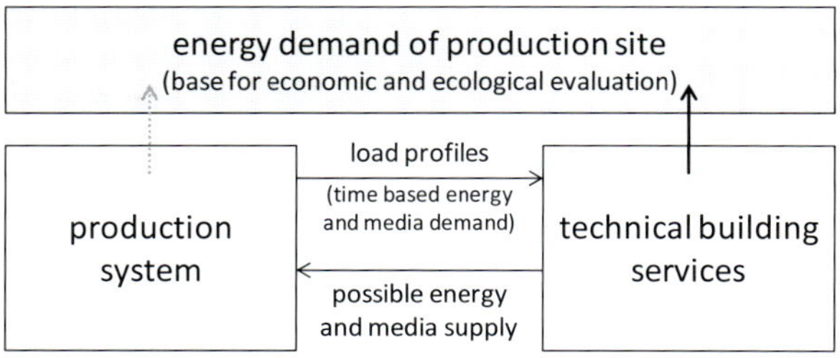

Figure 6: Simulation based interaction of production system and technical building services [17].

As unique approach these interdependencies can be consciously and realistically considered when designing and controlling production system as well as TBS. To reveal and balance conflicts of goals an integrated evaluation is available – typically variables for economic (energy costs including peak surcharges), environmental (energy consumption) and technical (production time) performance were used [16].

As mentioned the simulation approach enables to simultaneous consider production system and TBS. Measures for improvement can be derived and evaluated in both sub systems while considering interdependencies between them. To show the potentials two case studies are presented in the following. The first one is focusing on the improvement of TBS (dimensioning and control of compressed air system in weaving mill, based on [17]), the second on production system design.

4 CASE STUDY WEAVING MILL

The first case study considers a company running a large scale weaving mill to produce technical textiles which are being used for e.g. industrial purposes (e.g. supporting material for abrasive papers, printing industry). Please note that for reasons of confidentiality the values are just shown in relative values (in relation to the lowest value of this category as reference) in this case study. Table 2 shows the machine park of the weaving mill which basically consists of a total of 41 weaving machines based on four different basic machines types operating independently (no linkage, every machine with own production program) and almost continuously in a three shift system. Furthermore the necessary energy related input streams are depicted which are based on actual measurements. As shown, besides electricity a significant amount of compressed air is necessarily needed to run the machines. For both energy flows the consumption behaviour can be well described through two basic operating states (besides "off") – producing and standby (main switch and control on) – with sufficient accuracy. Each machine needs a system air pressure of at least 6.2 bar to be operated. For the generation of compressed air several huge compressors (each with 100 kW nominal power and potential generation of 1200m³/min compressed air, actual values for full or partial load were also measured) are available.

	Qty	power demand [kW]		compressed air [m³/h]	
		producing	standby	Producing	standby
Machine Type 1	17	4	1.2	106	5
Machine Type 2	5	6	1.4	88	10
Machine Type 3	14	5	0.9	109	6
Machine Type 4	5	5	1	100	6

Table 2: Machine park of the considered weaving mill (relative values).

For this study two questions are of major importance:

- What is the preferable configuration of the compressed air system (e.g. system volume/storage tanks, air pressure, amount of compressors)? Based on fundamentals presented before higher volume and/or higher nominal air pressure lead to higher energy inputs. Therefore both (directly connected) factors should be minimized. However, the risk of failures due to insufficient air pressure for operating has to be avoided as well to be able to tab the full production capacity.
- What is the actual relevance of compressed air induced electricity consumption within the weaving mill compared to direct electricity consumption of the machines?

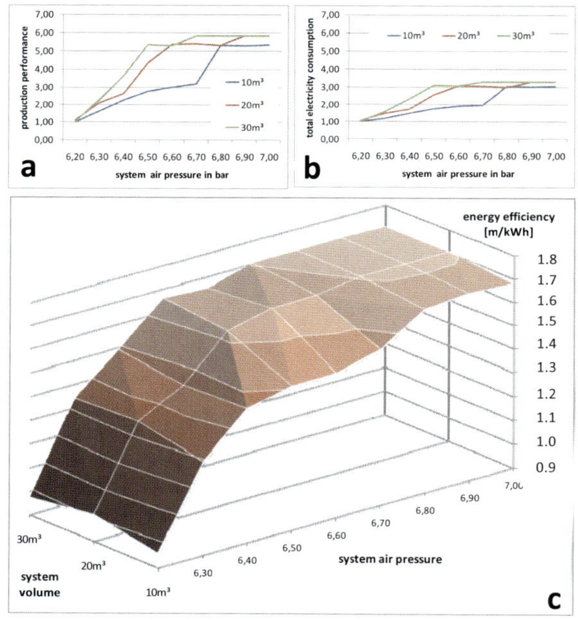

Figure 7: Graphical representation of simulation results (relative values) - a) production performance, b) electricity consumption and c) energy efficiency in relation to compressed air system volume and pressure with four compressors.

To conduct the analysis the weaving mill was modeled with the proposed simulation approach and diverse scenarios based on variation of relevant parameters were simulated (for one shift). The production machines consume electricity and compressed air based on the profile shown in Table 3. Compressed air is generated by compressors with the technical specifications also mentioned above. Additionally, the consideration also includes failure mechanisms of the production machines (based on actual MTBF – mean time between failures – data) and certain necessary time for repair (mean time to repair – MTTR, e.g. if compressed air is lacking) which causes further dynamics in the systems. As relevant target variables production amount (meters of textiles) and total electricity consumption are being considered. Energy costs are directly linked with consumption in this case (including peak surcharges). Figure 7 shows the results of the simulation runs for a system with four compressors. As expected, higher nominal system air pressure and/or higher system volume (e.g. through larger buffer tanks) lead to more stable production performance (Figure 7a) with fewer failures due to lack of pressure. Thus, total electricity consumption (Figure 7b) also increases. Energy efficiency is defined as the ratio of production performance and total electricity consumption and serves as main evaluation criteria here (Figure 7c). Simulation time was one shift, please note that due to continuous three shift production with very high production volumes even relatively small improvements can scale up to significant savings for e.g. a whole year.

The results confirm that volume and pressure are naturally closely connected – e.g. lower nominal air pressure can be balanced with higher system volume in order to still achieve a failure-free operation. As shown, the simulation approach allows the integrative evaluation of all target variables and a convergence to optimal system configuration (structure, parameters). While being operated with higher nominal pressure in reality, the simulation reveals that the maximum production volume can also be achieved with a air pressure of 6.7 bar (with 30m³ system volume) or 6.9 bar (with 20m³) in this case. As further simulation runs have shown, that the optimal number of compressors is four while failure-free operation cannot be guaranteed with just three compressors.

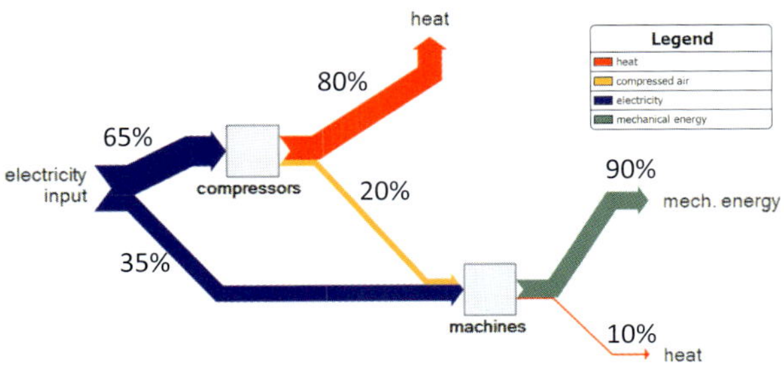

Figure 8: Simplified Sankey analyis of energy flows in weaving mill.

Further on, based on the simulation results and theoretical knowledge about energy losses (see section 2) the energy flows of the weaving mill can be depicted as a Sankey diagram (Figure 8). This kind of representation allows to clearly depicting coherences and losses within the system. In the case of the weaving mill, the major share of electricity input is being used for the generation of compressed air. However, most of this electricity gets lost through heat emissions (approx. 50% of total energy input). Altogether just approx. 44% of the inserted electricity can be actually used as mechanical energy for the weaving process. Based on this analysis the most promising measures for improvement of energy consumption/energy efficiency of the system are the recuperating of waste heat (for other internal purposes or even inserting in municipal district heat) and the critical review of the energy efficiency of used electric drives in the production machines.

5 CASE STUDY ALUMINIUM DIE CASTING

In the second case study the energy flow oriented simulation approach was applied to a simplified aluminium die casting process chain. In this case three casting cells with peripheral equipment serve two identical CNC milling machines for mechanical treatment before they are being transported automatically by a conveyor in a one piece flow into an abrasive blasting machine and finally to a palletizing device.

The simulation model that has been deducted from the process chain description is depicted in Figure 9. It is parameterized with real metered data like cycle times, (energy) load profiles depending on operation modes, process depended material efficiency et cetera and focuses on the consumption of electrical energy in this case.

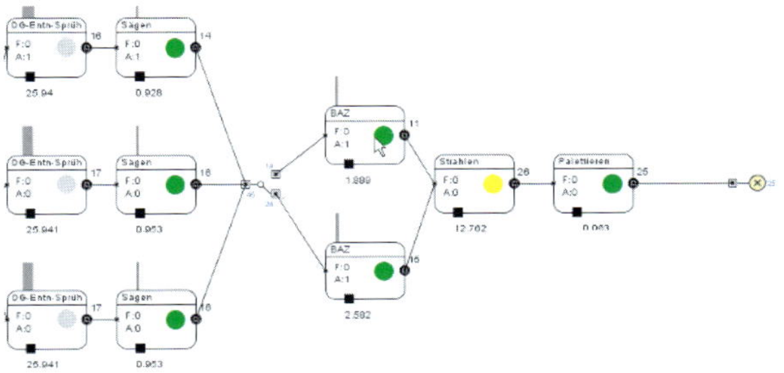

Figure 9: Simulation model based on case study process chain.

Figure 10 shows the results of simulation runs for different scenarios as well as the resulting electrical load profile for the base scenario (synchronous operation of all three die casting cells, production of at least 50 pieces). The analysis of the simulated load profiles of the single sub processes (base scenario) leads to the conclusion that the die casting cells with its peripheral equipment are the main drivers of energy consumption. Furthermore there is a high impact of the standby consumption of all machines especially of the abrasive blasting machine. Therefore three improvement scenarios were derived and simulated in order to evaluate their impact on the energy consumption and production performance. In scenario A the lot size for the abrasive blasting machine is changed from 1 to 10 pieces. Between these batches the blasting machine is switched off. In scenario B only two die casting cells are operating. The third cell and its peripheral equipment is switched off. In scenario A+B both prior scenarios are simulated simultaneously.

By comparing the simulation results is can be seen that conflicts of goal between energy consumption and production performance (e.g. time per piece) can occur. However for the given setting specifically in scenario B und A+B the specific energy consumption per piece and also in relation to the time per piece could be significantly improved while also saving capacity (running system) or investment (new system) of one die casting cell. Certainly, as this simulation uses simplifications the exact results cannot be directly transferred to a real industrial environment. However, the simulation study shows potentials of the proposed methodology as well as possible effects of technical and organizational measures on the energy efficiency of process chains. It gets clear that manifold evaluation criteria need to be considered. In the end it is a matter of the specific target system and priorities of the company.

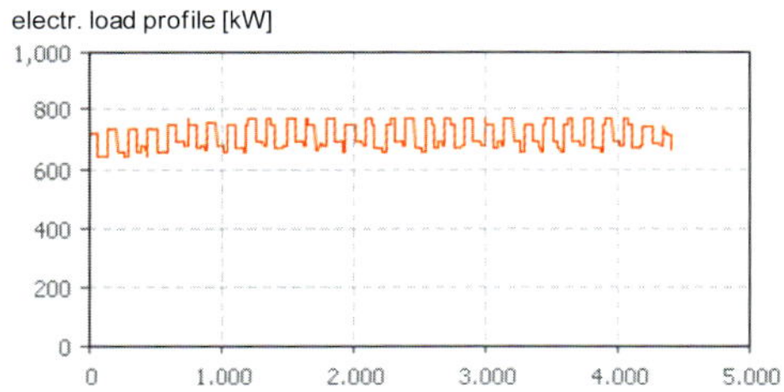

simulated electrical load profile of the process chain (base scenario)

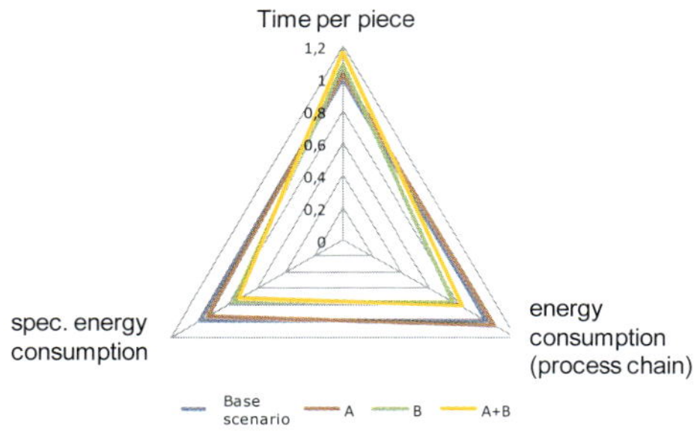

	Base scenario	A	B	A+B
produced output [pieces]	50	56	50	57
production time [sec.]	4422	5091	4775	5873
time per piece [sec./piece]	88	91	96	103
energy consumption (process chain) [kWh]	886	932	678	731
spec. energy consumption [kWh/piece]	17,71	16,7	13,6	12,8
spec. energy consumption [kWh/time per piece]	10,07	10,24	7,06	7,10

Figure 10: Simulation results from scenario simulation.

6 SUMMARY

Based on the strong need for improving the energy efficiency in manufacturing the paper underlines the necessity of a holistic system comprehension which considers all relevant energy flows in a production facility. In this context the conversion of energy sources like gas, oil or electricity to internally needed forms of energy like compressed air and process/space heat is of major importance. To enable the consideration of all interdependencies and an integrative evaluation with technical, economic and environmental target variables a simulation based solution approach is proposed. Its applicability and potentials are being proved with two case studies which underline that the simulation may help to derive an improved system configuration with better energy efficiency.

7 REFERENCES

[1] Environmental Science – Systems and Solutions, 4th edition, Jones and Bartlett, Sudbury, MA, ISBN 978-0-7637-4262-1.

[2] Viegas, J., 2005, Kinetic and potential energy: understanding changes within physical systems, 1st edition, Rosen publishing group, New York, ISBN 1-4042-0333-8.

[3] U.S. Energy Information Administration, 2009, Glossary, http://www.eia.doe.gov/glossary.

[4] Seefeldt, F.; Wünsch, M., 2007, Potenziale für Energieeinsparung und Energieeffizienz im Lichte aktueller Preisentwicklungen, Prognos AG.

[5] German Federal Ministry of Economics and Technology, 2009, Energy Statistics.

[6] Hesselbach, J., Herrmann, C., Detzer, R., Martin, L., Thiede, S., Lüdemann, B., 2008, Energy Efficiency through optimized coordination of production and technical building services; Proceeding of the 15th Conference on LCE, Sydney.

[7] Rebhan, E., 2002, Energiehandbuch - Gewinnung, Wandlung und Nutzung von Energie, Springer, Berlin.

[8] Radgen, P.; Blaustein, E., 2001, Compressed Air Systems in the European Union - Energy, Emissions, Saving Potentials and Policy Actions, ISBN 3-932298-16-0.

[9] Bierbaum, U.; Hütter, J., 2004, Boge Druckluft Kompendium, ISBN 3-935772-11-4.

[10] Gauchel, W., 2006, Energy-saving pneumatic systems, O+P Ölhydraulik und Pneumatik, Vol. 50, Nr 1.

[11] Gloor, R., 2000, Energieeinsparungen bei Druckluftanlagen in der Schweiz, www.energie.ch/themen/industrie/druckluft.

[12] Herrmann, C., Thiede, S., Heinemann, T., Ganzheitliche Ansätze zur Erhöhung der Energie- und Ressourceneffizienz in der Produktion, 10. Karlsruher Arbeitsgespräche Produktionsforschung 2010, Karlsruhe, Germany, 2010.

[13] Devoldere, T., Dewulf, W., Deprez, W., Willems, B., Duflou, J., 2007, Improvement Potential for Energy Consumption in Discrete Part Production Machines, Proceedings of the 14th Conference on LCE, Tokyo.

[14] Gutowski, T., Dahmus J., Thiriez A., 2006, Electrical Energy Requirements for Manufacturing Processes, in: Proceedings of 13th CIRP International Conference on LCE, Leuven, Belgium, 2006.

[15] Herrmann, C.; Thiede, S., 2009, Process chain simulation to foster energy efficiency in manufacturing, In: CIRP Journal of Manufacturing Science and Technology, Elsevier, ISSN 1755-5817

[16] Herrmann, C.; Thiede, S., 2009, Towards Energy and Resource Efficient Process Chains, In: 16th CIRP International Conference on LCE, Cairo, Egypt, pp. 303-309, ISBN 978-0-9783187-3-4

[17] Thiede, S., Herrmann, C., Simulation-based Energy Flow Evaluation for Sustainable Manufacturing Systems, 17th CIRP International Conference on LCE, Hefei, China, 2010.

Development of a Bionic Pump Based on the Sap-rising Principle of Trees

J. Theileis, O. Kopp, M. Stache, I. Rechenberg
Department of Bionics and Evolution Technique, Technische Universität Berlin, Germany

Abstract
Trees have the capability to transport water up to a height of 120 m. One of the driving forces is transpiration. At the stomata of the leaves, water is evaporated. Thus, suction is created that pulls the water through the xylem in the stem up into the leaves.
This paper introduces a suction pump, based on the sap-rising principles of plants. By condensing moist air, distilled water is gained. The pump has no moving parts and utilises only the sun as source of energy.
Combined with low maintenance and the use of solar energy or waste heat, this pump enables a sustainable method of delivering and purifying water. These advantages over ordinary pumps and purification systems lead to a wide range of possible applications such as water supply in areas with weak infrastructure, for cooling of buildings or irrigation in agriculture.
Keywords: Bionics, trees, pump, water purification

1 INTRODUCTION

This paper describes a pump based on the sap rising principles of trees. Therefore, it is necessary to understand the structure of tress and the transport systems for liquids in trees. Trees consist of three major parts: the crown, the trunk and the roots.

1.1 Mechanism of the water transport

The roots are the foundation of a tree. One major purpose is to fix the tree to the ground. Additionally they act as a reservoir for nutrition and enable the tree's water supply from the surrounding ground. The accumulated surface of the roots is about as large as the collective surface of the leaves of a tree. Water comes through the outer layers of the root system and reaches the xylem. A difference in the chemical potential between the root and the surrounding soil draws water into the roots and towards the xylem. The xylem, the long-distant sap transport system of the tree, hauls water from the roots into the trunk and to the leaves in the tree crown.

The xylem only transports against the direction of gravity. It consists of a complex, ligneous tissue. The xylem enables the sap rising from the roots to the leaves, to supply the tree with water and inorganic salts. There are two different kinds of xylem: tracheids and tracheas. They are between 0.3 mm (tracheids) and up to 11 m (tracheas) long and have a diameter of 20 to 700 µm [1]. Many of these tracheids or tracheas are inter-connected and arranged in bundles and form the xylem. These bundles reach from the roots to the ends of the leaves. Each of these vessels are single cells. The secondary walls of the cells are interspersed with pits. These pits enable the sap to flow from one vessel to another and in case an air embolism develops in one of the cells, they act like valves and seal the affected capillaries off.

The leaves are covered with a waxy coating, the cuticle. This waterproof layer protects the leave from unregulated evaporation. Embedded into this outer layer are stomata. These small openings enable gas exchange between the inside of the leave and the surrounding environment. The tree can open or close these small openings to regulate gas exchange. Moist air is emitted and CO_2 enters the leaves. The inside of each leave is filled with sponge-like parenchyma tissue. The vessels of the xylem end between the parenchyma cells. Due to a difference in the gas pressure between the inside of the xylem and the surrounding soft tissue water evaporates into the intercellular space of the soft tissue. From there the moist air diffuses to the stomata and reaches the outside environment. Convection effects and wind drive the moist air away from the leaves' surface and the stomata.

1.2 Fluid transport in the Xylem

The cohesion tension theory describes principles of the sap rising in the xylem of trees and why trees can pump or suck water up to heights of more than 100 m. J. A. Boehm first formulated the cohesion tension theory in 1893 [2]. He discovered that a cut twig connected to a mercury column by a water-filled capillary created a drag through evaporation that lifted the mercury up to a height of 760 mm (ambient pressure). He concluded that water strands connect the water evaporating cells in the leaves to the roots. The evaporation of water in the cells of the leaves create a negative pressure gradient in the xylem. Cohesive forces between the sap molecules and adhesive forces between the sap molecules and the xylem walls keep the water transport stable. Water from the roots and the xylem replaces the evaporated water, this leads to a water flow from the roots to the leaves against the direction of gravity.

The holding force compensating gravity and therefore preventing the water from flowing back, originates from the water´s surface tension (see equation (1)). The reference points for the surface tension force are pores with very small diameters (< 1 µm) between the xylem and the cell tissue in the leaves. The water evaporates through these pores into the voids between the surrounding cells. The small diameter of these pores together with the surface tension defines the holding force and subsequently the length of the water column. Knowing the diameter of these pores enables the calculation of the possible holding height (see equation (3)).

2 MATERIALS

Different materials were used as evaporation material for the experiments presented in this paper. The diameter of a pore is significant for the possible length of a water column this specific pore can hold. Materials with different pore diameters were tested [3, 4, 5]. At the beginning of the experiments hard plaster with a pore diameter of approximately 10 µm was used. Hard plaster is simple to process and can be formed into many different shapes before it gets hard. It is easy to use and widely available. However, hard plaster dissolves when constantly exposed to water. The water washes out the plaster and the diameter of the pores increase, which decreases the possible holding height (see equation (3)). Hard plaster is suitable for short term testing and the testing of different shapes. For analyzing long time series, a different material was used. Clay was found to fulfil all necessary requirements for a long lasting evaporation material. Fired clay is durable when exposed to water. It has smaller pore diameters than plaster, which lets it hold longer water columns.

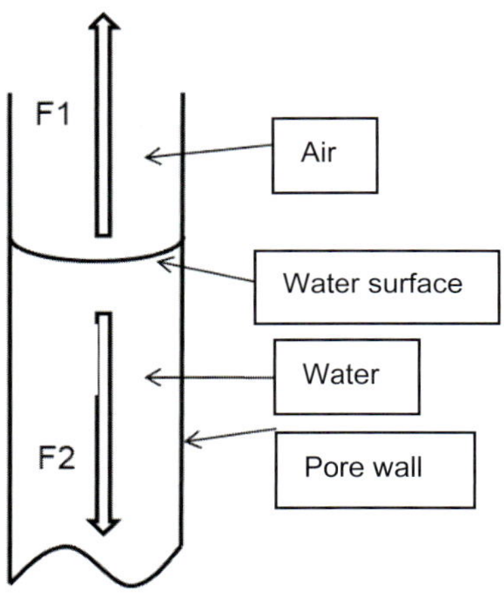

Figure 1: Pore with holding force (F1) and weight (F2) of a liquid in a pore

The average pore diameter is about 0.7 µm. Theoretical calculations showed that a fired clay evaporator can hold a water column of approximately 42 m (see figure 2).

$$F_1 = \sigma \times \pi \times d \quad (1)$$

$$F_2 = h \times \rho \times g \times \frac{\pi}{4} \times d^2 \quad (2)$$

where F_1 is the holding force from the surface tension σ, d the diameter of the pore, F_2 is the weight of the water column, h is the length of the water column, ρ is the density of water, g is the gravitational constant. Under the assumption that the water column does not move into any direction, F_1 and F_2 have to be equal. This leads to a new equation that describes the relation between the length of the water column and the size (diameter) of the pores.

$$h = \frac{4\sigma}{\rho g d} \quad (3)$$

Inter-cellular pores in tree leaves can hold a water column of more than 100 m length.

To illustrate equation (3) the theoretical delivery height in dependency of the pore diameter have been plotted in figure 2.

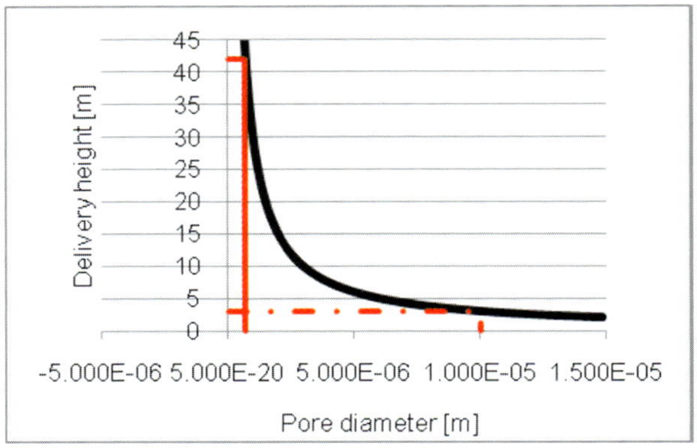

Figure 2: Delivery height in dependency of the pore diameter for water in a glass capillary at T=20°C. The red lines show fired clay (solid) and hard plaster (dashed)

3 STATE OF THE ART

Several approaches were followed to simulate the water ascent in trees.

J. A. Boehm used a porous ceramic body to describe his cohesion theory. Also using a porous ceramic body, a transpiration pump was patented by Emmerich and Milverton in 1993 [6]. Around the surface shell of a porous cylinder is a heating device. By utilizing a variable heater the volume flow can be controlled. This transpiration pump is used in low temperature applications to deliver liquid oxygen and hydrogen in nuclear technology.

Another approach is the use of transpiration pumps in micro fluidic systems. S. Vargo, E. P. Muntz, and G. Shiflett have developed a micro scale thermal-transpiration gas pump [7]. Instead of a porous ceramic body, they use a silica-aero-gel membrane that is arranged between two heating devices.

An experiment by T. D. Wheeler und A. D. Stroock aims to copy the sap-rising in plants and the water absorption in the roots of plants by building a synthetic tree [8]. They use hydro-gel as a porous substance. They model leaves and roots through two semipermeable membranes. By utilizing micro channels, produced with lithographic print they create a model of a synthetic tree, being able to transport water from the roots to the leaves.

All these pumps mimic the way liquids go in trees. The liquid is evaporated and after the pumping process available in its gaseous form. The pump presented in this paper aims at making available the pumped media in its liquid form. Therefore, the step of condensing water is introduced as a main part into the pumping process. The main goal of this pump is to deliver water in a certain height and to gain drinkable water.

4 DESIGN OF THE PUMP

The pump described in this paper is made out of three main components: the evaporating surface, the heat exchanger and the conduit casing. The casing is made out of pipes guiding the hot moist air from the porous body at the heated side to the heat exchanger where the air is cooled down and the moist air condensates (see figure 3).

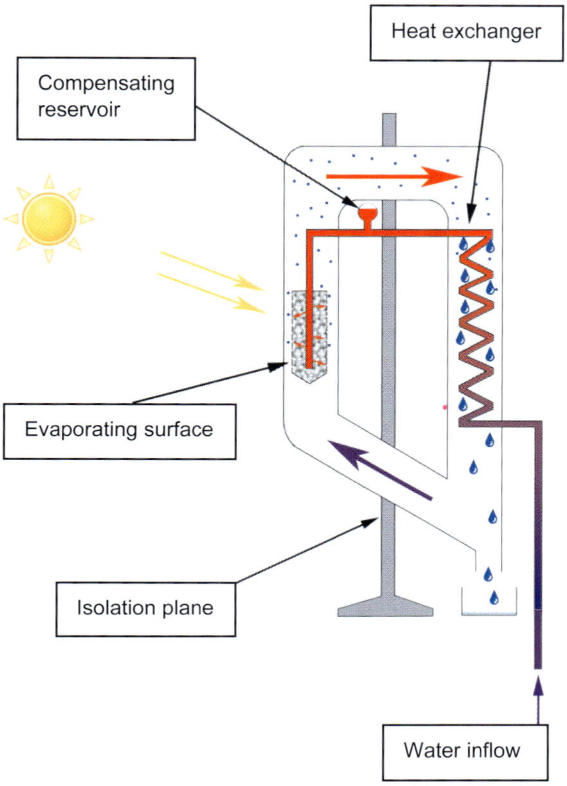

Figure 3 : Principle of a pumping system working with the sun as source of energy

Through the heating up and the cooling down of the air, natural convection is created. On the collector side a transparent casing is used. It enables the thermal radiation to shine onto the surface of the porous body. Water evaporates from the porous body and the moist air ascents inside the casing on the heated side (red arrow in figure 3).

On the other side, protected from the thermal radiation by an isolation layer, the heat exchanger is situated inside the flow channel. This heat exchanger is made out of a copper pipe spiral. Ascending cool water from the ground reservoir flows through this copper piping. A temperature difference is created between the heated moist air coming from the heated side and the cool ascending water inside the copper pipe. Thus, the moist air cools down and starts to condensate on the copper pipe.

The ascending water gains heat from the condensation process taking place along the copper pipe. Passing the heat exchanger the water flows directly into the porous body, where the suction pressure is created through evaporation at the porous surface. The porous surface is made out of fired clay and forms a hollow cylinder.

Since the water is pre-heated, it evaporates more easily, which also increases the energy efficiency of the pump.

The cool unsaturated air flows downwards on the heat exchanger side (see blue arrow in figure 3) and back into the heated side, where the dry air is heated up again and flows alongside the porous body. The porous body stays in contact with unsaturated warm air, which increases the evaporation rate.

To keep the pump running the hydraulic system has to be sealed airtight because the water is under a negative pressure gradient. A leak in the hydraulic system would lead to the development of an embolism, to loss of water and consequently to the failure of the pump.

5 EXPERIMENTS

The following section will give an insight in the different experiments conducted to validate the assumed pumping principles plants is utilize.

5.1 Materials and methods

For all the measurements water from the public water supply of Berlin was used.

The fired clay that was used has a porosity of 0.33 and the hard plaster has a porosity of 0.55. The porosity was calculated by measuring the volume of a material probe and measuring the volume of the water that the probe absorbed after 1 hour of watering.

To measure the volume flow, a Satorius weighing scale was used. Each experimental setup conducted for at least 3 days. In this period at least every 12 hours a measuring point was taken. By taking the arithmetic average of the measurements the average volume flow, which is displayed in figure 5, was calculated.

5.2 Suction pressure

To measure the suction pressure of a fired clay bowl, a hollow clay sphere with an outer diameter of 12 cm and an inner diameter of 10 cm is used. The bowl has an opening with a diameter of 10 mm. The clay bowl is filled with water. The opening of the bowl is joined to a 15 cm long plastic tube with an attached manometer. All these connections are sealed airtight, thus the pores of the clay bowl are the only way for mass exchange with the ambient surroundings.

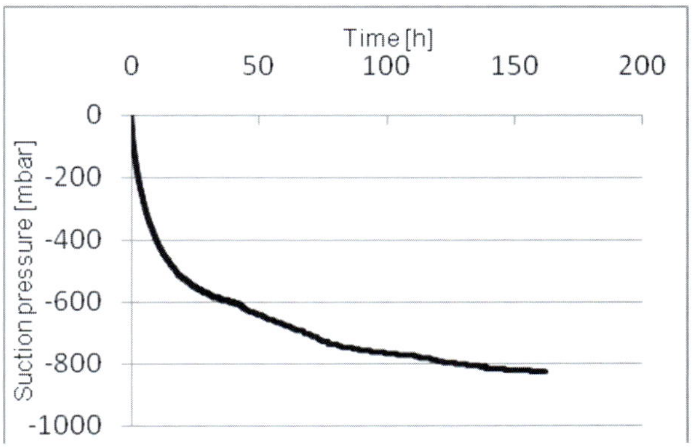

Figure 4: Increasing relative pressure of a porous clay bowl over time

The water evaporates from the clay bowl and builds up a negative pressure in the tube which can be measured (see figure 4).

The results of figure 4 can be interpreted as the maximum pressure difference a fired clay bowl can resists until it comes to an intrusion of air. The pores of the tested clay bowl can hold the water in the pores up to a maximum pressure difference of 825 mbar.

Subsequently this pressure difference translates directly into delivery height. Due to variability in the maximum pore size of different clay bowls caused by the fabrication process, the measured pressure is not as high as the achieved suction pressure of other clay bowls tested with a different method (compare with delivery heights in figure 5).

5.3 Delivery height and volume flow of a porous body

The possible delivery height of the porous body was measured using a simplified test set up. The same kind of hollow clay sphere (see section 5.1) is connected airtight to a pressure tube with a diameter of 8 mm. In the next step the sphere and the tube are filled with water, so that the inside of the porous clay bowl is in constant contact to water. The free end of the tube is connected to a water reservoir open to the ambient to allow for pressure compensation. After filling up the system the clay bowl is lifted up to a certain height. To measure the volume flow a pipette is connected to the water reservoir. With the help of the pipette the flow rate can be measured with an accuracy of ±0.1 ml. With this setup, the delivery height below 10 meters and the volume flow in dependency of the temperature of the surrounding air can be observed. The volume flow was calculated by measuring the weight of the water reservoir. The results for the fired clay bowls (circles and triangles) are presented in figure 5.

For researching the temperature dependency on the porous body a 500 W lamp is placed at 0.35 m distance of the clay bowl. By increasing the surface temperature, evaporation rates raise likewise leading to a higher volume flow rate. This can be seen by comparing the results in figure 5 of the irradiated clay bowls (circles) and the not irradiated clay bowls (triangles).

To compare the impact of the pore size, a hard plaster porous body has been used. Therefore a cone with a hard plaster plate with diameter 9cm as evaporation surface has been connected to a tube. The rest of the setup is like the setup for the fired clay bowls which were not irradiated. Compared to clay the hard plaster has bigger pores. In figure 5 there is a obvious hint, that bigger pores produce higher volume flows.The rates of hard plaster volume flow are much higher than the volume flow rates of the fired clay bowls.

5.4 Pump rate results

In a next step a test system was built to measure the volume flow rate of the complete pumping system. The conduit casing of the demonstrator was made out of PP-H plastic pipes with an inner diameter of 46.6 mm and a wall thickness of 1.8 mm. The irradiated part of the conduit circle had a length of 0.5 m. 0.12 m of the PP-H pipes have been replaced by a Plexiglas pipe with an diameter of 50 mm and a wall thickness of 5 mm. At this transparent part of the irradiated side, the porous body is placed in the middle of the conduit circle. The cylindrical porous body made out of fired clay, with an diameter of 30 mm and a length of 0.15 m had a wall thickness of 10 mm. Attached to the cylinder was an plastic tube with diameter of 6 mm and a wall thickness of 1 mm. The heat exchanger that is connected to the tube had a surface of 528 cm² and was made out of a copper pipe spiral with a diameter of 4 cm, a height of 0.2 m and a copper pipe with 6 mm diameter and 1 mm wall thickness. The heat exchanger is situated on top of the 0.5 m long non irradiated side of the conduit casing. Around the heat exchanger is a Plexiglas tube like the one used for the irradiated side to make the condensation process visible. Simulating the sun as source of energy a 500 W lamp is installed at a distance of 0.35 m from the evaporating surface. The calculated energy density on the surface amounts to 40 mW/cm².

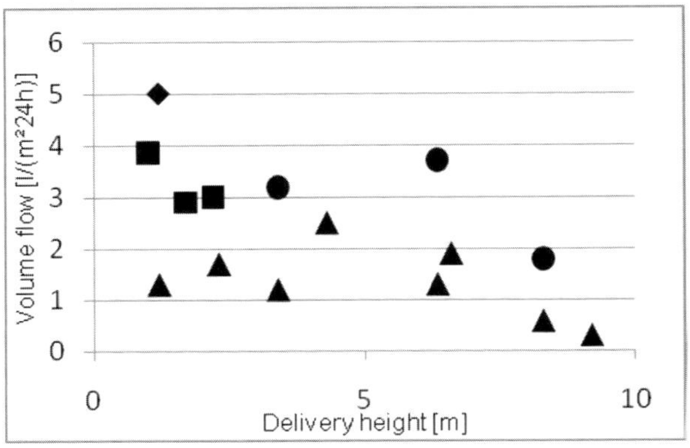

Figure 5: Volume flow in dependency of the delivery height of a porous clay bowl (triangle), a porous clay bowl irradiated by a 500 W lamp (circle), a hard plaster plate (square) and a pumping system irradiated by a 500 W lamp (rhomb).

The experiment is conducted with a delivery height of 1.2 m. In this setup, water is pumped from a water reservoir which is positioned on a weighing scale. With the results of the scale the volume flow rate of the pumped water is calculated. The gained water that was delivered into another reservoir was also measured with a scale.

In figure 5 the result of the pumping system at a delivery height of 1.2 m are shown (rhomb). In comparison to the flow rates of the clay bowls (triangle and circles) it can be seen that it is much higher. It is concluded that these different results are due to the construction of the different pumps. The generation of convection and the regaining of the condensation heat in the pumping system lead to higher volume flow rates (rhomb).

6 ENERGY CONSUMPTION AND EFFICENCY

The pump presented in this paper is able to run solely on solar power. Depending on the location, irradiation densities vary from 700 to 2200 kWh/m² per year. Because of the low energy density, the use of solar power is area intensive. The pump is best used in areas where enough space is available and conventional energy is expensive.

A main characteristic of pumps is their energy consumption, especially the energy consumption per liter of water delivered. The process to evaporate water from 20 °C consumes approximately 2443 kJ/l. For water delivery rates of 5 l/m² per day the energy consumption of the pump is approximately 6912 kJ/l. The energy efficiency, which we define as theoretically needed energy for evaporating 1 l of water divided by the total primary energy consumed, calculates to approximately 37 %. Some of the energy is recuperated and stays in the pumping system; it can be assumed that the actual efficiency is less. There exist no likewise pumps to compare the efficiency to but simple solar desalination units (including the phase change of water) have comparable consumption (2880 to 7020 kJ/l) and efficiency rates (< 50 %) [9, 10].

Under special conditions (good water supply, windy, high temperatures) plants can almost reach 100 % efficiency, i.e. they use nearly 100 % of the irradiated energy to transpire water. For longer periods they have efficiency rates between 40 and 70 % depending on the water supply and time of year (below 0 °C no water is delivered) [11].

6.1 Approaches to raise efficiency

Due to the fact, that the presented pumping system is only a model to show that the principle is working, there are several approaches to raise efficiency. The main points that would make the pump more efficient, is to raise the evaporation rate, to raise the condensation rate and to connect the two mechanisms as efficient as possible by preheating the inflow water with the use of the condensation heat.

A main point to raise the evaporation rate, is to raise the temperature at the evaporating surface. Operated with solar energy a setup similar to a tube collector system for solar heat should be used. Therefore the flow channel at the irradiated side should be build out of a double layer glass. By vacuuming the gap between the two glass layers heat loss due to convection can be reduced. Building a concentrating reflector around the tube would also raise the temperature at the evaporating surface. As well as the reflector would irradiate the back of the evaporating surface. The color of the evaporating surface should be black to absorb as much solar radiation as possible.To raise the evaporation a porous media with bigger pores can be used. This is suitable for applications where a small delivery height is sufficient.

On the not irradiated side the air should be cooled down as low as possible. To regain the condensation heat, the flow channel with the heat exchanger should be isolated and situated on top of the non irradiated conduit casing. In this section of the flow channel the saturated air should only condense at the counter flow heat exchanger where the inflow water is flowing through. Due to the fact, that the inflow water cannot absorb as much thermal energy as the saturated air has gained through evaporation, another cooling section is needed. This cooling section, situated under the counter flow heat exchanger, should cool down the flow of the saturated air to ambient temperature. For that a plate heat exchanger or a similar device can be used.

7 CONCLUSIONS

The described experiments on the running test system have proven the underlying concept and assumption to be correct.

Results such as a delivery height of up to over 9 meters are a promising first step. Yet being only a first test system of a bionic pump, significant increases in pumping performance can be obtained by elaborating on the process with a possible integration of further technologies.

A bionic pump combines low maintenance needs, due to the lack of moving parts in the concept with the inherent water purification through the water´s phase transitions. Another advantage is the continuous variability of the volume flow rate through controlling the input energy. High thermal input increases the evaporation rate and consequently the amount of gained water (see figure 5).

The pump could be interesting in a broad field of applications. It could be interesting as a cooling device or regaining energy from waste heat. Other than that, it could be used in agriculture as an irrigation device.

The next steps on this research project are to review delivery heights of more than 10 meters, to deliver water from wet grounds and building a long lasting and improved prototype to increase the volume flow rate. This also includes the review of ways to avoid the effects of contamination and calcination of the porous body. In addition, it should be explored whether the consumed energy is linked to the delivery height.

8 REFERENCES

[1] Bresinsky,A., Körner, C., Kadereit,J.W., Neuhaus,G., Sonnewald, U., 2008, Strasburger Lehrbuch der Botanik, Aufl. 36, Heidelberg-Spektrum Akademischer Verlag, ISBN 3-8274-1561-6

[2] Böhm, J.,1893, Berichte der Deutschen Botanischen Gesellschaft 11,

[3] Theileis, J., 2010, Entwicklung einer Kapillar-Tensiopumpe, diploma thesis, Department of Bionics and Evolutiontechnique, Berlin Inst. of Technology, Germany

[4] Messing, R., 2008, Entwicklung einer solaren Wasserförderungsanlage nach dem Vorbild der Pflanzen, student research paper, Department of Bionics and Evolutiontechnique, Berlin Inst. of Technology, Germany

[5] Kopp, O., 2010, Eine bionische Pumpe nach dem Vorbild der Transportmechanismen von Bäumen, student research paper, Department of Bionics and Evolutiontechnique, Berlin Inst. of Technology, Germany

[6] Hemmerich, J.L., Milverton, P.H.,1993, Transpirations Verdampfungspumpe, Deutsches Patentamt, DE 69402702T2.

[7] Vargo, S.E., Muntz, P.E, Shiflett, G.R., 2003 Thermal transpiration pump United States University of Southern California (Los Angeles, CA) 6533554 http://www.freepatentsonline.com/6533554.html

[8] Wheeler, T.D., Stroock, A.D, 2008 The transpiration of water at negative pressures in a synthetic tree, nature, vol. 455, doi:10.1038/nature07226

[9] Müller H.C., 2009, Theoretische und praktische Untersuchung einer mehrstufigen solarthermischen Kleinanlage zur Meer- und Brackwasserentsalzung, phd thesis, Rheinisch-Westfälische Hochschule Aachen

[10] Brendel T., 2003, Solare Meerwasserentsalzungsanlagen mit mehrstufiger Verdunstung, Ruhr Universität Bochum

[11] Guenther R., 2005, Zum Verhältnis von Strahlungsenergie und Verdunstung, 11. Gumpensteiner Lysimetertagung, Höhere Bundeslehr und Forschungsanstalt für Landwirtschaft pp.37-40

Sensitivity of CO_2 Emissions to Renewable Energy Penetration for Regions Utilizing Power and Water Cogeneration

P. Lin[1], A. Khalid[2], S. Kennedy[1,2], S. Sgouridis[2]

[1]Water and Environment Engineering Program, Masdar Institute of Science and Technology, Abu Dhabi, UAE

[2]Engineering Systems Management Program, Masdar Institute of Science and Technology, Abu Dhabi, UAE

Abstract

Use of renewable energy to produce electricity is generally viewed as a promising means to mitigate CO_2 emission. However, the substitution of a significant percentage of electricity generation from fossil fuel fired power plants with renewable energy, such as photovoltaic, concentrated solar power plant, and wind power, might not be as effective in CO_2 mitigation in regions utilizing power and water cogeneration. Due to a unique relationship between power cycle efficiency and the ratio of power to water production in the cogeneration system, counterintuitive results might appear as the penetration of renewable energy increases. Our analysis indicates that, without decoupling of the power and water cogeneration or active water conservation, generating electricity with renewable energy alone might be less effective in reducing CO_2 emissions than anticipated.

Keywords:

Cogeneration; Desalination, Power Generation; Renewable Energy; CO_2 Emission

1 INTRODUCTION

Water and electricity are cogenerated in dual-purpose power and desalination plants, also known as cogeneration plants, in many of the Gulf Cooperation Council (GCC) countries where freshwater resources are scarce and desalination is used as a means for producing fresh water supplies [1, 2]. Since desalination is an energy intensive process, cogeneration configurations are more economical in terms of fuel consumption than stand-alone thermal based desalination plants because the lower grade heat exhausted from the gas turbines can be recovered to produce steam for thermal desalination. The efficiency of a cogeneration plant is sensitive to the ratio of electricity to water production, referred to as "power to water ratio" (PWR). PWR determines the requirement of thermal energy for each process and the quantity of exhaust heat that can be recovered for distillation. At low electricity loads, the recovered heat may not be sufficient for water production. Therefore, an auxiliary boiler has to be fired to supply additional thermal energy. Large-scale cogeneration plants usually rely on fossil fuel such as natural gas or oil, and thus contribute significantly to carbon dioxide (CO_2) emissions.

For accounting purposes, it is often desirable to determine the relative impact of the two co-products, i.e. electricity or water, in terms of fuel use and emissions. However, due to the nature of cogeneration, the sharing of physical components and fuel exergy makes such separation difficult.

In practice, there is no single "correct" allocation method and several accounting methods have been proposed, including *energy content of product*, *exergy*, *incremental fuel consumption to electricity production*, *incremental fuel consumption to water production (reference cycle)*, *shared emission saving*, *economic value of products*, etc [3-6]. The choice of the appropriate method should be based on the configuration of the cogeneration plants and the objectives of the analysis.

A further complication in using allocation methods is that they are accurate for apportioning fuel and emissions for a specified operating state of the system and *marginal* changes. However, they cannot be reliably used for assessing the impacts of *substantial* changes in water and electricity demands from the existing system conditions due to the nonlinear relation between the PWR and plant efficiency. For example, if 60% of CO_2 emissions are allocated to electricity and 40% to water, this allocation ratio may no longer be valid if the electricity production is reduced substantially, while the water production remains constant. A particular allocation ratio is therefore only valid for a certain PWR. Because of this, a fixed allocation ratio cannot be used when estimating emissions or fuel savings due to a reduction in either water or electricity production. A "variable-PWR" approach is necessary to account for the change in PWR between the two operating states.

In an electricity and water production system without carbon capture and storage mechanisms, CO_2 emissions are directly related to the total fossil fuel consumption from all power and desalination plants in the system. It is possible that a combination of technologies coexist in such a system. For example electricity grids in GCC countries rely on conventional power generation options including natural gas or oil powered steam, gas or combined-cycle turbines. Non-fossil alternatives, such as nuclear energy power plants, concentrated solar power plants (CSP), photovoltaic and wind power are also being introduced. Similarly, on the desalination side, there are several technology options. Multi-Stage Flash (MSF) and Multi Effect Distillation (MED) are thermal based technologies that require heat for distillation and electricity for pumping water. Reverse Osmosis (RO) is the most popular membrane technology which only requires electrical power to mechanically separate salts and water.

In this paper, we will take Abu Dhabi Emirate in the United Arab Emirates (UAE) as a case study since more than 99% of Abu Dhabi's electricity production comes from cogeneration plants [7]. These plants are responsible for a significant portion of the emirate's fossil fuel consumption, primarily natural gas, and contribute proportionally to the CO_2 emissions in Abu Dhabi. Increasing the penetration of renewable energy in the electricity production is expected to reduce the reliance on cogeneration and mitigate CO_2 emissions. However, a high penetration of renewables, without water demand management or introduction of alternative water desalination methods, reduces the PWRs in the cogeneration system. At low PWRs auxiliary boilers need to be utilized to supplement the thermal distillation units with the steam needed to meet water production requirements. Therefore, high penetration of renewable might not necessarily result in CO_2 savings as high as expected. In this paper, the "variable-PWR" approach is used to analyze the potential for CO_2 reduction by renewable energy penetration in Abu Dhabi.

2 DEFINITIONS OF PWR, POWER CYCLE EFFICIENCY AND FUEL CONSUMPTION

2.1 PWR

The PWR of a cogeneration plant is defined as the ratio of net power output to water production of the plant. The net power output is the total power that is transmitted to the grid. It can be calculated by subtracting the plant auxiliary power consumption from the gross power generation. The system PWR is the sum of the net power of all cogeneration plants over the sum of the water production of all cogeneration plants. Power production usually has greater variation than water because power has to be provided instantaneously upon demand while water can be stored. Power demand in Abu Dhabi not only varies within a day because of peak and nonpeak difference, but also exhibits significant seasonal variation between summer and winter months due to variable air-conditioning requirements.

2.2 Power cycle efficiency

Power cycle efficiency is defined as the ratio of net electrical energy output to the thermal energy of the total fuel input for an individual cogeneration plant or the entire system. The fuel consumption, thus, can be estimated by dividing power demand by the power cycle efficiency. Based on literature and regression analysis against historical data, we have found that the cycle efficiency of any given cogeneration plant can be expressed as a nonlinear function of PWR. Therefore, for any change in aggregate power and water demands, we can estimate the corresponding change in fuel consumption based on the new power demand, PWR and cycle efficiency. Typical design PWRs and the corresponding power cycle efficiencies for cogeneration plants are shown in Table 1.

Combined Cycle Gas Turbine (CCGT) cogeneration is the dominant technology not only because it can achieve higher cycle efficiency at full load condition, but also because of the flexibility it provides in covering a wide range of PWR. When the electricity load is reduced, one or more gas turbines can be shut down, yet the steam flow to the Multi Stage Flash (MSF) desalination units can be kept constant by firing the auxiliary boiler to supplement heat to the Heat Recovery Steam Generator (HRSG). At an even lower electricity load, the steam turbine can bypass the steam directly from boiler to the distiller and sustain the same level of water production.

Table 1: Typical design values of power to water ratio (PWR) and power cycle efficiency for different cogeneration plants

Plant type	PWR (MW/MGD[1])	Cycle efficiency
Backpressure steam turbine + MSF	3.3-5.8	0.2-0.25
Extraction or condensing steam turbine + MSF	3.3-15.8	0.25-0.3
Gas turbine/unfired HRSG +MSF	5-10.8	0.25-0.35
Combined cycle gas and steam turbine / HRSG (CCGT) +MSF	7.5-15	0.35-0.45

Source: data adapted from [8]

1. MW: mega watt; MDG: million gallon day

2.3 The relationship between PWR and cycle efficiency

Historical values: single plant

Taweelah A1 power and desalination plant (TA1) in Abu Dhabi is taken as an example. This plant originally had three gas turbines and was equipped with bypass stacks so that they can be operated in high-pressure single or combined cycle modes. There were also four MSF desalination units. Taweelah A1 power and desalination plant has undergone a change in ownership in September 2000. After the transfer, the capacity of the plant had been upgraded from 255MW/29 MGD to 1014 MW/85 MGD, and the expansion was completed in 2002. Five new CCGTs and three backpressure steam turbines (BPST) (200 MW each) were added. Each CCGT has its associated HRSG equipped with supplementary firing. The steam from the HRSG is passed to the BPSTs, and the residual low pressure steam is then used in the desalination process. Each of the four existing MSF desalination units has been upgraded to 8 MGD and 14 new MED units (3.77 MGD each) have been added to make a total capacity of 85 MDG. Figure 1 shows a representative schematic diagram for the plant setting.

Figure 2 illustrates the plot of monthly aggregated data of cycle efficiency against PWR of the Taweelah A1 plant from 1999 to 2008. Monthly net power output, water production and fuel consumption were taken from the ADWEC 2008 statistical report [7]. It is clear from the figure that, after the upgrade, the cycle efficiency has improved greatly especially in the region of high PWRs which is close to the full power plant load. These data show that higher PWR would result in higher cycle efficiency, but this increment is not linear. Based on the regression model, the cycle efficiency (η_{cy}) can be expressed as a function of PWR as shown in Equation (1).

$$\eta_{cy} = -0.0009(PWR)^2 + 0.0876(PWR) \quad (1)$$

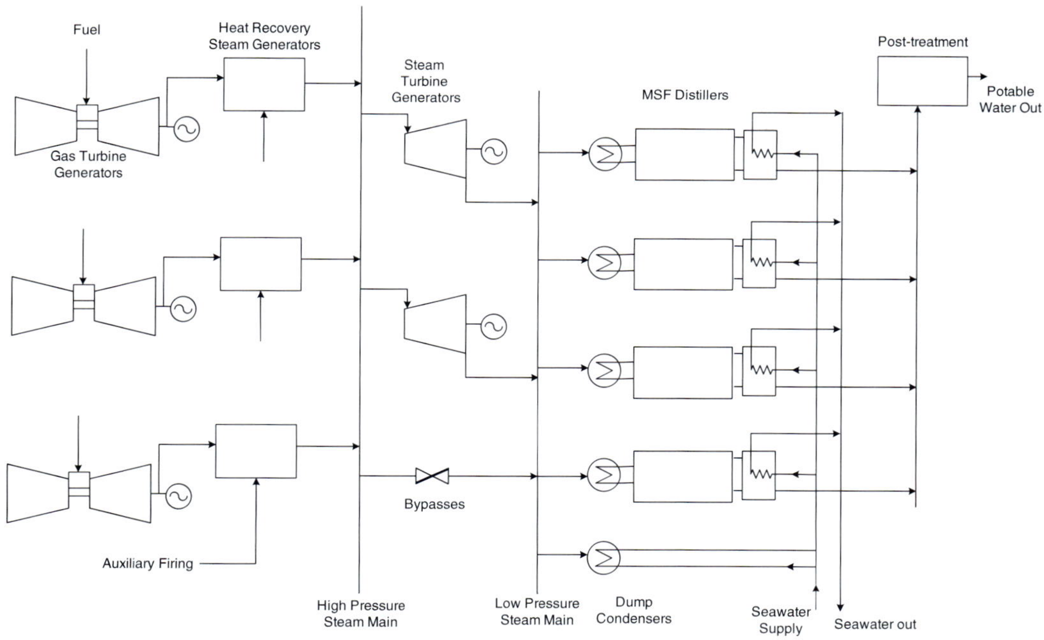

Figure 1: Schematic diagram of GGCT and MSF based cogeneration plant. (Source: [5])

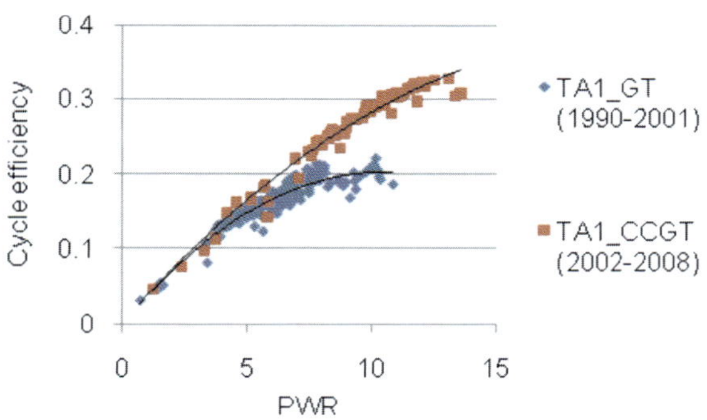

Figure 2: the relationship between PWR and cycle efficiency for Taweelah A1 power and desalination plant. After the upgrade, the plant has been changed from gas turbine based to combined cycle based cogeneration plant.

We can also compare the curve of PRW and cycle efficiency of TA1 with other CCGT cogeneration plants. For example, Taweelah A2 power and desalination plant (TA2) includes three CCGT, three HRSGs and two steam turbines (ST) in its power plant and four MSF distillers in the desalination plant. The power and generation capacities are 760 MW and 51 MGD respectively. The scale of TA2 is about half of TA1. Another plant, Shuweihat 1 (S1) has similar capacity to TA1. The power plant has five CCGTs, associated HRSGs, and two STs, providing a total power generation capacity of 1614 MW. On the other hand, the desalination plant has six MSF distiller units with a total capacity of 101 MGD. Table 2 summarizes the power and water generation capacities for the three CCGT based cogeneration plants [7]. It is expected that the power efficiency curves would show similar characteristics among these plants.

Table 2: Power and water generation capacities of three combined cycle based power and desalination plants.

Plant	TA1		TA2		S1	
Turbine type	GT	ST	GT	ST	GT	ST
Power capacity (MW)	813	601	540	220	1104	511
Water capacity (MGD)	85		51		101	

Figure 3 shows that the cycle efficiency of S1 is slightly higher than TA1 and TA2 for a given PWR. It might be due to the fact of technological improvement over time since S1 is a relatively new plant. Moreover, some data points that deviate from the curve representing lower efficiencies for given PWRs correspond to the early operation period of the plants. It is usually the case that newly commissioned plants require an adjustment/learning period until optimal efficiency operations are consistently achieved.

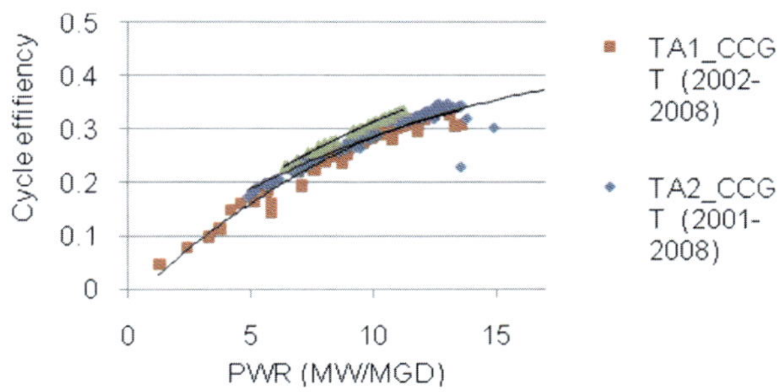

Figure 3: Comparison of the PWR/efficiency curve among different combined cycle based cogeneration plants in Abu Dhabi.

Similarly, we can also compare TA1 before its upgrade with other gas turbine cogeneration plant in Abu Dhabi. Al Mirfa power and desalination plant, belonging to Al Mirfa Power Company (AMPC), is equipped with four gas turbines, four waste heat recovery boilers (WHRB) and four additional auxiliary boilers in case the recovered heat cannot provide sufficient steam to the MSF distillers. This plant has total power and water generation capacities of 186 MW and 15 MGD (Table 3).

The scale of Al Mirfa plant is smaller than TA1. However, it shows better efficiency than TA1 (Figure 4). It might be a result of technology improvement since, compared to the pre-upgrade TA1, it is relatively new. When compared with CCGT cogeneration plants in Figure 3, as expected, the CCGT cogeneration plants shows higher efficiency especially at high PWR region.

Table 3: Power and water generation capacities of gas turbine power and desalination plants

Plant	Mirfa	TA1
Power capacity (MW)	186	255
Water capacity (MGD)	15	29

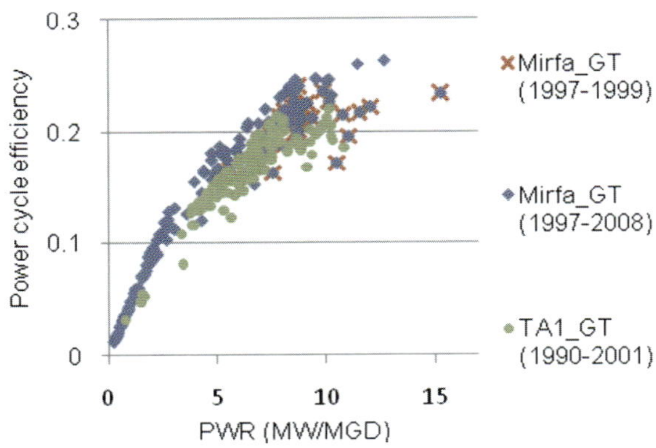

Figure 4: Comparison of the PWR/efficiency curve between two gas turbine based cogeneration plants in Abu Dhabi.

Historical values: system aggregated

As our objective is to estimate the impact of changes on a system-wide scale, we aggregated data from all the operational cogeneration plants from 1998 to 2008 in Abu Dhabi and plotted the cycle efficiency against the PWR (Figure 5). These plants include both GT and CCGT based cogeneration plants as well as ST based ones. The system-wide power cycle efficiency as a function of PWR curve was regression fitted to the function expressed in Equation (2).

$$\eta_{sy} = -0.001(PWR)^2 + 0.0372PWR \quad (2)$$

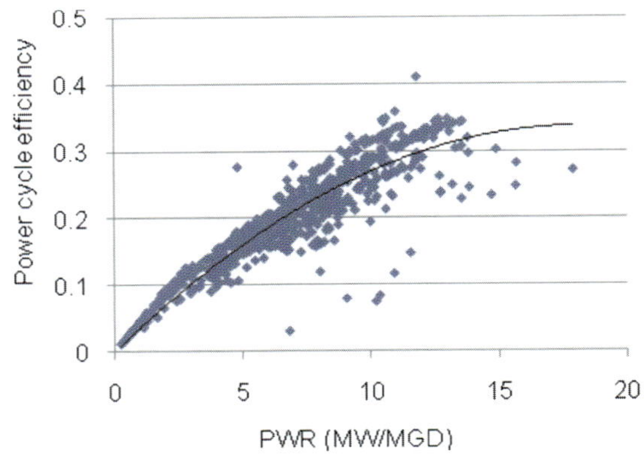

Figure 5: Monthly aggregated data of all the cogeneration plants in Abu Dhabi from 1998 to 2008.

Partial load PWR and cycle efficiency curve from literature

We also compared the historical data with some of the values of part-load performance of cogeneration plants from the literature [8]. The part-load performance analysis compared the fuel consumption for varying PWR for a given plant with constant water output but variable electricity output. Such a case is the norm as demand for electricity shows greater variability than water. Two type of cogeneration plants are chosen for the comparison: a combined cycle gas turbine, supplementary fired HRSG, extraction/condensing steam turbine (CCGT-EC), and a combined cycle gas turbine, supplementary fired HRSG, backpressure steam turbine (CC-BP). Both of them are operated at ambient temperature of 50 °C and the desalination units have a performance ratio (PR) of 8. PR is the ratio of distillate to steam in terms of mass, and it is a common indicator for the performance of a distiller. Low PR leads to high PWR and vice versa. It is shown in Figure 6 that the efficiency curves are similar between the historical monthly aggregated data and those presented in the literature.

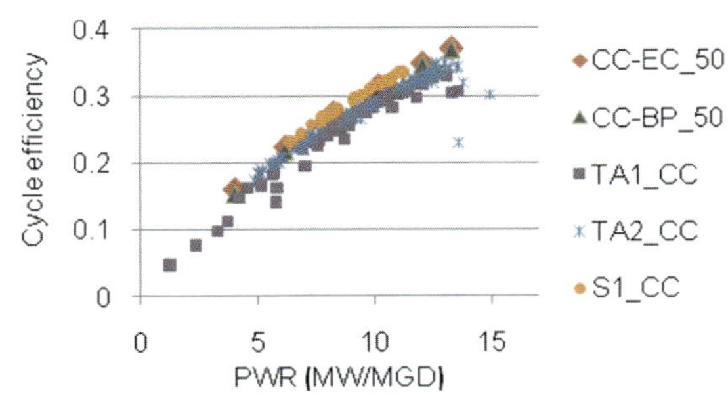

Figure 6: Comparison of the efficiency curve between the historical data and the values in the literature [8].

3 RENEWABLE ENERGY PENETRATION AND REDUCTION IN CO_2 EMISSION

Substitution of fossil-fuel powered electricity generation with renewables is expected to reduce CO_2 emissions, the primary greenhouse gas contributing to climate change. There are several means to generate electricity from renewable energy including photovoltaic, concentrated solar power, wind power, tidal, etc. However, in a system where almost all electricity and water are produced from cogeneration, moving the electricity production to renewable energy alone might not result in the expected CO_2 reductions (Figure 7 A). As mentioned earlier, the PWR, power cycle efficiency and fuel requirement are highly correlated in a cogeneration system. Penetration of renewable energy in electricity production alone will reduce the power generation requirement in the cogeneration system. However, the cogeneration plants may still have to operate to maintain the same level of water production. Therefore, with a lower PWR, the cogeneration plants are forced to operate at a less efficient mode. As a result, 5 % penetration of renewable in electricity generation, for example, may not give rise to a corresponding 5% CO_2 reduction.

In the current generation mix in Abu Dhabi, 96% of electricity is generated from combined cycle gas turbine (CCGT), 3% from steam turbine (ST) and 1% from gas turbine (GT) [7]. Equation 2 quantifies the relationship between PWR and cycle efficiency on a system-wide basis. If we assume that the cycle efficiency is independent of the PWR, a penetration of renewables from 0 to 80% would result in the same percentage of CO_2 reduction (Figure 7B). However, if we take the effect of variable efficiency into consideration, the savings are substantially lower than expected. For example, 10% of renewable penetration would only result in 2.46% emission reduction if water consumption and generation methods remain the same. The percentage of emission reduction also shows diminishing returns as when the penetration of renewable reaches 80%, the total system emissions are only reduced by 16.79%.

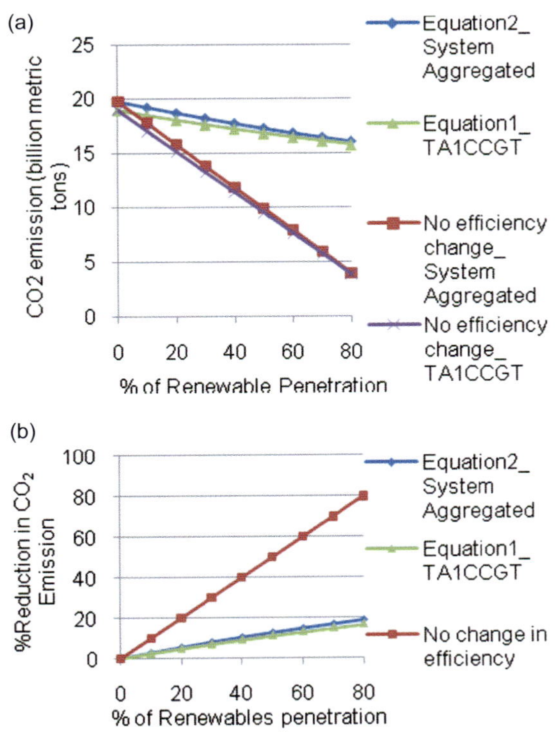

Figure 7: (A) Sensitivity of CO_2 emission to the penetration of renewable energy; (B) percentage reduction in CO_2 with regard to the penetration of renewables.

4 DISCUSSION

Abu Dhabi is in a unique region where cogeneration of electricity and water provides synergy to both entities. Cogeneration is more economical in terms of fuel savings because the process heat from power cycle still contains enough exergy to generate steam for the thermal distiller. Generating electricity with renewable energy is widely recognized as an effective approach to mitigate CO_2 emission. However, as demonstrated above, for regions that practice electricity and water cogeneration, moving the electricity generation forward to the renewable energy alone is not effective enough to achieve the expected emission reduction target. It has to be accompanied by either decoupling of electricity and water production or reduction in water demand. MSF needs both thermal energy (300 MJ/m^3) and electrical power (3-5 kWh/m^3) whereas RO only requires electricity (6-8 kWh/m^3) for mechanical separation of salts and water [9, 10].

Inefficient operation of a power system at a low PWR can be avoided by decoupling electricity and water production. Some of the highly energy intensive thermal desalination plants could be replaced by RO plants in order to maintain a high PWR among the remaining cogeneration plants. While RO adds to the electricity load of the system, at the same time, it reduces the burden of cogeneration, thus increasing PWR and cycle efficiency.

MSF and RO hybrid cogeneration plant is another configuration that can efficiently generate electricity and water. Fujairah emirate in the UAE has the world's biggest MSF-RO hybrid cogeneration plant which can produce water from MSF unit during peak electricity demand and switch to RO during the off peak period. At a small scale, solar thermal desalination may also be a viable option. Solar energy can be applied directly to produce distillate or collected by solar collectors to generate heat or electricity for conventional desalination techniques such MSF, MED and RO. On the demand side, water conservation policies, which have great potential in improving PWR, will be particularly useful in winter times when PWR is generally low.

Future work will include exploration of different electricity and water policy combinations and estimation of their potential and cost in emission reduction in regions utilizing electricity and water generation.

5 REFERENCES

[1] M. Al-Rashed and M. Sherif, "Water Resources in the GCC Countries: An Overview," *Water Resources Management*, vol. 14, no. 1, pp. 59-75, Feb. 2000.

[2] M. A. Dawoud, "The role of desalination in augmentation of water supply in GCC countries," *Desalination*, vol. 186, no. 1, pp. 187-198, Dec. 2005.

[3] S. P. Agashichev and A. M. El-Nashar, "Systemic approach for techno-economic evaluation of triple hybrid (RO, MSF and power generation) scheme including accounting of CO2 emission," *Energy*, vol. 30, no. 8, pp. 1283-1303, Jun. 2005.

[4] O. A. Hamed, H. A. Al-Washmi, and H. A. Al-Otaibi, "Thermoeconomic analysis of a power/water cogeneration plant," *Energy*, vol. 31, no. 14, pp. 2699-2709, Nov. 2006.

[5] N. M. Wade, "Energy and cost allocation in dual-purpose power and desalination plants," *Desalination*, vol. 123, no. 2, pp. 115-125, Oct. 1999.

[6] M. A. Rosen, “An exergy-based method for allocating carbon dioxide emissions from cogeneration systems,” *International Journal of Exergy*, vol. 6, no. 1, pp. 1 - 14, 2009.

[7] *ADWEC Statistical Report 1998-2008*. Abu Dhabi, UAE: Abu Dhabi Water & Electricity Company, 2008, p. 203.

[8] A. M. El-Nashar, “Cogeneration for power and desalination -- state of the art review,” *Desalination*, vol. 134, no. 1, pp. 7-28, Apr. 2001.

[9] N. H. Afgan, M. Darwish, and M. G. Carvalho, “Sustainability assessment of desalination plants for water production,” *Desalination*, vol. 124, no. 1, pp. 19-31, Nov. 1999.

[10] R. Raluy, L. Serra, and J. Uche, “Life cycle assessment of desalination technologies integrated with renewable energies,” *Desalination*, vol. 183, no. 1, pp. 81-93, Nov. 2005.

Optimizing the Design of a Hybrid Solar-Wind Power Plant to meet Variable Power Demand

K. Mousa, A. Diabat

Department of Engineering Systems and Management, Masdar Institute of Science and Technology, Abu Dhabi, UAE

Abstract

Not enough studies have been done on operating the two most available and renewable energy sources, sun and wind energy, alongside one another. A complementary relationship has been proven to exist between the two energy sources, and more research needs to be carried out in order to take advantage of it. This paper presents an optimal design for a hybrid solar-wind power plant, where several design parameters which are pertinent to the plant's performance are optimized, such as the number of photovoltaic modules, the wind turbine height, the number of wind turbines, and the turbine rotor diameter. With the goal of minimal design cost, our simulation and sensitivity analysis proves that the plant can reliably deliver energy throughout the year, to meet variable seasonal demand, by taking advantage of the sources' complementary nature.

Keywords

Hybrid;Renewable energy; Solar; Standalone;Wind turbines

1 INTRODUCTION

While a large number of actions have been found to have a negative effect on our environment, it is widely agreed that greenhouse gases (GHGs) have the largest of those effects. Nitrous oxide, methane, sulfur hexafluoride, hydro fluorocarbons, perfluorocarbons and carbon dioxide all make up the spectrum of GHGs. The earth's temperature is maintained at acceptable levels for organisms with the help of these gases, and a change in their levels could result in an increase/decrease in the earth's temperature. However, because these gases allow sunlight to enter the atmosphere while trapping the heat reflected against the earth's surface, an increase in the levels of GHGs increases the temperature of the earth, leading to global warming, to reach temperature levels that may be too high for organisms. An increase in natural disasters has also been linked to global warming by scientists, stating that oceanic and atmospheric patterns shift with a shift in earth's temperature.

In a global attempt to reduce GHG emissions, the Kyoto protocol was formed. It is part of the United Nations Framework Convention on Climate Change, and the protocol legally binds member nations to be committed in reducing their emissions. To make this easier on legally binded members, the protocol allowed for three "flexible mechanisms". Countries can gain emission credits using the Joint Implementation mechanism, which allows them to execute projects that would reduce emissions in other countries. Another option would be to use the Clean Development mechanism, and this would allow members to gain credits by carrying out emission reduction projects in developing countries. The final mechanism, Emissions Trading, pushes authorities to reduce their emissions by providing an incentive.

The push of governments towards renewable and clean energy does not only rest on the carbon-emissions caps and financial incentives provided. The rapid depletion of fossil fuels and their scarcity provides an even stronger incentive for governments to look for other sources of energy that would meet their demand. In addition, a large proportion of the world's population do not have proper access to electricity, since they live in rural/remoter areas which would be geographically-difficult to connect to the grid. Governments have been looking for creative, efficient and feasible methods to provide electricity to such areas, and renewable energy sources are thought to be the best-suited for such a problem.

Solar and wind energy are renewable, infinite and environmental friendly. This has attracted governments to renewable energy sources in general and solar/wind energy in specific. Clearly, those two energy sources are nature-dependent, which makes their use in applications vulnerable to unpredictable climatic drawbacks such as bad weather. However, both energy sources have shown a complementary nature, and can neutralize one another's drawbacks with their strengths. The hybrid solar-wind concept is brought up to take advantage of this complementary situation.

A hybrid system is one that utilizes two energy sources alongside one another. The concept itself has been revealed many years ago, but has only been in the focus of some studies a few years ago [1]. Hybrid power plants save fossil fuels, and are ideal for supplying power to geographically isolated areas [2] without any GHG emissions.

This means that a hybrid system that would utilize the advantages of solar and wind energy's complementary nature is a suitable solution to providing energy to both on-grid and rural areas without harming the environment. However, the complexity that arises from combining both energy sources into one power plant makes the hybrid system tougher to design and analyze. Hence, there is no wonder that a lot of work has gone into designing and modeling wind/solar power plants and other hybrid energy sources.

2 LITERATURE REVIEW

Dihrab and Sopian [1] studied the design of a grid-connected hybrid PV/wind system to be used in Iraq. The variables they kept constant were the land's meteorological data, PV and wind turbine sizes. They solved their method using MATLAB solver, and the system was simulated for cases of total power blackout in different locations. Their results showed that the system can be used to provide power to rural areas in need of power, as well as in times of total blackout, while also sending power to the grid. However, their study did not consider the power demanded by the load, or seasonal power demand.

Yang and Burnett [3] focused their research on analyzing the probability of power supply failure of a hybrid PV/wind system in Hong Kong that used a storage battery, while also assessing its reliability. First, using local weather data, the authors' simulation showed that solar and wind power truly possess complementary characteristics, and can make up for one another. They also set up equations and restraints for the loss of power supply probability (LPSP) to consider in their study, and their results showed that to maintain an LPSP of 1%, a battery bank with a storage capacity of 3 days is needed. To maintain a 0% LPSP, a battery bank of 5 days capacity would be needed. Their study did not consider PV and wind turbine design parameters, such as turbine height, turbine diameter, etc.

Shaahid and Elhadidy [4] studied the utilization of an off-grid hybrid PV/diesel power system with the use of batteries. The study was based on the load demanded by a residential building in Saudi Arabia, and the only factor considered in the design was the size of PV arrays. Average monthly solar irradiance values were used to test the system in three combinations of using PV alone, with a diesel back-up generator, and with the use of batteries. Their results showed that the system can meet the required power load with $225\ m^2$ of PV, along with a battery storage capacity of 12 hours, while the diesel generator provides 9% of the demand. Without a battery, the generator would have to provide 58% of demand. While this design does provide for the decreased consumption of fuels, the authors' system is not considered fully renewable due to the use of a consumable fuel.

Yang et al. developed [5] a model to optimize the design configurations of a hybrid PV/wind system with the use of batteries. The model was used to calculate the optimal configuration as well as the desired LPSP at minimal cost, and is based on a genetic algorithm. In the process, several design parameters were optimized except for the turbine rotor diameter. Good performance was found on the system, however it was only tested using data to simulate the hybrid system supplying power to a relay station. This means that the load demand was constant, and the authors did not take into account how the system would cope with seasonal changes in demand.

Onar et al. [6] presented a grid-independent hybrid system that consisted of a wind turbine, photovoltaic system, ultra-capacitor and fuel cell. The specifications of the PV and turbine components were pre-set, and the model was tested under various wind speeds and solar condition. Their model was used to find an optimal control strategy, and the simulation showed good performance of the system. However, as mentioned above, the various design parameters of the components of the systems (e.g. turbine height, PV number) were not considered in this study, and seasonal demand was not considered either.

A hybrid system model that incorporated fuel cell generation alongside wind and solar energy was developed by Ahmed et al. [7]. The main energy source came from the wind and the sun, while the fuel cells was utilized as a backup resource. The authors found that the system can be used as a reliable power source to supply the load, even in bad climatic conditions.

Kershman et al. [8] presented a model that optimized the design of a sea water reverse osmosis desalination plant. Power was supplied to the plant using a hybrid wind/solar hybrid system. The system was simulated as a plant on Libya's coast, to supply a village with drinkable water.

The technical and economic feasibility of an on-grid hybrid wind/solar system was studied by Bakos and Tsagas [9]. The system would be used to supply energy to a city in Greece through electrical and thermal energy.

Tina et al. [10] used a probabilistic approach to examine the long-term performance of a hybrid solar/wind system. The model was used to study the plant's performance for off-grid and grid dependent applications.

In this paper, we present an optimization of the design parameters for a hybrid stand-alone solar/wind power system. The system could be used to power various applications, including loads in rural areas. The design parameters optimized in our study are the number of wind turbines, the number of PV modules, the turbine rotor diameter, and turbine height. We simulate the model using seasonal demand patterns for the city of Abu Dhabi, United Arab Emirates, where demand is expected to peak during summer. Our model will optimize the design parameters of the system, with a primary objective of minimizing cost while meeting power demand.

3 THE MODEL

3.1 Notation: Parameters and Values

The Notation used in our model is summarized in Table 1. Parameter values are shown.

Table 1: Parameters and decision variables.

Parameter	*Value*	Description
N_w	Decision Variable	Number of wind turbine
C_{wm}	1000$	Annual maintenance cost for wind turbine
h	Decision Variable	Wind tower height (m)
r	Decision Variable	Radius of wind turbine (m)
C_{wf}	50000$	Installation + Fabrication cost of wind turbine (steel cost not included)
i	5%	Real Interest rate
ffY_{proj}	35 year	Project lifetime
N_s	Decision Variable	Number of solar cells
C_{sm}	500$	Annual maintenance + cleaning cost for solar panel
C_{sc}	5000$	Solar panel capital cost + installation cost
rho	1.225 kg/m^3	Air density
c_p	0.45	Coefficient of performance
V_w		Wind speed (m/s)
N_g	50%	Generator efficiency
N_b	95%	Gearbox bearing efficiency
Voc, V_{oco}		Voltage for open circuit
n	1< n <2	Ideality factor
K	$1.38*10^{-23}$ J/K	Boltzmann constant
q		Magnitude of the electron charge
R_s		Series resistance (ohm)
I_{sc}, I_{sco}		Short circuit current (A)
G, G_o		Solar radiation, W/m^2
T_o, T		Temperature under standard conditions (K)
α, β, γ		Constant parameters for PV module

2.2Equations

2.2.1 Objective Function

Our objective function is shown in Equation 1. It includes the costs incurred by the installation, use and maintenance of both wind turbines and solar arrays used. The aim of our model is to minimize this function. Interest rate is considered over the lifetime of the project, while costs are further elaborated on in section 2.2.3.

$$Cost(wind)+Cost(Solar) \quad (1)$$

2.2.2 Demand Constraint

Equation (2) warrants that the demanded power load is met by the power generated from the hybrid plant. The power generated from the wind turbines and solar arrays is accounted for separately. Seasonal weather is taken into consideration when the power of both sources is calculated. Power equations could be seen more clearly in section 2.2.4.

$$P_t(Wind)+P_t(Solar) \geq P_{demand} \quad (2)$$

2.2.4 Height And Radius Constraints

The following two equations are dimensional constraints for the wind turbines. We use Equation (3) to limit the height of the wind turbines to 50 meters, while equation (4) limits the rotor radius to 30% of the tower height.

$$h \leq 100 \quad (3)$$

$$r \leq 0.3*h \quad (4)$$

2.2.3 Overall Costs

The costs incurred from the installation, operation and maintenance of the wind turbines are given in Equation (5). The equation incorporates the costs of increasing the height of the wind turbine and the rotor diameter. These costs are a multiple of the number of wind turbines installed, *Nw*.

$$Cost(Wind)=N_w C_{wm} + \left(0.1\left[\frac{h}{12}-1\right]+1\right) N_w \left(2.448 r^{[illegible]} + C_{wf}\right)\left(\text{[illegible]}\right) \quad (5)$$

Equation (6) shows the costs incurred as a function of the design of the solar modules. The costs include capital, installation and maintenance costs. The function is a multiple of the optimal number of solar arrays *Ns*.

$$Cost(Solar)=N_s C_{sm}+N_s C_{sc}\left(\text{[illegible]}\right) \quad (6)$$

2.2.4 System Generated Power Output

The total power generated by the wind turbines is expressed in Equation (7). The turbine height and rotor diameter are both decision variables incorporated into the function.

$$P_o(Wind) = N_w(1+0.814\ln(h)-1.82) \\ *(0.6 \times rho \times \pi r^2 e_p V_h^3 N_g N_b) \quad (7)$$

Equation (8) gives the power output generated by the solar modules. The equation is based on a similar study by Yang et al. [11] in their look for an optimal design of a hybrid wind/solar system.

$$P_o(Solar) = N_s \left(\text{[illegible]} \right) \quad (8)$$

4 DISCUSSION AND RESULTS

An existing complementary relationship between solar and wind power has already been shown by Karim et al. [12] as well as many other studies. The model presented here was simulated using GAMS software, with the core objective of minimizing costs, while meeting the seasonal demand pattern across a year in Abu Dhabi. Our model was also used to find the optimal number of solar modules and wind turbines to be used in our design, as well as the rotor diameter and height of the wind turbines. We will also assess the feasibility of the designed plant to supply power, under seasonal weather conditions, to rural locations, such as a mini village or compound in the city of Abu Dhabi.

Our model was first tested without including weather or seasonal effects, meaning that the solar irradiation and wind speeds were constant over the year. The simulation results showed that our model would only install wind turbines under these conditions. This was expected, since wind turbines have a cheaper capital cost and would be cheaper to maintain, while they provide as much energy as the solar modules, meaning that the model would avoid the modules to minimize costs.

These results mean that the model went against our hybrid system design, and preferred installing the wind turbines only. However, this reinforces our idea that hybrid systems should be designed to take advantage of the sun and the wind's complementary characteristics, since weather conditions involving both sources are not constant over the years as was specified in our test. In some Middle Eastern countries, for example, solar radiation is higher in the summer months while the wind speed is low, and vice-versa in winter months. This means that solar modules may not be able to meet demand in winter on their own and will need wind turbines to generate that extra needed power, while in summer, there will not be enough wind for the turbines to generate sufficient power on their own.

We will now further test the ability of the system to cope with seasonal demand under varying weather conditions across a year. Actual weather data from the city of Abu Dhabi, is obtained from Islam et al.[13]. Solar radiation data is shown in Table 2.

Table 2:Solar radiation values for Abu Dhabi, given in W/m².

Jan	Feb	Mar	Apr	May	Jun
162.96	201.62	238.77	263.77	290.05	284.95
Jul	**Aug**	**Sep**	**Oct**	**Nov**	**Dec**
283.79	270	258.33	224.54	176.62	130.79

To have a clearer picture of Abu Dhabi's weather conditions, we plot the patterns of the two different sources. Figure 1 shows the solar radiation and wind speed as a percentage of the maximum attainable from each resource separately, over the duration of a year.

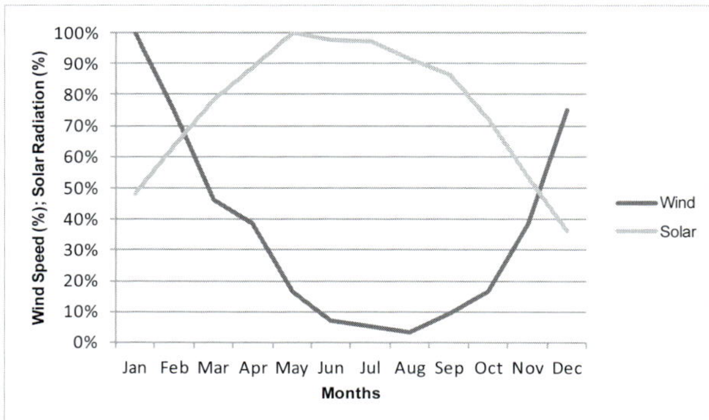

Figure 1: Resources as a percentage of their maximum over the year.

Figure 1 further reinforces the complementary relationship that exists between wind and solar radiations. The figure shows a negative correlation between the two resources.

After the weather data was input into our simulation, our model was used to calculate the power generated by the hybrid system's different components: the wind turbines and solar modules. Figure 2 shows the power output of the separate components. A line was drawn showing the demand load of a small and remote village or compound in Abu Dhabi, with a peak of 35 kW in summer. It should be noted that the figure shows the demand being met every month.

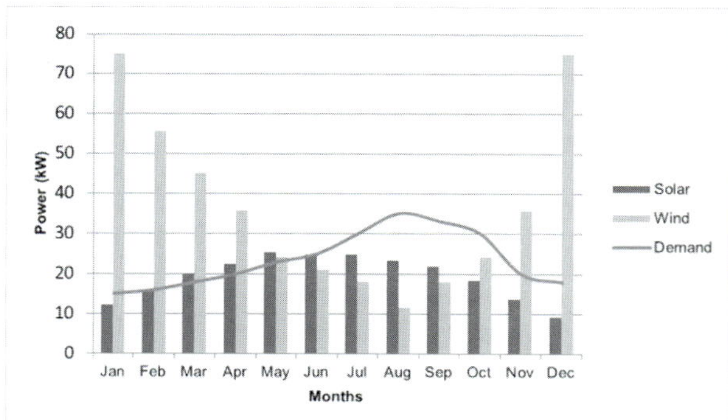

Figure 2: Power generated by the wind turbines and solar arrays over the duration of a year.

It could be seen from Figure 2 that the wind turbines will produce more power in Abu Dhabi during the winter times, due to higher wind velocities, while the solar modules generate very little power in winter. In summer however, the solar modules generate more power to make up for the reduction in power generated by the turbines.

The resulting design of our system is tested under two different varying seasonal weather conditions. In the first case, weather data from the city of Abu Dhabi is used. In the second case, we use data from Amman, Jordan. The varying data is useful, since it shows us how our model will react to opposing weather. In Amman, the electricity demanded increases in the winter time due to the use of heating equipment. On the other hand, Abu Dhabi experiences extremely higher power demand in the summer time due to the extreme usage of air conditioning systems. The resulting optimal design parameters for both cases are shown in Table 3.

Table 3: Optimal design parameters under different weather conditions.

City	Nw	Ns	h	r	Cost
Abu Dhabi	11	98	49.8m	14.9m	$140,314
Amman	7	29	49.7m	14.9m	$62,389

The results show that Abu Dhabi would need more wind turbines and solar arrays to meet its seasonal demand than Amman would. In other words, to achieve an optimal design to meet the varying seasonal demand, it would be cheaper to set up our proposed hybrid system in a city that needs heating devices in winter, rather than cities than need air conditioning in summer. This is logical, since cities which need heating do so in winter. Which means, that such cities witness peak demand in winter when there is plenty of wind at high velocities, sowind turbines, which cost cheaper than solar arrays to meet a load, can be installed to generate enough power at a lower price without the need of many solar arrays. In addition, not many solar arrays would be needed in summer, since electricity demand drops in summer when heating is not needed. Unlike a city such as Abu Dhabi, which witnesses peak demand in summer when the wind speeds are low, and more solar arrays are needed to meet the increasing demand, resulting in a higher cost for a hybrid system since solar arrays cost considerably higher than wind turbines to generate the same power.

It is visible in our results from Figure 2 that more power than is needed would be generated by our hybrid system if set up in Abu Dhabi, especially in the early and late months of the year. This shows that our design could be further improved with the addition of extra components such as batteries and inverters. Batteries could store the extra power generated, which would result in the need for less wind turbines and solar arrays to meet demand, which will also result in less costs. Adding these components will result in further improvement of the hybrid system's performance.

5 CONCLUSION

This paper presented a model that allowed for the optimal design of a hybrid solar-wind power plant. The aim of the model was to find the optimal design parameters which include the number of wind turbines, number of solar modules, turbine height and turbine rotor diameter. This aim was to be achieved by meeting varying seasonal demand of an application in a remote location, at minimal cost.

Taking advantage of the existing complementary characteristics between solar and wind energy, weather data of the city of Abu Dhabi was used to test how our model would perform, and to analyze the results of the presented optimal solution. The optimal design parameters of a hybrid plant were found, which would meet summer peak demand of 35 kW of a remote, off-grid application in the city.

Finally, to get a clearer understanding of the results, varying power demand of a different city (Amman) was used to compare our model's results. During summer, when there is plenty of solar radiation and little wind energy, the solar modules provide most of the needed energy. In winter, the wind turbines provide most of the energy due to the higher wind velocities and less solar radiation. Our model produced optimal design parameters for a hybrid-system that would cope with varying power demand across the period of a year, reinforcing the feasibility of such a project. The results also showed that setting up a hybrid plant to meet varying power loads would be cheaper in areas that witness peak demand in winter rather than in summer. This is due to the ability of wind turbines to supply the needed power at those times without the need of solar arrays which are more expensive. An area that has peak power demand during summer would depend largely on solar arrays than on wind turbines to generate that power, resulting in higher costs.

6 REFERENCES

[1] Dihrab, S.S. And Sopian, K.,2010, Electricity Generation of Hybrid PV/wind Systems in Iraq, Renewable Energy, 35:1303-1307.

[2] Kanase-Patil, A.B., Saini, R.P., and Sharma, M.P., 2010, Integrated Renewable Energy Systems For Off Grid Rural Electrification, Renewable Energy, 35: 1342-1349.

[3] Yang, H.X., Jurnett, B., and Lu, L., 2003, Weather data and Probability Analysis of Hybrid Photovoltaic-wind Power Generation Systems In Hong Kong, Renewable Energy, 28/11:1813-24.

[4] Elhadidy, M.A., and Shaahid, S.M., 2003, Opportunities for Utilization of Stand-alone Hybrid (wind + photovoltaic + diesel + battery) Power Systems in Hot Climates, Renewable Energy, 28:1741-1753.

[5] Yang, H., Zhou, W., Lu, L., and Fang, Z., 2008, Optimal Sizing Method For Stand-Alone hybrid Solar-Wind System With LPSP Technology By Using Genetic Algorithm", Solar Energy, 82:354.

[6] Onar, O.C., Uzunoglu, M., and Alam, M.S., 2008, Modeling, Control And Simulation Of An Autonomous Wind Turbine/Photovoltaic/Fuel Cell/Ultra-Capacitor

Hybrid Power System, Journal Of Power Sources, 185/2:1273-83.

[7] Ahmed, N.A., Miyatake, M., and Al-Othman, A.K., 2008, Power Fluctuations Suppression Of Stand-Alone Hybrid Generation Combining Solar Photovoltaic/Wind Turbine And Fuel Cell Systems, Energy Conversion and Management, Vol. 49:2711.

[8] Kershman, S.A., Rheinlander, J., Neumann, T., and Goebel, O., 2005, Hybrid Wind/PV and Conventional Power For Desalination In Libya – GECOL Facility For Medium And Small Scale Research At RasEjder, Desalination, 183/1-3:1-12.

[9] Bakos, G.C., and Tsagas, N.F., 2003, Technoeconomic Assessment Of A Hybrid Solar/Wind Installation For Electrical Energy Saving, Energy and Buildings,35/2:139-45.

[10] Tina, G., Gagliano, S., and Raiti, S., 2006, Hybrid solar/wind Power System Probabilistic Modelling for Long-term Performance Assessment, Solar Energy, 80:578-588.

[11] Yang, H., Zhou, W., and Lou, C.,2009, Optimal Design And Techno-economic Analysis Of A Hybrid Solar-Wind Power Generation System, Applied Energy, 86:163-169.

[12] Mousa, K., Alzu'bi, H., Diabat, A., 2010, Design of a Hybrid Solar-wind Power Plant Using Optimization, ICESMA.

[13] Islam, M.D., Kubo, M., and Alili, A.A., 2009, Measurement of Solar Energy Radiation in Abu Dhabi, UAE, Applied Energy, 86:511-515.

7 Sustainability Assessment

Sustainability is recognized as a driver for innovation. However, there is a critical need for improved sustainability evaluation methods to aid decision-making and selection of novel product and process designs for sustainable manufacturing. In the first section of this chapter Jawahir and Jayal present an overview of product and process sustainability evaluation methods and modeling techniques. In the second section Koho et al. introduce a toolset for designing and implementing competitive and sustainable manufacturing networks. In the third section Rabe et al. propose a conceptual framework for the modeling and simulation of value creation network which differentiates the effects of sustainability. In the fourth section Masui and Sonda show a method for quantifying the relational strength between two sustainability events by using their co-occurrence frequency in newspaper articles. In the fifth section Ammouri et al. initiate a carbon footprint calculator for construction projects for estimating the carbon footprint. In the sixth section Lu et al. present a framework for developing comprehensive product and process metrics for sustainable manufacturing. In the seventh section Dombrowski et al. evaluate the post-series supply strategies of electronic components towards achieving sustainability.

Product and Process Innovation for Modeling of Sustainable Machining Processes

I.S. Jawahir, A.D. Jayal
Institute for Sustainable Manufacturing, University of Kentucky, Lexington, USA

Abstract
Sustainability is recognized as the driver for innovation. However, there is a critical need for improved sustainability evaluation methods, as well as for improved predictive models and optimization methods, to aid decision-making and selection of novel product and process designs for sustainable manufacturing. In the case of machining, increased awareness of the need for sustainability in manufacturing operations has led to significant research in advancing with new and more sustainable processes such as dry, near-dry and cryogenic machining. This paper presents an overview of product and process sustainability evaluation methods and modeling techniques, including analytical, empirical and computational methods, as well as optimization procedures, developed for predicting the performance of sustainable machining processes and major sustainability elements in machined products. The paper also highlights the technological challenges involved, and the future work needed, in developing comprehensive predictive models and optimization techniques for sustainable machining.

Keywords:
Sustainable manufacturing, Machining, Modeling, Products, Processes

1 INTRODUCTION

The critical need to achieve general sustainable development and, in particular, sustainability in manufacturing operations is now well recognized. Sustainable manufacturing is defined as "*the creation of manufactured products that use processes that minimize negative environmental impacts, conserve energy and natural resources, are safe for employees, communities, and consumers and are economically sound*" [1]. Further, sustainable manufacturing includes the manufacturing of *sustainable products* as well as the sustainable manufacturing of *all* products [2]. Thus, the overall need for sustainability drives product and process innovation to yield benefits for all elements of the triple bottom line: environment, economy, and society. Due to the multiple and complex trade-offs involved in product and process design for sustainability, optimization, and hence modeling, becomes critical. Figure 1 presents a schematic view of the blueprint for model-based sustainable manufacturing, enabled by product and process innovation – i.e., model-based product and process design for sustainability, as shown in Figure 2, are essential to achieving overall model-based sustainable manufacturing. In the case of machining processes and machined products, which are the subject of this paper, increased awareness of the need for sustainability has led to significant research in advancing dry, near-dry and cryogenic machining as alternatives to conventional flood cooling. However, there is a lack of comprehensive models for dry, near-dry and cryogenic machining, as well as for sustainability elements of machined products.

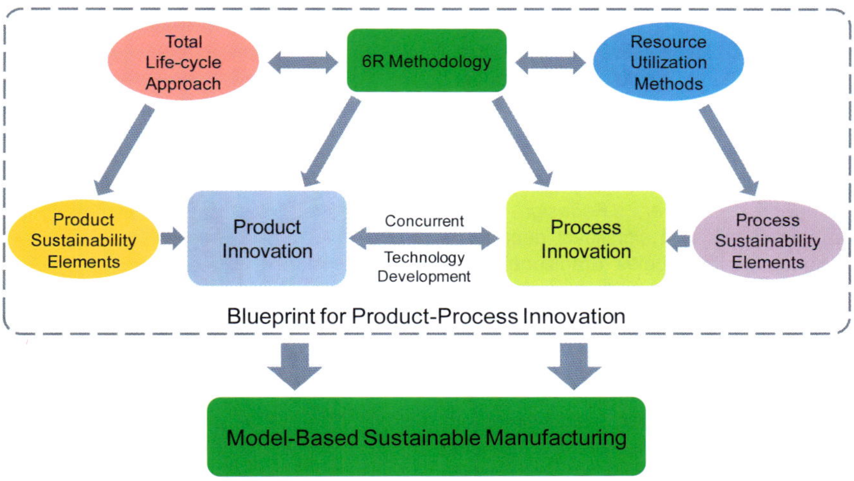

Figure 1: Blueprint for product and process innovation towards achieving model-based sustainable manufacturing.

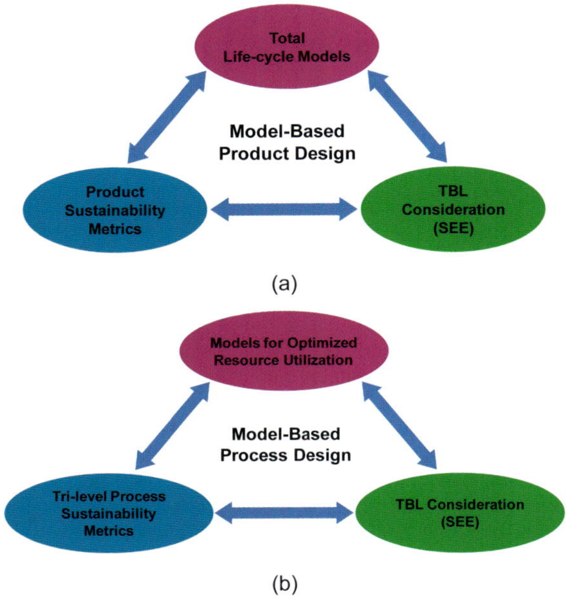

Figure 2: Integral elements of (a) model-based product design for sustainability, and (b) model-based process design for sustainability.

This paper presents an overview of product and process sustainability evaluation methods and modeling techniques, including analytical, empirical and computational methods, as well as optimization procedures, developed for predicting the performance of sustainable machining processes and major sustainability elements in machined products. The paper also underlines the challenges involved, and the future work needed, in developing comprehensive predictive models and optimization techniques for sustainable machining.

2 PRODUCT SUSTAINABILITY EVALUATION

Evaluation of product and process innovations aimed at improving sustainability must consider the effect on the total product life cycle due to the multiple energy and material flows involved in a product's life, and the effects of manufacturing processes on product performance and life. Graedel [3] has presented an extensive analysis of streamlined life-cycle analysis (SLCA) methods, including matrix approaches using target plots for five major product life-cycle stages: pre-manufacture; manufacture; product delivery; use; and recycling. In more recent work, a simplified total life-cycle of a manufactured product was assumed to consist of four key stages: pre-manufacturing, manufacturing, use and post-use [4]. Further, the "6R" approach [5] - *Reduce*, *Reuse*, *Recycle*, *Recover*, *Redesign* and *Remanufacture* – was introduced as a significant departure from current manufacturing methodologies (traditional and lean), and from the conventional 3R-based green concept (*Reduce*, *Reuse* and *Recycle*), as it enables the innovation-based transformation from an open-loop, single life-cycle paradigm to a closed-loop, multiple life-cycle paradigm. Developing upon the above-mentioned concepts, de Silva et al. [6] presented a simplified sustainability assessment of design alternatives for a commercial laser printer. The six major sustainability elements identified – environmental impact, functionality, manufacturability, recyclability and remanufacturability, resource utilization/economy, and societal impact (Figure 3) – were classified into 24 sub-elements, which were further divided into 46 influencing factors for the product (Figure 4). The overall product sustainability score was derived based on inputs from design engineers and a survey of manufacturers and consumers, which helped to not only quantify the individual impacts of different influencing factors, but also to determine their relative importance or weighting factors. This work was extended to develop a comprehensive rating system, or generic Product Sustainability Index (PSI), which is versatile enough to be applied to a wide range of products [4]. A (3x4) dimensional matrix, with three rows representing the components of sustainability, and four columns representing the four product life-cycle stages, is developed. A set of influencing factors are then identified and weighted based on their relative importance and company priorities and simple mathematical formulas are used to integrate the PSI across the rows and columns and the overall PSI is evaluated.

3 PROCESS SUSTAINABILITY EVALUATION

The critical challenge in sustainability assessment of machining processes is that only three of the six major sustainability elements of manufacturing processes [9] (Figure 5) can be modeled using analytical and numerical techniques because of their relatively deterministic nature. Modeling of the other three elements requires non-deterministic means, such as fuzzy logic. By extending the previously developed sustainability assessment methodology for machining processes [10], a hybrid model has recently been developed for comprehensive sustainability evaluation of machining processes [9]. Also, this model is extended to include an optimization module to provide the optimum sustainability level of the given machining process in the form of a sustainability index. Thus, as a first approximation, the overall sustainability score for a machining process is constructed as a combined function of the deterministic and non-deterministic sustainability elements:

$$S = (S_{SHE}, S_m) \quad (1)$$

where S_{SHE} is the sustainability component for safety, health and environmental issues assessed using fuzzy logic and S_m represents the deterministic component dealing with process-related elements such as machining cost, power consumption and waste management, which can be modeled analytically.

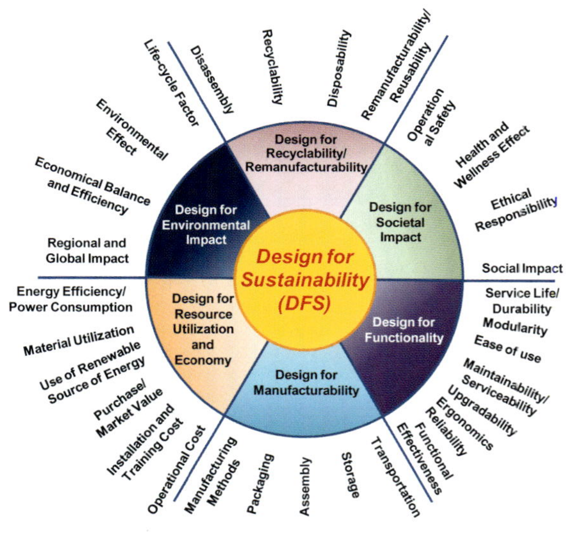

Figure 3: Product sustainability wheel [7].

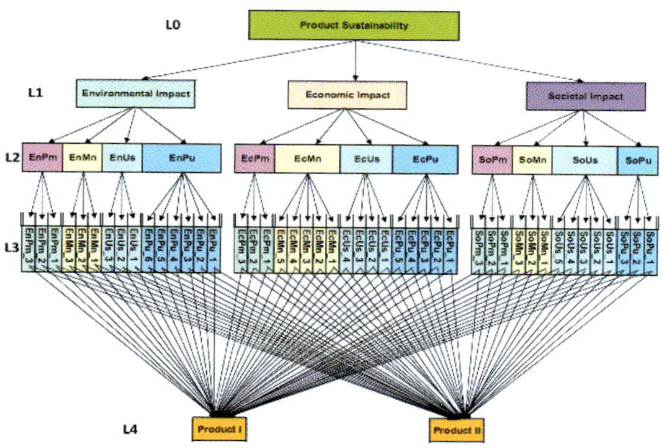

Figure 4: Product sustainability hierarchy [8].

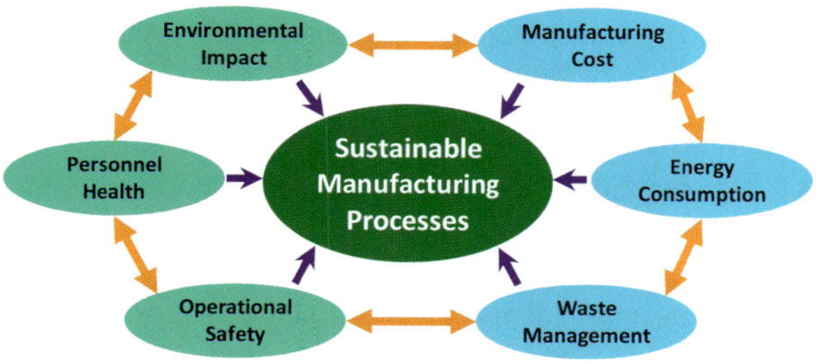

Figure 5: Six elements of sustainable manufacturing processes [9].

For assessing S_{SHE} several primary, secondary and tertiary level sub-elements are identified and classified according to linguistic variables, and then related through fuzzy rules. Finally, the area center method is employed to defuzzify the fuzzy set into a numerical value:

$$Z^* = \frac{\sum_{j=1}^{q} z_j \mu_c(z_j)}{\sum_{j=1}^{q} \mu_c(z_j)} \tag{2}$$

where z_j is the value of the jth element of the fuzzy set for the output variable and $\mu_c(z_j)$ is its membership grade. The deterministic component of the sustainability rating of the machining process, S_m, is evaluated by considering the machining operation cost, the power consumption and the waste generation. All these elements can be analytically modeled as a function of the cutting conditions. In this study, the generation of waste has been related to chip breakability since achieving good chips during the machining process is directly related to the reduction of waste generation as it increases their reusability and recyclability. *MC*, *PC* and *CB* denote the machining cost, power consumption and chip breakability, respectively. Corresponding constraints on these are assumed as MC_{min}, MC_{max}, PC_{min}, PC_{max}, CB_{min} and CB_{max}. The partial sustainability rating for this component of the machining model is constructed as:

$$S_m = C_{MC}\left(\frac{MC_{\max} - MC}{MC_{\max} - MC_{\min}}\right) + C_{PC}\left(\frac{PC_{\max} - PC}{PC_{\max} - PC_{\min}}\right) + C_{CB}\left(\frac{CB - CB_{\min}}{CB_{\max} - CB_{\min}}\right) \tag{3}$$

where each term is normalized by the user-provided information concerning machining operational sustainability requirements. The parameter S_m is expressed as a value between 0 and 1 and C_i (i = *MC*, *PC*, *CB*) are the weighting factors considered as the contribution coefficients of ith machining sustainability parameter to the value of the operation. Using the above procedure for calculating S_{SHE} and S_m, and by initially assuming that there is no significant interaction between the two, separate optimization techniques are applied to determine a global optimum for achieving maximum process sustainability.

4 DRY MACHINING

Research efforts in sustainable machining have mainly focused on: (a) dry machining, or machining without any coolants/lubricants [11]; (b) near-dry or minimum quantity of lubrication (MQL) machining, in which minute quantities of lubricants/coolants are employed in spray form [12]; and (c) cryogenic machining, which utilizes ultra-low temperature coolants (usually N_2 or CO_2) that are non-hazardous and easily evaporate leaving no residue [13]. This section presents brief summaries of recent modeling and optimization efforts, and the challenges involved, in dry machining.

4.1 Analytical modeling of turning operations

The difficulty in extending fundamental 2D models to practical 3D operations is the major challenge for analytical and computational modeling of dry machining. Our initial work to overcome this limitation includes the development of a comprehensive model for prediction of elemental primary forces (main cutting and thrust forces) and secondary (machine tool-oriented) forces, with the total forces evaluated as the sum of the 'edge' forces at the cutting edge and the 'area of cut' forces at the rake face [14]. Further, the equivalent toolface (ET) method [15] was developed for predicting cutting conditions and chip curl in turning with grooved tools by determining the equivalent flat-faced tool geometry that generates the same machining forces. Ghosh et al. [16] also developed a model for 3D chip curl by approximating the chip as a twisted 3D elastic beam and proposed a new failure criterion for chip breaking based on the calculated octahedral shear stress. In the case of fundamental 2D modeling, major research efforts have been undertaken at the University of Kentucky to develop new slip-line models for machining of ductile materials with rounded cutting edge restricted contact grooved tools [17]. Oxley's machining theory [18] was integrated with the new slip-line model to give predictions of cutting forces, chip thickness, chip curl radius, strain, strain-rates and temperatures in the shear zone and at the tool-chip interface [17].

4.2 Numerical modeling of turning operations

Recent work at the University of Kentucky involves the development of improved finite element models to study residual stresses in 2D dry machining [19] by: (a) employing a modified Johnson–Cook model to describe the material behavior as a non-Newtonian fluid; (b) using a remeshing scheme to simulate material flow around the cutting tool edge without the use of a separation criterion; (c) properly accounting for the unloading path; and (d) considering the thermomechanical coupling effect on deformation. Subsequently, the FE model was expanded to also include coolant effects for simulating near-dry and flood-cooled machining [20] (Figure 6).

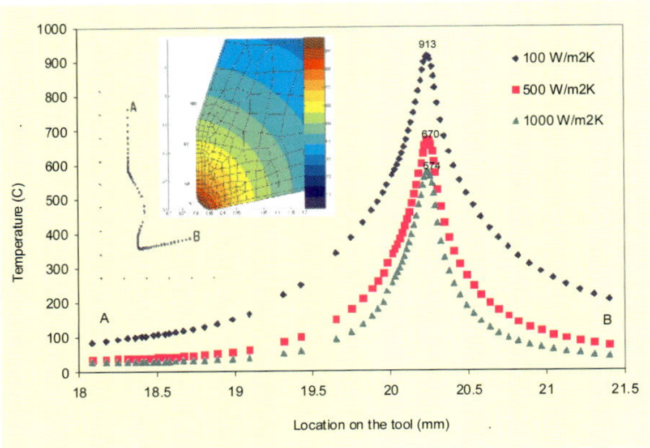

Figure 6: FE modeling of the temperature profile on the tool surface under different cooling conditions [20].

More recently, a hybrid empirical-FE model has been developed to investigate surface integrity phenomena – formation of white and dark layers in the surface and subsurface of hard machined parts – by including empirical equations for phase transformation, in combination with a hardness-based flow rule, in a FE model for orthogonal machining of hardened AISI 52100 steel [21].

4.3 Hybrid predictive modeling and optimization

Our recent research highlights the use of advanced computer-aided, fuzzy logic and genetic algorithms (GA) applications for single and multi-pass machining optimization in turning, milling and drilling using hybrid experimental and predictive models. For example, Wang and Jawahir [22] have employed GA methods for optimization of cutting conditions and tool selection in multi-pass turning with tool-wear by using predictive models for some machining performance measures, such as cutting forces, experimental databases for certain other measures, such as surface roughness, and a fuzzy logic-based assessment method for chip-form/chip breakability [23]. Predictable chip breaking is an essential requirement for automated machining and early work on the analysis of 3D cyclic chip formation and chip breaking cycles provides the foundation for subsequent analysis of chip breaking in machining with grooved tools [24]. Using the chip charts produced for different combinations of tools, work materials and cutting conditions, a new method for quantifying chip breakability was derived based on the fundamentals of fuzzy reasoning [23].

4.4 Tool-wear/tool-life prediction

Tool-wear/tool-life is one of the most important machining performance measures for process planning. The conventional methods of tool-wear measurement for flat-faced tools are inadequate for characterizing the multiple, concurrent and complex wear mechanisms in grooved tools. Jawahir et al. [25] presented a new methodology for the complex tool-wear in grooved tools, and developed a new method [26] to predict tool-life in turning with coated grooved tools, which includes the chip-groove effects (W_g) and the coating effects (W_C) as follows:

$$T = T_R W_g \left(\frac{V_R}{V} \right)^{W_C \frac{1}{n}} \tag{4}$$

where, T = tool-life, T_R = reference tool-life (1 minute), W_g = chip-groove effect factor, V_R = reference cutting speed (for 1 minute tool-life), V = cutting speed, W_C = coating effect factor and n = Taylor's tool-life exponent.

5 NEAR-DRY MACHINING (NDM)

5.1 Significance

NDM (or minimum quantity lubrication – MQL) provides a sustainable alternative to dry machining as it helps to reduce the use of cutting fluids significantly while the tool-life and performance requirements are maintained uncompromised. Our early work on spray cooling in machining of stainless steels showed significant improvement in surface roughness, cutting forces, chip forms and tool-life [27]. Attempts to quantify the performance improvements in NDM show a great potential for reducing the tool-chip friction coefficient and improving machining performance [20, 28-29].

5.2 Case studies in modeling of NDM

Near-dry turning of AISI 4140 steel

Experimental observations of NDM in turning lead to several conclusions [20]: (i) Flood-cooling produces approximately 10-18% higher cutting forces than oil-based NDM in finish-turning; (ii) Cutting forces and surface roughness are lower over time with NDM, while an increasing trend is observed in flood-cooling; (iii) Chip-form/chip breakability improves with proper application of NDM methods; (iv) In fine finish-turning with oil-based NDM the surface roughness is significantly lower than with flood cooling and water-based NDM; (v) Metallographic analysis shows reduced tool-chip interface friction in NDM compared to flood cooling; (vi) NDM is more effective in light machining operations as compared to heavy machining. The original tool-life equation by Jawahir et al. [25] was modified to include the fluid type, flow rate and nozzle(s) position with respect to the cutting tool and workpiece [28]:

$$T = T_R \frac{km}{f^{n_1} d^{n_2}} \left(\frac{V_R}{V} \right)^{\frac{1}{n} W_C \frac{1}{N_{NDM}}} \tag{5}$$

where N_{NDM} is the NDM effect factor, given by:

$$N_{NDM} = n_{mist} / n_c \tag{6}$$

where n_c is the coating effect factor and n_{mist} is the modified coating factor for NDM mist spray, defined as:

$$n_{mist} = \frac{\log V_1 - \log V_2}{\log(G_{F,W,N,M_{ZX},M_{ZY}}) - \log T_1} \tag{7}$$

where $G_{F,W,N,M_{ZX},M_{ZY}}$ is the new modified tool-wear using the NDM application, empirically derived while varying fluid type, flow rate and nozzle(s) position. The new tool-life equation provides significantly greater accuracy than the original equation proposed for dry machining.

Near-dry face milling of automotive alloy A380

The face milling of automotive aluminum alloy A380 was studied under four different lubrication/cooling conditions: dry, flood cooling, MQL (Oil), and MQL (Water). Empirical models of surface roughness and cutting forces were developed in terms of cutting speed, feed and depth of cut. A previously developed comprehensive optimization procedure for multi-pass turning [22] was extended to two-pass face milling to optimize the performance under different lubrication/cooling conditions based on a criterion integrating the effects of all major machining performance measures: surface roughness, cutting forces, tool-life and material removal rate [29]. Three combinations of lubrication/cooling conditions were studied for the two-pass face milling operation: Dry-Flood, Dry-MQL (Oil) and Dry-MQL (Water). Comparison between the finish passes of flood cooling, MQL (Oil) and MQL (Water) shows that the optimum point for finish pass with MQL (Oil) condition gives comparable results to that of the optimum point in finish pass with flood cooling. The surface roughness in case of flood-cooling is marginally lower than with MQL (Oil) but the cutting forces are nearly double of the MQL (Oil) condition.

6 CRYOGENIC MACHINING

6.1 Research activities in cryogenic machining

Research in cryogenic machining encompasses several areas, including: cryogenic properties of cutting tools [30]; cryogenically machined work surface properties [31]; and cryogenic machining performance in hardened materials and heat-resistant alloys [13]. Cryogenic machining has been shown to impart several improvements, including: (i) Environmentally-friendly and safe coolant: Nitrogen is an inert gas that constitutes 79% of atmospheric air; (ii) Increased productivity: Both hardness and toughness of cutting tools have been shown to increase under cryogenic cooling, allowing increased material removal rates; (iii) Improved part surface quality: Cryogenically machined parts were found to have improved surface integrity and fatigue resistance [31]; (iv) Process step reduction: Cryogenic cooling results in more economical machining in the hardened condition for many materials, allowing intermediate heat-treating steps to be eliminated [31].

6.2 Modeling and optimization of Inconel 718 machining

Inconel 718, a nickel-based aerospace alloy, is difficult to machine since it does not undergo significant thermal softening during cutting and its low thermal conductivity causes the heat generated to be concentrated near the cutting edge, leading to excessive tool-wear. Thus, the optimized selection of cooling/lubricating agents and cutting conditions is crucial for high-performance machining of Inconel 718. Pusavec [13] studied the effects of different cooling/lubrication methods (dry, near-dry, cryogenic, and combined cryogenic-MQL) on machining forces, surface roughness, tool-wear and chip breakability during turning of Inconel 718 under various conditions. The models developed through response surface methodology were employed to optimize the process using genetic algorithms. As shown in Figure 7, the combined application of cryogenic fluid from the flank and MQL application from the rake side yielded the best overall results.

7 CONTROL OF MACHINING FOR IMPROVED SURFACE INTEGRITY AND PRODUCT LIFE

An important aspect of product sustainability is the effect of manufacturing processes on the product's service life, which is often limited by failure through fatigue, corrosion, wear, etc., through their effect on the product's surface integrity. The product life is an important sustainability parameter since it determines the rate at which products end up in the landfill, or undergo remanufacture or recycling. Our recent work in this area includes experimental investigation and numerical modeling of the effects of cutting tool edge radius on microstructural changes and residual stresses in the surface and subsurface of machined parts [33] (Figure 8).

8 SUMMARY

This paper presents an overview of product and process innovation for sustainability and modeling techniques, including analytical, empirical and computational methods, as well as optimization procedures, developed for predicting the performance of sustainable machining processes and major sustainability elements in machined products, such as the surface integrity measures with critical influence on product performance and life. The paper also highlights the technological challenges involved, and the future work needed, in developing comprehensive predictive models and optimization techniques for sustainable machining.

9 ACKNOWLEDGMENTS

Contributions of previous PhD and masters students, as well as research collaborators and industrial partners, are acknowledged. The authors also thank the National Science Foundation, National Institute of Standards and Technology, US Department of Education, Kentucky Science and Engineering Foundation and companies such as Ford, General Motors, Toyota Motor Manufacturing, Semicon Associates, Air Products, Inc., Unist, Inc., Haas Automation, Inc., Sandvik, Seco, Stellram, and Kennametal for sponsoring various research projects in relevant areas.

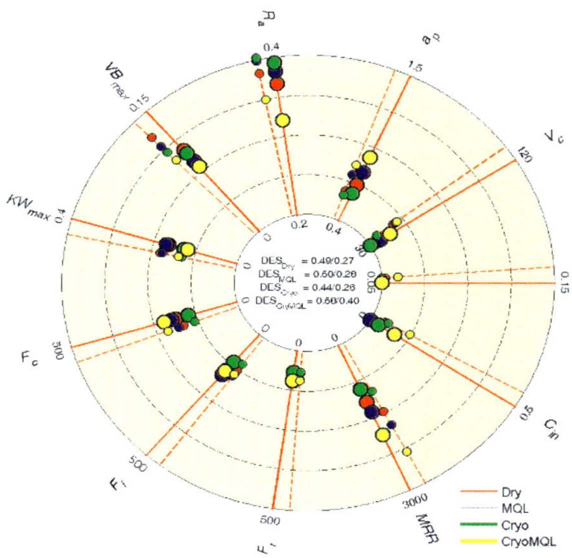

Figure 7: Optimum cutting conditions and corresponding desirability index (DES) with (solid lines) and without (dashed lines) inclusion of fuzzy-rule evaluation of chip breakability in the optimization procedure [13].

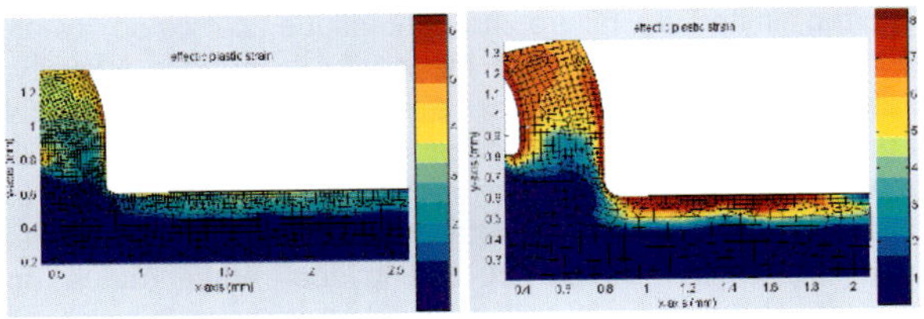

(a) (b)

Figure 8: FE simulation of plastic deformation produced in AISI 1045 steel by: (a) 50 μm and (b) 80 μm cutting edge radius tools (V = 175 m/min, f = 0.20 mm/rev) [33].

10 REFERENCES

[1] ITA, U.S. Dept. of Commerce, 2007, How Does Commerce Define Sustainable Manufacturing?, Available: http://www.trade.gov/competitiveness/sustainablemanufacturing/how_doc_defines_SM.asp.

[2] National Council for Advanced Manufacturing (NACFAM), 2009, Sustainable Manufacturing, Available: http://nacfam02.dev.web.sba.com/PolicyInitiatives/SustainableManufacturing/tabid/64/Default.aspx.

[3] Graedel, T.E., Streamlined life-cycle assessment, 1998, Prentice-Hall, New Jersey, USA.

[4] Jaafar, I.H., Venkatachalam, A., Joshi, K., Ungureanu, A.C., De Silva, N., Dillon, Jr., O.W., Rouch, K.E., Jawahir, I.S., 2007, Product design for sustainability: a new assessment methodology and case studies, Hdbk. Env. Conscious Mech. Des., (Ed.: M. Kutz), John Wiley & Sons, 25-65.

[5] Joshi, K., Venkatachalam, A., Jawahir, I.S., 2006, A new methodology for transforming 3R concept into 6R concept for improved product sustainability, IV Global Conf. Sust. Prod. Dev. & LCE, São Carlos, Brazil.

[6] Silva, N.D., Jawahir, I.S., Dillon, O.W., Russell, M., 2009, A New Comprehensive Methodology for the Evaluation of Product Sustainability at the Design and Development Stage of Consumer Electronic Products, Int. J. Sust. Manuf., 1/3, 251-264.

[7] Jawahir, I.S., Dillon, Jr., O.W., 2007, Sustainable manufacturing processes: new challenges for developing predictive models and optimization techniques, 1st Int. Conf. Sust. Manuf. (SM1), Montreal, Canada, 1-19.

[8] Gupta, A., Vangari, R., Jayal, A.D., Jawahir, I.S., 2010, Priority Evaluation of Product Metrics for Sustainable Manufacturing, Proc. 20th CIRP Des. Conf., Nantes, France.

[9] Granados, S., Jawahir, I.S., Fernandez, J., 2009, A comprehensive criterion for sustainability evaluation of machining processes, 7th Global Conf. Sust. Manuf., IIT Madras, Chennai, India, 385-391.

[10] Wanigarathne, P.C., Liew, J., Wang, X., Dillon, Jr., O.W., Jawahir, I.S., 2004, Assessment of process sustainability for product manufacture in machining operations, Global Conf. Sust. Prod. Dev. & LCE, Berlin, Germany, 305-312.

[11] Klocke, F., Eisenblatter, G., 1997, Dry Cutting, Annals of the CIRP, 46(2), 519-526.

[12] Weinert, K., Inasaki, I., Sutherland, J.W., Wakabayashi, T., 2004, Dry machining and minimum quantity lubrication, Annals of the CIRP, 53(2), 511-537.

[13] Pusavec, F., 2010, Modeling and characterization of sustainable cryogenic machining, PhD Dissert., University of Ljubljana, Ljubljana, Slovenia.

[14] Redetzky, M., Balaji, A.K., Jawahir, I.S., 1999, Predictive modeling of cutting forces and chip flow in machining with nose radius tools, 2nd CIRP Int. Wksp. Modeling of Machining Operations, Nantes, France, 160-180.

[15] Ghosh, R., Redetzky, M., Balaji, A.K., Jawahir, I.S., 1996, The equivalent toolface (ET) approach for modeling chip curl in machining with grooved tools, CSME Forum, 702-711.

[16] Ghosh, R., Dillon, O.W., Jawahir, I.S., 1998, An investigation of 3D curled chip in machining – part 1: A mechanics-based analytical model, Mach. Sci. & Tech., 2(1), 91-116.

[17] Wang, X., Jawahir, I.S., 2007, Recent advances in plasticity applications in metal machining: slip-line models for machining with rounded cutting edge restricted contact grooved tools, Int. J. Machining & Machinability of Materials, 2(3), 347-360.

[18] Oxley, P.L.B., 1989, Mechanics of machining: an analytical approach to assessing machinability, Ellis Horwood.

[19] Ee, K.C., Dillon, Jr., O.W., Jawahir, I.S., 2005, Finite element modeling of residual stresses in machining induced by cutting tool with a finite edge radius, Int. J. Mech. Sci., 47, 1611-1628.

[20] Wanigarathne, P.C., 2006, Experimental and analytical modeling of near-dry turning operations with coated grooved tools for improved sustainability, PhD Dissert., Univ. of Kentucky.

[21] Umbrello, D., Jayal, A. D., Caruso, S., Dillon, O. W., Jawahir, I. S., 2010, Modeling of white and dark layer formation in hard machining of AISI 52100 steel, Mach. Sci. & Tech., 14(1), pp. 128-147.

[22] Wang, X., Jawahir, I.S., 2005, Optimization of multi-pass turning operations using genetic algorithms for the selection of cutting conditions and cutting tools with tool-wear effect, Int. J. Prod. Res., 43(17), 3543-3559.

[23] Da, Z.J. and Jawahir, I.S., 1998, Optimal chip control in turning operations, Int. Sem. Improving Mach. Tool Perf., San Sebastian, Spain, Vol. II, pp. 607–618.

[24] Jawahir, I.S., 1990, On the controllability of chip breaking cycles and modes of chip breaking in metal machining, Annals of the CIRP, 39(1), 47-51.

[25] Jawahir, I.S., Li, P.X., Ghosh, R., Exner, E.L., 1995, A new parametric approach for the assessment of comprehensive tool-wear in coated grooved tools, Annals of the CIRP, 45(1), 49-54.

[26] Jawahir, I.S., Fang, X.D., Li, P.X., Ghosh, R., 1997, Method of assessing tool-life in grooved tools, US patent: 5,689,062.

[27] Um, J.-Y., Chow, L.C., Jawahir, I.S., 1995, An experimental investigation of the application of the spray cooling method in stainless steel machining, ASME-IMECE, MED, 2(1), 165-178.

[28] Marksberry, P., Jawahir, I.S., 2008, A comprehensive tool-life performance model in near-dry machining for sustainable manufacturing, Int. J. Mach. Tools & Manuf., 48(7-8), 878-886.

[29] Kardekar, A., 2005, Modeling and optimization of machining performance measures in face milling of automotive aluminum alloy A380 under different

lubrication/cooling conditions for sustainable manufacturing, MS Thesis, Univ. of Kentucky.

[30] Hong, Z., Hong, S.Y., 1992, Cryogenic properties of some cutting tool materials, J. Mat. Eng. & Perf., 1(5), 705-714.

[31] Zurecki, Z., Ghosh, R., Frey, J.H., 2003, Investigation of white layers formed in conventional and cryogenic hard turning of steels, Proc. ASME IMECE2003-42313.

[32] Outeiro, J.C., Kandibanda, R., Pina, J.C., Dillon Jr., O.W., Jawahir, I.S., 2010, Size-effects and surface integrity in machining and their influence on product sustainability, Int. J. Sust. Manuf., 2(1), 112-126.

Towards Manufacturing System Sustainability Assessment: An Initial Tool and Development Plans

M. Koho, H. Nylund, T. Arha, S. Torvinen
Department of Production Engineering, Tampere University of Technology, Tampere, Finland

Abstract
This paper focuses on sustainable manufacturing and is based on a research project that aims to assist in realizing competitive and sustainable manufacturing networks. In considering sustainability, environmental, social, economical, political and technological aspects are covered. A toolset for designing and implementing competitive and sustainable manufacturing networks that consists of a general level assessment and development tool and digital manufacturing solutions, e.g., simulation, is pursued. In this paper, an overview of the research project is provided, and then the focus turns to the general level tool for realizing sustainable manufacturing networks. Use and further development of TUTKA, a production system assessment tool that is based on best practices of lean and agile manufacturing, is proposed. An overview of the tool is presented, and its usefulness in assessing manufacturing system sustainability and the need for further development are considered by linking the tool to aspects of sustainability.

Keywords:
Assessment, Manufacturing, Manufacturing system, Sustainability

1 INTRODUCTION

This paper and the related research project are motivated by the increasing importance of sustainability and sustainable manufacturing. The work was initiated by the outlook that to succeed in the global competition and business environment, companies need to combine, in their decisions and actions, cost-effectiveness, customer orientation, and sustainable development. It has been pointed out that presently industrial systems are not sustainable in the long term because of their growing demands for non-renewable resources [1]. In the context of manufacturing, the need to consider sustainability and environmental effects during the design, implementation and operation of manufacturing systems and processes, rather than to focus on end-of-pipe solutions such as controlling and treatment of waste and emissions, has been emphasized [2]. Furthermore, to meet the goals of sustainable development, tools and measures for evaluating and measuring sustainability are required [1]. Thus, the authors' have initiated a research project that aims to assist manufacturing companies by developing tools and solutions for designing and realizing competitive sustainable manufacturing systems, and for assessing and improving the sustainability of existing manufacturing systems.

This paper reports the early phases of the research project on competitive sustainable manufacturing. Sustainability, sustainable development and competitive sustainable manufacturing are discussed and described (Section 2). Then, the main focus is on the toolset for designing, implementing and developing competitive sustainable manufacturing systems and networks. The toolset consisting of a general level tool and more detailed level solutions is presented in Section 3. Subsequently, the TUTKA-production system assessment tool is presented. Its ability to assist in assessing and improving manufacturing system sustainability is considered, and the need for further development is pointed out (Section 4).

With regard to research methods, information on sustainable manufacturing, design and operation of sustainable manufacturing systems and processes has been collected from literature, observations of state-of-the-art manufacturing systems and interviews. The results and ideas presented were generated by the analysis and synthesis of the collected information as well as outcomes from the authors' previous research projects.

2 COMPETITIVE SUSTAINABLE MANUFACTURING

The term Competitive Sustainable Manufacturing (CSM) has been presented and used, e.g. Jovane et al. [3] and the Ad-hoc Industrial Advisory Group [4]. The authors' perspective and description for CSM can be clarified by considering two aspects, competitiveness and sustainability. Competitiveness can be seen as the ability to profitably operate and compete in the global market. For manufacturing companies, the ability to produce and provide products that meet the needs of customers in a profitable way plays a key role in achieving and realizing competitiveness. Sustainability and sustainable development, on the other hand, are defined as actions and developments that meet the needs of the present without compromising the ability of future generations to meet their own needs [5]. Thus, competitive sustainable manufacturing requires that manufacturing systems, processes and outputs are competitive and sustainable, i.e. profitable and viable both for now and for the future. In principle, sustainability includes and presupposes competitiveness and economic viability, hence including the term competitiveness would not be necessary. However, combining competitiveness and sustainability aims to emphasise that efforts towards sustainable manufacturing, i.e. environmental friendliness, should not be regarded as additional costs, but as sources of competitiveness and competitive advantage.

2.1 Background and overview of CSM

The following figure presents the background and overview of CSM used in this research project. The approach adopted combines three production paradigms or approaches: lean manufacturing, agile manufacturing and sustainable manufacturing.

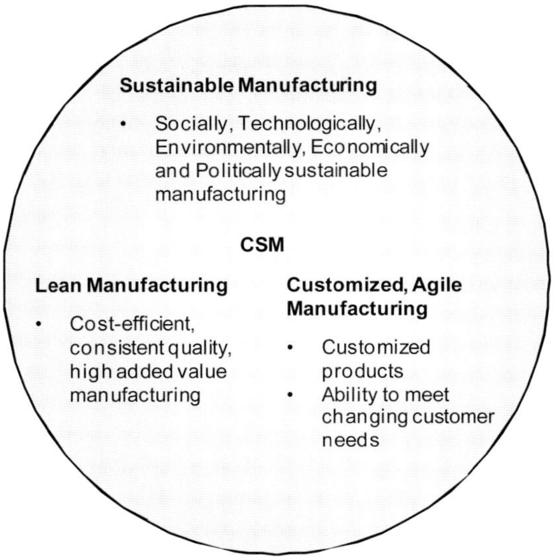

Figure 1: The background and overview areas of CSM.

The philosophy and practices of lean manufacturing aim at efficient manufacturing processes and at high, consistent quality by eliminating non-value adding operations and actions, standardising processes, and instituting continuous improvements [6]. By striving for cost-efficiency and reduction of waste such as unnecessary operations and use of materials, lean manufacturing contributes to both competitiveness and sustainability of manufacturing. Then, agility and agile manufacturing emphasise the ability to prosper in rapidly changing markets and to fulfil changing customer needs [7]. Means to achieve these objectives include customized products as well as flexible and reconfigurable manufacturing systems and processes (e.g. [8]). The ability to meet changing customer requirements, now and in the future, contributes towards competitiveness and sustainability. Furthermore, flexibility and adaptability lengthen the lifetime of manufacturing systems and hence support sustainability.

Finally, sustainable manufacturing, which the Lowell Centre for Sustainable Production [9] defines as "the creation of goods and services using processes and systems that are non-polluting, conserving of energy and natural resources, economically viable, safe and healthful for employees, communities, consumers and socially and creatively rewarding for all working people", focuses on the effects and contribution of companies and manufacturing on the environment, society and economy. In addition to sustainability, sustainable manufacturing and concepts such as the 3R approach (Reduce, Reuse, Recycle) can provide competitive advantage and improve competitiveness, for example by reducing energy and material usage and related costs.

The above discussion indicates that the three approaches, lean, agile and sustainable manufacturing, contribute towards both sustainability and competitiveness. Hence, they provide a good basis for the research project that focuses on realizing CSM.

2.2 STEEP framework

Commonly, sustainability is defined by three different aspects: the economy, society and the environment [5] [10]. Also the term "triple bottom line" or "the 3 P's", planet, people, and profit, has been presented and used to define sustainability [11]. Analogies within those are planet for the environment, people for the social aspects and profit for the economy. The aspect of economical sustainability focuses on securing economic viability in the short and long range. According to Jovane et al. [3] social sustainability can be achieved through people feeling that they can have a fair share of wealth, safety and influence. Then, according to Goodland [12], "environmental sustainability seeks to improve human welfare by protecting the sources of raw material used for human needs and ensuring that the sink for human wastes are not exceeded, in order to prevent harm to humans".

Jovane et al. [3] have introduced the ESET context by adding a technological aspect to the previously discussed ESE

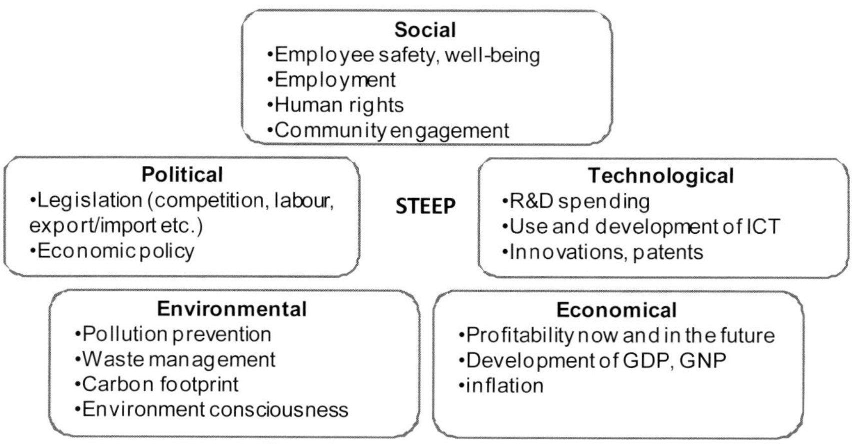

Figure 2 : The STEEP framework

context. Technology can offer huge potential in sustainable development [3] and it can be seen as an enabler and empowering aspect. To further elaborate the perspective on and definition of sustainability, the STEEP framework presents the political aspect that is related for example to national or international legislation. The STEEP framework is quite similar to the PESTEL framework [13] in which political and legal aspects have been separated.

Here, the STEEP framework is used to describe and to analyse sustainability. The framework and some examples of each aspect are presented in Figure 2.

3 OVERVIEW OF THE RESEARCH PROJECT ON MANUFACTURING SYSTEM SUSTAINABILITY

The research project focuses on and aims to assist companies in realizing competitive sustainable manufacturing. It aims to identify and gather state-of-the-art know-how and practices, and to develop a framework and toolset for designing and developing competitive and sustainable manufacturing systems and networks. The toolset can be divided into:

- A general level assessment tool that provides a quick overview of competitiveness and sustainability for manufacturing networks and highlights strengths and potential improvements.
- Digital manufacturing tools and solutions offering information rich models, and knowledge-based modelling and simulation solutions aiding in the design and development of sustainable manufacturing systems.

In practical assessment, development and design cases, the general level tool can be used for quick assessment and overview. On the other hand, the digital manufacturing tools enable more detailed evaluation and provide more detailed information. These two parts of the toolset can be used separately or together. The tools are briefly presented in the following subsections.

3.1 General level assessment tool

The general level assessment tool aims to enable a quick assessment and to provide an overview of competitiveness and sustainability of a manufacturing system or network. The tool is intended to assist in improving the sustainability of manufacturing networks by pointing out potential improvements and the means for their realization.

The tool will combine benchmarking, i.e. comparing the assessed manufacturing network with best practices of sustainable manufacturing and performance measurements focusing on aspects of sustainability. The benchmarking approach enables a quick overview of the network and sustainability, while the performance measurement data provides more detailed information on the performance and effects of the manufacturing network and processes.

The previously developed TUTKA production system assessment tool will be used as a starting point for developing the general level tool for assessing manufacturing systems or network sustainability. The applicability and usefulness of the TUTKA tool in assessing manufacturing system sustainability and directions for developing the general level tool for assessing manufacturing network sustainability are discussed in Section 4.

3.2 Digital Manufacturing Solutions

The research on digital manufacturing systems has no commonly agreed definitions but they share the idea of managing the typically isolated and separate product and production systems related activities as a whole, as an integrated system, by the evolving possibilities of Information and Communications Technology (ICT) [14]. The importance of ICT has also been recognized by the industry to improve the efficiency, adaptability and sustainability of production systems and their integration within agile business models and processes [3].

In this project, the possibilities of digital manufacturing systems are aimed to achieve more specific information related to the current situation as well as the potential and means for improvement found in the general level assessment. The structure of digital manufacturing systems can be divided into digital, virtual, and real parts, where the digital part represents the information and knowledge of a manufacturing system whilst the virtual part is a model of a real part existing physically [15]. The data gathered in the general level assessment are included in the digital part. It is used by humans, machines, as well as information systems and therefore is required to be formally presented. It is used in the decision making processes of existing real manufacturing systems as well as to create the virtual models of existing and possible future alternatives of manufacturing systems.

Based on the digital part, a more detailed assessment that utilizes modelling and simulation can be carried out and combined with the general level best practice based assessment. The simulation experiments can be carried out with both calculation and simulation models. The utilization possibilities can be divided into three levels, based on the needs of the participating companies i.e. calculation models built on e.g. spreadsheets for rough level analysis, rough level simulation focusing on the basic principles and structures of systems, e.g. layout planning, or detailed level simulation for controlling, and the planning and scheduling of production systems [16]. The simulation experiments make it possible to examine and evaluate the behaviour and characteristics of the system. The experiments can be focused on comparing the solution to best practices, to compare different solution alternatives as well as to validate the improved process itself.

3.3 Application Areas of the Toolset

The toolset can be applied in academic purposes and industrial case studies. From an academic perspective, several aspects of the toolset can be utilized to enhance the current and future education of university students. The education environment can also be used for the training of the personnel of the participating companies. The industrial case studies are focused on the specific development needs of the participating companies on the level of production networks, production systems or subsystems chosen from a production system. Several aspects of the toolset can also be tested in an academic research environment, which is a common development environment of several ongoing research projects and where any suitable research topics can be piloted.

4 TUTKA TOOL IN ASSESSING MANUFACTURING SYSTEM SUSTAINABILITY

The TUTKA production system assessment tool was developed in a research project that aimed to assist Finnish mechanical engineering companies in improving their production systems, production performance and competitiveness (see [17]). The tool and assessment focus on production system, i.e. one factory or factories of one company, and on make-to-order or assemble-to-order production approaches.

The TUTKA tool consists of key characteristics of a well-performing production system, methods for assessing a production system, and a scale for presenting the assessment results. The key characteristics, presented in Table 1 below, are mainly based on lean production philosophies and practices and they are seen to result in or at least contribute towards the achieving of good production performance in terms of lead time, quality and cost. The assessment of a production system is based on comparing the characteristics of the system with the key characteristics, and differences between these two are seen to indicate potential improvements. The pilot cases in which the tool was used and the feedback received demonstrate that the TUTKA tool is useful in assessing and improving production systems and production performance.

4.1 The TUTKA tool and the STEEP perspectives

To evaluate the usefulness and applicability of the TUTKA tool in assessing manufacturing system sustainability, the 33 key characteristics included in the tool were considered based on the STEEP framework and perspectives. The following table summarises the evaluation and findings by presenting the key characteristics and their linkage to Social (S), Technological (T), Economical (Ec), Environmental (En) and Political (P) sustainability. The checkmarks indicate the aspects of sustainability for which a key characteristic can have a positive effect and which can be considered in assessing the realisation of the key characteristic. For example, emphasising and ensuring occupational safety and the ability to work presented in the area of human resources improves employee safety and well-being and is hence connected to the aspect of social sustainability. Furthermore, the key characteristic focusing on the fit between production equipment and production objectives can contribute to and cover several aspects of sustainability, if, for example, environmental friendliness, energy consumption and safety are included in the objectives of production.

Due to the limited length of the paper, explaining all the key characteristics and their connections to the STEEP framework is not possible. Hence, only general explanation and logic of connecting the characteristics and aspects of the STEEP framework are presented.

With regard to the economical perspective, the main logic is that improving efficiency of production by reducing non-value adding activities, such as transportation and handling, and by supporting value adding work has a positive effect on the economical performance and sustainability. For example, the first three key characteristics of a production system structure aim at product-based layout and short distances within the production process, which compared to process-based layout reduces the need for transportation and hence shortens lead time and improves production efficiency. As the TUTKA tool is mainly based on lean manufacturing and seeks for cost-efficient production, the majority of the key characteristics are connected to and have a positive effect on the economical perspective.

The positive effects on and connection to environmental aspects are mainly related to characteristics of production equipment, improvements in production efficiency, and the ability to update products. With regard to equipment, considering, for example, energy consumption and emissions in selecting the equipment clearly contributes towards environmental sustainability. Then, improving production efficiency by reducing unnecessary activities, for example transportation within the manufacturing system, reduces the energy consumption. Furthermore, modular product structure and the use of platforms enable updating and modifying products to meet changing needs instead of producing new products. This for example reduces the need for and use of raw materials and energy.

Key characteristics that contribute towards and include aspects related to employee well-being, occupational safety and motivation are connected to societal sustainability. Occupational safety and ergonomic working positions are aimed at by several key characteristics in different areas, e.g. process and job descriptions, auxiliary tools and devices and personal safety equipment. As another example, the ability to participate in improving production systems and processes emphasised in the area of human resources can improve working environments and the motivation of employees.

Finally, the technological aspect is mainly related to solutions for achieving sustainability. Hence, key characteristics related to production equipment as well as information and communication systems are connected to that aspect.

4.2 Conclusions and need for further development

The evaluation of the TUTKA tool based on the STEEP framework shows that the tool includes key characteristics that are connected to and contribute towards aspects of sustainability. The tool and the characteristics mainly improve economical sustainability and reduce environmental effects by improving production efficiency and eliminating or reducing non-value adding work and activities. The tool can also indicate improvement needs related to societal aspects such as occupational safety. Thus, an assessment using the TUTKA tool can indicate potential ideas for improving sustainability for manufacturing systems and processes.

However, there clearly is a need and potential for further work and development. Table 1 shows that none of the key characteristics is connected or related to the political aspect. Furthermore, only a few checkmarks were placed within the S and T columns, indicating the need to elaborate on the aspect of social sustainability and to identify technologies that support and improve sustainability, for example environmental friendliness. Additionally, although key characteristics aiming to improve production efficiency can be connected to the environmental aspect, the need to present additional means for environmental issues such as pollution, prevention and waste management can be identified. Finally, as the TUTKA tool focuses on production systems in developing a general level assessment tool, the scope of assessment must be broadened to cover networks consisting of several production systems.

Table 1: Connections between the key characteristics of the TUTKA tool and the aspects of sustainability

Decision area	Key characteristic of well-performing production system	Sustainability aspect				
		S	T	Ec	En	P
Product architecture	Product architecture is modular			X	X	
	Product platforms are used			X	X	
	Product structure consists of predefined parts and subassemblies			X	X	
	Levels in product structure simplify the structure of the production system and support production control			X		
Production system structure	Production system consists of production units that are responsible for certain parts of a product			X	X	
	Production system structure corresponds to the structure and production process of products			X	X	
	Distances between production units and distances between production equipment are short			X	X	
	Production system structure makes it possible to observe the state of and the prerequisites for production			X		
	Layout and organisation of workplaces eliminate non-value-adding work and support value-adding work			X	X	
Production processes and management	Dissimilar main processes of production are identified and separated from each other			X	X	
	Processes of production units are defined	X		X	X	
	Processes and procedures between production units are defined	X		X	X	
	Plans and procedures for responding to production disruptions, problems and delays are in place			X	X	
	Cost-efficient and flexible processes are combined by late-point differentiation and appropriate positioning of order penetration point			X	X	
	Responsibility for production control is allocated to production units	X				
	Close cooperation between production units is supported and enabled			X	X	
Production equipment	Production equipment fits the requirements of the products and processes and the objectives of production	X	X	X	X	
	Changeover and set-up times of equipment are short			X		
	Production equipment enables integration and reduction of production phases			X	X	
	Production equipment is reliable and available for use			X		
	Production equipment is easy to use	X		X		
	Production equipment supports occupational safety and ability to work	X				
	Value-adding work, transportation and handling are supported and simplified by appropriate tools and auxiliary devices	X		X		
Information and communication	Information and communication support and enable decision-making		X	X		
	Information transfer and communication follow systematic and predefined principles and procedures			X		
	Information and communication systems used in the production system are compatible and integrated		X	X		
	Information and communication systems are reliable and available for use					
	Visual information and control systems are used in the production system			X		
Human resources	Teamwork and team organisation are used in production	X				
	Employees are cross-trained and skills of employees meet the requirements of work tasks and processes	X				
	Personnel policy and arrangement of working time support operational flexibility and reliable deliveries			X		
	Commitment to work and involvement in improving production system and production processes are promoted	X		X	X	
	Occupational safety and ability to work are emphasised and ensured	X				

5 SUMMARY

This paper presented an overview and background of a research project focusing on competitive sustainable manufacturing and then discussed the development of a tool for assessing manufacturing network sustainability. A previously developed TUTKA tool was presented and its use in assessing sustainability was evaluated. The evaluation indicated that the TUTKA tool can provide ideas and guidelines for improving sustainability of production systems and processes. Hence, it can be used as a starting point in developing a new assessment tool. The evaluation also pointed out guidelines and directions for further research and the development of an assessment tool that covers an entire production network and provides a sound overview of its sustainability.

6 ACKNOWLEDGMENTS

The authors wish to acknowledge the Finnish Funding Agency for Technology and Innovation, TUT Foundation and the participating companies for funding the research project on which this paper is based.

7 REFERENCES

[1] Krajnc, D., Glavic, P., 2003, Indicators of sustainable production, Clean Technology Environmental Policy, Vol. 5, pp. 279-88.

[2] O'Brien, C., 1999, Sustainable production – a new paradigm for a new millennium, International Journal of Production Economics, Vol.60-61, pp.1-7.

[3] Jovane, F., Westkämper, E. & D. Williams. 2008. The Manufuture Road: Towards Competitive and Sustainable High-Adding-Value Manufacturing, Springer.

[4] Ad-hoc Industrial Advisory Group, 2010, Factories of the Future PPP - Strategic Multi-annual Roadmap, 20th January 2010.

[5] WCED, 1987, Report of the World Commission on Environment and Development: Our Common Future, Oxford University Press, Oxford.

[6] Womack, J.P., Jones, D.T. & Roos, D., 1990, The machine that changed the world, New York: Rawson.

[7] Maskell, B., 2001, The age of agile manufacturing, Supply Chain Management, An International Journal, 6 (1): 5-11.

[8] Sharp, J.M., Irani, Z. & Desai, S., 1999, Working towards agile manufacturing in the UK industry, International Journal of Production Economics, 62 (1-2): 155-169.

[9] Veleva V., Ellenbecker M., 2001, Indicators of sustainable production: Framework and methodology, J. Cleaner Prod., Vol.9, No.6, pp.519-549,

[10] Martins, A.A., Mata, T.M., Costa, C. A. V., Sikdar, S.K., 2007, Framework for Sustainability Metrics, Ind. Eng. Chem. Res., Vol. 46, pp. 2962-2973.

[11] Elkington, J., Cannibals with Forks: The Triple Bottom Line of 21st Century Business, In: American Management Association. 2007. Creating a Sustainable Future: A Global Study of Current Trends and Possibilities 2007-2017.

[12] Goodland, R., 1995. The concept of environmental sustainability. Annual Review of Ecology and Systematic, Vol. 26, pp. 1–24.

[13] Walsh, P.R., 2005. Dealing with the uncertainties of environmental change by adding scenario planning to the strategy reformulation equation. Management Decision, 42 (1): 113-122.

[14] Nylund H., Andersson P., 2010, Simulation of service-oriented and distributed manufacturing systems. Robot Comput Integr Manuf, doi:10.1016/j.rcim.2010.07.009

[15] Nylund, H., Salminen, K., Andersson, P. 2008. Digital virtual holons – An approach to digital manufacturing systems. In: Mitsuishi, M., Ueda, K. & Kimura, F. (eds.) Manufacturing Systems and Technologies for the New Frontier, The 41st CIRP Conference on Manufacturing Systems, May 26.-28.2008, Tokyo, Japan pp. 103-106.

[16] Nylund, H., Koho, M., Torvinen, S., "Framework and toolset for developing and realizing competitive and sustainable production systems", The 20th International Conference on Flexible Automation and Intelligent Manufacturing, FAIM 2010, July 12.-14.2010, Oakland, CA, pp. 294-301.

[17] Koho, M., 2010, Production system assessment and improvement: a tool for make-to-order and assemble-to-order companies, TUT Publication 885.

Multi-perspective Modelling of Sustainability Aspects within the Industrial Environment and their Implication on the Simulation Technique

M. Rabe[1], R. Jochem[2], H. Weinaug[3]

[1]Chair IT in Production and Logistics, Technische Universität Dortmund, Germany

[2]Chair Quality Science, Technische Universität Berlin, Germany

[3]Department Enterprise Processes and Logistics, Fraunhofer IPK, Berlin, Germany

Abstract

Modelling is a broadly accepted analytic instrument and planning tool. Today, modelling is mainly applied for engineering and physics purposes and covers a short time horizon when compared to intergenerational justice. In parallel, sustainability is gaining increased importance for the industrial planning, because themes like global warming, child labour, and compliance with social and environmental standards have to be taken into account. Sustainability is characterized by comprehensively examining the three dimensions of economic, environmental, and social questions as well as their long-term perspective. Adequate indicators and an adaptation of modelling and simulation methods and procedures are necessary to evaluate the sustainability of industrial processes. A further challenge is the consideration of sustainability effects coming from the usual manufacturing process structure of a value creation network, in particular from cross-company cooperation and geographically distributed production. This article analyses the implications of those advanced tasks on the simulation technique and its application and proposes a conceptual framework for the modelling and simulation of such networks, which differentiates the effects of sustainability by the level of decision-making within enterprises.

Keywords:

Modelling; Simulation Technique; Sustainability

1. INTRODUCTION

'Making a development durable and viable for the future, means that the current generation meet their needs without threaten the ability of a future generation to satisfy their own needs' – in this way the Brundtland-Report from 1987 defines sustainable development [1]. Sustainability is seen as a process of change, by whom the use of resources, the structure of investments, the orientation of technological improvements and the institutional structures have to be adjusted and balanced between the current and future needs. Such needs are for example to ensure healthy food, availability of resources and a stable and safe social environment. In parallel to the long-term aim of an intergenerational justice the origin for a sustainable development was sharpened by the German Bundestag. In 1998, a committee of enquiry put sustainability on 'Three Columns' and pointed it out as a 'concept of enduring development viable for the future and related to the economic, environmental and social dimension of human existence' [2]. Sustainability can unfold its full potential just if it finds a way into a broad range of economy and society.

The modelling and simulation of technical and economic issues is a common analytic instrument and planning method. Presently, simulation models mostly focus on clearly delimited engineering and physics questions. Moreover, these questions are analyzed with a short-term time horizon compared with a period covered by intergenerational justice. If sustainability aspects of the issues should be analysed, then additionally the implications on the economic, environmental and social dimensions have to be taken into account. At the same time also a relation to the long-term development has to be facilitated. Appropriate indicators are requested to perform such combined analyses that bridge the local analysis content with all sustainability dimensions as well as the long-term orientation. Simulation methods and procedures have also to be adapted to the advanced content and time perspective. But, it is not that easy to gain a sustainability-integrating investigation of a part from a supply net compared with a pure material-flow-oriented analysis. This is especially a challenge because many manufacturing processes are organized across enterprise borders and are geographically distributed, often even globally. As an example, the choice of a 'sustainable' material for a product might induce an environmentally non-beneficial coating procedure in a later stage within the value creation chain or induce a higher scrap rate during the final manufacturing.

At the moment, discrete event simulation related to different planning horizons is applied for example for factory planning, for decision support at the manufacturing site and in operative control systems like production control centres and manufacturing execution systems. In these fields at all levels decisions are made that influence the sustainability. Whereas, decisions reached at a higher level (longer time perspective) limit the scope for decision making at lower levels. Taking into account these thoughts, the simulation of sustainability aspects requires a stronger integration of fields that are currently mostly analyzed independently, especially, along the different decision making levels within an enterprise as well as along a supply net crossing enterprise borders. Moreover, the product life cycle has an important influence on the sustainability. For a reasonable focus this last aspect is not taken into account for this paper.

This article examines in a first step the state of technique related to sustainability, to modelling of sustainability in the area of Quality Management and to simulation of value creation networks. Afterwards, a conceptual framework for the simulation of those networks is proposed, which differentiates the effects of sustainability by the level of decision making within enterprises.

2. STATE OF THE MODELLING AND SIMULATION OF SUSTAINABILITY IN THE INDUSTRIAL ENVIRONMENT

Correlated to the simulation of sustainability within the industrial environment, there is already content related to several research fields. In order to structure the related research, this chapter first examines which general sustainability concepts are available for single enterprises and a set of enterprises along value creation networks. Then, simulation approaches related to an individual production site and to complete value creation networks are reviewed.

With regard to the design and operation of sustainable industrial value creation networks, concepts and tools of environmental management and material flow management have to be taken into account and have to be integrated into a Sustainable Supply Chain Management concept. Regarding this topic accepted **norms and standards** are already defined internationally. The material flow management provides on the one hand solutions to reduce the environmental impact by the reduction of anthropogenic emissions and by substitution of materials. On the other hand, the environmental management solutions propose ways to improve the corporate environmental management (ISO 14001) and the operational environment performance (EMAS). The ISO 14001 is an environmental management system, which systematically anchors the environmental protection within the management in order to take environmental aspects into account in all daily tasks and in all strategic enterprise decisions [3]. The EMAS is based on the 'European Union Eco-Audit Regulation' by the European Commission from 1993. It regulates how all the effects of a production site on the environment have to be recorded, assessed, and documented [4]. In the field of social accountability, SA8000 is the world's first certifiable standard for the expansion of the company's reputation through socially responsible corporate governance. Based on the international convention of human rights and selected rules of the International Labour Organization (ILO), the SA8000 includes nine thematic areas. Among those issues SA8000 includes prohibition of child labour and forced labour, health and safety, freedom of association, prohibition of discrimination, limits on working hours, an adequate remuneration and management systems sufficient to guarantee compliance with these standard's conditions [5].

However, the degree of implementation of these standards is – compared with the ISO 9000 – neither comprehensive nor area-wide. Defined standards remain without significant effect, as long as they are not perceived and implemented. However, a study on environmental and social standards in the automotive industry showed in 2004 an overall positive attitude to questions of sustainability. 97% of the companies consider sustainability in general as important or very important, 92% classify sustainability for the automotive industry as important or very important and 79% consider environmental and social standards (ISO 14000, SA8000) as very helpful to assist and cultivate sustainability [6]. Often companies ostensible require compliance with standards such as ISO 9000, ISO 14001 and EMAS from their suppliers. Nevertheless, a holistic link and integration of environmentally-focused management concepts such as material flow management and environmental management with the supply chain management is not yet identified. First approaches to that can be seen in the EFQM Model [7]. In its Assessment Model it contains an integrated approach to describe and analyse indicators for different perspectives of an enterprise from customer satisfaction, employee satisfaction to environmental and social responsibility [8]. Other more operational approaches derived from lean management principles like "5S" (structurize, systematize, sanitise, standardize, self-discipline) provide checklists and data sheets to document and analyse sustainability indicators related to competitiveness, environmental and social responsibility in an integrated manner [27].

According to a study from the ETH Zurich the companies see the lack of appropriate software as the main obstacle to not have introduced a Sustainable Supply Chain Management. However, some companies already fail by the definition of appropriate metrics and their weightings, as well as the continuous measurement of the indicators [9].

With the title ManuFuture a **research roadmap** related to future production within the European Seventh Framework Program (2007 to 2013) has been defined. This roadmap is supported by Germany from the Fraunhofer Society, universities, associations and 200 companies. Beside the primary research tasks, the areas 'High Performance Manufacturing' and 'Digital Factory / Adaptive self-organizing cooperative processes / reconfigurability of machinery' appear as highlights. These are suggestions, mainly to accelerate production processes and design them more effective, by computer use. Anyway, it is not declared more precisely how to consider the operational, economic, environmental or social conditions and consequences. In the USA the theme 'Next Generation Manufacturing' focuses primarily on computer applications to increase the self-configuration of production facilities as well as a strengthening of life-cycle analysis of products and production facilities.

In the future, the existing techniques for simulation in production and logistics have to be brought together with the demands of sustainability, especially coming from the integration of economic, environmental and social perspectives. For this purpose a framework for the different simulation approaches is needed in a first step to be able to structure their relationships among each other and in the second step to model the required enrichment of sustainability needs. Here, both localized approaches (factory simulation) and simulation approaches for cross-company value creation networks have to be taken into account.

Simulative approaches for individual production sites, especially the simulation of specific aspects of factories up to their integrated consideration as 'digital factory', have been applied successfully since many years. Those simulation models are used in both conditions: the specific and unique use, as well as for the continuous employment in work processes. Further applications can be found in the fields of factory control and management. There, simulation is applied for the planning of portfolios as well as for operational planning (for example in production control). This field has been researched widely, and detailed descriptions of procedures, quality criteria and the validation of the resulting models are available.

The **simulation of value creation networks** – partly overlapping with the Supply Chain Simulation (SCS) – is an extra field in the simulation, which already exists for various applications. Some authors in this context use the term 'simulation' not in the technical sense, but simply in terms of analytical calculations and the comparison of different simulation scenarios [10, 11, 12, 13]. The discrete-event simulation is mostly used for cross-company networks [14, 15, 16] but also for internal supply chains [17, 18, 19]. A topic in the research is the distributed simulation by using the High Level Architecture (HLA) and its application for cross-company simulation [20, 21]. A particularly important issue currently is the structured and responsive derivation of simulation models from business process models and their integrative reuse, up to the implementation of solutions [22, 23, 24]. A simulative analysis of sustainability aspects, that means concretely an analysis beyond the pure technical and economic goals, but tackling a comprehensive study of value creation networks in terms of economic, social and environmental consequences, cannot be found currently at any of the listed authors.

Summarizing, there exists a wide variety of successful simulation applications. These have to be, however, more integrated with each other in terms of sustainability and to be extended for a consistent consideration of sustainability.

3. CONCEPTUAL FRAMEWORK FOR THE MODELLING AND SIMULATION OF SUSTAINABILITY ALONG VALUE CREATION NETWORKS

To define a framework that systematizes the existing simulation approaches and to examine this with respect to cause-effect relationships on sustainability, the authors see the following challenges:

1. Consideration of the long-term nature of sustainability in parallel to the relatively short-term industrial problems.
2. Evaluation of the impact of limited industrial problems on the economic, environmental, social aspects of sustainability.

In order to structure effects and potential solutions, the Cross-organizational Business Process Framework (CBP) can provide guidance. Originally, this framework was developed to support the modelling of cross-organizational business processes [22]. Similar to the interaction of several companies within the CBP framework, also to tackle the challenges of the simulation of sustainability aspects is forcing the cooperation between different actors in value networks and between different kinds of content.

Especially, the framework's differentiation of process modelling levels by using the requirements of different involved interest groups can be used as criteria also to structure the aspects of simulation of sustainability. The CBP framework differentiates between the modelling levels business, technology and execution.

Figure 1 highlights for each level the application aspects and the resulting implications for business process and simulation models.

	Application Aspects	Model Relevance
Business Level 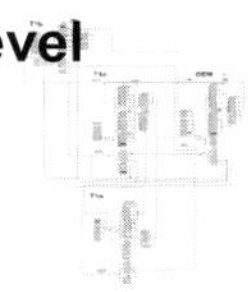	• Definition and management of several objectives (environmental, economic, social) • Different time horizons	Integration of different kinds of models; Transparency of workflow, information exchange. rules for modelling
Technical Level 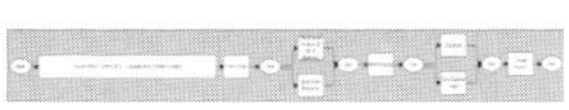	• Handling of distribute data and models • Feasibility check	Definition of technical processes by technical models (e.g. message transfer)
Execution Level 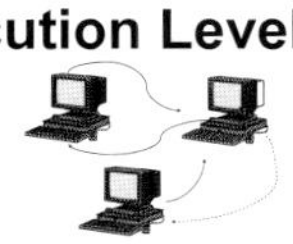	• Configuration and parameterization of processes • Process execution	Service definition, process flow, BPEL, ontology implementation

Figure 1: Conceptual framework: application aspects and model relevance at the simulation of sustainability.

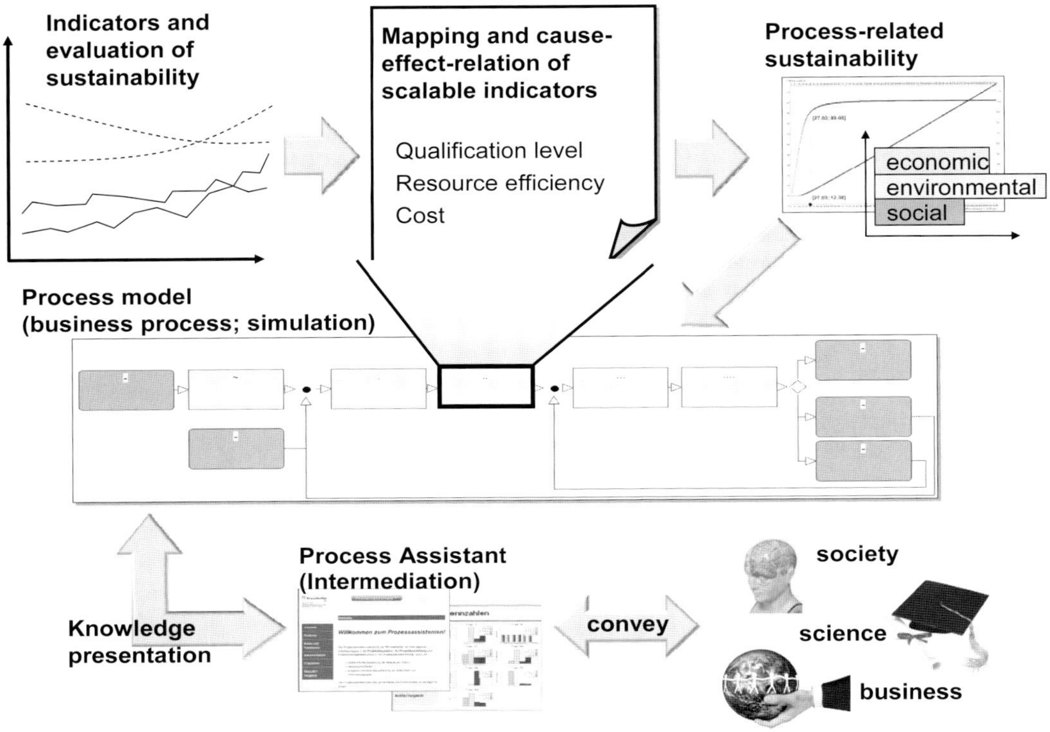

Figure 2: Example for the multi-perspective modelling: indicators, cause-effect-relations, processes.

The **business level** represents the co-operation and the interaction of overall aspects of the model content. This level is of particular importance because in this level the question has to be clarified how the sustainability aspects can be simulated in addition to the engineering and economic objectives. It is considered on the principle definition of sustainability as 'permanently sustainable development of economic, environmental and social dimension'. Economic goals for the simulation are, among others, the improvement of processing time, of stock level or the routing. Their direction and the measurements are usually clearly defined. However, the environmental objectives may overlap with the economic objectives, especially if they have monetary implications. Examples are the CO_2 reduction and the increase of the energy efficiency. As long as the legislature and the market provide indicators, a consideration in simulation models is easily possible. For environmental goals such as long-term availability of resources, recycling or efficient land use it is much more difficult to find the indicators to be used in the simulation. Moreover, the cause-effect relationships are not known directly between the simulated technical-economic issues and environmental objectives in terms of value, size and time. For example, what is the connection and correlation between optimal throughput time, maximum working time and energy efficiency? Even more serious, this occurs with the social objectives, such as social responsibility (for example child labour), to respect social standards, permanent education or the need for a stable and safe social environment. In addition, the attempt to follow different objectives of sustainability may lead to a conflict of aims, both within and between the dimensions economic, environmental and social. As an impression of the scope and complexity of the indicators for real products Finkbeiner et al. [25] gave the example of the Mercedes-Benz S-Class.

Summarizing, the presence of measurable and verifiable indicators is a prerequisite for the simulation of sustainability aspects. For technical and economic goals appropriate indicators are – as experiences show – available. For the environmental and social dimension, standards like ISO 14001, EMAS and SA8000 can provide a first set of indicators. However, these indicators represent individual aspects only, but not the sustainability of a certain decision as a whole. The relation between the indicators is not always clearly defined. A more promising approach with that respect is the EFQM Model. Here the relations between the indicators are defined and assessed in an integrated manner. But, the social and environmental indicators are often defined at a relatively high level of abstraction and related to long periods of time, compared to technical or economic metrics. A more concrete description of these indicators is provided in checklists of lean management approaches like "5S". However, also here the interdependencies of the different indicators are not clearly described.

Therefore, the key challenge will be to describe the objectives and indicators of the different dimensions in their relation to each other and to the business processes which are causing the values of the indicators. A further challenge is to downscale or to upscale the indicators to the level of individual enterprises and value-added networks (Figure 2).

Thus, the ways of modelling and describing sustainability indicators as well as their cause-effect relationships with processes and process models (business processes and simulation) have to be integrated formally to get a reliable simulation model. By using this integrated model, the company-internal and the cross-enterprise processes can be clearly described together with their alignment to the sustainability aspects. Beside this model-integrated description of sustainability aspects, a further challenging issue is the bi-directional intermediation of the modelled information with different actors. A web-based Process Assistant should provide various individually adaptable views of the model information to a large number of interested actors from society, business and science. Moreover, this community should be able to enhance the model with additional information by using the Process Assistant (figure 2).

The **technical level** of the conceptual framework is detailing the contents of the enterprise level down to the point of view of the control processes, including the exchanged messages. This happens regardless of specific platforms and software. At this level, the consideration of sustainability has particular influence on the volume and source of data and models to be included. The fact that in the same simulation model a variety of indicators from all dimensions of sustainability should be mapped will surely increase the amount of data. This will be increased by the need to look beyond an individual production site, taking into account the entire value creation network, at least for systems analysis. In addition, the effort to prepare the simulation model will increase, because now the cause-effect relationships and scaled indicators have to be described in detail. Therefore, it is helpful to reuse specific models or model components when they once have been implemented. Whenever possible, also external data sources should be used, such as comprehensive environment simulations, databases, social and environmental indicators, etc. This should be done in addition to the modelling and simulation of sustainability indicators in models of value creation networks themselves. If these data sources contain at a higher level of abstraction also process-related and scalable descriptions of the relevant indicators, then this could ensure a linkage of the data sources with technics and economics-oriented simulation models. This could enable a direct feedback of individual sustainability indicators and simplify the use of distributed data. Such implementation could use concepts and techniques of distributed simulation. The distributed simulation ensures that the data exchange occurs smoothly between the sub-components of the simulation scenarios on the technical level (message transfer).

At the **execution level**, the processes and messages are transformed and executed in the modelling language of specific IT systems, such as a business process engine or a simulation environment. The consideration of sustainability by simulation models has little influence to this level of the conceptual framework. At the technical level, the distributed simulation provides already mechanisms and tools, e.g. for the automatic configuration and parameterisation of simulation models and other data sources. Furthermore, it coordinates the execution of a common simulation environment.

The outlined three levels of the conceptual framework cannot be seen independently. Instead, each layer is directed to different levels of decision-making and actions within an enterprise. But, it is only achievable to get an effective planning tool to implement the measures if the stakeholders on all three levels act together. In this sense, the conceptual framework also points out that actors need to cooperate in modelling at the management level, during the design and simulation of their common business, the technical implementation in the workflow up to the derivation of models for the execution. The results of one level must be converted into the other.

Currently, it is planned to set up a 'demonstrator for sustainable value networks' to support the simulation of sustainability aspects in the industrial environment in terms of a more interactive knowledge transfer between all involved actors and to develop the simulation techniques [26].

4. SUMMARY

The issue of sustainability got internationally specific and sharp contours, especially since the common decision of the G20 countries to limit the global warming by 2050 to a maximum of 2°C. Derived from this demand, more practical objectives and arrangements at national level will follow soon. It is very likely that the economy will be obliged to implement this. Sustainability will therefore be an essential topic in the value creation networks and enterprises. An analysis of the state of the art of sustainability and simulation gave initial starting points.

However, there is the need to develop advanced concepts and tools for the different levels of simulation, especially to completely embed long-term sustainability – in the sense of intergenerational equity – into simulation scenarios. The conceptual framework described in this paper structures in three levels the effects and consequences resulting from the analysis of sustainability issues in value creation networks to the simulation technique.

For example, at the enterprise level the cross-enterprise definition and managing of sometimes divergent goals across all sustainability dimensions (economic, environmental and social) is one of the biggest challenges. This means, that the ways of modelling and description of (sustainability) indicators and their cause-effect relationships with processes and process models have to be systematically and formally combined. At the technical level, the size and complexity of simulation models will be increased by the consideration of sustainability indicators. In addition, a lot of data are not locally available, but obtainable only through external sources or by specific separate models. An implementation and execution of models under such conditions can be simplified by the use of concepts and techniques of distributed simulation.

Standards must be developed, enhanced and introduced as base to fully and systematically consider sustainability aspects in the industry. As base to plan sustainable value creation networks, in the medium term a unification has to be achieved in the following areas:

1. Scalable system of indicators to ensure consistent evaluation of sustainability.
2. Integration of different spheres of models (indicators, cause-effect relationships, processes).
3. Standardized interface definitions between simulation models and other sources (databases, control systems).

5. REREFENCES

[1] World Commission on Environment and Development (1987) Report of the World Commission on Environment and Development: Our Common Future. Transmitted to the UN General Assembly as an Annex to document A/42/427 - Development and International Co-operation: Environment. Oxford, web page http://www.un-documents.net/wced-ocf.htm, access April 2, 2010.

[2] Committee of enquiry of Bundestag (1998) Abschlussbericht der Enquete-Kommission des 13. Deutschen Bundestages. Berlin, Printing 13/11200.

[3] International Standardisation Organisation (2006) ISO 14044: Environmental Management – Life Cycle Assessment – Principles and Framework.

[4] Seidel, E.; Clausen, J.; Seifert, E.K. (1997) Umweltkennzahlen. Munich: publisher Vahlen.

[5] Social Accountability International (SAI) (2009) Social Accountability 8000 (SA8000) Standard. web page: www.sa-intl.org, access July 6, 2009.

[6] Seuring, S. (2005) Zur Umsetzung von Umwelt- und Sozialstandards in der Automobilindustrie – Ergebnisse einer Umfrage. Supply Chain Management Center, web page: www.uni-oldenburg.de/scmc, access April 27, 2005.

[7] EFQM, 'The Fundamental Concepts of Excellence', Brussels, Belgium, 2008

[8] EFQM, 'Assessoren Bewertungsbuch', EFQM, Brüssel, 2008

[9] Nienhaus, J. (2003) Trends im Supply Chain Management. ETH Zürich.

[10] Aigbedo, H. (2004) Analysis of parts requirements variance for a JIT supply chain. In: International Journal of Production Research. Band 42 (2004) 2 pp. 417-430.

[11] Hämmerling, A. (2000) Verdeckte Gewinne. Edscha AG bildet Supply Chain Netzwerk - Deutsche Telekom AG und Wassermann AG sind Partner. In: CYbiz Band 6 (2000), pp. 96-100.

[12] Nissen, V. (2000) Simulationsmöglichkeiten in SAPs Supply-Chain-Management-Werkzeug Advanced Planner & Optimizer. In: IM Information Management & Consulting. Band 15 (2000) 4, pp. 79-85.

[13] Zäpfel, G.; Piekarz, B. (2001) Interaktive Lehrmethoden im Supply Chain Management durch Planspiele und Simulation. In: Zeitschrift für Betriebswirtschaft. (Ergänzungsheft) Band 71 (2001) 3, pp. 135-161.

[14] Sihn, W.; Bischoff J.; Graupner, T. D.; et al. (2003) Ein prototypisches Werkzeug zur Planung und simulationsgestützten Bewertung von Versorgungsketten für die Automobilindustrie. In: Simulation und Visualisierung 2003. Proceedings, San Diego: Society for Modeling and Simulation International SCS-European Publishing House, pp. 59-67.

[15] Hickmann, J. (2003) OTD-SIM: Bewertung und Gestaltung der Order-to-Delivery-Prozesse in der Automobilindustrie. In: VDI-Berichte, Band 1744 Düsseldorf: VDI-Verlag, pp. 163-178.

[16] Nuttle, H.L.W.; King, R.E.; Hunter, N.A.; Wilson, J.R.; Fang, S.C. (2000) Simulation modeling of the textile supply chain, part I: The textile plant models. In: The Journal of the Textile Institute, part 2: Textile Economics, Management and Marketing. Band 91 (2000) 1, pp. 35-50.

[17] Bause, F.; Buchholz, P. (2006) Modellierung von Supply Chains unter Berücksichtigung variierender Organisationsstrukturen. In: Wenzel, S. (eds.): Simulation in Produktion und Logistik. pp. 53-62.

[18] Gupta, M.; Ko, H. J.; Min, H. (2002) TOC based performance measures and five focusing steps in a job shop manufacturing environment. In: International Journal of Production Research. Band 40 (2002) 4, pp. 907-930.

[19] Baumgarten, H.; Sommer, D. T. (2003) Efficient planning methods for reconfigurable logistics systems in recycling and disassembly networks. In: Colloquium e-ecological Manufacturing. Proceedings, pp. 19-23.

[20] Mertins, K; Jäkel, F.-W. (2003) Generierung von Schnittstellen für die verteilte Simulation am Beispiel Supply Chain. In: Datenmodelle in der Produktion. Fortschritt-Berichte VDI, Reihe 2: Fertigungstechnik. Band 633. Düsseldorf: VDI-Verlag, pp. 107-123.

[21] Rabe, M.; Jäkel, F.-W. (2004) Verteilte Simulation und Monitoring von Supply Chains. In: Weisbecker, A.; Renner, T.; Noll, S.: Electronic Business. Innovationen, Anwendungen und Technologien. Stuttgart: Fraunhofer IRB Verlag, pp.120-126.

[22] Greiner, U.; Jäkel, F.-W. (2007) Process Interoperabillity – From the Idea to Execution. In: Schulze, Th.; Preim, B.; et al. (eds.): Simulation und Visualisierung 2007. Proceedings, San Diego: Society for Modeling and Simulation International SCS-European Publishing House, pp. 141-153.

[23] Simulation Interoperability Standards Organization (SISO), SISO COTS Simulation Package Interoperability Product Development Group (2007) Standard for Commercial-off-the-shelf Simulation Package Interoperability Reference Models (SISO-STD-006-2007-DRAFT-1.0).

[24] Taylor, S. J. E.; Turner, S. J.; Strassburger, S. (2008) Guidelines for commercial Off-the-shelf Simulation Package Interoperability. In: Mason, S. J.; Hill, R. R.; Mönch, L.; Rose, O.; Jefferson, T.; Fowler, J. W. (eds.): Proceedings of the 2008 Winter Simulation Conference, pp. 193-204.

[25] Finkbeiner, M.; Hoffmann, R.; Ruhland, K.; et al. (2006) Application of Life Cycle Assessment for the Environmental Certificate of the Mercedes-Benz S-Class. In: International Journal of Life Cycle Assessment. Band 11 (2006) 4, pp. 240- 246.

[26] Rabe, M.; Jäkel, F.-W.; Weinaug, H. (2006) Supply Chain Demonstrator based on Federated Models and HLA Application. In: Schulz, T.; Horton, G.; Preim B.; Schlechtweg, S. (Eds.) Simulation und Visualisierung 2006, Proceedings. San Diego: Society for Modeling and Simulation International SCS- European Publishing House, pp. 329-338.

[27] Ho, Samuel. (2009) Integrated lean TQM Model for global sustainability and competitiveness. In: TQM Journal, Emerald Publisher, Vol. 22, Issue 2, page 143-158.

Relevance Analysis of keywords related to Sustainability

K. Masui, T. Sonda
National Institute of Advanced Industrial Science and Technology, Tokyo, Japan

Abstract
It is essential for all engineers to consider environmental influences caused by their products. Therefore education concerning to environmental issues, resource constrain, human daily life and so on, that is sustainability becomes more important. For this purpose, it is the first step to clarify the relationships among various events related to sustainability. In this research, we propose a method for quantifying the relational strength between two sustainability events by using their co-occurrence frequency in newspaper articles. Based on the results, the current status of sustainability is presented using multidimensional scaling (MDS) to visualize the data.

Keywords:
Sustainability, Relevance Analysis, Co-occurrence Frequency, Multidimentional Scaling

1 INTRODUCTION

It is essential for all engineers to consider environmental influences caused by their products. Therefore education concerning to environmental issues, resource constrain, human daily life and so on, that is sustainability becomes more important.

Although nowadays a lot of researches evaluating and forecasting sustainability are implemented [1-4], these researches' outputs seems to be useful for social scientists and government rather than for design engineers and process engineers due to their too big scope. For instance, "Limits to Growth – The 30-year Update" [1] published in 2004 treats various issues such as world population, world industrial production and human welfare concerning sustainability by system dynamics approach, and then United Nations pointed out that the further industrial development would affect food security from the viewpoint of sustainability [2]. These literatures are very important, however, it is necessary for sustainability education of engineers to develop some tools giving advices directly related to their ordinary tasks.

For this purpose, it is the first step to clarify the relationships among various events related to sustainability. Although ontology [5] is a well-known method for integrating much knowledge and illustrating the inter-relationships, much knowledge of many experts in various fields is required for structuring the ontology of sustainability. On the other hand, there have been many researches to determine similarities between items by using their co-occurrence frequency in text data (e.g. [6]). Authors proposed a method to determine relational strength between various events related to sustainability by using this technique [7]. In previous research, targeted keywords are mapped in two dimensions according to their similarity. The results showed several clusters of keywords formed on two-dimensional plane. However, when we consider many faces of sustainability, we further need to explore other aspects which can categorize various keywords.

In this research, we have two purposes achieved by clarifying the relationship among the events related to sustainability. One is to support decision-making for appropriate system boundary concerning sustainability study through visualizing the relation or similarity of sustainability keywords. The other one is to find conflict between the events related to sustainability. Although all the events strongly related each other do not come up against conflicts, we are trying to point out strongly related keywords as candidates of conflicts among enormous amount of relationship beyond various fields.

In this paper, we consider newspaper articles as a database of events that have happened in society. And then, we will clarify the relationships among the various events between manufacturing activities and the environmental protection without depending on experts and their knowledge (Fig. 1). Firstly, the events (keywords) related to sustainability are identified. Next, we try to quantify the relational strength between each two events by using their co-occurrence frequency in newspaper articles. Based on the result, an overview of the current status of sustainability is presented using multidimensional scaling (MDS) to visualize the data. In this instance, output of MDS is three-dimensional solution considering many faces of sustainability keywords dimension. We have also developed a tool which visualizes the relationship between each combination of individual events. This tool provides an analytical appearance frequency of an event by itself, co-occurrence frequency with other events, and also the results of cluster analysis based on co-occurrence frequency data.

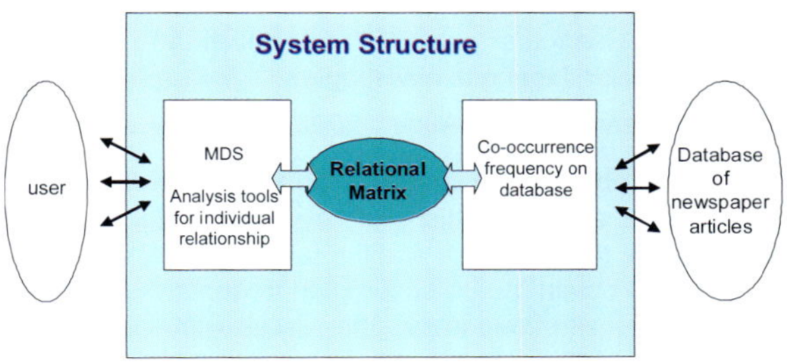

Figure 1: System structure of relevance analysis tool for sustainability

Table 1: Relational matrix showing the co-occurrence frequency between each pair of keywords

	crude oil	cereal grain	corn	soybean	gasoline	gold price	automobile	aluminum	chrome	server	sugarcane
crude oil	3003										
cereal grain	244	1044									
corn	229	454	1411								
soybean	229	413	557	1748							
gasoline	462	327	406	322	4122						
gold price	32	2	3	5	16	163					
automobile	200	31	59	19	380	5	7807				
aluminum	25	10	7	11	22	1	89	419			
chrome	7	0	0	0	0	0	21	5	148		
server	1	0	3	0	2	0	26	1	0	1228	
sugarcane	21	31	95	21	77	1	24	1	0	1	301
information	110	37	50	59	105	5	425	9	18	345	8
communication	17	8	11	4	40	1	138	9	0	77	1
silicon	1	1	1	1	2	1	26	15	0	2	1
stainless	12	2	0	2	4	0	42	10	29	1	0
diode	1	0	0	0	0	0	2	0	0	0	0
titanium	3	1	0	1	0	0	27	16	10	0	0
letter	6	1	5	6	12	0	20	2	1	1	3
videophone	0	0	0	0	0	0	3	0	0	5	0
naphtha	64	6	5	3	37	0	30	4	0	0	0
nickel	35	15	10	8	13	1	62	29	63	0	0
bioethanol	56	93	215	65	195	0	47	1	0	1	106
biofuel	83	134	263	132	189	1	69	4	0	1	91
pipeline	65	1	10	5	19	0	17	1	1	0	2
hybrid car	22	7	10	5	128	0	247	4	5	1	3
wind power generation	3	4	3	0	12	0	32	2	0	3	1
plastic	21	3	30	6	27	0	85	13	3	3	6
e-mail	7	1	9	10	39	1	45	2	2	132	3
travel	19	12	4	5	53	0	80	2	0	10	5

150 keywords (columns) / 150 keywords (rows)

2 METHOD FOR QUANTIFYING RELATIONAL STRENGTH

2.1 Identification of events related to sustainability

Firstly, the events (keywords) related to sustainability including the fields of the environment, resources, manufacturing and ordinary life are listed up in Japanese by brainstorming among the authors and collecting from journals to complement our knowledge. A total of 150 keywords were investigated in this preliminary study.

2.2 Acquisition of co-occurrence frequency

Secondly, the co-occurrence frequency of any two keywords in the Japanese newspaper database is obtained as an indicator of relational strength (similarity) between them. The search conditions are as follows:

Number of keywords: 150

Targeted information: four financial/industrial newspapers published by Nikkei and the four major Japanese newspapers

Search period: August 2007 to January 2008

We finally obtained a matrix (Table 1) expressing the relevance for all combinations of keywords. Note, however, that because the keywords and the targeted newspapers for searching to obtain the co-occurrence frequency among the keywords are all in Japanese, the results were obtained in Japanese and then translated for explanation. Next, we investigated the visualization of relevance because the relational matrix includes too many items and their combinations for an easy understanding.

3 VISUALIZATION OF RELATIONSHIP BY MDS

Multidimensional scaling (MDS) is a statistical technique often used in information visualization for exploring similarities of quantitative relationships in the data. MDS can produce a multidimensional geometric representation. For sufficiently small dimension, the resulting locations have been calculated in three-dimension and the following results are displayed in each two-dimension figures.

Figures 2 show the output of MDS applied to the above relational matrix to obtain an overview of similarity among all keywords. In these figures, the distance between keywords indicates their similarity. Figure 2(a) shows similarity of keywords on x-y plane, on the other hand figure 2(b) shows the similarity on x-z plane. In figure 2(a), it is easily found that similar keywords from the point of view of co-occurrence frequency make several clusters. For instance, the keywords related to "reduction of product weight" such as "lighter", "aluminum" and "carbon fiber" are located close to each other. The labels of axes are interpreted by analysts, who were the authors this time. The keywords related to environmental protection such as "biodiversity" and "carbon offset" are located on the left side of this figure, whereas the keywords such as "secondary cell" and "rare metal" on the right side indicate "development-orientation." It is also found that there are many keywords related to "government" on the upper side, and many keywords related to daily life on the lower side. In figure 2(b), x-axis expresses relationship of

"Protection-Oriented" versus "Development-Oriented" in common with x-axis of figure 2(a). On the other hand, the keywords related to "Production-Management" such as "supply chain" and "green purchase" are located on the upper side of this figure, whereas the keywords such as "automobile recycling law" and "acid rain" on the lower side indicate some emissions and their influences. When we look into the keywords related to "reduction of product weight" again, the cluster spreads in vertical direction and "carbon fiber" is located at the area of "Emission-Management". There are still few articles where "carbon fiber" and a mission-related keyword such as "recycling" and "waste" appear at the same time in the targeted newspapers for searching. Therefore location of carbon fiber seems to be influenced by the others strongly related to this keyword. As a result, combination of name of a keyword "carbon fiber" and name of a region "emission-management" indicates that emission of carbon fiber might become remarkable issue in the future.

As described above, the map shown in Fig. 2 is expected to yield objective criteria for assessing the system boundary of a more detailed simulation of sustainability, that is, the events located near a certain sustainability event on the map or the events located along the same axis should be considered in a more detailed investigation.

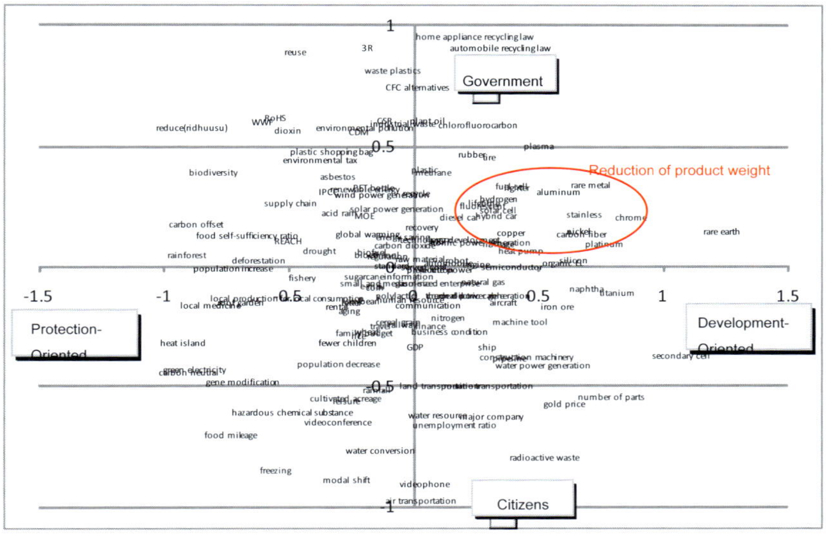

(a) Similarity of keywords on x-y plane

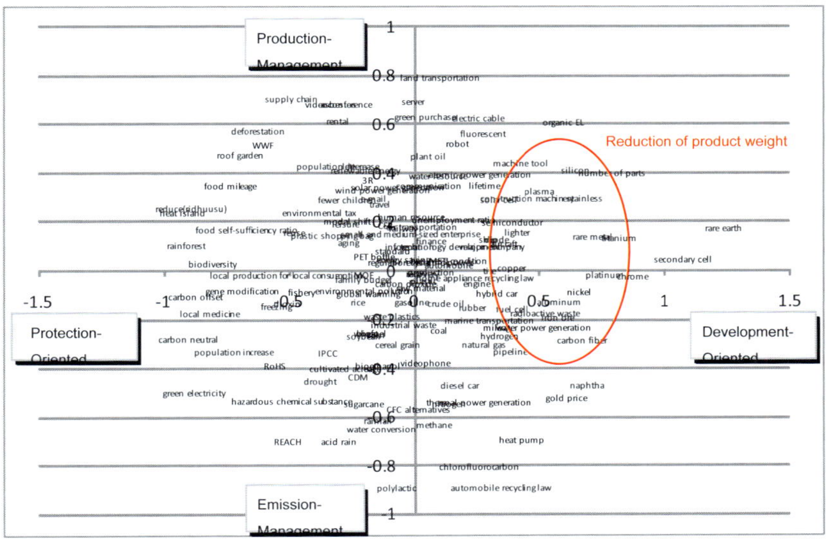

(b) Similarity of keywords on x-z plane

Figure 2: Overview showing similarity of keywords related to sustainability

4 VISUALIZING RELATIONSHIP BETWEEN INDIVIDUAL EVENTS

The output of MDS shown in Fig. 2 provides an overview of all keywords' similarity, but cannot provide information about the direct relationship between any two keywords. Hence, we developed a tool which can visualize the relationship between individual events and help engineers understand the concrete events related to their task.

4.1 Visualization method

We investigated the following two methods to visualize the relationship among individual events. Firstly, we investigated the method to indicate the relationship between each pair of keywords located on the circumference of the same circle by a link corresponding to their co-occurrence frequency. In the second method, a specified keyword which is focused on by a engineer is centrally located, and then the relationship between its keyword and another one is displayed in the same way. Here, the number of keywords and links displayed in the figures is adjusted by setting the threshold appearance frequency of a keyword by itself and the co-occurrence frequency between each pair of keywords in both methods described above. The size of a circle in the figures indicates the appearance frequency of a keyword by itself, displayed on a logarithmic scale. The color of a keyword indicates its category which the authors assigned in advance. All keywords are sorted into the seven categories of "material", "product", "technology", "system", "social", "nature", "environment" shown in Table 2 in this instance. Further, the keywords are also categorized by cluster analysis (furthest neighbor method) based on the information of co-occurrence frequency. This additional information will help engineers comprehensively observe the relationship among individual keywords.

4.2 Interpretation of Visualized relation

Figure 3 shows the relation between individual keywords visualized using the above method. In Figure 3(a) the links between items distributed on the circumference of the same circle indicate the existence of a co-occurrence relationship. On the other hand, in Figure 3(b), the event that users are focusing on is centered, and then a link is provided between its item and another which has a strong relation. In this example, "solar cell" is focused on here as a keyword. In both figures, only the items and links exceeding a certain threshold appearance frequency or co-occurrence frequency are displayed. The search period for counting frequency is from August 2007 to January 2008.

In Figure 3(a), the relationship between "crude oil" and "business condition (economy)" is shown to be strong by a link indicating high co-occurrence frequency. In Table 2, crude oil belongs to the category of "material" and business condition belongs to "social." That is, the result shows it is possible to find a linkage between two events belonging to different categories. Regarding the terms "soybean", "corn", "cereal grain", "gasoline", and "crude oil", they all belong to the same category of "material" in Table 2, as well as belong to the same cluster in the analysis based on co-occurrence frequency. Although the keywords of "engine", "automobile", "carbon dioxide", and "energy saving" belong to some different categories, the result of cluster analysis based on co-occurrence frequency shows that they form a cluster which could be interpreted as a "global warming problem." The keyword "automobile" is categorized as "product" in Table 2, but it should also be categorized as an environment-related keyword in a different context. We will further investigate this technique of combining co-occurrence frequency and cluster analysis so as to quantify the importance of each aspect of a multifaceted keyword.

Table 2: Categories and their keywords in individual analysis

material	product	technology	system	social	nature	environment
crude oil	automobile	pipeline	videophone	information	renewable energy	recycle
cereal grain	server	wind power generation	e-mail	communication	global warming	RoHS
corn	diode	thermal power generation	marine transportation	letter	cultivated acreage	3R
soybean	hybrid car	atomic power generation	air transportation	travel	deforestation	reduce(ridhuusu)
gasoline	construction machinery	lighter	ship	human resource	carbon dioxide	reuse
gold price	machine tool	solar power generation	railway	major company	rainfall	hazardous chemical substance
aluminum	aircraft	mileage	land transportation	recovery	biodiversity	heat island
chrome	plant oil	radioactive waste	finance	technology development	drought	CFC alternatives
sugarcane	diesel car	energy saving	family budget	leisure	acid rain	WWF
silicon	electric cable	water power generation	production	roof garden	rainforest	carbon neutral
stainless	secondary cell	plasma	electric power	population increase	fishery	carbon offset
titanium	fuel cell	heat pump	regulation	population de[illegible]		green electricity

In Figure 3(b), co-occurrence relations between centrally-located "solar cell" and the others under one set of conditions are shown. It is found that "solar cell" is related to "solar power generation", "carbon dioxide", "electric power", "silicon", and "semiconductor." By adding the result of cluster analysis here, it was also verified that there exist two topics of "power supply and the corresponding carbon dioxide emission" and "supply issue of silicon necessary for both solar cell and semiconductor manufacturing" in society.

5 SUMMARY

In order to investigate the "sustainability" of a complex society, firstly it is necessary to reveal the relationships among various events in the world. In this research, we considered newspaper articles as a database of social events and quantified the relational strengths between keywords by co-occurrence frequency of any two keywords in the articles, without depending on experts and their knowledge. For the purpose of supporting deciding system boundary of sustainability study, we applied multi-dimensional scaling (MDS) to the matrix finally obtained by the above method to express similarity among all keywords as distance on two-dimensional planes (x-y plane and x-z plane). We also developed a software tool which has a function to indicate the links between keywords directly and to provide the results of cluster analysis based on co-occurrence frequency information. This function is considered to be useful for engineers to find a conflict between the events related to sustainability. In the final step to identify whether the keywords which have strong relationship are in a conflict or not, human involvement is necessary. However, this tool could provide candidates of a conflict in automatic manner.

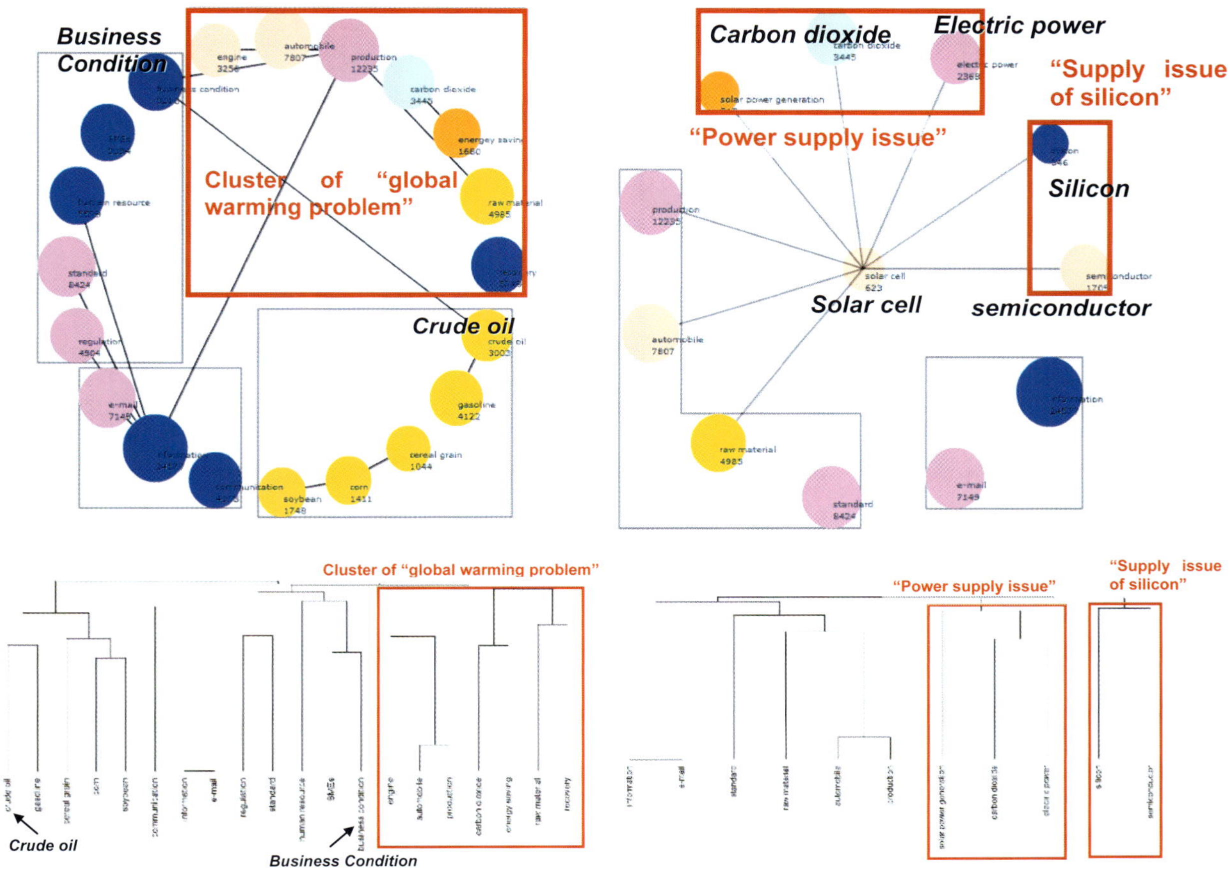

(a) Keywords appearing in more than 5,000 articles, or having a co-occurrence frequency of more than 450 times with another item. The furthest neighbor method is used in cluster analysis

(b) Keywords appearing in more than 7,000 articles or having a co-occurrence frequency of more than 50 times with "solar cell"

Figure 3: Examples of output from the developed tool for analyzing the relations between individual keywords

As a future work, we are planning to investigate the following issues:

1) Enhancing the completeness of events (keywords) related to sustainability
2) As ongoing work, we are also developing a method to observe time-series variation of relationship. Optimizing the search period to capture the changes of relationship appropriately should be investigated.

3) Introducing and developing other relational indices such as the Jaccard Coefficient and Simpson's Coefficient by combining the appearance frequency of a single keyword and co-occurrence frequency of two keywords.

6 REFERENCES

[1] Meadows, D., Randers, J., Meadows, D., 2004, Limits to Growth – The 30-year Update, Chelsea Green Publishing.

[2] United Nations, 2000, Sustainable bioenergy: A Framework for Decision Makers.

[3] Choucri, N., Mistree, D., Haghseta, F., Mezher, T., Baker, W. R., Ortiz, C.I., 2007, Mapping Sustainability, Springer.

[4] Nishiyama, T., Yamasaki, Y., Fukushige, S., Umeda, Y., 2007, Proposal of a Structural Description Method of Scenarios for Sustainable Society Scenario Simulation, Proc. of EcoDesign 2007 (CD-ROM).

[5] Kumazawa, T., Saito, O., Kozaki, K., Matsui, T., Mizoguchi, R., 2009, Toward Knowledge Structuring of Sustainability Science Based on Ontology Engineering, Sustainability Science, 4: 99–116.

[6] Ichise, R., Takeda, H., Ueyama, K., 2005, Exploration of Researchers' Social Network for Discovering Communities, JSAI 2005 Workshop, LNAI 4012, 458.

[7] Masui, K., Sonda, T., 2010, A Relevance Analysis Method to Find Conflicts between Events Related to Sustainability, 17th CIRP Inter. Conference on Life Cycle Engineering, pp. 417-421

Carbon Footprint Calculator for Construction Projects (CFCCP)

A.H. Ammouri[1], I. Srour[2], R.F. Hamade[1]

[1]Department of Mechanical Engineering, American University of Beirut, Lebanon

[2]Engineering Management Program, American University of Beirut, Lebanon

Abstract

A carbon footprint is defined as 'The total set of greenhouse gas emissions caused directly and indirectly by an individual, event, organization or product expressed as mainly CO_2 emissions [1]. Carbon footprints and their calculations have recently drawn considerable attention in order to limit Greenhouse Gas Emissions. Several carbon footprint calculators are available online to estimate the carbon dioxide emissions of individuals and households based on house, auto, and other consumption-related measures. On the other hand, there are a limited number of calculators that address construction projects. The importance of developing carbon calculators for construction projects is due to the fact that 13-18% of the total embodied carbon footprint of any construction project [2] and 100% of the total embodied carbon footprint of any landscape project is released the year the project is built or installed.

Therefore, the aim of this project, Carbon Footprint Calculator for Construction Projects (CFCCP), is to initiate a carbon footprint calculator that can be used in the market place for estimating the carbon footprint of construction projects. The CFCCP is capable of estimating the total carbon footprint of a construction project taking into consideration such variables as size, landscape, and material construction. This calculator produces better accuracy estimates compared with other existing construction carbon foot print calculators [3] that consider only the general inputs of a certain project and produce an estimate with ±25% accuracy. The cost of the accuracy is accomplished via quality library variables including: project design plan, detailed bill of quantities (BOM), and schedule timeline.

Keywords:
Calculator, Carbon footprint, Construction, Sustainability

1 INTRODUCTION

The considerable amount of carbon released upon the installation of construction projects requires quite an attention from environmentally concerned engineers. It is also crucial for those who consider 'green building' regulations in their project design. This is due to the fact that around 20% of the total embodied carbon footprint in construction projects is released the year the project is installed [2].

The remainder of the carbon footprint is the operational carbon released and the landscape carbon sequestered over the life of the project, typically 30 to 80 years [3]. This means that an average saving of 1% of the total carbon emissions during the construction phase of an 80 years old building is equal to the same amount of carbon emitted during an operation period of 12 months.

The Construction Carbon Calculator estimates embodied carbon. Embodied carbon is the carbon released when a product is manufactured, shipped to a project site and installed [3]. The proposed calculator will be capable of estimating the total carbon footprint of a construction project taking into consideration the size, landscape, and material construction. The calculator will be based on information gathered from any local Market. The calculator can be localized to any market based on the database of material and equipment that can be easily constructed with the calculator add-ons.

This calculator, unlike the existing ones, takes into consideration detailed project BOM and schedule timeline to produce a precise calculation of the total footprint. The cost of the accuracy will be paid by the need to input a library of variables. Although this will take more time than general calculators, the result will be a detailed report that describes the areas where the maximum carbon was emitted so that corrective measures can be taken to minimize these emissions.

2 METHODOLOGY

The following steps were pursued while building the carbon footprint calculator:

1. Data collection: this step includes the collection of data from the construction material suppliers.
2. Data analysis: for building the carbon footprint database of construction material using a special add-on for the calculator.
3. Programming: source code writing using NI LabVIEW.
4. Testing: using artificial data for a sample project.

2.1 Data collection

This step includes collecting data by deploying surveys in the local market where the calculator will be deployed. This is used in evaluating the performance of a certain supplier to know how much of carbon is emitted while producing a certain product.

The suppliers can be divided into two types. The first type is the importers only where no manufacturing/processing is carried out on the material. For this type of suppliers, we need to have information about the product manufacturing in its mother country along with the transportation method to be able to find its carbon footprint. For these suppliers, the estimate of the carbon footprint will be based upon information gathered from the exporting origin.

The other type of suppliers is the manufacturers in the local region. For these suppliers, the calculation of the carbon footprint is easier since we will have all the information we need about the manufacturing process from the factory.

The gathered information will be utilized to calculate the carbon emissions of the produced material. Once again we should also take into consideration that some raw material might be needed from a foreign source and thus information about the origin of the raw material should also be available.

Information is collected by surveys that will be deployed in the local market. A team for deploying the survey will be needed to contact the personnel responsible for the management of the production process in each material supplier company.

Sample survey questions that are used in the process of data collection are shown below:

- Company contact information.
- What is the product name and production unit?
- What is the geographic location of the factory/supplier (LAT/LON)
- How many workers/staff are there in the factory?
- What is the factory's yearly production capacity?
- What is the yearly energy consumption (electricity and fossil fuel)?
- Are there any raw material used in the production of this product?
- Where is its origin? (used to calculate travel distance)
- How is it transported to the factory? (land/sea/air)
- What is the embodied carbon of the raw material preparation?

The above collected data are used to calculate the production carbon footprint of the final product. Carbon emission of the transportation of this product will be accounted for when using the calculator.

2.2 Data analysis

Data collected from the survey are used to build a database of construction material for the local market.

To facilitate data entry and analysis, the graphical programming tool LabVIEW was used to program an add-on for the calculator named "CarbonAnalyzer". This tool is used in the construction of the database.

A sample snapshot of the main window of the "CarbonAnalyzer" is shown in Figure 1.

The figure shows a sample Cement product that will be saved in a database file for the Lebanese market called LebDB.tdms.

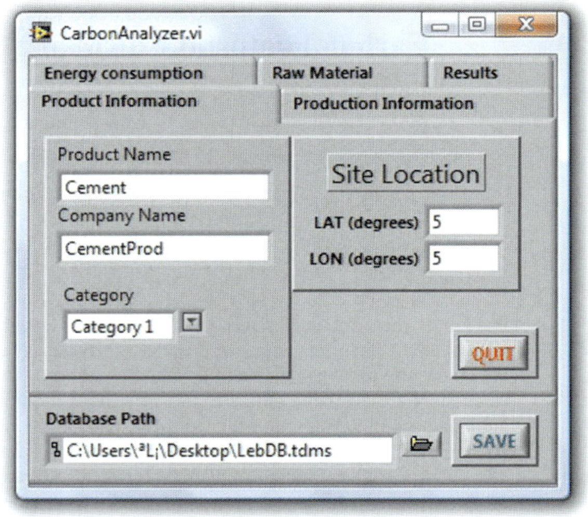

Figure 1: CarbonAnalyzer add-on

This tool works in the following manner:

1. Yearly energy consumption is transformed to corresponding carbon emissions
2. The total yearly carbon emissions is divided by the yearly production volume
3. The result is stored as production carbon emission
4. Raw material embodied carbon per unit along with its transportation carbon emitted is added to the production emission per unit
5. The resulting total carbon emission per unit production volume is then saved into the database under the name of the producing company under the category of the product.

Data conversions were gathered from the National Renewable Energy Laboratory (NREL) [4] who produced a database called the Lifecycle Inventory (LCI) database.

It should be noted that different countries might have different values for their material lifecycle. For example electricity carbon footprint differs from nuclear to hydraulic production processes.

These factors will lead to a slight increase/decrease in the total carbon footprint, unless taking into consideration the local carbon footprint of the energy sources available.

The output of the CarbonAnalyzer add-on is a database file with a .tdms extension (streaming data file).

This file will be used with the calculator along with the project properties to get the total carbon footprint of a project.

Database files can be viewed with another add-on called TDMS file viewer.

A sample database file is shown in the Figure 2. The file shows the LebDB.tdms database that includes different categories of construction materials. Each category contains different suppliers. A sample CEMENT product of the CementCO1 Company is shown selected with the corresponding product properties shown in the right pan of the TDMS file viewer add-on.

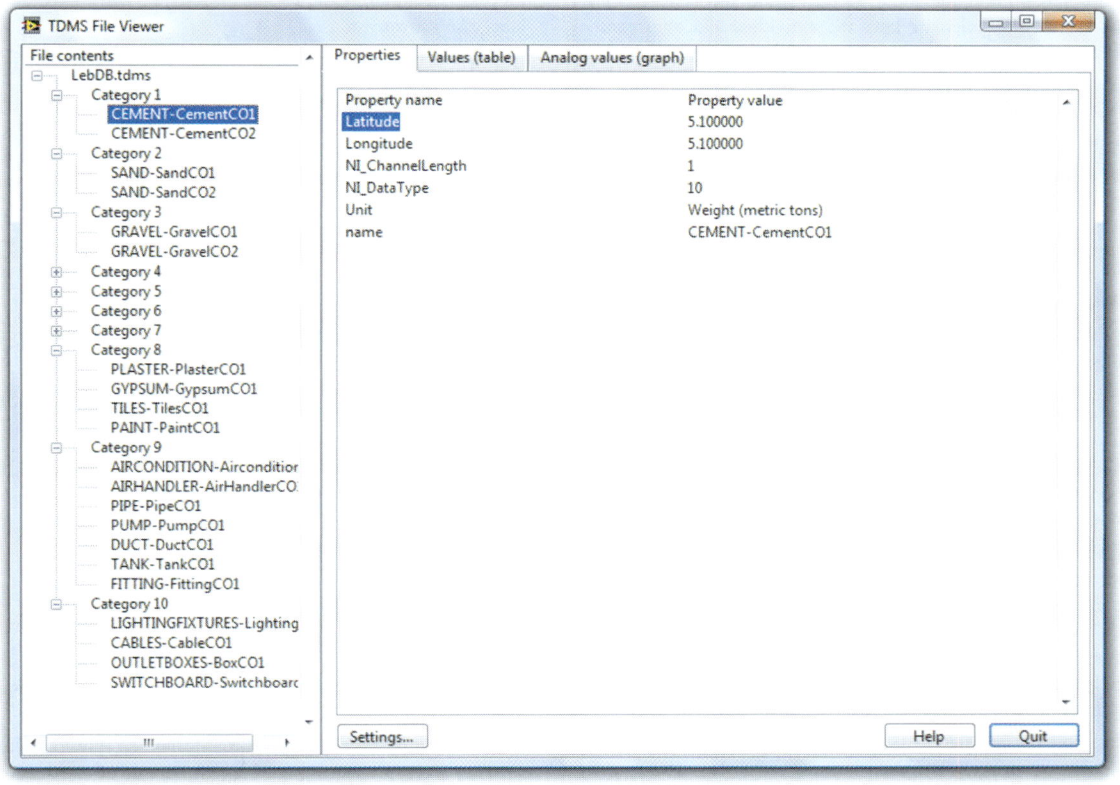

Figure 2: LebTDMS database file sample

2.3 Calculator programming

CFCCP is a very user friendly program that was programmed using the National Instruments LabVIEW virtual programming software.

Figure 3 shows a snapshot of the CFCCP showing the General Project Information tab. This tab is used to input the site information that includes its name, the consulting company, the site location, and the database file that is going to be used in the calculation of the carbon footprint.

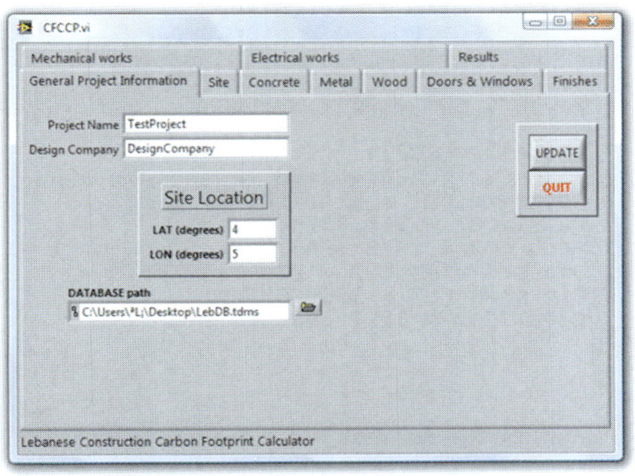

Figure 3: CFCCP main window

The calculator takes the database file constructed with the CarbonAnalyzer add-on along with project inputs to calculate the total carbon footprint of the project under consideration. The project inputs are divided into the following 9 categories:

1. General project information
2. Site works
3. Concrete works
4. Metal works
5. Wood works
6. Windows and doors
7. Finishes
8. Mechanical works
9. Electrical works

The General project information includes the information about the project name, Design Company and its geographic location (LON/LAT). The geographic information is used to calculate the approximate distance from the suppliers in the database built previously. The database file to be used is selected using the General project information tab. Future modifications to the calculator can include adding a routing option that can increase the accuracy of distance calculations.

The remaining 8 categories of the project works (from site to electrical) accounts for the equipment used, the quantities used, the duration of each task, the workforce on each task, and the transportation method of each material (including transport distance).

The information from the input variables and the local supplier database is used to calculate both the individual and total carbon footprint of the project task.

A sample site works (concrete) will be demonstrated to show the steps followed in the calculation of the carbon footprint of any site work.

The total carbon embodied in a certain work is divided into 3 categories:

1. Carbon embodied in the raw material
2. Transportation carbon footprint
3. Workforce carbon footprint

Since concrete is a mix of materials, each individual material should be selected from the available suppliers in the database file considered. Figure 4 shows an example of Concrete works with cement selected from CementCO1 supplier. Upon the selection of the material supplier, the distance indicator shows the geographic distance from the site. After inputting the quantity of cement and the corresponding truck load, the transportation and raw material embodied carbon can be calculated directly.

The embodied carbon is simply the quantity used multiplied by the carbon index of the raw material of the selected supplier. While the transportation carbon emissions is found by first dividing the quantity by the truck load (to find the number of truck deliveries) and then by the specific fuel consumption of each truck along the traveled distance. Indices for the traveled distance consumption of each truck can be found from the database file.

The last step in calculating the carbon footprint for cement will come from the work force of the cement team. This is found by multiplying the carbon index of the workers/trucks by the daily workload and the period of the site work.

The summation of all the sub-materials of the concrete works will result in the total concrete carbon emissions that is shown in the lower right side of the window in the concrete CO_2 indicator (Figure 4)

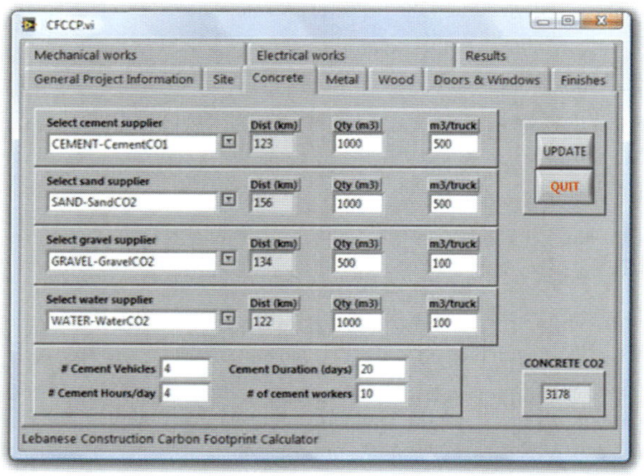

Figure 4: Concrete works in CFCCP

The same procedure is repeated for the other site works to conclude the total carbon footprint of the project being studied. The results are viewed in the results tab where it can be saved to a text file for comparison and research purpose.

A sample snapshot of the results is shown in Figure 5.

The figure shows the results on a pie chart with the label showing the detailed carbon emission of each task in the construction process. The total CO_2 emission in tons is shown below the chart label also shown.

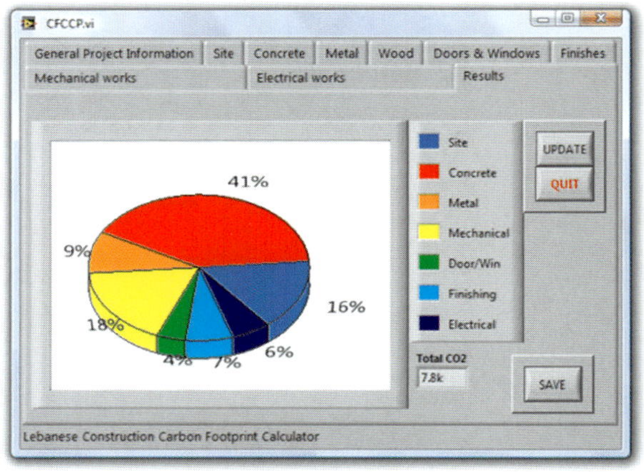

Figure 5: Results tab of the CFCCP

3 RESULTS

Results from any project can be saved by pressing the SAVE button in the results tab. The output will be a text file that contains the detailed description of the carbon footprint.

Figure 6 shows a sample saved text file for a virtual project taken into consideration.

The text file describes the project information, the total embodied carbon, and the detailed carbon emissions of each task while constructing it. It can be noticed that just playing with the project geographic location alone will change the carbon footprint dramatically.

These files are useful in comparing how changing material suppliers will alter the carbon emissions of the whole project. It can also be used to view the areas where emissions are higher than usual to try to find alternate scenarios with lower emissions.

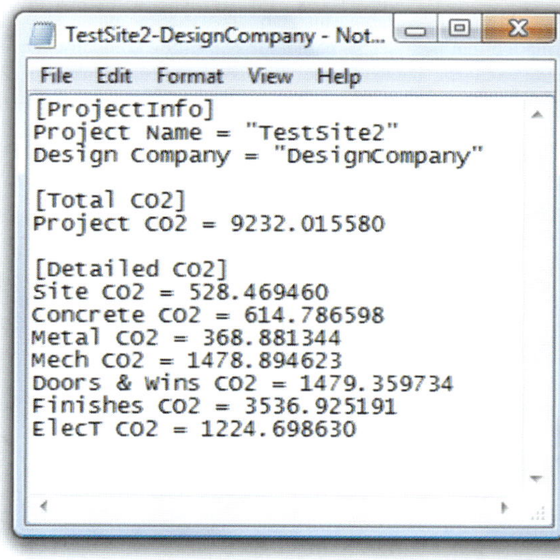

Figure 6: Sample output result file

4 CONCLUSIONS AND FUTURE WORK

A complete toolkit dubbed the 'Carbon Footprint Calculator for Construction Projects or CFCCP)' was built. This software tool is capable of building a construction material database of carbon emissions "CarbonAnalyzer", along with calculating the carbon footprint of construction projects.

Although the focus has so far been on construction projects, the utilization of CFCCP may also be extended to other project types if database files for the project material can be built. Therefore, future applications can involve the design of green buildings given that buildings typically utilize materials from distant sources resulting in noticeable increase in their initial carbon footprint. Since only about 25 percent (20 years of operation over 80 year projected service) of the carbon footprint of construction projects comes from the actual construction tasks then it is very important to reduce the initial carbon emission of construction projects. Such future work includes adding database for the Lebanese construction material suppliers, along with the tuning of the calculator constants and programming procedure to make it ready to be used on a commercial basis.

5 ACKNOWLEDGMENTS

The authors wish to acknowledge the financial support of the University Research Board (URB) of the American University of Beirut. The first author also wishes to acknowledge the support of Consolidated Contractors Company through the CCC Doctoral Fellowship in Manufacturing.

6 REFERENCES

[1] Carbon Trust,Carbon Trust website, 2009, What is a Carbon Footprint? [Online] [Cited: Nov 25, 2009.] http://www.carbontrust.co.uk/solutions/CarbonFootprinting/what_is_a_carbon_footprint.htm.

[2] United Nations Environment Programme, 2007, Buildings and Climate Change: Status, Challanges and Opportunities, UNEP 2007, ISBN: 978-92-807-2795-1 http://www.unep.fr/pc/sbc/documents/Buildings_and_climate_change.pdf

[3] Build Carbon Neutral, 2009, About the Construction Carbon Calculator.[Online] [Cited: NOV 25, 2009.] http://buildcarbonneutral.org/about.php

[4] National Renewable Energy Laboratory NREL, US life cycle inventory database, 2009, http://www.nrel.gov/lci

A Framework of Product and Process Metrics for Sustainable Manufacturing

T. Lu[1], A. Gupta[1], A. D. Jayal[1], F. Badurdeen[1], S.C. Feng[2], O. W. Dillon, Jr. [1], I. S. Jawahir[1]

[1]Institute for Sustainable Manufacturing, Department of Mechanical Engineering, University of Kentucky, Lexington, USA

[2]National Institute of Standards and Technology, Gaithersburg, USA

Abstract

This paper presents a framework for developing comprehensive product and process metrics for sustainable manufacturing, using machined products and machining processes examples, and addressing all three aspects of the triple bottom line – environment, economy and society. The need for developing standardized metrics is discussed for the wider use of these metrics by different manufacturers. The occurrence of similar measurements in some of the metric categories indicates the potential and need for data sharing between product and process metrics. The differences, relationships, and potential interactions between the product and process metrics are discussed from the viewpoint of their applications.

Keywords:

Sustainable manufacturing, Metrics, Products, Processes

1 INTRODUCTION

The U.S. Department of Commerce defines sustainable manufacturing as *"the creation of manufactured products that use processes that minimize negative environmental impacts, conserve energy and natural resources, are safe for employees, communities, and consumers and are economically sound*" [1]. It is also stated that sustainable manufacturing includes both the manufacturing of "sustainable" products as well as the sustainable manufacturing of all products [2].

Sustainable manufacturing should consider the economical, environmental, and societal impacts, usually addressed as the triple bottom line. The impacts of the sustainability elements of a manufactured product and its manufacturing processes also need to take its entire life-cycle into consideration, which includes the four life-cycle stages: pre-manufacturing, manufacturing, use, and post-use. The analysis can be carried out on the product level, process level or system level. The innovation based 6R (*reduce, reuse, recycle, recover, redesign,* and *remanufacture*) approach allows for a significant transformation from a cradle-to-grave concept to multiple life-cycle consideration for a specific product [3]. The major product sustainability sub-elements identified, shown in Figure 1, consider all the 6R components.

Product and process metrics for sustainable manufacturing are necessary for evaluating the performance of a product or a manufacturing process considering the sustainability aspect. Aside from the basic application of proper evaluation of products and processes, the ultimate goal of developing product and process metrics for sustainable manufacturing is to provide improved decision-making criteria when optimizing product design and process design for sustainable manufacturing. Current metrics or indicators focus primarily on company, regional, national and global levels. Highly technical methodology of assessing sustainability performance of products and processes has not been fully addressed and there is a critical need for developing improved, comprehensive, and useful metrics for sustainability evaluation of products and processes [4].

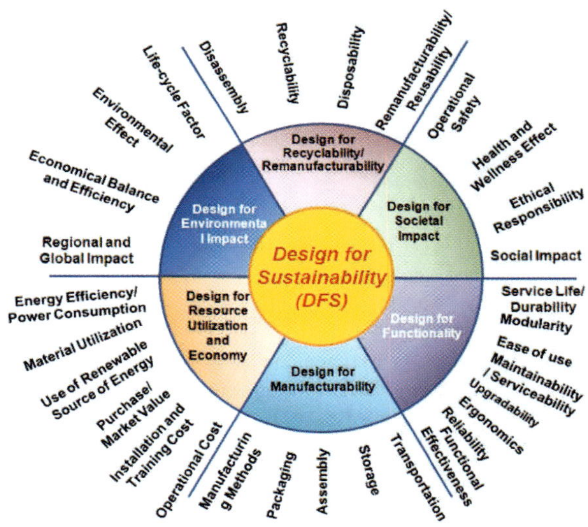

Figure 1: Product sustainability wheel [3]

This paper presents a framework for developing comprehensive product and process metrics for evaluating sustainable manufacturing of discrete products. Their relationships are also discussed from the total life-cycle viewpoint.

2 PREVIOUS WORK

Achieving sustainable manufacturing starts with the proper assessment of relative performance concerning sustainability in manufacturing. Ultimately, sustainability performance assessment requires commonly accepted standardized metrics. There have been many attempts that try to build such a system, and Feng and Joung [5] presented a comprehensive list of previous approaches including the indicator sets by Global Reporting Initiative (GRI) [6] and core indicators by the Organization for Economic Corporation and Development [7].

Sustainable products are generally defined as those products providing environmental, societal and economical benefits while protecting public health, welfare, and environment over their full lifecycle, from the extraction of raw-materials to final disposal [8]. Previous work performed in the area of product sustainability has produced various indicators and metrics to evaluate the sustainability content. Fiksel et al. [9] developed a comprehensive list of product sustainability indicators and categorized these under environmental, societal and economic aspects.

Significant work has been done at the University of Kentucky in developing indicators and metrics for product sustainability. Jawahir and Wanigarathne [10], in their early work, identified six major sustainability elements and corresponding sub-elements in manufactured products, incorporating environmental, societal, and economic impacts. The broadly identified sustainability elements are as follows: environmental Impact, societal Impact, functionality, resource utilization and economy, manufacturability, and recyclability and remanufacturability.

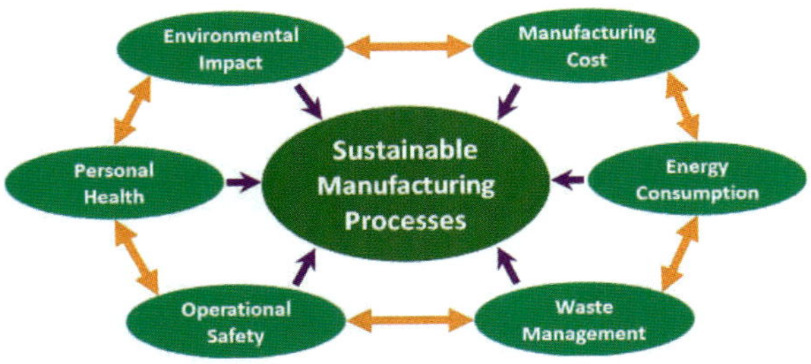

Figure 2: Six elements of sustainable manufacturing processes [11]

Based on the triple bottom line considerations, a set of sustainability elements for sustainable manufacturing is proposed by Jawahir and Dillon [11], as shown in Figure 2.

Among these, the manufacturing cost, energy consumption and waste management are considered as deterministic elements, and the environmental impact, personnel health and operator safety are non-deterministic elements.

Seven guidelines for choosing an appropriate set of measurements in industrial applications were proposed by Fiksel et al. [12]. These are: comprehensiveness, controllability, cost-effectiveness, manageability, meaningfulness, robustness and timeliness. Feng et al. [4] also indicated seven characteristics of the sustainability performance indicators, requiring the metrics to be measurable, relevant and comprehensive, understandable and meaningful, manageable, reliable, cost-effective in data access and timely. In a recent General Motors report on metrics for sustainable manufacturing [13] the following five criteria that the metrics have to satisfy are proposed: (1) the metrics need to address the need of all stakeholders, (2) facilitate innovation and growth, (3) harmonize business units of different geographical locations, (4) be compatible with current value-adding business systems and (5) the related measurement needs.

Several researchers have investigated the environmental impact of manufacturing [14-18]. However, the ideas discussed mainly focus on environmental aspects only, even though elements of the economical and societal aspects are occasionally mentioned.

When performing manufacturing process optimization, defining the objective function is a critical issue. Conventionally, maximizing the productivity or minimizing the manufacturing cost is the usual target, subject to fulfilling quality requirements [19-21]. Preliminary work on machining process optimization with total sustainability as the target involves setting up a methodology to account for a partial set of indicators generated according to both the deterministic and non-deterministic elements of sustainable manufacturing processes [22]. However, a critical need remains for developing a comprehensive set of metrics for sustainability evaluation of products and manufacturing. The metrics should include a vocabulary and objective functions for sustainability optimization [22].

This paper focuses on presenting a framework for developing comprehensive and practical product and process metrics for sustainable manufacturing, and presents the interrelationships and potential interactions among the metrics.

3 PRODUCT METRICS

In addition to considering environmental, societal and economic aspects and incorporating the 6R (Reuse, Reduce, Recycle, Remanufacturing, Redesign, and Recover) approach, a major emphasis is given to categorize the developed indicators and metrics into four stages in a product life-cycle. The four key stages of a manufactured product in a closed loop system are represented as follows: pre-manufacturing, manufacturing, use, and post-use [3]. Jawahir et al. [23] evaluated the sustainability content of a product using a generic product sustainability index (PSI), incorporating the three major components of sustainability (economy, environment, and society), over all four life-cycle stages. Numerous influencing factors are identified and categorized appropriately. The weights assigned to the influencing factors are arbitrary numbers based on their relative importance and company priorities. Gupta et al. [24] used the analytic hierarchy process (AHP) to determine the relative importance of different influencing factors and compare the sustainability content of two similar products

Table 1: Product metric clusters

	ENVIRONMENT	ECONOMY	SOCIETY
METRICS CLUSTERS	Residues	Cost	Education
	Energy Use and Efficiency	Innovation	Customer Satisfaction
	End-of-Life Management	Profitability	End-of-Life Management
	Material Use and Efficiency	Product Quality	Product Safety and Societal Well-being
	Water Use and Efficiency	Investment	Employee Safety and Health

Based on the six major sustainability elements and corresponding sub-elements of manufactured product metrics, as shown in Figure 1, a new product metrics system is being developed at the University of Kentucky. The metrics are grouped under a range of metrics clusters to make them more structured. The proposed metric clusters are defined for the environmental, economic and societal aspects, as shown in Table 1. Furthermore, several metrics have been identified and defined under different clusters to make the metrics system comprehensive. Some example metrics are shown

Table 1: Examples of product metrics in different clusters, and the life-cycle stages to which they apply

Metrics Clusters	Example Metrics	Unit (D/L: dimensionless)	PM (pre-mfg.)	M (mfg.)	U (use)	PU (post-use)
Residues	Emissions Rate (carbon-dioxide, sulphur-oxides, nitrous-oxides, etc.)	mass/unit	√	√	√	√
Energy Use and Efficiency	Remanufactured Product Energy	kWh/unit		√	√	√
	Maintenance/ Repair Energy	kWh/unit			√	
Product End-of-Life Management	Design-for-Environment Expenditure	$/$ (D/L)	√			
	Ease of Sustainable Product Disposal for End Users	$/unit				√
Material Use and efficiency	Restricted Material Usage Rate	mass/unit	√	√		√
Water Use and Efficiency	Recycled Water Usage Rate	gallons/unit	√	√		√
Cost	Product Operational Cost	$/unit			√	
Innovation	Average Disassembly Cost	$/unit				√
Profitability	Revenue	$/unit			√	
Product Quality	Defective Products Loss	$/unit		√		
	Warranty Cost Ratio	$/unit			√	
Education	Employee Training	Hours/unit	√	√		√
Customer Satisfaction	Repeat Customer Ratio	(D/L)			√	
	Post-Sale Service Effectiveness	(D/L)			√	
Product Safety and Societal Well-being	Product Processing Injury Rate	incidents/unit	√	√		√
	Landfill Reduction	mass/unit	√	√	√	√

(along with the clusters in which these metrics occur) in Table 2. All metrics are categorized across the life-cycle stages of a product to have a detailed understanding of the influence of a particular metric. An interesting observation while categorizing metrics across life-cycle stages is that some of these metrics have presence across multiple life-cycle stages. This provides an opportunity to organize metrics at different levels; for example, top level metrics can be the ones that are present across all four life-cycle stages. Further, the priorities derived through analytical techniques, such as the previously mentioned AHP, can be combined with this to obtain a system of levels which is more science-based. The ongoing work involves defining measurement methods to determine each metric quantitatively.

4 PROCESS METRICS

4.1 Process sustainability metrics

When evaluating a manufacturing process with respect to sustainability, each input and output needs to take into account the total life-cycle. A simple input/output chart of a manufacturing process, using machining as an example, is shown in Figure 3.

Metrics for a sustainable machining process are proposed by carefully examining the inputs and outputs of a machining process based on the six elements of sustainable manufacturing processes, shown in Figure 2. Examples of the metrics are shown in Table 3.

Use, reuse, recycle, and disposal aspects are considered whenever these life-cycle phases take place within the manufacturing process. For example, manufacturers buy new cutting tools and coolants and use these for a certain amount of work. After that, they may regrind the cutting tool and filter the coolant to make these applicable for further use. When the functionality is lost and/or goes beyond certain acceptable level, they will be disposed, either by sending to land-fill or by delivering to dedicated recycling plants. The cutting tools and coolants may have multiple life-cycles within the manufacturing plant in this sense, and the measurements represent the influence of the multiple life-cycles.

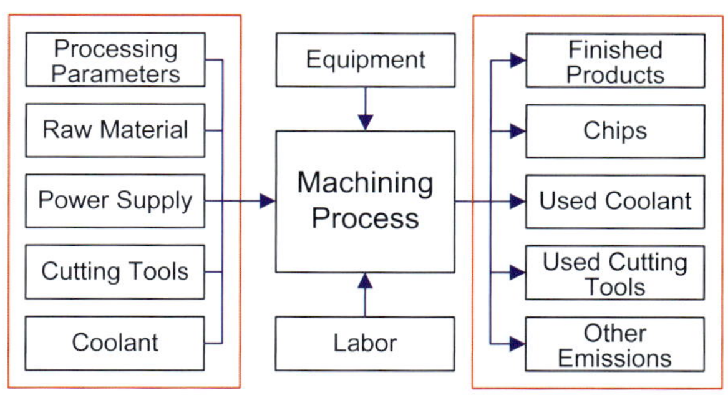

Figure 3: Input/output diagram for a machining process

For measurements of deterministic elements, the data can be collected onsite, experimentally measured, empirically predicted, or analytically calculated. For measurements of non-deterministic elements, by using fuzzy-logic techniques and creating a corresponding system of linguistic rules, the measurements can be quantitatively evaluated.

4.2 Hierarchy structure of the process metrics

A hierarchy structure of the process metrics is proposed. A sample is shown in Figure 4. In this hierarchy structure, the measurements are categorized into three levels: the manufacturing cell/line/plant level, the workstation level and the operation level. In a previous literature [4], the manufacturing system within a plant is considered as factory/line level, work cell level, machine tool level, and process level.

With this hierarchy-based structure of metrics the widely accepted/standardized indices can be fit into the metrics properly. These indices can be either a high level measurement or a value summarized from the metrics. Furthermore, when current industrial applications fail to take the measurements in such detail at the operation level, or even workstation level, because of current routine or difficulty of application they might be able to collect data from higher levels. The data collected will be able to represent the influence of its sub-measurements on the indicators. The disadvantage is that the investigator may have difficulty utilizing presently available scientific models to build correlations between the input parameters and the higher level measurements, as the outputs of scientific models usually correspond to operation level measurements. The metrics proposed in Section 4.1 are actually a summary of line level measurements. It gives a guideline and overview of all the aspects which need to be evaluated.

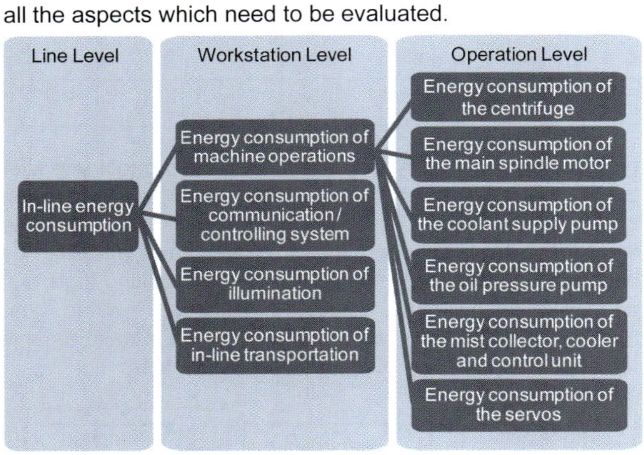

Figure 4: Hierarchical structure of an example process metric, listed in Table 3

At the operational level, all measurements focus on single process operations. The machine is doing a specific job with certain tools and materials under some particular set of operating conditions. In the case of machining, a single operation can be a face turning step or a hole drilling operation. The workstation level measurements focus on one single machine doing one or more operations or an accessory equipment providing some specific function. The plant/line/cell level includes the measurements inside the whole manufacturing unit. It can be a mass production flow-line, a manufacturing cell or a machine-shop plant, depending on the organization of the manufacturing facility.

All measurements at higher levels are composed of corresponding measurements at lower levels. In other words, with sufficient data from lower level measurements, we should be able to integrate them into the higher level measurements. Some measurements might appear under different sustainability elements, showing their influences on multiple aspects of sustainability. When calculating different indicators concerning these sustainability elements, different weighting ratios should be assigned to the repeated measurements as their importance and contribution towards that particular indicator would be different.

5 DISCUSSION

5.1 Relationships between the product metrics and process metrics from the life-cycle viewpoint

It is not surprising to see that some of the measurements in the product metrics and the process metrics are closely interconnected. For example, the manufacturing cost is considered in both the product and the process metrics. Furthermore, under the waste management element of the sustainable manufacturing process metrics, the proposed measurements take use, reuse, recovery and recycling into consideration, while similar measurements are considered in the sustainable product metrics.

The fundamental reason for the repeated measurements is that, a manufacturing process occurs in the "manufacturing" phase when considering the life-cycle of a product. On the other hand, if the manufacturing consumables or the production equipment is taken as the product under investigation, the "use" phase of their life-cycle occurs within the manufacturing processes. For the recycled and remanufactured products, the recycling and remanufacturing process is their "manufacturing" phase. The overall assessment therefore might be totally different. It suggests that the target under investigation has to be clarified at the beginning.

From the viewpoint of the products, manufacturing processes are a series of small periods in their life-cycles. Choosing the "correct" manufacturing processes, which is usually

Table 3: Examples of process metrics for sustainable machining

Environmental Impact	Energy Consumption	Cost
GHG emission from energy consumption of the line (ton CO_2 eq./unit) Ratio of renewable energy used (%) Total water consumption (ton/unit) Mass of restricted disposals (kg/unit) Noise level outside the factory (dB)	In-line energy consumption (kWh/unit) Energy consumption on maintaining facility environment (kWh/unit) Energy consumption for transportation into/out of the line (kWh/unit) Ratio of use of renewable energy (%)	Labor cost ($/unit) Cost for use of energy ($/unit) Cost of consumables ($/unit) Maintenance cost ($/unit) Cost of by-product treatment ($/unit) Indirect labor cost ($/unit)
Operator Safety	**Personal Health**	**Waste Management**
Exposure to Corrosive/toxic chemicals (incidents/person) Exposure to high energy components (incidents/person) Injury rate (injuries/unit)	Chemical contamination of working environment (mg/m^3) Mist/dust level (mg/m^3) Noise level inside factory (dB) Physical load index (dimensionless) Health-related absenteeism rate (%)	Mass of disposed consumables (kg/unit) Consumables reuse ratio (%) Mass of mist generation (kg/unit) Mass of disposed chips and scraps (kg/unit) Ratio of recycled chips and scraps (%)

considered as process planning, involves selecting the optimized routine according to different criteria. These criteria can be conventional economics-oriented criteria, such as maximizing manufacturability, minimizing manufacturing cost and achieving best product functionality/quality, or relatively innovative sustainability-oriented criteria, such as minimizing environmental impact or maximizing societal benefits. The sustainability assessment of manufacturing processes provides a comprehensive criterion for total-sustainability-oriented process design.

5.2 Driving forces for data sharing

The nature of overlaps among the product and process metrics indicates the possibility and need for data sharing when carrying out sustainability assessment of a product or a manufacturing process. Judging from the measurements themselves, it is essential that the sustainability assessment data of manufacturing supplies, when considered as the product under investigation, should feed into the process metrics of the manufacturing process in which they are used. For example, the exact content of some specific coolant and the associated environmental impact will be used in the process metrics to judge the environmental impact of the used coolant. Similarly the sustainability data for a manufacturing process should feed into the product metrics of the product being manufactured by the process as the source for the "manufacturing" phase data.

Furthermore, as the data sources are often outside the domain under the investigators' control, the data collection of input materials as products can be difficult for the user. On the other hand, the product designer or the product sustainability evaluator carrier might have limited knowledge of the exact manufacturing process of the product. Also, taking the same measurements repeatedly by different investigators for similar purpose of using them would be a waste of time and effort. All of these point towards the need for data sharing between the product metrics and the process metrics.

However, there are problems when sharing the data for sustainability assessment. A very common problem is that different companies or industries use different business languages generated from their own company culture. Further, there are very few commonly accepted data formats. Even worse, interpretations of same words in different companies, industries, or countries can be very different. These problems are major barriers for data sharing among manufacturing organizations.

One of the important purposes of sustainability assessment of a product and its manufacturing processes is to identify areas with opportunity for improvement. It requires comparing different plants, companies or industries. A single set of measurements might be applicable when evaluating similar products manufactured with similar processes. However, it might be hard to compare the data measured for products coming out of different manufacturing processes or different geographical locations. The situation gets worse when comparing different products even when they may have similar functionality.

Both the data sharing and the comparison between different products require comparable data measured on correctly corresponded points. Thus, the sustainability metrics and data format for information exchange need to be standardized. The definition of measurements and correspondence of the data need to be addressed clearly.

5.3 Interface of analysis

The interface of analysis between the product and process metrics needs to be emphasized. As we can see in Figure 5, based on the product requirements, the product design phase would address a series of manufacturing output requirements. These requirements become the constraints of the process design. Though the ultimate goal is to implement the product performance expectations, the direct output of the process design is evaluated by the criteria provided by product design.

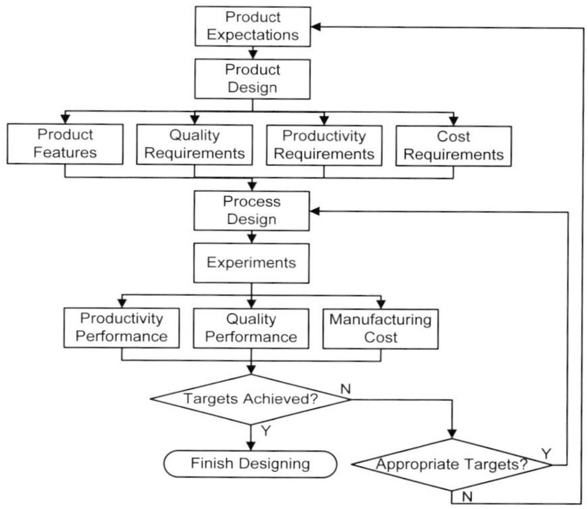

Figure 5: Relationship between the product design and process design

The sustainability assessments of a product, and its corresponding processes, have different emphases. The manufacturing processes, serve to implement a product design, and their constraints are decided by the current product design. To evaluate a manufacturing process, its fulfillment of product design features and requirements need to be considered. For sustainability assessment of a product design, the overall product sustainability performance is the ultimate criteria and the process assessment is only one of the sub-elements. To be specific, the sustainability assessment of a process would not cover the other phases of the manufactured product's life-cycle. The product assessment usually covers broader aspects than the process assessment, such as the entire life-cycle and 6R aspects.

An optimized manufacturing process routine does not necessarily mean that the product is optimal concerning its sustainability performance. On the other hand, to achieve optimal overall sustainability performance when designing a product, the corresponding manufacturing processes need to be optimized based on some sustainability criteria.

6 SUMMARY

This paper has presented a framework for developing comprehensive product and process metrics for sustainable manufacturing. The interactions among the two sets of metrics are discussed in view of the need for proper application of sustainability assessments. Potential need for information exchange is also briefly introduced.

7 ACKNOWLEDGMENTS

The work presented in this paper is part of an on-going project at the University of Kentucky, funded by the National Institute of Standards and Technology (Award No.: 60NANB10D009).

8 DISCLAIMER

No approval or endorsement of any commercial products by the National Institute of Standards and Technology (NIST) is intended or implied. Certain company names are identified in this paper to facilitate understanding. Such identification does not imply that their products are necessarily the best available for the purpose of sustainable products or manufacturing.

9 REFERENCES

[1] International Trade Administration, U.S. Department of Commerce, 2007, How Does Commerce Define Sustainable Manufacturing?, Available: http://www.trade.gov/competitiveness/sustainablemanufacturing/how_doc_defines_SM.asp.

[2] National Council for Advanced Manufacturing (NACFAM), 2009, Sustainable Manufacturing, Available: http://nacfam02.dev.web.sba.com/PolicyInitiatives/SustainableManufacturing/tabid/64/Default.aspx.

[3] Jawahir, I.S., Dillon, O.W., Rouch, K.E., Joshi, K.J., Venkatachalam, A., Jaafar, I.H., 2006, Total Life-cycle considerations in product design for sustainability: A framework for comprehensive evaluation, 10th Int. Research/Expert Conf., TMT 2006, Barcelona, Spain, 1-10.

[4] Feng, S.C., Joung, C., Li, G., 2010, Development Overview of Sustainable Manufacturing Metrics, Proceedings of the 17th CIRP International Conference on Life Cycle Engineering 2010, Hefei, China.

[5] Feng, S.C., Joung C.B., 2009, An Overview of a Proposed Measurement Infrastructure for Sustainable Manufacturing, Proceedings of the 7th Global Conference on Sustainable Manufacturing, Chennai, India, 355-360.

[6] Staniskis, J. and Arbaciauskas, V., 2009, Sustainability Performance Indicators for Industrial Enterprise Management, Environmental Research, Engineering and Management, 2/48:42-50.

[7] OECD, 2009, Policy Brief: Sustainable Manufacturing and Eco-innovation: Towards a Green Economy, http://www.oecd.org/dataoecd/34/27/42944011.pdf

[8] Sustainable Products Corporation, Available: www.sustainableproducts.com/susproddef.html.

[9] Fiksel, J., McDaniel, J., Spitzley, D., 1998, Measuring Product Sustainability, Journal of Sustainable Product Design, 6:7-19.

[10] Jawahir, I.S., Wanigarathne, P.C., 2004, New Challenges in Developing Science-based Sustainability Principles for Next Generation Product Design and Manufacture, (Keynote Paper), Proceedings of TMT, Neum, B&H, 1-10.

[11] Jawahir, I.S., Dillon, O.W., 2007, Sustainable Manufacturing Processes: New challenges for developing predictive models and optimization techniques, Proceedings of First International Conference on Sustainable Manufacturing, Montreal, Canada, 1-19.

[12] Fiksel, J., McDaniel, J., Mendenhall, C., 1999, Measuring Progress towards Sustainability Principles, Process, and Best Practices, Greening of Industry Network Conference Best Practice Proceedings, available: http://www.economics.com/images/Sustainability%20Measurement%20GIN.pdf.

[13] Dreher, J., Lawler, M., Stewart, J., Strasorier, G., Thorne, M., 2009, General Motors: Metrics for Sustainable Manufacturing, available: http://actionlearning.mit.edu/files/slab_files/Projects/2009/GM,%20report.pdf

[14] Kondo, T., 1997, Environmentally Clean Machining in Toyota, International Journal of the Japan Society for Precision Engineering, 31/4:249-252.

[15] Munoz, A.A., Sheng, P., 1995, An Analytical Approach for Determining the Environmental Impact of Machining Processes, Journal of Materials Processing Technology, 53:736-758.

[16] Choi, A.C.K., Kaebernick, H., Lai, W.H., 1997, Manufacturing Processes Modelling for Environmental Impact Assessment, Journal of Materials Processing Technology, 70:231-238.

[17] Dahmus, J. B., Gutowski, T. G., 2004, An Environmental Analysis of Machining, ASME International Mechanical Engineering Congress and RD&D Expo, IMechE2004, Anaheim, CA, USA.

[18] Gutowski, T., Murphy, C., Allen, D., Bauer, D., Bras, B., Piwonka, T., Sheng, P., Sutherland, J., Thurston, D., Wolff, E., 2005, Environmentally Benign Manufacturing: Observations from Japan, Europe and the United States, Journal of Cleaner Production, 13:1-17.

[19] Boothroyd, G., Rusek, P., 1976, Maximum Rate of Profit Criteria in Machining, Journal of Engineering for Industry Transactions of the ASME, 98/1:217-220.

[20] Shin, Y.C., Joo, Y.S., 1992, Optimisation of machining conditions with practical constraints, International Journal of Production Research, 30:2907-2919.

[21] Jawahir, I.S., Balaji, A.K., Rouch, K.E., Baker, J.R., 2003, Towards Integration of Hybrid Models for Optimized Machining Performance in Intelligent Manufacturing Systems, Journal of Materials Processing Technology, 139/1-3:488-498.

[22] Granados, S., Jawahir, I.S., Fernández, J., 2009, A Comprehensive Criterion for Sustainability Evaluation of Machining Processes, Proceedings of the 7th Global Conference on Sustainable Manufacturing, Chennai, India, 385-391.

[23] Silva, N.D., Jawahir, I.S., Dillon, O.W., Russell, M., 2009, A New Comprehensive Methodology for the Evaluation of Product Sustainability at the Design and Development Stage of Consumer Electronic Products, International Journal of Sustainable Manufacturing, 1/3:251-264.

[24] Gupta, A., Vangari, R., Jayal, A. D., and Jawahir, I. S., 2010, Priority Evaluation of Product Metrics for Sustainable Manufacturing, Proceedings of the 20th CIRP Design Conference, Nantes, France.

Evaluation of Post-series Supply Strategies in Regard of Sustainability

U. Dombrowski, S. Schulze, S. Weckenborg

Institute for Production Management and Enterprise Research, Technische Universität Braunschweig, Germany

Abstract

The use of electronic components is rising in industries like the automobile industry since several decades. The characteristics of these parts are different to these of mechanical components, especially with regard to the spare parts management. Today economical aspects and technological restrictions are in the main focus of the planning process for the post-series supply. However the ecological impact of the post-series supply is not considered in this process. Often the economic and the ecologic preferable strategy are conflictive. The consideration of these two aspects with regard to the technological feasibility is a complex problem. One possible solution to consider these aspects simultaneously and to increase the sustainability of the products and processes is the design of electronic components according to the requirements of post-series supply.

Keywords:

Post-series Supply, Spare Parts Management, Product Development

1 INTRODUCTION

In order to guarantee customer satisfaction it is necessary to be able to provide spare parts at reasonable prices during the whole usage phase of the product [1], [2], [3]. Spare parts are parts or components, which are determined to replace broken, worn out or missing parts or components in a product. [4] Spare parts management is meant to guarantee the availability and comprises all activities of a company concerning spare part supply, e. g. spare parts manufacturing or the distribution of spare parts [1]. Within the spare parts management electric and electronic components become more and more important [5]0. The reason is the constantly increasing value share of electronic and electrical components in many long-living primary products like automobiles or machine tools. As innovations are more and more driven by electronics this trend will continue in the future [5]. Until the year 2015 almost a third of the value creation in automotive engineering will for example arise from electric and electronic components (figure 1) [6].

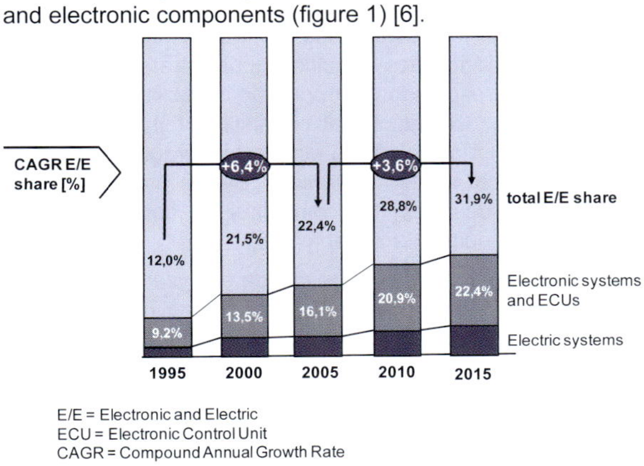

Figure 1: Rising share of electronic and electric [6]

However, the increased application of electric and electronics components poses new challenges for spare parts management [1]. The short product life cycles of electronic components, for instance, entails problems concerning the long-term availability. The components are mainly developed and produced for the consumer electronic and information technology. These branches are characterized by short product life- and innovation-cycles so that components are produced only during a short time frame. The automobile industry only has a market share of 7% of the global semiconductor market [7], which means that only in few cases the market power is sufficient to enforce an extended production of semiconductors. The discontinuation of electronic parts is therefore a constant problem. Furthermore, the demand forecasts for long supply periods of up to 25 years in the automobile industry have a high uncertainty as there are no empirical values concerning the failure performance of newly developed electronic control units (ECU) [1], [8]. Further challenges of planning are technological restrictions like the unknown suitability for storage of parts or components. After long time storage the components might not be processable anymore due to corrosion or the operability might be affected [9]. Therefore, the planning of long-term spare parts supply is a complex problem for many companies from different branches.

The use of electric and electronic components leads furthermore to a big challenge in terms of sustainability [10], [11]. As shown in figure 2 sustainability has three dimensions. The social, the ecological and the ecological impact of business decisions have to be considered. For each dimension there are several key figures to evaluate the impact. For the social dimension exist for example the human development index, the global compact or the Gini Index [10]. In this paper the focus is on the ecological impact, but also in this area is a wide range of key figures, which try to evaluate the impact. They evaluate the used resources, the pollution connected with the production and the use of the products or the toxicity [11]. A single key figure to evaluate the ecological impact does not exist.

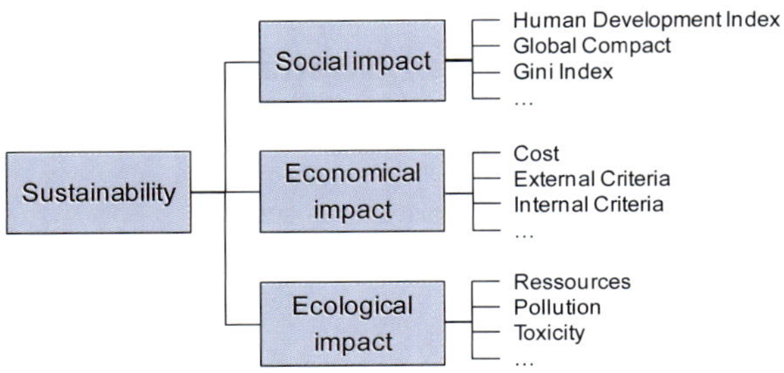

Figure 2: Sustainability [11]

The production of electronics requires a high effort in energy and resources. Moreover, often toxic substances are used in the production which means that the recycling of these materials requires knowledge and technical equipment [12]. According to Williams et al. the production of a single 32 MB DRAM Chip needs 1.600 g fossil fuel, 72 g chemical inputs and 32 liters of water [13]. These facts have also been recognized by the international law which gives strict guidelines for the production of electronics for example by directives such as the WEEE (Waste Electrical and Electronic Equipment) or the RHOS (Restriction of Hazardous Substances) in the European Union. In the USA or Japan similar settlements exist [14], [15]. The efforts for sustainability, however, may not end with the series production but have to consider also the post-series supply. In this article will be shown how ecological aspects can be taken into account in the process of planning the post-series supply.

2 STATE OF THE ART SPARE PARTS PLANNING

During the serial production of the product in most cases the necessary spare parts are manufactured on the regular serial facilities. [3] After the End of Production (EOP) a strategy change is necessary as the facilities are designed for mass production and cannot be run economically given the small lot sizes of spare parts during the long supply period.

2.1 Supply Strategies

The planning goals of the spare parts management are deduced from the business objectives [16]. Planning goals may be costs, risks, flexibility and customer relationship. In industrial practice the post series supply after the EOP is guaranteed on a long term basis by means of six different strategies:

1. Compatible successive product generations,
2. storing of a final lot,
3. periodical internal production,
4. periodical external production,
5. reuse of used components and
6. repair of used components.

A detailed description of the supply strategies and their corresponding advantages and disadvantages is given in [1], [3], [4] and [17]. Main points in regard to the ecological sustainabilty is that the first four strategies will lead to the recycling of used components, meanwhile the supply strategies reuse and repair comprise the remanufacturing of used components.

Due to changing general conditions during the supply period, e. g. a decreasing demand or a higher forecast confidence, it is often necessary or economically worthwhile to change between different strategies. In this way several strategies are combined to a supply scenario in order to cover the whole supply period. [1]

2.2 Planning Process

The long period of the post-series supply and the multitude of influencing factors create a high planning complexity. Furthermore interdependencies between products exist. Is a manufacturing facility just used for the periodical production of one type of spare parts, the resulting machine hour rate is high. If other spare parts are also produced on this facility the overall costs may be lower, even if for the other spare parts a supply strategy like the repair would be preferable. In order to control this complexity a systematic planning process for the development of strategies for post series supply and for its control is necessary [16]. Figure 3 depicts the current planning process, which can be divided into six steps.

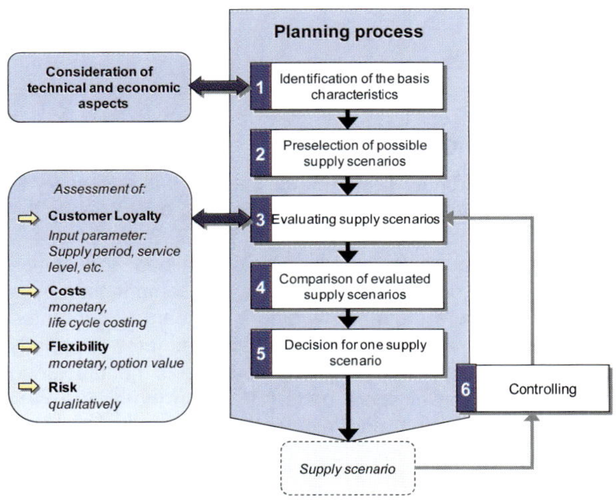

Figure 3: Planning Process [18]

During the first step the basis characteristics related to the component are identified. In industrial practice technical and economical aspects are focused as for example suitability for storage, market data, an estimation of the supply period and the prognosis of future demands. With the information about the basis characteristics specific supply strategies can be evaluated with respect to their principle suitability in step two. If, for instance, the functional capability of the assembled component is not given after a long-term storage the strategy "storing of a final lot" is not feasible. By the pre-selection of the possible scenarios the complexity of further planning steps can be reduced remarkably. [18]

The remaining supply scenarios can be analyzed in detail and evaluated within the detailed planning during the third step. Hereby, a cost as well as a risk evaluation (e. g. probability of a component discontinuation notice or probability of non-availability) is carried out. Furthermore, the influence of possible scenarios on customer satisfaction and thereby customer loyalty are considered. Eventually, flexibility is a further important aspect for the decision for a supply scenario. How fast a change to another scenario can be carried out and how much costs will arise from this action. [1], [18]

The supply scenarios evaluated are compared to each other by means of a benefit calculation in the fourth step. After a comparison of all alternatives the most promising scenario is determined and implemented (step 5). Due to the long supply periods decisions made at this point in time – as for example the scrapping of production lines and testing facilities, which are no longer in use - might have negative impacts on the economical efficiency of the company several years later. However, the selection is not a static process, which is only carried out once. Due to the multiplicity of constantly changing internal and external influences the reasonability of the chosen scenarios is controlled continuously (step 6). [18] By this means the economical efficiency and long-term availability of spare parts can be guaranteed. External influences like component discontinuation notices or changed forecasts of future demand have to be taken into account immediately during the planning process. However, also internal influences like for example damaged production lines or changes in cost structures have to be considered within the control of the post series supply.

3 SOLUTION APPROACHES

Up to now there are no systematic approaches taking into account the ecological impact in spare parts management. On the one hand, there is the need for a reactive approach for products already available in the market, which are already in the focus of spare parts management or will be in near future. The range for solution is strictly limited. Because of the commitment to the customers that spare parts are available for years, this problem will be current for a long time. On the other hand, a proactive approach for future parts that are still in the developing phase is necessary. For this the product development needs to take care of the requirements for a sustainable post series supply. The range for solution is much larger since many determining factors for the post series supply are defined during product development as for example the type and characteristics of employed electronic parts.

3.1 Reactive Approach

As pointed out in paragraph 2.2 the process of planning and control of the post series supply comprises exclusively economical and technical restrictions. Due to the high ecological impact of electronics it would be necessary to consider aspects of sustainability already during the identification of basis characteristics (step 1) in order to take them into account in the pre-selection. These criteria would be for example the resource, energy and water consumption during the manufacturing processes in contrast to the usage during the remanufacturing process. A further point is the formation of harmful substances during soldering or etching. According to the characteristics of a specific part, some supply strategies may be excluded early in the preselection (step 2).

During the detailed planning and evaluation of the supply scenarios (step 3) a thorough analysis of the environmental impact of the strategies can be done. For this purpose there are several established methods as for example the carbon footprint or the global warming potential. In addition to the costs and risks implied by the supply scenario chosen the sustainability of the approach can be taken into consideration (step 4). The criteria for the selection of a supply scenario therefore should be:

- Cost
- Risk
- Flexibility
- Customer Satisfaction
- Ecological Impact

For a systematic consideration it is useful to develop a standardized process, in which, for example, values for the achievement of the objectives are given and an objective and comprehensible decision for a supply scenario can be guaranteed (step 5). The continuous control for necessary adjustment (step 6) should be done with regard to the ecological impact.

3.2 Proactive Approach

Instead of reactive measures in planning and control of post-series supply a product design taking into account sustainability of post series supply is a promising approach. In product development only a relatively small value of about 5-10% of the lifecycle costs arises but up to 75% of the cost in later life cycle phases are set [19]. Taking into account the ecological damage by the products this percentage is estimated to be much higher [12]. However, not only the costs for the product are set during the product development to a large extent, but also the impact on the environment.

This results in the approaches of the Design for Environment and Design for Recycling [20]. In this approaches the impact of the product on the environmental sustainability are taken into account systematically. However, the post-series supply is often not considered. Due to the increasing share of electronic, the spare parts are gaining importance and the product development should regard the requirements of the post-series supply.

The objective of the product development for post-series supply is the early definition of a framework for an effective and economic spare parts management [5]. To reach this goal the aspects of the post series supply should be considered using the degree of freedom given by the product specification. Main points are the continuous regard of the defined requirements and the consideration of the chances and risks resulting of the product draft.

The preferred supply strategies have different requirements to the product development. It is important to identify a supply strategy for a product to be developed, which is not only based on economical aspects and technological restrictions but also offers potentials for the ecological sustainability. To support the product development general effects of each post-series supply strategy have to be identified.

4 ECOLOGICAL IMPACT OF SUPPLY STRATEGIES

The six supply strategies will be described accordingly to their environmental impact, their specific advantages and disadvantages and the derived requirements for the product development. The schematic evaluation of post series supply is a first step towards the evaluation of sustainability. This does not imply an evaluation of the advantages of supply strategies in general. It needs further detailed product specific analyses comprising an objective evaluation of product characteristics for every application.

4.1 Compatible successive product generations

This supply strategy implies that the next product generation can be used as spare parts for the product generation before. Therefore the spare parts can be easily manufactured during serial production of the post processor. This is mostly very economical, the investments and the risk are low. A major disadvantage is the restriction of innovative changes between the product generations. In terms of environmental impact it has to be stated, that a reuse or repair of used components is generally preferable. Otherwise the used parts have to be recycled, which will also consume energy and water.

Requirements for the product development

The post processor has to fit in the installation place of the considered part. Furthermore the interfaces (e.g. power supply, communication) must be identical or compatible between the product generations. Otherwise an adapter has to be used. A possible solution approach is the division in strictly defined modules that can be easily exchanged. There has to be the possibility to switch additional function of the next product generation off, when the part is used as spare part.

4.2 Storing of a final lot

In this strategy the estimated demand for the whole supply period is produced at once and stored. The essential difference from the other strategies is the fact that demands are estimated for a long period of time. This implies a higher uncertainty and a lower confidence for the prognosis. This might result in excessive inventories, which could lead to disposal of surplus components. Of course the disposal is a major disadvantage. Furthermore the storing conditions of electronic components are particular concerning temperature or humidity. Certain products can only be stored in a nitrogen atmosphere or need a regular power supply in order to remain functional. To establish these conditions high energy consumption is necessary.

Requirements for the product development

The product development has to use parts with a guaranteed functionality after long-term storage. Electronic parts like aluminum electrolytic capacitor, flash memories etc. should be avoided.

Furthermore the requirements during the storage period should be low. For example plastics often discolor or get breakable if the storage conditions are not optimal. Especially for plastics seals this is important. Often it is necessary to test the parts before distribution, if the functionability is given. Therefore an option for testing the devices is required.

4.3 Periodical internal or external production

Spare parts can also be manufactured in small batches demand-driven. This can be done internal or external. In regard to the ecological impact that does not make any difference therefore both strategies are summarized in this chapter. In relation to the above described strategies the internal or external production requires to keep the manufacturing and testing facilities.

Requirements for the product development

The used electronic components must have a long-term availability i. e. standard components should be preferred. The manufacturing and test facilities must be available for long periods. Specialized facilities that were used for a component during the serial production cannot be used effectively due to the low utilization in the post serial production. It is therefore necessary that the components are designed in a way, that standardized manufacturing and test facilities can be used.

4.4 Reuse of used components

Reuse means that the electronic components are recovered from the primary product after the usage phase. The components are transported afterwards to a facility where the functionality is tested, the component is cleaned, packaged and stored until delivery. By means of the reuse of the product during the post series ecological advantages arise as no resources, component as well as production resources, need to be provided. Negative aspects are the packaging and the transportation after the recovery. Especially if the parts are distributed worldwide this may lead to a negative environmental impact.

Requirements for the product development

This strategy has not much requirements for the product development. Solely the testing of the used components without opening of the housing should be possible to detect and separate the faulty components.

4.5 Repair of used components

The repair of products for post series supply is basically affected by the same general conditions as reuse. Deviating from reuse the strategy of repair entails further ecological disadvantages resulting from the repair process itself. Production expenses for e. g. energy, water and auxiliary material for the brazing process and components are to be taken into account. As not every redelivered part is reparable, either from an economical or a technological point of view, many redelivered components are scrapped at that point in time. Hence, the expenses for the return transport were useless. In case the production is accomplished in low-wage countries because of a high percentage of manual work there might be lower environment regulations so that production might release harmful substances to the atmosphere.

Requirements for the product development

For the repair process an easy opening of the housing is necessary. Therefore a housing, that can be closed with screws or a snap fit are preferable against gluing or sealing the housing. The detection of the error and the derivation of the concerned electronic elements should be easily conducted without opening the housing. In that way, every irreparable component can be excluded immediately. Often electronic components are glued for the purpose of heat transfer or to safeguard against humidity. Of course this will prevent any repair action.

4.6 Conclusions

The two strategies reuse and repair show considerable parallels. Current and impending product take-back legislation supports these two strategies. While this type of legislation is often negative for manufacturers, who prefer the recycling or disposal of their products, remanufacture provides a potentially profitable end-of-life alternative. Especially the environmental impact of both strategies is in general lower than the other supply strategies. [21], [22]

5 SUMMARY

The spare parts management of electronic parts gains more and more importance for manufacturing enterprises due to a high profitability and a constant demand. The increased use of electronics poses new challenges in order to guarantee a long-term availability. Today the supply scenario for spare parts is often determined at the end of serial production and in the decision process mainly technological and economical aspects are considered. The ecological impact of the supply strategies is not in the focus of the enterprises. Due to the high ecological impact of electronic components it becomes necessary to establish a spare parts planning with consideration of sustainability. For existing products reactive measures in the actual planning process like a rough evaluation of the strategies and analysis of the environmental impact like the carbon dioxide emission are necessary.

For products that are still in the development process proactive measures should be initiated. The large influence of the product development in terms of costs and sustainability of the products leads to the approach of a product development according to the demands of post-series supply. In the paper the environmental impact and the main requirements for several supply strategies are described.

6 REFERENCES

[1] Bothe, T., 2003, Planung und Steuerung der Ersatzteilversorgung nach Ende der Serienfertigung, Dissertation Technische Universität Braunschweig.

[2] Rustenburg, W., van Houtum, G., Zijm, W., 2000, Spare parts management for technical systems: resupply of spare parts under limited budgets, IIE Transactions, 32:1013-1026.

[3] Inderfurth, K., Mukherjee, K., 2008, Decision support for spare parts acquisition in post product life cycle, Central European Journal of Operations Research, 16: 17-42.

[4] DIN 24220 Teil 1, 1976, Ersatzteillisten – Allgemeines, Beuth Verlag, Berlin.

[5] Dombrowski, U.; Weckenborg, S.; Schulze, S., 2010, Method-supported product development for post-series supply, 20th CIRP Design, Nantes, France.

[6] Bauer, P., 2006, Wie Elektronik die Automo-bilindustrie verändert, Gottschalk, B., Mastering the Automotive Challenge, Sueddeutscher Verlag, München, 269-284.

[7] ZVEI-FV, 2010, Electronic Components and Systems, ZVEI-Mitteilungen 6/2010.

[8] Kennedy, W., Wayne Patterson, J., Fredenhall, L., 2002, An overview of recent literature on spare parts inventories, Int. J. Production Economics 76:201-215.

[9] Dittmann, K., 2007, Alterungsmechanismen und optimales Einlagern von Ersatzteilen, Plus 6.

[10] Lye S.W., Lee S.G., Khoo M.K., 2002, ECoDE – An Environmental Component Design Evaluation Tool, Engineering with Computers, 18.

[11] Finkbeiner, M., 2007, Nachhaltigkeitsbewertung von Produkten und Prozessen, Tagungsband Produktionstechnisches Kolloquium, 12:123–136.

[12] Nagel, M., 2003, Managing the environmental performance of production facilities in the electronics industry, More than application of the concept of cleaner production, Journal of Cleaner Production, 11:11–26.

[13] Williams, E. D.; Ayres, U. A.; Heller, M., 2002, The 1,7 Kilogram Microchip, Energy and Material Use in the Production of Semiconductor Devices, Journal of Environmental Science and Technology, v. 36, n. 24: 5504-5510.

[14] Ciocci, R.; Pecht, M., 2006, Impact of environmental regulations on green electronics manufacture, Microelectronics International Volume 23, 2:45–50.

[15] Seliger, G.; Kernbaum, S.; Zettl, M., 2006, Remanufacturing Approaches Contributing to Sustainable Engineering, Gestao & Producado, v.13, n.3:367-384.

[16] Dombrowski, U.; Bothe, T.2003, Ersatzteilversorgung nach Ende der Serienfertigung, Industrie Management 19, 4:19-22.

[17] Dombrowski, U., Schulze, S., Wrehde, J., 2007, Efficient Spare Part Management to Satisfy Customers` Need, Proceedings of the International Conference on Service Operations and Logistics and Informatics, 304-309.

[18] Hesselbach, J.; Dombrowski, U.; Bothe, T.; Graf, R.; Wrehde, J.; Mansour, M., 2004, Planning Process for the Spare Part Management of Automotive Electronics, WGP Production Engineering XI/1:113-118.

[19] Wiendahl, H.-P., 2008, Betriebsorganisation für Ingenieure, 6. vollständig überarbeitete Auflage, Carl Hanser Verlag München.

[20] Herrmann, C., 2003, Unterstützung der Entwicklung recyclinggerechter Produkte, Vulkan-Verlag Essen.

[21] Williams, J., Shu, L. H., 2001, Analysis of Remanufacturer Waste Streams Across Product Sectors, Annals of the CIRP Vol. 50/2.

[22] Hansen, U., 2000, Reverse Logistics is the key for remanufacturing and a sustainable development. IEEE International Symposium on Electronics and the Environment, 238-242.

8 Logistics andGreen Supply Chain Management

Logistics and supply chain management include activities required by companies in a value creation network to bring a product or service to market. The design of supply chain networks must not only seek economic benefits for shareholders and value for customers, but also must assure environmental protection and societal impacts affecting all stakeholders. In the first section of this chapter Shuaib et al. present a product life-cycle based framework for sustainable supply chains. In the second section Straube and Doch integrate a view on ecological footprint analysis and decision-making in logistics. In the third section Mohammed and Sadique introduce the concept of lean value stream manufacturing approach. In the fourth section Tracht et al. investigate the effects of non-homogeneous demand and the resulting queue in the repair shop. In the fifth section Helmig et al. present a dynamic and quantitative approach for evaluating the sustainable application of logistic concepts in networks. In the sixth section Al Dhaheri and Diabat introduce a multi-product capacitated inventory-location model with risk pooling and CO_2 emissions considerations. In the seventh section Bushi focuses on a cradle-to-grave life cycle assessment that evaluates the potential environmental impacts of the magnesium front end auto parts over its full life cycle. In the eighth section Schuh and Wienholdt develop a holistic model consisting of all elements for the subsystems of a production system in spare parts industry. In the ninth section Schuh et al. outline detailed results of designing a methodological framework for implementing an integrative reverse supply chain in case of manufacturing companies.

Design and Performance Evaluation of Sustainable Supply Chains: Approach and Methodologies

M. Shuaib[1], H. Metta[1], T. Lu[1], F. Badurdeen[1,2], I.S. Jawahir[1,2], T. Goldsby[3]

[1]Department of Mechanical Engineering, University of Kentucky, Lexington, USA

[2]Institute for Sustainable Manufacturing, University of Kentucky, Lexington, USA

[3]Gatton College of Business & Economics, University of Kentucky, Lexington, USA

Abstract:
Supply chains are networks of companies engaged in activities required to bring a product (or service) to market. The design of products, processes and systems in sustainable supply chains (SSCs) must not only seek economic benefits for shareholders and value for customers but also manage environmental and societal impacts affecting all stakeholders. Realizing this goal mandates considering activities that span the entire life-cycle of a product which, in reality, are conducted by many companies in the supply chain. This paper presents a product life-cycle based framework for SSCs and examines the modeling capabilities necessary to achieve sustainability goals. On-going research to develop models for SSC design and management are also presented.

Keywords:
Sustainable supply chains; Ontology; Performance metrics; Coordinated design

1 INTRODUCTION

Traditionally, the design and management of activities across the supply chain has been focused solely on increasing economic gains. To ensure triple bottom line [1] benefits economic, environmental and societal implications of supply chain activities to all stakeholders must be considered. Therefore, sustainable supply chains (SSCs) require a much broader focus that integrates a number of different aspects.

If companies operated independently without consideration for activities taking place upstream or downstream in their supply chains, it will only lead to sub-optimization of sustainability benefits across those supply chains. Therefore, a holistic systems-based approach is necessary for SSCs where all the SC partners' activities are integrated. One approach to achieve integration is adopting a product life-cycle perspective. A product goes through four stages during its life-cycle: (1) pre-manufacturing, (2) manufacturing, (3) use, and (4) post-use [2]. Different companies across the supply chain engage in the activities in each of these four stages. Thus, sustainability benefits cannot be achieved in SSCs unless all the four stages are considered collectively when making decisions relevant to activities in any one of the stages. In conventional supply chain management the focus has been limited to the first three stages. Even within these three stages companies operate independent of each other without much consideration to what happens to their inputs and outputs upstream or downstream in the supply chain respectively. Sustainability in supply chain operations cannot be achieved with such an open-loop approach. Instead, a closed-loop material flow-based, integrated approach that balances activities across all the four life-cycle stages is required to achieve true sustainability in SCs.

Product life-cycle management (PLM) tools seem to have the capability necessary to enable the coordination and integration between SC partners. Current PLM practices, however, consider only a single product life-cycle as opposed to the multiple life-cycle perspective needed to enable recovering a product at its end-of-life to extract materials, components and possibly reusing the product itself, with refurbishing as needed. One approach to sustainable manufacturing that facilitates multiple life-cycle closed-loop material flow is the 6R methodology. This methodology extends the previous 3R's of reduce, reuse and recycle to include recover, redesign, and remanufacture [3]. The 6R's provide a framework to implement closed-loop flow and connect the activities and information sharing across all the four life-cycle stages.

Thus to achieve the desired performance in SSCs, there is a need to integrate activities across the four product life-cycle stages using a framework such as the 6Rs methodology as illustrated in Figure 1. Based on this product life-cycle based approach SSC management can be defined as the *"the planning and management of sourcing, procurement, conversion and logistics activities involved during pre-manufacturing, manufacturing, use and post-use stages in the product lifecycle in closed-loop through multiple lifecycles with seamless information sharing about all product lifecycle stages between companies by explicitly considering the social and environmental implications to achieve a shared vision"* [2].

2 MODELING SUSTAINABLE SUPPLY CHAINS

The challenge to developing SSCs arises partly due to the need to integrate the multiple facets described above. Even addressing each factor (TBL benefits/impacts, total product life-cycle emphasis or closed-loop multiple life-cycle flow) independently to model the SC and assess the impacts is quite complex. This complexity increases exponentially when multiple aspects (say, modeling TBL impact across four product life-cycle stages) are integrated. Often this leads to many difficult to quantify aspects being overly simplified or completely overlooked. Moreover, when it comes to relevant SC sustainability concepts and criteria, a common understanding appears to be lacking; different enterprises tend to have varied interpretations of what factors are important to promoting sustainability, how they are related or how/whether addressing environmental and/or societal considerations influence economic performance or vice versa.

Such misunderstandings are a drawback to coordination, integration and interoperability between companies in the SC which is vital to adopting the total life-cycled based approach in SSCs. Thus, one of the primary requirements to promoting SSCs is a common knowledge framework that can increase integration and avoid sub-optimization of performance.

Ontology could help to overcome misunderstandings about the concepts and structure of SSCs, and to develop a comprehensive framework that integrates both the concepts and the knowledge and information across the SSC. While ontology promotes a common understanding, in any domain what gets measured gets managed. Therefore, comprehensive performance metrics that enable enterprises to manage operations to achieve the sustainability goals is another requirement to promote SSCs. It is also necessary to design products, processes and the systems that provide the capabilities to achieve desired performance as measured through the metrics. Predictive models to assess the impact of product and process design on SC design and operations, to identify SC partner capabilities and to evaluate the impact of numerous risks on SSC behavior are also required. This paper identifies capabilities necessary to develop SSCs, presents findings of on-going research in the above areas and further research necessary.

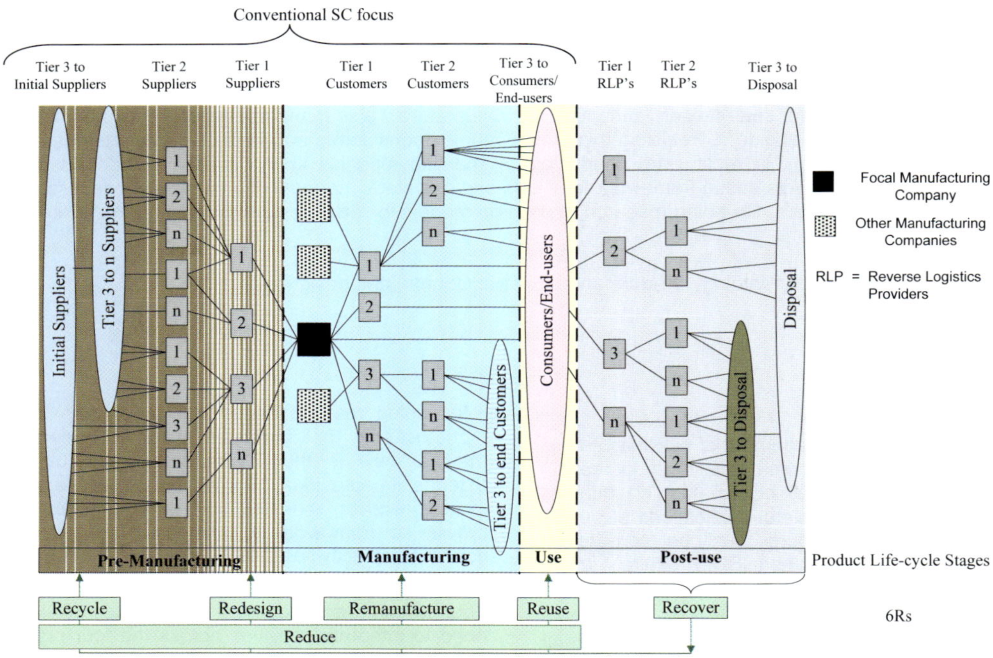

Figure 1: Integrated Framework for SSCM [2]

3 SUSTAINABLE SUPPLY CHAIN ONTOLOGY

The globalization of markets has lead to a significant increase in the size and complexity of SCs, the range of stakeholders affected by them and the nature of activities and systems that must be monitored and managed. As such, integration and interoperability is essential to promote sustainability in SCs; establishing standardized and consistent interfaces within and between enterprises in the SC, timely and comprehensive information and knowledge sharing and balance between multiple objectives are all important in this regard.

Unfortunately, a comprehensive framework for SSCs that integrates all the relevant concepts (for example, the TBL criteria, 6R methodology, all stages across product life-cycle, SC processes, product and process dimensions, etc.,) does not exist. The development of such a framework would promote sustainability in SCs and would enable integration, interoperability and information sharing across today's complex SC networks.

3.1 Ontology to Achieve Interoperability

Ontology is "an explicit specification of conceptualization" [4]. It can be described as "semantic primitives that specify a particular domain of knowledge". It is not a product, but a tool to provide an input to decision-making and management processes [5]. Ontologies are developed for complex domains to [6]:

- Share a common understanding of the structure of information within a group
- Enable the reuse of domain knowledge to integrate or reapply existing ontologies in different domains
- Make explicit, easy to update domain assumptions
- Separate the domain knowledge from the operational knowledge to enable the application of a single ontology to multiple domains
- Analyze domain knowledge based on declarative specification of the terms used in an ontology

Developing an ontology as a tool to achieve these objectives will enhance integration and interoperability between enterprises in SSCs. Developing such an ontology requires: 1) Comprehensive review of literature, practices, regulations, etc. to define classes/sub-classes of concepts relevant to SSCs; 2) Developing a taxonomic hierarchy based on review, brainstorming, consulting, etc.; 3) Defining slots (properties and attributes) and describing allowed values for these slots; 4) Validation of the model with industry partners.

In an effort to develop ontology for SSCs, comprehensive reviews of existing SC management models and the assimilation of concepts relevant to SSCs was conducted initially [2,7]. This helped develop a preliminary classification based on the key concepts and identify the following main classes/sub-classes for the SSC ontology: TBL areas of economy, environment and society; SC performance drivers within each TBL area (these become sub-classes within the TBL areas); the product life-cycle stages; SC management processes and the enterprises that constitute the SC itself [8]. The proposed domain and main concepts for the ontology are presented in Figure 2.

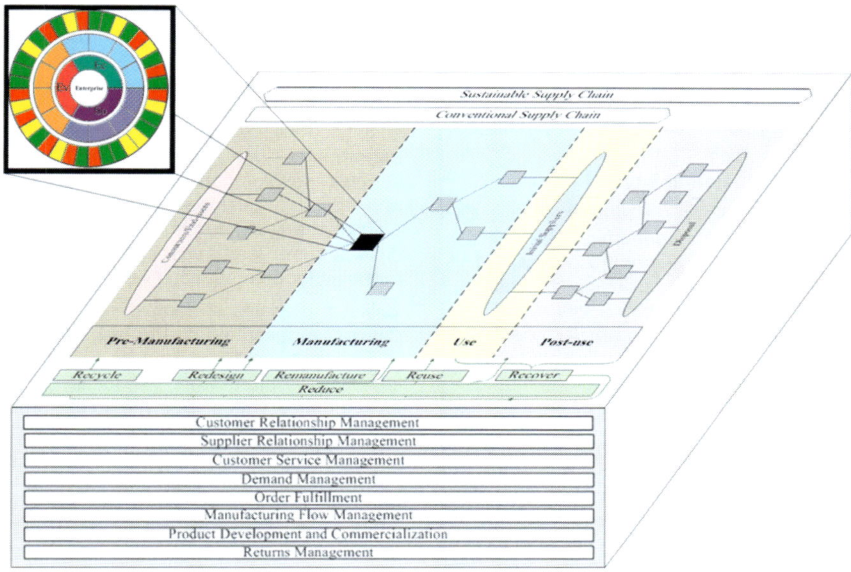

Figure 2: Representing Relations in SSCM ontology

Work is on-going to refine the class and sub-class structure, defining properties/attributes and permissible values for them and to expand its focus from the higher conceptual level to the product and process levels. While numerous ontology development tools are available, Protégé for was chosen for this research due to its ease of use, support for developing large ontologies and building end-user applications [9,10].

Validation of the SSC ontology is planned subsequently with support from industry partners. This will enable verifying and evaluating the ontology use for SSC performance evaluation, predictive modeling and to improve communication and integration.

4 SUSTAINABLE SUPPLY CHAIN PERFORMANCE METRICS

Performance metrics play a very important role in assessing a system's performance, ensuring that progress is in the right direction and provide an opportunity for continuous improvement. Therefore, for successful SSC management there is a need for developing effective performance metrics that assess the overall SC performance. While the importance of comprehensive metrics for SSC performance is widely discussed such metrics at the SC level are still lacking. In order to effectively measure the performance of SSCs there is a need to develop a hierarchical framework of metrics that can enable measurement at the process, product, enterprise and SC levels. The aggregation of metrics at the lower levels (say process) must enable assessing performance at the higher levels (enterprise and SC) and vice versa. Also, because each enterprise in the SC will have a portfolio of products, it is also necessary to be able to aggregate and segregate metrics along these dimensions as illustrated in the sustainability metrics hierarchy in Figure 3.

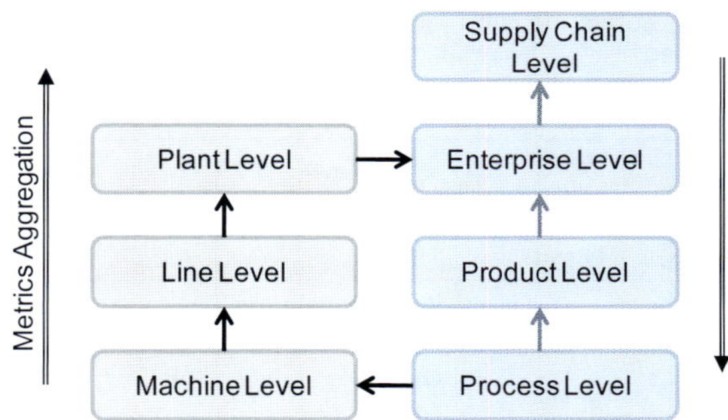

Figure 3: Sustainability Metrics Hierarchy

The development of process and product level sustainability metrics, following the systems based approach discussed in section 1, is presented in a separate paper [11]. These metrics must then be extended to define comprehensive metrics for the enterprise level and aggregated, as appropriate, for the SC level. Thus, as the first step towards developing SSC metrics, here we present the approach and sample metrics for enterprise level sustainability evaluation.

4.1 Enterprise Level Performance Metrics

Corporate reporting of sustainability performance has increased in recent years with more companies publishing either a corporate social responsibility (CSR) or a sustainability report in addition to the annual report. However, a review of such reports quickly reveals the large variation in the metrics used, making an objective comparison of performance across companies nearly impossible. One of the widely used is the Global Reporting Initiative (GRI) framework which has 70 indicators [12]. There are also numerous other frameworks such as the Dow Jones Sustainability Index criteria or the OECD core indicators. For an objective enterprise performance evaluation that will then show the way to necessary improvements at the product, process and systems level the sustainability metrics must be based on an integrated framework; the framework must be consider all relevant concepts and their relationships, as captured, for example, through an ontology. Thus, metrics at the enterprise level must be an integration of those at the product level (assimilated across multiple products or at plant level and aggregated for all plants). Initial efforts to develop enterprise level metrics (metrics are more quantifiable than indicators) that are based on a comprehensive conceptual framework that evolves from the concepts identified in the ontology and that enables predictive performance modeling are presented in this section.

4.2 Metrics for Economic Sustainability

A large number of economic metrics are already in use. Of these, the most relevant metrics for economic sustainability evaluation must be singled out. The balance score card [13] criteria of the customer, internal business processes, financial, and learning and growth provide a well accepted framework that can be used to organize economic sustainability metrics. We identify multiple metrics along each of these criteria and formulas to calculate each of the metrics.

4.3 Metrics for Environmental Sustainability

For sustainability evaluation, the environmental impact has been the single most focused on aspect. This is partly due to the relative ease quantifying environmental impacts. In order to identify comprehensive metrics, a comprehensive review of existing frameworks, various life-cycle assessment (LCA) tools and ISO standards were reviewed. Three broad criteria –resources, energy and residues—were chosen to organize the enterprise level sustainability metrics. For each of these metrics corresponding formulas for computations were developed.

4.4 Metrics for Societal Sustainability

Societal metrics are the least developed and reported within any of the existing sustainability measurement systems. One reason is the absence of any definite standards to measure societal sustainability; the fact that they are less quantifiable has also made it difficult to define objective metrics. While guidelines such as ISO 26000 and IBM supplier conduct principles/guidelines facilitate companies to be socially responsible [14], well-defined metrics for enterprise societal sustainability evaluation are still lacking. Here the ISO 26000 guidelines were used as a basis to develop quantifiable societal metrics. First, relevant criteria to evaluate performance were identified. They are: anti-corruption, supplier development and training, employee development and training, customer satisfaction, customer awareness, compliance with policies, employee well-being, community development and diversity. Metrics and corresponding formulas for computations were then developed.

Table 1 presents a sample set of metrics (notations used cannot be described due to lack of space). The complete list of TBL metrics [15] cannot be presented here due to space limitations. Future work will focus on linking the enterprise level metrics with the lower level—product, plant, process, etc., as in Figure 3—metrics and extending them to develop supply chain level metrics. It must be noted that to achieve the desired performance in SSCs developing metrics alone is not sufficient. There is a need for identifying metrics that influence sustainability performance and modify the product, process and system designs appropriately to achieve the desired performance.

Table 1: Sample Metrics for Quantifying TBL Performance

TBL Aspect	Performance Criteria	Metric	Formula to Measure Metric	Equation	Desired Direction	Metric Number
Economic	Internal Business Process	BTO index	$\frac{\text{Number of products that are built-to-order}}{\text{Total number of product configurations}}$	$\sum_{a=1}^{F} \frac{BTO_a}{PC_a}$	↑	EC9
	Financial	Inventory turnover ratio	$\frac{\text{Total cost of sales}}{\text{Average Inventory}}$	$\sum_{a=1}^{T} \frac{CS_a}{AI_a}$	↑	EC21
Environmental	Residues	Gaseous emissions rate	$\frac{\text{Total emissions}}{\text{Total number of product types}}$	$\frac{\sum_{a=1}^{N} EM_a}{N}$	↓	EN1
	Energy	Energy savings rate	$\frac{\text{Total regenerated energy}}{\text{Total energy consumption}}$	$\frac{E_r}{E_t}$	↑	EN13
Societal	Supplier Dev. and Training	Supplier training hours ratio	$\frac{\text{Total number of supplier training hours}}{\text{Total number of suppliers}}$	$\frac{\sum_{a=1}^{T} S_{ha}}{\sum_{a=1}^{T} S_a}$	↑	SC3
	Employee Dev. and Training	Employee contribution ratio	$\frac{\text{Number of employee suggestions implemented}}{\text{Total ideas offered}}$	$\frac{\sum_{b=1}^{P} E_{sib}}{\sum_{b=1}^{P} E_{sb}}$	↑	SC7

5 COORDINATING SUSTAINABLE PRODUCT AND SUPPLY CHAIN DESIGN DECISIONS

Coordination between product and SC design decisions is one of the factors that contribute to improving SSC performance. Nearly 80% of the product's cost is determined during its design [16]. All the costs (both forward and reverse loop) incurred across the SC, too, are dependent on the product design. These SC costs are driven by the SC configuration (number and location of SC partners, their capabilities and capacities).

Therefore, decision-support models for coordinated sustainable product and SC design (CSD) that can evaluate the impact of a given set of product designs on their corresponding SC configurations can help managers identify the best product design that will bring maximum sustainability benefits to the SSC. While there has been emphasis on the importance of CSD models, very limited literature is available on how this can be approached [17,18]. A model developed to address the CSD problem is presented in this section.

SCs encompass a large number of entities with conflicting objectives spanning many geographic boundaries. Optimization even to maximize economic benefits in such a system is challenging due to the complexities involved. When performance must be optimized from multiple aspects (TBL), as in this case, the mathematical models quickly become NP-hard and are difficult to solve in real time. To overcome this challenge, a hierarchical approach is adapted here to solve the CSD problem. Thus, an economic optimization model is first developed, which selects for each alternate (varying in terms of materials used, product life, post-use capability, etc.,) product design a closed-loop SC configuration that maximizes the profit. Then, for every product design-SC configuration combination (PDSCC), a multi life-cycle economic analysis (if product life-cycle = 8 yrs, analysis done for, say, 20 yrs) is performed. All combinations are then ranked based on profit potential. The outcome at this point is a collection of PDSCCs with highest profit potential across the period considered. In the next step, a multi life-cycle environmental analysis is conducted for the PDSCCs to identify those with minimal environmental impact. In the final stage, a multi life-cycle societal analysis evaluates the PDSCCs (with best economic and environmental performance) to select from among them the best PDSCC with maximum societal performance. This CSD model provides a basis for selecting one PDSCC from among those economically viable that also enable meeting environmental and societal performance expectations. The hierarchical approach is illustrated in Figure 4 and explained below.

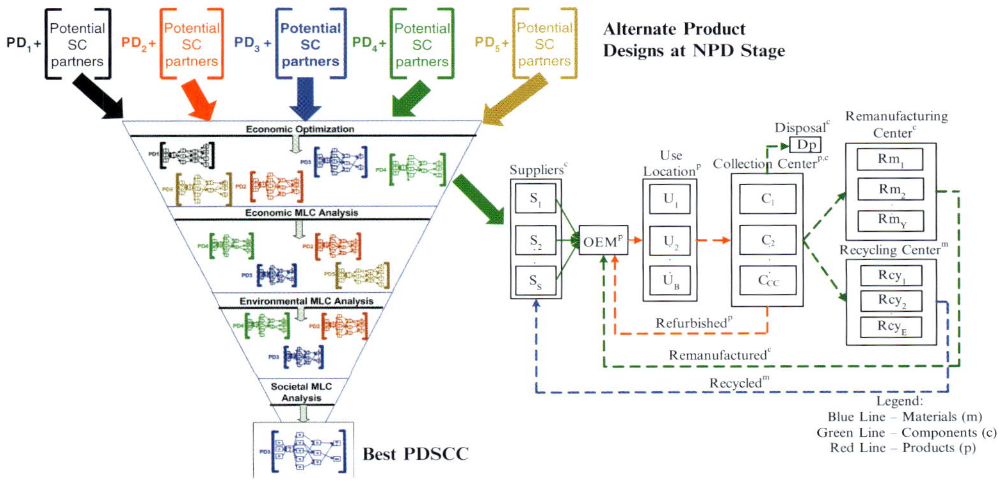

Figure 4: Hierarchical Approach for CSD Model

5.1 Economic Optimization

A detailed overview of the economic optimization model is presented in [18]. The model is formulated as a Mixed Integer Linear Programming (MILP) problem and solved using the IBM ILOG CPLEX software with an objective to maximize the profit. Some of the key aspects of this model include consideration of 6R concept of SM, different prices for new and refurbished/remanufactured products, consideration of capital and recurring costs such as assembly, processing, inventory, transportation, etc. The model is subjected to capacity, capability, balanced flow, binary and non-negativity constraints based on the demand and manufacturing requirements. This model selects for each alternate product design an optimal SC configuration that maximizes profit. Figure 5(a) illustrates the locations of all possible SC partners (and transportation costs on arrows) for a given product design. The SC configuration chosen using the MILP model is shown in Figure 5(b).

5.2 Economic Multi Life-cycle Analysis

In the economic multi life-cycle analysis, for each PDSCC, corresponding SC costs are computed and analyzed over multiple years. The output of this analysis is to identify the best PDSCCs with maximum cumulative profit.

5.3 Environmental Multi Life-cycle Analysis

In the environmental multi life-cycle analysis, the top PDSCCs from the economic multi life-cycle analysis are assessed in terms of material usage, energy consumption and carbon dioxide emissions. The output of this analysis is to select all the combinations that have minimal cumulative environmental impact.

5.4 Societal Multi Life-cycle Analysis

Here the best combinations that have maximum profit and minimal environmental impact are analyzed from a stakeholder's perspective. Factors such as employee and supplier training and development and customer satisfaction are considered in evaluating the PDSCCs. From this analysis, the best PDSCC with minimal societal impact is identified.

Therefore, decision support tools for CSD such as above can provide managers with a holistic picture of the impact of their product and SC design decisions both in short-term and long-term and also identify the best product design with highest TBL benefits. On-going research in this area focuses on applying this model for a refrigerator case-study to select the best refrigerator design that provides highest TBL benefits. Because many factors considered in the CSD model are highly variable, extensive sensitivity analysis is also being conducted to evaluate the vulnerability to such factors.

6 OTHER CAPABILITIES NEEDED TO DEVELOP SSCS

While the above sections address a number of modeling and decision making capabilities needed to develop SSCs, there are numerous other issues that must also be dealt with. The exposure to unexpected risks increases the vulnerability of operations often leading to financial losses and negative environmental and societal implications (for example, the recent BP experience with the Deepwater Horizon oil rig explosion). Therefore, comprehensive predictive modeling for risk management in SSCs is another requirement [8]. Most such risks propagate through SC partners, bringing about the need for better approaches for supplier sustainability evaluation and selection. Many other opportunities, such as potential transition to product service systems, also bring about the need for predictive models to evaluate the impact on SSCs and more research is needed to develop such modeling capability.

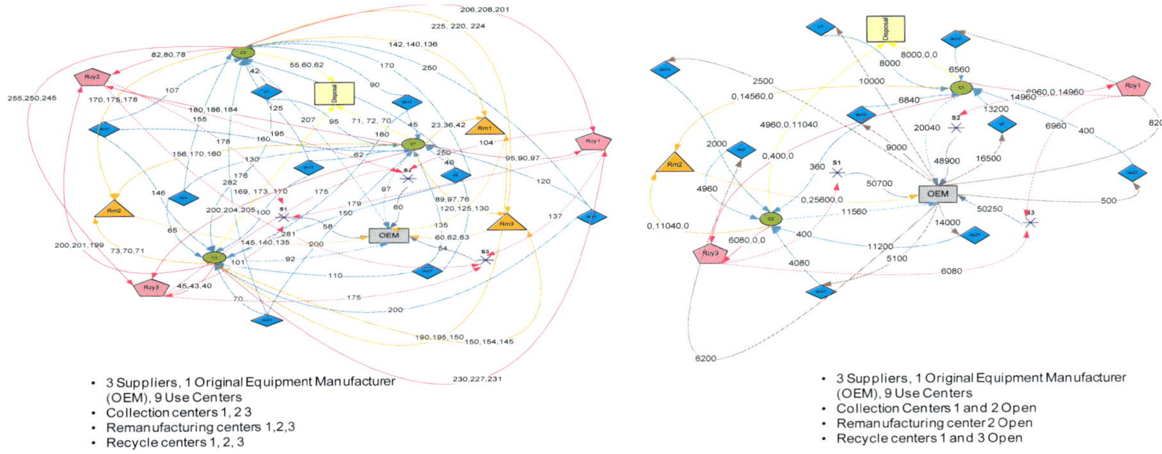

(a) Possible SC partners (b) Optimal SC partners

Figure 5: Results from Economic Optimization

7 CONCLUSIONS

This paper presented a comprehensive approach to SSCs and some on-going research to develop various modeling capabilities necessary for improve SSCs performance along the TBL areas. Further research must be conducted to continue work in the above areas and also develop models that are capable of capturing the complexities in closed-loop SSCs. These models must also be applied to real life test scenarios to validate their usefulness.

8 REFERENCES

[1] Elkington, J., 1998, Cannibals with Forks: The Triple Bottom Line of the 21st Century Business, New Society Publishers, USA.

[2] Badurdeen, F., Iyengar, D., Goldsby, T.-J., Metta, H., Gupta, S., Jawahir, I.-S., 2009, Extending total life-cycle thinking to sustainable supply chain design, International Journal Product Lifecycle Management, 4/1/2/3:49-67.

[3] Jaafar, I.H., Venkatachalam, A., Joshi, K., Ungureanu, A.C., De Silva, N., Dillon, Jr., O.W., Rouch, K.E., and Jawahir, I.S., 2007, Chapter 2: Handbook of Environmentally Conscious Mechanical Design, (Ed. Kutz, M.), John Wiley & Sons, pp. 25-65.

[4] Gruber T.R., 1993, A translation approach to portable ontology specifications, Knowledge Acquisition, 5/2:199-220.

[5] Ahmad, A., Mollaghasemi, M., Rabelo L., 2004, Ontologies for Supply Chain Management, Proceedings 13th Annual Industrial Engineering Research Conference, Houston, 2004.

[6] Noy, N., Mc Guiness, D.L., 2001, Ontology Development 101: A Guide to Creating Your First Ontology, Knowledge Systems Laboratory, Stanford University.

[7] Badurdeen, F., Metta, H., Gupta, S., 2009, Taxonomy of Research Directions for Sustainable Supply Chain Management, Proceedings of the 18th Industrial Engineering Research Conference, May 19-21, Miami, FL.

[8] Badurdeen, F., Shuaib, M. A., Jawahir, I. S., 2010, Ontology Development for a Systems Approach to Sustainable Supply Chains, Proceedings of the 4th International Conference on Sustainable Energy & Environmental Protection, June 29 – July 2, 2010, Politechno di Bari, Italy.

[9] Ahmad, A., Mollaghasemi, M., Rabelo, L., 2004, Ontologies for Supply Chain Management, Proceedings 13th Annual Industrial Engineering Research Conference, Houston, 2004.

[10] The Protégé Ontology Editor and Knowledge Acquisition System, http://protege.stanford.edu.

[11] Lu, T., Gupta, A., Jayal, A. D., Badurdeen, F., Dillon, Jr., O. W., Jawahir, I.S., 2010, Product and Process Metrics for Sustainable Manufacturing, submitted for the 8th Global Conference on Sustainable Manufacturing, November 21-24, Abu Dhabi.

[12] Global Reporting Initiative (GRI), http://www.globalreporting.org/Home

[13] Kaplan, R. L., Norton, D. P., 1996, Using the Balanced Scorecard as a Strategic Management System, Harvard Business Review, January-February 1996, pp. 75-85.

[14] Kastenhofer, K., Rammel, C., 2005, Obstacles to and potentials of the societal implementation of sustainable development: a comparative analysis of two case studies, Journal of Sustainability: Science, Practice, and Policy, 1/ 2:5-13.

[15] Badurdeen, F., Metta, H., Gupta, A., Stovall, C., Goldsby, T.J. and Iyengar, D., 2010, Metrics and framework for corporate sustainability assessment: The Enterprise Sustainability Assessment Triangle (ESAT), University of Kentucky Working Paper.

[16] Boothroyd, G., Dewhurst, P., Knight, W., 1994, Product Design and Manufacture for Assembly, New York, Marcel Dekker.

[17] Rungtusanatham, M., Forza, C., 2005, Coordinating product design, process design, and supply chain design decisions. Part A. Topic motivation, performance implications, and article review process, Journal of Operations Management, 23/3–4:257–265.

[18] Metta, H., Badurdeen, F., 2009, A framework for coordinated sustainable product and supply chain design, Proceedings of the 2nd International Conference on Value Chain Sustainability (ICOVACS), October 19-21, 2009, Louisville, KY, pp. 20-25.

A contribution to Sustainable Logistics and Supply Chain - conceptual design to evaluate ecological and economical cause-effect relations in logistics planning processes

F. Straube[1], S. Doch[2]
[1]Department of Logistics, Technische Universität Berlin, Germany
[2]International Transfer Center for Logistics GmbH, Berlin, Germany

Abstract
In the recent years environmental protection is getting to the top of the agenda in politics, society and customer behaviour. In literature the basic principles of sustainability are known for a long time. Nevertheless, actual research does not cover adequately the practical facets of ecological sustainability in logistics. Our objective is to present an integrated view on ecological footprint analysis and decision-making. Starting with an exploratory empirical survey study the actual objectives, challenges and developments of ecological sustainability in logistics were identified. Based on a brief literature review a blueprint of the methodology was developed. In an action research approach this concept was prototypical used for the calculation of current emissions of major logistics process configurations and the development of improvement possibilities out of the calculation results.

Keywords:
Green Logistics, Green Supply Chain Management, Ecological Sustainability, Green Logistics Assessment, Carbon Footprint

1 INTRODUCTION

Due to climate change, increasing environmental degradation and the depletion of natural resources, the public is developing an increased awareness for the importance of environmental and resource protection. Thus the Stern Review, published by Nicholas Stern in 2006, on the Economics of Climate Change transparently shows the economic consequences of climate change (Stern 2007). It is also clear that possible climate change would cost trillions and early prevention is more economically viable. Companies are, therefore, required to design sustainable products and services, as economic demands on account of various stakeholder groups and development are constantly gaining in importance. The same applies to the processes of value creation of a company which have to support environmentally sustainable procurement, production, distribution, use and recycling of products (Singh 2004).

The main influencing factors for the topic of “environment and resource protection” are the stakeholders within the economic environment of a company and especially the cost increasing factors, such as rising energy and commodity prices, partly resulting from the global shortage of raw materials. Recent studies show that in transport and logistics 30 percent of the total annual energy expenses of 180 billion euro are incurred in Germany (EyeforTransport 2007). Since logistics processes and transport are particularly identified with somewhat high fuel consumption, an exceptional challenge for logistics can be found here.

On the political level, this fact has been taken into account in the form of the appropriate climate change objectives, which have a direct or indirect impact on companies. The starting point for European policy on climate protection is the Kyoto Protocol, which prescribes the EU to reduce the six main greenhouse gases by 8% in the period from 2008 to 2012 in comparison to the levels of 1990. Although a continuation of international climate protection programmes after the failure of the conference in Copenhagen is uncertain, the EU continues to hold fast to its reduction targets (EU 2007). Based on a council decision, The EU has set the following binding targets for itself in March of 2007: reduction of greenhouse gas emissions by 20% compared to 1990, increasing the share of renewables for end user energy consumption to 20% and reducing primary energy consumption by 20% in regard to a projected business-as-usual scenario.

It is, therefore, necessary to meet these requirements of the stakeholders and analyse the complete life cycle of products and their entire value creation chain with regard to environmentally sustainable aspects as well as to organise them according to the wishes of the customer (Beamon 1999). Currently, it should be noted, however, that environmentally sustainable logistics services are usually limited to the measurement of CO_2-emissions and these services only occur sporadically. Accordingly, customers have not become aware of these service offers yet on a broad level and have also not demanded them. Therefore, environmental aspects currently only play a role for one third of customers for purchasing common consumer goods such as textiles or toys. However, the protection of resources has increasingly developed into a major selling point for end customers and employees as they are being sensitised to these topics through the media.

It is, therefore, the objective of this paper to enable companies to a more sustainability oriented action taking. It will be shown that it is necessary to analyse logistics processes regarding their actual emission level and to systematically identify optimization possibilities. This process of measurement and improvement has to be supported by appropriate decision support methods and systems. The main purpose in this context is to identify the factors that influence a successful implementation of ecological sustainability concepts.

This implementation affects almost all business sectors and particularly logistics. An essential requirement for logistics for example is to reduce greenhouse gas emissions. Based on estimates, up to 75% of the so-called carbon footprint, for example harmful CO_2-emissions of companies, can be attributed to the transport and logistics industry and clearly demonstrates the current need for action to increase the environmental sustainability of the logistics industry. With regard to the total CO2-emissions which, in the context of the Stern report, are broken down into the sectors of energy (24%), industry (14%), transportation (14%), buildings (8%), land use (18%), agriculture (14%), waste (3%) and miscellaneous (5%), the great responsibility of the logistics industry becomes apparent (Stern 2007). Logistics processes are not just responsible for a relevant share of transport-related emissions, but also for a part of the emissions caused by industry, buildings and waste.

Knowing about this macroeconomic share of transport-related emissions, companies are looking for hints on their individual share on total emissions. An accurate calculation of the actual emission level for individual companies and single logistics processes can lead to transparency for logistics management on that issue. Therefore ecological measurement methods and tools are on the actual agenda of most companies. While companies are able to report their current emission balance by that, little support is given for the identification of optimization potentials. Therefore, the design of a methodology that goes beyond reporting purposes is the ultimate goal of this paper. The research approach is oriented on a qualitative and applied research process: Starting with an exploratory empirical survey study the actual objectives, challenges and developments of ecological sustainability in logistics will be identified (section 3). Based on a brief literature review (section 2) a blueprint of the methodology will be developed (section 4). In an action research approach this concept will be prototypical used for the calculation of current emissions of major logistics process configurations and the development of improvement possibilities out of the calculation results in a multinational industrial company (section 5). The insights out of that practical concept application will be aggregated in a generalised methodology to measure, evaluate and improve the ecological sustainability in logistics.

2 ECOLOGICAL SUSTAINABILITY IN LOGISTICS

One of the most well-known definitions of sustainability is the one of the World Commission on Environment and Development (WCED) initiated by the UN General Assembly in 1983 (United Nations 1983). According to the commission's final report, a development is to be recognised as sustainable if it meets the needs of the present without compromising the ability of future generations to meet their own needs (United Nations 1983).

The definition, in line with the principle of distributive justice and widely used in the past, does convey some idea about the concept of sustainability, but it is too broad and unspecific for use in a micro-economic context. It is for this reason that organisations have often found it difficult to find their individual role in this overall macro-economic context.

The triple bottom line developed by Elkington provides a more detailed overview over possible characteristics of the sustainability of companies (micro-economic context). This model requires companies to implement their value creation in such a manner that capital preservation is not compromised. In regard to capital, there is a distinction between economic, natural and social capital. Accordingly, this model divides the sustainability of companies into three dimensions: social, environmental and economical sustainability (Elkington 1998).

In literature the basic principles of sustainability like the triple bottom line, the relation of sustainability and competitive advantage or the role of ethics and regulations in sustainability are known for a long time (Srivastava 2007). In logistics a wide range of sustainability topics such as design for green logistics, remanufacturing and reverse logistics, green logistics network design, sustainable procurement, sustainability KPI's or green routing and location optimization are discussed (Zhu et al. 2008). In addition, latest publications focus on interorganizational and collaborational aspects of sustainability (Vachon & Klassen 2006). In literature three different theoretical concepts do influence the discussion on environmental sustainability in logistics. General supply chain theory includes important aspects such as collaboration, process orientation and total costs but is not yet aware of the performance dimension of environmental sustainability. On the other hand environmental concepts like material flow management (Wagner & Enzler 2005) and environmental management do have that focus on the environmental perspective. Environmental sustainability can, therefore, also be understood to be the implementation of environmental and resource protection. The company management must make decisions not only based on economic considerations but also on the basis of social and environmental requirements, so a balance may ensue while meeting these requirements. While the logistics of companies since their development was geared toward efficiency and increasing economic performance and, due to social requirements, for a long time especially toward employees, the third dimension has gained considerably in importance in the recent past. Environmental sustainability as a target dimension for logistics as well as the discussion about a so-called Green Supply Chain Management has evolved since the early 90s as part of recycling management and reverse logistics for example. In the recent past, this topical area has once again gained in importance. Green supply chain management is often understood in literature as the sum of environmental management and traditional supply chain management concepts. Green supply chain management serves to reduce risks and implement process innovations. Green supply chain management also includes aspects for the design of supply and customer relationships, which result in an expansion of approaches to sustainability in the supply chain. Darnall, Jolly and Handfield researched and empirically evaluated the relationship between environmental management systems and green supply chain management. The results showed that companies using environmental management systems tend to more readily implement green supply chain management systems than those that do not use environmental management systems (Darnall et al. 2008). It was stressed here that environmental management systems in supply chain management do not affect environmental sustainability within a company but cover the entire supply chain of the company. In his discussion of current literature on green supply chain management, Srivastava (2007) comes to the conclusion that this has not yet been sufficiently established in the companies. Also, it is noted that a concise systematisation of the subject area is required to promote an increasing recognition and integration of green supply chain management systems. Other authors point to the increasingly observable degree of effort to establish green supply chain management approaches (Hoffren & Apajalahti 2009, Darnall et al. 2008, Halldorsson et al. 2009). The concept of green supply chain management is becoming an all-encompassing environmentally relevant initiative in the areas of procurement, logistics, production, distribution logistics and disposal logistics with the involvement of suppliers, service companies, distributors and end users. As key leverage for green supply chain management, cross-company co-operation with the overall goal of conserving resources is mentioned. This view is based on the fact that potentially more and more effective solutions to environmental challenges are developing through customer and supplier integration and are implementable, as it is the case when taking an isolated, business related viewpoint. It is assumed that cross-company partnerships or projects with customers result in higher quality and greater flexibility while those with suppliers primarily promise efficiency gains. A comparison between supply chain management, as a holistic optimisation and management concept, and the environmentally focused concepts of material flow and environmental management represent approaches to reciprocal integration (see figure 1).

The most significant commonality of the three management approaches is the focus on increasing the competitiveness of the company. Additionally, a consideration of resources and products takes place, and a process optimisation to improve performance is encouraged. Cooperation and communication for optimising interfaces are essential for implementation, whereby an environmentally oriented supply chain management emerges, which regards the logistical goods and information flows from an environmental viewpoint. To achieve economic and environmental objectives, a holistic view of the cost, service and environmental performance the total supply chain is needed. Environmentally sustainable logistics must act as a bridge between the resource-intensive processes in the company and the environmental demands of stakeholders (McIntyre et al. 1998). Therefore, there is already a demand during the design of a product for example and, thereby, during the designation of volume and weight, to achieve an agreement and a coordination with logistics in order to guarantee an optimal procurement and distribution (Design for Logistics) (Aronsson et al. 2006).

In total, earlier works have a limited focus and narrow perspective (Zhu et al. 2008, Darnall et al. 2008, Halldorsson et al. 2009, Aronsson et al. 2006). They do not cover adequately the practical facets of ecological sustainability in logistics and do not have the goal to present an integrated methodology to measure, evaluate and improve the ecological sustainability in logistics. Our objective is to present an integrated view on ecological footprint analysis and decision-making. This leads to the following research question: What elements and process steps should a methodology consist of, that supports the measurement, evaluation and improvement of the ecological sustainability in logistics?

3 PRACTICAL RELEVANCE AND IMPLEMENTATION STATUS OF ECOLOGICAL SUSTAINABILITY IN LOGISTICS

The majority of consumers would like companies to provide information on the causes of environmental effects (75 percent) and also to measure and record these in a well-founded and systematic manner (73 percent) (Manget et al. 2009). Despite the pronounced ecological awareness of the end users, they are not the prevailing external influencing factor affecting the environmental protection activities of companies. Rather, the companies are reacting to the specific demands of external partners and influential players. Companies view the expectations of consumers as comparatively less important (Ballas & Richert 2009).

A survey by the industrial branch association BME elucidates the current value placed on ecological sustainability in the logistics field. Accordingly, 70 percent of companies questioned attribute major importance to the acquisition and reduction of CO_2-emissions. The rationale and motivation behind environmental protection activities is for the most part concerned with the creation of a positive image for the company (85 percent) and the increasing environmental awareness of customers (75 percent). More than half of the companies (56 percent) expect a continued increase in costs for CO2-emissions and use this as a rationale to already identify potential reductions today (BME et al. 2009). 81 percent of the companies aim to steadily continue their activities in this area and some even intend to expand them. This data indicates that ecological sustainability has a weighty significance for companies even in critical times. The results of a survey of leading German industrial and consumer goods companies also confirms this (Ballas & Richert 2009)

Besides measures taken within their own companies, many businesses focus on improving the ecological sustainability of their supply-chain partners. Studies show that, in regard to the assessment of ecological sustainability of their supply chain, the most successful companies do not merely focus on the "last mile" of production, but possess an integrated view over the entire chain of production processes (Wright et al. 2009). In a survey, 45 percent of participants declared that they take the degree of environmental friendliness of logistical service providers into account when tendering for services. However, only 23 percent of those surveyed are currently willing to accept higher costs and prices for ecologically friendlier transports (Straube & Borkowski 2008).

Companies which meet these challenges with a proactive strategy are able to profit in numerous ways. On the one hand, the acquired knowledge can potentially be utilized to advise partners and other companies; on the other hand, the expansion of one's own expertise may increase company sustainability rankings on the financial markets, when the environmental topic is thematised more strongly (PricewaterhouseCoopers et al. 2009). Apart from reactive reasons, such as the prevention of risk and damage to a company's image, differentiation advantages are the main driving force behind those companies which implement ecological sustainability (ATKearney 2007 and Seuring & Müller 2008). Thereby, the perception of the demands of ecological sustainability may change: these demands may no longer be seen as a burden but as a potential competitive advantage (Srivastava 2007 and ATKearney 2007).

On the whole, studies show that mostly small companies fail to recognise the value of disclosing their CO_2-balance sheets (Straube & Borkowski 2008, BME et al. 2009, Straube & Pfohl 2008). However, as part of the delivery chain, this data will have to be provided in the future. The differences in implementation status also become apparent when examining the availability of green logistics products: while globally active companies, who do not have their own transportation fleets, are able to offer green products to 34 percent of their clients, smaller haulage contractors can only offer this to 25 percent of clients (PricewaterhouseCoopers 2009).

Besides investing in technology, such as in the regeneration of transportation fleets, the most common features are currently an increased pooling of transportation, the provision of training for drivers, the optimisation of route planning and the deployment of telematics systems. On an operative level, measures like these are relatively cost-efficient, as they quickly redeem themselves by generating savings. The transportation shift towards other carriers, such as railway or barge shipping, is practiced only rarely at the moment (BME et al. 2009). As described earlier, the environmental protection activities implemented are mostly limited to the operative level at present; only a few companies have strategic concepts for a "green" corporate orientation (Ballas & Richert 2009).

A study of trends by the German logistics association shows how difficult such a strategy formulation still is in the practical reality of logistics (Straube & Pfohl 2008): although more than 76 percent of the major enterprises questioned had established policy guidelines and visions in company policies and strategy for the protection of the environment and resources, only 53 percent of major enterprises have strategically defined environmental and resource protection for their logistics. Even fewer companies (26 percent) have operationalised environmental and resource protection objectives in their logistics systems of objectives or KPIs and defined them according to value. The main reason for this relatively small degree of operationalisation stems from the difficulty of making binding promises regarding certain environmental measurement parameters in the field of logistics. Since cost-benefit assessment procedures do not exist and no adequate measurement procedures are available the effects of logistical decision-making on measurement parameters of ecological sustainability are not transparent. Experts deem it probable that, in the future, a measurement and source-specific correlation of emissions will be established for logistic procedures, and that clients will bear the resulting expenditure for the ascertainment and designation of emissions (PricewaterhouseCoopers et al. 2009). Furthermore, experts expect that the ascertainment of CO_2-emissions is only a first step (Brown 2008). Apart form other forms of emissions, such as nitric oxide and noise pollution, environmental effects relating to resource and surface area consumption are coming increasingly into focus. The capability of disclosing extensive information will be a basic prerequisite in procuring commissions in the future (PricewaterhouseCoopers et al. 2009).

29 percent of companies view the cause-oriented measurement and allocation of CO_2-strains to be of a high and very high significance for products, but only 18 percent have so far taken up measures to implement such a measurement. The lack of standards and difficulties arising from the methodology of measurement are identified as being the reasons for the obstruction. Investigations into the relation between the realisation of specific measures and the resulting sustainability success of companies indicate that the ex-act measurement and cause-oriented allocation of emissions is of even greater importance for the practice of logistics than previously assumed (Ballas & Richert 2009). Ultimately, the cause-oriented allocation of emissions forms the basis for the product-specific designation of the CO_2-balance sheet as well as for the precise planning, governance and controlling of measures to increase sustainability on the product level. Even if presently there are numerous initiatives that push the topic forward, a lack of internationally valid and acknowledged standards for measurement and evaluation remains. Until these standards are reached and defined, various methods in the framework of studies and pilot projects need to be developed and tested in regard to their suitability (Halldórsson et al. 2009).

In total four different areas of issues can be summarized that avoid a more advanced practical implementation of ecological sustainability in logistics (see figure 2). The first one refers to missing decision support methods that enable companies to operationalize their strategic sustainability objectives to detailed KPIs and methods on the bottom-line. Therefore an assessment method should support the evaluation and improvement process rather then being limited to the measurement of the current emission status. A second reason is related to the practical implementation issues of ecological sustainability methods in logistics. An appropriate measurement method has therefore to take practical issues like missing planning data availability and compatibility issues to other functional measurement areas into consideration. Regulatory uncertainty concerning reporting standards, trading schemes and sustainability labels can be summarized as the third area of requirements. This missing standardisation and regulation requires a high methodological flexibility that an appropriate measurement method has to fulfil. The consequence of the fourth area of missing standardization can be seen in a high level of requirements on transparency and flexibility of measurement methods. An open measurement approach and the possibility of evaluating the influence of each calculation factor on the quality of the calculation result are two relevant solutions dealing with this issue.

4 REVIEW OF CURRENT MEASUREMENT METHODOLOGIES OF ECOLOGICAL SUSTAINABILITY IN LOGISTICS

Many measurement models for evaluating emissions in the field of logistics share a common focus on the processes of transport logistics. Often they have their origin in scientific institutions or organisations and have been designed for non-commercial purposes. A short synoptic view of their principle characteristic will be presented below.

GEMIS 4.5 - The Global Emissions Model of Integrated Systems is a computer programme for the comparative study of the environmental effects of energy provision and utilisation (Fritsche & Schmidt 2008). It offers a database with possibilities for equilibration and analysis for life cycles of energy, matter and transport processes as well as any desired combination of these. GEMIS was developed from 1987 to 1989 by the Ecology Institute e. V. and the University of Kassel. The model lacks the integration of traffic and street conditions into the evaluation of emission factors. Furthermore, the weight of the vehicles and their load capacity cannot be arbitrarily altered, and the energy consumption of turnover processes merely focuses on the branch of industrial suppliers.

TREMOVE 2.7b - TREMOVE is a transport and emission simulation model which has been developed by the Catholic University of Leuven and by Transport & Mobility Leuven for the benefit of the European Commission (De Ceuster et al. 2007). The model estimates the transport demand, the modal split, the emissions of air pollutants and the level of driving comfort, taking various political scenarios into account. It is a uniform simulation model which has been developed for the strategic analysis of costs and effects for a broad range of political instruments and measures. TREMOVE is applicable to local, regional and European transportation markets and is able to cover both the transport of passengers and freight of 31 countries within the time period from 1995 to 2030. A disadvantage of the TREMOVE model is its focus on the macro level. Thus, the stated data refers to European traffic as a whole and not to individual vehicles. The fact that national emissions vary further expounds the problem of an EU-wide evaluation. Furthermore, the focus is on the evaluation of emissions while considering political scenarios. Thus, political and not entrepreneurial aspects are the focal point of interest.

COPERT 4 - COPERT 4 is a software programme which evaluates air pollution emissions caused by street traffic. The technological development of COPERT is financed by the European Environment Agency (EEA), a programme within the framework of the activities of the European Topic Center on Air and Climate Change (Gkatzoflias et al. 2007). Since 2007 the European Commission's Joint Research Center has coordinated the further scientific development of the model. In principal COPERT was designed for use by national experts to estimate the traffic emissions for official, annual, national emission registries. The TREMOVE model has to a large extent inherited the vehicle emissions structure from COPERT. Consequently, the same deficits are present. Furthermore, there is a lack of factors pertaining to the surface area consumption of a vehicle.

HBEFA 2.1 - The environmental agencies in Germany (UBA), Austria (UBA) and Switzer-land (BUWAL) have for many years promoted various research projects and measurement sequences to evaluate the extent of air pollution caused by traffic and appropriate measures for pollution reduction (Pischinger 2002). The Manual for Emission Factors is a synthesis of results for these projects and provides data for specific emissions, for example emission data for individual vehicles with a high degree of differentiation. For the first time the current version 2.1 introduces a joint version for all three countries. Despite the high degree of differentiation of the HBEFA-model, it invariably assumes the weight of the vehicle and the load capacity to be based on a full load of freight. Especially the weight of the vehicle has a substantial influence on fuel consumption and the resulting emissions. Thus, this restriction is not sufficient to illustrate real transport processes. Information concerning surface area consumption is not provided.

TREMOD 4 - The emission evaluation model TREMOD (Transport Emission Model) de-scribes the motorised street, railway, ship and air traffic in Germany in matters pertaining to traffic and driving performance, energy consumption as well as the related air pollution emissions for the time period of 1960 to 2030 (IFEU 2005). It was created by The Institute for Energy and Environmental Research in 1993 and has been continually developed since that time. TREMOD is used by the Federal Environmental Agency and the various federal agencies for the preparation of legislation submitals, political decisions and environmental reporting. In the "street traffic" sector, TREMOD operates in conjunction with the Manual for Emission Factors of Street Traffic (German: HBEFA). This means that the emission factors of the HBEFA are used in TREMOD for evaluation purposes and that the HBEFA uses the calculated fleet composition to as-certain Germany's average emission factors. As a consequence of this classification, the model exhibits similar gaps. For street characteristics and traffic conditions, only one average value is assumed. There are only three categories to choose from when evaluating the segmentation of vehicle loading conditions. A more extensive segmentation would be conducive to a more precise emission evaluation. Figure 3 summarises the deficits of the individual models.

In summary, one may affirm that measurement models like GEMIS, TREMOVE, COPERT, HBEFA and TREMOD exhibit the following deficits:

- Depictions of cause and effect correlation between logistical influencing factors (for example use to full capacity, the stockpile factor of palettes) and measurement parameters are insufficient.
- There is no exhaustive consideration of all emission types and a lack of comparability or overall evaluation of all emission types.
- There is no illustration of sensitivities for environmental measurement parameters and, closely related to this point, no holistic system for substantiated derivation of tangible activity fields and measures.
- A link between environment, process and financial key figures to evaluate performance and cost for active measures and their environmental effects is also missing.
- Many measurement models are isolated and were not designed with integration into existing (logistics) controlling systems in mind.

Drawing a comparison of these deficits with the implementation issues of ecological sustainability in logistics shows the major requirements on an ecological measurement framework. A measurement methodology has to support cause and effect analyses and sustainability KPIs to overcome the operationalization gap shown in section 3. Methodological flexibility and compatibility can be concluded as the major success factors for the implementation of ecological measurement systems for decision support.

5 DEVELOPMENT OF A METHODOLOGY TO MEASURE, EVALUATE AND IMPROVE THE ECOLOGICAL SUSTAINABILITY IN LOGISTICS

5.1 Relevant emissions for Green Logistics Assessment

The point of origin for the development of a methodology for measurement, evaluation and enhancement of ecological sustainability for the field of logistics is formed by defining those environmental factors which need to be integrated. Environmental influences are understood to be the environmental strain caused by pollutant emitters. Another form of emission is noise emission. It mainly affects humans and not the environment as a whole. Significance must also be attributed to the environmental effects of surface area consumption through industry and traffic and to the effects of the fragmentation of natural landscapes. Ecological and ethical considerations reveal another form of influence, namely the use of limited, natural resources and energy sources. Pollutants are generally regarded as substance emissions (of natural and anthropogenic origin) which damage an organism or system. In this case these may be substances which previously have never or seldom been present in a system, but they may also be emissions which disturb the balance of a system through an increased influx of a pre-existing substance (Fonger 1993). The main focus of logistics is on the problems arising from air pollution of energy conversion through combustion engines, factories and production. The most important substances in this context are: the nitric oxides (NOx), carbon dioxide (CO_2), dust and soot particles, and ozone near the ground (as a photochemical product of other pollutants) as well as the fleeting organic compounds (VOC), caused by incomplete combustion. The main sources of nitric oxides can be traced back to traffic (yet with a greatly diminished tendency since the introduction of catalytic converters), other mobile sources (including especially ship transportation), power generation and agriculture (fertilisers, animal farming). Besides the tropospheric water vapour, carbon dioxide (CO_2) is the most important climatically relevant trace gas. The natural, dynamic balance is increasingly being disturbed by the combustion of fossil fuels. These anthropogenic CO_2-emissions are above all made responsible for the climatic changes which are summarized as the greenhouse effect. The main sources of dust and soot particles are the energy harnessing industry and traffic (combustion processes). In recent years it has been possible to reduce particle emissions - yet it was only during this development that the severe threat of fine dust particles has been recognised. Until 1990 the main cause of VOCs was assumed to be traffic (incomplete combustion, but also the evaporation of fuels), but today, on account of the substantial reduction of traffic emissions, the most essential source categories are the use of solvents and fertilisers as well as the emissions from production processes and small combustion facilities. Noise is defined as a sonic sound of an intensity perceived to be disruptive. In this context the sentience of noise has a strong subjective variance and results in a basic problem of evaluation. Among the main noise sources are traffic and specifically freight traffic alongside the producing industries and construction. In the framework of surface area consumption, one may distinguish between primary landscape consumption as a sealing of areas and the complementary landscape consumption as a consumption and usage of adjacent areas or the fragmentation of landscapes. Generally speaking, the external costs of surface area consumption and the visual impairment of the overall appearance of the landscape are somewhat marginalised in comparison with the other external effects. Concerning the aspect of finite resource consumption, one must affirm that an economic evaluation of consumption proves to be difficult, since neither the existing quantity nor the prospective demand can be determined with certainty. Moreover, the technological innovations and substitution possibilities for the respective natural resource can hardly be predicted in the long-term.

5.2 The measurement process of Green Logistics Assessment

The developed model for measuring and evaluating emissions of logistical processes serves to ascertain and compare environmentally relevant emissions, such as CO_2, NOx, VOCs, fine dust particles, surface area consumption and noise. The model is applicable to any inbound and outbound process and permits assertions about the value of total emissions and also about the distribution of emission types in the process sequences of logistics.

The developed evaluation procedure is based on the fundamental assumption that basic logistical processes, such as transport and transhipment, are characterised by the physical movement of freight, requiring the intentional deployment of energy. The cause for the demand on energy and, thereby, the propeller of emissions is the expenditure to overcome various resistances of movement whose specific value imminently depends upon the driven route, the mass to be moved, the type of vehicle and the speed of movement within the framework of transport processes. The overall measurement process is shown in figure 4.

In order to reduce the evaluation expenditure, various simplifications of the evaluation procedure become necessary. Regarding the ascertainment of relevant influential factors, these simplifications can be realised in various ways:

- Simplification via the formation of factors for certain influential factors (for example co-efficients of roll and air resistance for average acceleration)
- Linear convergence via the formation of average values (for example route profile, age of vehicle fleet)
- Convergence via cluster and category formation (for example average speeds and acceleration proportions, clustered according to various street types and day time zones)

Within the framework of the evaluation and the designation of various emissions, average and/or threshold value related conversions of energy consumption or consumptions for utilised energy carriers are used. It must be noted that the evaluation takes direct and indirect emissions into consideration, but not emissions from previous chains (for example the infrastructure for power generation). The detailed calculation model for road transport is shown in figure 5.

5.3 Evaluation techniques in Green Logistics Assessment

In general environmental improvements can be seen on the level of the product, the structure of the logistics systems, the planning and management of the logistics processes and on the operational and technological level (Straube & Pfohl 2008). An appropriate evaluation technique has to be chosen for each of these areas of environmental improvement. For showing the influence of single operational improvements like better transport routing or the use of environmentally friendly technologies the analysis of the sensitivity of relevant input factors can be a useful technique. Knowing that for example the transport distance has an influence above average on the emission level, relevant improvements can be prioritized. For the selection of appropriate planning and management approaches more complex evaluation methods are needed. Redesigning the production or transport planning systems can influence the ecological sustainability in different ways. By that the evaluation of different planning approaches has to be done by the comparison of different process configurations. The same evaluation technique has to be taken into consideration for the evaluation of different supply chain structures. On an even higher design level like the overall strategy or the product design even more possible trade-offs have to be taken into account. For most operational and management improvement an influence on the different emissions in the same direction can be observed. When comparing different transport modes, supply chain structures or product configurations the effects on the different emissions can be oppositional. In order to weight the various types of emissions according to their environmental effect and, thereby, achieve comparability, the method of ecological scarcity is implemented. The term ecological scarcity was coined in 1974 on the basis of ecologically sustainable substance fluency (BUWAL 2009). The objective is not only to record industrial effects on the environment, but also to enable these effects to be managed using environmental policies. The basic principle behind this is the concept that the environment is able to endure a certain amount of strain (emissions, resource depletion etc.); this strain must be distributed across the various sources of pollution. By use of ecological factors the various environmental effects are converted into environmental damage points, which permit a comparability of environmental strains for various emissions on the one hand and varying logistics processes or configurations on the other. The evaluation of ecological factors takes the current environmental situation into account, the present standardised environmental situation as it pertains to the referential size and the target situation, pursued by environmental policies (governmental environmental policy priorities). The fit of the different evaluation techniques and the targeted areas of improvement are shown in figure 6.

5.4 A case study on improvement opportunities out of the Green Logistics Assessment results

The presented Green Logistics Assessment methodology was successfully used in different ecological sustainability projects. One practical case study of the Green Logistics Assessment application will be described. The starting point of the case study company, a large multinational industrial goods producer, was the objective to gain insights into the management of ecological aspects in logistics by starting optimization projects on single logistics processes. One of the selected processes was the distribution process of finished industrial goods from European factories to the port of Hamburg. From there, the goods were shipped to an industrial building site in China. The total sales volume of the selected project was 50.000.000 €, consisting of 250 tonnes of goods. At the beginning the logistics process was analysed in detail. The process included a packaging process near the factory, the road transport to a warehouse in the port area, the containerization of the goods, the transport of the packed containers to the port terminal and the loading on board of the container vessel. For each process step the used logistics equipment, the quantity structure and detailed material flow was analysed in detail by structured interviews. Based on these input information the calculation of the emissions for the actual process configuration was done. The calculation results show a total energy consumption of 65.000 kWh differentiated in minor 50 kWh energy for the packaging and transhipment process, 62.350 kWh for the transport process, 150 kWh for the containerization and warehousing process, 950 kWh for the transport from the warehouse to the port terminal and 1500 kWh for the loading on board. Using the presented emission factors the different emissions and the total emission was calculated. In the evaluation phase of the project the presented evaluation technique of sensitivity analysis was applied to develop a cause-and-effect relationship model for possible environmental improvements. A summary of this analysis is shown in figure 7.

The different improvement potentials were discussed at a management workshop, including representatives of the shipper, the logistics service providers and the terminal and port company. The realization opportunity for each option were analysed and financially evaluated. The major ecological improvements the management workshop agreed on are a modal shift to rail transport, the early containerization of the goods at the beginning of the distribution process, and the postponement of the shipping to the latest date to avoid the warehousing process near the port. For this alternative configuration a new Green Logistics Assessment calculation was proceeded. The environmental improvement in terms of CO_2-emissions as well as a comparison of logistics costs and transport lead times are presented in figure 8.

6 SUMMARY

In total, the described case study shows the usefulness of an integrated methodology for the measurement, evaluation and improvement of the ecological sustainability in logistics. The presented Green Logistics Assessment methodology is oriented on a process companies are choosing for their green logistics projects. The methodology supports companies on each single step in decision-making and does not end up with a calculation result of the actual process configuration. By including all relevant emissions in the methodology, one single approach is appropriate for most company requirements. The approach offers methodological flexibility by using a physical and open-source calculation logic as well as open calculation factors and data. The methodology can be individually customized on company specific data availability and objectives as well as on changing regulatory standards. Decision support is realized by orientating the design of the calculation method on the logistics planning and management logic and not on macroeconomic reporting practices. While logistics planning is demanding for detailed insights into the cause-and-effect relations of logistics practices and ecological emissions, reporting purposes have to deal with a much higher level of aggregation. The research on the interoperability of ecological reporting and ecological planning support systems should therefore be on the agenda of further research. In the area of decision support Green Logistics Assessments enables companies to build up ecological KPIs. While single optimization decisions can be argued with these KPIs, complex network related issues need holistic systems of metrics rather than single KPIs. In the next steps of the research the focus should be on the evaluation of the influence of emissions on logistics costs and service quality. A long-time objective can be seen in having objective sustainability KPIs in logistics that allow a comparison of ecological performance to cost and service levels. By that process and structure alternatives can be evaluated and logistics management can force top management decisions on what investments are needed to reach a certain ecological performance level. The case study has shown that ecological sustainability is one planning dimension of the logistics process. Having an integrated decision-making process that not only includes emissions but also logistics costs and service levels will be even more beneficial for companies and can boost the implementation of sustainable solutions in logistics. The key success factor for that is the integration of evaluation and decision support methods into the measurement methodology itself. Research should focus on this issue by evaluating the capabilities of measurement methodologies and processes under the logistics planning rather than the reporting point of view.

7 REFERENCES

Aronsson, H.; Brodin, M. H. (2006): The environmental impact of changing logistics structures, The international Journal of Logistics Management, Vol. 17, No. 3, p. 394-415.

ATKearney (2007): Chain reaction. Your firm cannot be "sustainable" unless your supply chains become sustainable first.

Ballas, K.; Richert, J. (2009): Green Supply Chain Management. Nachhaltige Wertschöpfung in Handel und Konsumgüterindustrie, J&M Management Consulting AG and DGMF.

Beamon, B. M. (1999): Designing the green supply chain, Logistics Information Management, Vol 12, No. 4, p. 332-342.

BME; DHBW Lörrach (2009): Green-Logistics - hohe Bedeutung auch in Krisenzeiten, http://www.bme.de/fileadmin/bilder/PDF/AuswertungGreenLog.pdf.

Brown, D. (2008): Environmentally friendly credentials are influencing business outsourcing decisions, Strategic Outsourcing: An International Journal, Vol. 1, No. 1, p. 87-95.

BUWAL (2009): Die Methode der ökologischen Knappheit, http://www.bafu.admin.ch/publikationen/publikation/01031/index.html?lang=de&download=NHzLpZig7t,lnp6I0NTU042l2Z6ln1acy4Zn4Z2qZpnO2Yuq2Z6gpJCGdoF9fWym162dpYbUzd,Gpd6emK2Oz9aGodetmqaN19XI2IdvoaCVZ,s-.pdf.

Ceuster, G.; Herbruggen, B.; Ivanova, O.; Carlier, K.; Martino, A.; Fiorello, D. (2007): Service contract for the further development and applica- tion of the transport and environmental TREMOVE model Lot 1 (Improvement of the data set and model structure), http://www.tremove.org/documentation/Final_Report_TREMOVE_9July2007c.pdf.

Darnall N.; Jolly J.; Handfield R. (2008): Environmental Management Systems and Green Supply Chain Management: Complements for Sustainability?, Business Strategy and the Environment 2008, Vol. 17, No. 1, p. 30-45.

Endres, A.; Bertram, R.; Finus, M.; Rundshagen, B. (2007): Umweltökonomie; Verlag Kohlhammer; Stuttgart 2007.

Elkington, J. (1998): Cannibals with Forks: The Triple Bottom Line of the 21st Century. Stoney Creek.

EyeforTransport (2007): Green Transportation & Logistics, Summary and analysis of eyefortransport's survey, London, 2007.

Fritsche, U.; Schmidt, K. (2008): Global Emission Model of Integrated Systems (GEMIS): Öko Institut, Darmstadt, http://www.oeko.de/service/gemis/files/doku/g45_manual.pdf.

Fonger, M. (1993): Gesamtwirtschaftlicher Effizienzvergleich alternativer Transportketten (multimodaler Verkehr). Göttingen: Vandenhoeck & Ruprecht.

Gkatzoflias, D. ; Kouridis, C. ; Ntziachristos, L. ; Samaras, Z. (2007): COPERT 4 : Computer programme to calculate émissions from road transport, http://lat.eng.auth.gr/copert/files/tech01.pdf.

Halldórsson, A.; Kotzab, H.; Skjøtt-Larsen, T. (2009): Supply chain management on the crossroad to sustainability: a blessing or a curse?, Logistics Research, Vol. 1, No. 2, p. 83–94.

Hoffren, J.; Apajalahti, E.-L. (2009): Emergent Eco-Efficiency Paradigm in Corporate Environment Management, Sustainable Development, Vol. 17, Nr. 1, S. 233-243.

IFEU (2005): Energieverbrauch und Schadstoffemissionen des motorisierten Verkehrs in Deutschland 1960-2030, http://ifeu.de/verkehrundumwelt/pdf/IFEU(2005)_TREMOD_Grundlagenbericht_Zusammenfassung.pdf.

Manget, J.; Roche, C.; Münnich, F. (2009): Capturing the Green Advantage for Consumer Companies. The Boston Consulting Group. http://209.83.147.85/impact_expertise/publications/files/Capturing_Green_Advantage_Consumer_Companies_Jan_2009.pdf.

McIntyre, K.; Smith, H.; Henham, A.; Pretlove, J. (1998): Environmental performance indicators for integrated supply chains: the case of Xerox Ltd. Supply Chain Management, Vol. 3, No. 3, p. 149-156.

Pischinger, R. (2002): Update of the Emission Functions for Heavy Duty Vehicles in the Handbook Emission Factors for Road Traffic, http://www.hbefa.net/e/documents/DACH_SNF_Endbericht.zip.

PricewaterhouseCoopers (2009): Land unter für den Klimaschutz? Die Transport- und Logistikbranche im Fokus.

PricewaterhouseCoopers; Supply Chain Management Institute (2009): Transportation & Logistics 2030. Volume 1: How will supply chains evolve in an energy-constrained, low-carbon world?

Seuring, S.; Müller, M. (2008): From a literature review to a conceptual framework for sustainable supply chain management. Sustainability and Supply Chain Management. Journal of Cleaner Production, Vol. 16, No. 15, p. 1699–1710.

Singh, M. (2004): A Review of the Leading Opinions on the Future of Supply Chains; Supply Chain 2020 Project Working Paper. MIT, Boston, Massachusetts.

Srivastava, S. K. (2007): Green supply-chain management: A state-of-the-art literature review. In: International Journal of Management Reviews, Vol. 9, No. 1, p. 53–80.

Stern, N. (2007): The Economics of Climate Change - The Stern Review; Cambridge University Press; Cambridge.

Straube, F.; Borkowski, S. (2008): Global logistics 2015+. How the world's leading companies turn their logistics flexible, green and global and how this affects logistics service providers, Berlin.

Straube, F.; Pfohl, H.-C. (2008): Trends und Strategien in der Logistik. Globale Netzwerke im Wandel. Bremen.

United Nations (1983): Process of preparation of the Environmental Perspective to the Year 2000 and Beyond. General Assembly Resolution 38/161. New York.

Vachon S.; Klassen R. D. (2006): Extending green practices across the supply chain: The impact of upstream and downstream integration. International Journal of Operations and Production Management, Vol. 26, No. 7, p. 795-821.

Wagner, B.; Enzler, S. (2005): Material Flow Management: Improving Cost Efficiency and Environmental Performance. Physica,.Heidelberg. 2005.

Wright, J.; Jones, D.; Hoyle, S. (2009): The Sustainable Supply Chain. Balancing cost, customer service and sustainability to achive a high-performance supply chain, Accenture. http://www.accenture.com/NR/rdonlyres/A67BA82D-EC74-4B71-A039-481B4C308351/0/Accenture_The_Sustainable_Supply_Chain.pdf.

Zhu Q.; Sarkis J.; Lai K. (2008): Green supply chain management implications for “closing the loop”. Transportation Research Part E, Vol. 44, No. 1, p. 1-18.

Figure 1: Supply Chain Management, material flow management and environmental management

	Supply Chain Management	**Material Flow Management**	**Environmental Management**
Objective	Efficiency and effectiveness of the Supply Chain in terms of costs, quality and service	Reduction of environmental impact by lowering anthropogenic material flows and the substitution of materials	Implementation and continuous improvement of the environmental management system (for example ISO 14001) and improvement of the corporate environmental performance (for example EMAS)
Strategic base	Fundamental logistics principles like total cost orientation, process integration, etc.	Environmental management principles	Environmental management principles
Object	Material, information and financial flows	Those Physical material and energy flows that have an environmental impact	Environmental governance, strategy, objectives, management systems and reports
Actors	Focal company, suppliers and customers	One or more companies and their stakeholders	One company or company's site and the stakeholders
Cooperation	Horizontal and vertical	Horizontal and vertical	Not specified
Methods	Process analysis, optimization methods, etc.	Material flow reports and calculation	Environmental audits and reports

Figure 2: Summary of the review of current measurement methodologies

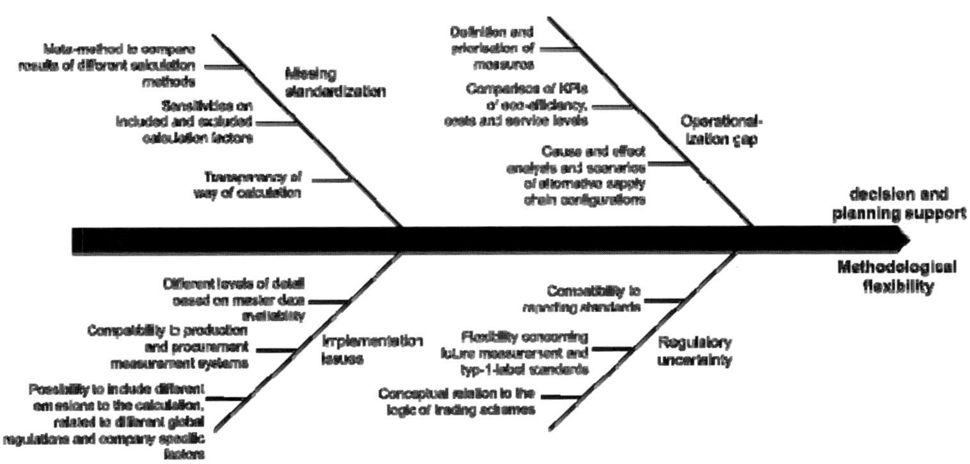

Figure 3: Summary of the review of current measurement methodologies

Measurement methodology	Summary of the review
GEMIS 4.5	• No consideration of dynamic variables like traffic situations or transport routings • Empty weight of vehicles and their payload no variable input factors in the calculation • Energy consumption of warehousing processes not clearly documented and transparent • Systematic evaluation of the measurement results and deduction of improvement potentials not objective of the methodology
TREMOVE 2.7b	• Macroeconomic measurement model with limited applicability for single logistics processes and companies • Based on political instead of management related goals and objectives • Systematic evaluation of the measurement results and deduction of improvement potentials not objective of the methodology
COPERT 4	• Macroeconomic measurement model with limited applicability for single logistics processes and companies • Not all relevant emissions part of the model • Based on political instead of management related goals and objectives • Systematic evaluation of the measurement results and deduction of improvement potentials not objective of the methodology
HBEFA 2.1	• Empty weight of vehicles and their payload no variable input factors in the calculation • Not all relevant emissions part of the model • Systematic evaluation of the measurement results and deduction of improvement potentials not objective of the methodology
TREMOD 4	• No consideration of dynamic variables like traffic situations or transport routings • Empty weight of vehicles and their payload no fully variable input factors in the calculation • Systematic evaluation of the measurement results and deduction of improvement potentials not objective of the methodology

Figure 4: Measurement process of Green Logistics Assessment

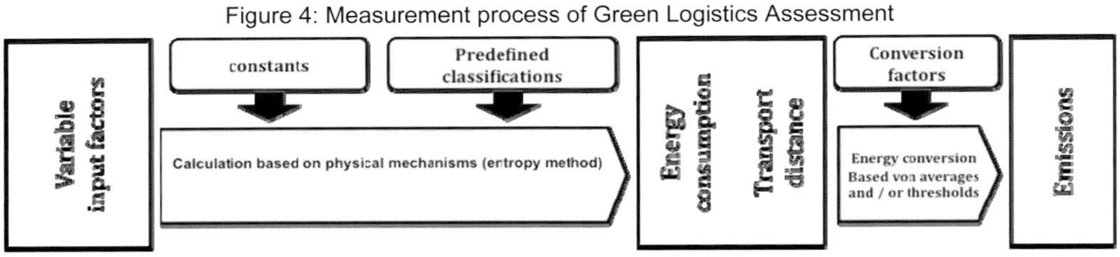

Figure 5: Calculation model for road transports

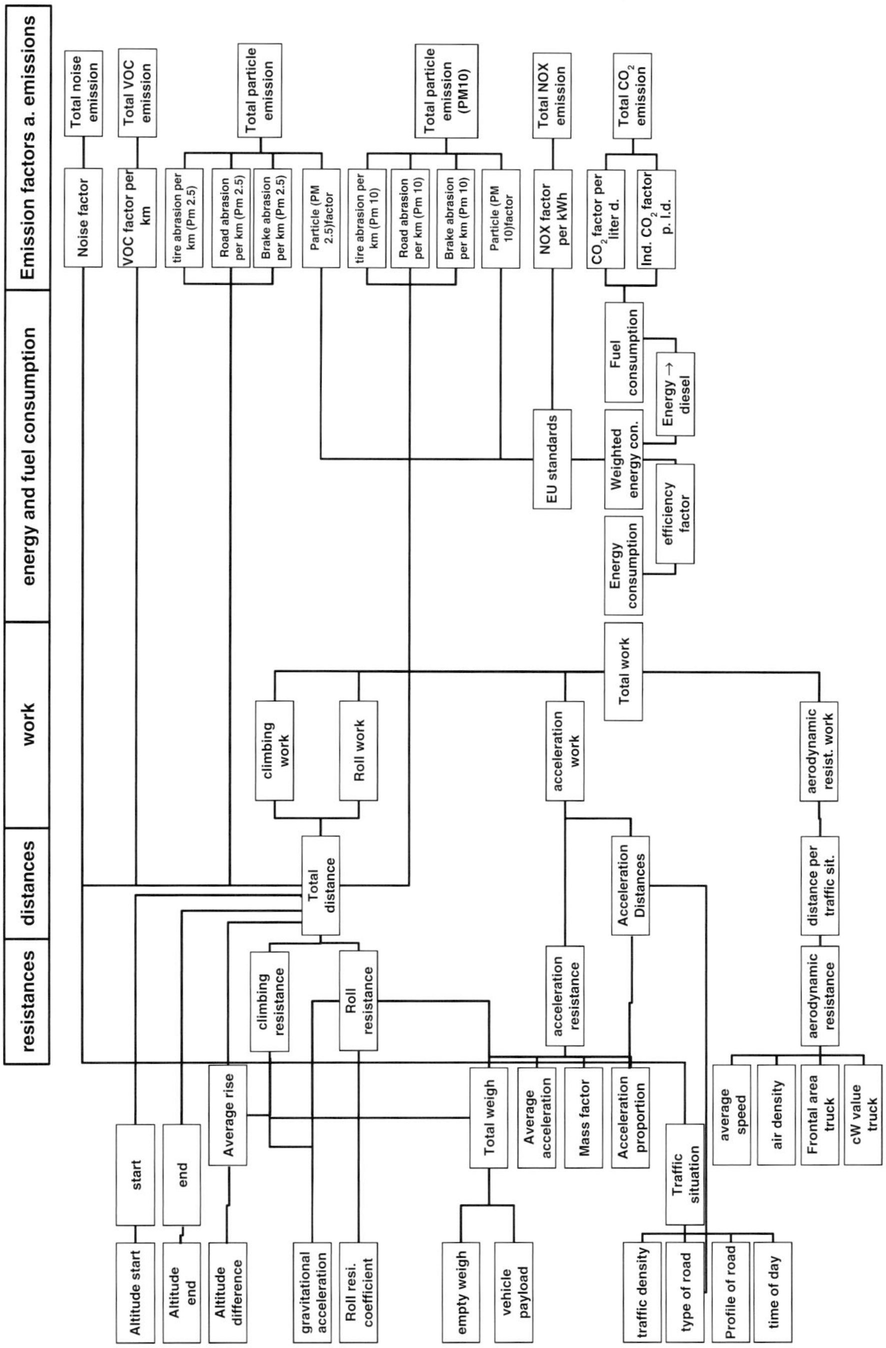

Figure 6: Fit of used evaluation techniques and targeted area of improvement

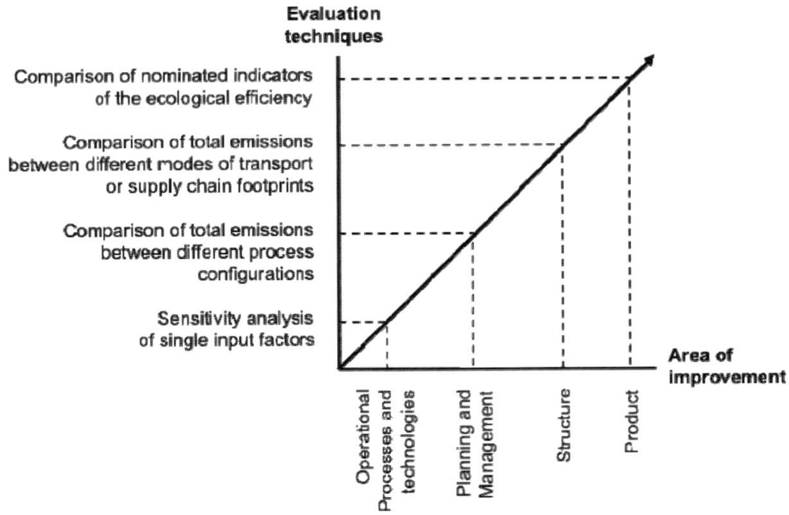

Figure 7: Cause-and-effect analysis in the different improvement dimensions

Improvement dimension	**Cause**	**Effect**
Technology	- Age of trucks - Alternative actuation - Alternative fuel - Electrification	- Energy efficiency - Resistances - Emission factors
Driver's behavior	- Economic driving	- Lower speed - Less acceleration
Mode of transport	- Alternative modes of transport	- Emission factors - Trade-offs between degree of utilization, transport distance, transport emissions
Supply chain structure	- Fewer process steps - Decupling point	- Fewer handling and warehousing processes
Transport management	- Delivery frequency - Degree of utilization - Distances	- Number of transports - Volume and Weight - Distances
Loading equipment	- Packaging	- Weight - Fewer repackaging processes

Figure 8: Comparison of emissions, logistics costs and transport lead time

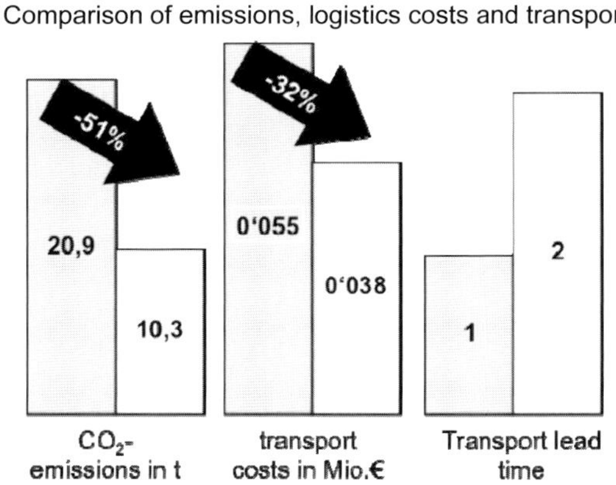

Lean Value Stream Manufacturing for Sustainability

J. Mohammed[1], A. Sadique[2]
[1]The Petroleum Institute, Abu Dhabi, UAE
[2]UX Bay Inc., Ohio, USA

Abstract
Most people are familiar with the old axiom "what gets evaluated and measured gets done"; the primary responsible people who measure in any organization are the finance and accounting department, which makes them have a powerful influence on organization strategy and decisions. Traditionally these fields have only tried to measure the one leg of the three legged sustainability stool (where economy, environment, and society are the three legs of the sustainability stool). In this paper, authors would like to introduce the concept of Lean Value Stream Manufacturing (LVSM) approach where the concept of lean manufacturing and value adding process is integrated. The value stream approach would be used to quickly identify and eliminate wastes that may well include water consumption, energy consumption, landfill avoidance, chemical consumption and much more

Keywords:
Lean value stream mapping, Sustainability, Clean Value stream

1 INTRODUCTION

One could argue that the only way to compete and win in today's manufacturing world is by adopting a methodology that will increase productivity, reduce cost & resource needs at the same time eliminating wasteful inefficiencies. Successful companies around the world have embraced Lean manufacturing to achieve the above objective. In any organization, most of the accounting systems are heavily focussed on trading which is actually based on the amount of money it makes, which makes it a mismatch between the manufacturing system and accounting system.

In this paper, authors would like to introduce the concept of Lean Value Stream Manufacturing (LVSM) approach where the concept of lean manufacturing and value adding process are integrated. The value stream approach would be used to quickly identify and eliminate wastes that may well include water consumption, energy consumption, landfill avoidance, chemical consumption in addition to many other considerations. The LVSM approach focuses on finding quick hit projects that can undertaken with very little time, effort, and investment which produce immediate savings and benefits. This paper would also discuss the steps involved in implementing the LVSM approach. The LVSM approach consists of seven steps which include, financial assessment, initial education, benchmarking, sustainability strategy, action plan, team launch, and, continuous improvement.

2 LEAN VALUE STREAM MANUFACTURING

2.1 Lean manufacturing

The core idea of Lean manufacturing is to maximize customer value while minimizing all type of wastes. In simple terms, lean means creating more value for customers, both internal and external with fewer resources. True lean organization understands what customer value is and focuses its processes with a view to increasing it continuously. The ultimate goal is to provide perfect value to the customer through a perfect value creation process that has zero waste.

Lean organization always eliminates waste along the entire value streams, instead of at isolated points. This creates processes that need less human effort, less space, less capital, and less time to make products and services at far less costs and with much fewer defects, compared with traditional business systems. Lean Organizations are able to respond to changing customer desires with high variety, high quality, low cost, and short time.

Lean applies to every business and every process. It is not a short term cost reduction program, but a way of thinking for the entire organization. All types of businesses including healthcare and governments are using lean methodology as the way they think and do business. Lean transformation is often used to characterize a company moving from a traditional old way of thinking to lean thinking. This requires a complete transformation on how an organization conducts business.

2.2 Value stream

Value stream is a process for planning and linking lean initiatives through systematic data capture and analysis. Value stream process supports the transformation into a lean enterprise by providing a structure to ensure that the lean implementation team functions effectively. In any proven process can fail to achieve results if the people do not apply it properly or if they lack fundamental understanding of the nature of the process.

2.3 Sustainability

There are many definitions that are proposed for sustainability, the Brudtland report defines sustainable development as the development that meets the needs of the present without compromising the ability of the future generations to meet their own needs. Sustainability relates to economics, society, and environment. All the three are considered as the three legged stool model [1]. The increase or decrease in the size of the length makes the system dysfunctional and disproportional. The common fact in that links the three legs is the generation of the waste

Sustainability is not a methodology; it is actually how the people make the changes. The challenge of sustainability is that it is neither technical not rational. It has a lot to do with one's behaviours and attitude. It also has to do a lot with the social and fundamental issues in one's society. With the increasing diversification population in terms of behaviours, attitudes, spending and purchasing habits one thing that would be sensible and common would be the money factor.

In any process (for example consider a manufacturing process) where there are seven types of wastes [2] that are involved and they are shown with the help of the pie diagram in the figure 1. The type and amount of waste depends upon

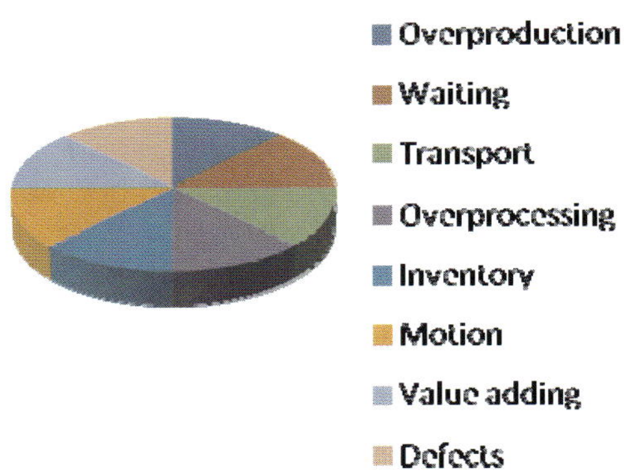

Figure 1: The seven types of wastes in manufacturing

There are two stages of the waste situation, first is the generation of the waste and the second would be to establish policies for the further reduction of the waste that is already created. Both involve money. From the economic stand point, large amounts of money are spent hauling, transporting and disposing waste, which could be used more efficiently in other activities; which could help check generation of wastes thereby supporting and sustaining efficient utilization of lean techniques. From the social point of view, waste is a health hazard regarding the environment, waste would produce serious unwanted effects such as flow of unwanted material into ground water which could bring about devastating environmental degradation by contaminating the ground water resources.

3 LEAN VALUE STREAM MANUFACTURING

The lean value stream manufacturing concept is the integration of lean value concepts and design for sustainability. Sustainability was mostly considered as environmental issues. Most of the efforts focused on creating profits by increasing production then later moved into clean technology or green technology production producing eco friendly products. The next phase was the production impacts, which takes into account the complete life cycle of the product. The impacts could be made into two categories: one would be from the design of the product to the life of the product and the other would be the end of life of the product. As discussed earlier, sustainability relates to economics, society, and environment. Lean value stream concept to sustainability goes well beyond the environmental issues; the concept helps to attain the consumer needs in terms of value and quality. Lean value stream manufacturing would be considered as a life cycle based efforts on improving the life cycle of the product by having a better awareness of the product, improving the efficiency, environmental concerns, and sustainable innovations. With lean value stream manufacturing to sustainability, the introduction of the sustainable product comes into the cycle, as shown in figure 2.

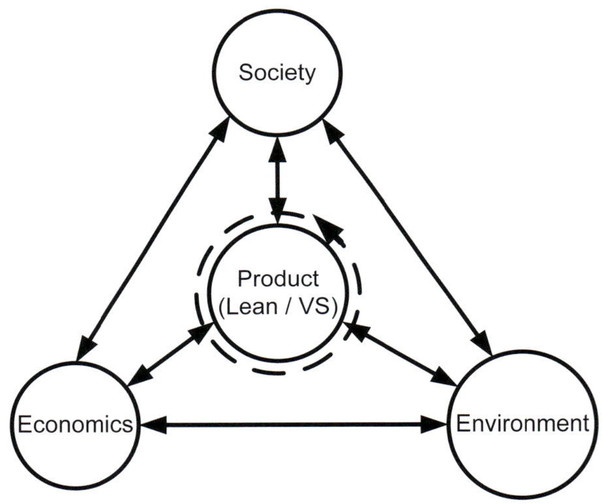

Figure 2: Lean value stream manufacturing concept

The lean value stream manufacturing is a strategy for an organization to manage by fact and data. This approach is based on the financial data to in order to identify where lean value stream concept could be applied. Unlike the quality control initiatives like six sigma, total quality management, sustainable lean value stream manufacturing has also the business objective based on profitability from the business and also the environmental responsibility. Organizations have attempted to achieve their business objectives through their functional hierarchy. Not many organizations have looked at the complete life cycle of the product, sustainable lean value stream manufacturing looks at the complete life cycle of the product and assesses based on the product based on its design, process, and end of life of the product.

The lean value stream manufacturing approach consists of seven steps, which are shown by the flow diagram illustrated in figure 3.

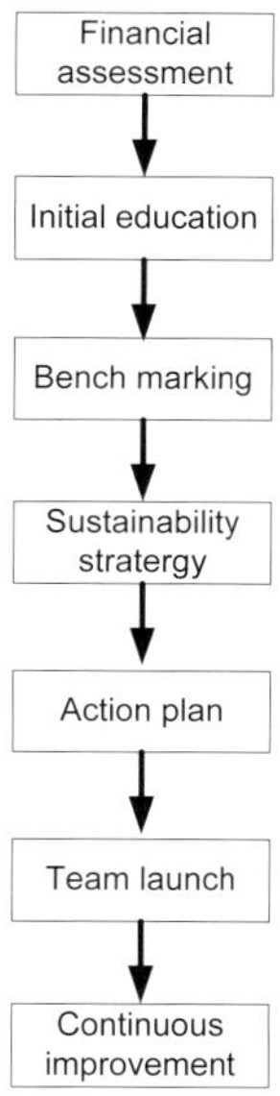

Figure 3: Steps in lean value stream manufacturing

The foremost important factor in implementing lean value stream manufacturing for sustainability is documentation or effective managing of the data in the value stream. Effective data management process should be implemented throughout the value stream. There are various tools that are utilized for capturing and processing the data right from 3D visualization of the process, simulating the process, capturing the bottle necks and other waste generation in the process. Once the lean is implemented the next focus should be on the sustainable development of the process and product throughout the product life cycle. Industry strives to attain continuous improvement in the process as well as with the product that is economical in the long run by caring for the environment. The lean value stream value stream manufacturing approach focuses on establishing a framework to identify objectives internally and externally with supply chain partners and then on embedding problem solving functionality throughout the value stream.

3.1 Financial assessment

The first and the foremost to apply the LVSM approach is the financial assessment. Understanding the big picture is very important to accurately identify and eliminate waste in the system. Financial analysis will help to understand the entire system better and also help to quantity the amount of waste in the system. The first step in financial analysis is to identify the process you want to study and draw an initial process diagram for the entire organization with all the inputs and outputs.

Figure 4: Organisational flow

The key here is INPUT MUST BE EQUAL TO OUTPUT.

Everything that goes into the organization must come out in some form or another. The output can be tangible things like solid waste and sometimes intangible like some type of energy going out. This process diagram will help to see clearly all the physical flows around the system. The accuracy in calculating the amount of inputs and outputs is important.

After identifying the inputs and outputs it is time to quantify the amount of waste in the system. Using the formula

INPUT = FIRST TIME QUALITY PRODUCT+WASTE

From the basic formula one can quantify the amount of waste that comes out of the organization. The waste could be in any form, it could be in the form of materials, labour, energy, or any form that does not translate to the first quality product. Financial assessment would help to identify the areas to pursue the lean concepts.

3.2 Initial education

Depending on the location, product, and awareness in the society; the strategy of initial education varies. The concept of sustainability may or may not be well appreciated. There might be places in the developing world where the product innovation is by nature, problematic, characterized by poor business and government rules. But one thing which always takes the lead is the money factors. All organizations irrespective of where they are located, would keep on using techniques on increasing the profit of the product and this is where Lean Value stream play its role.

The initial education and the focus needs to come from two perspectives, from top down as well as bottom up in the organizational hierarchy in order to find the problems and opportunities.

3.3 Bench marking

Many institutions and organizations consider bench marking as a process or a practice to keep their products or service in par with the industry needs. Bench marking typically would involve redesigning or designing a new production with the intention on improving quality, time, and cost. In most scenarios the improvement directions are received from the competitor's products or comparable products elsewhere. Benchmarking involves management identifying the best firms in a similar industry where a similar product is manufactured or where a similar process is used in the manufacture of the product.

Benchmarking is used as a performance indicator, either in terms of cost, productivity, defects, or in terms of measured parameter. Considering the LVSM approach benchmarking would involve different strategies either it could be applied in the design level, considering the percentage of reusing and recycling the product or it could be used in the process benchmarking where the process is compared to the best practices in minimizing the unwanted actions. Once a process or the designed is benchmarked, the organization tries to achieve the same, and it is often treated as a continuous process which the organization seeks to improve their practice in designing and manufacturing the product.

3.4 Sustainability strategy

Continuous improvement across a value stream relies on basic principles of lean which always had been minimizing waste, lowering costs, improving productivity, enhancing revenue generation through better use of assets and resources. One way to do is to map the complete process like the value stream mapping method. The Value Stream Mapping method is a visualization tool used in Lean Manufacturing. Primarily it is used as a strategic planning tool and a management change tool. The goal of Value stream mapping is to identify, expose and decrease Waste in the process. In lean manufacturing waste is any activity that does not add value to the final product. Value Stream mapping method visually maps the flow of materials and information from the time products come in the back door as raw material, through all manufacturing process steps, and off the loading dock as finished products.

Mapping out the activities in the processes including all the inputs goes into each step of the process like, energy usage, water usage, and other information's like cycle times, down times, in-process inventory, material moves, helps to visualize the current state of the process activities and guides towards the future desired state. Here are the five steps for developing a value stream map for the sustainability strategy.

- Identify similar target product, product family, or service.
- Draw a current state value stream map, which shows all the inputs like, water usage, electricity usage, delays, etc
- Assess the current state value stream map in terms eliminating waste by reducing the amount of environmental waste
- Draw a future state value stream map.
- Develop an action plan to work toward the future state condition

Figure 5 shows systematically where the points of generation of waste are. Value stream mapping would provide more macro information on the process. By looking at the figure one could easily tell that the procedure is either sustainable or easily redesignable.

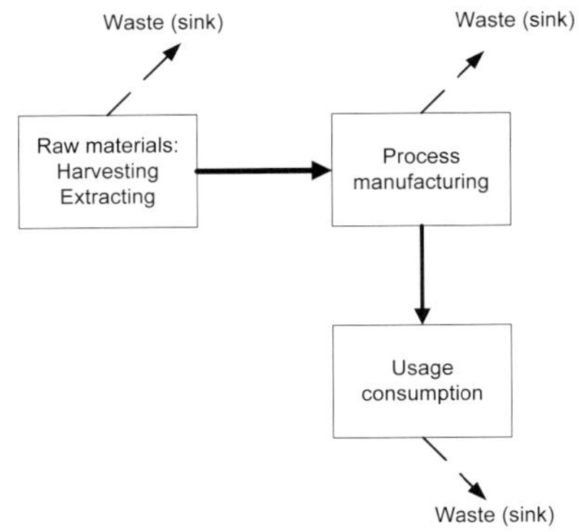

Figure 5: Linear path of generating waste

Several Life cycle Assessment methodologies are used currently for developing the sustainability strategy [4, 5]. For example, potential impacts due to green house gases (GHG) could be considered as one of the sustainability strategy. Selecting a strategy would depend upon where the type of product or process under consideration.

The environmental analysis also plays a crucial role in developing the sustainability strategy. The purpose of the environmental analysis helps the designers understand the implication of alternative design choices that would be beneficial for the future generations. Figure 6 shows the environmental foot prints that is involved with the development of a product, it provides the information of the areas that needs to be considered while designing the product and process. The system becomes more complex and dynamic when coupled with the life cycle assessment of the product.

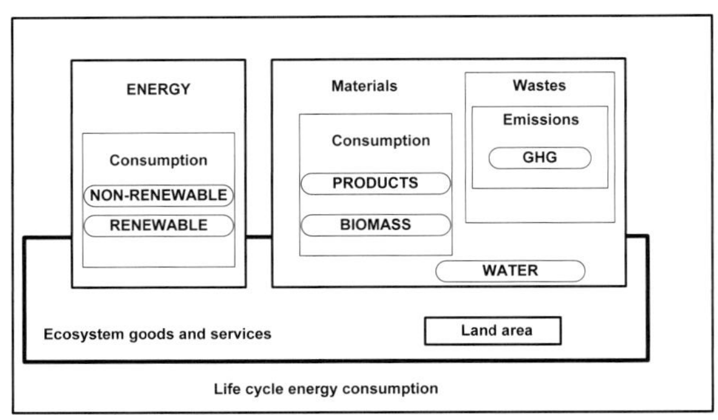

Figure 6: Environmental foot print chart

The International Organization for Standardization (ISO) has incorporated series of standards focusing on environmental management. Most of the standards are included on the ISO 14040 series. The standards on the Life cycle assessment is as follows

- ISO 14040 – life cycle assessment, principles and framework
- ISO 14041- life cycle assessment, Goal and scope definition and life cycle inventory analysis

- ISO 14042 – life cycle assessment, life cycle impact assessment
- ISO 14043 – life cycle assessment, life cycle interpretation.

The international standard organization has been developing standards for life assessment as a part of its 14000 series standards on environmental management. Theses standards address both technical and conceptual organization of the life cycle assessment.

Life cycle assessment is a technique that is used to assess the environmental aspects and potential impacts associated with a product, process, or service by compiling an inventory of relevant energy and material inputs and environmental releases, evaluating the potential impacts associated with identified inputs and releases; and interpreting the results to make a more informed decision. The functional unit is the central concept in the life cycle assessment. The goal and scope definition is designed to obtain the required specification based on the product and process for the life cycle assessment study.

The sustainability strategy is the most important step that is used while designing a new product, or redesigning an existing product based on its design or process.

3.5 Action plan

Prioritization Matrix is a useful technique for sorting the diverse set of projects into an order of importance. It also enables the relative importance of projects to be identified by deriving a numerical value for the three legs of sustainability (economy, environment, and society). For example an item with a score of 190 is clearly far more important than one with a score of 18, but is not much more important than one with a score of 185.

To be able to compare with one another in this way, each item is scored against a set of key criteria derived from the three legs of sustainability, and the scores for each item are then multiplied to get the rank. For example, a potential project to reduce power usage will get a high score on the criterion of, Low power cost, but might get low score on improving stake holder's morale. Selecting good criterion is very important say as it reflects key goals and enables objective measurements to be made.

When there are multiple criteria, it may also be important to take into account the fact that some criteria are more crucial than others. This can be implemented in the prioritization process by allocating weights to the criteria. After weighting the criteria those weights will be included in the multiplication to come up with the final rank. The weights recommended are 1-10 scale with spacing between the different weights like 1-3-7-10.

In addition to the performance rating against each of the levels an overall performance rating is then developed again assigning a weighting factor based on their importance given the nature of the product that is been developed with the intention of improving the sustainable characteristics.. This performance rating is then compared to the competitor's product which would determine the urgency in the improvisation of the product.

3.6 Team Launch

Many organizations launch products in different direction, it can be redesign of a product, development of a new product, or it can be also of a reverse engineering product. A substantial amount of time and money is often spent: on the new product, in re-engineering certain aspects of the product for developmental reasons, all of which calls for better training and use of incremental expertise in regard to the implementation strategies used to work. This structured launching of the work should enable the organization to quickly develop an action plan for improving the development process.

3.7 Continuous Improvement

The financial person assigned to work with the project would work with the team and would be listed with the project charter. Without this involvement from the personnel finance, engineers, and people who are involved with the project, the declared savings would be simply not credible. Aside from the built-in bias involved in calculating the benefit created from one' own project, there is an issue of qualification. The best qualified people to look into the calculation of benefits would be the people who understand the financial matters and the life cycle of the product and the process it goes through. This is not to imply that the finance expert's calculations should go unchallenged. The organizations should also be able to formally define the different categories of savings. Savings for the categories are placed in two categories, namely hard saving and soft savings..

Hard savings are the actual reductions in the amount of money that is spent on the product developments and manufacture. Soft savings are projected reductions that should result from the project. For example, applying the sustainable design and calculating the life cycle assessment of the product and calculating the overall benefits of the project. It is important that the savings be integrated into the business system of the organization. If the institutional framework and commitment does not change the savings could eventually be lost. Therefore there is a need to continuously visit the problem for redesigning the product and process. What worked yesterday may fail tomorrow; the organization must be well prepared to meet the new and changing demands from customers.

Building the shareholder value especially depends on the company's ability to innovate, improve, and learn. Innovation and learning are the two important areas that are to be addressed in continuous improvement. Both of them are driven based on the customer requirement and financial balance sheet. The performance measure for learning and innovation addresses mainly three areas, namely employee involvement and competency, technology, and corporate culture.

4 SUMMARY

This paper summarized the sustainable lean values stream manufacturing concepts and the various steps involved in the successful implementation of the same. Sustainable lean value stream manufacturing benefits claimed for every project are to be claimed by the experts in accounting, finance, engineers, and environmentalist working for the organization and the community it serves.

5 REFERENCES

[1] Dawe, N., Ryan, K., 2003, The faulty three legged stool model of sustainable development, conservation biology, Vol. 17, No. 5, pp. 1458-60

[2] Tapping, D., Luyster, T., Shuker, T., Value Stream Management, 2002, Productivity press, New York.

[3] Fiksel, J., Design for environment – A guide to sustainable product development, 2009, Second edition, McGraw hill publication, New York.

[4] Bieda, B., The use of life cycle assessment (LCA) conception of Mittal steel Poland SA energy generation – Krakow plant case study, 2007, First international conference on energy and sustainability, Wessex institute of technology press.

[5] Munier, N., Introduction to sustainability – road to a better future, 2005, Springer publications

Queueing of Seasonally Demanded Spare Parts in a Repair Shop of a Closed-Loop Supply Chain

K. Tracht[1], M. Mederer[2], D. Schneider[1]
[1] Bremen Institute for Mechanical Engineering (bime), University of Bremen, Germany
[2] m² hycon, Oberolm, Germany

Abstract
Spare parts in the aviation industry are repaired after removal from the aircraft and led back to stock, forming a closed-loop supply chain. Typically the demand of spare parts is assumed to follow a homogeneous Poisson distribution. The repair times in the shop with limited capacity are taken to be exponentially distributed. This paper investigates the effects of non-homogeneous demand and the resulting queue in the repair shop due to limited capacity in periods of high demand.

Keywords:
Logistics, Simulation, Queueing, Closed-Loop Supply Chain

1 INTRODUCTION

1.1 Problem Description

Dealing with problems in closed-loop supply chains, becomes more important, because of rising awareness of environmental issues and increasing number of recycling activities. Taking into account feedback effects, inventory forecasting is more difficult in closed-loop supply chains than in one way supply chains.

Industrial waste can be reduced by repairing spare parts and reusing them. Spare parts supply in the aviation industry is realized with a closed-loop supply chain, repairing parts after removal from the aircraft and putting them back into stock. The necessary number of spare parts in stock is usually calculated based on homogenous Poisson distributed demand and constant throughput times in the shops. As many spare parts have seasonal, lumpy demand, the assumptions of Poisson distribution and constant times in the repair shop are inappropriate. Results of the analytic solution become obsolete when limited capacity causes queueing of spare parts in the repair shop and rising throughput time. In addition to high demand, longer throughput times diminish the spare parts in stock.

1.2 Course of Investigation

The paper starts with a literature review of forecasting methodologies for intermittent demand, inventory planning and queueing theory. It is followed by an introduction to the supply chain of spare parts in the aviation industry and a presentation of the simulation model. The succeeding chapter compares the total costs of providing a spare part for aircrafts in different scenarios. In order to cope with seasonal demand, stock level and numbers of workers are analyzed.

2 LITERATURE REVIEW

The problem addressed in this paper is a combination of different research areas that have been focused on in the past: forecasting intermittent demand, inventory planning, queueing theory, and simulation of supply chains.

Different authors concentrate researches on forecasting intermittent demand in the aviation industry. Because of its importance for economical supply of spare parts, it has been a topic of interest for decades. Campbell first analyzed forecasting of intermittent demand in the aviation industry, finding a correlation between flight hours of aircrafts and the failure rate of spare parts [1]. The results are not accurate for all parts, because the failure rate of the landing gear, for instance, depends on the number of cycles and not flight hours [2]. One cycle consist of take off, flight and landing.

As Ghobbar and Friend found out in a survey with maintenance, repair, and overhaul providers and airlines, many of them do not use forecasting methods but rely on experience and average values when predicting future demand and necessary inventory levels [3].

Forecasting intermittent demand is a challenge not only limited to the aviation industry. Willemain, Smart and Schwarz present different approaches for dealing with intermittent demand in general and compare these approaches for their accuracy [4]. Syntetos and Boylan also compare forecasting methods [5][6]. They furthermore categorize different demand patterns in order to choose the adequate forecasting method, since the categorization is an essential element of inventory control management systems [7].

Other authors conducted research in queueing problems and supply chain simulations. Kleinrock explains analytical calculations of queues and different arrival processes of customers in a shop [8]. His work, published in 1975, still serves as a basis for today's researchers working with queueing phenomena. Cruz, Duarte, and Woensel address queueing and buffering problems with computational and simulation approaches [9]. Ashaheri and Lemmes simulate how enhanced inventory and demand planning in an organization's division impact the whole organization's revenue [10]. Terzi and Cavalieri review more than 80 articles about supply chains and analyze difficulties arising during simulation model building, because due to confidentiality companies along the supply chain do not share all the relevant parameters [11]. Tracht, Schneider and Schuh simulated a closed-loop supply chain as a feedback control system [12].

None of the cited authors considers the queueing of spare parts in a repair shop within a closed-loop supply chain. This paper will present a newly developed simulation model in order to determine cost optimal parameter settings when queueing influences the supply chain performance.

3 CLOSED-LOOP SUPPLY CHAIN

3.1 Spare part supply in the aviation industry

The broad variety of spare parts in the field of aviation and the difficulties in forecasting spare part demand are challenging for economical management of the closed-loop supply chain.

When a part fails unexpectedly, it has to be replaced in order to comply with flight safety regulations. Keeping downtimes of aircrafts to a minimum by quick spare part supply, avoids revenue loss for airlines.

The supply chain under investigation consists of three parts: the warehouse, the aircraft and the repair shop. In case of a spare part demand, the part is taken from stock and shipped to the location of the aircraft. A worker installs the part in the aircraft, replacing the broken one.

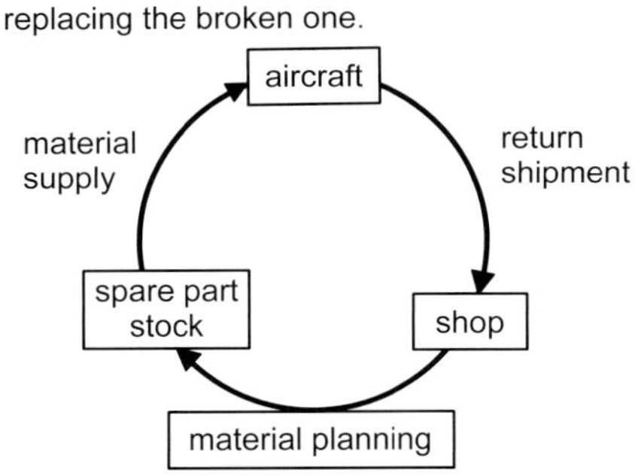

Figure 1: Supply chain in the aviation industry.

The broken part is sent to the repair shop, repaired, and put back into stock (see Figure 1). Watching the demand and the remaining spare parts in stock, the material planning buys new parts if necessary.

3.2 Seasonal Demand

For homogenous Poisson distributed demand the mean number of spare parts being repaired in the shop during a time interval equals the mean number of spare parts being removed from the aircrafts. The number of spare parts kept in stock and the total quantity in the supply chain will be sufficient, if calculated with standard methods of inventory forecasting. Queues are short and variation of their length only depends on stochastic impacts of the demand.

Rising demand due to seasonal effects will lead to insufficient spare part quantities in the supply chain, if seasonality is not considered in planning. In addition, stochastic nature of the throughput time will increase queueing effects. Spare parts queue in the shops during periods of high demand and diminish the number of parts in stock.

4 SIMULATION MODEL

4.1 Description

The difficulties of dealing with queues of parts in the repair shop and the lack of approaches addressing queues in closed-loops supply chains require development of a simulation model.

The model generates demand for a spare part with exponentially distributed time intervals between two requests. Every time a demand occurs, the stock in the warehouse is diminished by one. If there is a demand but no spare part in stock, the part will be loaned from a competitor.

The parts taken out of stock, queue in the repair shop and are repaired when a work space turns vacant. The repair time is assumed to be exponentially distributed. After leaving the repair shop, the part is put back into stock. Transportation time from the warehouse to the aircraft and the time elapsing before the broken part is shipped from the location of the aircraft to the repair shop, are neglected.

4.2 Cost Functions

The total costs c_t, which are subject of investigation in this paper, consist of capital costs c_c, costs of workers in the repair shop c_r, and loan costs c_l.

$$c_t = c_c + c_r + c_l \qquad (1)$$

The capital costs c_c depend on the purchase price c_p of a spare part and the number of spare parts in the supply chain n_p.

$$c_c\ (c_p,\ n_p) \qquad (2)$$

c_c = capital costs
c_p = purchase price of spare part
n_p = number of spare parts in the supply chain

The repair shop costs vary with the number of workers and workers' wages.

$$c_r\ (c_{w,}\ n_w) \qquad (3)$$

c_r = repair shop costs
n_w = number of workers
c_w = costs per worker (wages)

The longer the time spent in the repair shop due to queueing the fewer spare parts are in stock, worsening the delivery reliability in periods of high demand. The system time t_s, which is the sum of queueing time of spare parts t_q and repair time t_r, has a direct impact on supply chain performance. Actions taken to shorten queueing time, raise delivery reliability and lower loan costs c_l.

The loan costs c_l are a function of the purchasing price c_p, the number of spare parts n_p, the number of workers n_w the demand d, and the repair time t_r.

$$c_l\ (c_p,\ n_p,\ n_w,\ d,\ t_r) \qquad (4)$$

c_l = loan costs
d = demand
t_r = repair time for one spare part

Smaller values of d and t_r and raising n_p and n_w, diminish the loan time t_l as well as loan costs c_l.

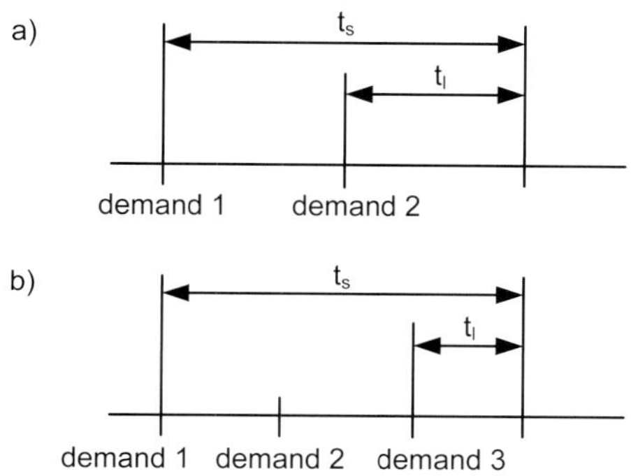

Figure 2: Loan time t_l.

Assuming n_p =1, the part is taken out of stock, when demand 1 occurs (see Figure 2 (a)). If a second spare part demand occurs during t_s of the first part, the demand has to be satisfied by loaning. The remaining time for the first part to get back to stock is the loan time t_l. Assuming n_p=2, a part has to be loaned, when demand 3 occurs (Figure 2 (b)). The remaining t_l is on average shorter than in case of n_p=1.

The loan time t_l is a function of t_s and n_p. The greater n_p and the smaller t_s are, the smaller t_l becomes.

$$t_l = \frac{t_s}{n_p + 1} \tag{6}$$

5 SIMULATION RESULTS

In this chapter impacts of seasonal demand on the supply chain are investigated. Variation of the number of spare parts n_p and the number of workers n_w while keeping other parameters constant, shows cost optimal solutions As introduced above, the function used for calculating c_t is:

$$c_t = c_c\ (c_p,\ n_p) + c_r\ (n_w,\ c_w) + \ c_l\ (c_p,\ n_p,\ n_w,\ d,\ t_r)$$

Ten year simulation runs generate the results discussed in the following. For every different n_p or n_w the simulation runs are repeated ten times, delivering an expected mean value of total costs c_t and system time t_s as well as a confidence interval with a confidence coefficient of 0.95. Calculating these values, the first year of every simulation run is neglected to allow for settling time of transient effects. The mean and standard deviation of the exponentially distributed repair time t_r=3 days. Demand is 40 parts a year.

5.1 Homogenous Poisson Distributed Demand without Seasonal Pattern

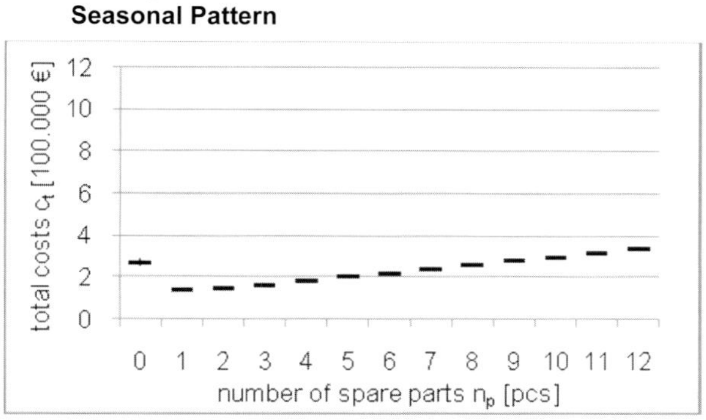

Figure 3: Total costs c_t with one worker, non-seasonal.

As a reference the total costs for homogenous Poisson distributed demand of 40 pieces a year are analyzed. Mean queueing time of the spare parts and necessary inventory levels for this scenario could also be calculated with standard approaches presented in [8].

Figure 3 shows the change of total costs c_t depending on the total number of spare parts n_p in the supply chain. If n_p=0 every spare part demanded has to be loaned from a competitor. The figure includes the mean value and the upper and lower limits of the confidence intervals for every n_p. For n_p>2 the confidence interval narrows to almost the mean value. The optimal total costs $c_{t,min}$ incur, if n_p=1.

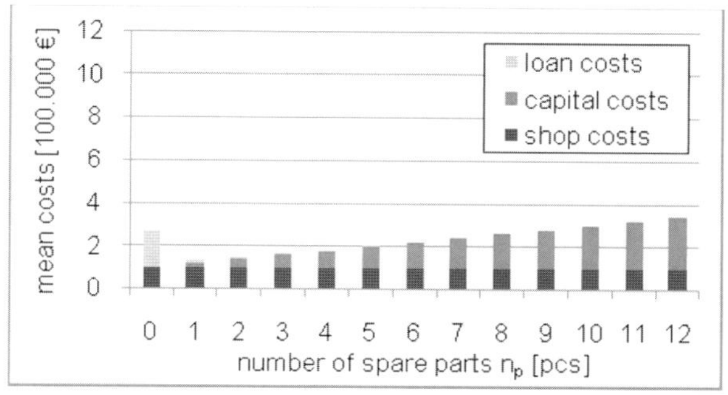

Figure 4: Proportion of cost types, one worker, non-seasonal.

The proportions of different cost types included in total costs for the scenario of one worker and non-seasonal demand are displayed in Figure 4. Repair shop costs c_r stay constant and capital costs c_t rise, when n_p increases. Loan costs are only significant if there is no part in the system (n_p=0).

5.2 Poisson Distributed Demand with Seasonal Pattern

Contrary to the homogenous distribution discussed above, the demand considered here has a seasonal pattern, the majority of the 40 parts demanded accumulating at the end of the year. Compared to the first quarter, demand rises three-fold in the 3rd quarter and fifteen fold in 4th quarter (see Table 1).

Table 1: Seasonal demand.

quarter	demand [pcs]
1	2
2	2
3	6
4	30

One Worker

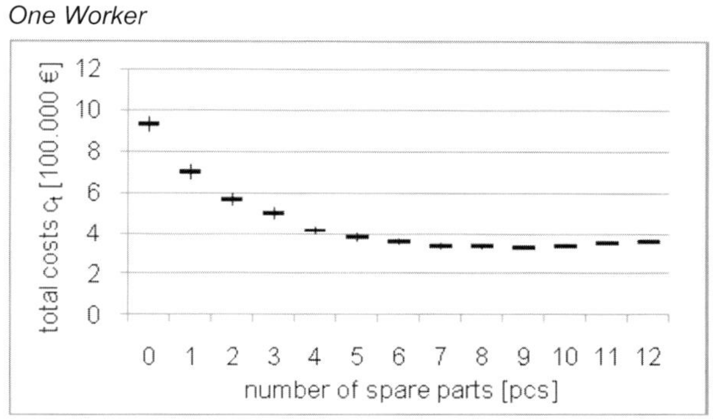

Figure 5: Total costs c_t with one worker, seasonal demand.

Figure 5 shows the total costs for seasonal demand with one worker in the repair shop over the number of spare parts. Minimal total costs $c_{t,min}$ can be realized with n_p=9. A hundred simulation runs were necessary for this scenario in order to narrow the confidence interval and achieve satisfying accuracy.

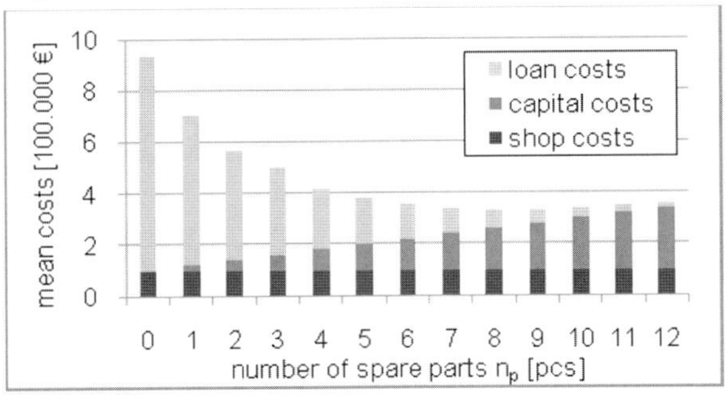

Figure 6: Proportion of cost types, one worker, seasonal.

Figure 6 specifies the proportions of different cost types included in total costs for one worker and seasonal demand. Loan costs c_l decrease with increasing n_p values, because of diminishing loan probability. They have the greatest share of total costs, if n_p<5.

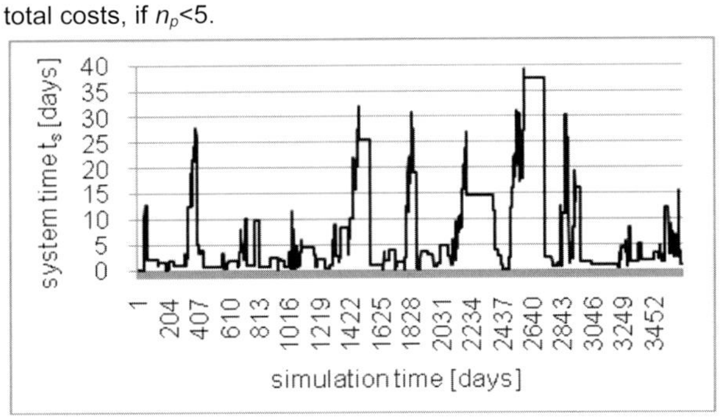

Figure 7: System time t_s of spare parts in the repair shop.

The system time t_s of the scenario with one worker and seasonal demand is plotted in Figure 7. In periods of high demand the worker's repair rate is lower than the rate of spare parts arriving at the shop. As t_s>15 days in many cases, additional workers seem useful to reduce queueing time.

Two workers

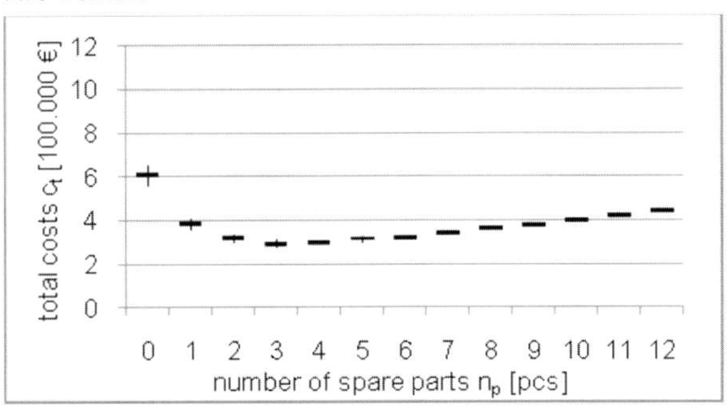

Figure 8: Total costs c_t with two workers, seasonal.

Figure 8 shows the total costs c_t with two workers over the number of spare parts. The lowest total costs $c_{t,min}$ arise with n_p=3.

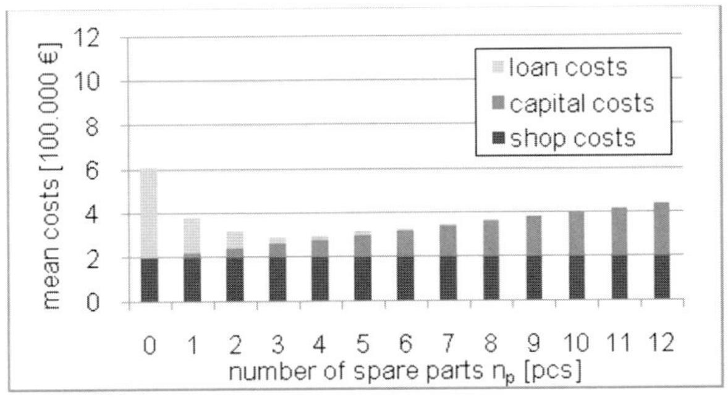

Figure 9: Proportion of cost types, two workers, seasonal.

The proportion of different cost types for two workers and seasonal demand are displayed in Figure 9. Loan costs c_l have a significant share of total costs only if n_p<3. If 1<n_p<10, repair shop costs incur more than half of the total costs.

Three workers

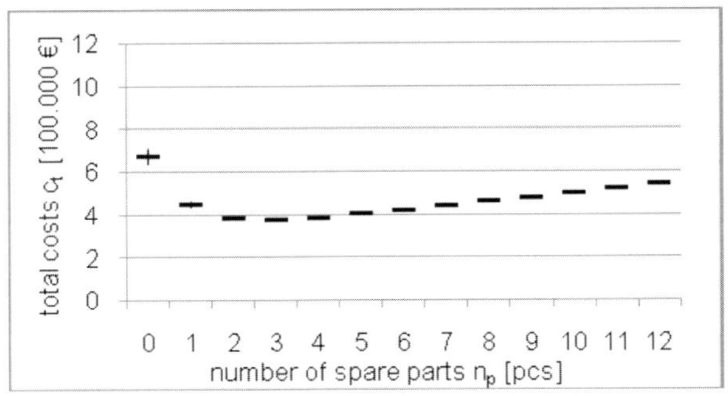

Figure 10: Total costs c_t with three workers, seasonal.

Figure 10 displays the total costs c_t with three workers over the number of spare parts n_p. The optimal n_p is the same as in the scenario with two workers. The total costs c_t, however, are lower if only two workers are hired.

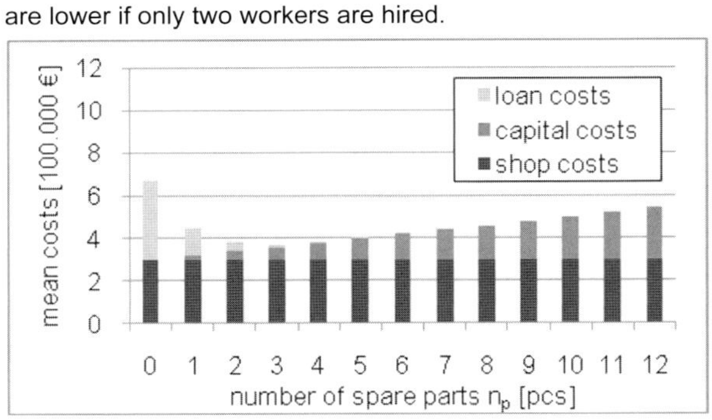

Figure 11: Proportion of cost types, three workers, seasonal.

Figure 11 presents proportions of cost types for three workers. The results are similar to the scenario with two workers. The greater repair shop costs c_r due to the additional worker hardly decrease loan costs c_l.

System Time t_s

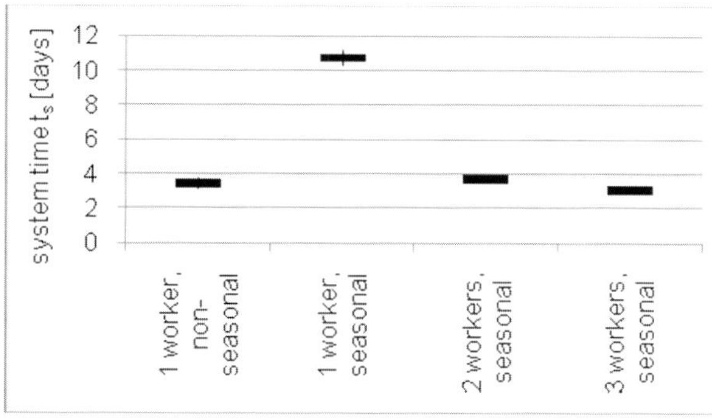

Figure 12: System times t_s of the cost minimal scenarios.

Figure 12 shows the mean system time t_s of the spare parts for the case of minimal total costs c_{tmin} of every scenario. For n_w=1 and non-seasonal demand the mean t_s=3.45 days. Taking into account that mean repair time t_r=3 days, the parts only spend 0.45 days queueing on average. With seasonal demand and one worker, the mean system time t_s=10.73 days within a wider confidence interval. Hiring a second worker lowers the mean of t_s to almost the mean it has with non-seasonal demand. As a third worker hardly influences t_s, the savings in loan costs do not compensate higher repair shop costs c_r.

Optimized Total Costs

Table 2 gives an overview of the mean total costs c_t depending on the number of spare parts n_p and the number of workers n_w. The grey shaded cells highlight the cost optimum of each column, the dark grey shaded highlights the overall optimum. Lower and upper limits of the confidence intervals are omitted for simplicity. The overall minimum total costs $c_{t,min,all}$= 293,659 € are realized with n_w=2 and n_p=3. Comparing the minimum of non-seasonal and seasonal demand with one worker, one can see that seasonal demand raises costs by 187,324 € or 136 %. Optimizing the supply chain with a second worker lowers costs by 30,676 € (9.5%), but costs are still significantly higher than without seasonality.

Table 2: Mean total costs c_t [€].

n_p	1 worker Non-seasonal	1 worker seasonal	2 workers seasonal	3 workers seasonal
0	270,079	1,007,638	608,284	670,542
1	137,011	685,507	383,753	446,997
2	143,152	565,291	316,561	383,443
3	161,331	497,077	293,659	372,274
4	179,830	440,353	298,745	384,751
5	199,840	448,092	315,644	403,255
6	219,745	345,722	321,199	420,086
7	239,702	338,307	340,473	439,702
8	259,660	324,335	359,785	459,660
9	279,617	332,052	379,617	479,617
10	299,575	337,044	399,575	499,575
11	319,532	348,459	419,532	519,532
12	339,489	359,335	439,489	539,489

6 CONCLUSION

In this paper a closed-loop supply chain in the aviation industry with limited repair shop capacity is analyzed and solutions to cope with seasonal demand are proposed. As calculation methods and analytical approaches known so far, do not deal with queueing phenomena in closed-loop supply chains, a simulation model was developed to forecast total costs in case of inhomogeneous demand.

The simulation model considers repair shop capacity and calculates total costs of providing spare parts during one simulation run. The total costs include capital costs, worker wages, and loan costs that incur when a part has to be loaned because of empty stock. Stochastic influences on demand and repair time are also taken into account. Due to queueing during periods of high demand, system time in the repair shop (queueing and repair time) and total costs rise and make optimization necessary.

In the initial scenario one worker repairs spare parts, which arrive at the shop. During peak periods the low repair rate increases system time of spare parts in the shop to more than ten days and causes queues to grow long. Hiring an extra worker reduces the mean system time of the spare parts in the shop by almost seven days. Shortened system time allows for reduction of the number of spare parts in the system from eight to three. Optimization by hiring an additional worker and by reduction of inventory level, lowers total costs by 9.5 %.

The presented simulation model is a first approach to optimizing closed-loop supply chains with queueing effects. As environmental issues become more important, an increasing number of parts will be recycled and led back to the original equipment manufacturer. Rising costs for disposing of used parts, will make reuse of low cost products more attractive. Developed for cost forecasting of expensive high-tech spare parts, the simulation model can also estimate costs of repairing low-cost items and their impacts on queues in repair shops.

7 REFERENCES

[1] Regattieri, A., Gamberi, M., Gamberini, R., Manzini, R., 2005, Managing Lumpy Demand for Aircraft Spare Parts, Journal of Air Transport Management, 11: 426-431.

[2] Ghobbar A.A., Friend, C.H., 2002, Sources of Intermittent Demand for Aircraft Spare Parts within Airline Operations, Journal of Air Transportation Management, 8: 221-231.

[3] Ghobbar, A. A., Friend, Chris H., 2004, The Material Requirement Planning System for Aircraft Maintenance and Inventory Control, Journal of Air Transport Management, 10: 217-221.

[4] Willemain, T.R., Smart, C.N. Schwarz, H.F., 2004, A new Approach for Forecasting Intermittent Demand for Service Parts Inventories, International Journal of Forecasting, 20: 375-387.

[5] Syntetos, A. A, Boylan, John E., 2005, The Accuracy of Intermittent Demand Estimates, International Journal of Forecasting, 21, 303-314.

[6] Syntetos, A.A. Boylan, John, E., 2006, On the Stock Control Performance of Intermittent Demand Estima-

tors, International Journal of Production Economics, 103: 36-47.

[7] Syntetos, A.A. Boylan, John, E., Croston, 2005, On the Categorization of Demand Patterns, Journal of the Operational Research Society, 56: 495-503.

[8] Kleinrock, L., 1975, Queueing Systems, John Wiley & Sons.

[9] Cruz, F., Duarte, A. van Woensel, T., 2008, Buffer Allocation in General Single-Server Queueing Networks, Computers and Operations Research, 35, 3581-3598.

[10] Ashayeri, J., Lemmes, L., 2006, Economic Value Added of Supply Chain Demand Planning: A System Dynamics Simulation, Robotics and Computer-Integrated Manufacturing, 22: 550-556.

[11] Terzi, S., Cavalieri, S., 2004, Simulation in the Supply Chain Context: a Survey, Computers in Industry, 53: 3-16.

[12] Tracht, K., Schneider, D. Schuh, P., 2010, A Feedback Control System for Spare Parts Logistic Networks in the Field of Civil Aviation, 7th CIRP International Conference on Intelligent Computation in Manufacturing Engineering, Capri, Italy.

8 ACKNOWLEDGEMENTS

The results presented in this paper were developed in the research project dLP – dynamic LRU planning (03CL02H). dLP is part of the Aviation Cluster Metropolitan Region Hamburg funded by the German Federal Ministry of Education and Research (BMBF).

Sustainable Cooperation in Networks

Evaluating the Sustainable Implementation of Logistic Concepts in Networks

J. Helmig, J. Quick, H. Wienholdt, T. Brosze

Research Institute for Operations Management, RWTH Aachen University, Germany

Abstract

Since the millennium the business conditions for companies changed significantly. Rising energy cost and various legal regulations concerning environmental aspects like CO_2-emissions foster the need of an ecological-economical integration in entrepreneurial decisions. In addition, companies have to handle a high variety of parts, distribution channels and variants. These circumstances require the establishment of a network between the partners in the supply chain. Because of the specific current economic situation companies finally have to face a tight liquidity situation.

Supply Chain Management (SCM) can address these changes. It includes different approaches for logistic concepts which characterize the logistics between customers, suppliers and logistic service providers. Sustainability matters need to be included in SCM in order to guarantee ecological-economical integration.

In this context choosing the best combination of cooperation concepts and services for the network is an urgent question for companies. This paper presents a dynamic and quantitative approach for evaluating the sustainable application of logistic concepts in networks. The result is a practical decision support which enables companies to identify adequate alternatives based on different scenarios.

Keywords:

Dynamic; Evaluation; Logistic Service Provider; Supply Chain Management; Sustainability

1 COMPANIES FACE THREE MAJOR CHANGES

At the beginning of the 21st century the world markets are dominated by a deregulation of trade relations, political and economical integration as well as a trend to a globalization of the corporate activities [1]. Despite great efforts in reducing the variety of parts, distribution channels and variants are still increasing [2]. Companies face tightened market requirements with short reaction times, high delivery reliability, increasing cost pressure and shorter product life cycles. As a consequence, it is necessary to respond with an entire and intact market entity [3].

In case companies build up such an entity, it is possible for them to react on the described changes with specific measures. One possibility to face the variant circumstances is to establish production networks with locations all over the world [4]. The main targets of those networks are cost reduction, inventory reduction, shorter reaction times and benefiting from synergy effects [5].

Beside those economic arguments for cooperations between different companies in networks, ecological questions become more important because of the public discussions in the 21st century concerning the sustainability of economic activities.

Based on general reflections in the 1970s regarding the economic growth limit [6] a discussion concerning the involvement of ecological aspects in economic questions and decisions came up [7].

During the following years, the relevance of ecological aspects in economical decisions rises. The raw material stocks will continue to decrease because of the accelerated growth of the emerging markets like China and India [8]. Furthermore the competitions for bottleneck resources like fossil fuels or iron ore for steel production led to worldwide price increases [9]. Additionally, various governmental activities created new policies like e.g. the CO_2 emission based taxation of vehicles. Companies face certain legislative regulations, which require a clear definition of the sustainability of their products. This becomes mandatory for products and services by so called Environmental Product Declarations (EPD´s). Those policies have a great impact on economical questions.

Beside those ecological aspects it is necessary to include also the special economical situation at the end of the 2000s. Such a negative environment in the financial market has a destructive impact on the investments in real economy. On the one hand declining payment behaviour and rising risks for bad debt losses make the collaboration even with established customers very difficult. On the other hand insolvencies of key suppliers endanger the material availability. The protection of the own liquidity as well as those of the partners in the network is an essential measure for economic survival [10].

Hence, environment of companies during the last decade is affected by three main impacts. The first impact deals with the established challenges in economy. Examples for this group are the increasing cost pressure or the rising number of product variants. Those challenges are supplemented with the second impact concerning the ecological aspects like the availability of resources and the emission production. The third influence involves the current business situation with a low degree or even absence of liquidity.

2 ARISING LONGTERM CHALLENGES

Companies can execute different measures regarding the established requirements of the millennium. Those steps are for example analysing the structure of the value creation or setting up a network with other companies based on cooperation concepts. It is problematic in this context that the organizational structures in companies normally do not support taking such a strategic decision. Furthermore those decisions often require investments which are regarded critical in the current situation. Based on the established challenges (like increasing cost pressure or variety) it is problematic to identify the best cooperation concept for the specific network. Companies lack resources for analyzing the situation holistically and are furthermore not willing to take the investment risk.

In scientific discussion different approaches for handling the specific economic situation came up. The main focus is to ensure liquidity not at least due to the difficult availability of credits on the financial markets. The result is that only companies with an appropriate liquidity are capable of acting or even surviving [11]. As a consequence the quick identification of activities that ensures liquidity gain in importance.

The restructuring of the supply chain and the assignment of services to external providers are seen as possible approaches. Generally, the effect is to adapt the depth of added value, which increases in turn the performance of the own company [12]. In reference to liquidity, companies have to take the decision to what extent they can allocate their logistic services and which cooperation concept should be applied.

It should be kept clearly in mind that the discussion concerning in- and outsourcing in real economy still existed prior to the financial crisis. Depending on different reasons the outsourcing decision event at those times was not always a successful one. Companies have to consider long-term consequences as well instead of focussing on quick fixes concerning liquidity. Quick decisions might increase the short-term liquidity but maybe also exert negative influence on the competitiveness.

Regarding the ecological aspects the main challenge is integrating the relevant ecological parameters into the economical decision making process. The basic problem is risen by the reduction of resource availability because of low market power on the supply side for most companies. This leads the affected companies to accept the exogenous pricing process. Concentrating on the efficiency of using resources is an appropriate measure in this situation.

Especially small and medium sized entities (SME) have limited possibilities of lobbying concerning the regulations for emissions. In fact they have to implement the new legislative provisions (like trading with CO_2 certificates or emission badges) without having an impact on the changes. Consequently also emission reduction measures come to the fore. The fact is that small and medium sized companies are not able to influence the supply side prices or the legislative conditions. This is why SME´s often pose the question: Which internal measures take ecological as well as economical parameters into account?

The integration of economical and ecological aspects in the company's system of performance indicators defines sustainable action. Social components of sustainability will not be contemplated in this approach. The described current situation led companies to three relevant problems. They have to make decisions and implement measures for solving those problems.

The answer to face the established requirements like a high level of cost pressure or the increasing variants is to form networks. Appropriate cooperation concepts are able to support this process. Second, it is indispensable to incorporate the long-term consequences while looking at short-term measures for protecting liquidity. Therefore it is necessary to check the benefit of logistic service providers supporting the execution of cooperation concepts and the short- and long-term effects. Third, ecological and economical parameters have to be taken into account. This leads to the requirement for sustainable performance indicators.

On the one hand, the initial situations like ensuring the liquidity and on the other hand the way of handling problems like the introduction of cooperation concepts are part of a complex interdependency. It is essential to consider the whole field of interest for an integral examination. There are internal relations between the different dimensions in the initial situation as well as in the group of the possible fields of action. As a consequence it is not possible to regard exclusively the need to form networks without looking at the liquidity availability or the ecological-economical integration. Furthermore, the introduction of cooperation concepts has an influence on the analysis of a possible outsourcing decision and on the sustainable performance indicators.

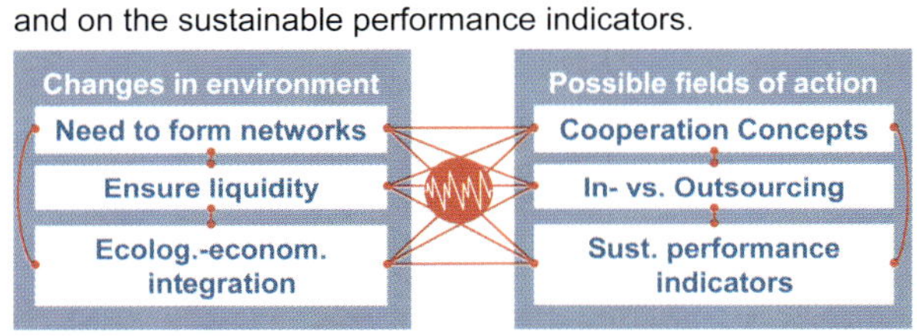

Figure 1: Changes in environment and possible fields of action.

In addition, there are dependencies between the initial situation elements and the possible field of action elements. The theoretical measures for choosing a cooperation concept might be influenced by all of the identified initial situation elements. The need to form a network, ensuring the liquidity and the ecological-economical integration has an impact on the outsourcing decision to a logistic service provider.

The result is a wide variety of influences and dependencies that must be considered. Besides regarding the context relations between the relevant elements the chronological aspect is important as well. In general the measures concerning the cooperation concepts and the sustainable performance indicators have a long-term view. The implementation of a cooperation concept is designed for several years. Also the integration of sustainable performance indicators into the company's system of performance indicators is a long-term process.

In contrast, ensuring liquidity seems to be a quick and short-term measure. The downside is that making quick decisions concerning the outsourcing of services can show negative consequences in the long run. As a result, short term and long term effects of an outsourcing decision have to be considered simultaneously. All three elements of the possible field of action are relevant in regard to the company's strategic orientation. Therefore they have to be considered in a holistic and simultaneous way. The chronological inspection taking into account the dependencies concerning the context leads to a dynamic conjunction. Figure 1 summarizes this dynamic conjunction.

The depicted complex interactions between the system elements have to be detected by applying an appropriate modelling method. At least this model has to be approachable for a scenario analysis.

3 EXISTING APPROACHES TO MEET THE CHALLENGES

Based on the outlined problems, this chapter depicts different existing approaches to meet the arising challenges. Special attention is paid to approaches which consider sustainability explicitly and those which cover dynamic system theory.

3.1 Sustainability

The idiom sustainability originates from forestry. It means in this context that regenerative and living resources are only used to such an extent that allows a natural replacement [13]. Meadows used the expression sustainability in 1972 in economic terms for the first time. During the 1980´s the notion of sustainable development was established by the UN committee for environment and development (Brundtland-Committee) as an objective for global development in a non-scientific context [14]. Meanwhile many different definitions of sustainability have been developed. Most of them refer to the same core ideas, which are in general to preserve the future and to ensure safe paths of future development

Besides the different aspects of several definitions, there is one aspect which is usually shared in a common agreement in terms of defining sustainability. It is argued in general, that sustainability provides the three dimensions of economical, ecological and social sustainability which are intensely connected to each other and therefore need to be coordinated in balanced way in the long run.

Especially the integration of economical and ecological aspects of sustainability is of importance for the further development of this paper. Companies have a direct impact on ecological issues like for instance the use of natural resources by deciding on economical questions like investments, product developments, changes in production or education. They have an additional indirect impact by promoting certain lifestyles through commercials or by supporting lobbyists who influence political frame conditions or decisions [15].

Sustainability is defined for the use in this paper as the integration of aspects of the mentioned ecological and economical dimension. Social concerns are not considered. The objective of this paper is to develop a methodology to evaluate the sustainable implementation of logistic concepts. Therefore sustainability as the integration of ecological and economical aspects means, that companies ought to consider classical performance indicators as well as ecological parameters. For example, turnover, growth or capital employed are as equally important as CO_2 emission or the carbon footprint of the distributed products [16].

3.2 Systems Thinking and System Dynamics in Supply Chain Management

We have elaborated above, that there are a couple of different factors in a company´s environment which are strongly connected to each other. In order to analyze the impact of a logistic concept on relevant parameters, a holistic view of the different factors needs to be implemented. Such holistic approaches are covered by the fields of Systems Thinking and System Dynamics in research literature. At first we will have a closer look at the background of these relevant idioms and second some already existing applications of these approaches to Supply Chain Management.

Systems Thinking

In 1958 Jay W. Forrester published at Massachusetts Institute of Technology (MIT) research results, which dealt for the first time with interchanging dependencies between factors of an industrial production environment [17]. His work founded the approach of Systems Thinking at the interface between engineering and business science. The idea was developed further during the following 40 years, so that John D. Sterman defines Systems Thinking as follows: Systems Thinking is the ability to understand and to see the world as a whole. It is essential to realize, that all considered elements are linked to each other. This means that an action at one spot leads to consequences in other spots [18]. Senge shares this point of view. In his work, Senge describes certain disciplines, which establish a learning organization. Systems Thinking is according to Senge the fifth discipline, which continuously improves the systems performance [19].

System Dynamics in Supply Chain Management

There are different means in research literature which enable the transfer of the theoretical approach of Systems Thinking into practical life. One example is the System Dynamics Method which offers different tools for analyzing dynamic systems.

Scientific literature provides many different sources that describe a standardized process for implementing the System Dynamics approach. This process usually comprises 5 to 7 steps [20]. We will focus on the 5-step process further on.

The first step is the problem articulation, followed by dynamic hypothesis about the connection of the several parameters. At this point, the problem is usually split up into sub-problems which leads to sub-model-structure in order to facilitate the further process. Third, the whole structure needs to be designed in terms of a simulation model. This means that the quantitative relations are identified e.g. by defining default values, which are afterwards transferred into a simulation software. The simulation model is tested on its validity in the fourth step. The fifth and last step is to derive consequences for a decision situation. This part develops a decision support methodology which supports the user in terms of the dynamic problem.

There is a variety of different models of System Dynamics concerning Supply Chain Management [22]. An/Ramachandran provide an extensive overview of various approaches which show remarkable similarities [23]. These models usually deal with logistical cooperation concepts between two companies, especially regarding parameters like return, time or cost.

4 A NEW METHODOLOGY TO EVALUATE THE SUSTAINABLE IMPLEMENTATION OF LOGISTIC CONCEPTS

The existing approaches to describe the dynamic situation of companies in their environment consider up to now the interface between two companies and a focussed selection of cooperation concepts. With regard to the arising challenges, some more crucial aspects need to be considered. It is not sufficient to deal only with a selection of cooperation concepts. The entire variety needs to be covered. The idea of sustainability in terms of the integration of economical and ecological parameters is not treated either in the current discussion. Hence, in this chapter a new methodology for the dynamic evaluation of the sustainable implementation of logistic concepts is developed. The development comprises several steps that are now described in detail.

4.1 Analysis of network-modules

The first step in order to develop the methodology is to define the elements that need to be analyzed. Since the focus on networks, the relevant elements of such a network need to be identified. These components can be structured to modules which finally constitute a network.

This step enables companies to draft the relevant part of their network. Possible examples for these network elements can be Sources, Drains and Service Providers. The specific activities are detailed in this step.

4.2 Analysis of network exchange relations

Once the several elements of the network are defined, every element shows exchange relations to another element. There are in general three possible flows which can be exchanged between the partners. The flows are depicted by their certain characteristics. First there is the material flow, which comprises certain products and goods. These are usually characterized by product specifications like size, weight, lot sizes or delivery times.

The partners exchange besides mere materials second also information. That information can be e.g. data about planned promotions, demand forecasts or data about planned restructuring of plant locations. As a third aspect, ecological flows need to be considered. Every transport causes for instance a certain amount of carbon dioxide. In terms of a sustainable evaluation, these aspects need to be considered as well. Step 1 and 2 form the basis for the description of a network. Not every identified module or exchange relation will be relevant for every specific network. Important parts can be chosen by the user and used according to strategic frame conditions like already existing partnerships to other companies or certain objectives.

4.3 Analysis and structuring of logistic concepts

Based on possible exchange relations between partners, there are several concepts in the scientific discussion which shape the theoretically possible exchange relations according to certain guidelines. These concepts are a combination of inter-organizational cooperation concepts like for instance Just in Time or Vendor Managed Inventory. Companies usually combine the concepts with certain aspects of a logistic service provider. There are many different logistic concepts in scientific discussion. This leads to a mere undistinguishable complexity of concepts for companies. In order to reduce this complexity, possible cooperation concepts need to be clustered. This is done in this step so that the following steps use the condensed clusters of concepts for the further development of the approach.

4.4 Identification of relevant performance indicators

The goal of the methodology is to evaluate the sustainable implementation of logistic concepts. In order to be able to evaluate, relevant performance indicators need to be identified. By analyzing the different values of a relevant indicator for the specific logistic concept, the impact of this concept on each parameter becomes clear and forms the basis for a later decision.

By identifying these parameters, it is important to take classical economical elements like ROCE and ecological aspects like Carbon Footprint into account. This step provides a pool of different parameters, which are not necessarily arithmetically convertible into each other. Such a quantitative test will follow in a later section in the field of System Dynamics. The indicators should be used for every part in the network locally, so that a clear comparison of a logistic concept´s impact on every partner in the network is possible.

4.5 Analysis of dependencies between exchange relations and performance indicators

After the different elements of a network, possible logistic concepts and relevant indicators are described, it is possible to study the dependencies between the different aspects. It would be most interesting to announce the relation between logistic concepts and performance indicators directly. Unfortunately such a direct link is too difficult, since the comparison of goal indicators and general concepts is not specified enough. Due to this fact, we will at first have a look at the dependencies between exchange relations in a network and the performance indicators.

To clarify the dependencies, a matrix is set up, which lists all possible exchange relations on the one hand and all identified performance indicators on the other hand. In this matrix the connection between every indicator and every characteristic of exchange relation is analyzed. If there is a dependency identified, this will be indicated by a filled circle (see Figure 2). Once every dependency is marked, it is possible to study the kind of connection that has just been indicated. This will be done by using the System Dynamics approach. With the help of this approach the way in which the different parameters are linked to each single characteristic of the exchange relations can be described.

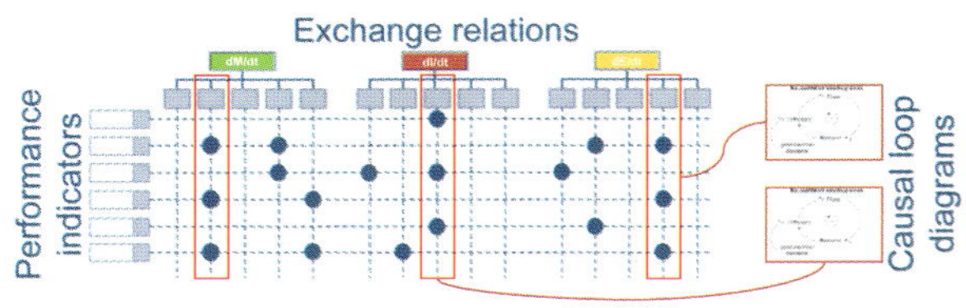

Figure 2: Causal matrix between performance indicators and exchange relations.

By using System Dynamics, dependencies between elements can be described in a standardized manner. In turn of this analysis, all relevant performance indicators per characteristic are marked and examined towards their kind of connection. The kind of dependency can be depicted by seven basic types of dynamic behaviour. This can be achieved with the help of so called causal loop diagrams. A basic structure enables during the ongoing process a simulation of the specific dependencies.

4.6 Analysis of dependencies between exchange relations and specific classes of logistic concepts

The previous step described the dependencies between exchange relations and performance indicators. This step is going to indicate the connection between exchange relations and classes of logistic concepts. In turn of this step, we will use again a matrix in which on the one hand exchange relations and on the other hand structured classes of logistic concepts are listed (see Figure 3).

Using the exchange relations in both matrices enables us to establish a link between the chosen classes of cooperation concepts and performance indicators in the following step. The exchange relations are used as a help to link the generic logistic concepts to the performance indicators.

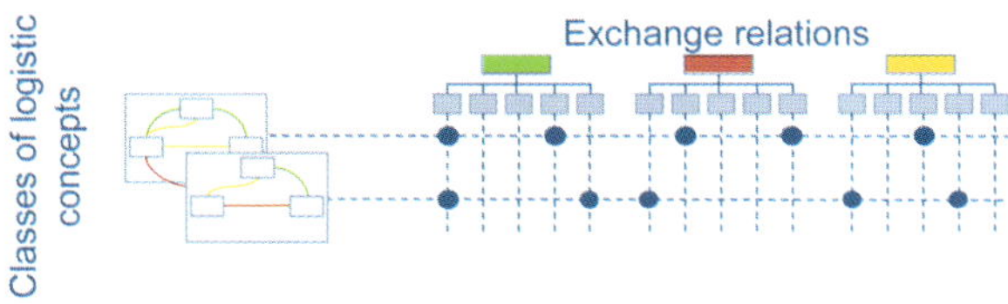

Figure 3: Causal matrix between classes of logistic concepts and exchange relations.

4.7 Derivation of a causal-model for specific classes of logistic concepts

The results of both previous steps are now combined and form the basis for a holistic causal-model for every class of logistic concepts.

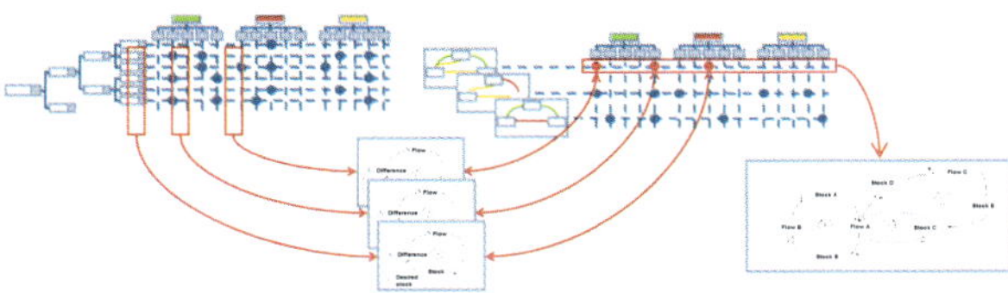

Figure 4: Derivation of causal model for specific classes of logistic concepts.

The first causal matrix delivers a causal loop diagram for every characteristic of the exchange relations. Examples of such diagrams are shown in the middle of Figure 4. This information is "deposited" in the second causal matrix on the right hand side of the Figure . It becomes clear with the help of the second causal matrix which causal loop diagrams play a role for which class of cooperation concept. By using these marks, all relevant causal loop diagrams for a class of logistic concepts are identified. This selection of diagrams for the chosen classes is now put together to another, bigger causal loop diagram that describes the dependencies between the single causal loop diagrams. Such a consolidated diagram finally describes the connection among relevant performance indicators for the chosen class of logistic concepts (see Figure 4).

4.8 Designing a simulation model

The causal loop diagrams for the specific classes of logistic concepts provide the basis of a quantification of the considered elements. This means that e.g. differential equations have to be set up, default values for relevant parameters have to be defined or decision rules should be implemented. Identifying the correct approaches for a quantitative description is a challenging task. At this point, the 5 step method, elaborated earlier will be used in order to create a valid model.

Once the model has been designed, it needs to be tested on its validity. Especially tests regarding robustness and sensitivity are important and will be conducted in this stage. At that point, an evaluation of the sustainable implementation of logistic concepts is possible.

4.9 Deriving a decision support mechanism for the evaluation and selection of logistic concepts

Finally, the scientific results in shape of the simulation model need to be transferred to a practical decision support mechanism that can be applied by companies. This last step is an important one, since it makes the general scientific results accessible for companies. Thus, a clear four-step approach is designed to guarantee the practicability of the research results.

At first the relevant part of the network is described and schematically depicted. A selection of customer, vendor and service provider relations are included. Second, relevant performance indicators and specific current and planned values for the performance indicators are recorded. The planned values are the basis for the later evaluation of the different scenarios. In a third step, these values are entered into the simulation model. Accordingly specific results from the simulation for relevant indicators are retrieved. These results form the basis for the scenario analysis. They are compared to the planned values and provide the opportunity to decide, which class of logistic concepts is the most suitable one. This comparison is conducted in the fourth step and reveals the logistic concept which fulfils best the planned values.

An example for the implementation of this method can be for instance the following one: A vendor and a customer are linked in their supply chain by a logistic service provider via Just-in-Time delivery. The customer sets up a new sustainability strategy which is why ecological performance indicators gain actually in importance in his local view. So the customer uses the proposed simulation model and checks the adequacy of the Just-in-Time delivery concept in terms of his new sustainability strategy. He finds out, that a Vendor Managed Inventory serves his purposes better, which is why he then shifts his logistic cooperation from a Just-in-Time delivery to a Vendor Managed Inventory.

5 SUMMARY

Changed business conditions like the rising importance of ecological factors lead to the need to evaluate possible logistical cooperations with regard to sustainable aspects. A dynamic approach based on System Dynamics is presented to address this need and to come up with a practical decision support mechanism for companies. This support tool composes 4 steps and is free to use. By using the decision support, companies can additionally realize some advantages. So it is possible that the enabled transparent evaluation of a logistic concept leads to a quick beneficial change in liquidity of the company. The application of the decision support tool does not use a lot of resources and might reveal beyond that additional, unknown potentials in the company. Finally, companies which use the decision support tool get an insight in the way other companies work, since the methodology shows a cross-section of company relevant issues.

6 REFERENCES

[1] Webster, P. (2008): Principles and Tools for Supply Chain Management. McGraw-Hill, Boston, MA, USA.

[2] Melzer-Ridinger, R. (2007): Supply Chain Management – Prozess- und unternehmensübergreifendes Management von Qualität, Kosten und Liefertreue. Oldenbourg Verlag, München/Wien.

[3] Bretzke, W.-R. (2006): SCM: Sieben Thesen zur zukünftigen Entwicklung logistischer Netzwerke. In: Supply Chain Management 6 /2006:7-15.

[4] Sydow, J. (2007): Management von Netzwerkorganisationen – Zum Stand der Forschung. In: Sydow, J. (Ed.): Management von Netzwerkorganisationen. Beiträge aus der „Managementforschung". 4th Edition. Gabler Verlag, Wiesbaden:387-472.

[5] Ijoui, R.; Emmerich, H.; Ceyp, M.; Diercjs, W. (2008): Supply Chain Event Management als strategisches Unternehmensführungskonzept. In: Ijoui, R.; Emmerich, H.; Ceyp, M. (Ed.): Supply Chain Event Management, Physica-Verlag, Heidelberg:3-15.

[6] Meadows, D. H.; Meadows, D. L.; Randers, J.; Behrens III, W. W. (1972): Limits to growth. Universe, New York.

[7] Meadows, D.; Randers, J.; & Meadows, D. (2004): Limits to growth, The 30-years update. White River Junction, VT, Chelsea Green.

[8] Meyer, B. (2009): Wie muss die Wirtschaft umgebaut werden? Fischer Taschenbuch Verlag, Frankfurt am Main.

[9] Bartelmus, P. (2008): Quantitative Eco-Nomics, Springer, London et al.

[10] Schuh, G. (2009): „Für Schönrederei ist der Wellengang zu groß". In: Industrieanzeiger 24/25 (2009):14/15.

[11] Haarman, H.; den Exter, E.; van der Schaar, J.; van Heijst, W. (2009): Crisis dominates the supply chain agenda in 2009. URL: http://www.de.capgemini.com/m/de/tl/Crisis_dominates_the_supply_chain_agenda_in_2009.pdf (Retrieved on 27.08.09).

[12] Wildemann, H. (2008): Entwicklungslinien der Logistik. In: Nyhuis, P. (Ed.): Beiträge zu einer Theorie der Logisitik, Springer Verlag, Berlin et al.:19-42.

[13] Ott, K.; Döring, R. (2008): Theorie und Praxis starker Nachhaltigkeit. Marburg, Metropolis-Verlag.

[14] Grundwald, A.; Kopfmüller, J. (2006): Nachhaltigkeit. Frankfurt am Main, Campus Verlag.

[15] Grundwald, A.; Kopfmüller, J. (2006): Nachhaltigkeit. Frankfurt am Main, Campus Verlag.

[16] Grundwald, A.; Kopfmüller, J. (2006): Nachhaltigkeit. Frankfurt am Main, Campus Verlag.

[17] Forrester, J. W. (1958): Industrial Dynamics. In: Harvard Business Review, July/August 1958:37-66.

[18] Sterman, J.D. (2000): Business Dynamics. Irwin McGraw Hill, Boston et al.

[19] Senge, P.M. (1990): The Fifth Discipline: the Art and Practice of the Learning Organization. Random House, London.

[20] Warren, K. (2002): Competitive Strategy Dynamics. John Wiley & Sons, London.

[21] Göttlich, S.; Herty, M. (2008): Dynamic Models for Simulation and Optimization of Supply Networks. In: Ijoui, R.; Emmerich, H.; Ceyp, M. (Ed.): Supply Chain Event Management, Physica-Verlag, Heidelberg:249-265.

[22] An, L.; Ramachandran, B. (2006): System Dynamics Model to Understand Demand Conditioning Dynamics in Supply Chains. In: Proceedings of IEEE International Conference on Services Computing (SCC'06), Chicago, USA:277-293.

An Integrated Supply Chain Problem with Environmental Considerations

N. Al Dhaheri, A. Diabat
Engineering Systems and Management, Masdar Institute of Science and Technology, Abu Dhabi, UAE

Abstract
In this paper, a novel formulation for a multiple product capacitated inventory-location supply chain model with risk pooling and carbon emission considerations is proposed. Various products are shipped from a single plant to retailers with stochastic demand via a network of capacitated warehouses in the presence of environmental regulations. The locations and inventory policies of the warehouses are chosen so as to minimize the sum of fixed facility location, transportation, inventory carrying costs, and emissions-related costs. The warehouses retain safety stock so as to maintain appropriate service levels in the face of uncertain demand at the retailers for multiple products. Computational results are also presented.

Keywords:
Green Supply Chains; Location-Inventory; Mixed Integer Non-Linear Programming; Multiple Products.

1 INTRODUCTION

The design of efficient supply chains has become all the more challenging in view of current environmental regulations such as the Kyoto protocol. The designer must make decisions on facility location, technology selection, inventory management, and distribution while considering the environmental repercussions of these decisions, and also while taking into account the additional flexibility allowed by the trading of carbon emissions credits. The designer must balance the oftentimes conflicting requirements of customers, investors and environmental agencies. In this challenging design environment, the importance of integrating the strategic, tactical, and operational decisions to minimize costs and emissions and to maximize the satisfaction of all concerned parties cannot be underestimated. Much of the supply chain literature considers strategic, tactical and operational level decisions separately; few articles deal with optimizing the different levels jointly, and even fewer involve multiple products with distribution centers (DCs) of limited capacity and environmental constraints.

In this paper, we study a multi-product capacitated inventory-location model with risk pooling and CO_2 emissions considerations, which considers the impact of tactical and operational level decisions (e.g., inventory management and distribution) on strategic decisions (e.g., facility location). Our model considers a three-tiered supply chain network consisting of suppliers; distribution centers (DCs) and retailers. This model integrates the traditional location problem with the inventory problem for multiple products, while considering capacity constraints, the risk pooling effect and the flexibility allowed by trading carbon emissions credits, as shown in Figure 1.

The remainder of the paper is organized as follows: Section 2 reviews the literature on multi-commodity location problems and sustainable supply chain design; Section 3 introduces our multi-product capacitated inventory-location model with risk pooling and carbon emissions considerations; Section 4 provides computational results and a sensitivity analysis; and Section 5 concludes with directions for future research.

2 LITERATURE REVIEW

Traditionally, the supply chain management literature has considered strategic, tactical and operational level decisions separately. Strategic level decisions have tended to be addressed by the location theory literature, while operational level supply chain decisions have tended to be addressed by the inventory theory literature. Since our work is very much related to location theory, inventory theory, integrated inventory-location models, and green supply chain designs we provide a short review of the literature in each of these areas.

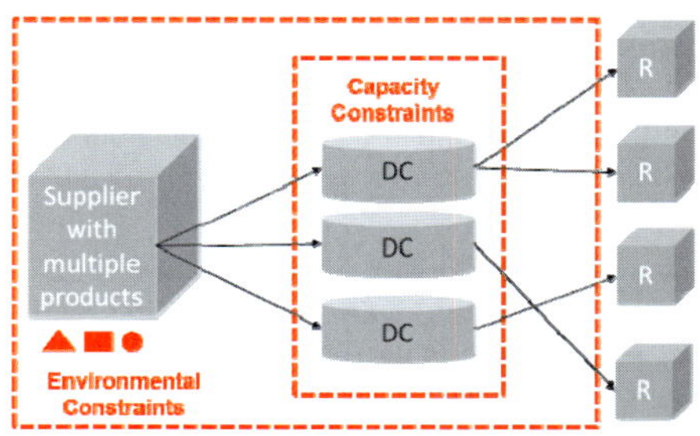

Figure 1: Constraints faced by the supplier and the DCs

The main objective of location theory is to optimize the number of DCs needed to satisfy the demand, identifying their locations and assigning retailers to them, while minimizing the fixed facility location and transportation costs. Prior to the early 1990's, most location models that appeared in the literature failed to consider inventory-related costs. Overviews of the location theory literature are provided by Daskin [1] and Drezner [2]. Adding capacity constraints that limit the quantity of product that each facility can handle gives the capacitated facility location problem (CFLP). The classical CFLP problem, as considered by, for example, Sridharan [3],

Tragantalerngsak et al. [4], Cortinhal and Captivo [5], and Liu et al. [6], is formulated as an integer linear programming model. The CFLP literature has also tended to overlook inventory-related costs.

The main objective of inventory theory is to optimize order times and order sizes to as to minimize costs. An overview of the inventory literature in the context of multi-echelon supply chains is provided by Clark [29]. The concept of risk pooling represented an important advance in inventory theory. Eppen [8] quantified the benefits of risk pooling, assuming uniform or negligible transportation costs. This work was later generalized by Federgruen [9].

More recently, integration of the location and inventory problems for single products has attracted much attention. Baumol and Wolfe [10] were among the first to consider incorporating inventory costs into location models by adding a square root term to the objective function of the uncapacitated facility location problem (UFLP). An overview of the single-commodity location-inventory theory literature is provided by Shen et al. [12] and Kaijun et al. [6]. The majority of these models either ignores the nonlinear inventory costs or approximate these costs; exceptions are the work of Erlebacher and Meller [13] and Daskin and Shen [14].

The multi-commodity location problem was considered by Warszawski [15] and Warszawski and Peer [16]; this work was later extended in a number of ways [11,17,18]. These models considered fixed location costs and linear transportation costs, and assumed that each warehouse stocks at most one commodity. Geoffrion and Graves [19] considered a capacitated version of the multi-commodity location problem which imposes capacity constraints on the plants and on the DCs. These models commonly assumed single sourcing, in which each customer must be served by a single DC or plant. Hind and Basta [20] relaxed the single sourcing assumption, and solved the resulting model using a branch-and-bound method. Lee [21] and Mazzola and Neebe [22] studied a multi-product capacitated facility-location problem with choice of facility type. Shen [23], Daskin et al. [14] and Shen et al. [12] studied a single-product location model with risk pooling (LMRP) for the case of stochastic demand. Shen [24] considered a multi-commodity version of the LMRP without a capacity constraint. Liu et al. [6] studied a capacitated version of the LMRP for a multi-channel supply chain and suggested a Lagrangian-relaxation-based algorithm to solve it.

More recently, international concerns over reducing carbon dioxide emissions have driven increasing interest in green supply chains, that is, supply chains that are designed taking into account environmental considerations. Srivastava [25] provides an overview of green supply chains. Ramudhin et al. [26] studied the impact of transportation, subcontracting, and production activities on the design of a green supply chain. They considered a multi-supplier, multi-plant, multi-product, and multi-retailer problem, but assumed that the facility locations and sizes are determined in advance. Diabat and Simchi-Levi [27] studied a carbon-capped supply chain problem in which the throughput and storage capacity of the manufacturing site and distribution centers are considered as decision variables.

The work of Ozsen et al. [28] is more specifically related to our work, since they study a capacitated LMRP that incorporates tactical inventory management decisions into strategic facility location decisions. Our model can be considered as the multi-product version of their capacitated model; it thus inherits the property of their model that a warehouse can order more frequently so as to lower the inventory levels at the warehouse, thus loosening the capacity constraints. In addition, we consider the environmental impact of the location and inventory decisions, and also allow for the possibility of trading carbon emissions credits.

In this article, we extend the capacitated warehouse location model with risk pooling of [28] to accommodate multiple products and the flexibility allowed by trading carbon emissions credits; this is one of the first green supply chain design models for capacitated warehouses that considers multiple products, economies of scale, and risk pooling. This model also captures the interdependence between inventory levels and capacity limitations at the DCs in the presence of emissions considerations. We describe the model and its formulation in the next section.

3 MODEL FORMULATION

In this section we introduce our multi-product capacitated inventory location model with carbon emissions considerations. This model integrates the traditional location problem with the inventory problem for multiple products, taking into account capacity constraints, the risk pooling effect and the flexibility allowed by carbon trading.

This work models the storage and the transportation of different products from a supplier to a set of distribution centers, which supply the demands of a set of retailers. We assume that the locations of the supplier and the retailers are known and that the supplier has the capacity to meet an arbitrarily large demand for each product. We further assume that the different products can be stored in the same warehouse within the available space depending on their specific volume, and that each retailer is supplied by a single DC via direct shipment. In addition, we consider the daily demands to follow a Poisson process [12,14,28]. Our problem can be stated as: given a set of candidate distribution centers, a set of retailers and a set of different products, determine the location of the DCs and their inventory policy – the reorder time, reorder quantity and safety stock – to serve retailers' demands for different products at minimum cost and while keeping emissions below the carbon cap, given a certain price for carbon emissions credits.

In this framework, for each DC to meet the customers' stochastic demand, it holds two different types of inventory: working inventory and safety stock. The working inventory is determined by the ordering policy at each DC, while the safety stock is determined by a constraint that keeps the probability of a stock-out below a specified threshold α.

We assume that carbon emissions come from two sources:

1. Emissions at the warehouses: the emissions are proportional to the volume of the warehouse.
2. Emissions due to the distribution of products: the emissions are proportional to the distance between facilities.

We define the following decision variables:

$$X_j = \begin{cases} 1 & \text{if a DC is opened at candidate site } j \\ 0 & \text{otherwise} \end{cases}$$

$$Y_{ijp} = \begin{cases} 1 & \text{if the DC at location j serves retailer i's demand for product p} \\ 0 & \text{otherwise} \end{cases}$$

Q_{jp}: the reorder quantity at DC j for product p

We use the following notation:

P	≜	set of products or product categories
I	≜	set of retailers
J	≜	set of candidate DC locations
f_j	≜	fixed annualized cost of opening a DC at location j
F_j	≜	fixed ordering cost from DC j to the supplier
g_j	≜	fixed order shipping cost from the supplier to DC j
a_{jp}	≜	per-unit shipment cost of product p from supplier to DC j
h_p	≜	annual holding cost for product p
v_p	≜	volume of product p
C_j	≜	inventory capacity at DC j (units of volume)
L_j	≜	lead time for deliveries from supplier to DC j
ϕ_p	≜	cost of transporting product p per unit distance
μ_{ip}	≜	mean daily demand of customer i for product p
β	≜	weight factor associated with transportation costs
θ	≜	weight factor associated with inventory costs
Ω_p	≜	CO_2 transportation emissions factor for product p (weight of CO_2 per unit distance)
ω	≜	CO_2 emissions factor of each warehouse (weight of CO_2 per unit volume)
CAP	≜	current carbon cap for the company
$CO2$	≜	Total CO_2 emissions from all activities
∞	≜	expected market price of emissions credits
κ_{ij}	≜	distance between retailer i and DC j
λ_j	≜	distance between DC j and the supplier
χ	≜	number of days in a year (= 365)
α	≜	Threshold on the probability of stock-outs
z_α	≜	$P(Z \le z_\alpha) = \alpha$ for Z a standard Normal random variable

We begin by analyzing the cost components and CO_2 emissions associated with the supply chain; these are shown in Figure 2.

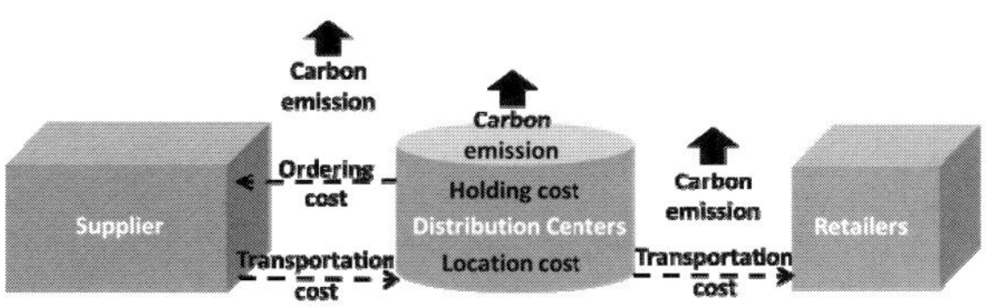

Figure 2 : Cost components and CO_2 emissions components

According to our assumptions, CO_2 emissions are produced by warehouses and by the distribution of products. The emissions level at a DC j is found by multiplying the volume of that DC, C_j, by the CO_2 emission factor ω, for each DC that is open. The distribution-related emissions level due to the demand for product p at a retailer i served by DC j is obtained by multiplying the demand for product p at retailer i, μ_{ip}, by the distance over which the product is transported, $\kappa_{ij} + \lambda_j$, and by the emissions factor for product p, Ω_p. Summing over all products, retailers, and DC's gives the total CO_2 emissions:

$$CO2 = \sum_{j \in J} \left(\omega C_j X_j + \sum_{i \in I} \sum_{p \in P} \Omega_p (\kappa_{ij} + \lambda_j) \mu_{ip} Y_{ijp} \right) \tag{1}$$

We assume that each DC uses a (Q, r) inventory policy with type I service level [14,28]. To determine the inventory policy we adopt the two step approach used by Ozsen et al. [28]. The first step is to determine the order quantity (Q) in the EOQ model, and the second step is to determine the reordering point (r). This approach yields an inventory policy that has a maximum relative error of .125 as compared to the optimal (Q,r) policy [28], well within the acceptable range.

Assuming that the daily demands at each customer are uncorrelated over time and across customers, and approximating the Poisson demand distribution with a Normal distribution, which is a good approximation when the demands are sufficiently large [28,29], we find that the safety stock of product p needed to ensure that stock-outs occur with a probability of α or less at a DC j which supplies a set of retailers $S_j \subseteq I$ is:

$$z_\alpha \sqrt{L_j \sum_{i \in S_j} \mu_{ip}} = z_\alpha \sqrt{L_j \sum_{i \in I} \mu_{ip} Y_{ijp}}$$

Given that our model deals with different products of possibly different volumes, we introduce a parameter v_p, which is the volume that one unit of product p occupies at the DC. At any point in time the total volume of products held at any DC should not exceed the capacity of that DC. In the worst-case this is the sum of the order quantity, the safety stock amount and the expected demand during the lead time. Thus the capacity constraint for each DC j becomes

$$\sum_{p \in P} v_p \left[Q_{jp} + z_\alpha \sqrt{L_j \sum_{i \in I} \mu_{ip} Y_{ijp}} + L_j \sum_{i \in I} \mu_{ip} Y_{ijp} \right] \le C_j, \forall j \in J \tag{2}$$

This expression generalizes the capacity constraint in [28] by using product volume as the basis for assessing utilization of capacity in distribution centers that hold multiple types of products.

To facilitate testing of the relative importance of transportation and inventory costs as compared to the fixed facility location costs, we introduce parameters β and θ, which are weight factors associated with transportation and inventory costs, respectively.

Having calculated the total CO_2 emissions, the safety stock term, and the capacity constraint for the DCs, we can now formulate our model:

Minimize

(3)

$$\sum_{j\in J}\{ f_j X_j + F_j \sum_{p\in P} \frac{\chi \sum_{i\in I}\mu_{ip}Y_{ijp}}{Q_{jp}} + \sum_{p\in P}\theta \frac{h_p Q_{jp}}{2}$$

$$+\beta g_j \sum_{p\in P}\frac{\chi\sum_{i\in I}\mu_{ip}Y_{ijp}}{Q_{jp}} + \theta\sum_{p\in P} h_p z_\alpha \sqrt{L_j \sum_{i\in I}\mu_{ip}Y_{ijp}}$$

$$+\beta\chi \sum_{i\in I}\sum_{p\in P} a_{ip}\mu_{ip}Y_{ijp} + \beta\chi\sum_{i\in I}\sum_{p\in P}\kappa_{ij}\phi_p\mu_{ip}Y_{ijp}$$

$$+\infty\left(\left[\sum_{j\in J}\omega\, C_j\, X_j + \sum_{j\in J}\sum_{i\in I}\sum_{p\in P}\Omega_p(\kappa_{ij}+\lambda_j)\mu_{ip}Y_{ijp}\right] - CAP\right)$$

Subject to:

$$\sum_{j\in J} Y_{ijp} = 1 \qquad \forall i \in I, p \in P \tag{4}$$

$$Y_{ijp} \le X_j \qquad \forall i \in I, p \in P, j \in J \tag{5}$$

$$\sum_{p\in P} v_p \left[Q_{jp} + z_\alpha \sqrt{L_j \sum_{i\in I}\mu_{ip}Y_{ijp}} + L_j \sum_{i\in I}\mu_{ip}Y_{ijp}\right] \le C_j \quad \forall j \in J \tag{6}$$

$$Q_{jp} \ge 0 \qquad \forall j \in J, p \in P \tag{7}$$

$$Y_{ijp} \in \{0,1\} \qquad \forall i \in I, p \in P, j \in J \tag{8}$$

$$X_j \in \{0,1\} \qquad \forall j \in J \tag{9}$$

The objective function (3) includes the fixed cost of locating distribution centers, the cost of transporting goods from the supplier to the DCs, the cost of transporting goods from DCs to retailers, the safety stock cost, the holding costs, the fixed-order costs at the DCs and the CO_2 emissions cost or credit. Constraints (4) require that each customer be assigned to exactly one DC. Constraints (5) require that customers can be assigned only to open DCs. Constraints (6) are the capacity constraints at the DCs. Constraints (7) require the decision variables Q_{jp} to be nonnegative, while Constraints (8) and (9) require the decision variables Y_{ijp} and X_j to be binary. This model has nonlinear terms in both the objective function and the constraints, unlike the UFLP or the LMRP. Also, unlike the LMRP, we must explicitly solve for the order quantity variables Q_{jp}.

4 COMPUTATIONAL ANALYSIS

We used the BARON solver in the General Algebraic Modeling System (GAMS) version 23.3 to test our formulation. The experiment was designed to investigate the effect of DC capacity on carbon emissions levels. We ran our experiments with three different numbers of products: a single product, two products, and four products. We used a 5-node data set consisting of the five nodes with the highest demand from Daskin's 49-node data set [2; pp. 480–482], which consists of the capitals of the lower 48 United States plus Washington, DC. Except for node five, which represents the location of the supplier, each node represents a retailer location and also a candidate DC location. The demands for the first product are those associated with the four selected nodes, while the demands for other three products are those associated with nodes that are selected randomly from the 49-node data set. The resulting mean demands for each node and for each product, and the fixed costs for each node are shown in Table 1 below.

Table 1: Demands of 4 products and the associated fixed costs of 4-node data set (μ_{ip})

i/j	$\mu_{ip=1}$	$\mu_{ip=2}$	$\mu_{ip=3}$	$\mu_{ip=4}$	*Fixed Cost*
1	297.6002	118.81643	77.30188	60.16425	115800
2	179.9046	114.30602	66.28637	55.44159	101800
3	169.8651	108.47115	64.78216	51.17073	72600
4	129.3793	92.95297	61.87358	48.91769	72400

We also require the distance between the supplier and each candidate DC location j, and also the distance between each candidate DC location j and each retailer i. These distances, which were calculated based on the great circle distance between the locations, are tabulated in Tables 2 and 3 below.

Table 2: Distance between retailer i and DC j (κ_{ij})

	R_1	R_2	R_3	R_4
DC_1	0	2454.8	717.3	1419
DC_2	2454.8	0	1745	1379.2
DC_3	717.3	1745	0	937.7
DC_4	1419	1379.2	937.7	0

Table 3: Distance between DC j and the supplier (λ_j)

	DC_1	DC_2	DC_3	DC_4
Supplier	77.6	2395.2	666.3	1342.6

The values of the other parameters are shown in Table 4.

Table 4: Values of other parameters

Parameter	value	Applicability
F_j	10	$\forall j \in J$
g_j	10	$\forall j \in J$
a_{ip}	5	$\forall p \in P, \forall j \in J$
h_p	1	$\forall p \in P$
v_p	1	$\forall p \in P$
ϕ_p	2	$\forall p \in P$
Ω_p	2	$\forall p \in P$
ω	3	-
χ	1	-
z_α	1.96	-

Note from Table 4 that χ was set to 1, and the differences between the daily parameters and the yearly parameters are realized through the weights β and θ. Also, z_α was set to 1.96 to achieve a service level of 97.5%.

4.1 Emissions versus DC Capacity

To study the dependence of CO_2 emissions on the capacity of the DCs, we considered seven different DC capacities, denoted c1 – c7; these capacities, shown in Figure 3, are in decreasing order, where c1 corresponds to the highest capacity and c7 to the lowest capacity.

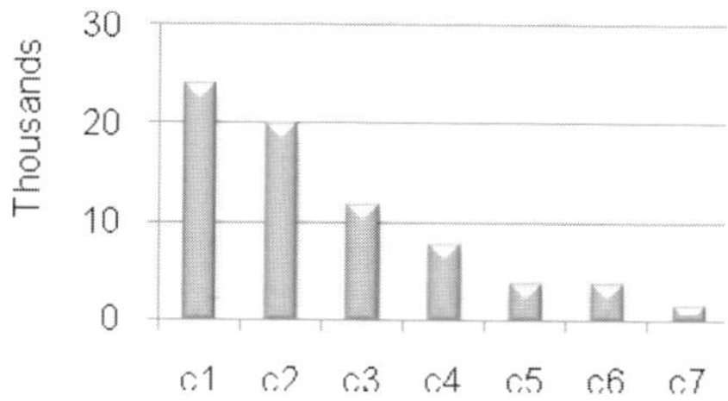

Figure 3: DC capacities for seven test cases

For each of these DC capacities, we used our model to estimate the resulting CO_2 emissions.

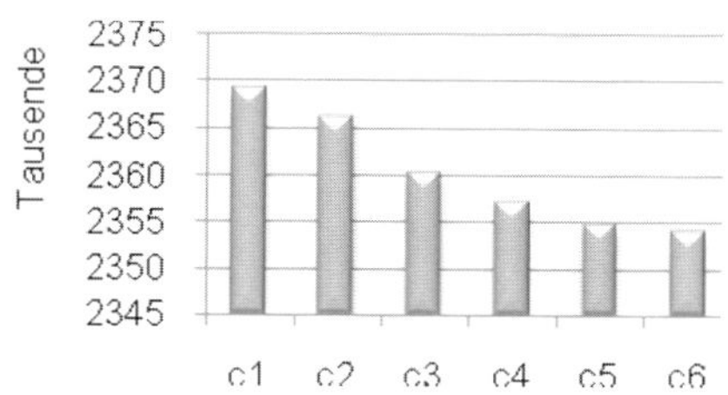

Figure 4: CO_2 Emissions – Cap for six DC capacities

We observed that as the DC capacity decreases, the CO_2 emissions and the total costs both decrease, as shown in Figure 4. This behavior is most likely due to the influence of the first term in Equation (1), which is directly proportional to the DC capacity. However, once the capacity is decreased below a certain threshold, the CO_2 emissions and the total costs both increase dramatically, as shown in Figure 5. A likely explanation for this behavior is that when the capacity of each DC is small, more DCs need to be opened to hold the products, and this will increase the transportation-related CO_2 emissions and costs.

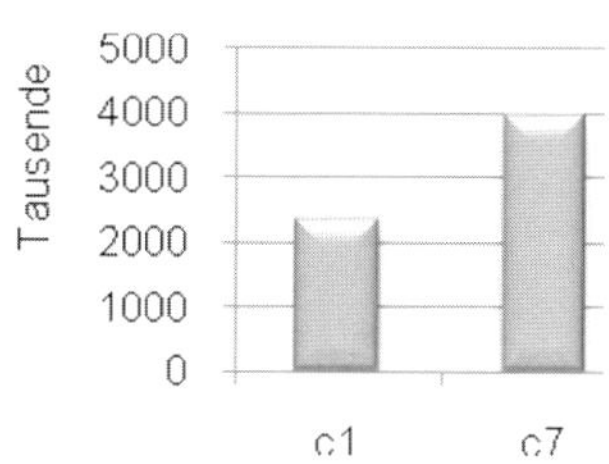

Figure 3 : CO_2 Emissions – Cap for the largest (c1) and smallest (c7) DC capacities considered

5 CONCLUSION

We introduced a multi-product capacitated inventory-location model with risk pooling and CO_2 emissions considerations. The model determines the location of the DCs and their inventory policy – the reorder time, reorder quantity and safety stock – to serve customers' demands for different products at minimum cost. We studied the dependence of CO_2 emissions and total costs on DC capacity, and observed that the CO_2 emissions and total costs both decease as the DC capacity is decreased, but once the DC capacity is decreased below a certain threshold, both CO_2 emissions and total costs increase dramatically.

In the near future, we plan to extend our model by adding certain service constraints. Also, we plan to solve this problem using heuristic evolutionary search algorithms such as Genetic Algorithms.

6 REFERENCES

[1] Daskin, M. S., 1995, Network and discrete location: models, algorithms, and applications, Wiley, New York.

[2] Drezner, Z., 1995, Facility location: A survey of applications and methods, Springer, New York.

[3] Sridharan, R., 1995, The capacitated plant location problem, European Journal of Operational Research 87 (2), pp. 203–213.

[4] Tragantalerngsak, S., Holt, J., Ronnqvist, M., 1997, Lagrangian heuristics for the two-echelon, single-source, capacitated facility location problem, European Journal of Operational Research 102 (3), pp. 611–625.

[5] Cortinhal, M. J., Captivo, M. E., 2003, Upper and lower bounds for the single source capacitated location problem, European Journal of Operational Research 151 (2), pp. 333–351.

[6] Liu, K., Zhou, Y., and Zhang, Z., 2010, Capacitated location model with online demand pooling in a multi-channel supply chain, European Journal of Operational Research 207, pp. 218–231.

[7] Clark, A. J., 1994, Multi-echelon inventory theory – A retrospective, International Journal of Production Economics, Vol. 35, Issues 1-3, pp. 271-275.

[8] Eppen, G. D., 1979, Effects of centralization on expected costs in a multi-location newsboy problem, Management Science 25 (5), pp. 498–501.

[9] Federgruen, A., Zipkin, P., 1984, Approximations of dynamic, multi-location production and inventory problems, Management Science 30 (1), pp. 69–84.

[10] Baumol, W. J., Wolfe, P., 1958, A Warehouse-Location Problem, Operations Research 6 (2), pp. 252–263.

[11] Karkazis, J. and Boffey, T. B, 1981, The multi-commodity facilities location problem, The Journal of the Operational Research Society, 32, pp. 803–814.

[12] Shen, Z. J. M., Coullard, C., Daskin, M. S., 2003, A joint location-inventory model, Transportation Science 37 (1), pp. 40–55.

[13] Erlebacher, S. J., Meller, R. D. s. I., 2000, The interaction of location and inventory in designing distribution systems, IIE Transactions 32, pp. 155–166.

[14] Daskin, M. S., Coullard, C., and Shen, Z.-J. M., 2002, An inventory-location model: Formulation, solution algorithm and computational results, Ann Oper Res, Vol. 110, pp. 83-106.

[15] Warszawski, A., 1973, Multi-dimensional location problems, Operational Research Quarterly, pp. 165–179.

[16] Warszawski, A. and Peer, S., 1973, Optimizing the location of facilities on a building site, Operational Research Quarterly, 24, pp. 35-44.

[17] Klincewicz, J.G., Luss, H. and Rosenberg, E., 1986, Optimal and heuristic algorithms for multiproduct un-capacitated facility location, European Journal of Operational Research, 26, pp. 251–258.

[18] Klincewicz, J. G. and Luss, H., 1987, A dual-based algorithm for multiproduct un-capacitated facility location, Transportation Science, 21, pp. 198–206.

[19] Geoffrion, A. M. and Graves G.W., 1974, Multi-commodity distribution system design by Benders

decomposition, Management Science, 20, pp. 822–844.

[20] Hindi, K. S. and Basta, T., 1994, Computationally efficient solution of a multiproduct, two-stage distribution-location problem, The Journal of the Operational Research Society, 45, pp. 1316–1323.

[21] Lee, C. Y., 1991, An optimal algorithm for the multiproduct capacitated facility location problem with a choice of facility type, Computers and Operations Research, 18, pp. 167–182.

[22] Mazzola, J. B. and Neebe, A. W., 1999, Lagrangian-relaxation-based solution procedures for a multi-product capacitated facility location problem with choice of facility type, European Journal of Operational Research, 115, pp. 285–299.

[23] Shen, Z. J., 2000, Approximation algorithms for various supply chain problems. Ph.D. thesis. Evanston: Northwestern University, Department of Industrial Engineering and Management Sciences.

[24] Shen, Z. J. M., 2005. A multi-commodity supply chain design problem, IIE Transactions 37 (8), 753–762.

[25] Srivastava, S., 2007, Green supply-chain management: A state-of-the-art literature review, International Journal of Management Reviews 9, 53–80.

[26] Ramudhin, A., Chaabane, A., Kharoune, M., Paquet, M., 2008, Carbon Market Sensitive Green Supply Chain Network Design, IEEE International Conference on Industrial Engineering and Engineering Management, IEEM 2008, pp. 1093–1097.

[27] Diabat, A., Simchi-Levi, D., 2009, A carbon-capped supply chain network problem, IEEE International Conference on Industrial Engineering and Engineering Management IEEM 2009, pp. 523–527.

[28] Ozsen, L., Coullard, C., Daskin, M. S., 2008, Capacitated Warehouse Location Model with Risk Pooling, Naval Research Logistics, pp. 295-312.

[29] Simchi-Levi, D., Kaminsky, P. and Simchi-Levi, E., 2003, Designing and managing the supply chain: Concepts, strategies, and case studies. Boston, MA: McGraw-Hill/Irwin.

The Impact of the Upstream Supply Chain and Downstream Processes to the Cradle-to-Grave Environmental Profile of Mg Lightweight Front End Auto Parts

L. Bushi
LCA Consultant, Toronto, Canada

Abstract
This paper focuses on a cradle-to-grave life cycle assessment (LCA) that evaluates the potential environmental impacts of the magnesium (Mg) front end auto parts over its full life cycle and highlights the impact of the upstream processes along the average supply-chain, auto parts manufacturing and assembly, and the downstream processes following the use stage and end-of-life value chain.
Pathways for improving sustainability of magnesium use in automobiles through greening the supply chain processes, technology improvements including the end of life chain are also discussed.
Keywords:
Cradle-to-grave analysis; Downstream processes; Mg front end auto parts; Upstream supply-chain processes

1 INTRODUCTION

In the framework of a Canada-China-US collaborative project titled Magnesium Front End Research and Development (MFERD), which was initiated in 2007, key enabling technologies for structural applications of magnesium in the automotive industry are developed [1].

The environmental aspects and potential impacts associated with the application of magnesium in the North American auto sector are captured by a three-year Phase I LCA task.

To achieve a sustainable production of metal front end auto parts, the environmental implications of the whole supply-chain of the auto parts, the use stage, and end-of life (EOL) processing- in other words, the entire life cycle from 'cradle-to-grave' must be considered.

This paper focuses on the results and 'hot spots' of the LCA of magnesium front end auto parts for a 2007 GM-Cadillac CST- considered an alternate light structure scenario to the standard steel-based design. Magnesium designs while maintaining the same safety performance are lighter and thus improve the fuel economy.

This LCA study is conducted in compliance with international standards ISO 14040:2006 and ISO 14044:2006 [2,3].

2 SUPPLY CHAIN OF MAGNESIUM FRONT END

2.1 Supply chain life cycle model

LCA is an internationally standardised method that addresses the environmental aspects and potential environmental impacts throughout a product's life cycle from raw material acquisition through to production, use, end-of-life treatment, recycling and final disposal [3].

The LCI modelling framework is based on a purely descriptive documentation of the product system under analysis, without implying a decision-context that would account for potential additional consequences on other systems. Furthermore, the existing interactions with other systems are included in the LCI model by considering recycling benefits or avoided production for co-products. This type of 'accounting' LCI modelling is known as the archetypal goal Situation C1 [4].

In Figure 1, the life cycle modelling depicts existing supply-chain of Mg front end auto parts, by connecting the single processes within the techno sphere along the flow of mass, energy, and services.

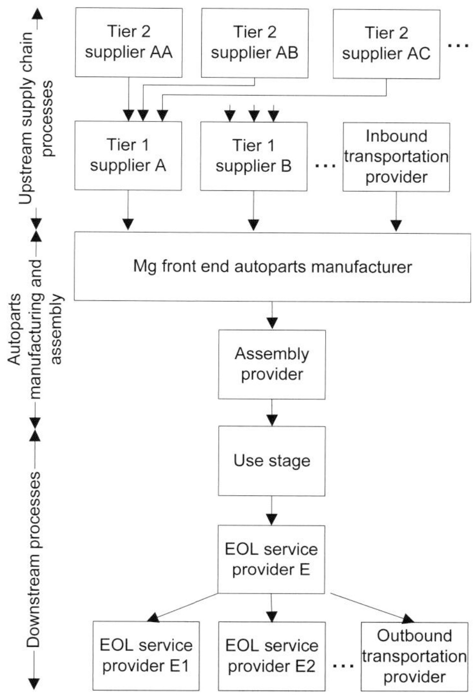

Figure 1: A simplified schematic supply-chain life cycle model of magnesium front end auto parts.

For this accounting type modelling, technology-specific data of the supply-chain are used for the foreground system and average market consumption mix data for the background system.

2.2 Identifying processes in LCI modelling

It's ultimately the goal and scope of this work that defines the principle system boundaries and the following included life cycle stages: upstream supply chain processes, manufacturing and auto parts assembly and downstream processes. The specific processes of Mg front end systems are identified step-wise starting from the foreground system (auto parts manufacturing and assembly) and following the process-chain and supply-chain as well as use-stage stepwise upstream and downstream.

Table 1 shows the processes attributed on a life cycle stage basis to the analysed system.

Table 1: Identified processes on a life cycle stage basis

Life cycle stages	Processes
Upstream supply chain processes	Mg primary production
	Mg alloy ingot production
	Al primary production (for Mg alloy)
	Inbound transportation
Auto parts manufacturing and assembly	Die-Casting
	Stamping
	Extrusion
	Pre-consumer scrap recycling
	Assembling
Downstream processes	Use stage
	End-of-life value chain
	Outbound Transportation

3 MAGNESIUM FRONT END PRODUCT SYSTEM

3.1 Description of automotive front end

It is calculated by the North American auto parts manufacturers that the advanced design of Mg front end have a weight of 45.2 kg with only 27 parts and a 45% weight reduction compared to conventional stamped steel [5]. These magnesium alloy parts provide a service equivalent to 82.2 kg with 79 parts made of carbon steel, used in 2007 GM-Cadillac CTS.

Die-casting, stamping, and extrusion technologies are used for Mg front end parts manufacturing in the following weight ratios of 67%, 9%, and 24%, respectively. The corresponding alloys for these technologies are AM60B (6% Al, 0.22% max Zn, and 0.25% min Mn, balance Mg), AZ31 (3% Al, 1% Zn, balance Mg), and AM30 (3% Al, 0.40% Mn, balance Mg), respectively. As per standard cut-off criteria of this LCA study, any mass flow less than 0.5% of the cumulative mass of reliable estimates of inputs into the life cycle inventory (LCI) model can be excluded, providing its environmental relevance is not a concern. The amounts of Zn and Mn in magnesium alloys are estimated to contribute respectively 0.24% and 0.26% of the cumulative mass and are therefore excluded from this analysis.

3.2 Upstream supply chain processes

The analysis captures the recent primary magnesium market trends with the consideration of a rate 4:1 between China's thermal reduction process (the Pidgeon process) and electrolytic process [6].

China is the world's largest primary magnesium producer. The Pidgeon process starts with the mining of dolomite ($CaCO_3.MgCO_3$) followed by its calcination process. The resulting mixed oxide of calcium and magnesium is reduced at high temperature under vacuum by ferrosilicon. The Ramakrishnan, S. and Koltun, P. 2004 publication is used as a source of the secondary LCI data on Mg primary production- as it was found to be the most transparent and complete work conducted in this field [7]. The base case data are conservative, as they do not include technological improvements made recently in China on the Pidgeon process. The Chinese national average electricity grid from 2005 is applied [8]. It is also assumed that sulphur is used to produce the cover gas in order to protect the magnesium from oxidation. Since part manufacturing is assumed to occur in North America, transportation of alloy ingots from the Chinese plant to port by rail (750 km) and from the Chinese to the US port via cargo ship (8,000 km) is being considered, with the final shipment from the port to the auto parts manufacturer occurring by truck (300 km).

The electrolytic process involves the preparation of anhydrous magnesium salt. The primary magnesium is produced by the electrolysis of a molten anhydrous salt solution at about 400°C. Generic LCI data for the electrolytic process is applied, based on the ecoinvent database v 1.3, SimaPro 7.1 LCA Software. Based on IMA's 2008 statistics, the major world magnesium producers as per electrolytic process are: US (32%), Israel (22%), Brazil (9%), Russia (23%) and Kazakhstan (14%) [9]. In this analysis, the 2008 world average production mix of the electrolytic magnesium is considered to enter the North American market. National electricity grid published data for US, Israel, Brazil, Russia and Kazakhstan were applied to calculate the weighted average grid mix for world production of electrolytic magnesium, which resulted to coal (43%), natural gas (18%), nuclear (6%), hydro (23%), others (10 %)- divided 50/50 fuel oil and renewable energy. In order to protect the molten magnesium from oxidation, sulphur is used to produce the cover gas. An average transportation distance of 7,000 km is assumed by cargo ship to the US. Furthermore, 300 km average transportation distance is assumed by truck– from the default US port to the auto parts manufacturer.

3.3 Auto parts manufacturing and assembly

Technology-specific primary data from 2008 have been collected and provided confidentially by North American auto parts manufacturers on the 'foreground' system processes including the three fabrication technologies of the Mg front end parts: die-casting, stamping and extrusion.

The die-casting fabrication process included melting and die-casting steps. For the stamping fabrication process, data was collected for slab casting, sheet forming and stamping. The extrusion fabrication technology consists of two processes (1) direct-chill billet casting (including melting and casting) and (2) the extrusion process (including sawing, scalping, billet pre-extrusion, feedstock preheat, extrusion and finishing). As part of the 'foreground' system, primary data were collected on the pre-consumer scrap recycling as well. This scrap (known also as prompt scrap) is clean and generated during the manufacturing phase. It's typically recycled in a relatively short time and in a 'close loop' – and it avoids the primary magnesium production. For assembling process, generic sector specific LCI data are applied following a modification to reflect the US average electricity grid [10].

3.4 Downstream processes

Use phase

The use stage is a critical aspect of the cradle-to-grave life cycle of magnesium front end parts. The lightweight advanced designs such as magnesium, aluminium and high strength steel while maintaining the same safety performance are lighter and thus improve the fuel economy.

This analysis applies the Kloffler et. al 2010 method to estimate the weight-induced fuel consumption for a steel reference Cadillac CTS 2007 and the mass-induced fuel savings (with powertrain modifications) to achieve equal driving performance with the reference vehicle [11]. To quantify the vehicles fuel consumption, EPA CFE- EPA combined fuel economy (55% city FTP-75/ 45% highway HWFET) driving cycle is considered which consists of the following segments for the FTP-75: cold start phase, transient phase and hot start phase [12]. The following are basic parameters of the cycle: 11.04 miles (17.77 km) distance travelled, 1,874s duration and 21.2 mph (34.1 km/h) average speed. EPA Highway Fuel Economy Cycle (HWFET) characteristic parameters of the cycle are: 765s duration, 10.26 miles (16.45 km) total distance and 48.3 mi/h (77.7 km/h) average speed.

The total mass-induced energy demand (W_{sum}) for the EPA's CFE is calculated as follows [11]:

$$W_{sum} = W_R * 0.85 + W_A = m * \left(0.85 * g * f_R * C_{WR} + C_{WA}\right) \quad (1)$$

where,

W_R work to overcome the 'rolling resistance'

WA work to overcome the 'acceleration resistance'

m curb weight of 2007 Cadillac CTS (1,595 kg)

g gravitation constant (m/s^2)

f_R rolling resistant coefficient (dimensionless)

C_{WR} characteristic value (17,198 m) for the EPA CFE driving cycle (55% city / 45% highway) [11]

C_{WA} characteristic value (2,221 m^2/s^2) for the EPA CFE driving cycle (55% city / 45% highway) [11]

The work to overcome the 'aerodynamic resistance' (W_L) is deemed negligible in order to calculate the energy needed to move a certain weight through the driving cycle profile since it is independent of the vehicle's mass [11].

For 100 km and 100 kg, the mass-induced energy demand therefore is:

$$W_{sum\ (100kg,100km)} = \left(\frac{100}{17.176}\right) * 100 * \left(0.85 * 9.81 * 0.01 * 17{,}198 + 2{,}221\right) = 2.128\,\text{MJ} \quad (2)$$

The mass-induced fuel consumption $V_{100kg,EPA\ CFE}$ for naturally aspirated gasoline engines is calculated [11]:

$$V_{100kg,EPA\ CFE} = W_{sum(100kg,100km)} * 1.02 + \eta_{diff,g} = 2.128 * 1.02 * 0.073 = 0.158 \frac{l}{100km * 100kg} \quad (3)$$

where,

$\eta_{diff,g}$ differential efficiency for naturally aspirated gasoline engines (0.073 l/MJ).

By multiplying the Cadillac CTS 2007's curb weight by the value of $V_{100kg,EPA\ CFE}$ and the nominal life time driving distance (200,000 km), we get the weight-induced fuel consumption for a steel reference Cadillac CTS 2007 of 5,040 l.

It's estimated with magnesium front end substitution vehicle curb weight would reduce from 1,595 kg to 1,547 kg, (assuming a secondary weight savings factor of 0.3 [13]) resulting in fuel economy improvements from 0.1120 l/km to 0.1105 l/km. Thus 300 l mass-induced fuel savings (with powertrain modifications) can be achieved for the new Mg front end lightweight design of the Cadillac CTS 2007 over the life of the car (200,000 km). It should be emphasized that 300 l fuel savings do not only apply to Mg front end auto parts but instead to the full lightweight vehicle concept design. **Mass-induced fuel consumption** of Mg front end auto parts over the life time driving distance of 200,000 km is estimated to be 138.5 l (= 4,740*45.2/1,547).

Energy use and emissions to air estimates during the use stage are based on the North American year 2007 mix of conventional and reformulated gasoline fuel using the Greenhouse Gases, Regulated Emissions, and Energy Use in Transportation (GREET) model developed by Argonne National Laboratory Emissions [14]. Estimates for the use of the baseline vehicle are derived using US Environmental Protection Agency's MOBILE6 model where emission standard levels for each pollutant are adjusted to reflect expected in-use performance over the vehicle lifetime [15].

End of life value chain including transportation

The recycling scenario is based on the current situation for end-of-life vehicles in North America [16]. This analysis assumes that 94% of vehicles are collected at the end-of-life phase. A recovery rate of 95% at the recycling step is applied by assuming that 5% of the magnesium is lost in shredding operations. The end-of-life value chain includes dismantling, shredding, and separation of old scrap. About fifty percent of sorted magnesium also contains the aluminium rich metal fraction. This aluminium scrap is used to produce A380 casting alloy. As this alloy has strict requirements on the maximum concentration of magnesium, the level of magnesium is reduced by the addition of chlorine. As a consequence, magnesium is lost. The other fifty percent of the magnesium scrap is shipped to China for hand sorting using the returning empty shipping containers. This magnesium is used for steel desulfurization and avoids the primary production of magnesium.

4 LCA FRAMEWORK

This LCA study is conducted in compliance with international standards ISO 14040:2006 and ISO 14044:2006 and uses the 'cradle-to-grave' approach by including upstream supply chain processes, auto parts manufacturing and assembly, and use phase and end-of-life value chain of auto parts.

The goal of this LCA study is to evaluate the potential environmental impacts of magnesium based front end part used in a GM-Cadillac CTS, built and driven for 200,000 km in North America. It includes performance at 5-star frontal, offset and side impact crash ratings.

This study is not a comparative study in and of itself; however, it is intended to enable comparative studies in this field.

The intended audiences of this LCA study are governments, the North American car industry including auto parts manufacturers, magnesium primary producers, auto recycling industry and academia.

The reference flow is 45.2 kg magnesium front end that provide a service equivalent to 82.2 kg front end parts made of carbon steel.

The scope of the manufacturing and auto parts assembly stage and downstream processes is limited to North America. The scope also reflects that eighty percent of the world's magnesium production is occurring in China using the Pidgeon process and the remaining twenty percent is from the electrolytic processes.

This analysis is focused on the total primary energy (TPE) and the greenhouse gas emissions (GHGs) such as carbon dioxide, methane, dinitrogen oxide etc for the entire value chain. The TPE covers the non-renewable, renewable, and feedstock energy sources. TPE is expressed in MJ. Higher heating value (HHV) of primary energy carriers is used to calculate the primary energy values used in this study. The methodology and global warming potentials (GWPs) metrics of the International Panel on Climate Change (IPCC) are used for calculation of the potency of GHGs relative to CO_2 to estimate the climate change impact category. The 100-year time horizons recommended by the IPCC and used by the US for policy making and reporting are adopted within this impact category.

SimaPro 7.1 LCA software package designed for modelling and analysing the environmental potential impact of products and services was used for the life cycle assessment [17].

5 LCA RESULTS

Table 2 shows the breakdown into each stage of the cradle-to-grave life cycle of 45.2 kg magnesium front end auto parts: upstream supply chain processes (USCP), manufacturing and auto parts assembly (MAA) and downstream processes (DP).

Table 2: Selected LCIA results for magnesium front end

Life cycle stages	USCP	MAA	DP	Total
TPE (MJ)	20,191	3,797	-1,455	22,534
Climate change (kg CO_2-eq)	1,936	208	-221	1,923
TPE (%)	90%	17%	-7%	100%
Climate change (%)	101%	11%	-12%	100%

The upstream supply chain stage is the most energy intensive, and has the most significant impact in this cradle-to-grave life cycle analysis. Downstream processes amount to -7% and -12% of the TPE and climate change, respectively. The negative values at the DP stage show the net environmental benefit attributed to the avoided magnesium primary production. Manufacturing and auto parts assembly play a minor role to the overall impact. Use phase (mass-induced fuel consumption of 45.2 kg Mg front end auto parts) accounts for 24% of TPE and 23% of climate change of the total results, with the EOL value chain processes contributing -31% and -35% respectively. The 'hot spot' phase of the cradle-to-grave magnesium front end is upstream supply chain stage that contributes 90% and 101% of TPE and climate change impact categories.

6 IMPROVING THE SUSTAINABILITY OF MAGNESIUM

In this section, pathways are presented which can improve the sustainability of magnesium use in automobiles through greening the supply chain along with technology improvements.

6.1 Technological improvements for the Pidgeon process in China

The reduction of the energy consumption and alteration of fuels e.g. replacing coal with producer gas, coke oven gas (COG) or natural gas, are the most promising strategies to reduce the total GHG emissions of the primary Mg production as per Pidgeon process.

A sensitivity analysis according to ISO 14040 series of standards is conducted to evaluate the impact of the technological improvements in the climate change impact category by comparing the base case Mg front end scenario with the 'COG' scenario as described in Gao, F et al. 2009 [18]. Both the base case and the 'COG' scenario assume the use of sulphur as a cover gas.

In Table 3, Scenario 1, the results for the upstream supply chain processes and cradle-to grave system show that climate change can be decreased by 13% and 6%, respectively - as a result of technological improvements in the Pidgeon process.

6.2 Improvements on the EOL value chain.

Based on the state-of-the-art recovery technology, magnesium base case scenario assumes that per 1 kg primary magnesium auto parts, around 0.37 kg end-of-life secondary magnesium is generated which will replace the same amount of primary magnesium ingot.

As the quantity of magnesium will increase with enhanced penetration of structural magnesium in automotive applications, there will be an economic incentive to have an additional sorting and to segregate the aluminium and magnesium alloys. Magnesium chunks can be directed to aluminium can production and used as an alloying element.

For sensitivity analysis purposes, the assumption is being made here that the application of the new separation technologies in the near future will increase the amount of the secondary magnesium up to 0.48 kg per kg of magnesium auto parts. The results in Table 3, Scenario 2 show the increase of the Mg recovery by around 11% (from 37% to 48%) will result in a decrease of the primary energy and climate change by around 9 to 10%.

7 SUMMARY

In the recent years, the world auto industry in general and the North American auto sector in particular, have shown an increased interest in developing a better understanding of the sustainability of magnesium applications in automobiles. A LCA task, under the MFERD Canada-China-US collaborative project, was engaged to assist in evaluating the environmental impacts of magnesium front end auto parts through a detailed Life Cycle Assessment (LCA).

A LCA study conducted according to ISO 14040 series of Standards is an important step in developing a more strategic approach for considering sustainability attributes (both risks and opportunities) of magnesium structural parts and applications in the auto industry.

The goal of the study was to evaluate a 'cradle-to-grave' Life Cycle Assessment of magnesium front end auto parts for a 2007 GM-Cadillac CST, build and driven for 200,000 km in North America. The 'cradle-to-grave' scope includes upstream supply-chain processes, auto parts manufacturing and assembly, and the downstream processes following the use stage (mass-induced fuel consumption of 45.2 kg Mg auto parts) and end-of-life value chain.

The primary finding of the study is that the upstream supply-chain processes drive the cradle-to-grave environmental profile of Mg front end auto parts. The manufacturing and auto parts assembly phase has a minor impact to the overall results. EOL value chain processes contribute to a net environmental benefit (reduction of TPE and climate change) attributed to the avoided magnesium primary production.

At the end, technology improvements of the Pidgeon primary magnesium production process and improvements of the EOL value chain, are effective strategies that will strongly improve the sustainability of magnesium and increase its applicability in the auto industry.

8 ACKNOWLEDGMENTS

The author wishes to thank the LCA task members: Dr. Alain Dubreuil, Natural Resources Canada; Dr. Sujit Das, Oak Ridge National Laboratory, US and Dr. Ambalavanar Tharumarajah, Australia Commonwealth Scientific and Industrial Research Organization (CSIRO), Casting & Solidification Technology Cooperative Research Centre (CAST-CRC) as well as the MFERD project sponsors: Natural Resources Canada, United States Department of Energy (DOE), United States Automotive Materials Partnership (USAMP) and Chinese Ministry of Science and Technology.

9 REFERENCES

[1] Nyberg, E.-A., Luo, A.-A., Sadayappan, K., Shi, W., 2008, Magnesium for future autos, Advanced Materials & Processes, 166 (10): pp. 35-37.

[2] ISO 14040, 2006, Environmental management – Life cycle assessment – Principles and framework.

[3] ISO 14044, 2006, Environmental management - Life cycle assessment - Requirements and guidelines.

[4] EC JRC IES, 2010, ILCD Handbook - General guide for LCA- Detailed guidance, © European Union.

[5] Luo, A, 2007, AMD 603-Magnesium front end design and development, Automotive lightweighting materials, FY 2007 Annual Report.

[6] Aghion, E., Bartos, S. C., 2008, Comparative review of primary magnesium production technologies as related to global climate change, 65th Annual World Magnesium Conference, Warsaw, Poland.

[7] Ramakrishnan, S., Koltun, P., 2004, Global warming impact of the magnesium produced in China using the Pidgeon process, Resources, Conservation and Recycling, 42 (1): pp. 49-64.

[8] Xianzheng, G., Zuoren N., Zhihong, W., 2006, Research and development of Chinese LCA database, Beijing University of Technology, Eco-materials Institute Center for Materials LCA, presentation.

[9] International Magnesium Association, 2010 http://www.intlmag.org/files/yend2008r.pdf.

[10] Koltun, P., Tharumarajah, A., Ramakrishnan, S., 2005, Life cycle environmental impact of magnesium automotive components, Magnesium Technology, pp. 43-48.

[11] Koffler. Ch., Rohde-Brandenburger, K., 2010, On the calculation of fuel savings through lightweight design in automotive life cycle assessments, International Journal of Life Cycle Assessment 15: pp. 128–135.

[12] EPA's FTP-75 and HWFET driving cycles

http://www.epa.gov/oms/emisslab/methods/ftpdds.gif

http://www.dieselnet.com/standards/cycles/hwfet.html

[13] Geyer, R., 2007, Life cycle greenhouse gas emission assessments of automotive materials- The example of mild steel, advanced high strength steel and aluminium in body in white applications, Methodology report.

[14] Argonne National Laboratory, The Greenhouse Gases, Regulated Emissions, and Energy Use in Transportation (GREET) Model, On-line version from September 2009,

http://www.transportation.anl.gov/modeling_simulation/GREET.

[15] U.S. EPA, MOBILE6 Vehicle Emission Modelling Software, On-line version from September 2009.

http://www.epa.gov/otaq/m6.htm.

[16] Gesing, A. J., Dubreuil, A., 2008, Recycling of post-consumer Mg scrap, 65th Annual World Magnesium Conference, Warsaw, Poland.

[17] PRé, 2008, SimaPro 7.1 LCA software, Netherlands.

[18] Gao, F., Nie, Z., Wang, Z., Gong, X. and Zuo, T., 2009, Life cycle sssessment of primary magnesium production using the Pidgeon process in China, International Journal of Life Cycle Assessment, 14 (5): pp. 480-489.

Table 3: Summary of the sensitivity checks to improve the sustainability of magnesium for use in auto industry

Scenario 1- Technological improvements for the Pidgeon process in China				
Climate change impact category (kg CO2-eq)	**Base case (80% Pidgeon: 20% electrolytic)**	**Altered assumption - COG scenario**	**Deviation - in absolute basis**	**Deviation - in %**
Upstream supply chain only	2,144	1,874	-270	-13%
Cradle-to grave	1,923	1,801	-122	-6%
Scenario 2- Improvements on the EOL value chain (cradle-to-grave)				
Impact category	**Base case (80% Pidgeon: 20% electrolytic)**	**Altered assumption - EOL value chain**	**Deviation - in absolute basis**	**Deviation - in %**
TPE (MJ)	22,534	20,500	-2,034	-9%
Climate change (kg CO_2-eq)	1,923	1,725	-198	-10%

Designing the Spare Parts Supply Chain in the Wind Energy Industry

G. Schuh[1, 2], H. Wienholdt[1]

[1]Research Institute for Operations Management, RWTH Aachen University, Germany
[2]Laboratory for Machine Tools and Production Engineering, RWTH Aachen University, Germany

Abstract
In a subproject of the cluster of excellence "Integrative Production Technology for High-Wage Countries" at RWTH Aachen University a configuration logic is under development that enables companies to configure their production system according to the dynamic requirements of the market. As a result of this project, a holistic description model for production systems has been defined. With numerous attributes in the sub-models a detailed characterization of the production system is possible.
The sub-model for the design of the supply chain will be depicted in detail in this paper. Representative for the design of a supply chain, the spare parts logistics of the wind energy industry is analyzed in depth. Designing this supply chain is not only one of the most challenging tasks in logistics. Only a responsive but also cost efficient design of the spare parts supply chain guarantees high productivity, extended life spans of the wind turbines as well as the expected profit for all companies in the supply chain.
Design decisions for the spare parts supply chain are within one of the three design fields: network design, cooperation concepts (e.g. with logistics providers, customers, suppliers) and inventory management. There are many interdependencies that have to be taken into account, but the complexity among all interdependencies can hardly be understood. Therefore it is necessary to reduce the complexity of design decisions by focusing on the most important design elements according to the logistical requirements of different spare part categories and to develop a classification of spare parts by their key characteristics. For different spare part categories only a smaller set of design elements and their interdependencies has to be taken into account. The reduced number of key design elements per spare part category can be analyzed and understood in depth. Thus a system dynamics approach is used to allow a better configuration of the network design, cooperation concepts and inventory management in spare parts supply chains on the basis of specific logistics requirements of different spare part categories.
The depicted research has been funded by the German Research Foundation DFG as parts of the Cluster of Excellence "Integrative Production Technology for High-Wage Countries".

Keywords:
Spare Parts Logistics, Supply Chain Design, Configuration Logic

1 WIND ENERGY INDUSTRY AND ITS AFTER-SALES-CHALLENGES

The use of wind power as an energy source is growing worldwide. From 1995 to 2009 the installed capacity of wind turbines has increased from 5.000 megawatts to 160.000 megawatts, respectively [1]. Moreover, the recent world economic crisis seemed to have negligible effect on the wind energy sector. During 2008 and 2009, in the midst of crisis, the annual increase remained at 30 per cent [2]. It is expected that this growth will continue despite global economic setback, aided by public subsidies, as in the US, by the green energy act in Canada, or by similar support programs in Asia [2,3].

The increasing installed base of wind energy plants draws more and more attention towards after-sales activities. After installing new facilities, focus is set on securing plant availability. Maintenance and servicing become major fields of activity, which draw attention towards spare parts management. However, the management of spare parts in the wind energy sector faces several challenges, deriving from the market's dynamics caused by the described rapid growth of the installed base, as well as from the fact that many companies have not yet implemented the required service strategies. The installed base is characterized e.g. by a regional distribution, site-dependent wear behavior and the diversity of variants. Besides, as an emerging market, wind energy industry lacks the availability of historic data and experience.

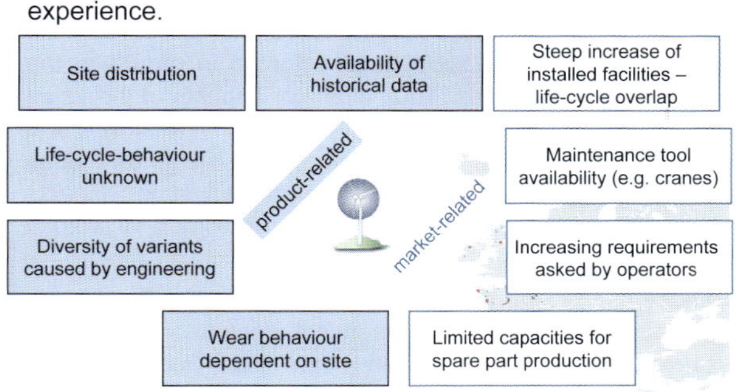

Figure 1: Product-related and market-related challenges

The major challenges of wind energy industry are depicted in Figure 1. They can be distinguished into product-related and market-related challenges. Product-related challenges include the fact that preferred site-locations are mostly found in regions with poor infrastructure, the life cycle of the installed plant, the diversity of product variants, site-dependent wear behaviour and the absence of historical data sets. Market-related challenges comprise life cycle overlap by the rapidly increasing installed base and resulting bottleneck situations in maintenance tools, missing capacities for spare parts production and increasing customers requirements as to availability.

The research depicted in this paper therefore puts special emphasis on the management of spare parts in wind energy industry. It has been carried out within the framework of the research project "Integrative Production Technology for High-Wage Countries" at RWTH Aachen University. Its objective was to develop technologies and methods to optimize production systems in high-wage countries. Production industry companies in these countries are facing an increasing competitive pressure caused by globalised markets. To withstand this pressure, they have to concentrate on the production of customer-specific and high-quality goods, while keeping production cost as low as possible. This results in the polylemma of production, depicted in Figure 2. It is set up by the interplay of economies-of-scale vs. economies-of-scope on the one axis and individual, flexible, value oriented production vs. planning oriented production on the other axis [4]. The reduction and finally the resolution of the polylemma of production can only be achieved by the implementation of new production systems, focused on the requirements of high-wage markets.

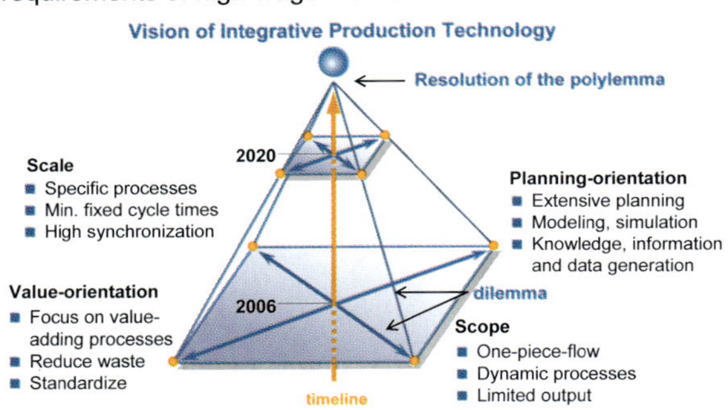

Figure 2: Resolution of the polylemma of Production [4]

In the research project a framework for the configuration of production systems has been developed. It is based on the five phases of the SCOR-model: Source, Make, Deliver, Plan and Return [5], which have been adjusted in regard to the project's main objectives. Thus all major aspects of the value chain (procurement (Source), production (Make), and distribution (Deliver)) are incorporated in the model as well as further sub-models which describe additional major aspects like quality management, production technology and product architecture.

Based on the framework, configuration logic has been developed to support the dynamical design of production systems. The supply chain in spare parts logistics, with special emphasis on wind energy industry, will be depicted in this paper as an example for the make and deliver sub-model. Therefore the special challenges and characteristics of spare parts management have to be analyzed first and will end in a classification of spare parts according to their characteristics. Finally based on this classification the configuration logic for the spare parts supply chain will be developed.

2 OPPORTUNITIES AND CHALLENGES IN SPARE PARTS MANAGEMENT

In a survey Boone [6] identifies several key challenges for spare parts business, whereas the lack of a holistic perspective was named as the most relevant one. Holistic in this context signifies the understanding of interdependencies while designing the spare parts logistics in the three major design fields: Network Design, Cooperation Concepts and Inventory Management. In Figure 3 these three design fields are depicted. Nevertheless, Figure 3 depicts also another key element for the design of the spare parts logistics: the ability to forecast the demand precisely.

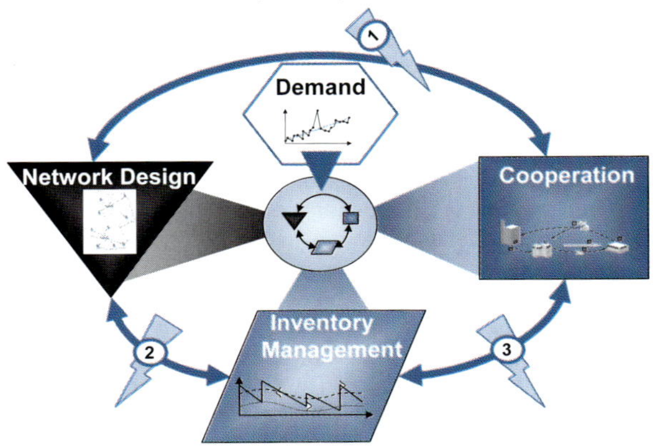

Figure 3: Interdependencies among design fields

As already mentioned above, understanding the interdependencies is a key challenge in spare parts logistics. Three examples will support this statement (see Figure 3):

1. While designing the distribution network for spare parts, decisions about the location of warehouses have to be made. Several algorithms exist to support these decisions [7], [9]. Nevertheless, while deciding about methods and ways of cooperation with suppliers, customers or logistic service providers, the location of their sites has also to be taken into account. Thus the warehouse location decision in Network Design is influenced by a decision about the way of cooperation with all possible partners in the supply chain and vice versa.

2. By using cluster algorithms, different scenarios for the number of necessary warehouses in the network can be generated. Afterwards the algorithm for the warehouse location problem (WLP) minimizes the distances to the customers for each warehouse in the preferred scenario [8], [9]. This can be supported by software tools. In Inventory Management, the inventory level that is necessary to achieve a certain service level is getting lower the more customer demands can be consolidated. This is due to the reduction of variance in the total demand by so called pooling effects [10]. Thus, from inventory cost perspective one central warehouse is optimal to reduce inventory costs, whereas from network design perspective a distribution of many warehouses is beneficial to reduce distances and therefore transportation

time and costs. This dilemma between inventory costs and transportation costs and time has to be quantified to achieve the best configuration for the spare parts logistic.

3. In Inventory Management, decisions about service level and necessary stock levels are made. According to prevalent formulae used to calculate safety stocks and reorder stock levels, inventory costs increase exponentially with the increase of the desired service level [11]. Thus, inventory and warehousing costs depend on the decision about the desired service level. For make-or-buy decisions in inventory management these costs have to be taken into account, depending on the desired service level. Furthermore, integrating a supplier managed inventory (SMI) concept [12] at a customers production site, decentralizes the inventory for the spare parts supplier. As mentioned above, the more decentralized the inventory is, the less pooling effects can be achieved and higher inventory costs or lower service levels are the inevitable consequence.

The described examples illustrate the interdependencies between decisions regarding the design of the spare parts supply chain. Often a decision in one design field has direct consequences on decisions in other design fields. The cause-effect relations in designing the spare parts supply chain are therefore not unidirectional. Mathematically spoken, the configuration steps for designing the spare parts supply chain are in many cases not topologically ordered.

To address this "chicken-egg-dilemma" a categorization approach is used. Different categories of spare parts are identified in order to reduce the complexity of the decision problem. The underlying assumption is that categories of spare parts with the same characteristics exist. In each category many design decisions are already determined by these characteristics and the configuration problem can be focused on certain sub-problems within the categories.

3 CATEGORISATION OF SPARE PARTS

Spare parts have specific characteristics that complicate planning their respective demands. Detailed knowledge about these characteristics is necessary to configure the spare parts supply chain [13]. The following paragraphs will, based on a literature review, show that understanding the difficulties caused by the characteristics of spare parts can be advantageous for the configuration of the spare parts supply chain.

Criticality

The criticality of an item is one of the first named special characteristics of spare parts. The criticality of a part is related to the consequences caused by its failure in case a replacement is not readily available [13]. Therefore critical and non-critical items can be distinguished [14]. A critical item directly causes a disorder of the primary product, which e.g. can be a wind turbine interrupted in operation. On the contrary, a non-critical item does not have direct consequences on the primary functions of the turbine.

Wear behavior

Spare parts show different types of wear behavior. If wear behavior is continuous, the abrasion depends on the usage time of the item. The wear behavior can therefore be planned by considering the time of usage. On the other hand, spontaneous defects can occur for certain items. This sporadic wear behavior cannot be planned. Thus, spare parts can be distinguished in spare parts with continuous wear behavior and sporadic defects [15].

Installed base

Another key characteristic of spare parts is that their demand is not directly dependent on the customer's behavior. Instead, it is delayed by the life span of different items in the primary product. Therefore the demand for spare parts is mainly caused by the number of primary products that have been sold to the market – the installed base – as well as the age of these products. Modern forecasting techniques for spare parts take this into account and perform much better than usual forecasting techniques [16,17].

While the installed base directly influences the total demand of certain spare parts, criticality and wear behavior have obviously a strong impact on the design of the spare parts supply chain. The criticality defines whether the replacement of the part has to be done immediately or if a longer time to repair or replace the item is possible. The impact on the supply chain strategy can directly be deduced, either being extremely fast and flexible or focusing on efficiency and cost savings. The wear behavior directly influences the planning character within the supply chain. Sporadic defects cannot be foreseen and planning is very difficult. On the other hand continuous wear behavior allows much better planning of replacements and repairs.

Combining these results, a matrix can be spanned which distinguishes spare parts in critical, non-critical items on the one axis and predictable and non-predictable on the other axis (Figure 4). The four possible combinations can be used to define norm strategies for the configuration of the spare parts supply chain. A third dimension, which includes the installed base and the forecasted long-term demand respectively, might be useful as well, but will not be discussed in this paper.

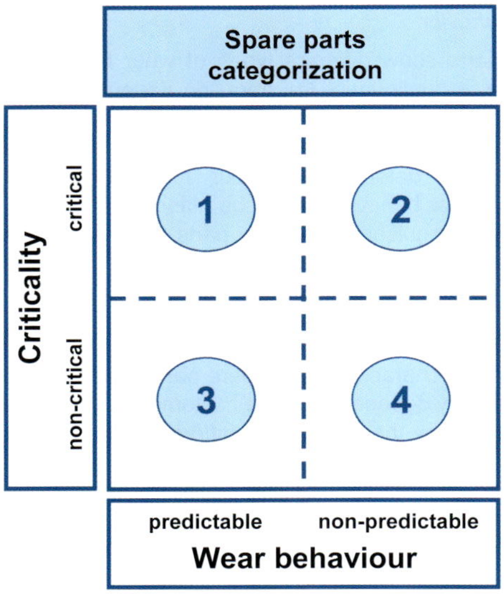

Figure 4: Spare parts categorization

4 NORM STRATEGIES FOR THE CONFIGURATION OF THE SPARE PARTS SUPPLY CHAIN

In Supply Chain Management norm strategies are used to generally define tendencies for the configuration of the supply chain. Fisher [18] was one of the first to distinguish Responsive and Efficient Supply Chains for innovative and functional products. Other authors [19,20] distinguish between Lean and Agile Supply Chains. Lean Supply Chains – comparable to Efficient Supply Chains – focus on efficiency and cost reduction. In Agile Supply Chains the focus – comparable to Responsive Supply Chains – is set on flexibility and the reduction of reaction times. This is caused by the customers' expectation and behavior. Lee [21] advances Fisher's approach by adding the perspective of volatility of the supply side. To reduce procurement-risk caused by instable supply markets, Lee adds the Risk-Hedging Supply Chain and consequently proposes three different norm strategies: Responsive, Flexible and Risk-Hedging.

Obviously the norm strategies for supply chain management depend on different characteristics, e.g. of the product, the customer's behavior or the supply side. In chapter 3 we described an approach to categorize spare parts by criticality and wear behavior. This classification is usable for an application of the norm strategies to support the configuration of spare parts logistics.

Responsive Spare Parts Supply Chain for critical items

As mentioned before, critical spare parts have to be replaced immediately to reduce down-time of the primary product. For critical spare parts the main focus therefore lies in the reduction of reaction time. While configuring the spare parts supply chain, this yields to higher service levels and shorter reaction times by a decentralized network design, etc. As a consequence, the supply chain in total has to be responsive.

A distinction can be made by taking the wear behavior into account. Predictable wear behavior reduces complexity in designing the Responsive Supply Chain. Designing a Responsive Supply Chain for critical spare parts with non-predictable wear behavior is the biggest challenge in spare parts logistics design.

Efficient Spare Parts Supply Chain for non-critical items

Non-critical items do not demand as short reaction times as critical items. The supply chain for non-critical items can therefore be designed very efficient. This is comparable to Fishers approach for functional products. Non-critical and non-predictable spare parts have nevertheless to be stocked with certain safety stocks. Yet, the safety stock can be reduced by applying a very centralized network design. Non-critical and predictable items can be stocked with lower inventory levels. The predictable and non-critical item's supply chain is the less critical and should therefore be the most efficient supply chain.

5 SYSTEM DYNAMICS APPROACH TO DESIGN THE NORM STRATEGIES

By using the categorization of the spare parts to define norm strategies, the configuration of the spare parts supply chain becomes a less complex problem. Fewer interdependencies have to be taken into account, as the norm strategies already give tendencies for certain design fields (e.g. de-centralized network design for responsive spare parts supply chains). Nevertheless, the configuration problem remains difficult to solve. Interdependencies still exist and the sporadic demands make it difficult to design the supply chain in detail. Therefore current research aims at defining the implications of the different spare part categories as detailed as possible. For each of the four spare parts categories (Figure 4) a norm strategy will be defined in detail. System Dynamics techniques help to understand the remaining interdependencies in depth by applying causal-loop-diagrams [22]. Finally, decision support systems will be generated by using stock-and-flow-simulations in which logistics costs and performance will be measured.

In the following case a stock-and-flow-model is used to design the supply chain of a critical item with predictable wear behavior for a gear producing company in the wind energy industry. The produced gear boxes are critical components in regard to the operation of the wind turbines. The gear producer has an individual service division, which maintains the gear boxes as an own profit center. In the recent process, broken gear boxes are sent back to the factory site and repair is done by replacement and refurbishment of broken parts. After repair, the gear boxes will be sent back to their site of operation. The main problem is the inventory planning for the items to be replaced. High variety in the installed base of gear boxes leads to a very sporadic demand for the different items. Figure 5 illustrates this by showing an ABC-Analysis for the number of movements of certain spare parts. As depicted, roughly 80% of parts are very slow moving with a historical demand of less than 10.

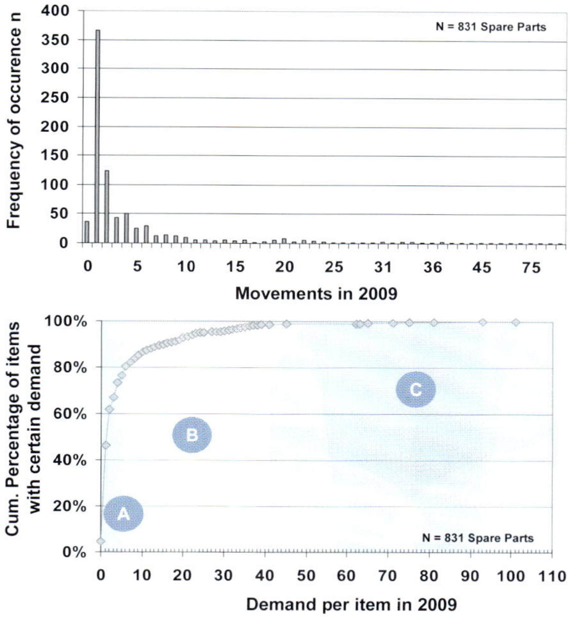

Figure 5: ABC-analysis to identify fast and slow-movers

The dilemma between customer's demand for quick repair and company's need for low inventory seems obvious. As the gear box' damage can be classified as critical but predictable (see Figure 4), a Responsive Supply Chain approach should be used. Nevertheless, the predictability of the wear behavior also demands inventories to be as low as possible.

Thus the company decided for a supply chain with a kanban system to control inventory. The kanban system is used for the main runner gear boxes as well as for the different spare parts to be replaced. Figure 9 depicts the implemented system as a stock-and-flow-model. A central kanban inventory of very few gear boxes satisfies the demand. Once a gear box breaks down at a customer's site, it can be replaced by one gear box of the kanban stock. The broken gear box will be repaired by taking the necessary spare parts from the kanban stock of spare parts. Finally the repaired gear box refills the kanban stock of gear boxes. The stock-and-flow-model (Figure 6) helps to configure the whole system, depending on the forecasted customer's demand as well as considering the varying lead times in supply and production. The performance is measured by a combination of total inventory costs and opportunity costs if stock-outs occur and penalties have to be paid. Scenario techniques can be applied to the model to investigate future behavior of the system for early adaption of the parameters in the supply chain (e.g. kanban stock level).

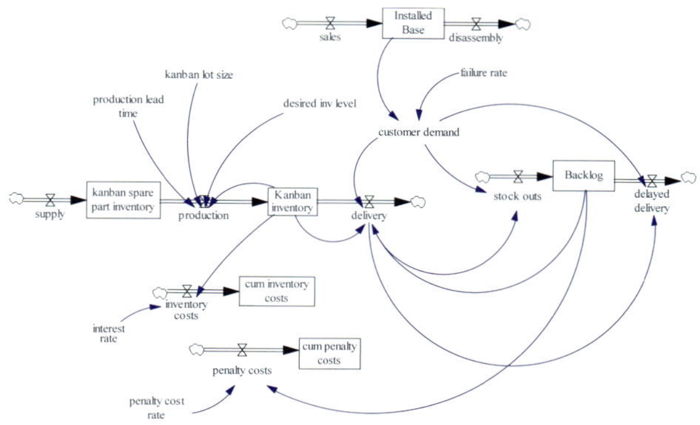

Figure 6: Stock-and-Flow-simulation of a responsive Supply Chain

6 SUMMARY

In this paper, we have shown the major problems companies in wind energy industry have to face regarding after-sales and spare parts management. While the rapid increase of wind energy plants offers various opportunities, even under the circumstances of economic crisis, it also creates new challenges for maintenance and servicing of existing facilities. To cope with these challenges, we developed a holistic description model for production systems. The model consists of all describing elements for the subsystems of a production system in spare parts industry. The partial model of the supply chain and an approach to configure the supply chain for spare parts has been worked out.

The difficulty in designing the spare parts supply chain, not solely in wind energy industry, is caused by interdependencies among the design elements. Firstly, a categorization of spare parts into four categories by combining the characteristics criticality and wear behavior has been developed. According to norm strategies in Supply Chain Management, norm strategies for the spare parts supply chain can be derived. Responsiveness and efficiency are the key distinguishing elements. Performance and costs of the norm strategies have been investigated further by applying system dynamics methods to the norm strategies. As a simple case study shows, system dynamics can be applied to support the decision process for implementing norm strategies into the spare parts supply chain. Thus, companies involved in spare parts management of the wind energy industry are enabled to improve their logistical performance and efficiency.

7 ACKNOWLEDGMENTS

The illustrated polylemma of production and the derived configuration logic for production systems are research results of the major research project "Integrative Production Technology for High-Wage Countries", wherein the development of the configuration logic is a sub-project. The depicted research has been funded by the German Research Foundation DFG as part of the Cluster of Excellence "Integrative Production Technology for High-Wage Countries". More information about the Cluster of Excellence can be found on the webpage (www.production-research.de) or by contacting the authors of this publication.

8 REFERENCES

[1] Bundesverband Windenergie e.V. (BWE): Welt: Entwicklung der installierten Leistung. 2010, www.wind-energie.de, downloaded 15.07.2010.

[2] N.N.: USA and China experience wind energy boom. Renewable Energy Focus, 2009, www.renewableenergyfocus.com, downloaded 17.07.2010.

[3] Ontario Sustainable Energy Association (OSEA): www.greenenergyact.ca, downloaded 17.07.2010.

[4] Schuh, G., Kreysa, J. und Orilski, S.: Integrierte Produktionstechnik. Published in: Schuh, G.; Klocke, F.; Brecher, C.; Schmitt, R. (Hrsg.): Excellence in Production. Aachen: Apprimus, pp. 29-53, 2007.

[5] Supply Chain Council: SCOR Model Reference 9.0. http://www.supply-chain.org. downloaded 03/2010.

[6] Boone, C.: Managing Service Inventory in the Suppy Chain, 2006.

[7] Domschke, W., Drexl, A.: Standorte. Oldenbourg, München, 1996.

[8] Domschke, W., Drexl, A.: Einführung in Operations Research. Berlin: 6., revised edition Springer, 2005.

[9] Klose, A., Drexl, A.: Facility location models for distribution system design. European Journal of Operational Research Nr. 162, pp. 4-29, 2005.

[10] Alicke, K. 2003: Planung und Betrieb von Logistiknetzwerken, Springer-Verlag, 2003.

[11] Silver, E. A., Pyke, D.F. und Peterson, R.: Inventory management and production planning and scheduling, Wiley, New York, NY, 1998.

[12] Werner, H.: Supply Chain Management, Betriebswirtschaftlicher Verlag Dr. Th. Gabler | GWV Fachverlage GmbH, Wiesbaden, 2008.

[13] Huiskonen, J.: Maintenance spare parts logistics: Special characteristics and strategic choices. International Journal of Economics, No. 71, pp. 125 – 133, 2001.

[14] Dekker, R., Kleijn, M. und de Rooij, P.: A spare parts stocking policy based in equipment criticality. International Journal of Production Economics, Vol. 56 - 57, pp. 69 – 77, 1998.

[15] Meidlinger, A.: Dynamisierte Bedarfsprognose für Ersatzteile bei technischen Gebrauchsgütern. Lang, Frankfurt am Main, 1994.

[16] Abernethy, R.B.: The new Weibull handbook, R.B. Abernethy, North Palm Beach, Fla., 2006.

[17] Rinne, H.: The Weibull distribution, Chapman und Hall/CRC, Boca Raton, Fla., 2009.

[18] Fisher, M.L.: What is the right supply chain for your product? Harvard Business Review, Vol. 75, No. 2, pp. 105-116, 1997.

[19] Mason-Jones, R., Naylor, B. und Towill, D.: Lean, agile or leagile? Matching your supply chain to the marketplace. International Journal of Production Research, Vol. 38, No. 17, pp. 4061 – 4070, 2000.

[20] Naylor, J.B., Naim, M.M. and Berry, D.: Leagility: Integrating the lean and agile manufacturing paradigms in the total supply chain. Int. J. Production Economics, Vol. 62, pp. 107-118, 1999.

[21] Lee, H.L.: Aligning supply chain strategies with product uncertainties. California Management Review, Vol. 44, No. 3, pp. 93-102, 2002.

[22] Sterman, J.D.: Business dynamics. Irwin/McGraw-Hill, Boston, 2000.

Rebound Logistics: An Integrative Reverse Supply Chain for Multiple Usage Products

G. Schuh ,T. Novoszel, M. Maas
Research Institute for Operations Management, RWTH Aachen University, Germany

Abstract
Today, the management of product returns has become a crucial issue for manufacturing companies. While the impact of green laws, legislative provisions and the increasing demand for a sustainable production management have been among the dominant drivers forcing companies to design and manage the reverse supply chain actively, recently also the opportunity for enormous profit generation and for improving the competitive advantage of manufacturing companies offered by product returns has gained increasing recognition. This paper outlines detailed results of designing a methodological framework for implementing an integrative reverse supply chain in case of manufacturing companies based on a Reverse Supply Chain Reference Model. The reference model was inductively derived from a series of industrial case studies and represents the four basic business processes collecting, selecting, reprocessing and reintegration as well as specific sub-processes. Furthermore the paper describes a two-dimensional typology of product returns initiated by OEMs. The development of the product return types is based on characteristics of motivation, intention and purpose of multiple usages of the returned products.
Keywords:
Supply Chain Management, Reverse Supply Chain Management, Product Returns, Reverse Logistics, Reference Model, Typology

1 INTRODUCTION

Traditionally manufacturing companies are primarily concerned with offering new products and value added services to the market. Product returns have mostly been considered a nuisance and uneconomical; hence their legacy is a reverse supply chain (RSC) that has been designed to minimize costs and effort. Over the last decade, the recognition of the increasing value of products at the former end of product life-cycle has forced companies to engage in managing their RSC actively. Today, the flow of product returns is becoming a significant concern for many manufacturing companies. In this research area, three fundamental aspects of product returns need to be taken into consideration: First, companies become increasingly aware of the fact that product returns may offer an opportunity for enormous profit generation and for improving the competitive advantage of a manufacturing company when taking into account the accretive value of the products and technology. Second, the impact of green laws, legislative provisions and the increasing importance of a sustainable production management due to marketing aspects force companies to design and manage the RSC actively. Third, the importance of managing the RSCs effectively will be enforced by the currently volatile economic climate.

From the point of view of a manufacturing company, the main drivers for managing consciously a RSC can principally be classified as exogenous and endogenous factors. In the last decades, an increase in environmental debates clearly caused the environmental consciousness of consumers to rise. This led to an increase in the interest towards sustainable manufacturing systems. The challenges imposed by the global increase of resource scarcity and the introduction of eco-political laws, like the European Directive for Waste of Electrical and Electronic Equipment (WEEE), on the one hand and the importance of ecological buying criteria in customer's expectations towards companies are the main endogenous factors rendering the harmonization of economic as well as ecological target sizes a requirement and an opportunity for companies at the same time [1,2,3].

Economic and market strategy aspects are some of the key exogenous driving forces that influence the handling of redeemed products. This is especially true in industries where redemption of used products is already at a high level, or if products may be sold at a high price on the primary market and after redemption and refurbishment on the more price-sensitive secondary market. The increasing importance of all these aspects leads to a rise in the complexity of the supporting planning and logistics processes, and subsequently to an increase in the importance of reverse supply chain management as a discipline.

2 LITERATURE REVIEW

2.1 Environmental sustainability

Savitz and Weber suggest that business interests and the interests of the environment and society interact in every firm's operation [5]. Thus, environmental sustainability is only possible if economic decision making starts incorporating the presumption that natural resources are limited. For a long time, producers have neglected the economic and ecologic potential underlying their used goods [6] as outlined in chapter 1. This position has changed over the last 10 to 15 years. Although there is no encompassing theory of Reverse Supply Chain Management (RSCM) yet, there are some useful points of departure for our research framework in the literature. In this chapter, we will present the status of current research regarding the topic of our paper.

2.2 The Modularization of reverse logistics: collection and recovery options

A number of authors have focused on the conceptualization and empirical analysis of the main activities within reverse logistics, which can roughly be divided into those who refer to product collection and those who refer to product recovery activities.

Beulens et al. study the costs of strategic management decisions for the transport of used products from the customer back to the producer and describe the essential steps of the collection process [7]. The authors identify ideal types of collection systems and assess the possibilities of different routing options considering the opportunities by integrating transport capacities of the traditional Supply Chain. Similar approaches focus on the role of the agent of product returns [8,9], the price policy [10], the level of integration [11], inventory management [12], the spatial disposition of collection facilities [13] and different types of collection systems [14] for the network design and profitability of the collection process.

In the last years, the research focus has mainly been on the topics of remanufacturing and product recovery. Thierry et al established the concept of different product recovery options [6]. These recovery options are differentiated according to their level of disassembly, the purpose of the recovered product and the respective quality requirements. In the context of the design of supply chain management networks, the existing research approaches have also been extended by a number of authors to account for the influence of product recovery activities in the design of a production or distribution network [15,16].
Seliger looks at the underlying process focusing on the product as such with all its potentials and characteristics [17]. The author provides a sound analysis of sustainable product design, starting with a complete life-cycle approach, as well as the planning and the implementation of possible processing or disposal activities. Next to the numerous preparatory efforts undertaken in the area of remanufacturing, there are several other works that deal with the diverse forms of product-value-preserving actions [6], different product recovery measures within certain industries [18,19] or the optimal implementation of value-preserving activities [20,21].

2.3 The Closed-Loop Supply Chain

The concepts of closed-loop supply chain and closed-loop recycle network are described in detail by Morana [22]. The author is using an extensive definition, systematization and analysis of systems that are characterized by a closed material cycle. Furthermore, the research investigation addresses the steps and actors needed to account for product refluxes in a traditional supply chain.
Schröter investigates closed-loop supply chains under the premise of an optimal supply of spare parts [23]. He details on legal and economic conditions that affect product returns, in order to develop a decision support concept for the implementation of closed-loop supply chains for spare parts. Seitz, however, describes closed-loop recycle networks primarily in terms of sustainability, he thus elaborates on remanufacturing and its integration into existing structures [24]. Starting from a general description of a closed-loop system, he models an automotive manufacturer's business processes based on the product processing operations.

2.4 Summary

In summary, the literature review conducted revealed that there are already some ready-to-use approaches focusing on parts of the field of reverse supply chain management. Product technological remanufacturing processes as well as closed-loop logistics are analyzed using both descriptive studies and studies of a structuring nature. While the processes of product collection and recovery have been studied from a variety of analytical perspectives, the process of the reintegration of the returned and recovered products has been widely neglected. As such there is a lack of theoretical-conceptual and empirical understanding of the process of reintegration, its challenges, opportunities and network design options, which will be discussed in chapter 3 and 4. The consideration of a holistic supply chain through an integrated approach of "forward" and "reverse" relationships is usually only marginal or absent altogether. None of the authors analyzed a holistic approach in terms of a complementary, vertical diversification of the business model from the perspective of a manufacturing company. Although there are already some best-practice solutions regarding the design and implementation of some sub-process steps in a RSC, the existing pieces of work can only partly be transferred to other firms and industries due to each particular business' situation. To enable a uniform view on both the traditional supply chain and the RSC, it is necessary to use a scientifically sound approach. The aim is to examine this "advanced" and integrated supply chain in order to identify the business potential, as well as to derive decision-making and design options.

3 REVERSE SUPPLY CHAIN MANAGEMENT

3.1 The role of RSCM

The main role of RSCM can be derived from the definition of Prahinski and Kocabasoglu, who consider RSCM as "the effective and efficient management of the series of activities required to retrieve a product from a customer and either dispose of it or recover value" [25,26]. According to this definition, reverse logistics and product recovery are the two essential components of RSCM. Reverse logistics is defined by the „Council of Logistics Management" as „the process of planning, implementing and controlling the efficient, cost-effective flow of raw materials, in-process inventory, finished goods and related information from the point of consumption to the point of origin for the purpose of recapturing or creating value or for proper disposal" [11].

The second main topic of RSCM, i.e. product recovery, deals with the recovery, reprocessing or remanufacturing of products that are returned by the consumer to the manufacturer at the preliminary end of their lifecycle. The terms recovery, reprocessing or remanufacturing are often used synonymously and generally refer to all activities, which aims at making a used product ready for further usages.
The current research focuses mostly on models for the design of the supply chain directed to business customers. These models describe the holistic value creation process of a product, from raw material suppliers to the end consumer. Schönsleben interprets the role of the classic supply chain management as the coordination of strategic and long-term cooperation between companies within a supply chain [27].

The focus of supply chain management is on methods and concepts for planning, managing and monitoring the delivery of value adding processes. In addition to the primary consideration of material and information flows, increasing attention is paid to money flows, product support services, or repatriation (e.g. recycling of end products) [28,29,30].

The drivers for RSCM as outlined in chapter 1 force companies to reconsider its existing logistics and production processes, especially since product returns from customers nowadays play an increasingly important role. This development imposes some challenges to companies. First, they should evaluate their specific ability of integrating the RSC. Second, they need to derive their individual RSC design possibilities. An initial design advice is given by Hompel and Beck [31], who argue that SCM and RSCM need to be evaluated holistically. The closed-loop supply chain as mentioned above can be idealized as an existing form of an integrated RSC.

3.2 Core processes and network design structure

The typical network structure of a RSC is comprised of a number of nodes of different agents which regulate the flow of materials. As the flow of materials is returned from the customer through retailers back to the producer, the logistic network takes on a convergent structure with many nodes at the customer side and only a few nodes at the producer side. As the returned products are recovered and the flow of materials is led on to its subsequent point of consumption the network structure, similar to traditional logistic networks, becomes divergent as the material is transported from the few points of recovery to many points of consumptions while passing through a number of retail nodes [32,33,34].

As indicated in figure 1, the main tasks of RSCM can also be structured according to a process sequence of generic RSCM phases. In the literature, which conceptually describes RSC from such a process theoretical point of view, at least three phases can be distinguished [6,32]:

- Product Acquisition
- Product Recovery
- Product Reintegration

While reverse logistics tasks are mainly concluded in the first and the last phase, the product recovery phase evidently includes activities related to product recovery tasks as outlined above. The challenges and main tasks of each RSCM phase shall be described below.

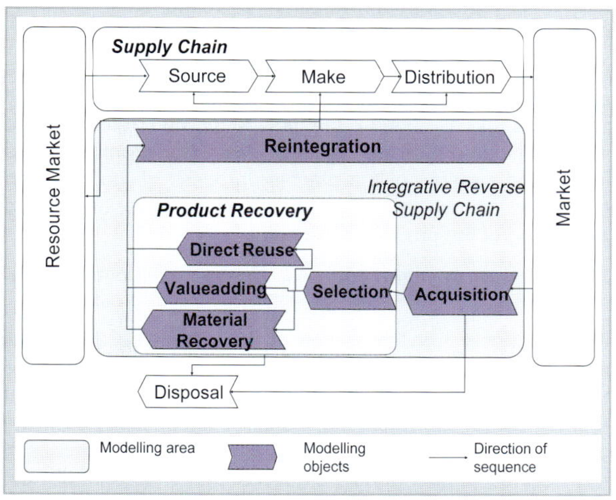

Figure 1: Integrative Reverse Supply Chain

3.3 Acquisition

Product acquisition describes the process of returning the used products from the customer to the producer and making them available for the subsequent recovery process. The primary market is the source of the input stream for the RSC, which is composed of the products that have been utilized according to their mode of operation and at this point are considered as end-of-use or end-of-life products [19].

The acquisition phase is comprised of collection, selection, stocking and transport activities. The management of product acquisition aims at coordinating and regulating the flow of returned products in order to avoid an unnecessary accumulation of returned products while meeting the demand for recovered products on the market. The main challenge consists of regulating and coordinating an input stream, which varies to a significant degree in terms of quantity, quality and timing. As opposed to a Forward Supply Chain the product returning actor usually has only little knowledge about how many products are returned on what date and in which condition [6,7,32,34].

The network design of the acquisition process is dependent upon the type of returning actor, the collection system as well as the degree of centralization and integration of the logistical resources and facilities [8,22]. Returning actors can be the OEM, a syndicate of OEMs or third-party logistics provider. Two different collection systems can be identified: The Bring system relies on the customer to take the used products back to the producer or to respective collection points. Within Pick-up Systems the to-be returned product is directly collected from the customer and brought to respective collection and selection facilities [22]. The degree of centralization describes the spatial disposition of the collection facilities and is dependent on the collection system. Collection facilities concentrate the multiple flows of returned products in order to aggregate their relative low volume. Additionally they can conduct selection and inspection services [33]. The level of integration represents the extent to which capacities and resources, mainly transport and stocking capacities, of the Forward Supply Chain are simultaneously used for acquisition purposes of the RSC [35].

3.4 Product Recovery

After the used products have been returned they are made available for recovering their economic and functional value, which has been decreased while being used, or to remanufacture the product for another functional purpose [6,36]. This phase includes the activities of cleaning, inspecting, selecting and disassembling [19,34]. Cleaning is considered as a prerequisite for the subsequent inspection and for recovering the functional value as well as the optical impression for its further usage [19]. The inspection is necessary due to the unknown quality of condition of the returned product, which again is an important criterion for the decision regarding the further usage of the returned product. The returned products are then, based on the findings of the inspection process, grouped together according to their further purpose of usage and hence selected for different recovery options [37,38]. Disassembly activities aim at systematically separating the product components at least into two units. Disassembly activities are, thus, a necessary step for all recovery options except for the direct reuse of the returned product [39]. Two forms of disassembly can be distinguished. The non-destructive disassembly separates the product components without affecting its functional value. The destructive disassembly separates the products components and destroys the functional value of the product, hence, making only resources or energy available for recovery [40].

According to Thierry et al. and depending on the condition and the type of the product, recovery operations mainly take

on six different forms [6]: repair, refurbishing, remanufacturing, cannibalization, recycling and direct reuse.
The aim of repair is to make a used good again fully functional. This may include the replacement or the repair of some of the product's items. Reconditioning is attempted to make the product meet a specified quality level. This quality level is usually below that of a new product. To this end, the product is disassembled into modules, which are inspected and, if necessary, repaired or replaced.
The purpose of remanufacturing is to raise the quality level of a used product to the quality level of a new product. For that purpose, the product is completely disassembled, inspected and extensively restored. If necessary, damaged components are replaced by a newer component version. The term cannibalization refers to the complete decomposition of a product and the subsequent reuse of parts of its components into other goods. The fifth and final element of product recovery is recycling, whereby the main objective is the recovery of valuable materials that the product consists of.
Moreover, there is the possibility of using the products directly as spare parts without giving them one of the five treatments outlined above. Part of this option is to sell the product on the so-called "second market" [6].
The recovery options can further be categorized according to their level of disassembly, condition and purpose of recovery into direct reuse, value adding and material recovery. Direct reuse resembles the same recovery option by Thierry et al. Value adding encompasses refurbishing, remanufacturing and repair activities, which are characterized by a low to high level of disassembly (depending on the condition of the product) for the purpose of using the product or product components again for the same or a different function. Material recovery, on the other hand, aims at destroying the functional value of the product for recovering its material and using them for the production of new products. The recycling recovery option is related to this recovery category [19].

3.5 Reintegration

Product Reintegration takes place after the returned product has been recovered and directs the flow of recovered products to different phases of a traditional Supply Chain. This phase encompasses all steps to sell, stock and transport the recovered product to its new point of usage, it thus resembles the distribution process of the Forward Supply Chain [41].

The point of reintegration is dependent on the purpose for which the used product has been returned and recovered [6]. If the product has been recovered for a direct reuse, it will be reintegrated into the distribution process of a Forward Supply Chain. If individual product components are to be used for the same or a different purpose, these parts will be reintegrated either into the production process or into the sourcing process of a Forward Supply Chain. Additionally recovered material can be reintegrated on the market and made available for the Supply Chain of other companies, which is commonly referred to as an Open-Loop Supply Chain [42,43,44].

4 RESEARCH FRAMEWORK

In principle, two strategies can be used to develop a reference model, the first being the development of an entirely new model. Another strategy may as well be the reuse or an adaption of an existing model [34]. Both strategies are widely used and accepted for the development of both practically and theoretically oriented reference models. The selection of the used strategy is commonly a result of the modelling requirements and is not based on an explicit choice.

Based on a holistic view, the hierarchical reverse supply chain reference model consists of the key processes collecting, selecting, reprocessing and redistribution.

On the first level (Selection), the core process of the RSC is selected. The core processes chosen in this paper can be found in a large range of other pieces of work [6,45,46,33,47,32,11,47,19]. The second level consists of several categories of core processes that enable a company to configure its supply chain and to introduce an operative strategy. The third level is used to develop and adjust the operative strategy by modelling the relevant processes. This last level can be used generically and is thus valid for all companies active in the RSC segment. Next to these hierarchically structured levels, the model usually encompasses a fourth level that is used for company specific processes or activities. Note that this level is not illustrated in the reference model outlined in Figure 1.

In the following paragraphs, the hierarchically structured levels of the reference model will be outlined in detail. In addition, a notation that will increase the comprehensibility of the model is introduced.

4.1 Notation of the Reverse Supply Chain Reference Model

The RSC reference model is characterized by a set of standard notations, which can still be found through the entire model.

To represent the core processes at the first level, these are identified by single letters. "P" describes plan processes, "C" describes collecting processes, "S" describes selecting processes, "RP" describes reprocessing and "RE" reintegration processes.

The notation system on the 2nd level of the reference model adds consecutive numbers to the letters representing the respective core processes due to the variety of execution processes on the configuration level (e.g. C1, C2 etc.). If the letter "P" precedes the notation, it refers to a planning process of the respective core process.

On the 3rd level of the hierarchically structured reference model, the notation is expanded by another number, which is separated by a "." from the first letter and refers to the individual element on the 3rd level. PS1.2, for example, refers to the 2nd element of the planning process of the selection core process, which is the identification, assessment and monitoring of the product resources.

4.2 Structure of the Reverse Supply Chain Reference Model

The five basic core processes of the original SCOR model (plan, source, make, deliver and return) are not part of the modelling environment as originally designed, thus, not represented in the reference model. It rather shifts from the processes of the traditional Supply Chain to the processes of the Reverse Supply Chain as described in chapter 3. Hence, the processes plan, collection, selection, reprocessing and reintegration are modelled as new core processes of the reference model.

Furthermore, the reference model is hierarchically structured and encompasses three levels of processes. The first level is related to the individual core process of the RSC. Specific configurations and design options for companies are introduced on the 2nd level. Descriptions of the sequential activities of each process are found on the 3rd level of the reference model.

Level 1: Selection

The first level (Selection) is comprised of the five core processes of a RSC (plan, collecting, selecting, reprocessing, and reintegration), which together form the basic structure of the model and thus delimit the content of the model [32].

Level 2: Configuration

On the second level (Configuration), the core processes can be further configured by the two different types of processes: "planning" and "execution". The process type "planning" designates the coordination of the process resources with the requirements on the respective processes. Based on these, a consistent basis is developed, taking into account a consistent planning horizon of a strategic approach to the core process. The process type "execution" includes all activities that are directly related to the handling of the repatriated used products, and therefore change the status of these products. The underlying activities basically include the scheduling and flow control, the workmanship on the used product, and transporting the products to the next process. The activities of this type of process thereby determine the processing time of the product.

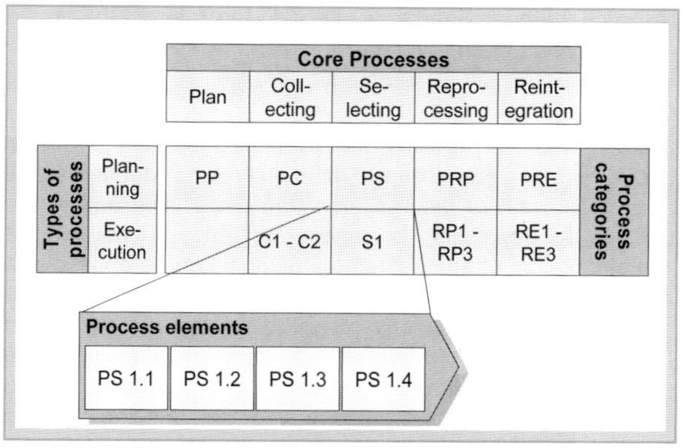

Figure 2: Hierarchical structure of the reference model

Level 3: Element

The third and final level of the model (Element) includes the preparation and coordination of an operational strategy. In addition, the process categories from the second level are further decomposed. The relevant processes or process elements are modelled on the third level.

5 A REFERENCE MODEL OF RSCM

The designed reference model consists of the mentioned 5 core processes, 5 planning process types respectively, 9 execution process types, which represent different process categories on the second level and allow for a company specific design of the RSC, as well as 74 process elements on the third level. Each core process along with its different configurations on the second level shall be described below. The elements on the third level of a specific category of the reprocessing process are exemplified in chapter 5.4.

5.1 Core Process: Plan

The core process "plan" includes activities that match the drivers of a product take-back with existing resources and develop a strategic approach that meets the requirements of the core processes. The reference model represents one configuration of the plan process for each core process on the second level. Each plan configuration is thus concerned with the planning of the respective RSCM process.

5.2 Core Process: Collection

The core process "collecting" is composed of all the activities that are necessary for the purchase of used goods from consumers. It represents the interface between the market and the reprocessing phase and illustrates the point of return. Two system configurations are represented in the reference model, which refer to the collection systems described in 3.3: A bring system (C1) and a pick-up system (2). All must meet the challenges imposed by the uncertainty of quantity and timing, which characterizes the flow of material of RSC'.

5.3 Core Process: Selection

The core process "selecting" describes the processes that contribute to the acquisition of information about the nature and quality of the used product. Furthermore, within this core process, the recirculated used products are classified in order to select the most appropriate and economically valuable reprocessing procedure. It encompasses all activities for the cleaning and inspection, if necessary disassembly, of the returned product in order to identify the potential of multiple usage. Based on the results of the inspection, the returned product is being selected for the appropriate reprocessing option. The system context of the selection process is characterized by the uncertainty of quality of the returned product, which sometimes makes the disassembly of the product for the purpose of the inspection a requirement. The reference model only considers one configuration of the selection process which is common to every RSCM (S1).

5.4 Core Process: Reprocessing

The core process "reprocessing" involves various methods by which the re-circulated used products are reprocessed. All activities to disassemble, to recover the functional and economic value of the re-circulated used product or to remanufacture it for another functional purpose take place in the reprocessing phase. Depending on the quality of the product, cutting costs and the degree of possible reuse three different reprocessing options are considered by the reference model, which represent different configurations and cover all reprocessing methods as identified by Thierry et al. [6,35,24,39]: Added Value Recovery (RP2) encompasses the recovery options remanufacturing, refurbishing and repair. The process configuration Material Recovery (RP3) aims at recovering the material of the returned product and includes the recycling recovery option. The configuration Direct Reuse (RP1) aims at recovering the original functional value of the returned product when its condition does not require any further disassembly or value adding activities and resembles the recovery option direct reuse by Thierry et al.

To illustrate the role of process elements within the reference model, the direct reuse category of the reprocessing process is illustrated on the third level (RP1.1-1.6). The first element consists of the reception of the product (RP1.1). After the returned product has been cleaned and checked for its functionality (RP1.2), it is upgraded to be directly reused (RP1.3). It is then being checked again (RP1.4) and prepared for its further transport (RP1.5). The transport itself (RP1.6) is the last process element of the direct reuse reprocessing category.

5.5 Core Process: Reintegration

Last but not least, the core process "Reintegration" is characterized by the activities encompassing the management of orders, management of transportation and distribution of products that were re-made marketable in the reprocessing process. Hence, the reintegration process is concerned with the redistribution of the reprocessed products to different points of entry of traditional Supply Chains according to their purpose of multiple usages, which allows for the generation of profit within RSCM. The reference model represents three different configurations according to their respective point of entry. The Return-to-Source configuration (RE1) reintegrates the reprocessed products into the source process of the Forward Supply Chain. The Return-to-Make configuration (RE2) directs the flow of reprocessed products to the production process of traditional Supply Chains. Finally, the Return-to-Market configuration (RE3) reintegrates the reprocessed products into the market phase of the traditional Supply Chain.

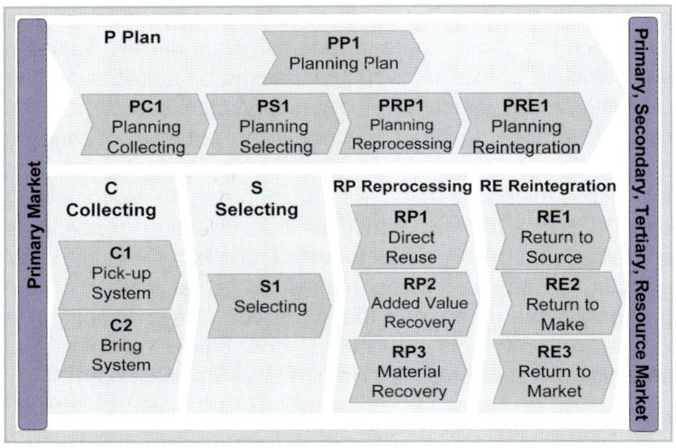

Figure 3: Process configurations

6 DISCUSSION AND FURTHER DEVELOPMENT OF THE RESEARCH FRAMEWORK

The Reverse Supply Chain Reference Model outlined in this paper enables users to address, improve and communicate business activities among all interacting participants. It is a process based reference model for reverse supply chain management, spanning from the end-user to the original equipment manufacturer (OEM).

By describing a reverse supply chain network using process elements, the reference model can be used to describe intuitive RSC' as well as complex networks using a common set of definitions. However, the high variety of RSC situations calls for an adequate specification level of the generic reference model. A typology of ideal types of reverse logistics companies is considered for developing respective best practices and network design recommendations for each type. The typology is based on characteristics of motivation, intention and purpose of multiple usages of the returned products. The motivations or intentions of reverse logistics companies refer to the existing principle incentive systems of economic and environmental drivers, as mentioned in chapter 1.

Moreover, all types of commercial product returns should be considered using a multi-dimensional typology of purpose of multiple usages. Applying a morphologic analysis, types of purposes are categorized based on the dichotomous idea, that returned products can either be re-processed and re-integrated for the same purpose of the original product or for a different purpose. In order to configure an integrative RSC in a mid-term perspective, the underlying principle incentive systems of the individual companies as well as the purpose of multiple usages are assumed to be a constraint for integrating an efficient RSC from the manufacturer's points of view. It will be shown that the application of the analysis leads to specific clusters of manufacturing companies dealing with certain type of products and diverse strategies. Furthermore, findings can be interpreted in light of types specific process selection, process configuration and design recommendations towards a specific reference model for each rebound type. This research work addresses a holistic approach in terms of a complementary, vertical diversification of the business model from the perspective of a manufacturing company.

7 CONCLUSION

In this survey, we addressed the reverse logistics of industrial reuse of products. We have first outlined a number of endogenous and exogenous drivers for RSCM. Within the same topic, we reviewed the literature on environmental sustainability, specific business processes and the Closed Loop Supply Chain, thereby outlining our white spot on the research agenda. Chapter 3 specified the role of RSCM, the structure and general processes of RSC, which serves as the theoretical-conceptual foundation of the designing of the reference model.

Chapter 4 and 5 presents our research framework, which is based on a Reverse Supply Chain Reference Model, as well as a discussion of the research framework. We outline the difficulty of developing a generic reference model for the implementation of an integrated RSC that can be applied to a large number of different types of companies. We show that a multi-dimensional typology of rebound types, based on different incentive systems for reverse supply chain management as well as a number of types of products would significantly improve the process selection and the process configuration by delivering specific design recommendations.

The illustrated Polylemma of Production and the derived Configuration Logic for Production systems are research results of the major research project "Integrative Production Technology for High-Wage Countries", wherein the development of the Configuration Logic is a subproject. The depicted research has been funded by the German Research Foundation DFG as part of the Cluster of Excellence "Integrative Production Technology for High-Wage Countries". More information about the Cluster of Excellence can be found on the webpage (www.production-research.de) or by contacting the authors of this publication.

REFERENCES

[1] Dhanda, K. K., Hill, R. P., 2005, The role of information technology and systems in reverse logistics: a case study, International Journal of Technology Management, 3/1/2:140-151.

[2] Georgiadis, P., Vlachos, D., 2004, The effect of environmental parameters on product recovery, European Journal of Operational Research, 157:449-464.

[3] Scholz-Reiter B., de Beer C., Freitag M., 2008, Dynamik logistischer Systeme, Beiträge zu einer Theorie der Logistik, Springer, Berlin.

[4] Straube, F., Rösch, F., Klaar, K., 2008, Neue Strategien in der Logistik, Ruprecht, R. (ed.), 9. Karlsruher Arbeitsgespräche Produktionsforschung - Spitzentechnologie für den Wirtschaftsmotor Produktion, FZKA, Karlsruhe, pp. 314-323.

[5] Savitz, A. W., Weber, K., 2006, The triple bottom line: How today's best-run companies are achieving economic, social and environmental success – and how you can too, Jossey-Bass, San Francisco, CA.

[6] Thierry, M., Salomon, M., Van Nunen, J., Van Wassenhove, L., 1995, Strategic Issues in Product Recovery, California Management Review, 37/2:114-135.

[7] Beullens, P., Van Wassenhove, L., Van Oudheusden, D., 2002, Collection and vehicle routing issues in reverse logistics, Dekker, R., Fleischmann, M., Inderfurth, K., Van Wassenhove, L.N. (ed.), Reverse Logistics - Quantitative Models for Closed-Loop Supply Chains, Springer, Berlin, 96-134.

[8] Spicer, A. J., Johnson, M. R., 2004, Third-party demanufacturing as a solution for extended producer responsibility, Journal of Cleaner Production 12/1:37-45.

[9] Meade, L., Sarkis, J., 2002, A conceptual model for selecting and evaluating third party reverse logistics providers. Supply Chain Management: An international journal 7/5:283-295.

[10] Savaskan, R. C.,Van Wassenhove, L. N., Reverse Channel Design: The Case of Competing Retailers,. Management Science 52/:1-14

[11] Krumwiede, D. W., Sheu, C., 2002, A model for reverse logistics entry by third-party providers, Omega The international Journal of Management Science, 30/5:325-333.

[12] Toktay, L. B., Wein, L. M., Zenios, S. A, 2000, Inventory Management of Remanufacturable Products, Management Science, 46/11:1412-1426.

[13] Srivastava, S. K., Srivastava, R. K., 2006, Managing product returns for reverse logistics, International Journal of Physical Distribution & Logistics, 36/7:524-546.

[14] Guide, D., Van Wassenhove, L. N., 2000, Managing Product Returns for Remanufacturing.

[15] Alicke, K., 2003, Planung und Betrieb von Logistiknetzwerken, Springer, Berlin.

[16] Beamon, B. M., Fernandes, C., 2004, Supply-chain network configuration for product recovery, Production Planning & Control, 15/3:270-281.

[17] Seliger, G., 2007, Sustainability in Manufacturing. Berlin, Heidelberg: Springer, Berlin.

[18] Seitz, M., Wells, P. E., 2006, Challenging the implementation of corporate sustainability. Business Process Management Journal, 12/6:822-836.

[19] Stölting, W., 2006, Lebenszyklusorientierte strategische Planung von Remanufacturing-Systemen für elektr(on)ische Investitionsgüter, VDI-Verlag, Düsseldorf.

[20] Bey, N., 2004, Application of LCA and Eco-Design in Industry – A Successful Example of Putting Life Cycle Thinking into Practice, Proceedings Global Conference on Sustainable Product Development and Life Cycle Engineering, Berlin.

[21] Seliger, G., Zettl, M., 2008, Modularization as an enabler or cycle economy, CIRP Annals – Manufacturing Technology, 57/1:133-136.

[22] Morana, R., 2006, Management von Closed-Loop Supply Chains, Dt. Univ.-Verl., Wiesbaden.

[23] Schröter, M., 2006, Strategisches Ersatzteilmanagement in Closed-Loop Supply Chains, Dt. Univ.-Verl., Wiesbaden.

[24] Seitz, M., 2006, Remanufacturing, Supply Chain Management and Sustainability, Fachhochschule Hof, Hof.

[25] Prahinski, C. K. C., 2006, Empirical research opportunities in reverse supply chains, Omega The international Journal of Management Science, 34/6:519-532.

[26] Guide, D. R., Van Wassenhove, L. N., 2002, The Reverse Supply Chain, Harvard Business Review, 80/2:25-26.

[27] Schönsleben, P. D. P., 2007, Integral Logistics Management- planning and control of comprehensive Supply Chains. St. Lucie Press. Zürich.

[28] Busch, A., Dangelmaier, W., 2004, Integriertes Supply Chain Management – Ein koordinationsorientierter Ansatz, Busch, A., Dangelmaier, W. (ed.), Integriertes Supply Chain Management: Theorie und Praxis effektiver unternehmensübergreifender Geschäftsprozess, Gabler Verlag, Wiesbaden.

[29] Dyckhoff, H., Souren, R., 2008, Nachhaltige Unternehmensführung: Grundzüge des industriellen Umweltmanagements, Springer, Berlin.

[30] Fettke, P., 2007, Supply Chain Management: Stand der empirischen Forschung, ZfB Zeitschrift für Betriebswirtschaft 77/:417-461.

[31] Hompel, M., Beck, M., 2008, Materialfluss und Energieeffizienz, 9. Karlsruher Arbeitsgespräche Produktionsforschung - Spitzentechnologie für den Wirtschaftsmotor Produktion, pp. 300-301.

[32] Fleischmann, M., 2001, Reverse Logistics Network Structures and Design, Research Paper from Erasmus Research Institute of Management (ERIM).

[33] Fleischmann, M., Krikke, H. R., Dekker, R., Flapper, S. D. P., 2000, A characterisation of logistics networks for product recovery, Omega The International Journal of Management Science, 28/6:653-666.

[34] Schmid, E., 2008, Koordination im Reverse Logistics, PhD thesis at the University of Braunschweig, Wiesbaden.

[35] Fleischmann, M., Beullens, P., Bloemhoff-Ruwaard, J. M., Van Wassenhove, L. N., 2001, The Impact of Product Recovery on Logistics Network Design,10/2:156-173.

[36] Geyer, Roland, Jackson, Tim, 2004, Supply Loops and Their Constraints: The Industrial Ecology of Recycling and Reuse, California Management Review, 46/2:55-73.

[37] Brüning, R., 2003, Möglichkeiten und Grenzen der erneuten Verwendung elektr(on)ischer Geräte und Komponenten, VDI-Fortschritt Berichte 1775:57-84.

[38] Steinhilper, R., 1999, Produktrecycling: Vielfachnutzen durch Mehrfachnutzung, Fraunhofer IRB, Stuttgart.

[39] Dinge, A., 2000, Demontage komplexer Produkte in einer Kreislaufwirtschaft, Josef Eul Verlag, Lohmar.

[40] Walther, G., 2005, Recycling von Elektro- und Elektronik-Altgeräten, Dt. Univ.-Verl., Wiesbaden.

[41] Rinschede, A. ; WEHKING, Karl-Heinz ; JÜNEMANN, Reinhardt: Entsorgungslogistik III. Berlin: Schmidt, 1995

[42] Luger, T., Herrmann, C., Steinborn, J., Walther, G., Spengler, T., 2008, Wertschöpfung durch Mehrfachnutzung, Zeitschrift für wirtschaftlichen Fabrikbetrieb 103/9:602-606.

[43] Mertens, Jan-Peter, 2005, Dynamische Aspekte der Reverse Logistics in Closed-Loop Supply Chains, Grin, München.

[44] Barros, A.I., Dekker, R., Scholten, V., 1998, A two-level network for recycling sand: A case study, European Journal of Operational Research, 110/2:199-214.

[45] Fleischmann, M., Bloemhof-Ruwaard, J. M., Dekker, R., Van der Laan, E., van Nunen, J. A. E. E., Van Wassenhove, L. N., 1997, Quantitave models for reverse logistics: A review, European Journal of Operational Research, 103/1:1-17.

[46] Dowlatshahi, S., 2000, Developing a Theory of Reverse Logistics, Interfaces, 30/3:143-155.

[47] Guide, V. D. R. J., Harrison, T. P. and Van Wassenhove, L. N., 2003, The Challenge of Closed-Loop Supply Chains, Interfaces, 33/6:3-6.

Printing and Binding: Stürtz GmbH, Würzburg